2nd ATLAS *of the* BREEDING BIRDS
of MARYLAND *and the*
DISTRICT *of* COLUMBIA

2nd ATLAS *of the* BREEDING BIRDS *of* MARYLAND *and the* DISTRICT *of* COLUMBIA

Edited by WALTER G. ELLISON ❦ Foreword by Chandler S. Robbins

THE JOHNS HOPKINS UNIVERSITY PRESS
Baltimore

© 2010 The Maryland Ornithological Society
All rights reserved. Published 2010
Printed in China on acid-free paper
9 8 7 6 5 4 3 2 1

The Johns Hopkins University Press
2715 North Charles Street
Baltimore, Maryland 21218-4363
www.press.jhu.edu

Library of Congress Cataloging-in-Publication Data

Second atlas of the breeding birds of Maryland and the District of
Columbia / Walter G. Ellison, editor ; foreword by Chandler Robbins.
      p.   cm.
  "A project of the Maryland Ornithological Society, the Maryland
Department of Natural Resources, and the U.S. Geological Survey"—
Preliminaries.
  Includes bibliographical references and index.
  ISBN-13: 978-0-8018-9576-0 (hardcover : alk. paper)
  ISBN-10: 0-8018-9576-6 (hardcover : alk. paper)
  1. Birds—Maryland. 2. Birds—Washington (D.C.). 3. Bird populations—
Maryland. 4. Bird populations—Washington (D.C.). 5. Birds—Maryland—
Geographical distribution—Maps. 6. Birds—Washington (D.C.)—
Geographical distribution—Maps. I. Ellison, Walter G. II. Maryland
Ornithological Society. III. Maryland Dept. of Natural Resources.
IV. Geological Survey (U.S.). V. Title: Atlas of the breeding birds of
Maryland and the District of Columbia.
  QL684.M3S43 2010
  598.09752—dc22   2009043633

A catalog record for this book is available from the British Library.

Special discounts are available for bulk purchases of this book. For more information, please contact Special Sales at 410-516-6936 or specialsales@press.jhu.edu.

The Johns Hopkins University Press uses environmentally friendly book
materials, including recycled text paper that is composed of at least 30
percent post-consumer waste, whenever possible. All of our book papers
are acid-free, and our jackets and covers are printed on paper with recycled
content.

To the county coordinators, who were the backbone of this volume

CONTENTS

FOREWORD

This atlas gives an up-to-date account of every wild bird species that nests in the state of Maryland and the District of Columbia. Each species account includes a color photograph, dates of seasonal occurrence and nesting, maps of current distribution and change during the past two decades, and a graph of trend in abundance. The accounts also describe habitat requirements, threats, and conservation needs. This book is based on five years of intensive fieldwork by professional and amateur observers who searched every 5 × 5–km (3 × 3–mi) tract of land in Maryland and the District. Through comparison with the results of a similar survey 20 years ago, it shows where changes in bird populations are occurring, how these changes are related to ongoing changes in the environment, and what steps are needed to prevent further loss of nesting species.

This breeding bird atlas is unique in its multiscale approach, which permits comparisons by county, by physiographic regions, and for Maryland as a whole. Thanks to the North American Breeding Bird Survey it also allows comparison with eastern North America and the entire continent. All modern grid-based atlases map the distribution of each species during a specified period (generally five years), and most use a grid of about 5 or 10 km (3 or 6 mi) on a side. A few (including this one) incorporate special techniques to estimate changes in abundance as well as changes in distribution. As in the previous atlas, the Baltimore-Washington corridor was subdivided into quarterblocks (2.5 × 2.5 km [1.5 × 1.5 mi]) to increase the ability to detect changes in bird distribution in the counties experiencing the most change in human activities. The southernmost county and the northwestern-most were also quarterblocked to facilitate the detection of bird distribution changes that might result from global climate change.

The basic fieldwork was conducted by amateur and professional observers who first explored the 5 × 5–km (3 × 3–mi) block in which they lived, searching for breeding evidence of each species they found. As the fieldwork progressed, many workers shifted their activities into other blocks that had not been as well sampled. In the final years the Atlas Committee solicited help from experts to visit undercovered blocks in order to attain adequate coverage of every block in Maryland and the District of Columbia.

In addition to the original observers who provided basic coverage, expert birders were trained to run 15-stop miniroutes in alternate 5-km (3-mi) blocks throughout the state for comparison with results from miniroutes run during the previous atlas. Special efforts were also made to detect nocturnal and crepuscular birds.

The Atlas Committee met almost monthly from February 2002 to June 2009, planning coverage, tracking results, planning illustrations, solving coverage problems, and tracking progress with the text and illustrations.

Nearly one-fifth of Maryland's breeding bird species were confirmed in fewer than 10 atlas blocks, raising concerns that some of them might disappear before the next atlas survey is conducted. In recent decades we have observed how natural catastrophes have caused sudden declines in certain species: nesting Purple Martins that starved during the prolonged heavy rains from Hurricane Agnes in June 1972, migrating Chimney Swifts swept into the middle of the Atlantic Ocean by Hurricane Wilma in late October 2005, sleeping Northern Bobwhites trapped under an impenetrable crust of ice in the winter of 1977, House Finches infected with conjunctivitis since January 1994, and several species affected by West Nile virus in the present decade.

We have also observed how increases in human population and changes in people's lifestyle are taking a greater and greater toll on bird populations every year. Maryland leads the nation in passing legislation to protect prime nesting habitat of forest interior–dependent species from fragmentation. It is now the responsibility of every citizen to seek ways to further reduce our cumulative impact on avian populations. For example, more Maryland counties should enact cats indoors legislation. U.S. cities should follow the Toronto City Council's lead in mandating bird-friendliness in the construction of new high-rise and nonresidential low-rise buildings and reduction of nighttime glare and light trespass. And we must all strive to reduce the avian casualties at clear and reflective glass windows, which are estimated to take the lives of a billion birds annually in our country. We must drastically reduce the hazards that we construct along migratory corridors. We should all slow down when driving on narrow rural roads in the early morning. Every landscaper and home owner should plant native trees and shrubs that are preferred by our native birds rather than exotic species that are of little use to them. And finally, we must wean ourselves away from the vast biological wastelands called lawns that are now the major crop in our suburban counties; they provide neither food nor shelter for any of the birds we would like to protect.

Chandler S. Robbins

PREFACE

I was first made aware of breeding bird atlases by a talk given by Massachusetts breeding bird atlas coordinator Deborah Howard in June 1975 at a bird conference in Vermont. A year later my father and I visited my first atlas block in the steep heavily wooded hills of east-central Vermont during the pilot year of the first Vermont bird atlas. Atlas surveying encouraged my father and me to explore new territory and was a crash course on bird nesting behavior. We made discoveries large and small during each day of atlas work. For instance, we found many more Louisiana Waterthrushes and Mourning Warblers than we had expected. Atlas surveying was fascinating and fun. Three decades later I have worked on my fifth atlas project and I still enjoy atlasing.

I had the pleasure of coordinating the second statewide breeding bird atlas of Maryland and the District of Columbia over the past eight years (2002–2009). This role required much more responsibility of me than did my prior atlas experiences. But the talents and steadfastness of the people who worked with me on this project allowed me to delegate many responsibilities, and I am humbled and proud to have worked with them. An atlas project is ultimately powered by talented volunteers, and this project was no exception. The bird enthusiasts of the Maryland Ornithological Society, the staff and leadership of the Wildlife and Heritage Division of the Maryland Department of Natural Resources, and the staff of the Biological Resources Division of the U.S. Geological Survey at the Patuxent Wildlife Research Center were all instrumental in making this atlas project a success.

Nearly a thousand observers, largely under the leadership of the county coordinators, compiled more than 190,000 records of 206 bird species in 1,284 atlas blocks. These numbers are impressive, but the maps in the species accounts that form the core of this book are even more impressive for the breadth of information conveyed about the nesting birds of Maryland and DC. In the process of compiling these data, we met many landowners who showed an interest in and concern for the wildlife on their property. They were generous in giving us permission to survey their land and sharing their wealth of knowledge about the birds that lived there. These data will provide an invaluable store of information for land use planners, environmental agencies, bird researchers, conservationists, and adventurous birders looking for new places to explore.

Although several bird species have increased their ranges since the 1980s and many others have maintained the same distribution, far too many have lost ground. Some species, such as Northern Bobwhite and Barn Owl, have declined precipitously. Two new breeding species were added to the Maryland tally: Common Mergansers proved surprisingly well established along the Potomac and Monocacy rivers, and summering Ruddy Ducks finally produced well-documented offspring at two locations. Once again we have proved that bird ranges are dynamic entities reflecting environmental change. As important agents of environmental modification, witting and unwitting, humans need to assess how birds and other organisms are reacting to such ongoing change. The data we present here should be more than sufficient to convince future generations to repeat atlas projects often. Let us all fervently hope that those future generations will continue to find as diverse an array of nesting birds as this project's participants had the good fortune to document.

ACKNOWLEDGMENTS

Successful completion of the *Second Breeding Bird Atlas of Maryland and the District of Columbia* was the result of close collaboration by state, federal and county employees, private individuals, and the various conservation organizations within our borders.

The Atlas Committee benefited greatly from the collective experience acquired during the original Maryland-DC Atlas Project in 1983–1987. Chair Jane Coskren and the majority of her committee had served on the original advisory board. Committee members who served throughout the second atlas period were Jane Coskren, Walter G. Ellison, Janet Millenson, Karen Morley, Robert F. Ringler, Chandler S. Robbins, Joanne K. Solem, Glenn D. Therres, and Mark Wimer. Others who served initially were Marcia Balestri, Mark Johnson, Kathy Klimkiewicz, and John Malcolm.

The Maryland Ornithological Society (MOS) was again the primary sponsor of the Atlas Project. MOS provided funding through its annual budget, special fund-raising events at annual conferences, and matching grants. Financial contributions also came from MOS chapters and individual members. The MOS presidents who served during the fieldwork and publication preparation gave their full support to the project: Karen Morley, Paul Zucker, Janet Millenson, Marcia Balestri, and Wayne Bell.

The Maryland Department of Natural Resources (DNR) was the other driving force and the project's primary financial supporter. DNR contributed $225,000 during the life of the project. The money was State Wildlife Grant funds provided by the federal government to the states for wildlife diversity conservation, including birds. The expertise of DNR personnel was also valuable in obtaining and analyzing information on many bird species.

Data collection, analysis, and presentation were greatly facilitated by the Biological Resources Division of the U.S. Geological Survey (USGS) at the Patuxent Wildlife Research Center (PWRC), which used the second Maryland-DC Atlas Project as a model for establishing a nationwide atlas network to monitor changes in bird distribution. Data management provided by USGS PWRC was supported financially by the National Biological Information Infrastructure. Mark Wimer, Bruce Peterjohn, and assistants Anna Ott, Naoko Griffin, Jessica Sushinsky, and Allison Sussman were extremely helpful in developing a system that allowed atlas observers to enter and edit data. The same USGS staffers worked diligently with the Atlas Committee to develop maps for this publication that would effectively display key information such as changes in distribution, both statewide and among physiographic regions, and atlas status. Keith Pardieck graciously helped us contact Breeding Bird Survey (BBS) observers to seek their data contributions to the atlas. John Sauer provided invaluable aid and guidance in producing the Maryland BBS graphs accompanying the species accounts. Danny Bystrak provided the file of miniroute observations from the first Maryland-DC Atlas Project along with valuable advice on planning and running miniroutes. Jane Fallon contributed her expertise to produce the relative abundance maps derived from the miniroute data.

County coordinators were essential for recruiting, training, and encouraging field-workers, directing them to particular habitats and undervisited areas, retrieving and editing observations, and assisting with data entry. They are listed here by county: Allegany—Ray Kiddy; Anne Arundel—Sue Ricciardi; Baltimore—Elliot Kirschbaum, Debbie Terry, Paul Kreiss, and Scott Crabtree; Calvert—Arlene Ripley; Caroline—Wanda Cole; Carroll—Robert F. Ringler; Cecil—Laura Balascio and Marcia Watson; Charles—George Jett and George Wilmot; District of Columbia—Rob Hilton; Dorchester—Lynn Davidson; Frederick—David R. Smith; Garrett—Fran Pope; Harford—Dennis Kirkwood and Bill Russell; Howard—Joanne K. Solem and Bonnie Ott; Kent— Wayne Bell; Montgomery—Jim Green; Prince George's— Fred Fallon; Queen Anne's—Glenn Therres; Somerset— Charles Vaughn; St. Mary's—Kyle Rambo and Patty Craig; Talbot—Stan Arnold and Amy Bourque; Washington—Anna Hutzell and Judith Lilga; Wicomico—Don and Carol Broderick; Worcester—Mark Hoffman.

The relative abundance surveys, or miniroutes, were organized by Nancy Martin, who planned coverage, recruited observers, and entered observations into the database. She also provided invaluable assistance to the atlas coordinator throughout the project, especially as material was prepared for publication. Robert Solem took on a large portion of miniroute data entry and also made available quarterblock records from the first two Howard County atlas projects.

Species account authors researched each species, explained the history of the species in the state, and provided a narrative account. The account authors are Gwenda L. Brewer, David F. Brinker, T. Dennis Coskren, Lynn M. Davidson, Deanna K. Dawson, Walter G. Ellison, James M. Mc-Cann, Robert F. Ringler, and Glenn D. Therres. Color photographs add visual interest, beauty, and whimsy to species accounts and habitat descriptions. The MOS held a photo contest and encouraged photographers of all skill levels to contribute to the effort. Bill Ellis and Les Eastman gave generously of their time and skill to facilitate the complex judging process. The photographers are Robert Ampula, Stan Arnold, Scott Berglund, Melissa Boyle, David Brenneman, Jim Brighton, David Brinker, Philip Brody, Steve Collins, T. Dennis Coskren, Ralph Cullison III, Walter Ellison, Monroe Harden, Mark Hoffman, Bill Hubick, George M. Jett, Craig Koppie, John Landers, Charles Lentz, Rick Mandelson, Nancy Martin, Sean McCandless, Karen Morley, Danny Poet, Evelyn Ralston, Fran Saunders, Kurt Schwarz, Bill Sherman, Eric Skrzypczak, Gary Smyle, June Tveekrem, Gary Van Velsir, and David Ziolkowski. Graphics support for atlas logo items came from Sue Probst.

Financial support came from a range of organizations and individuals. The Maryland Department of Natural Resources and the Maryland Coastal Bays Program supplied generous grants. Other contributions came from individuals, MOS chapters, and organizations that share a concern for birds and conservation.

DONORS: PEREGRINE LEVEL

Anne Arundel Bird Club
Anonymous
Audubon Naturalist Society
Baltimore Bird Club
Jane Coskren and T. Dennis Coskren
Frederick Bird Club
Robert Hilton Jr.
Howard County Bird Club
Maryland Coastal Bays Program
Maryland Department of Natural Resources
Elliott J. Millenson
Montgomery Bird Club
Fran Pope
Chandler Robbins, in memory of Eleanor Robbins
SAIC, Inc.
Joanne and Robert Solem

DONORS: MERLIN LEVEL

Allegany & Garrett Counties Bird Club
Henry T. Armistead
Stan Arnold
Dennis and Linda Baker
Harry and Maud Banks
Dr. and Mrs. Robert Batchelor
Gwen Brewer
Anne Brooks
Catherine Carroll
Martha Chestem
John Davidson
Lynn Davidson
Suzonne Davidson
Joan DeCarli
Carol Ghebelian
Nelse Greenway
Harford Bird Club
William H. Hildebrandt and Mabel A. Quinto
Emmalyn Holdridge
Anna and Doug Hutzell
George Jett
Kent County Bird Club
Debi and Mike Klein
Elise and Paul Kreiss
Ellen M. Lawler
Annette and James Livengood, in honor of Fran Pope
Ann Lucy
Nancy Magnusson
Marietta Community Foundation, in memory of Rick Blom
Howard McIntyre
Elayne and Jeff Metter
Paul O'Brien
Helen Patton
Patuxent Bird Club
Ron and Susan Polniaszek
Susannah F. Prindle
Chandler Robbins, in memory of Kathy Klimkiewicz
James and Anne Robinson Foundation
Gertrude and Joseph Rogers
Fran and Norm Saunders
Kurt Schwarz
John and Pamela Smith
Gary Smyle
Joanne and Robert Solem, in memory of Rosamond Munro
Eva Sunell
Tri-County Bird Club
Washington County Bird Club
Frank G. Witebsky

DONORS: KESTREL LEVEL

Anonymous (3)
Richard Adams
Dianne Aguilera
Frances and George Alderson
Joyce and Wayne Bell
Lisa Bellamy
Dana and Scott Berglund
Mary-Jo Betts
Monika Botsai
D. H. Michael and Joy Bowen
L. Jeanne Bowman
Fred and Sarah Bradbury
Tom and Sharon Bradford
Carol and Don Broderick
Joseph Byrnes
Eileen Clegg

Edward and Anne Clucas
Renie Collmus
Cliff Comeau
Barry Cooper
John and Adelina Cornette
Patty Craig
Frances Curnow
Julie Daniel
Nancy Dunn
Walter Ellison
Mark England
Helen and John Ford
Shirley Ford
Sam Freiberg
Jeff Friedhoffer
Jean and Larry Fry
Helene Gardel
Garrett County Memorial
    Hospital, in honor of
    Fran Pope
Shirley Geddes
Kay Gibbons
Deborah and Keith Gingrich
David and Helen Gray
Jim Green
Jane and Joseph Hanfman
David Harvey
Maureen Harvey
Susan Henyon
Charles and Janet Hesler
Barbara Holloman
Kim Hudyma
Betty and Frank Hughes
Parke John
Carey and Beverly Johnston
Kay and Vince Jones
Mary and Robert Keedy
Lynn Kenney
Margie King
Jeanne Lacerte
Tim Larney
Candita and Rick Lee
Jeanne Leroy
Victor A. Levi
John and Judith Lilga
Susan Lin
Cyndie Loeper
Tom Loomis
Grace and Hugh Mahanes
John Malcolm
Andy Martin
Jule McCartney
Daniel and Georgia
    McDonald

Janet Millenson
Mike and Laurie Miracle
Gorham Miscall
Ann Mitchell
John and Margie Mogarvero
Karen Morley, in memory
    of Kathy Klimkiewicz
Harvey and Marion Mudd
Lou Nielsen
Elizabeth Nobbe
Wendy Olsson
Doug and Nancy Parker
Michael Parr
Michael Pearl
Mark Pellerin
Betsy Perlman
Evelyn Phucas
Paul Pisano
Elizabeth Pitney
Marie Plante
David Powell
Wafi Rains
Jim Rapp
Sue Ricciardi
George and Grace-Louise
    Rickard
Bob Rineer
Arlene Ripley
James Ripley
Laurie Robertson
Johnna Robinson
Brian and Sara Rollfinke
Les Roslund
Barbara Ross
Sally Rowe
Ken Sander
Lydia Schindler
Douglas Schmenner
Robert Schreiber
Janet Shields
Don Simonson
Alan and Carol Sleeper
Joanne and Robert Solem,
    in memory of Kathy
    Klimkiewicz
Southern Maryland
    Audubon Society
Charles R. Stirrat
Nancy and Rick Sussman
Talbot Bird Club
Todd Taylor, Bollard Yachts
Patricia Tice
Kathy Tinius
June Tveekrem

Pat Valdata
Charles Vaughn
Sherrye Walker
David Wallace
Mike Walsh
Marcia Watson
Sandy Webb

Hal Wierenga
John Williamson
Erika Wilson
Ann Wing
Martha Witebsky
Patricia Wood
Paul and Sherry Zucker

Nearly a thousand individuals devoted many hours to collecting essential field data. Without these dedicated fieldworkers and other volunteers, the Atlas Project would not have been possible. People whose names are followed by an asterisk (*) also conducted miniroute surveys.

ATLAS VOLUNTEERS

George Adams
Phil Adkins
Andy Aguilera
Dianne Aguilera
Terry Allen
Jaime Alvarez
Frank Ammer
Laura Anderson
Brad Andres
Justin Armetta
Henry T. Armistead*
Elaine Arnold
Stan Arnold*
Jean Artes
Fred Atwood
Bob Augustine
Linda Ault
Glenn Austin
Zach Baer*
Shirley Bailey
Linda Baker
Laura Balascio
Neil Baldacchino
Erin Baldwin
Marcia Balestri*
Lisa Balmert
Tom Bancroft
Tony Barbour
Chris Barnard
Miles Barnard
Bonnie Barnes
Marty Barron
Polly Batchelder
Robin Batcheller
Denise Bayusik
Tom Beal

Brenda Belensky
Tyler Bell*
Wayne Bell*
Nancy Bellaire
Debby Bennett
Janice Bennett
Tom Benton
Doris Berger
Peter Bergstrom
Lance Biechele
Denny Bingamin
Anne Bishop
J. Catherine Bishop
Alex Bitzel
Ray Bivens
John Bjerke
Debbie Blake
Peter Blank
Rosemary Bliss
Caroline Blizzard
Rick Blom
Bill Blum
Karan Blum
Ken Board
David Bohaska
Paula Bohaska
Craig Bolyard
Dan Boone
Jeff Bossart
June Bourdat
Orley Bourland
Peg Bourland
Amy Bourque
D. H. Michael Bowen*
Deborah Bowers
Bob Boxwell

Ginger Boyce
Ed Boyd
Ron Boyer
W. H. Boyer
Alicia Boyers
George Boyers
Frank Boyle
Melissa Boyle
Tom Bradford
Sharon Bradford
Bill Braerman
Hugh Brandenburg
Mike Braun
Martin Brazeau
Robert Brennan
Gwen Brewer*
Bill Bridgeland
Jim Brighton
David Brinker
Jim Brinkley
Marilyn Brinkley
Christina Brinster
Jim Britt-Baker
Sheryl Britt-Baker
Carol Broderick
Don Broderick
Keith Broderick
Cathy Brodo
Doris Brody
Ed Bromley
Anne Brooks
Andy Brown
Cathy Brown
Kathy Brown
Roy Brown
Sarah Brown
Mike Bryan
Shelly Buhlman
Andrew Bullen
Elaine Bullock
Robert Burch
Mike Burchett
Judy Burdette
Rod Burley
Susan Buswell
Betty Butler
Bryce Butler
Brent Byers
Mary Byers
Lori Byrne
Joseph Byrnes
Danny Bystrak*
Paul Bystrak
Kevin Caldwell*

Mike Callahan
Kathy Calvert*
Brian Campbell
Paul Canner
Steve Cardano
Vicki Carlson
Pat Caro
Catherine Carroll
Kyle Carstensen
Larry Cartwright
Linda Cashman
Annalea Catulle
Scott Catulle
Centreville Middle School
George Chase
Chesapeake Marshlands
    NWR Complex
Martha Chestem
Frank Chetelat
Mary Chetelat
Teri Christensen
J. B. Churchill*
Bob Churi
Carol Churi
Ed Clark
Eileen Clegg
Ken Clements
Charlotte Clive
Lisa Colangelo
Joyce Colbert
Judy Cole
Ken Cole
Wanda Cole*
Steve Collins
Kathy Colston
Thomas Congersky
Andy Cooper
Barry Cooper
Donald Cooper
Jane Coskren
T. Dennis Coskren
Keith E. Costley*
Meg Cowenhoven
Josephine Cox
Scott Crabtree
Patty Craig*
Randee Craig
Jack Cremeans
Marty Cribb
Steve Croker
Virginia Croker
C. Crout
Steve Crout
Jean Crump

Ruth Culbertson
Jeff Culler
Ralph Cullison III
David Curson*
Judy Dance
Julie Daniel
Linda Daniel
Jamie Darcy
Karen Darcy
Lonnie Darr
Lynn Davidson*
Barbara Davis
Lynn Davis
Michael Davis
Phil Davis
Donna Daviss
Deanna Dawson
Frances Dawson
Lou DeMouy
Eric Decker
Patsy Decker
Hal Delaplane
Richard Demmitt
John Dennehy
Bob Depuy
Jan Depuy
Deidre DeRoia
Richard DeRycke
Curtis Dew
Tina Dew
Allen Deward
Jim Dewing
Janelle Dietrich
John DiTomaso
Robert Dixon
Kevin Dodge*
Marian Dodson
Chris Dominick
Mike Donovan
Tim Dowd
Barbara Dowell*
Cynthia Downs
Ken Drier
Cindy Driscoll
Sam Droege*
John Drummond*
Billie Durham
Dyke Marsh Breeding Bird
    Survey
Sam Dyke
Carl Dyson
Georgia Eacker
Mark Eanes
Susan Earp

Wesley Earp
Les Eastman
Jane Ebert
Ward Ebert
Darius Ecker
Scott Edie
Jane Edwards
Tom Edwards
Jeff Effinger
Coleman Eldridge
Bill Ellis
Walter G. Ellison*
Mark England*
Courtney Englar
Kirsten Enzinger
Tracy Eve
Erin Eve
Fred Fallon
Jane Fallon
Cheryl Farfaras
Mark Farfaras
David Farner
Chris Feaga
Tom Feild
Jim Felley
Valerie Fellows
Bill Fiege
Tray Fiege
Bruce Field
John Field
Linda Field
John Finerty
Harry Fink
Chuck Finley
Emanuel Fisher
Emma Fisher
Nolley Fisher
Reuben Fisher
Shalom Fisher
Vernie Fisher
Kathy Fleming
Jerald J. Fletcher
Carol Flora
Jim Flynn
Charlotte Folk
Robert Foor
B. Forbus
Betsy Ford
Diane Ford
Shirley Ford
Steve Ford
Doug Forsell
Susan Fowler
John Fram

Franciscan Friary
Adriana Frangos
Gail Frantz
Jeff Friedhoffer
Linda Friedland
Norman Friedman
Alexa Lee Fry
Jean Fry
Larry Fry
Tammy Fuehrer
Lynette Fullerton
Mary Beth Furst
Tony Futcher
Barbara Gaffney
George Gaffney
Helene Gardel
Rod Gardner
Lynda Garrett
Kurt Gaskill
Donna Gates
Ed Gates
Barbara Gearhart*
Dale Gearhart
Erma Gebb
Shirley Geddes
George Gee
Aelred Geis
Zach Gent
Ned Gerber
Ross Geredien*
Jane Geuder
Ralph Geuder
Carol Ghebelian
Kay Gibbons
Denise Gibbs
Rob Gibbs
Thomas Gibson
Doug Gill
Sue Gleichauf
Ricardo Gonzalez
Hilmar Gottestall
Greg Gough
Kevin Graff
Katherine Grandine
David Gray
Helen Gray
Jim Green
Nancy Green
Matthew Grey
Todd Gribling
Beth Griffin
Dave Griffin*
Naoko Griffin
Phyllis Grimm

Trish Gross
John Groutt
Jim Gruber
Shiras Guion
Donna Gur
Dot Gustafson
Mary Gustafson
J. Guyader
Matt Hafner
Paul Hagen
Charles Hager
Kate Halla
Joe Halloran
Sue Hamilton*
Jane Hanfman
Joseph Hanfman
Erika Hanner
Hugh Hanson
Michael Haramis
Monroe Harden
Jocelyn Harding
Tyler Harding
George Harrington
Clive Harris
Karen Harris
Randy Harrison
Tom Harten
Bill Harvey
David Harvey
Maureen Harvey
Michael Harvey
Allan Haury*
Kevin Hazlett
Gene Healy
Richard Hearn
Glen Hedelson
Kevin Heffernan
Jane Heim
Leland Hein
David Hemphill
Elaine Hendricks*
John Herder
Susan Herder
Donna Hershberger
Wil Hershberger
Norman Hershey
Charles Hesler
Janet Hesler
Robin Hessey
Bill Hill
Gayle Hill
Susan Hill
Dirk D. Hillegass
Steven Hillyer

Rob Hilton*
Larry Hindman
Steve Hinebaugh
Amy Hoffman
Mark Hoffman*
Deanna Hofmann
Hans Holbrook
Robert Holbrook
Emmalyn Holdridge
Linda Holley
Lynn Holley
Don Hollway
David W. Holmes*
Aaron Holochwost
Angela Holocker
Lenore Holt
Rob Holtz
Dave Holyoke
Judy Holzman
Phil Hoover
John Horger
Edward Horner
Helen Horrocks
Marshall Howe
John Hubbell
Bill Hubick*
Dave Hudgins
Mary Huebner
Stephen Hughes
Sara Hulbert
Stephen Hult
Jay Hurry
Anna Hutzell
Doug Hutzell
Marshall Iliff
Hillar Ilves
Connie Imler
Leroy Imler
E. B. James
Gayle Jayne
Carol Jelich
Simone Jenion
Kye Jenkins*
Barbara Jensen
Kristopher Jensen
George Jett*
Don Jewell
Parke John
Diane Johns
Anne Johnson
Dale Johnson
Heather Johnson
Jen Johnson
Jeremy Johnson

Mark S. Johnson
Phil Jones
Richard Jones
Sharon Jones
Terry Jordan
Jennifer Jowdy
Emily Joyce
Paul Jung
Fran Juriga
John Juriga
Athena Kahler
Harry Kahler
Romona Kasdan
Tina Katsampis
Lois Kauffman
Sylvan Kaufman
Dave Keane
Robert Keedy
Cherry Keller
Linda Keller
Wendy Keller
Robin Kendall
Lynn Kenney
Jessica Kerns-McClelland
Ray Kiddy
David Kidwell
Caitlin Kight
Margie King
Dennis Kirkwood
Elliot Kirschbaum
Nancy Kirschbaum
Val Kitchens
Greg Klein
Susan Klein
Kathy Klimkiewicz*
Janet Klingamen
Molly Knazek
Cathy Knight
Penny Knobel-Besa
Tom Knot
Beatrice Kondo
Jane Kostenko
Russell Kovach
Kimberly Kraeer
Shirley Kramer
Walter Kraus
Wen Kreider
Elise Kreiss*
Paul Kreiss*
Millie Kriemelmeyer
David Kubitsky
Charlie Kucera
Bill Kulp
Kathie Lambert

John Landers
Rhyan Lange
Brad Lanning
Kathleen Lathrop
Margaret Laughlin
Ellen M. Lawler
Candita Lee
Lauri Lee
Rick Lee
Shashi Lengade
Bronnen Lerner
Jeanne Leroy
Henry Leskinen
John Leskinen
Mike Leumas
Peter Lev
Victor A. Levi
Nate Levy
Frank Lewis
Lloyd Lewis
Larry Lewman
Nancy Lewman
Judith Lilga*
Letty Limbach
Dana Limpert
Roland Limpert
Delores Line
Larry Line
Bill Link
Suzanne Lisiewski
Doug Lister
Lisa Lister
Meta Little*
Kathy Litzinger
Cyndie Loeper
Rob Long
Sue Lorentz
John Lorenz
Glen Lovelace III
Felicia Lovelett
Woody Luber
Rosa Lubitz
Ann Lucy
Bob Lukinic
Brigitte Lund
Mikey Lutmerding*
Janet Lydon
Kevin Mack
Gail Mackiernan
Hugh Mahanes
John Maloney
Lauren Maloney
Peter Mann

Thomas Margevicius
Kathy Mariano
Nacole Marquess
Peter Marra
Andy Martin*
Elwood Martin
Joel Martin
Nancy L. Martin*
Peter Martin
Jeff Marx
Paul Mauss
Julie Maynard*
Colin McAllister
Maureen McAllister
Nancy McAllister
Sean McCandless
Jim McCann
Jane McConnel
Carol McDaniel
Joe McDaniel
Georgia McDonald*
Karen McDonald
Liz McDowell
Anne McEvoy
Frank McGilvrey
Howard McIntyre
Janet McKegg
John McKitterick
Dolores McLean
Taylor McLean
George McNabb
Diane Mecham
Don Messersmith
Elayne Metter
Jeff Metter
Helen Metzman
James Meyers
John Milbourne
Sarah Milbourne
Janet Millenson
Martin Miller
Mary Miller
Thomas W. Miller
Ann Mitchell
Lynn Moore
James Morrin
Georgia Morris
Brian Morrison
Mark Moxley
Mr. and Mrs. Moxley Jr.
Brian Moyer
David Mozurkewich
Rodney Mulchansingh

Bob Mumford
Dotty Mumford
Gil Myers
Cecily Nabors*
Lou Nabors
Diane Nagengast
Ian Nagengast
Jean Neely
Neighborhood Nestwatch
  Program at the
  Smithsonian Migratory
  Bird Center at the
  National Zoo
Gary Nelson
Lyn Nelson
Skip Nelson
Sue Neri
Br. Brian Newbigging
Cheryl Newcomb
Bea Newkirk
Carol Newman
Dean Newman
Kathy Newman
Melissa Newsome
Wayne Newsome
Paul Nicholson
Lou Nielsen
Nancy Elizabeth Nimmich
Paul Nistico
Randy Nixon
Paul Noell
Phil Norman
Steve Noyes
Marianna Nuttle
Trisha Nyland
Holliday Obrecht
Paul O'Brien
Doug Odermatt
Dave Oktavec
Amo Oliverio
Beth Olsen
Daryl Olson
Chris Ordiway
Richard Orr
Peter Osenton
Anna Ott
Bonnie Ott*
Tammy Otto
Carol Pace
Jay Pape
Keith Pardieck
David Pardoe
Elaine Pardoe

Tom Pardoe
Doug Parker
Jason Parker
Nancy Parker
Dipal Patel
David Patrick
James Patrick
Oliver Pattee
Edward Patten
Michael Patterson
Helen Patton
Patuxent Wildlife Research
  Center Volunteers
John Paul
Mary Paul
Karla Pearce
Patsy Perlman
Tony Perlman
T. Perrell
Dave Perry*
Sherry Peruzzi
Bruce Peterjohn
James Peters
Katharine Peters
Susan Peterson
William Pfingsten
Susan Phipps
Anita Picco
Jane Picot
Mary Piotrowski
Paul Pisano
Renee Piskor
Elizabeth Pitney
Bart Pitts
Danny Poet
Dawn Poholsky
Ron Polniaszek
Susan Polniaszek
Madonna Pool
Bill Pope
Fran Pope*
Edward Potter
David Powell
Frank Powers
Phil Powers
Megan Preston
Michael Price
Michelle Price
Niles Primrose
Suzanne Probst
Suzanne Procell
Diann Prosser
Nick Pulcinella

Sharon Pulcinella
John Putnam
Mike Quinlan*
Mark Raab
George Radcliffe
Gemma Radko*
Kyle Rambo*
Jim Rapp
Joanna Rawlings
Tim Ray
Joe Razes
Dave Rea
Joyce Rea
Karen Readel
Sherald Reagle
Donald Reed
Betsy Reeder*
Jan Reese
Kim Reichert
Bobbi Reichwein
Virginia Reynolds
Dusty Rhodes
Frank Rhodes Sr.
Robyn Rhudy
Sue Ricciardi*
Emily Rice
Derek Richardson
Kate Ricks
Bob Rineer
Robert F. Ringler*
Arlene Ripley*
John Rizzello
Julie Rizzello
Chandler Robbins*
Eleanor Robbins
Jane Robbins
Randy Robertson
Clyde Robinette
Anne Robinson
Ray Robley
Arthur Rogers
Brian Rollfinke
Les Roslund
Barbara Ross
Katherine Rowe
Mike Rudy
Ron Runkles
Luann Rushing
Bill Russell
Denise Ryan
Jack Saba
Ellen Salmon
Steve Sanford

Joanna Santarpia
John Sauer
Fran Saunders
Norm Saunders
Mark Scallion
Eugene Scarpulla
Shannon Schade
Robert Schaefer
Mark Schilling
Lydia Schindler
Bill Schreitz
Margaret Schultz
Bob Schutsky
Tammy Schwaab
Kurt Schwarz
Sharon Schwemmer
Don Schwikert
Bill Scudder
Harry Sears
Fred Shaffer
Lisa Shannon*
Ann Sharp
Bob Sharp
Chuck Sharp
Denise Sharp
Martha Shaum
Steve Sheffield
Jeff Shenot*
Jay Sheppard
Janet Shields
Rick Shilling
Craig Sholley
L.T. Short
Ann Sidor
Hugh Simmons
Teresa Simons
Ed Sipes
Mary Sipes
Susan Sires
Sisters of Bon Secours
Connie Skipper*
Adam Smith
David R. Smith*
Dawn Smith
John Smith
Kevin Smith
Lori Smith
Michael Smith
Pamela Smith
Philip Smith
Romayne Smith
Gary Smyle*
Chris Snow

Jean Snyder
Stephanie Snyder
Mary Sokol
Joanne Solem
Robert Solem
Duvall Sollers
Emily Solomon
Don Soubie
Dan Southworth
Linda Southworth
Ed Soutiere
Amanda Spears
Paul Speyser
Pat Sporn
Eric Sprague
Andy Sprenger
Bobbie Stadler
Randy Stadler
Mark Staley
Pam Stanitski
Rob Stansbury
Hank Stanton
Jennifer Stanton
Chris Starling
Leslie Starr
Jim Stasz
Nancy Steen
Richard Steffan Jr.
Charles Stegman
Wade Stephen
Barbara Stephens
Jim Stephens
James Stevenson
Karen Stewart
Michelle Stewart
Barry Stimmel
Charles R. Stirrat
Ralph Stokes
Roger Stone
Scott Streib
Raymond Strickroth
Tom Strikwerda
Carolyn Sturtevant
Sigrid Stiles
Susan Sullivan
August Sunell
Eva Sunell*
Allison Sussman
Rick Sussman
Albert Swann
Chris Swarth
John Swartz
Donald Sweig

Byron Swift*
Jim Swift
Brian Sykes
Nita Sylvester
Nancy Szlasa
Jerry Tarbell
Laura Tarbell
Pat Tate
David Taylor
Evelyn Taylor
John P. Taylor
John W. Taylor
Marilyn Taylor
Randy Taylor
Deborah Terry*
Sallie Thayer
Glenn Therres*
George Thomas
Amy Thompson
David Thorndill
Cindy Thornton
Kevin Thorpe
Robin Todd*
Thomas Trafton
Ashley Traut
Jen Traver
Todd Treichel
George Tsiourmar
Helen Tuel
John Tuel
Kate Tufts
Craig Tumer
Ken Tyson
Kermit Updegrove
Anna Urciolo
Bill Urspruch
Judy Van Dyke
Carol Vangrin
Joe Vangrin
Gary Van Velsir
Kate Vaugh
Charles Vaughn
Susan Venturella
Steve Vincent
Richard Vine
Kellie Vlahos
Van Vogel
David Walbeck*
Beverly Walker
Ginny Walker
Jake Walker*
Warren Walker
David Wallace*

Mark Wallace
Rose Wallace
Mike Walsh
Kyle Warfield
Paula Warner
Washington College Center
    for Environment and
    Society
Marcia Watson*
Donald Waugh
Martha Waugh
David Webb
Peter Webb
Colleen Webster
Ben Wechsler
Sally Wechsler
Dave Weesner*
Linda Weir
Matilda Weiss
Mike Welch

Mr. and Mrs. James Welling
Bill Wells
Steve Westre
Jean Wheeler
Joy G. Wheeler
Jessica White
Lisa Whitman
Hal Wierenga
Joanne Wilbur
Carol Wilkinson
Jim Wilkinson
Guy Willey
Levin Willey
Dwight Williams
John Williamson
Terry Willis
Ernest Willoughby
George Wilm
Dave Wilmot
George Wilmot

Becky Wilson
Dave Wilson
Mark Wimer
Bob Wolf
Patty Wolf
Betty Wolfe
Patricia Wood
Shirley Wood
Bruce Woodford
Paul Woodford
Jean Woods
Kathy Woods
Loretta Woods
Jean Worthley
Martha Wright
Michele Wright
Tommy Wright
Karin Wuertz-Schaefer
Carol Yates
Roxann Yeager

Barbara Yoder
Irene Yoder
Samuel Yoder
Bill Young
Dave Young
Orrey Young
Howard Youth
Joseph Zadjura
Beth Zang*
Helen Zeichner
Peter Zerhusen
Donna Zile
David J. Ziolkowski Jr.*
Harry Zoller
Larry Zoller
Paul Zucker*
Sherry Zucker

ACRONYMS AND ABBREVIATIONS

ac	acre(s)
AOU	American Ornithologists' Union
BBA	breeding bird atlas
BBS	Breeding Bird Survey
C	Celsius
cm	centimeter(s)
CRP	Conservation Reserve Program
DC	District of Columbia; Washington, DC
DDT	dichlorodiphenyltrichloroethane
DNR	Maryland Department of Natural Resources
DOD	Department of Defense
F	Fahrenheit
ft	feet
ha	hectare(s)
IBA	Important Bird Areas
in	inches
km	kilometer(s)
m	meter(s)
mi	mile(s)
MNRF	Maryland Nest Record File
MOS	Maryland Ornithological Society
NPS	National Park Service
NWR	national wildlife refuge
PWRC	Patuxent Wildlife Research Center
sq km	square kilometer(s)
sq mi	square mile(s)
USFWS	U.S. Fish and Wildlife Service
USGS	U.S. Geological Survey
UTM	Universal Transverse Mercator
WMA	wildlife management area
WWTP	waste water treatment plant
yd	yard(s)

2nd ATLAS *of the* BREEDING BIRDS *of* MARYLAND *and the* DISTRICT *of* COLUMBIA

Introduction

THE FIRST BREEDING BIRD ATLAS OF MARYLAND AND DC

In 1996 the University of Pittsburgh Press published Robbins and Blom's *Atlas of the Breeding Birds of Maryland and the District of Columbia*, which was the result of an intensive survey of the distribution and relative abundance of all species of wild birds nesting in Maryland and the adjacent District of Columbia in 1983–1987. The distribution of each of the 209 species was mapped on a grid of atlas "blocks," each measuring approximately 5 × 5 km (3.1 × 3.1 mi) and representing one-sixth of a 7.5-minute USGS topographic map. Each of the 1,256 blocks was visited multiple times during the five-year period by one or more observers in an effort to detect all nesting species and to upgrade its status from "possible" or "probable" to "confirmed." This was the first attempt in Maryland to map bird distribution on a fine scale throughout the state. It required the dedicated work of 797 field-workers and numerous authors, cartographers, and editors.

In addition to a current distribution map for each species, there was a map showing the location of observations in 1958, a map showing relative abundance from 15-stop miniroutes in 1983–1989, a summary of the number and percentage of blocks for each status category, and a trend graph showing changing abundance in Maryland from 1966 to 1989 from the North American Breeding Bird Survey. The text described status and nesting requirements in DC and throughout the state, using the extensive MOS nest record file to summarize nest locations and contents and provide Maryland dates of occurrence for each species. Each species was illustrated by an original pen-and-ink drawing created specifically for the book.

THE PURPOSES OF THE SECOND ATLAS PROJECT

Recognizing that many species of native birds no longer nested in parts of Maryland where they had been breeding in the mid-1980s, Maryland's Atlas Committee, together with committees from Pennsylvania, New York, Massachusetts, and Ontario, decided it would be important to repeat their original breeding bird atlases early in the twenty-first century to simultaneously document the changes over a wide geographic scale.

By documenting the changes early and determining the reasons for them, committee members hoped remedial action might be taken before it was too late. Many environmental changes that affect bird populations are taking place simultaneously, including natural habitat succession; fragmentation of formerly extensive forests and fields; an increase in human population, land values, vehicular traffic, and the number and height of communication towers and wind turbines; nest parasitism by cowbirds; predation from house cats; habitat degradation from deer and invasive exotics; and, of course, global climate change.

Unlike the North American Breeding Bird Survey, which detects bird population changes on a state, regional, or continental scale, the atlas work documents changes at the neighborhood level, changes that affect the individual. All of us live and work and spend our leisure time in atlas blocks.

The History of Grid-based Ornithological Atlases

PRE-ATLAS

The concept of mapping bird distribution dates back to January 1928, when Jane S. Elliott of the U.S. Bureau of Biological Survey began the daunting task of mapping the breeding and winter distribution of every North American bird species that breeds north of Mexico. Using the millions of published and unpublished records from Dr. C. Hart Merriam's bird distribution and migration card file that dates back to the 1880s, Mrs. Elliott, who was a member of the American Ornithologists' Union (AOU) and an accomplished artist, began with the grebes and meticulously lettered on a large map every locality for which she found nesting or wintering evidence. She continued updating these maps until her retirement in 1939. These maps were the chief source of information for the range descriptions in Bent's *Life Histories of North American Birds* and for the ranges given in various issues of the *AOU Check-list of North American Birds*. Earl Godfrey, at the National Museum of Canada, maintained a similar series of maps showing the distribution of birds within Canada, and used these to produce maps for nearly all species in *The Birds of Canada* (Godfrey 1966).

The first state bird books to include distribution maps were Florence Merriam Bailey's *Birds of New Mexico* (1928) with 59 distribution maps, Edward Howe Forbush's *Birds of Massachusetts and Other New England States* (1925–1929) with 1 winter and 29 summer maps for Massachusetts, and Robert E. Stewart and Chandler Robbins' *Birds of Maryland and the District of Columbia* (1958) with breeding range maps for 70 species. Robert E. Stewart's (1975) *Breeding Birds of North Dakota*, although not called an atlas, was in fact the first grid-based atlas in North America. His grid was the 9.6 × 9.6–km (6 × 6–mi) township, and from 1960 to 1973 he personally visited every township in North Dakota in search of nesting evidence. That is nearly two thousand 9.6-km blocks. Stewart used four map symbols, solid or open squares for confirmed nesting within or prior to his study period (1950–1972), and solid or open triangles for territorial males during or prior to his study period.

Next on the scene was Palmer David Skaar of Bozeman, inventor of the latilong. Again, he did not use the word atlas, but he set out to map Montana bird distribution by its 47 latitude-longitude blocks, and published the results as *Montana Bird Distribution: Preliminary Mapping by Latilong* (1975). This was followed by similar cooperative efforts in nearby states: *Colorado Bird Distribution Latilong Study* (28 blocks) edited by Hugh Kingery and Walter Graul (1978), *Wyoming Avian Atlas* (28 blocks) by Bob Oakleaf et al. (1979), *Utah Bird Distribution Latilong Study* (23 blocks) by Robert Walters et al. (1983), and *Idaho Bird Distribution* (27 blocks) by Daniel Stephens and Shirley Sturts (1991).

THE FIRST EUROPEAN ATLASES

The Botanical Society of the British Isles published the first grid-based natural history atlas, the *Atlas of the British Flora* (Perring and Walters 1962), a monumental 10-year effort by 1,500 botanists to map the distribution of about 2,000 species of plants on the 10-km (6.2-mi) national grid. This was the great-grandfather of the hundreds of natural history grid-based atlases that were to follow in the next few decades as the atlas movement swept over the face of the earth.

Scientists at the British Trust for Ornithology (BTO) were intrigued by the idea of a corresponding atlas of breeding birds, but were a bit hesitant about adding such an enormous project to their growing list of active projects for amateurs. Nevertheless, a trial project using the 10-km national grid was undertaken in three central counties of England in 1966–1968, resulting in the publication of *Atlas of Breeding Birds of the West Midlands* (Lord and Munns 1970). As fieldwork neared completion, the BTO collaborated with the Irish Wildbird Conservancy and launched a plan to conduct a five-year atlas for the British Isles in the period 1968–1972 using the same 3,862 10-km blocks as the botanical atlas. Enthusiastic birders in several of the British counties saw great benefits in more intensive atlasing at the county level and soon initiated county atlases at the tetrad (2-km [1.2-mi]) scale. The *Atlas of Breeding Birds in Britain and Ireland* (Sharrock 1976) set the format pattern with its drawing of each bird, life history information, distribution map with symbols for possible, probable, and confirmed records, habitat information from BTO files, a statistical summary, and appropriate references to the literature.

The International Bird Census Committee, meeting in various European countries every year or two starting in 1966, became very interested in atlas methods and results, and adopted standard international terminology and procedures for atlas projects. The Monks Wood Environmental Information Centre at Abbots Ripton, Huntingdon, Cambridgeshire, which became the nerve center for atlas studies of all forms of animal and plant life in Britain, did much to encourage atlas research in Europe and beyond. In 1971 a European Ornithological Atlas Committee was formed to promote atlas projects in other European countries, and in 1976 delegates from 18 countries participated in a census/atlas meeting in Szymbark, Poland. This was the year in which national atlases were published for France (Yeatman) and Denmark (Dybbro). In the next few years, national atlases were published for West Germany (Rheinwald 1977, 1982), Belgium (Lippens and Wille 1972), the Netherlands (Teixeira 1979), Switzerland (Schifferli et al. 1980), Finland (Bergman et al. 1983), Czechoslovakia (Stastny, Randik, and Hudec 1987), Belgium (Devilliers et al. 1988), Continental Portugal (Rufino 1989), Latvia (Priednieks et al. 1989), Sweden (Risberg et al. 1990), East Germany (Nicolai 1993), Austria (Dvorak et al. 1993), all of Germany (Rheinwald 1993), Estonia (Renno 1993), Norway (Gjershaug et al. 1994), Romania (Weber et al. 1994), Southwest Iceland (Skarphédinsson et al. 1994), Slovenia (Geister 1995), and Spain (Purroy 1997). Spain had also published substantial regional atlases for Rioja (de Juana 1980), Navarra (Elosegui 1985), Catalunya and Andorra (Muntaner et al. 1984), and Salamanca (Carnero and Peris 1988), Their Huesca atlas (Woutersen and Platteeuw) was published in 1998. Italy has produced more than 50 regional atlases including Campania 1983–1987 (Fraissinet and Kalby 1989), Provincia di Varese, Lombardia 1983–1986 (Guenzani and Saporetti 1988), and Napoli (Fraissinet and Caputo 1995). All the European nations collaborated in a mammoth project to produce the seven-pound *EBCC Atlas of European Breeding Birds* (Hagemeijer and Blair 1997) in 13 languages using a 50-km (31-mi) grid.

TETRAD ATLASES

A tetrad is a group of four, so a tetrad atlas has a 2-km (1.2-mi) grid, which is only slightly finer than the quarterblock grid (2.5 km [1.5 mi]) used in parts of Maryland. Tetrad atlases in Britain began concurrently with the BTO's 10-km (6.2-mi) *Atlas of the British Isles* (Sharrock 1976). They were undertaken initially in counties that were flush with birdwatchers in order to produce detailed distribution maps. The first British tetrad atlases to appear were those for London (Montier 1977), Bedfordshire (Harding 1979), Kent (D. Taylor et al. 1981), and Hertfordshire (Mead and Smith 1982). Others published before the end of the century were Devon (Sitters 1988), Cheshire and Wirral (Guest et al. 1992), Shropshire (Deans and Sankey 1992), Hampshire (Clark and Eyre 1993), the second atlas for Hertfordshire (K. W. Smith et al. 1993), Cambridgeshire (Bircham et al. 1994), Northumbria (J. Day et al. 1995), Lancaster (K. Harrison 1995), Berkshire (Standley and Swash 1996), Essex (M. Dennis 1996), Leices-

tershire and Rutland (Warrilow 1996), and Southeast Scotland (Murray et al. 1998). Others have been published since 2000.

Other fine-scale atlases include those for West Berlin (1-km [0.62-mi] grid; Bruch Elvers, Luddecke, et al. 1984; Bruch, Elvers, Pohl, et al. 1978), the High Valley of the Orbe, Switzerland (1-km grid; Glayre and Magnenat 1984), Lower Saxony, Germany (5-km [3.1-mi] grid; Heckenroth 1980), Province of Brescia, Italy (Brichetti and Cambi 1985), Tenerife in the Canary Islands (A. H. Martin 1987), Famenne, Belgium (1-km grid, Jacob and Parquay 1992), and Canton of Fribourg, Switzerland (Cercle de Ornithologique de Fribourg 1993).

NORTH AMERICAN ATLASING

In North America atlases started in the Northeast and quickly became established throughout the continent. (They also spread into parts of Africa and to Australia and New Zealand.)

The first atlas project in the New World was a pilot study in two suburban counties of central Maryland (Klimkiewicz and Solem 1978), published together with progress reports from two other Maryland counties. Fieldwork for the Montgomery County (1971–1973) and Howard County (1973–1975) atlases was actually completed before the British atlas was published.

Five-year statewide atlas projects began shortly after 1978. Vermont was the first state to publish an atlas (Laughlin and Kibbe 1985), followed by the province of Ontario (Cadman et al. 1987), Maine (Adamus 1987), New York (Andrle and Carroll 1988), Ohio (Peterjohn and Rice 1991), Michigan (Brewer et al. 1991), Pennsylvania (Brauning 1992), Canada's Maritime Provinces (Erskine 1992), the province of Alberta (Semenchuk 1992), Rhode Island (Enser 1992), New Hampshire (Foss 1994), and Connecticut (Bevier 1994). All the Atlantic coastal states from Maine to Virginia had completed fieldwork for their first atlas by 1990, as had several inland states and provinces. Some states experienced unexpected delays between the end of their fieldwork and publication of the results, and their atlases were not published until years later. The fieldwork for Maryland and DC's first atlas was conducted in 1983–1987 but the book was not published until 1996.

SECOND ATLASES

Europeans took the lead in producing second atlases to show the changes taking place in breeding bird populations. The first of these to come off the press were those for West Germany (Rheinwald 1982), West Berlin (Bruch et al. 1984), Netherlands (SOVON 1987), Belgium (Devilliers et al. 1988), Hertfordshire (K. W. Smith et al. 1993), Britain and Ireland (Gibbons et al. 1993), France (Yeatman-Bertelow and Jarry 1994), and Czech Republic (Stastny, Bejcek, and Hudec 1995).

Procedures

COORDINATION

Planning for the second Maryland-DC atlas began in 1987, the final year of fieldwork for the first atlas. The actual start date would depend in part on when neighboring states began censusing. The atlas committee believed that demonstrating changes in bird populations would be most effective if adjacent or nearby states and provinces conducted fieldwork in the same years.

The second Maryland-DC atlas fieldwork began in 2002. Early in 2002 Atlas Committee members and volunteers prepared packets for each block. Each packet included a map of the block, a copy of the atlas "Handbook," field cards, a copy of results from that block in the previous atlas, a documentation form, a sample letter of introduction to landowners, and other helpful information. An experienced atlas coordinator was hired in 2001, before the beginning of the fieldwork, to oversee the day-to-day operations, work with the national atlas coordinator at the USGS Patuxent Wildlife Research Center developing data entry and review procedures and designing maps to show records of occurrence and change, provide guidance and stimulation to the county atlas coordinators, train and inspire participants, take minutes at monthly meetings of the MOS Atlas Committee, prepare progress reports for the Administrative Board of the Maryland Ornithological Society, and at the completion of the fieldwork write the species accounts. The atlas coordinator, county coordinators, and atlas committee members provided field-worker training in each county.

Most of the county coordinators had worked on the first Maryland-DC atlas. Their duties included soliciting observers, providing material, training, assisting new observers in the field, and helping observers locate key habitats within their assigned atlas blocks. The county coordinators edited data for their respective counties and gave progress reports at chapter meetings. Some county coordinators prepared progress reports for some or all species within their county in an effort to encourage observers to fill in gaps.

THE MARYLAND AND DC BREEDING BIRD ATLAS PROJECT HANDBOOK

A 22-page "Second Maryland-DC Breeding Bird Atlas Project Handbook" (Maryland Ornithological Society 2001) was prepared for the use of all participants. The "Handbook" included a brief history of grid-based atlas projects, purposes and scope of the project, explanation of the grid system, and definition of the codes, as well as examples of their use. Recommendations were provided on atlas terminology, blockbusting (forming task forces to cover areas that are falling through the cracks), time management, how to upgrade observations, and how to use the forms. A map of Maryland and DC and a list (fig. 1) identified each of the 241 7.5-minute quadrangles used for the project by name and number and showed the numbering system for the six blocks in each quadrangle and for quarterblocks. Land access, atlasing ethics, atlas calling cards, permission and thank-you letters, fund-raising, and sources of additional data were covered briefly. A table gave the status for each species in each of the six major regions of Maryland, the preferred breeding habitat, and safe dates (dates within which the occurrence of transients was unlikely). A brief list of references along with telephone numbers and addresses of Atlas board members and county coordinators was included, as was a sample of the documentation form.

Figure 1. Maryland USGS Quadrangles. Atlas blocks are identified by the name of the 7½ minute USGS quadrangle, the quadrangle number and the position within that quadrangle: NW(1), NE(2), CW(3), CE(4), SW(5), and SE(6).

001	Friendsville	013	Clear Spring	025	Fawn Grove	037	Patterson Creek	049	Westminster
002	Accident	014	Mason Dixon	026	Delta	038	Oldtown	050	Hampstead
003	Grantsville	015	Hagerstown	027	Conowingo Dam	039	Paw Paw	051	Hereford
004	Avilton	016	Smithsburg	028	Rising Sun	040	Big Pool	052	Phoenix
005	Frostburg	017	Blue Ridge Summit	029	Bay View	041	Hedgesville	053	Jarrettsville
006	Cumberland	018	Emmitsburg	030	Newark West	042	Williamsport	054	Bel Air
007	Evitts Creek	019	Taneytown	031	Sang Run	043	Funkstown	055	Aberdeen
008	Flintstone	020	Littlestown	032	McHenry	044	Myersville	056	Havre de Grace
009	Artemas	021	Manchester	033	Bittinger	045	Catoctin Furnace	057	North East
010	Bellegrove	022	Lineboro	034	Barton	046	Woodsboro	058	Elkton
011	Hancock	023	New Freedom	035	Lonaconing	047	Union Bridge	059	Oakland
012	Cherry Run	024	Norrisville	036	Cresaptown	048	New Windsor	060	Deer Park

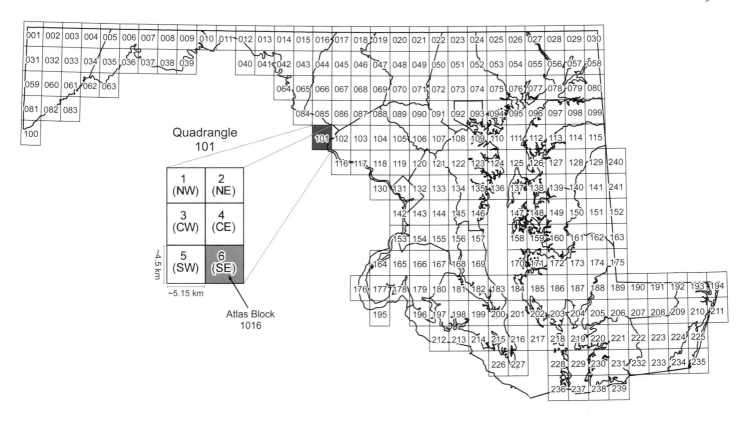

061	Kitzmiller	098	Galena	135	South River	172	Cambridge	209	Ninepin
062	Westernport	099	Millington	136	Annapolis	173	East New Market	210	Berlin
063	Keyser	100	Davis	137	Kent Island	174	Rhodesdale	211	Ocean City
064	Shepherdstown	101	Waterford	138	Queenstown	175	Sharptown	212	Stratford Hall
065	Keedysville	102	Poolesville	139	Wye Mills	176	Widewater	213	St. Clements Island
066	Middletown	103	Germantown	140	Ridgely	177	Nanjemoy	214	Piney Point
067	Frederick	104	Gaithersburg	141	Denton	178	Mathias Point	215	St. Marys City
068	Walkersville	105	Sandy Spring	142	Alexandria	179	Popes Creek	216	Point No Point
069	Libertytown	106	Clarksville	143	Anacostia	180	Charlotte Hall	217	Richland Point
070	Winfield	107	Savage	144	Upper Marlboro	181	Mechanicsville	218	Bloodsworth Island
071	Finksburg	108	Relay	145	Bristol	182	Broomes Island	219	Deal Island
072	Reisterstown	109	Curtis Bay	146	Deale	183	Cove Point	220	Monie
073	Cockeysville	110	Sparrows Point	147	Claiborne	184	Taylors Island	221	Princess Anne
074	Towson	111	Swan Point	148	St. Michaels	185	Golden Hill	222	Dividing Creek
075	White Marsh	112	Rock Hall	149	Easton	186	Blackwater River	223	Snow Hill
076	Edgewood	113	Chestertown	150	Fowling Creek	187	Chicamacomico River	224	Public Landing
077	Perryman	114	Church Hill	151	Hobbs	188	Mardela Springs	225	Tingles Island
078	Spesutie	115	Sudlersville	152	Hickman	189	Hebron	226	St. George Island
079	Earlville	116	Sterling	153	Mount Vernon	190	Delmar	227	Point Lookout
080	Cecilton	117	Seneca	154	Piscataway	191	Pittsville	228	Kedges Straits
081	Table Rock	118	Rockville	155	Brandywine	192	Whaleysville	229	Terrapin Sand Point
082	Gorman	119	Kensington	156	Lower Marlboro	193	Selbyville	230	Marion
083	Mount Storm	120	Beltsville	157	North Beach	194	Assawoman Bay	231	Kingston
084	Harpers Ferry	121	Laurel	158	Tilghman	195	King George	232	Pocomoke City
085	Point of Rocks	122	Odenton	159	Oxford	196	Colonial Beach North	233	Girdletree
086	Buckeystown	123	Round Bay	160	Trappe	197	Rock Point	234	Boxiron
087	Urbana	124	Gibson Island	161	Preston	198	Leonardtown	235	Whittington Point
088	Damascus	125	Love Point	162	Federalsburg	199	Hollywood	236	Ewell
089	Woodbine	126	Langford Creek	163	Seaford West	200	Solomons Island	237	Great Fox Island
090	Sykesville	127	Centreville	164	Indian Head	201	Barren Island	238	Crisfield
091	Ellicott City	128	Price	165	Port Tobacco	202	Honga	239	Saxis
092	Baltimore West	129	Goldsboro	166	La Plata	203	Wingate	240	Marydel
093	Baltimore East	130	Falls Church	167	Hughesville	204	Nanticoke	241	Burrsville
094	Middle River	131	Washington West	168	Benedict	205	Wetipquin		
095	Gunpowder Neck	132	Washington East	169	Prince Frederick	206	Eden		
096	Hanesville	133	Lanham	170	Hudson	207	Salisbury		
097	Betterton	134	Bowie	171	Church Creek	208	Wango		

FORMS

An atlas field card, front and back, is shown in figures 2 and 3. The chief change in the field card since the first atlas project was additional space to record the origin of the records: from an assigned block, incidental records, blockbusting, a miniroute, or Breeding Bird Survey. Space was also provided for entering the date of first confirmation for each species. Because the safe dates were published in the "Handbook," observers were not required to report the date of each observation for possible or probable observations. Nest record cards were provided to observers who wished to use them, but training sessions stressed that finding nests is not necessary for effective atlasing. The Atlas Board wanted to discourage nest hunting in order to protect the birds from disturbance.

MAPS

Three complete sets of 7.5-minute U.S. Geological Survey topographic maps were purchased. One set was divided among the county coordinators for planning purposes; another set was cut into sixths (atlas blocks) for use by fieldworkers. The atlas coordinator kept the third set.

BLOCK SIZE

The traditional size of atlas blocks has been 10 × 10 km (6.25 × 6.25 mi), and from the beginning of the atlas movement, the European Ornithological Atlas Committee has strongly recommended that all atlas projects use 10 km or a multiple of that (such as 5 km in Denmark and the Netherlands, 50 km for the atlas of Europe). In actual practice, the boundaries have been determined on the basis of the most readily available suitable maps. Some projects have used the worldwide Universal Transverse Mercator (UTM) grid; others have used a national metric grid; and still others have used a grid based on latitude and longitude.

For the Maryland-DC project we continued to use one-sixth of the USGS 7.5-minute topographic map as an atlas block for the following reasons:

1. Standard topographic maps (scale of 1:24,000) were readily available.
2. These maps carry tick marks in the left and right margins for dividing them into thirds, and a scale at the bottom for dividing them in half; only the top margin needed to be measured to divide the map into six equal parts (2.5 minutes of latitude by 3.75 minutes of longitude).
3. The resulting subdivisions at the latitude of DC measure about 4.6 km (2.8 mi) north-south by 5.4 km (3.3 mi) east-west, for an area of 24.7 sq km (9.5 sq mi), which is within 1.2% of the area of a 5 × 5–km block.
4. All surrounding states, in fact, almost all states in the eastern United States, also use one-sixth of a 7.5-degree block, making them all comparable and compatible.
5. Although the UTM grid is shown in the margin of recent Maryland topographic maps, using it would have required extensive cutting and pasting to construct each atlas map. Furthermore, the UTM grid, being a rectangular grid superimposed on a spherical surface, has a column of odd-size blocks every 6 degrees of longitude, so blocks would not have been all the same size.

In many states, including New York, Pennsylvania, and Delaware, observers atlased all blocks in the state. In others, such as Virginia and West Virginia, they targeted only one block out of six for coverage (the one in the southeast corner of the quad). Vermont and New Hampshire randomly selected one block in each quad. Maryland covered all blocks, and, in addition, designated the northwest block in each quadrangle for quarterblock coverage.

QUARTERBLOCK COVERAGE

Quarterblocks are unique to the Maryland-DC atlas. In the original atlas of Montgomery County from 1971 through 1973 (Klimkiewicz and Solem 1978), 32 of the 135 breeding species were found in all 60 of the 5-km blocks. This meant that if the Montgomery atlas were repeated in the future, it would be impossible to document any increase in these species and difficult to show changes in many of the other species that were found in a high percentage of the blocks. As long as any suitable habitat for a species remained anywhere within the 25-sq-km block, a declining species might still be present and the decline not detected. So, for the Howard County atlas from 1973 through 1975 (Klimkiewicz and Solem 1978), each of the 34 blocks was divided into quarters. A separate species list was kept for each of the quarterblocks, but the highest atlas breeding category achieved in any of the quarterblocks within a block was applied to all other quarterblocks in that block in which the species was recorded. Because it was not necessary to look for confirmation or even probable status in each quarterblock, little additional time and effort were required beyond normal good coverage of all sections of the block, although additional record keeping was necessary. The original quarterblock effort was highly successful. Of the 39 species found in all 34 blocks in Howard County, only 7 were found in all 136 quarterblocks.

After considerable discussion of the pros and cons of quarterblock coverage, the committee decided to atlas DC and the most rapidly changing portions of Maryland at the quarterblock level (fig. 4). This included all of Montgomery, Prince George's, Howard, and Baltimore counties, and the two southern quadrangles in Carroll County. Because a number of Maryland's breeding species are restricted to Garrett County in the far west, the blocks in that county were also divided into quarters, as were those in Somerset County

FIELD CARD

Maryland/DC Breeding Bird Atlas Project
2002 through 2006

Welcome to the Breeding Bird Atlas Program of the Maryland Ornithological Society, Inc.

This card is for your use in the field and for computer entry. For each species please enter the code for the highest category observed, and if CONFIRMED give the date of that observation. Also, if your block is subdivided into quarter-blocks, mark each quarter-block in which the species was found. Please return this card to your county coordinator by September 1.

BREEDING CRITERIA & CODES

POssible
- **O** Observed but not in breeding habitat
- **X** Heard or seen in breeding habitat

PRobable
- **A** Agitated behavior or anxiety calls
- **P** Pair seen
- **T** Bird holding territory
- **C** Courtship or copulation
- **N** Visiting probable nest site
- **B** Nest building by wrens or woodpeckers

COnfirmed
- **DD** Distraction display
- **NB** Nest building
- **UN** Used nest
- **FL** Recently fledged young
- **FS** Parent with fecal sac
- **FY** Parent with food for young
- **ON** Parent leaving/entering nest site or on nest
- **NE** Nest with eggs
- **NY** Nest with young

Quarter-block Numbering: 1 2 / 3 4

ADDITIONAL SPECIES	PO	PR	CO	OB	DATE	CO

COVERAGE (optional)

DATE	LOCATION	START	END	HOURS

TOTAL HOURS THIS YEAR

RARE AND LOCAL SPECIES

The following species breed in Maryland, but are expected in fewer than 20 blocks. Reports of any of these species from outside of known nesting areas (see Table 1 of Handbook) must be accompanied by completed verification forms.

Eastern Shore Specialties

Brown Pelican
Double-cr. Cormorant
American Bittern
Great Egret
Snowy Egret
Little Blue Heron
Tricolored Heron
Cattle Egret
Black-cr. Night-Heron
Glossy Ibis
Gadwall
Black Rail
Am. Oystercatcher
Wilson's Plover
Piping Plover
Laughing Gull
Herring Gull
Gt. Black-backed Gull
Gull-billed Tern
Royal Tern
Common Tern
Forster's Tern
Black Skimmer
Swainson's Warbler

Western Maryland Specialties

Sharp-shinned Hawk
Upland Sandpiper
N. Saw-whet Owl
Yellow-bel.Sapsucker
Alder Flycatcher
Red-breast. Nuthatch
Winter Wren
Golden-cr. Kinglet
Nashville Warbler
Northern Waterthrush
Mourning Warbler
Dark-eyed Junco

Rare Throughout Maryland

Pied-billed Grebe
Yellow-cr. Nt.-Heron
Hooded Merganser
Northern Harrier
Loggerhead Shrike
Sedge Wren
Henslow's Sparrow
Dickcissel

The following 16 species have been recorded breeding in Maryland at only a few localities, or are anticipated breeders. Extensive details are required. Nesting attempts by these species or any not on this card must be reported immediately.

Green-winged Teal
Ruddy Duck
Mississippi Kite
Northern Goshawk
Purple Gallinule
American Coot
Sandwich Tern
Roseate Tern
Long-eared Owl
Short-eared Owl
Olive-sided Flycatcher
Swainson's Thrush
Bachman's Sparrow
Lark Sparrow
White-throated Sparrow
Pine Siskin

Block information for computer entry

QUADRANGLE	
Quad name	Quad no

YEAR

BLOCK (circle one)

NW	NE	CW	CE	SW	SE
1	2	3	4	5	6

Observer's Name:

Address:

Phone:

Fax:

E-mail

Source of Records (circle one)
- A Assigned block
- B Block busting
- I Incidental
- M Miniroute
- S BBS

Total hours this year: _____

(Circle one)   Actual   Estimated

Figure 2. Atlas field card (front).

Quarter-block Numbering

1	2
3	4

SPECIES	PO	PR	CO	OB	DATE CO
Bittern, Least					
Heron, Great Blue					
Green					
Vulture, Black					
Turkey					
Goose, Canada					
Duck, Wood					
American Black					
Mallard					
Teal, Blue-winged					
Osprey					
Hawk, Cooper's					
Red-shouldered					
Broad-winged					
Red-tailed					
Kestrel, American					
Pheasant, Ring-necked					
Grouse, Ruffed					
Turkey, Wild					
Bobwhite, Northern					
Rail, Clapper					
King					
Virginia					
Moorhen, Common					
Killdeer					
Willet					
Sandpiper, Spotted					
Woodcock, American					
Tern, Least					
Dove, Rock					
Mourning					
Cuckoo, Black-billed					
Yellow-billed					
Owl, Barn					
Eastern Screech-					
Great Horned					

SPECIES	PO	PR	CO	OB	DATE CO
Owl, Barred					
Nighthawk, Common					
Chuck-will's-widow					
Whip-poor-will					
Swift, Chimney					
Hummingbird, Ruby-thr.					
Kingfisher, Belted					
Woodpecker,Red-headed					
Red-bellied					
Downy					
Hairy					
Flicker, Northern					
Woodpecker, Pileated					
Pewee, Eastern Wood-					
Flycatcher, Acadian					
Willow					
Least					
Phoebe, Eastern					
Flycatcher, Gt. Crested					
Kingbird, Eastern					
Vireo, White-eyed					
Yellow-throated					
Blue-headed					
Warbling					
Red-eyed					
Jay, Blue					
Crow, American					
Fish					
Raven, Common					
Lark, Horned					
Martin, Purple					
Swallow, Tree					
N. Rough-winged					
Bank					
Cliff					
Barn					

SPECIES	PO	PR	CO	OB	DATE CO
Chickadee, Carolina					
Black-capped					
Titmouse, Tufted					
Nuthatch, White-breast.					
Brown-headed					
Creeper, Brown					
Wren, Carolina					
House					
Marsh					
Gnatcatcher, Blue-gray					
Bluebird, Eastern					
Veery					
Thrush, Hermit					
Wood					
Robin, American					
Catbird, Gray					
Mockingbird, Northern					
Thrasher, Brown					
Starling, European					
Waxwing, Cedar					
Warbler, Blue-winged					
Golden-winged					
Northern Parula					
Yellow					
Chestnut-sided					
Magnolia					
Black-throated Blue					
Black-throated Green					
Blackburnian					
Yellow-throated					
Pine					
Prairie					
Cerulean					
Black-and-white					
Redstart, American					
Warbler, Prothonotary					

SPECIES	PO	PR	CO	OB	DATE CO
Warbler, Worm-eating					
Ovenbird					
Waterthrush, Louisiana					
Warbler, Kentucky					
Yellowthroat, Common					
Warbler, Hooded					
Canada					
Chat, Yellow-breasted					
Tanager, Summer					
Scarlet					
Towhee, Eastern					
Sparrow, Chipping					
Field					
Vesper					
Savannah					
Grasshopper					
Saltmarsh Sharp-tailed					
Seaside					
Song					
Swamp					
Cardinal, Northern					
Grosbeak, Rose-breast.					
Blue					
Bunting, Indigo					
Bobolink					
Blackbird, Red-winged					
Meadowlark, Eastern					
Grackle, Common					
Boat-tailed					
Cowbird, Brown-headed					
Oriole, Orchard					
Baltimore					
Finch, Purple					
House					
Goldfinch, American					
Sparrow, House					

Figure 3. Atlas field card (back).

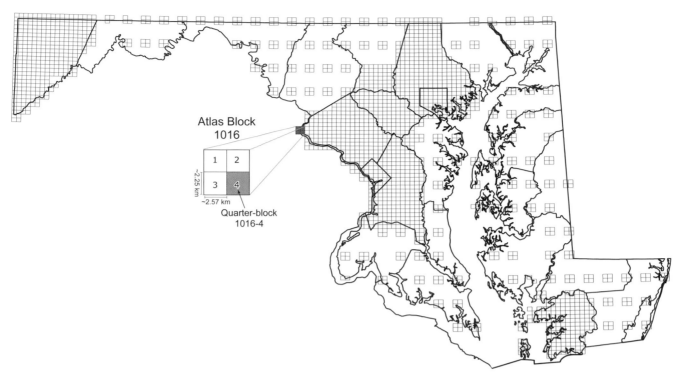

Figure 4. Quarterblocks.

on the lower Eastern Shore. The northwest block in all other quadrangles was designated for quarterblock coverage. (See fig. 9 in "The Environment" for a map of Maryland's counties.)

STATUS CODES

One primary objective of atlas programs is to define the degree of certainty that each species observed is actually nesting within the specified block of territory. For that reason four status categories are recognized worldwide: observed, possible, probable, and confirmed. The codes adopted for the Maryland-DC project are essentially those of the European Ornithological Atlas Committee and are the same as those adopted at the Northeastern Breeding Bird Atlas Conference (Laughlin 1982), with the following exceptions: we used separate codes of FY (food for young) and FS (fecal sac) instead of combining them under the code AT (attending young); we considered carefully identified eggshells on the ground under a nest as a nest with eggs (NE) instead of used nest (UN); and we restricted use of the code for physiological evidence (PE), which includes incubation patch and egg in the oviduct, for bird banders who examine birds in the hand. We did not publish the PE code in the "Handbook" for fear of misuse by nonbanders.

The Maryland-DC definitions included more explanation than those published by the atlas conference. The only code that might be interpreted more loosely in the "Handbook" is the code for territorial display (T). In the conference proceedings, the observations must be made at least a week apart to qualify for this code; in the "Handbook," the period between observations is not defined, in part because migrants were thought to be eliminated by the safe dates. Early training sessions urged caution in using the code and suggested that at least a week should pass between observations, especially for uncommon or rare species. The three major categories, possible, probable, and confirmed, have generally been accepted worldwide. Examples of use of the various codes were published in the "Handbook" (table 1).

ATLAS SAFE DATES

The table of "safe dates" (appendix A) in the "Maryland-DC Breeding Bird Atlas Project Handbook" indicated the period between migration seasons when each species can be assumed to be on territory. Many species present in Maryland throughout the year are considered safe for only a few months or less because they wander or because additional birds from the north are present. Barn Owls, for example, nest here every month of the year, but birds from farther north migrate into or through the state without nesting. Blue Jays begin nesting here in April, but migrants, especially one-year-old birds, are still passing northward through the state as late as 10 June; in one year the migration continued to the extraordinary date of 2 July (Robbins 1967). Although Blue Jays confirmed as nesting in April are good for atlas purposes, their mere presence, even in pairs, is not acceptable evidence unless they are observed within safe dates. Prob-

Table 1.   Status Codes Used in the Maryland-DC Atlas Project

Observed

O = Species observed in block but not in breeding habitat. This code was primarily for birds that do not breed in a block, for example, the thousands of Laughing Gulls in plowed fields on the lower Eastern Shore or the subadult Ring-billed Gulls that spend the summer in Maryland. Flyovers, such as soaring Turkey Vultures, were also in this category. Any species seen within safe dates (see appendix B) with no further evidence of breeding were recorded as O.

Possible

X = Species heard or seen in breeding habitat within safe dates. Outside of safe dates observers were urged to seek evidence of Probable or Confirmed breeding, as these were the only categories that would be accepted at those times.

Probable (always a single-letter code)

A = Agitated behavior seen in or anxiety calls heard from an adult. Parent birds respond to threats with distress calls or by attacking intruders. This did not include response to "pishing" or tape playing.

P = Pair observed in suitable breeding habitat within safe dates. Caution was urged in using this code.

T = Territorial behavior or singing male present at the same location on at least two different days. Territoriality was presumed from defensive encounters between individuals of the same species or by observing a male singing from a variety of perches within a small area.

C = Courtship or copulation observed. This included displays, courtship feeding, and mating.

N = Visit to probable nest site observed; primarily applied to cavity nesters. This code was applied when a bird was observed visiting the site repeatedly but no further evidence was seen.

B = Nest building by wrens or excavations by woodpeckers observed. Both groups build dummy nests or roosting cavities at the same time they are building one for nesting, but because an unmated male will exhibit the same behavior, nest building by these groups is coded probable.

Confirmed (always a two-letter code)

NB = Nest building (except by wrens and woodpeckers) or adult carrying nest material observed. Carrying sticks is part of the courtship ritual (code C) of some species, so caution was urged in using this code.

DD = Distraction display observed. This behavior included such behavior as feigning injury to distract an observer away from a nest. The Killdeer is well known for this behavior.

UN = Used nest found. Because nests are difficult to identify, extreme caution was applied in using this code. Nests were not collected for further identification because federal and state permits are required. This code was particularly useful after the leaves had fallen.

FL = Recently fledged young or downy young seen or heard. This included only dependent young. Some species range widely soon after fledging, so caution was encouraged. Dead fledglings or nestlings on the road also confirmed breeding. Young Brown-headed Cowbirds begging for food confirmed both the cowbird and the host species.

FS = Adult bird seen carrying fecal sac. Feces of nestlings of many species are contained in a membranous sac that parents carry away from the nest.

FY = Adult carrying food for young. This code was used with caution because a few species feed young long after they have wandered from the nest site, carry food a long distance, or engage in courtship feeding.

ON = Occupied nest presumed by activity of the parents: entering a nest hole and staying, parents exchanging incubation responsibilities, and so on. This category was primarily intended for cavity nesters and nests too high for the contents to be seen.

NE = Nest with eggs or eggshells seen on the ground. These had to be carefully identified. Brown-headed Cowbird eggs in nests confirmed both the cowbird and the host species.

NY = Nest with young seen or heard. A Brown-headed Cowbird chick in a nest confirmed both the cowbird and the host species.

---

lems developed for a few species for which the safe dates were not safe enough: Blue-winged Teal, Spotted Sandpiper, cuckoos, Common Raven, and some of the swallows.

VERIFICATION OF UNUSUAL RECORDS

Observers were required to submit a documentation form for each unusual record, detailing the exact location, date and time, full details of the breeding evidence and behavior, a description of the habitat, and contact information for all observers. No reports of birds out of season or out of normal breeding range were accepted without accompanying forms.

DATA PROCESSING

The USGS Patuxent Wildlife Research Center at Laurel, Maryland, used the Maryland-DC second breeding bird atlas project as a pilot for creating an electronic repository for atlas data. Our atlas project relied on the resulting web-based data management system for managing data entry and review as well as for data storage and retrieval throughout the project.

County coordinators used the website to add observers and assign them to their atlas blocks. Data from field cards were entered directly into the online data entry system by the field-workers or county coordinators. Information included block identifier, year, observer name, species, breeding code, dates if breeding was confirmed, and quarterblock data. The data entry screen was programmed to flag data as they were entered, warning users about gross errors, incorrect breeding codes, birds outside of safe dates, and rare or unexpected species. This saved time in the editing process. County coordinators then reviewed each field document, marking sightings as approved for inclusion or for exclusion if documentation was insufficient. Once coordinator review was completed for a field card, the data became part of the publicly available reports, such as listings of the best species evidence for each block and the highest evidence in each block for a species. The database accepted multiple records for each species within a block, but the highest evidence was selected for these reports.

Once all data for the year were entered and reviewed, a master data file was sent to the project coordinator for further examination and analysis. The reporting and summary tools were developed over the course of the atlas project, aiding coordination more in the last two field years. Mapping tools, which were not added until after fieldwork was completed, aided in data review and preparation for publication.

At the conclusion of fieldwork, a more careful review of the data was conducted, and statistical summaries of block coverage and species status were prepared. An interim map was generated for each species for editing purposes.

Miniroute data for relative abundance were collected and entered into a separate database for convenience, and those data were edited and reviewed along with the breeding evidence data during preparation for publication. Over the course of the project, data from the first atlas were assembled for comparison and measures of change between atlases. While data were being reviewed and edited, programs were written to handle the mapping and data summaries for the final publication, including maps showing change from the first to the second atlas.

Technology Overview

The atlas data management system was built using Microsoft SQL Server as the relational database management system, using the ColdFusion web-scripting language to control the website. Interim copies of data, as well as summary queries and reports, were developed using Microsoft Access. Spatial information used in the atlas, including county and physiographic boundaries, and block, quarterblock, and quadrangle boundaries, were stored in an ESRI Geographic Information System. Maps for the website and interim review maps were created using Mapserver, open-source mapping software developed at the University of Minnesota.

The atlas database, the same dataset used in this publication, is housed on servers maintained by the USGS. The Maryland Ornithological Society has a duplicate copy. Requests for data should be made to USGS Patuxent Wildlife Research Center, Laurel, Maryland, via their breeding bird atlas website (www.pwrc.usgs.gov/bba). If these data are used extensively in a publication, please contact the MOS for permission and attribute MOS, PWRC, and Maryland DNR as the source.

Coverage

The chief goals of the second atlas were to repeat satisfactory coverage of every Maryland-DC atlas block and to repeat coverage of each miniroute in the state and to have the results available in electronic and book form.

Most Maryland atlas blocks were presumed to still contain 90 to 100 breeding species; exceptions were those in urban and agricultural areas and those containing a large percentage of open water. Maryland adopted the same goal as many other eastern states: to locate 75% of the number of species estimated to occur in the block. As a general rule, observers were encouraged to try for 70 species in rural blocks and 40 species in urban areas. Minimum time suggested for the coverage of a single block was 20 hours, but coordinators emphasized that thorough coverage of all habitats in the block was more important than the number of hours involved.

Some states have put considerable emphasis on confirmations and have attempted to confirm half the species detected in each block. Because many species require hours of effort to confirm, Maryland chose to put more emphasis on raising all of the common species to at least probable status, suggesting 25% confirmed, 50% probable, and a maximum of 25% possible as a valid goal for each block. Coordinators also requested special effort to confirm rare and locally unexpected species.

The atlas board also believed that an understanding of the breeding status of widely distributed birds was not a function of the number of confirmations (see Red-eyed Vireo, for example). A high percentage of confirmations for common species would occur in the normal course of fieldwork, and additional time and effort spent attempting to confirm birds such as Mourning Dove, Carolina Chickadee, and Tufted Titmouse could detract from the effort to find and establish the breeding status of rarer species.

Special efforts were necessary in some areas of the state, especially in counties without MOS chapters. These efforts included county blockbusting trips, all-night owl trips, and special atlas trips during MOS conferences. For two seasons, the Atlas Board hired experienced observers to provide coverage in more remote blocks. The Maryland DNR undertook several surveys of colonial waterbirds and marsh birds to coincide with the breeding bird atlas effort. During the final season, county coordinators and other interested observers were provided with MP3 recordings of nocturnal birds and encouraged to spend extra time looking for these secretive species.

The Environment

LOCATION AND SIZE OF MARYLAND AND THE DISTRICT

Maryland is located midway along the Atlantic coast of the United States between north latitudes 38° and 40° and west longitudes 75° and 79°30'. The total area of Maryland is only 27,394 sq km (10,577 sq mi), of which the total land area is 25,478 sq km (9,837 sq mi). DC, which lies near the geographic center of Maryland, is crossed by latitude 39° and longitude 77°. Despite Maryland's small size, it occupies a strategic position. In addition to stretching from the Ohio River drainage west of the Allegheny Divide to the Atlantic Ocean, Maryland boasts one of the world's great estuaries, the Chesapeake Bay, which has a longer tidal shoreline than the entire state of California (fig. 5).

The Mason-Dixon Line separates Maryland from Pennsylvania on the north; the Potomac River, which lies entirely in Maryland, forms most of the southern boundary. Delaware forms the eastern boundary, except for the southernmost 50 km (31 mi), which are on the Atlantic Ocean. Garrett, the westernmost county, is bounded on its western and southern borders by West Virginia.

SOILS

Consolidated sedimentary rocks from acid shale and sandstone are the parent materials from which all soils in western Maryland were derived. This includes all of Maryland from the western boundary east through the Hagerstown Valley in Washington County. It also includes the Monocacy Valley from the eastern base of Catoctin Mountain east to Union Bridge, Woodsboro, and Buckeystown in Frederick County, and the western tip of Montgomery County from Dickerson and Martinsburg east to Seneca Creek.

Basic igneous and metamorphic rocks formed the chief parent material for soils of the Maryland Piedmont and also of Catoctin and South Mountains, the Middletown Valley, Sugarloaf Mountain, and Maryland's small share of the Blue Ridge Mountains.

The entire Coastal Plain is comprised of soils formed from unconsolidated Coastal Plain sediments. Within the Coastal Plain, the most distinctive soil subdivisions are the tidal marshes, sand dunes, and coastal beaches, and the greensand belt that extends from DC to Annapolis and Mayo in Anne Arundel County.

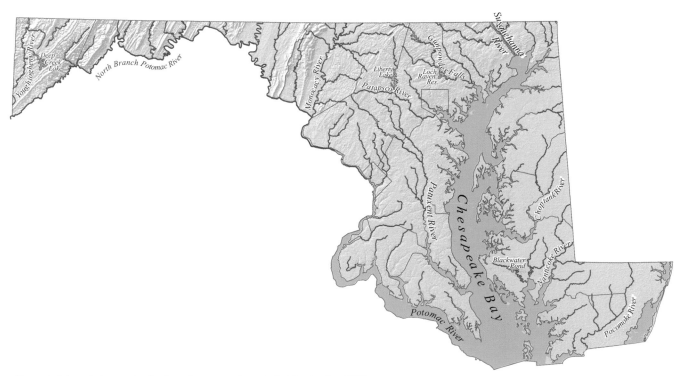

Figure 5. Selected rivers, other bodies of water, and terrain in Maryland and DC.

PHYSIOGRAPHIC REGIONS

The six physiographic regions (fig. 6) are the same as those adopted by Stewart and Robbins (1958): Allegheny Mountain, Ridge and Valley, Piedmont, Upper Chesapeake, Eastern Shore, and Western Shore. These are modified slightly from those proposed by Braun (1950) for classifying the regions of the eastern deciduous forests of North America.

The Allegheny Mountain section, in the far west, extends westward from the base of the Allegheny Front in western Allegany County. It consists primarily of the Allegheny Plateau, which is crossed by several higher ridges. The western half of Garrett County lies in the Mississippi drainage basin. Deciduous trees now dominate the Allegheny Mountain section.

The Ridge and Valley section includes the Catoctin ridge, the Middletown Valley, the South Mountain ridge, the Hagerstown Valley, and the series of heavily wooded ridges between there and the base of the Allegheny Front to the west. The Piedmont and the Ridge and Valley sections are in Braun's (1950) oak-chestnut forest region, from which the once typical American chestnut has all but disappeared. Flowering dogwood was still an ever-present understory tree during this atlas period, but because of disease it may not long remain an important indicator for this region.

The Piedmont section extends eastward from the base of Catoctin Mountain, which is the northern extension of the Blue Ridge Mountains of Virginia, to the fall line. The fall line, which runs northeastward from DC through Baltimore to Wilmington, Delaware, separates the Piedmont from the Coastal Plain province.

The southeastern half of Maryland lies on the Atlantic Coastal Plain, which is divided into three sections not recognized by Braun (1950): the Western Shore (west of the Chesapeake Bay), the Eastern Shore (east of the Chesapeake Bay and north to southern Kent County and central Caroline County), and the Upper Chesapeake (which includes the flat necks on the northwestern side of the Chesapeake Bay and the portion north of the Eastern Shore section on the eastern side). The Coastal Plain lies in the oak-pine forest region and occupies slightly more than half the state. Loblolly pines are characteristic of the Eastern Shore section, whereas Virginia pines are prominent in much of the Western Shore section. The Upper Chesapeake is characterized primarily by the absence of pines.

ELEVATIONS

The Allegheny Mountain section in western Maryland averages about 760 m (2,500 ft), with the ridges about 152 m (500 ft) above the rolling valleys. The highest point in Maryland is at 1,024 m (3,360 ft) on Backbone Mountain, which is within 3 km (2 mi) of the southern tip of Garrett County. The Ridge and Valley section ranges from 152 to 244 m (500–800 ft) in the valleys to 610 m (2,000 ft) on the higher ridges. The Piedmont is largely between 91 and 244 m (300–800 ft); most

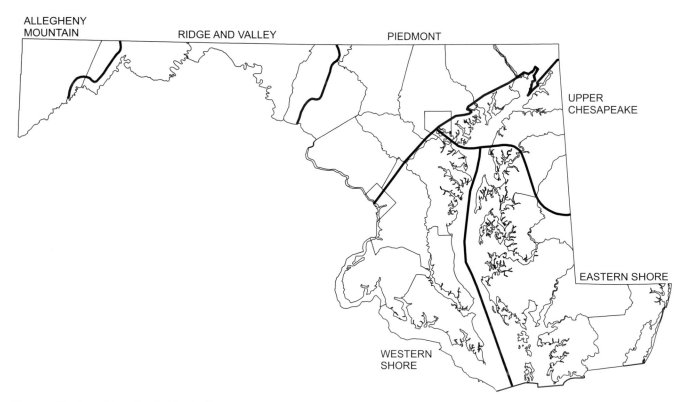

Figure 6. Physiographic regions in Maryland.

of the Coastal Plain west of the Chesapeake Bay ranges between 30 and 91 m (100–300 ft) in elevation. The Eastern Shore, which is low and flat, is composed of three terraces, all below 30 m (100 ft) elevation.

CLIMATIC CHANGES

The chief climatic change affecting breeding birds has been the gradual increase in mean temperature over the past hundred years (see fig. 7). Many nesting species are showing indications of nesting earlier in the season, as revealed by earlier nest-with-eggs and nest-with-young dates submitted by atlas observers for various birds (e.g., American Goldfinch). Some northern species' breeding ranges may be retreating northward or upslope in the Allegheny Mountains (e.g., Baltimore Oriole). Southern species appear to be showing similar upslope movement (e.g., Carolina Wren, Northern Mockingbird) or expanding their ranges to northern portions of the state (e.g., Brown-headed Nuthatch, Blue Grosbeak). At the same time, certain boreal species (e.g., Hermit Thrush, Yellow-rumped Warbler, Pine Siskin) have increased

or expanded their range. Most of the latter are resident or short-distance migrants, which may be responding to milder winter conditions or declines in neotropical migrants. The rise in sea level and increase in frequency of severe storms associated with climate change are also having a detrimental effect on nesting colonial waterbirds and other species that breed on islands in Chesapeake Bay. Those islands are eroding at an accelerated rate.

CHANGES IN LAND USE

The current land use map is an exciting tool for explaining the breeding distribution of many species (fig. 8). The Marsh Wren and the rails are restricted to the marshlands; Horned Larks and Vesper Sparrows are concentrated in the most extensive agricultural regions; and Worm-eating and Hooded warblers are restricted to the more extensive woodlands. Birds are highly dependent on the environment for feeding and for nesting, so as land use changes we can expect bird distribution to change with it.

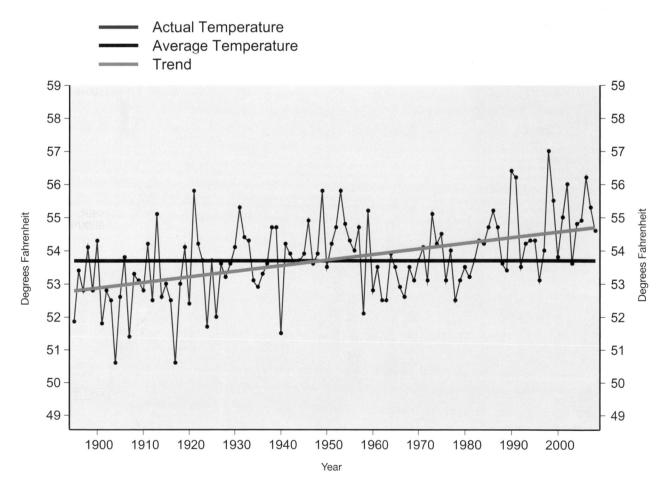

Figure 7. Average annual temperature in Maryland, 1895–2008. Annual 1895–2008 average temperature = 53.77°F; annual 1895–2008 trend = +0.17°F per decade.
Source: National Oceanic and Atmospheric Administration / National Climate Data Center.

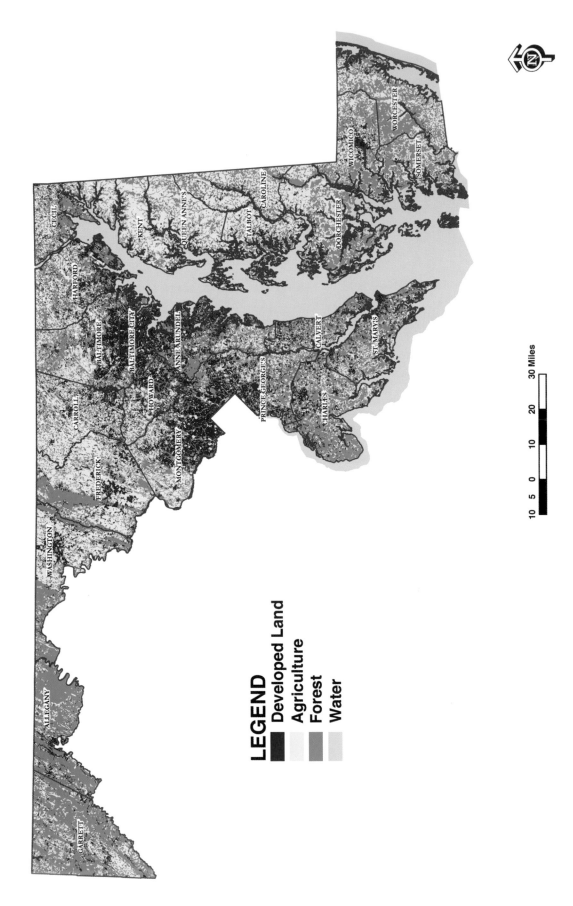

Figure 8. Land use in Maryland, 2002.
Source: Maryland Department of Planning.

Forest Cover

Environmental changes, as indicated by changes in forest type in Maryland (table 2), are taking place more rapidly than one might have predicted. The state continues to lose forest land to development for housing and business. Among conifers, there has been a relatively small loss of loblolly / shortleaf pine acreage, but a 50 percent decline in acreage of red and white pine since the first atlas. There has been a significant increase in spruce and fir forest, which remains a minor component of Maryland's forests critical to several species that are largely restricted to higher elevation forests in Garrett County. In the hardwood categories it is no surprise that species associated with early forest regeneration such as red

maples, elms, and ashes showed the greatest increase at the expense of oaks and hickories, but the greatest decline was in oak/gum/cypress on the lower Eastern Shore. The aspen/birch category, which is not well represented in Maryland, showed a high percentage decline, indicative of the maturation of forests in the Alleghenies.

Agriculture

The number of farms and the acreage of farmland in Maryland continues to decline, as has the acreage of most crops (table 3). Acreage planted to the three main crops—corn, wheat, and soybeans grown primarily as feed for poultry and other livestock—varies from year to year depending

Table 2. Change in area of Maryland timberland between inventories by forest type

Forest type	Thousands of acres		
	1986	1999	Change (% change)
White / red pine	49.5	24.7	−24.8 (−50.1)
Spruce/fir	0.0	3.2	+3.2 (n/a)
Loblolly / shortleaf pine	292.2	282.6	−9.6 (−3.3)
Oak/pine	243.4	229.6	−13.8 (−5.7)
Oak/hickory	1,502.3	1,387.9	−114.4 (−7.6)
Oak/gum/cypress	147.6	124.8	−22.8 (−15.4)
Elm/ash / red maple	95.6	108.2	+12.6 (+13.2)
Northern hardwoods	188.7	209.1	+20.4 (+10.8)
Aspen/birch	3.0	1.8	−1.2 (−40.0)
All types	2,522.3	2,371.9	−150.3 (−6.0)

Source: 1999 Maryland Forest Inventory, Northeastern Forest Inventory and Analysis (Newtown Square, PA: USDA Forest Service, 1999).

Table 3. Change in number of Maryland farms, crop acreage, and livestock

	1900	1987	1997	2007
Number of farms	46,012	14,776	12,084	12,834
Average acreage per farm	112	162	178	160
Total farm acreage	5,170,075	2,396,629	2,154,875	2,051,756
	Number of acres			
Corn	658,010	432,409	405,451	524,116
Wheat	634,446	146,081	199,351	166,713
Barley	n/a	59,268	47,405	34,288
Oats	44,625	n/a	5,611	2,338
Hay	357,224	255,676	223,014	223,390
Soybeans	n/a	405,170	509,683	386,604
Sorghum	n/a	n/a	14,600	5,903
Vegetables	78,801	38,238	35,958	33,447
Tobacco	42,911	10,780	7,939	423
Orchards	n/a	17,982	5,251	4,542
Commercial fertilizer applied to acreage	n/a	1,135,625	1,132,052	1,085,396
	Number of livestock			
Cattle and calves	170,842	308,052	261,324	190,504
Hogs and pigs	n/a	197,214	80,850	n/a
Poultry	n/a	55,500,461	50,659,636	65,977,169

Source: USDA, National Agricultural Statistics Service, Census of Agriculture, available at www.agcensus.usda.gov.

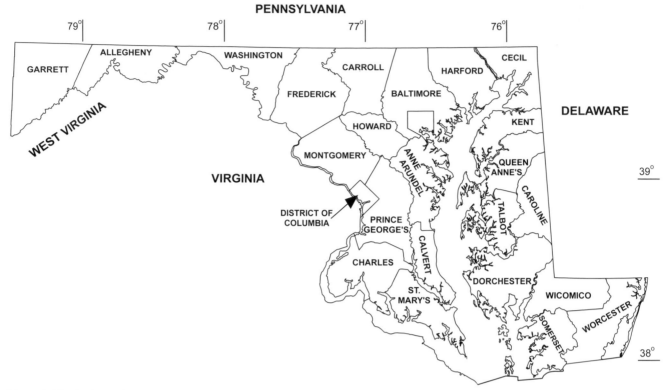

Figure 9. Maryland counties.

on anticipated market demand. Dairy cattle numbers have declined dramatically, leading to fewer pastures and forage-crop fields. Across much of the state, agricultural practices have changed dramatically since the mid-twentieth century. Intensive row-crop agriculture often results in removal of hedgerows, reducing diversity of nesting habitat available for breeding bird species. No-till farming, implemented to protect water quality in Chesapeake Bay, has increased the use of chemical herbicides. Reduction of weedy plants alters habitat structure for grassland birds and reduces abundance of prey. The U.S. Department of Agriculture's Conservation Reserve Program (CRP), begun in 1985, has encouraged some land owners and managers to keep portions of their farms out of production, benefiting certain grassland species (e.g., Northern Bobwhite, Grasshopper Sparrow). Others have taken the extra step of encouraging nesting birds by planting CRP fields with native prairie grasses and reducing use of agricultural chemicals.

Human Population

The population of Maryland continues to increase, which results in more intensive environmental alteration (fig. 9, tables 4 and 5). Since the 1983–1987 atlas, the greatest population increases have taken place in Calvert, Howard, Frederick, and Charles counties. There was a major decline in population in Baltimore City and substantial declines in Allegany County,

as well as in the District of Columbia. The suburban areas in Howard, Carroll, Frederick, Prince Georges, and Montgomery counties grew as the population of associated cities fell. Rockville in suburban Montgomery County supplanted Frederick as Maryland's second largest city. As a commuting lifestyle increased in the 1990s and 2000s, development expanded even farther beyond major cities and is reflected in the rapid population growth of Charles and Frederick counties and the continued growth of Washington County.

Table 4. Maryland Human Population Increase

Year	Population (% increase)
1900	1,188,044 (—)
1910	1,295,346 (9.0)
1920	1,449,661 (11.9)
1930	1,631,526 (12.5)
1940	1,821,244 (11.6)
1950	2,343,001 (28.6)
1960	3,100,689 (32.3)
1970	3,922,399 (26.5)
1980	4,216,975 (7.5)
1990	4,781,468 (13.4)
2000	5,296,486 (10.8)
2008	5,633,597 (6.4)

Source: U.S. Census Bureau.

Table 5. Population Growth in Maryland and DC

Jurisdiction	1980 census		2006 estimate	
	Population	People per sq mi	Population	People per sq mi (% change)
Allegany	80,548	188	72,831	170 (−9.6)
Anne Arundel	370,775	886	509,300	1,217 (+37.4)
Baltimore Co.	655,615	1,097	787,384	1,317 (+20.1)
Baltimore City	786,775	9,793	631,366	7,859 (−19.8)
Calvert	34,638	162	88,804	415 (+156.4)
Caroline	23,143	72	32,617	101 (+40.9)
Carroll	96,356	213	170,260	376 (+76.7)
Cecil	60,430	168	99,506	277 (+64.7)
Charles	72,751	161	140,416	311 (+93.0)
Dorchester	30,623	52	31,631	54 (+3.3)
Frederick	114,792	173	222,938	336 (+94.2)
Garrett	26,498	40	29,859	45 (+12.7)
Harford	145,930	326	241,402	539 (+65.4)
Howard	118,572	472	272,452	1,085 (+129.8)
Kent	16,695	60	19,983	72 (+19.7)
Montgomery	579,053	1,169	932,131	1,882 (+61.0)
Prince George's	665,071	1,366	841,315	1,728 (+26.5)
Queen Anne's	25,508	69	46,241	125 (+81.3)
St. Mary's	59,895	161	98,854	266 (+65.1)
Somerset	19,188	57	25,774	77 (+34.3)
Talbot	25,604	99	36,062	139 (+40.8)
Washington Co.	113,086	249	146,748	323 (+27.1)
Wicomico	64,540	170	91,987	242 (+42.5)
Worcester	30,889	65	48,866	103 (+58.2)
Maryland	4,216,975	429	5,615,727	571 (+33.2)
DC	638,432	10,398	581,530	9,471 (−8.9)

Source: U.S. Census Bureau.

Representative Maryland Habitats

FORESTS AND WOODLANDS

Hemlock Forest

On the northern Piedmont this type of forest is restricted to ravines; in the Ridge and Valley and Allegheny Mountains it is found increasingly at upper elevations, usually on north-facing slopes. It is dominated by hemlock and often has scattered hardwoods, such as black birch and various oaks, and white pine (to the west). Some characteristic birds found in such forests are Barred Owl, Pileated and Hairy woodpeckers, Acadian Flycatcher, Veery, and Louisiana Waterthrush (along brooks). In the western mountains such birds as Blue-headed Vireo, Hermit Thrush, and Black-throated Green and Blackburnian warblers also occur in hemlock forest. Hemlock wooly adelgid infestation may reduce the extent of this habitat in the future. *(Photograph by Nancy L. Martin)*

Loblolly Pine Woodland

Found on both the Eastern and Western Shore from southern Kent County southward, loblolly pine woodland is moist to wet and often found near tidewater; it is medium to tall open woodland dominated by loblolly pine with attendant broad-leaved trees and shrubs especially willow oak, sweet and black gums, and American holly. Birds characteristic of this habitat include Chuck-will's-widow, Brown-headed Nuthatch, and Yellow-throated and Pine warblers. Red-cockaded Woodpecker formerly occurred. Bald Eagles often place their huge stick nests in tall open-crowned loblolly pines. *(Photograph by Nancy L. Martin)*

Mature Moist Upland Hardwood Forest

This is a widespread forest type found largely from the Upper Chesapeake westward through the Ridge and Valley. Common trees include beech, several oaks (especially white oak), several hickories, tulip-poplar, and white ash. Birds frequently found in these haunts include Red-bellied and Pileated woodpeckers, Eastern Wood-pewee, Tufted Titmouse, Acadian Flycatcher, Red-eyed Vireo, White-breasted Nuthatch, Wood Thrush, Ovenbird, and Scarlet Tanager. *(Photograph by Karen D. Morley)*

Dry Upland Oak-Hickory Forest

Found on well-drained soils, widespread but best developed on dry ridge tops in the Ridge and Valley, the forest's understory varies; luxuriant mountain laurel is seen in this image. Some expected bird species include Eastern Wood-Pewee: Red-eyed Vireo; Wood Thrush; Black-and-white, Worm-eating, and Hooded warblers; and Scarlet Tanager. Mature ridge top oak-hickory forest is an important Cerulean Warbler habitat. *(Photograph by Karen D. Morley)*

FORESTS AND WOODLANDS *(cont.)*

Moist Riparian Hardwood Forest

Found along rivers and streams in much of Maryland and DC (Rock Creek Park), the habitat's common trees include tulip-poplar, beech, sweet gum, and elm. American holly, flowering dogwood, and spicebush are often in the understory. Typical birds include Red-bellied and Hairy woodpeckers, Acadian Flycatcher, Great Crested Flycatcher, Red-eyed Vireo, Carolina Chickadee, Carolina Wren, Blue-gray Gnatcatcher, Veery (on Piedmont), Wood Thrush, Northern Parula (vines and near water), Kentucky Warbler, and Scarlet Tanager. *(Photograph by Walter G. Ellison)*

WETLANDS

Cattail Marsh on Small Pond

Of Maryland's wetland habitats, among the most common are those found on small ponds, including mitigation ponds such as the one shown here. These habitats are dominated by emergent plants such as cattail, as well as sweet flag, irises (flags), and common reed. Such wetlands often host Canada Goose, Mallard, Wood Duck (especially if boxes are provided), Green Heron, Killdeer, Common Yellowthroat, Song Sparrow, and Red-winged Blackbird. Colonial herons often feed on small ponds. An American Coot was confirmed at a small farm pond during the second atlas project. *(Photograph by Nancy L. Martin)*

Freshwater Impoundment

Impoundments, whether built for wildlife management or for other purposes, are important habitats for waterbirds and wetland songbirds. Two such wetlands are found at McKee-Beshers Wildlife Management Area in Montgomery County and Lilypons Water Gardens in Frederick County. Among the birds found in such places are Wood Duck, Hooded Merganser, Green Heron, Least Bittern (rare), Virginia Rail, Sora (rare), and Common Moorhen (rare). Colonial herons come to feed, as do swallows. Nesting songbirds include Willow Flycatcher, Warbling and Yellow-throated vireos, Tree Swallow, and Yellow and Prothonotary warblers. *(Photograph by Philip Brody)*

Tidal Freshwater Marsh

These marshes feature deep water channels, tall emergent vegetation, and pools (often opened by muskrats). Dominant vegetation is diverse and may include cattails, big cordgrass, wild rice, bur-reeds, pickerelweed, tuckahoe, and button-bush. Drier areas and the edge of the marsh feature shrubs such as willow, dogwoods, and elderberry. Birds that may be found include Wood Duck, Green Heron, Least Bittern, Virginia Rail, Willow Flycatcher, Marsh Wren, Common Yellowthroat, and Red-winged Blackbird. Among the birds of the shrubby edges are Eastern Kingbird, Gray Catbird, Yellow Warbler, and Orchard Oriole. *(Photograph by Karen D. Morley)*

High Salt Marsh

High salt marsh is generally only shallowly flooded except at highest tides with high-salinity shallow ponds called pannes. Dominant plants are saltmeadow cordgrass, seashore salt-grass, and saltmeadow rush, with smooth cordgrass along creeks and on some panne margins. Other wetland plants such as narrow-leaved cattail and common reed invade disturbed areas and roadsides and reduce habitat quality for specialty organisms. Several rare Maryland birds nest in high salt marshes, including Black Rail, Black-necked Stilt, and Saltmarsh Sparrow. Other birds include herons and ibis feeding in pannes, American Black Duck, Virginia and Clapper rails, Willet, Seaside Sparrow, Eastern Meadowlark (drier sites), and Boat-tailed Grackle (nesting in scattered shrubs at margin). *(Photograph by Nancy L. Martin)*

WETLANDS *(cont.)*

Coastal Plain Wooded Swamp

Shown here is an early spring aspect of a lush riparian swamp with shaded pools, a habitat frequently associated with upper parts of tidal streams on the Coastal Plain. Common constituent plants include black and sweet gum, red maple, American holly, wetland oaks such as water oak and pin oak, sweet pepperbush, highbush blueberry, sweetbay, skunk cabbage, and lizard's tail. Vines, which are abundant, include trumpet vine, greenbriers, poison ivy, Virginia creeper, and grapes. The diverse birdlife of this habitat includes Wood Duck, Yellow-billed Cuckoo, Barred and Eastern Screech-owls, Red-bellied and Pileated woodpeckers, Acadian Flycatcher, Great Crested Flycatcher, Yellow-throated Vireo, Carolina Wren, Blue-gray Gnatcatcher, Prothonotary Warbler, Worm-eating Warbler (in dry patches), Louisiana Waterthrush, and Summer Tanager. *(Photograph by Danny Poet)*

Open Glade and Boreal Shrub Swamp

Pictured here is an upper elevation sedge-grass meadow with forbs such as goldenrod and joe-pye-weed in the foreground. In midground lies an alder-willow shrub swamp; in the background is red maple-cherry-elm-spruce wet woods. The distant slopes are clothed in northern hardwoods with frequent conifers on the ridges. Some of the birds of this habitat are Black-billed Cuckoo, Alder and Willow flycatchers, Veery, Golden-winged Warbler, Nashville Warbler (rare), Yellow Warbler, American Redstart, Swamp Sparrow, and Baltimore Oriole. *(Photograph by John Landers)*

Hay Meadow

This is a widespread agricultural habitat dominated by non-native forage grasses such as meadow fescue (seen here), timothy, orchard grass, and smooth brome. Forbs are frequent; in this field there are chicory and daisy fleabane. Birds regularly seen in this habitat include Savannah and Grasshopper sparrows, Bobolink, and Eastern Meadowlark. Upland Sandpipers also nest in very small numbers in hay meadows in Garrett County. Frequent cuttings can reduce nesting success and some farmers are replacing hay fields with crop fields for producing silage. *(Photograph by Sean McCandless)*

Beach and Dunes

Barrier islands feature a transition from beach, to fore dune, across an open sandy storm overwash zone, to shrubs or woodland, eventually passing to salt marsh on the inshore side. Seen here are dunes and open sand of a beach with sand flats and trees beyond. On the dunes is a sparse cap of wax myrtle, sea rocket, and beach grass. Piping Plovers and Least Terns often nest on flats behind dunes, with plovers feeding with chicks on the beach. Also feeding on the beach are gulls, American Oystercatchers, Willets, and Boat-tailed Grackles. A few songbirds such as Eastern Kingbird and Song Sparrow may nest in grasses, shrubs, and small trees. Common Nighthawks and Horned Larks are scarce nesters on dry sand flats.

(Photograph by Fran Saunders)

Fallow Field

With regular three-crop rotation and conversion to a hay-alfalfa rotation, fields left fallow are becoming less frequent. Some landowners set aside open lands as Conservation Reserve grasslands and plant fields to warm-weather prairie grasses. The field shown here is a more traditional fallow meadow with a mix of non-native grasses and forbs. Some of the birds nesting around fields such as this are Northern Bobwhite, Common Yellowthroat, Grasshopper Sparrow, Indigo Bunting, Blue Grosbeak, Dickcissel (this field hosted a singing male), Eastern Meadowlark, and Red-winged Blackbird. *(Photograph by Nancy L. Martin)*

Abandoned Industrial Site

Even abandoned and active industrial sites where soil banks have been exposed and weedy open ground exists attract a few birds. This is especially true of birds that dig nests in bluffs or exploit used burrows and crevices, such as Belted Kingfisher and Northern Rough-winged and Bank swallows. Killdeer, Horned Larks, and Vesper Sparrows may also nest on the thinly vegetated flats. *(Photograph by June Tveekrem)*

Reclaimed Surface Mine

After coal has been removed from an open surface or strip mine, a mining company is required to revegetate and resurface the site. After landscaping, a site is usually planted with grasses and forbs. As the site recovers it eventually becomes an extensive grassland with a variety of open-land grasses, forbs, and shrubs, often remaining in this state for more than a decade. Reclaimed mines are frequent in the coal mining areas of western Allegany and Garrett counties. Such places host most of Maryland's remaining Henslow's Sparrows, as well as Northern Harrier; Vesper, Savannah, and Grasshopper sparrows; Bobolink; and Eastern Meadowlarks. The most recent nesting records for Long-eared and Short-eared owls in Maryland were from a Garrett County reclaimed surface mine. *(Photograph by George M. Jett)*

Shrub Meadow

When crop fields or other exposed ground is abandoned long enough, late successional grasses, forbs, and shrubs invade. The species that invade vary across Maryland and DC. On the Coastal Plain (shown here) they include sweet gum, various oaks, cherry, sumac, and sassafras. Forbs include goldenrod, asters, tickseed sunflower, and legumes such as pencil flower. Grasses include fescue, redtop, switchgrass, broomsedge, and poverty grasses. Many birds are well adapted for nesting in shrublands, including Northern Bobwhite, American Woodcock, White-eyed Vireo, Gray Catbird, Brown Thrasher, Cedar Waxwing, Prairie Warbler, Yellow-breasted Chat, Eastern Towhee, Field Sparrow, Blue Grosbeak, Indigo Bunting, and American Goldfinch. In western Maryland shrub succession specialists include Golden-winged, Chestnut-sided, and Mourning warblers. *(Photograph by Danny Poet)*

Wooded Suburban Neighborhood

Suburban housing tracts, which can host a suite of widespread bird species, have more diverse birdlife if large trees were left by the developer or have matured from ornamental plantings. Homeowners who provide feeders and nest boxes also increase bird numbers and diversity. Characteristic birds include Mourning Dove, Ruby-throated Hummingbird, Red-bellied and Downy woodpeckers, Eastern Wood-Pewee, Eastern Phoebe, Blue Jay, chickadees, Tufted Titmouse, White-breasted Nuthatch, Carolina and House wrens, American Robin, Gray Catbird, Northern Mockingbird, European Starling, Cedar Waxwing, Chipping Sparrow, Northern Cardinal, Baltimore and Orchard orioles, House Finch, and House Sparrow. Cooper's Hawk and Pileated Woodpecker have increasingly come to nest in well-wooded suburbs. *(Photograph by Nancy L. Martin)*

Results

The current atlas, like its 1983–1987 predecessor, is a multiscale project designed to map the current distribution of every breeding bird species in Maryland and the District of Columbia in sufficient detail so that past and future changes in the distribution of each species can be documented, each species' habitat requirements can be understood, and conservation plans can be made to retain each native bird species far into the future.

COVERAGE

Thanks to the extraordinary efforts of participants, every atlas block in Maryland and the District of Columbia was visited, most of them multiple times. Our goal of finding at least 40 species in each urban block was far exceeded and the goal of at least 70 species in each full-size rural block was reached in most instances (fig. 10). See appendix B for the species totals for each atlas block in the two atlas projects. Nearly two hundred blocks bordering Chesapeake Bay or the Atlantic Ocean included substantial amounts of tidewater, thus limiting the amount of land habitat and reducing the number of breeding species in those partial blocks. In the previous atlas, birds found in atlas blocks that were 95% or more water were plotted in adjacent blocks. In the current atlas an effort was made to visit each block no matter how little land area existed and to plot each bird in the block where it was found. Fifteen blocks partly in Maryland along the Delaware state line were covered by the first Delaware atlas project run concurrently with the 1983–1987 Maryland-DC atlas. These data were inadvertently omitted from Robbins and Blom (1996). These blocks were included in the second Maryland-DC atlas as they were not scheduled to be covered by a second Delaware atlas (begun in 2008).

SCALE

The basic scale employed statewide is one-sixth of the area covered by the standard USGS topographic map, or the equivalent of a 5 × 5–km square (about 10 sq mi). Every bit of dry land in Maryland and DC is included in one of these 1,284 atlas blocks.

In the District of Columbia and central Maryland (Montgomery, Prince George's, Howard, Baltimore, and southern Carroll counties, and Baltimore City), where environmental changes are taking place most rapidly, each 5 by 5 atlas block was subdivided into four quarterblocks for both atlas projects. Garrett, Maryland's westernmost county, and Somerset, the southernmost, were also designated for complete quarterblock coverage. Throughout the rest of Maryland, only the northwestern block in each 7.5 minute quadrangle was selected for quarterblock coverage. No effort was made to count or estimate the number of individual birds in each block or quarterblock. We were simply documenting nesting of each species in each block and quarterblock.

At the block level, we achieved a remarkable set of statewide quantitative data in the form of 15-stop miniroutes, which were run by experienced observers to provide estimates of quantitative population changes. In the initial Maryland-DC atlas, miniroutes tended to be clustered close to where observers lived (fig. 11). For the second atlas, miniroute coordinator Nancy Martin made a point of spreading the pattern evenly across the state (fig. 12).

And finally, courtesy of the USGS Breeding Bird Survey program, we have included in the species accounts many references to statewide, national, and continental trends in the breeding populations of individual species. The Breeding Bird Survey (BBS) routes are about 40 km (25 mi) long, so each one extends through 4 to 6 atlas blocks.

EFFORT

Atlas field-workers were encouraged to record the amount of time they spent surveying their blocks, although not all of them did so. This recorded effort, which totaled 61,363 hours of fieldwork (fig. 13), was spread out fairly evenly over the five-year period, but it varied greatly from block to block depending on whether observers lived in their assigned block and whether they had easy access to a wide variety of habitats. In the final years of the project, observers from other areas were encouraged to visit blocks that appeared to be undercovered. Howard County is recognized for achieving the most effort hours, and along with that, the most atlas confirmations per block.

Fifty or more species were found in nearly all the full-size blocks (see fig. 10), and most counties averaged more than 70 species per block. Observers achieved excellent species tallies in even heavily developed regions. In the Baltimore East quadrangle the block species totals ranged from 65 to 79, and in Baltimore West, from 79 to 88 species except for a low tally of 60 in the northeast block. In Washington West, tallies ranged from 63 to 73, and in Washington East, from 58 to 92

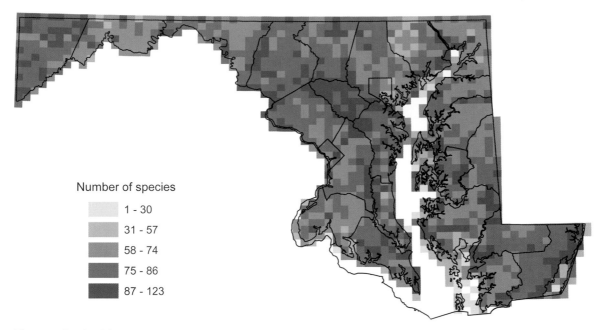

Figure 10. Species richness.

Number of species
- 1 - 30
- 31 - 57
- 58 - 74
- 75 - 86
- 87 - 123

Figure 11. Miniroute distribution, 1983–1989.

species. In Annapolis, blocks ranged from 63 to 73 species; in Hagerstown, from 57 to 68; and in Salisbury, from 69 to 83 species.

As in the previous atlas, counts of 100 species or more came primarily from Garrett County, but not necessarily from the same blocks as before. The highest count this time, 121 species, came from Frostburg CW, the block that includes Finzel Swamp and Carey Run Sanctuary. Outside of Garrett County, the highest species counts were 112 species at Lonaconing CE in Allegany County; 107 at Bristol SW in

Prince George's County; 101 species at Relay NE in Anne Arundel and Baltimore counties; and 102 and 100 species, respectively, at Laurel NE and CE in Anne Arundel and Prince George's counties.

Block totals of 90 species or more tended to be found along the Potomac and Patuxent rivers and were relatively scarce along the Pennsylvania state line and on the Eastern Shore. Surprisingly no atlas blocks within the Pocomoke drainage system had more than 86 species recorded.

Figure 12. Miniroute distribution, 2003–2007.

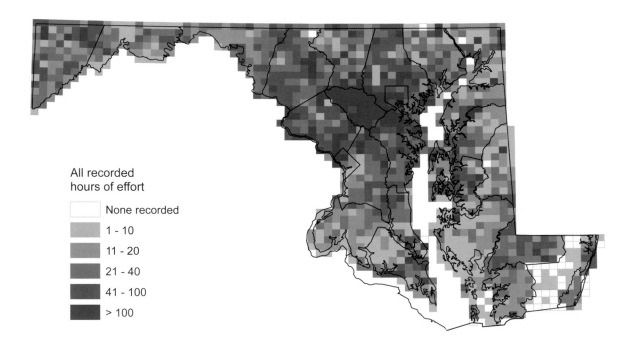

All recorded
hours of effort

	None recorded
	1 - 10
	11 - 20
	21 - 40
	41 - 100
	> 100

Figure 13. Recorded hours of fieldwork for the second atlas.

CHANGES IN BIRD DISTRIBUTION

The species maps show where in Maryland each species is most likely to be found and the extent to which its distribution has changed over the past two decades. We knew that changes were taking place in the distribution of various species, but never before has it been possible to view these changes from a statewide perspective, to relate these changes to changes in land use, and to consider conservation measures that might help retard or reverse bird population trends. Gross changes in distribution, such as the spread of the Wild Turkey and loss of the Northern Bobwhite became apparent during the first year of the second atlas project, as did the retreat of Eastern Meadowlarks and Field and Grasshopper sparrows from many haunts at the edge of creeping urbanization. Observers in the mountains of Garrett County detected Carolina Wrens and Northern Mockingbirds in places where they had not been found previously. Observers in central Maryland searched in vain for Broad-winged Hawks; they found Cooper's Hawks instead. In tidewater regions the Bald Eagle had become a common sight.

See appendix C for maps displaying changes in distribution at the 2.5 km by 2.5 km quarterblock level over the three atlasing periods in Howard County (1973–1975, 1983–1987, and 2002–2006). Also included are maps for 12 species in Garrett County during the two statewide atlas projects (1983–1987 and 2002–2006). Quarterblock maps for the entire state for all species are available online at the USGS Patuxent Wildlife Research Center, Laurel, Maryland, via their breeding bird atlas website, www.pwrc.usgs.gov/bba.

CHANGES IN ABUNDANCE

Increases or decreases in bird populations may be occurring without being readily apparent on regular atlas maps. One cannot estimate changes in abundance by merely noting increases or decreases in the number of blocks in which a species is found. One spot on an atlas map may represent a single pair of birds or it may represent 500 pairs. In order to measure changes in abundance, the normal atlas effort was supplemented by a network of 15-stop miniroutes that were run throughout Maryland during each atlas project by specially trained observers. Each of the 15 stops, selected as representative of the habitats in each atlas block, was visited for exactly 3 minutes in the early morning, during which time the observer concentrated on detecting all the species present—not counting the number of robins or starlings but seeking to detect every species that might be seen or heard. Because miniroute efforts were identical during each atlas period, always 3 minutes times 15 stops, those 512 miniroutes that were run in both projects give an index of population change from one period to the next. Table 6 shows

Table 6. Species showing 10% or greater change in relative abundance on Maryland miniroutes, 1983–1989 and 2003–2007

Birds increasing			*Birds decreasing*		
Species	Number of routes in first atlas	Increase in second atlas (% change)	Species	Number of routes in first atlas	Decrease in second atlas (% change)
Wild Turkey	104	223 (214)	Yellow-throated Warbler	96	10 (10)
Mute Swan	22	45 (205)	Grasshopper Sparrow	371	37 (10)
Cooper's Hawk	85	165 (194)	Yellow Warbler	311	34 (11)
Bald Eagle	41	76 (185)	Blackburnian Warbler	16	2 (13)
Great Blue Heron	18	33 (183)	Red-headed Woodpecker	125	18 (14)
Canada Goose	165	273 (165)	Hooded Warbler	201	31 (15)
Tree Swallow	152	250 (164)	Yellow-breasted Chat	436	80 (18)
Common Raven	36	28 (78)	Swamp Sparrow	51	9 (18)
Black Vulture	137	105 (77)	Virginia Rail	47	9 (19)
Brown-headed Nuthatch	37	22 (59)	Willet	21	4 (19)
Osprey	98	45 (46)	Canada Warbler	21	4 (19)
White-breasted Nuthatch	282	127 (45)	Black-and-white Warbler	234	48 (21)
Savannah Sparrow	49	21 (43)	Eastern Meadowlark	448	92 (21)
Cedar Waxwing	311	131 (42)	Prairie Warbler	296	73 (25)
Purple Finch	21	8 (38)	Black-billed Cuckoo	85	23 (27)
Blue-headed Vireo	33	11 (33)	Marsh Wren	75	22 (29)
Pileated Woodpecker	347	96 (28)	American Kestrel	301	93 (31)
Wood Duck	256	69 (27)	Ruffed Grouse	57	18 (32)
Northern Rough-winged Swallow	188	50 (27)	Least Bittern	33	11 (33)
Bobolink	32	8 (25)	Brown Creeper	48	17 (35)
Black-throated Blue Warbler	21	5 (24)	Cerulean Warbler	71	25 (35)
House Finch	374	83 (22)	King Rail	30	11 (37)
Red-shouldered Hawk	261	53 (20)	Blue-winged Warbler	60	22 (37)
Barred Owl	268	54 (20)	American Black Duck	91	35 (38)
Northern Parula	237	45 (19)	Kentucky Warbler	354	140 (40)
Mallard	369	61 (17)	Vesper Sparrow	151	62 (41)
Magnolia Warbler	24	4 (17)	Broad-winged Hawk	162	69 (43)
Ruby-throated Hummingbird	436	42 (10)	Northern Bobwhite	457	202 (44)
Warbling Vireo	135	13 (10)	American Woodcock	170	87 (51)
Eastern Bluebird	450	46 (10)	Northern Harrier	26	14 (54)
			Eastern Whip-poor-will	192	108 (56)
			Golden-winged Warbler	46	31 (67)
			Common Nighthawk	52	38 (73)
			Ring-necked Pheasant	187	147 (79)

changes of 10% or greater in statewide populations along these miniroutes, sorted from greatest increases to greatest declines. Most species are detected on 3-minute miniroute stops as only single individuals, so for these species the number of stops is the same as the number of birds. For the more common species the number of stops serves only as an index to the actual abundance.

Although quite a few species showed a substantial increase in numbers over the 20-year period many populations remained stable (less than 5% change) between the two atlas periods: Blue-winged Teal, Green Heron, Clapper Rail, Killdeer, Mourning Dove, Yellow-billed Cuckoo, Eastern Screech-Owl, Chuck-will's-widow, Chimney Swift, Belted Kingfisher, Red-bellied Woodpecker, Downy Woodpecker, Northern Flicker, Eastern Wood-Pewee, Acadian Flycatcher, Least Flycatcher, Eastern Phoebe, Great Crested Flycatcher, Eastern Kingbird, White-eyed Vireo, Yellow-throated Vireo, Red-eyed Vireo, Blue Jay, American Crow, Fish Crow, Horned Lark, Cliff Swallow, Barn Swallow, Carolina Chickadee, Tufted Titmouse, Carolina Wren, House Wren, Blue-gray Gnatcatcher, Veery, Wood Thrush, American Robin, Gray Catbird, Northern Mockingbird, Brown Thrasher, European Starling, Chestnut-sided Warbler, Black-throated Green Warbler, Ovenbird, Louisiana Waterthrush, Common Yellowthroat, Eastern Towhee, Chipping Sparrow, Field Sparrow, Song Sparrow, Scarlet Tanager, Northern Cardinal, Rose-breasted Grosbeak, Indigo Bunting, Red-winged Blackbird, Common Grackle, Brown-headed Cowbird, Baltimore Oriole, American Goldfinch, and House Sparrow.

NEW BREEDING SPECIES

Three large, conspicuous species that were not confirmed on the first atlas were found nesting in Maryland this time.

The Double-crested Cormorant is a colonial fish-eating species whose population has increased under protection; it was first recorded as nesting in 1990 after being recorded only as observed in 54 blocks in the previous atlas. The second newcomer, the Common Merganser, is a fish-eating duck whose nesting range had previously extended south only as far as Pennsylvania. The third is the Ruddy Duck, a prairie-nesting species never before proven to have nested in Maryland. Two other large species absent from the 1983–1987 atlas were found nesting during the 2002–2006 atlas: both the Long-eared Owl and the Short-eared Owl had previously nested in Maryland, but neither was observed in the first statewide atlas. Red-breasted Nuthatch was confirmed in the Pennsylvania portion of a Maryland atlas block in 1983, but a Maryland nesting confirmation was not documented until April 1990. Although there had been circumstantial evidence of nesting by Pine Siskins in Maryland before 1987, a nest was not found until the year after the 1983–1987 atlas. Several other records of nesting by Red-breasted Nuthatch and Pine Siskin have been documented, including during the 2002–2006 atlas.

SPECIES LOST

Northern Shoveler, Wilson's Plover, and Bewick's Wren, all of which were found during the first atlas, have been lost, probably permanently in the case of the Bewick's Wren. See the short accounts following the main accounts for a list of other former nesting species.

Conservation

Bird conservation in Maryland is accomplished through a variety of means, especially through federal and state laws that protect the species and through land acquisition, management, and protection that provides habitat. The first major action taken in the United States to protect birds was passage of the Migratory Bird Treaty Act of 1918. The act, which has been amended several times and is still in force, implemented various treaties and conventions between the United States and Canada, Japan, Mexico, and the former Soviet Union to protect migratory birds. Under the act, taking, killing, or possessing migratory birds is unlawful except during legal hunting seasons or with special permits. The vast majority of birds in Maryland are migratory and, thus, protected by this law.

Certain species have benefited from the federal Endangered Species Act. The Brown Pelican, Bald Eagle, and Peregrine Falcon are species whose populations have fully recovered, in part as a result of the Endangered Species Act. The Piping Plover is still protected by the act, as are two former Maryland breeding birds, the Roseate Tern and Red-cockaded Woodpecker.

The state of Maryland has its own endangered species act, called the Nongame and Endangered Species Conservation Act. Enacted in 1971 and revised in 1973, the law authorized the Maryland Department of Natural Resources (DNR) to establish a list of threatened and endangered species that are at risk of disappearing from the state and to develop programs for the conservation of all wildlife species (Therres 1998). There are three categories of listing: endangered, threatened, and in need of conservation. Twenty-seven species of birds are state-listed (table 7). Including the state-listed species, the Maryland Department of Natural Resources (2005) has identified 141 species of birds as species of greatest conservation need and will be focusing conservation actions on these species and their key habitats over the next 10 years.

The state law prohibits direct taking of listed species but does not clearly provide habitat protection. Threatened and endangered species habitat does receive protection through other environmental laws that mandate such considerations for state-listed species. State laws that include certain levels of habitat protection for threatened and endangered species include the Nontidal Wetlands Protection Act, the Maryland Forest Conservation Act, and the Chesapeake Bay Critical Area Law. The critical area law was the first legislation in Maryland to mandate local jurisdictions to protect habitat for threatened and endangered species within 304.8 m (1,000 ft) of mean high tide of all tidal waters of the state (Therres, McKegg, and Miller 1988). This law also mandates conservation of colonial waterbird nesting sites, waterfowl concentration and staging areas, and forest interior breeding bird habitat. Technical guidance is provided by state DNR biologists to other government agencies to ensure that appropriate protection or conservation measures are incorporated into these mandated programs (Therres 1998).

PROTECTED LANDS

Thousands of acres of conservation lands in Maryland and the District of Columbia provide habitat for birds and other wildlife. Many of these areas are managed primarily for wildlife; some areas are managed for other natural resources but provide suitable habitat for birds in the process. In addition to public lands, many conservation areas owned and managed by private conservation organizations provide habitat for birds as well.

Federal Lands

The two major federal land management agencies that own conservation lands in Maryland and the District of Columbia are the U.S. Fish and Wildlife Service (USFWS) and the National Park Service (NPS). The U.S. Department of Defense owns military lands that secondarily provide habitats for many species of birds.

There are three large national wildlife refuges in Maryland and several other properties managed under the Chesapeake Marshlands National Wildlife Refuge Complex by the USFWS. The largest refuge is Blackwater National Wildlife Refuge on the Eastern Shore near Cambridge. The refuge includes more than 10,900 ha (27,000 ac) composed mainly of rich tidal marsh characterized by fluctuating water levels and varying salinity. Other habitat types include freshwater ponds, mixed evergreen and deciduous forests, and small amounts of cropland and managed impoundments that are seasonally flooded for waterfowl use. The 925-ha (2,285-ac) Eastern Neck Island National Wildlife Refuge, an island refuge at the mouth of the Chester River on the Eastern Shore, is a major feeding and resting place for migrating and wintering waterfowl. This refuge provides breeding habitat for forest, field, and wetland birds. The Patuxent Research Refuge is unique among national wildlife refuges. Established in 1936 by executive order of President Franklin D. Roosevelt, the

Table 7. Bird species listed by the state of Maryland as endangered, threatened, or in need of conservation

Species	State status	Federal status
American Bittern	in need of conservation	—
Least Bittern	in need of conservation	—
Bald Eagle	threatened	delisted
Northern Goshawk	endangered	—
American Peregrine Falcon	in need of conservation	delisted
Black Rail	endangered	—
Common Moorhen	in need of conservation	—
Wilson's Plover	endangered	—
Piping Plover	endangered	threatened
Upland Sandpiper	endangered	—
Gull-billed Tern	endangered	—
Royal Tern	endangered	—
Least Tern	threatened	—
Black Skimmer	endangered	—
Short-eared Owl	endangered	—
Alder Flycatcher	in need of conservation	—
Olive-sided Flycatcher	endangered	—
Bewick's Wren	endangered	—
Sedge Wren	endangered	—
Loggerhead Shrike	endangered	—
Nashville Warbler	in need of conservation	—
Blackburnian Warbler	threatened	—
Swainson's Warbler	endangered	—
Mourning Warbler	endangered	—
Henslow's Sparrow	threatened	—
Saltmarsh Sparrow	threatened	—
Coastal Plain Swamp Sparrow	in need of conservation	—

Source: Maryland Department of Natural Resources 2007.

Patuxent Research Refuge is the nation's only national wildlife refuge established to support wildlife research. Located along the Patuxent River near Laurel, this 5,196-ha (12,841-ac) refuge houses the Patuxent Wildlife Research Center operated by the U.S. Geological Survey. Much important bird research has been conducted there over the decades.

The National Park Service owns much of the remaining natural habitat in the District of Columbia and several properties in Maryland. Assateague Island National Seashore is a barrier island bordering the Atlantic Ocean. More than half of the 19,425 ha (48,000 ac) is composed of near-shore and estuarine waters; the other portion is sand dunes, maritime forest, and tidal wetlands. It supports several species of plants and animals found nowhere else in Maryland, including nesting Piping Plovers. Catoctin Mountain Park and C&O Canal National Historic Park are two other major NPS properties in Maryland. Catoctin Mountain Park lies within the mountains west of Frederick. This 2,351-ha (5,810-ac) park supports habitat for hardwood forest breeding birds. The C&O Canal parallels the Potomac River for 297 km (184.5 mi) from Washington, DC, to Cumberland, Maryland. It provides breeding habitat for many birds, but is best known for its birding opportunities during migration. Rock Creek Park, the largest forested area within the District of Columbia, supports most of the species of birds that breed in the nation's capital. Several historic parks and other NPS properties in Maryland and the District of Columbia support breeding birds.

Department of Defense (DOD) properties that provide many hectares of habitat for breeding birds in Maryland include Aberdeen Proving Ground; Patuxent River Naval Air Station; Indian Head Division, Naval Surface Warfare Center; and the U.S. Navy's Bloodsworth Island. Other DOD facilities provide limited habitat for birds and other wildlife in Maryland and the District of Columbia.

State Lands

The Maryland DNR is the state agency that owns and manages the majority of the state-owned conservation lands in Maryland. More than 182,000 ha (450,000 ac) of conservation lands are scattered throughout all 23 of Maryland's counties (Maryland Department of Natural Resources 2008). Most of these lands are managed as state forests, state parks, or wildlife management areas. The creation of the public system of lands began in 1906 with a donation to the state of 775 ha (1,917 ac) of forest land in Garrett County by John and Robert Garrett to establish a forest reserve (R. Bailey 2006). These lands are now part of Potomac-Garrett State Forest.

Now there are more than 80,900 ha (200,000 ac) of state forests, most of which are in western Maryland or on the Eastern Shore. The three large state forests in the mountains are Green Ridge, Potomac-Garrett, and Savage River. On the Eastern Shore, the 23,876-ha (59,000-ac) Chesapeake Forest is scattered among the lower five counties. Pocomoke State Forest on the lower Eastern Shore provides breeding habitat for several southeastern forest species.

The Maryland Park Service operates 48 state parks, totaling more than 37,600 ha (93,000 ac), and 31 natural resources management areas and natural environmental areas, totaling more than 12,100 ha (30,000 ac). Some state parks, such as Gunpowder Falls, Patapsco Valley, and Patuxent River, provide valuable riparian forest habitats in the Baltimore-Washington metropolitan area. Others, such as Assateague, Calvert Cliffs, Janes Island, and Soldiers Delight, provide unique habitats, while still others, such as Point Lookout and Washington Monument state parks, are important for migrating birds.

Wildlife management areas are state-owned lands managed primarily for wildlife and wildlife-oriented recreation, including hunting and birdwatching. More than 40,450 ha (100,000 ac) of such lands have been designated in Maryland. Several large wildlife management areas on the Eastern Shore, including Deal Island, Fishing Bay, and South Marsh, provide important tidal wetland habitats. Important forested wildlife management areas include Indian Springs, McKee-Beshers, Millington, Mount Nebo, and Dan's Mountain.

DNR's land acquisitions are made possible by Program Open Space funds. Those funds, derived from a percentage of the state real estate transfer tax, are used for land acquisitions not only by DNR but also by local governments for parks and recreation areas. Many of these county and municipal parks provide habitat for birds and other wildlife.

Private Lands

In addition to acquiring land in fee simple, the State, through the Maryland DNR and the Maryland Environmental Trust, purchases or receives donated conservation easements as a way of protecting private properties from development. DNR has purchased conservation easements on more than 25,500 ha (63,000 ac) through Rural Legacy and Forest Legacy programs. As of 2008, the Maryland Environmental Trust had permanently protected 48,387 ha (119,569 ac) on 983 properties statewide.

Several private conservation groups also own and manage land in Maryland for wildlife and natural resources. The Maryland Ornithological Society owns 9 sanctuaries located throughout the state from Garrett County in western Maryland to Somerset County on the lower Eastern Shore, protecting a total of 885 ha (2,187 ac). They range in size from 3.2 ha (8 ac) to 635 ha (1,570 ac) and provide woodlands, fields, swamps, and tidal marshes for birds and other wildlife. The newest sanctuary is in Prince George's County. Acquired in 2008, it is 33 ha (82 ac) of mostly deciduous forest.

The Nature Conservancy owns and manages 27 properties in Maryland for biodiversity conservation. Most of these sites were purchased to protect rare species or unique natural communities but also provide habitat for a variety of birds. Some of these properties especially important to birds are Cranesville Swamp, Finzel Swamp, Nanjemoy Creek Preserve, and Nassawango Creek Preserve.

The National Audubon Society also owns and manages conservation lands in Maryland. Its two largest properties are Pickering Creek Audubon Center and the Jean Ellen duPont Shehan Audubon Sanctuary, both in Talbot County. Audubon is coordinating the Important Bird Areas (IBA) Program, a global effort to identify and conserve IBAs throughout the United States. As of 2009, there were 27 designated IBAs in Maryland.

Scattered throughout the state are other conservation lands operated by organizations such as the Izaak Walton League of America, DuPont's Chesapeake Farms, Chesapeake Wildlife Heritage, Wildfowl Trust of North America, and Ducks Unlimited.

Using the Species Accounts

SPECIES PHOTOGRAPHS

Each nesting species is illustrated with a photograph that was, with a few exceptions, selected by the Atlas Board from among pictures submitted in a contest. Nearly all of the winning photos were taken in Maryland or the District of Columbia. Each photo carries the photographer's name.

ATLAS DISTRIBUTION, 2002–2006

The green squares on the large map show the current distribution of the species against a background of topographic map boundaries. If readers are familiar with the USGS 7.5-minute topographic maps they will be able to identify each quadrangle and block by reference to figure 1. The four categories of nesting evidence are explained in table 1. The total number of blocks with reports of the species and a count of blocks for each nesting category are presented in the inset table. Records coded as observed are pale gray rather than green and are not counted in the block total. The highest category of evidence reported in any of the five years is the one displayed for that block. Note that some block codes may be obscure where the black outline of the coast is complex and the color is pale (e.g., possible or observed blocks).

CHANGE IN ATLAS DISTRIBUTION, 1983– 1987 TO 2002–2006

The change map compares the 2002–2006 map with the 1983–1987 map in the previous atlas and has a green circle for every atlas block that lacked the species in the first atlas and gained it in the second. For every block that lost the species from the first atlas to the second there is a red triangle.

CHANGE IN TOTAL BLOCKS BETWEEN ATLASES BY REGION

The color-coded table at the bottom left shows the number of blocks in which the species was detected in each atlas project in each of the six physiographic provinces together with the change and percentage change in each. Note that the numbers of blocks reflect blocks covered in the two atlas projects and those that did not receive coverage in both projects were excluded from the comparison. However, comparisons did use data from the 1983–1987 Delaware atlas for the 15 Maryland-Delaware state-line blocks covered solely by Delaware atlas fieldwork at that time and not published in the 1983–1987 Maryland atlas.

RELATIVE ABUNDANCE FROM MINIROUTES, 2003–2007

This map shows the percentage of 3-minute early morning miniroute stops at which the species was detected in each atlas block during the current atlas. A similar miniroute map for 1983–1989, using the same percentage categories, was published in the first atlas (Robbins and Blom 1996).

BREEDING BIRD SURVEY RESULTS IN MARYLAND

The North American Breeding Bird Survey was designed to track population changes in North American birds by a network of 50-stop roadside counts that are repeated each year at the height of the breeding season. Since its inception in 1966, Maryland has been sampled annually by about 50 BBS routes per year, or at a density of 16 routes per 1-degree block of latitude and longitude. This is double the sampling density of any other state except Delaware, so Maryland's trends are well documented.

For species with sufficient BBS data, graphs of yearly estimated populations in Maryland are presented with the 95% Bayesian confidence limits which approximates a margin of error around the estimated abundance index. Vertical gray lines demarcate the two five-year Maryland-DC atlas project periods (1983–1987 and 2002–2006). In addition to the graph of change, reference is frequently made in the text to long-term (1966–2007) estimates of percentage yearly change, or trends, detected in the various physiographic regions that include Maryland and to national and continental trends as well. Trends referenced in the text, along with associated probability levels that show the significance of the trends, were obtained from the USGS BBS website's data analysis pages (www.mbr-pwrc.usgs.gov/bbs/bbs.html). For further discussion of the calculation of BBS abundance indices and trends, see page 37 in Robbins and Blom (1996).

NOTES

A list of acronyms and abbreviations precedes the main text. Abundance terms used in the species accounts are defined in appendix D. Scientific names for nonavian fauna mentioned in the text are given in appendix E, and those for plants are in appendix F.

Species Accounts

Canada Goose
Branta canadensis

Canada Geese have been present in the Chesapeake Bay region for thousands of years. No doubt Native Americans marveled at the flocks of honking geese as the birds arrived in October to winter in Maryland, leaving in March to return to their breeding grounds in northern Canada and start the cycle anew. Although migratory Canada Geese are still a notable seasonal phenomenon, they have been present year-round in Maryland for more than 60 years. Female Canada Geese are highly philopatric, breeding in the area where they were raised and returning to the same breeding area each year. When, in 1946, a small flock of wild birds was captured, pinioned, and released, their offspring remained to breed in Maryland (R. Stewart, Cope, et al. 1952). Subsequent releases from private landowners, hunt clubs, and federal and state wildlife agencies led to a slow increase in resident breeding populations during the 1970s and 1980s and a rapid increase in the 1990s (Costanzo and Hindman 2007). Owing to multiple releases of captive-raised birds and to translocations, resident Canada Geese in the eastern United States represent a mixture of several of the large-bodied subspecies (Mowbray et al. 2002)

Canada Geese have adapted well to human alterations of the landscape, making use of constructed wetlands for nesting, and feeding on crops, waste grain, and planted grass. Nests are almost always located very close to water in upland sites, on islands, or on artificial elevated structures; they may also be placed over water on dense vegetation or on muskrat or beaver lodges (Bellrose 1980). Nests are constructed of nearby vegetation and previous nest sites may be reused. Broods require easy access to water, plant food (short grasses or sedges, or semiaquatic plants), and open mud areas (Mowbray et al. 2002). Females first breed when they are 2–4 years old, typically incubating 5 eggs for about 28 days. Nest success can be very high, especially on islands (87.7%; Costanzo and Bidrowski 2004) and in urban populations where predation rates are reduced and competition for resources is low (Conover 1998; Cline et al. 2004). Young are accompanied by both parents, who are highly aggressive in defense of their broods. Parents molt their wing feathers and become flightless late in the brood-rearing period. Goslings fledge at about 7–8 weeks and families remain together through about the first year (Mowbray et al. 2002).

Stan Arnold noted female Canada Geese on nests in Maryland during the 2002–2006 atlas period as early as 5 March 2002, and other observers found them on nests as late as 1 July 2002 (Polly Batchelder, Mount Vernon, Somerset County) and 13 July 2002 (Jim Rapp, near Catchpenny, Wicomico County). Dennis Kirkwood saw downy young as early as 20 March 2002, and Bob Ringler noted downy or fledged young as late as 12 August 2005; Shirley Wood observed others on 27 August 2002. During the atlas period, 117 nests were located and 738 broods with downy or fledged goslings were reported. The number of blocks with probable and confirmed breeders increased dramatically from the 1983–1987 atlas (49 and 198, respectively; Meritt 1996b) to the 2002–2006 atlas period (124 and 776, respectively), for an overall 187% increase in blocks with at least possible breeders. Compared to the earlier atlas period, nesting Canada Geese have increased in all parts of the state, although they are still absent from areas where there are few suitable wetlands.

BBS data also show an increasing trend in breeding Canada Geese in Maryland (overall increase of 14.6%), the United States (7.3%), and North America (8.5%). In the Chesapeake Bay region, the average number of Canada Geese per BBS route increased from 0.651 (1966–1970) to 48.2 (2001–2005), although short-term trends indicate a weaker increase and an estimated average 24,273 breeding pairs from 1993 to 2005 (Costanzo and Hindman 2007). Growth of resident Canada Goose populations in Maryland appears to be slowing as available habitat is saturated and as harvest has increased during special September hunting seasons designed to reduce resident but not migrant populations. Given continued urbanization in Maryland and the large number of resident Canada Geese nesting in the mid Atlantic region, we can expect this species to continue as a common breeding bird in the state.

GWENDA L. BREWER

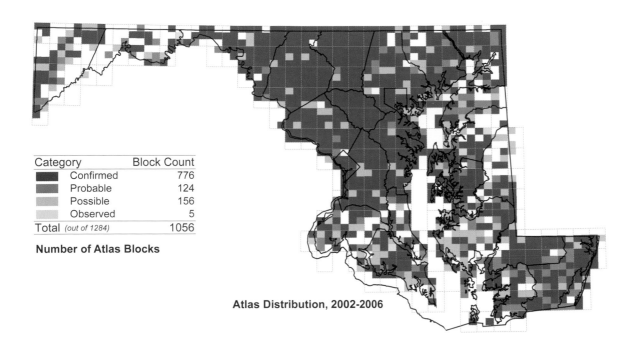

Category	Block Count
Confirmed	776
Probable	124
Possible	156
Observed	5
Total (out of 1284)	1056

Number of Atlas Blocks

Atlas Distribution, 2002-2006

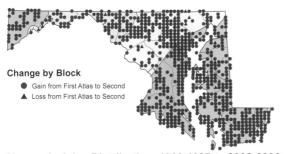

Change by Block
- Gain from First Atlas to Second
▲ Loss from First Atlas to Second

Change in Atlas Distribution, 1983-1987 to 2002-2006

Percent of Stops
- 50 - 100%
- 10 - 50%
- 0.1 - 10%
- < 0.1%

Relative Abundance from Miniroutes, 2003-2007

Atlas Region	1983-1987	2002-2006	Change No.	%
Allegheny Mountain	11	50	+39	+355%
Ridge and Valley	35	110	+75	+214%
Piedmont	155	301	+146	+94%
Upper Chesapeake	24	86	+62	+258%
Eastern Shore	66	287	+221	+335%
Western Shore	75	218	+143	+191%
Totals	366	1052	+686	+187%

Change in Total Blocks between Atlases by Region

Breeding Bird Survey Results in Maryland

Philip Brody

Mute Swan
Cygnus olor

The large and stately Mute Swan, with its knobbed orange and black bill, has adorned paintings, china, tapestry, dinnerware, and many a palace pond through the ages. In the mid-1800s Mute Swans from Europe were introduced to private estates, city parks, and the wild in the eastern United States. Reese (1996a) chronicles the introduction and spread of this species into the Chesapeake Bay, starting from an escaped group of five birds in 1962. Mute Swans also likely escaped from other sources in nearby states to coastal areas and other parts of Maryland; free-living birds have been sighted since 1929 in Pennsylvania, 1954 in Delaware, and 1957 in Virginia (Ciaranca et al. 1997). Population sizes in Maryland grew relatively slowly until the mid-1980s. From 1986 to 1992 the Mute Swan population increased at an annual rate of about 23% (Hindman and Harvey 2004). Breeding Bird Survey data show a significant increase of about 21% over the past 25 years. The 2002–2006 USFWS survey data show still greater gains. Midwinter surveys estimated an increase from 2,145 to 4,345; breeding waterfowl surveys reported a rise from 1,216 to 2,167 over the same period (DNR, unpubl. data).

Mute Swans are sedentary or undergo short-distance movements outside of the breeding season. Birds first breed when they are 3–4 years old, setting up territories that they defend from other waterfowl. In their native range of Europe to central Asia, Mute Swans inhabit steppe lakes, rivers, and marshes (Ciaranca et al. 1997). In the Chesapeake Bay, Reese (1980) found that large nests of mounded aquatic vegetation were built in marshland near the shoreline. Fe-

males deposited 4–10 eggs (mean 6.2, n = 151 nests) and incubated them for 36 days, with first eggs observed 8 March and downy young first observed on 28 April. Hatching success varied from 27% to 75% (mean 49%) with high hatching to fledging success (mean 82%) (Reese 1980). Causes of egg failure included disappearance (probably from predation) and high tides. Second clutches replaced eggs lost before mid-May; observers noted no eggs past 23 June (Reese 1980). In the 2002–2006 atlas period Harry Armistead noted a female on a nest as late as 1 July 2005 on Ferry Neck, Talbot County, and Lynn Davidson observed small cygnets as late as 12 August 2006 near Blackwater NWR, Dorchester County. In general, cygnets fledge at 120–150 days; they sometimes remain with their parents through their first winter (Ciaranca et al. 1997).

Breeding bird atlas data demonstrate an impressive increase in the distribution of Mute Swans between the first and second atlas periods. The number of blocks with possible, probable, or confirmed breeders went from 57 in 1983–1987 to 181 in 2002–2006, an increase of 216%. Mute Swans continue to occupy their former area of concentration along the east-central Chesapeake Bay shoreline. From this area, the distribution of breeding birds has primarily expanded south to eastern Dorchester and Somerset counties, north to the Upper Chesapeake Bay, and westward across the bay from Harford County to southern St. Mary's County. Observers noted additional records from the lower Potomac River and tributaries, middle and lower Patuxent River, back bays of Worcester County, and scattered locations associated with large reservoirs as far west as Frederick County.

Mute Swan behavior and use of resources have raised concerns about impacts on native species. Defense of territories averaging 5.2 ha (13 ac) (Reese 1996a) excludes other waterfowl from using an area during the breeding season and beyond if pairs defend a territory year-round. Territory defense can be so vigorous that birds and mammals entering Mute Swan territories have been killed (see Ciaranca et al. 1997). Exclusion from resources as well as other impacts can occur when large flocks of nonbreeders congregate. In Maryland, a large flock of molting Mute Swans (600–1,000 individuals) caused the abandonment of nesting sites and probably loss of young for state-threatened colonial nesting birds because of intensive use of favored oyster shell bars and beaches for loafing (Therres and Brinker 2004). These

large flocks and sometimes even family groups dislodge and consume large quantities of submerged aquatic vegetation, negatively impacting percent cover, shoot density, and canopy height (Tatu et al. 2007). To limit the negative impacts of Mute Swan aggression and resource use, control measures for Mute Swan populations in the Chesapeake Bay have been carried out in recent years (Hindman and Harvey 2004; Costanzo and Hindman 2007). These efforts have slowed the expansion of the non-native Mute Swan in Maryland and are likely to reduce its breeding distribution in the near future.

GWENDA L. BREWER

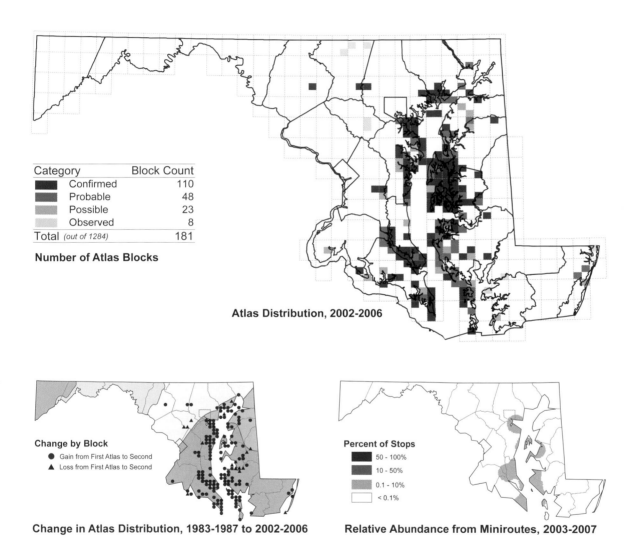

Category	Block Count
Confirmed	110
Probable	48
Possible	23
Observed	8
Total (out of 1284)	181

Number of Atlas Blocks

Atlas Distribution, 2002-2006

Change by Block
● Gain from First Atlas to Second
▲ Loss from First Atlas to Second

Change in Atlas Distribution, 1983-1987 to 2002-2006

Percent of Stops
- 50 - 100%
- 10 - 50%
- 0.1 - 10%
- < 0.1%

Relative Abundance from Miniroutes, 2003-2007

			Change	
Atlas Region	1983-1987	2002-2006	No.	%
Allegheny Mountain	0	0	0	-
Ridge and Valley	0	0	0	-
Piedmont	3	4	+1	+33%
Upper Chesapeake	4	21	+17	+425%
Eastern Shore	46	90	+44	+96%
Western Shore	4	65	+61	+1525%
Totals	57	180	+123	+216%

Change in Total Blocks between Atlases by Region

Charles Lentz

Wood Duck
Aix sponsa

The most common of the three cavity-nesting ducks in Maryland is the Wood Duck. In wooded wetlands across the state, it is not uncommon to hear the plaintive, drawn-out *oo-eek oo-eek* call of the female and the much softer, buzzing *ji-ihb* whistle of the male as a pair takes flight from a tree perch. What the male lacks in the complexity of his call, he makes up in the richness of his plumage, being one of the most striking male ducks. We are fortunate to see this species year-round in Maryland on a wide variety of wetland habitats, with peak numbers during fall migration in October and early November (Haramis 1991). Wood Ducks breed across much of the United States and extreme southern Canada; wintering birds are primarily found in the southeastern United States (Bellrose 1980).

Wood Ducks were once less common in the state and across their breeding range in the United States and southern Canada because of unregulated hunting and the loss of nesting cavities and wetlands as swamp forests were harvested and drained (Hepp and Bellrose 1995). Passage of the Migratory Bird Treaty Act of 1918, regrowth of forests in the eastern United States, enforcement of wetlands regulations, expanding beaver populations, and active nest box programs over past years have helped this beautiful but shy duck increase across its entire range.

Breeding Wood Ducks are present in Maryland in good numbers by late March and early April. Wood Duck nests are located close to or over water in mature forest or for-ested wetlands, with broods also using fresh tidal marshes and brackish sections of estuaries (Bellrose 1980). In addition to nest boxes, Wood Ducks nest in cavities in dead or live trees. The female lines the cavity with down and incubates the eggs for about 30 days. After hatching, ducklings leap, sometimes from great heights, to the ground in response to their mother's encouraging calls. In Maryland, McGilvrey (1969) found that females accompanied broods for about 5 weeks and 47% of ducklings survived from hatching to flight stage at about 70 days. Wood Ducks regularly produce two broods in one breeding season in the southern parts of their range, including Maryland (summarized in Hepp and Bellrose 1995); they are the only North American ducks to have second broods.

In the 2002–2006 atlas period, D. Perry found the earliest nest with eggs on 8 March 2002, one week later than the early date for eggs reported in Robbins and Bystrak (1977). Charlie Vaughn found the latest occupied nest on 25 July 2006. Other observers found downy young as early as 30 March 2006 near Centreville, Queen Anne's County (new early date; Lori Byrne) and found them as late as 28 July 2005 (DNR), Bob Ringler reported a late date for dependent young on 13 August, compared to a latest date of 2 September, noted 53 years before by Stewart and Robbins (1958).

Wood Ducks increased markedly in Maryland from the 1983–1987 to the 2002–2006 atlas period. Breeding Wood Ducks were found in all parts of the state, with a lower number of blocks in Garrett and Allegany counties. Observers found these ducks in a net total of 149 more blocks, a 25% increase over the previous atlas period. Although observers noted new blocks with Wood Duck records in all counties, areas in the Hagerstown Valley, southern Prince George's, and central Charles appear to have decreased Wood Duck distribution, possibly because of development pressures. In comparison with the 1983–1987 atlas, reports indicated fewer blocks with Wood Ducks along the Potomac River in Garrett County, where access can be difficult.

Many reports confirmed Wood Ducks breeding. Observers recorded 419 females with broods or young Wood Ducks (some may have been in the same block) and 82 records of nests with eggs or young, or with a laying or incubating female. Increases in Maryland breeding records for Wood Ducks mirror long-term trends from BBS data (increase of 5.9% in Maryland). Costanzo and Hindman (2007) estimate

that there were on average about 15,000 breeding pairs in the Chesapeake Bay region from 1993–2005. Although Wood Duck populations have made great progress, these authors caution that risks remain as forested wetlands are lost or degraded because of increasing human development pressure. In order for a breeding population of Wood Ducks to remain in Maryland, riparian areas need to be managed for mature and old growth forest, and small wetlands and riparian forest need to be protected in addition to large bottomland and riverine marshes (Haramis 1991).

GWENDA L. BREWER

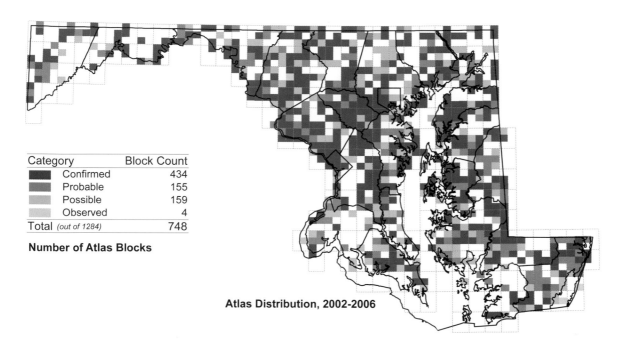

Category	Block Count
Confirmed	434
Probable	155
Possible	159
Observed	4
Total (out of 1284)	748

Number of Atlas Blocks

Atlas Distribution, 2002-2006

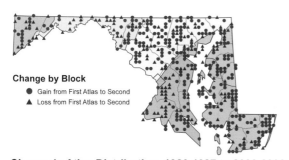

Change by Block
- ● Gain from First Atlas to Second
- ▲ Loss from First Atlas to Second

Change in Atlas Distribution, 1983-1987 to 2002-2006

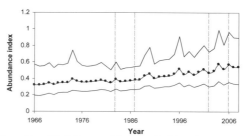

Percent of Stops
- 50 - 100%
- 10 - 50%
- 0.1 - 10%
- < 0.1%

Relative Abundance from Miniroutes, 2003-2007

Atlas Region	1983-1987	2002-2006	Change No.	Change %
Allegheny Mountain	31	31	0	0%
Ridge and Valley	90	84	-6	-7%
Piedmont	161	218	+57	+35%
Upper Chesapeake	59	75	+16	+27%
Eastern Shore	115	190	+75	+65%
Western Shore	140	147	+7	+5%
Totals	596	745	+149	+25%

Change in Total Blocks between Atlases by Region

Breeding Bird Survey Results in Maryland

Mark Hoffman

Gadwall
Anas strepera

The coming of fall brings a familiar sound to numerous bodies of water in Maryland: the nasal croak of male Gadwalls. It is this frequent sound that earned Gadwalls the Latin name *strepera,* meaning "noisy." In late August, Gadwalls start to arrive to overwinter, primarily in Maryland's coastal areas. Their active courting parties, where males compete for females through visual and vocal displays, are soon in evidence, as Gadwall males are the first among North American ducks to begin pairing up with females (LeSchack et al. 1997; Fox 2005). Numbers continue to increase to about mid-October, when most Gadwalls have arrived in the mid Atlantic region. By April most Gadwall pairs have left Maryland to journey to the prairie potholes of the United States and Canada, where the female returns to a familiar breeding site. A small number, however, breed in Maryland's fresh and tidal marshes.

On the northern prairies, where Gadwalls are a common nester, females nest on the ground in dense vegetation, often far from water. In general, 5–13 eggs are incubated by the female for about 25 days. Nest success is high and most young fledge by about 50 days (Bellrose 1980). An unusual predator on ducklings (noted in impoundments at Deal Island WMA) is the blue crab (L. Hindman, pers. comm.). Maryland egg dates range from 4 May to 19 July, with broods documented from 18 May into August (Stewart and Robbins 1958; MNRF).

Nesting Gadwalls were first noted in Maryland in 1948 (Springer and Stewart 1950), heralding the dramatic increase that occurred in the eastern United States in the 1950s when Gadwalls colonized freshwater impoundments in brackish areas (Henny and Holgerson 1974). In the 1983–1987 atlas period, observers confirmed Gadwalls in 8 blocks, found them probable in 10, and possible in 7 (Meritt 1996h). Confirmed and probable records were concentrated in lower Dorchester County, Deal Island WMA, and Chesapeake Bay islands to the south, and in lower Worcester County, with 2 possible records on the lower Western Shore. Results from the 2002–2006 atlas period show that the Gadwall remains a rare nester in Maryland. In the recent atlas period, Gadwalls occurred as possible breeders in 6 blocks, probable breeders in 11 blocks, and confirmed breeders in only 2 blocks. Most records continue to be in the tidal and freshwater marshes of lower Dorchester and western Somerset counties, the lower Chesapeake Bay islands, and the coastal bay marshes of Assateague Island. Observers reported no Gadwalls in southern Maryland during the breeding season, although they had been possible breeders in the previous atlas period. A probable breeder was reported from south of Baltimore and a possible breeder from Kent Island. Although observers documented breeding Gadwalls in the same general areas during the two atlas periods, they found this duck in fewer blocks in lower Dorchester County in the 2002–2006 atlas period. Loss of marsh habitat in Blackwater NWR may have contributed to this change.

BBS trend data show a slight increase for Gadwalls across North America, but note few individuals per route. Data from the midwinter waterfowl surveys for Maryland indicate that Gadwall estimates are generally increasing, although numbers declined during 2005 and 2006 (USFWS Migratory

Bird Data Center, http://mbdcapps.fws.gov). Gadwall trends from 1955–2008 in the core northeastern and midwestern breeding areas also show an overall increase, although there was a decline during the 2002–2006 atlas period (Zimpfer et al. 2008) that was probably caused by unfavorable weather conditions. In the Chesapeake Bay region, habitat loss from sea level rise and other causes of marsh and island loss could especially impact nesting Gadwalls into the future, as females in other areas have shown high site fidelity to breeding areas (D. Johnson et al. 1992). Although Gadwalls can benefit from the creation of artificial habitats, they do not appear to tolerate disturbance and are susceptible to increasing use of these areas by humans (Fox 2005).

GWENDA L. BREWER

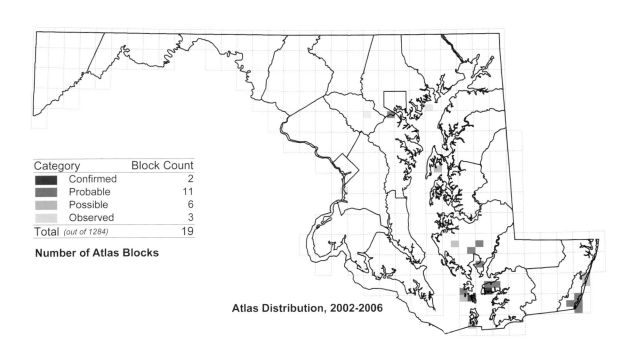

Category		Block Count
▉	Confirmed	2
▉	Probable	11
▉	Possible	6
▢	Observed	3
Total (out of 1284)		19

Number of Atlas Blocks

Atlas Distribution, 2002-2006

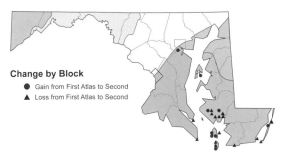

Change by Block
- ● Gain from First Atlas to Second
- ▲ Loss from First Atlas to Second

Change in Atlas Distribution, 1983-1987 to 2002-2006

Atlas Region		1983-1987	2002-2006	Change No.	%
	Allegheny Mountain	0	0	0	-
	Ridge and Valley	0	0	0	-
	Piedmont	0	0	0	-
	Upper Chesapeake	0	0	0	-
	Eastern Shore	23	18	-5	-22%
	Western Shore	2	1	-1	-50%
Totals		25	19	-6	-24%

Change in Total Blocks between Atlases by Region

American Black Duck

Anas rubripes

American Black Ducks both winter and breed in Maryland, with peak numbers in the state from early November to mid-March. The breeding range of American Black Duck includes wooded wetlands and bogs in the northern United States and eastern Canada south to the brackish coastal marshes of North Carolina (Longcore et al. 2000). The Chesapeake Bay region once harbored one of the largest southern breeding populations, with the largest concentrations and densities on bay islands and the marshes of Tangier Sound (Stotts and Davis 1960; Haramis et al. 2002).

In our region, observers found American Black Duck nests in woods, fields, brush piles, and shrub thickets, in elevated structures such as duck blinds and tree trunks, and in marshes, especially low salt marsh in some areas (Stotts and Davis 1960; Haramis et al. 2002). Male black ducks defend a territory during the breeding season, including against Mallards, until the female is about halfway through the 26-day incubation period (Stotts and Davis 1960; Longcore et al. 2000). Mean clutch sizes for 360 Maryland Chesapeake Bay nests were 10.9 for early nests (peak, April 20) and 7.5 for later nests (last hatch, early August); 38% of 574 nests hatched at least one duckling (Stotts and Davis 1960). Haramis et al. (2002) noted great variability in hatching success of nests on Smith Island over a two-year study (0%, n = 8; 80%, n = 5), with both studies noting predation as the major cause of nest loss. Young fledge at about 60 days and depend on shallow wetlands with emergent vegetation or scrub-shrub vegetation, new or reflooded beaver meadows, and brackish marshes interspersed with shallow ponds (Longcore et al. 2000).

American Black Ducks were found in 31% fewer blocks during the 2002–2006 atlas period (total 163) compared to the 1983–1987 atlas period (total 235). In the earlier atlas period, probable and confirmed records were located around the Chesapeake Bay, especially on bay islands, lower Dorchester and Somerset counties, and coastal Worcester County. Records were also noted along the Patuxent River and along the Potomac River south of Washington, DC, with scattered records from other locations, including western Maryland. During the 2002–2006 atlas period, nesting black ducks were much less widely distributed along the Patuxent River and were almost absent from the Potomac River. A few scattered records were still found in the Allegheny Plateau region, and new locations were documented in Cecil County. Declines in the number of blocks were seen in the Ridge and Valley, Piedmont, and much of the Upper Chesapeake region.

BBS data show declining but not significant trends for American Black Ducks (Maryland: 4.3% per year; Upper Coastal Plain 2.8% per year), possibly due to low numbers of breeders along few roadside routes. Data from Atlantic Flyway breeding waterfowl plot surveys, 1993–2005, show a significant decline in the Chesapeake Bay region, but these data are limited in assessing trends due to the small number of breeding black ducks noted on the surveys (Costanzo and Hindman 2007). Data from midwinter surveys in the eastern United States and Canada show a decrease in American Black Ducks from 1955 to 2005, but a slight increase in breeding populations, with fluctuations from 1990 to 2008 (USFWS 2008). Costanzo and Hindman (2007) estimated an average of 3,658 breeding pairs in the Chesapeake Bay region from 1993 to 2005.

Concerns in the early 1980s about continued declines and disappearances from former breeding areas led to harvest restrictions and an increase in exploration of black duck biology and conservation (Black Duck Joint Venture 2008). Early hypotheses suggested that the release of captive-reared Mallards in traditional black duck wintering and breeding areas could lead to increased competition and hybridization. Further study indicated that interactions between these two species may negatively affect black ducks, although McAuley, Clugston, and Longcore (2004), for example, did not see evidence that Mallards were excluding American Black Ducks from resources through direct competition during the breeding season in Maine. Bolen et al. (2002) assert that sensitivity to human disturbance by black ducks is more likely to explain black duck declines than competition with Mallards. Manke et al. (2004) found a dramatic reduction in the genetic differences between Mallards and black ducks in post-1998 samples compared to pre-1940 ones. But the impact of hybridization on black duck populations is still not well understood and captive Mallard releases are much less common. Losses in the quality and quantity of breeding and wintering habitats and overharvest are other factors that may also influence black duck populations (Conroy et al. 2002).

In Maryland, loss of submerged aquatic vegetation appears to affect wintering black ducks in the Chesapeake Bay, and breeding habitat has been lost because of human distur-

bance, development, habitat degradation, and sea level rise impacts on islands and marshes (Stotts 1987; Krementz 1991; Erwin, Sanders, et al. 2006). Predators, especially on nesting islands, are also a concern. Increased development since the last atlas period in southern Maryland, the Piedmont, and other areas may have resulted in fewer blocks with breeding black ducks through increases in direct disturbance and habitat degradation. Conservation of this species in Maryland and across its range depends on minimizing human impacts to wintering and breeding habitats and maintaining remaining nesting islands as productive breeding habitats.

GWENDA L. BREWER

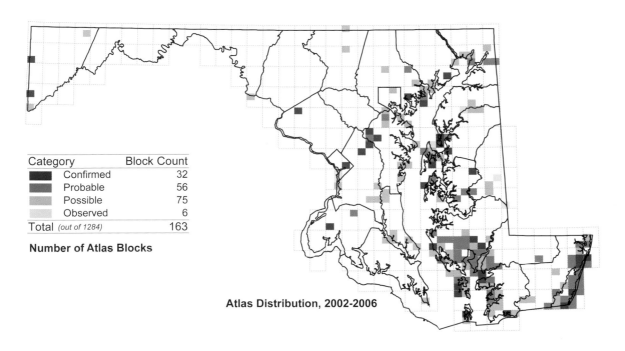

Category	Block Count
Confirmed	32
Probable	56
Possible	75
Observed	6
Total (out of 1284)	163

Number of Atlas Blocks

Atlas Distribution, 2002-2006

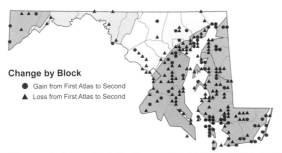

Change by Block

● Gain from First Atlas to Second
▲ Loss from First Atlas to Second

Change in Atlas Distribution, 1983-1987 to 2002-2006

Percent of Stops

■	50 - 100%
■	10 - 50%
■	0.1 - 10%
□	< 0.1%

Relative Abundance from Miniroutes, 2003-2007

Atlas Region	1983-1987	2002-2006	Change No.	Change %
Allegheny Mountain	4	4	0	0%
Ridge and Valley	6	1	-5	-83%
Piedmont	16	10	-6	-38%
Upper Chesapeake	27	13	-14	-52%
Eastern Shore	120	104	-16	-13%
Western Shore	60	29	-31	-52%
Totals	233	161	-72	-31%

Change in Total Blocks between Atlases by Region

Breeding Bird Survey Results in Maryland

Gary Smyle

Mallard
Anas platyrhynchos

The metallic green head of the male Mallard and the patterned brown of the female are an increasingly familiar sight across the state, from natural wetlands to neighborhood ponds. Mallards breed across the United States and Canada, migrating short to medium distances to wintering areas or remaining sedentary (Bellrose 1980). The Mallard is surely one of the most widespread and successful ducks across its range in North America because of its great adaptability to human-created and altered environments. Releases of captive-reared birds for hunting into the eastern and northeastern states have also hastened the spread of Mallards. In Maryland, Hindman et al. (1992) estimate that more than 260,000 Mallards were released under the Maryland Waterfowl Stamp Program between 1974 and 1987 to supplement hunter harvest and increase local breeding stocks. While state releases were discontinued in 1993, birds are still released in regulated shooting areas, and a study by D. B. Smith (1999) indicates that some of these birds traveled to other areas in Dorchester County.

The Mallard breeds in a great variety of habitats, from marshes and grasslands to forested areas with nearby rivers or ponds. In Maryland, L. Adams et al. (1985) documented the use of stormwater management ponds by breeding Mallards in the Columbia area, finding that they preferred shallow ponds to deep ponds or lakes. Females form a shallow nest bowl in dense cover in grasslands, marshes, roadside ditches, cropland, riverine floodplains, and forests (Drilling

et al. 2002), as well as in a variety of artificial structures, including docks, boats, and flower boxes in urban/suburban areas (Figley and VanDruff 1982). Incubation lasts for 28 days and young fledge at 50–70 days (Drilling et al. 2002). MNRF indicates that clutch sizes range from 1 to 40 (likely the result of dump nesting), with an average of 9 for 175 presumed nondump nests, and brood sizes range from 1 to 12, averaging 6.8 for 53 nests (Meritt 1996e). Data from midwestern prairie nests indicate that nest predation rates can be as high as 72% and brood survival rates vary widely (summarized in Drilling et al. 2002). Renesting is common if eggs are destroyed or the nest is abandoned. Urban Mallards, unlike wild Mallards, will frequently renest even when broods are lost (Figley and VanDruff 1982). Factors likely to affect nest success and brood survival in Maryland include predation and weather during brood rearing. Survival of adult Mallards in Maryland is influenced by predation (especially of the female while incubating) and hunter harvest, less so by pesticides and contaminants, and by collisions with vehicles and stationary structures.

During the 2002–2006 atlas period Jane Coskren noted Mallard females on nests as early as 17 March 2002, and Lynn Davidson found a late nest on 12 August 2006 at Blackwater NWR. Dave Wilson first saw downy or fledged young on 15 April 2004; Connie Skipper recorded the last on 26 August 2005. Earlier egg dates and downy young were noted in the 1983–1987 atlas period (12 March and 25 March, respectively; Meritt 1996e). Although Mallards bred widely across the state during both atlas periods, the number of blocks with probable breeders increased from 192 to 290, and the number of blocks with confirmed breeders increased from 493 to 518. Possible records remained similar: 208 for 1983–1987 and 215 for 2002–2006. Observers noted many records of broods (554) and 96 nests during the 2002–2006 atlas period. Most new block records were in the Allegheny Plateau, Piedmont, Western Shore, and Eastern Shore regions of the state. Highly forested areas in Charles County and Worcester County, and areas in Allegany and Garrett counties with few wetlands were the only places in Maryland where nesting Mallards were not seen during either atlas period.

BBS trends show increases in Maryland, the United States, and North America, although recent increases are smaller in magnitude than in previous years. Data from the upper Midwest and Canadian breeding ranges indicate a stable or slowly increasing population during the 2002–2006 atlas period (Zimpfer et al. 2008). Costanzo and Hindman (2007) estimated that an average of 43,575 pairs nested annually in the Chesapeake Bay region 1993–2005 and that short-term trends are relatively stable. The halt of the state release of captive Mallards, more liberal hunting seasons since 1995, and density-dependent population regulation have all contributed to the leveling off of breeding population size in Maryland. Given their great adaptability to environments that include human-dominated landscapes, Mallards are expected to remain very common breeders in the state.

GWENDA L. BREWER

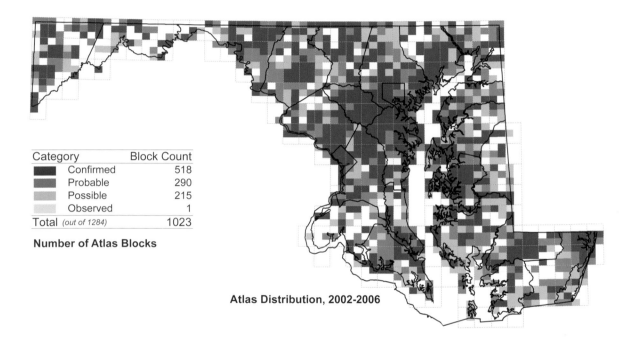

Category	Block Count
Confirmed	518
Probable	290
Possible	215
Observed	1
Total (out of 1284)	1023

Number of Atlas Blocks

Atlas Distribution, 2002-2006

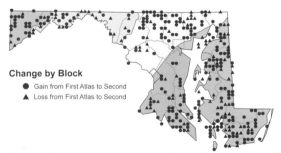

Change by Block
- ● Gain from First Atlas to Second
- ▲ Loss from First Atlas to Second

Change in Atlas Distribution, 1983-1987 to 2002-2006

Percent of Stops
- 50 - 100%
- 10 - 50%
- 0.1 - 10%
- < 0.1%

Relative Abundance from Miniroutes, 2003-2007

Atlas Region	1983-1987	2002-2006	Change No.	Change %
Allegheny Mountain	40	60	+20	+50%
Ridge and Valley	107	105	-2	-2%
Piedmont	243	283	+40	+16%
Upper Chesapeake	87	88	+1	+1%
Eastern Shore	233	279	+46	+20%
Western Shore	179	205	+26	+15%
Totals	889	1020	+131	+15%

Change in Total Blocks between Atlases by Region

Breeding Bird Survey Results in Maryland

Blue-winged Teal
Anas discors

Sightings of the Blue-winged Teal, with its flash of blue and quick flight, are a welcome but rare occurrence during the breeding season in Maryland. They are present only in relatively small numbers in winter as well, as most birds winter in southern Mexico and northern South America after a 26-hour flight across the Gulf of Mexico. These long-distance migrants are among the first to leave in fall and the last to return in spring (Rohwer et al. 2002). In Maryland, the majority of Blue-winged Teal do not arrive until early April and most birds have left by mid-September (Iliff, Ringler, and Stasz 1996). Their primary breeding range is in the north-central United States and adjacent Canadian provinces, although they are known to breed in good numbers outside of this area when conditions permit (Bellrose 1980). The creation of freshwater impoundments and a growing number of beaver-created wetlands have probably contributed to the expansion of Blue-winged Teal into new breeding areas in eastern North America (Rohwer et al. 2002).

Meanley (1975) reports the first recorded sighting of a Blue-winged Teal female and brood in Maryland along the Blackwater River in 1929. Blue-winged Teal prefer shallow ponds with abundant invertebrate food, nesting in grassy or herbaceous cover nearby (summarized in Rohwer et al. 2002). In Maryland, this habitat has primarily been represented by small pothole ponds formed from muskrat eat-out activity in the midst of brackish salt marsh meadows. Nesting cover, in addition to appropriate water levels, appears to determine

breeding area selection. In the upper Midwest, management that reduced residual cover, such as burning or mowing, resulted in lower nest densities in subsequent years (Livezey 1981). Male mates of nesting females are intolerant of other Blue-winged Teal and defend a territory until late in incubation (R. Stewart and Titman 1980). Females typically lay 10–11 eggs in a shallow ground nest lined with grasses and down. The incubation period lasts for 23–24 days and young fledge at about 40 days (Evarts 2005). Egg dates for Maryland range from 25 April to 12 July, with broods documented from 25 May to 11August (Stewart and Robbins 1958; Meritt 1996f). Nest success can be quite variable, with low success primarily resulting from mammalian predators and flooding (Bellrose 1980).

BBS trend data show a very slight decrease for Blue-winged Teal across the United States and North America, but observers noted few individuals on few routes. Population estimates for Blue-winged Teal from 1955 to 2008 in the core upper midwestern breeding areas show a great degree of fluctuation, with high population levels from 1995 to 2000 that then decreased during the 2002–2006 atlas period (Zimpfer et al. 2008), probably as a result of weather conditions. Recent breeding surveys in the core of their range show an increase in Blue-winged Teal.

Historically in Maryland, Blue-winged Teal nested primarily in tidewater areas of the lower Eastern Shore although records also exist for the Western Shore and Upper Chesapeake (Stewart and Robbins 1958). Consistent with this information, in the 1983–1987 atlas period, 3 probable and 7 confirmed blocks were noted in lower Dorchester and western Somerset counties, with a number of possible records (10 blocks) located nearby (Meritt 1996f). Two blocks in Worcester County had probable records, and 1 block on Smith Island had a confirmed record. In the 2002–2006 atlas period, however, Blue-winged Teal were confirmed as breeding in only 1 block, located in southwestern Queen Anne's County. A total of 4 blocks held probable breeders and 2 blocks, possible breeders. Only half of these blocks (2 probable and 1 possible) were found in areas identified as strongholds for this species in Maryland in the previous atlas, the tidal marshes of lower Dorchester and western Somerset counties (Meritt 1996f). One possible and 2 probable blocks were identified on the Western Shore, in St. Mary's, Anne Arundel, and Harford counties. Habitat loss and possibly

vegetation alteration in the marshes of lower Dorchester County, particularly along the Blackwater River, may explain the loss of records from 9 blocks between the two atlas periods. Records from the Western Shore in the most recent atlas period may reflect the propensity for this species to take advantage of favorable conditions. Loss of marsh habitat remains the greatest threat to Blue-winged Teal populations, especially in traditional stronghold areas, but increases in beaver-created wetlands and the management of freshwater impoundments could favor a wider and more consistent distribution of breeding Blue-winged Teal in Maryland.

GWENDA L. BREWER

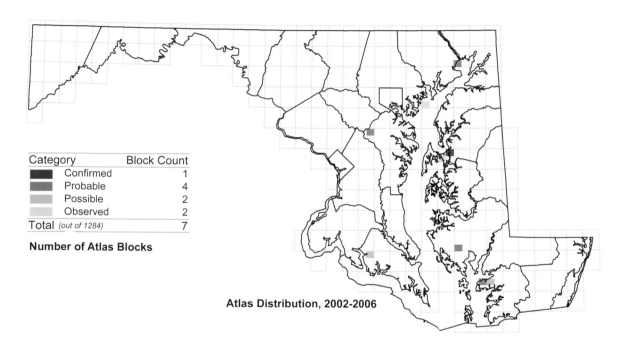

Category	Block Count
Confirmed	1
Probable	4
Possible	2
Observed	2
Total (out of 1284)	7

Number of Atlas Blocks

Atlas Distribution, 2002-2006

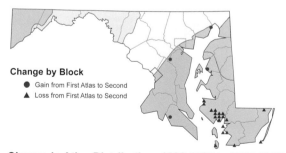

Change by Block

● Gain from First Atlas to Second
▲ Loss from First Atlas to Second

Change in Atlas Distribution, 1983-1987 to 2002-2006

Atlas Region	1983-1987	2002-2006	Change No.	Change %
Allegheny Mountain	0	0	0	-
Ridge and Valley	0	0	0	-
Piedmont	0	0	0	-
Upper Chesapeake	0	1	+1	-
Eastern Shore	22	4	-18	-82%
Western Shore	0	2	+2	-
Totals	22	7	-15	-68%

Change in Total Blocks between Atlases by Region

Hooded Merganser
Lophodytes cucullatus

What a nice surprise when a crested female with a thin bill and white wing patches leaves a Wood Duck box. The Hooded Merganser is the only other cavity-nesting duck in Maryland that is small enough to make use of these boxes. Its cousin, the Common Merganser, is the largest breeding duck in the state. The Hooded Merganser breeds exclusively in North America, generally migrating only short distances between its breeding and wintering areas (Bellrose 1980). Although found throughout the year in Maryland, most individuals are present from mid-November to mid-April (Iliff, Ringler, and Stasz 1996). During this period, the male shows off his fan-shaped crest as part of elaborate courtship behavior. Breeding habitat consists of forested wetland systems, while wintering habitat includes a wider range of freshwater and brackish areas (Dugger et al. 1994). Winter food of Hooded Merganser in the Chesapeake Bay was found to be quite varied, including fish, crayfish, mud crabs, dragonfly nymphs, and other aquatic invertebrates (P. Stewart 1962).

Hooded Merganser females deposit 7–13 almost spherical white eggs with unusually thick shells in a cavity over or near water (Dugger et al. 1994). In other parts of their range, eggs are incubated for about 32 days and young birds fledge at about 70 days (Bellrose 1980). Hooded Mergansers will lay eggs in the nests of other females of their own species and in Wood Duck nests, with parasitism rates in Minnesota as high as 45% and 38% of nests, respectively (Zicus 1990). Data from a 2006 effort to document Hooded Merganser nesting on Maryland public lands found this species in 32 Wood

Duck nest boxes (of 577 in areas where Hooded Mergansers were found) and 2 natural cavities, primarily on the Western Shore (Maryland Wood Duck Initiative 2006). The percent of ducklings hatched (n = 150) compared to the number of eggs laid (n = 227) was 66%, comparable to that of Wood Ducks in the same year.

Data from the 2002–2006 atlas period show occupied nests as early as 5 March 2002 (Edward Clark, Greenbelt, Prince George's County) and nests with eggs as early as 9 March 2003 (D. Perry, Patuxent Wildlife Research Center, Anne Arundel County). The latest records of nests with eggs are 21 June 2003 by Bill Harvey (Anne Arundel County: Riverwood, Bayard; Prince George's County: Wells Corner). The earliest nests with young were noted in Anne Arundel County on 8 April 2002 and the earliest downy young were seen in Maryland at McKee-Beshers WMA (Montgomery County) by Jim Green on 24 April 2003. New late brood records for the state are 28 June 2003 (Mike Welch, Lilypons, Frederick County; Jeff Shenot, Upper Marlboro, Prince George's County).

Recent data show that Hooded Mergansers have increased dramatically in the state as a nesting species. Hooded Merganser breeding records increased by 42 blocks between the 1983–1987 and 2002–2006 atlas periods, a 525% increase. In the earlier atlas period observers found Hooded Mergansers to be possible breeders in 2 blocks, probable in 1 block, and confirmed in 12 blocks (Wilmot 1996a). Distribution of breeding Hooded Mergansers was widely scattered, with confirmed and probable records from Garrett County, sites along the Patuxent River, and Charles County. Observers from the later atlas period found possible breeders in 16 blocks and confirmed breeders in 34 blocks. Confirmed breeding records increased along the Patuxent River, with new locations scattered along the Patapsco River, in the Piedmont region including along the Potomac River, along major rivers of the upper Eastern Shore, and in coastal Worcester County. Breeding locations appear to have shifted somewhat in Garrett County. Preliminary data from the 2004–2008 Pennsylvania breeding bird atlas project indicate a similar large increase in confirmed records, with many new records in southern Pennsylvania along river courses (http://bird. atlasing.org/Atlas/PA/). The BBS and the aerial waterfowl breeding survey are not well suited for measuring trends for this species. BBS data show an increase, but low numbers of birds are noted per route and few routes include counts of Hooded Mergansers.

Like other cavity-nesting ducks in Maryland, Hooded Mergansers are expected to increase as more forests attain ages over one hundred years, providing more nest sites as the number of cavity-producing trees increases. Nest box programs for Wood Ducks are also likely to continue to assist breeding Hooded Mergansers. To support current populations and sustain future increases, riparian forested corridors and other wetlands must be conserved, and water quality levels that support a sufficient food base must be maintained.

GWENDA L. BREWER

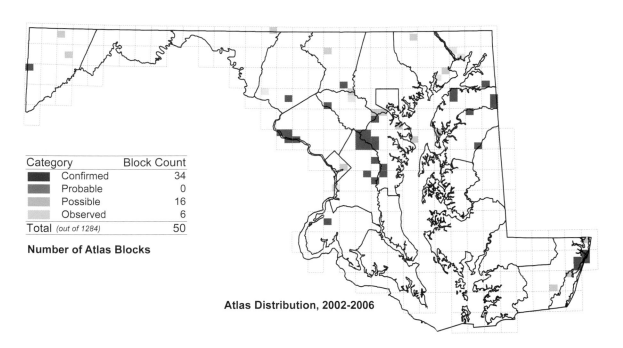

Category	Block Count
Confirmed	34
Probable	0
Possible	16
Observed	6
Total (out of 1284)	50

Number of Atlas Blocks

Atlas Distribution, 2002-2006

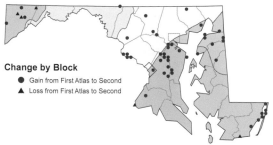

Change by Block
- ● Gain from First Atlas to Second
- ▲ Loss from First Atlas to Second

Change in Atlas Distribution, 1983-1987 to 2002-2006

Atlas Region	1983-1987	2002-2006	Change No.	%
Allegheny Mountain	3	3	0	0%
Ridge and Valley	0	1	+1	-
Piedmont	1	10	+9	+900%
Upper Chesapeake	0	8	+8	-
Eastern Shore	1	8	+7	+700%
Western Shore	3	20	+17	+567%
Totals	8	50	+42	+525%

Change in Total Blocks between Atlases by Region

Common Merganser
Mergus merganser

The "sawbill," so named for its elongated and serrated bill, breeds primarily in the northern forests of the United States and Canada where these habitats include large lakes and rivers (Bellrose 1980). Its specialized bill assists in capturing and holding fish, its main prey. This large diving duck, a top predator in aquatic food chains, has served as an indicator for contaminants and the impacts of acid precipitation in northern systems (Mallory and Metz 1999).

The Common Merganser was noted as a vagrant species during the 1983–1987 atlas period (Robbins and Blom 1996), and Stewart and Robbins (1958) cited records of wintering birds and casual summer vagrants in Maryland. During the 2002–2006 atlas period, Common Merganser was recorded as a breeding species in the state for the first time. Paul Zucker first confirmed Maryland nesting on 3 July 2004 when he saw a brood of eight large young in the Monocacy River at Legore Bridge, Frederick County. In 2005 females with young were observed at two other locations in Frederick County in the Monocacy watershed (18 May, Ed Miller; 8 June, Paul Zucker). In addition, there are several records of females with broods or ducklings in the Potomac River. It is not known if these females nested on the Maryland side of the Potomac, as females are known to move broods up to 8 km (5 mi) away from their nest sites (Mallory and Metz 1999). However, the most likely sites are on river islands that are in Maryland waters. On 23 June 2002 James McCann and Dan Feller observed a female Common Merganser with five downy young on the Potomac River near Paw Paw, WV (Ringler 2003b). A pair of Common Mergansers had been observed by these same two birders the previous day on the river near Oldtown. Dave Czaplak observed a female Common Merganser fly from the crown of a tree on the riverbank of the Potomac River in Montgomery County on 4 May 2003. The following year, he observed a lone downy chick in the Potomac River sliding down the rapids near Violettes Lock. On 14 May 2005 Dorothy Tella photographed a hen with young near Watkins Island where downy young had been observed the year before by Christie Huffman on 8 June. On 18 June 2006 Byron Swift saw a female with 11 young near Cohill, Washington County, and Hugh Mahanes confirmed fledged or downy young 25 June 2006 near Seneca Creek. After the atlas fieldwork had been completed, Fran Pope found downy young on Buffalo Run, Garrett County, on 12 May 2007.

Historical egg records show that Common Mergansers bred in West Virginia and Tennessee; other records show nesting in North Carolina and Virginia, including Dyke Marsh, as recently as 1965 (Kiff 1989). Nesting requires forest mature enough to have tree cavities of appropriate size, although cavities formed between rocks, spaces among tree roots, and holes in banks that are close to the water may be used in some areas (summarized in Mallory and Metz 1999). Broods are cared for in large lakes or rivers, where ducklings older than 7 days may amalgamate into groups of up to 40 young tended by one or more females (Mallory and Metz 1999). Females typically accompany broods 30–50 days after hatching, and young fledge at about 65 days (Erskine 1971).

Peak numbers of migrating Common Mergansers occur in Maryland mid-February to mid-March and early December to mid-January. It is not known when breeding Common Mergansers arrive in or leave Maryland or if these few birds remain in the general area throughout much of the year. In general, Common Merganser is one of the last waterfowl to leave for winter and one of the first to return north to breed. Observations of ducklings now extend from 6 April to 3 July

2004. These dates indicate nesting at least by early March and possibly as late as early May, assuming a 45-day interval from start of egg laying to duckling hatch. Pennsylvania records for Common Merganser indicate that ducklings were seen from 10 May to 25 July with a median date of 21 June (Reid 1992a).

Breeding Common Mergansers were documented in 6 blocks as confirmed, 2 blocks as probable, and 3 blocks as possible during the 2002–2006 atlas period. Preliminary results from the 2004–2008 Pennsylvania breeding bird atlas project (http://bird.atlasing.org/Atlas/PA/) show a substantial increase in the number of blocks with confirmed

breeders compared to the previous Pennsylvania atlas period (Reid 1992a), in part because of a southward spread along major river systems including the Susquehanna. BBS data show an increase of more than 5% in the United States in recent years. The dramatic expansion of Common Merganser to the south in the eastern United States and its presence as a breeding bird in Maryland has likely been influenced by improved water quality of large rivers and availability of nesting cavities as forests age. As long as these favorable conditions remain in our state, we can expect Common Merganser to be a regular, if uncommon, breeding bird here in Maryland.

GWENDA L. BREWER

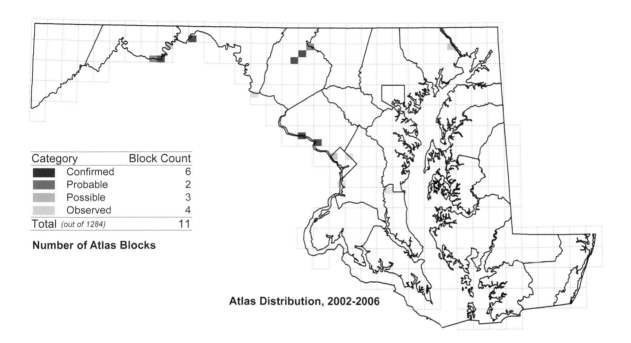

Category	Block Count
Confirmed	6
Probable	2
Possible	3
Observed	4
Total (out of 1284)	11

Number of Atlas Blocks

Atlas Distribution, 2002-2006

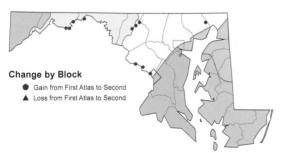

Change by Block

● Gain from First Atlas to Second
▲ Loss from First Atlas to Second

Change in Atlas Distribution, 1983-1987 to 2002-2006

Atlas Region	1983-1987	2002-2006	Change No.	%
Allegheny Mountain	0	0	0	-
Ridge and Valley	0	5	+5	-
Piedmont	0	6	+6	-
Upper Chesapeake	0	0	0	-
Eastern Shore	0	0	0	-
Western Shore	0	0	0	-
Totals	0	11	+11	-

Change in Total Blocks between Atlases by Region

Steve Collins

Ruddy Duck
Oxyura jamaicensis

Best known in Maryland for its compact wintering flocks of hundreds or even thousands, the Ruddy Duck arrives here beginning in September clothed in its drab gray plumage. It remains wherever there is open water on lakes or sheltered coves of major rivers and the Chesapeake Bay. The ruddy is one of very few duck species to molt into a breeding plumage in the spring. The male transforms from the little gray gnome of winter to the jaunty sprite of the breeding season with its sky blue bill and white cheeks offsetting the chestnut body and black crown, its stiff tail held upward. Most leave the region by May but some stay through the summer.

The Ruddy Duck is a very rare breeder in eastern North America. It is adapted to the prairies where it can hide its nest among the tall vegetation of potholes, lakes, and ponds. In Maryland it finds similar habitat among stands of phragmites or cattails in man-made lakes and impoundments. No nests have been found in Maryland. Another unusual aspect of the Ruddy Duck is the size of its eggs; they are quite large, similar in size to those of the Common Merganser. Young Ruddy Ducks are able to dive for food shortly after leaving the nest; the young of other diving species remain feeding on the surface for weeks. The nesting season for the Ruddy Duck is longer than for most ducks; downy young have been seen into October in some parts of its range (Bent 1925).

The enigmatic history of breeding Ruddy Ducks in Maryland begins in the 1970s when an unverified report of young seen at Druid Lake in Baltimore was received (reported by Peggy Bohanan). Unfortunately details of this report can no longer be found. In 1979 Gordon and Sally Paul reported a female with 4 young in a protected cove on Deep Creek Lake, Garrett County, on 20 July. There was no follow-up to this report either. The next report came 20 years later, in 1999, when David Mozurkewich observed an adult with 5 downy young at Deal Island WMA, Somerset County, on 11 July. There were no further sightings of these particular birds. In 2001 Carol Erwin counted 8 downy young in a flock of 18 Ruddy Ducks at the Hurlock WWTP, Dorchester County, on 2 July. Though ruddies continued to be seen here throughout the summer no one else was able to identify juvenile birds in the flock.

In adjacent states there are equally few breeding records, which are also restricted to birds with dependent young. In Pennsylvania, broods were seen in 1969 at Pymatuning Lake, and in 1997, 1998 (McWilliams and Brauning 2000). and 2000 at Glen Morgan Lake, Berks County (R. Keller 1997, 1998, 2000). In Delaware, broods were seen in 1961 and 1979 at Little Creek Wildlife Area (Hess et al. 2000). The Berks County site was slated for development and access was barred there. Water level was discovered to be a limiting factor at these sites. During drought years there was no available breeding habitat for the ducks.

During the present atlas there were 2 confirmations (both with young observed), 1 probable, and 22 blocks classed as observed with no evidence of breeding. The latter number shows the propensity for nonbreeding Ruddy Ducks to spend the summer in Maryland on quiet backwaters, impoundments, and lakes. The Ruddy Duck is the most frequently observed species of nonbreeding duck seen here in summer. There are June, July, and August records throughout the state that would not pertain to migrants. The probable report for the atlas was from the Hurlock WWTP, where

Ruddy Ducks were seen every year during the atlas. The first of the 2 confirmations was at Druid Lake where Ed Smith found two males, a female, and three downy young on 1 July 2004. The young continued to be seen by other observers through 20 July. The second confirmation was at the Hart-Miller Dredged Material Containment Facility in the Chesapeake Bay waters of Baltimore County; Robert Ringler and party photographed and videotaped a female with a begging fuzzy-headed juvenile on 23 September 2006.

As all the breeding sites in Maryland have been artificial and maintained by humans, the future of Ruddy Ducks as nesters in the state is dependent on human control. If water levels in managed impoundments are kept at satisfactory levels, opportunistic Ruddy Ducks will continue to breed in the state at irregular intervals.

ROBERT F. RINGLER

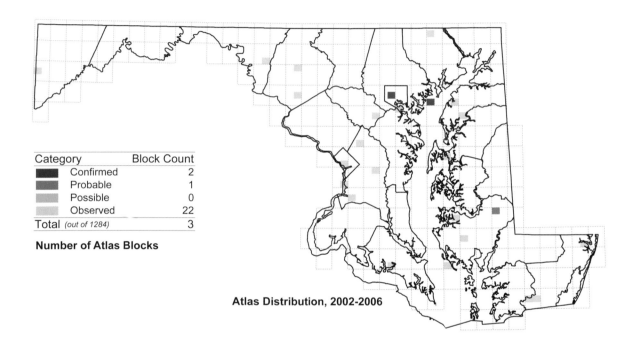

Category	Block Count
Confirmed	2
Probable	1
Possible	0
Observed	22
Total *(out of 1284)*	3

Number of Atlas Blocks

Atlas Distribution, 2002-2006

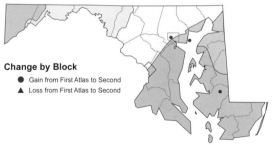

Change by Block
- ● Gain from First Atlas to Second
- ▲ Loss from First Atlas to Second

Change in Atlas Distribution, 1983-1987 to 2002-2006

			Change	
Atlas Region	1983-1987	2002-2006	No.	%
Allegheny Mountain	0	0	0	-
Ridge and Valley	0	0	0	-
Piedmont	0	1	+1	-
Upper Chesapeake	0	1	+1	-
Eastern Shore	0	1	+1	-
Western Shore	0	0	0	-
Totals	0	3	+3	-

Change in Total Blocks between Atlases by Region

Northern Bobwhite
Colinus viginianus

The sweet hollow voices of male bobwhites calling across brushy fields were characteristic sounds in pastoral Maryland as recently as 20 years ago. The Northern Bobwhite is no longer as common or widespread as it once was. It is still a favorite legal game bird and is often stocked by sportsmen, although the practice has not maintained this quail's distribution west of the Coastal Plain. They are still widespread but declining on the Coastal Plain.

Northern Bobwhites inhabit mixed farming landscapes that combine woodland edges, overgrown fields, and cropland with hedgerows. Grassy areas near cover are required for nesting, and waste grain near cover helps ensure survival through the winter months (Roseberry and Klimstra 1984). Winter climate can affect abundance. Harsh winters, such as the one of 1977–1978 (Robbins, Bystrak, and Geissler 1986), have caused declines, and quail are scarce in the higher elevations of western Maryland. Dispersal distance varies, but most bobwhites probably disperse no more than 1 km (0.62 mi) from their hatching place. Dispersal distance appears to be influenced by how much suitable adjacent habitat exists (Brennan 1999). Loss of farmland and the need for remaining farms to increase their efficiency in order to survive has led to considerable loss of habitat for the bobwhite.

These quail are resident year-round in Maryland and DC with an annual peak in numbers in early autumn. Under normal conditions high winter mortality is balanced by the fecundity of nesting females. Clutch size is highly vari-

able but generally exceeds a dozen eggs (Brennan 1999). Hens may produce up to three broods in a season (Guthery and Kuvlesky 1998). In one study 30% of a population attempted a second brood after hatching a first (L. Burger et al. 1995). Nests are on the ground in tall grass. Maryland egg dates range from 21 April to 14 October (Mathews 1996b; MNRF).

The Northern Bobwhite was mentioned because it is a useful game animal in seventeenth-century accounts of Maryland and the Delmarva Peninsula (Mathews 1996b; Hess et al. 2000). Historic bird references generally refer to this quail as common (Coues and Prentiss 1883; Kirkwood 1895), although Eifrig (1904) noted that serious declines took place in response to harsh winters in western Maryland. Stewart and Robbins (1958) called them common on the Coastal Plain, fairly common west to the Allegheny Front, and uncommon west of the Front. They were still widespread during the 1980s with records from more than a thousand atlas blocks from 1983 to 1987 (Mathews 1996b). They were scarce in Garrett and Allegany counties, and there were noticeable holes in the northern Piedmont and around DC.

The population losses evident during the 1980s (Mathews 1996b) continued over the next two decades and contributed to what have become wholesale losses in breeding range as seen on the 2002–2006 atlas map. Northern Bobwhites are still scarce in Garrett and Allegany counties, but declined less than in other regions, perhaps they are being sustained somewhat by regular restocking by hunters. From eastern Washington County east through the Piedmont declines have been steep, 229 blocks on the Piedmont alone. They were also lost from most blocks in southern Cecil County, and show severe thinning on the Western Shore, especially in western Prince Georges and Calvert counties. Although they are still widespread on the Eastern Shore, the breeding confirmation rate has plummeted, suggesting that there are now far fewer broods to find there.

The Northern Bobwhite is declining on BBS routes throughout its range at rates of 3% per year in the entire range and 3.6% per year in the east. In Maryland they have declined at a rate of 5.1% per year since 1966, and at an accelerated 7.3% per year since 1980. These declines are largely attributable to habitat lost to suburban sprawl and to elimination of hedgerows and fallow fields on intensively farmed cropland. Effects of restocking are uncertain but pen-raised birds are neither as wary nor as hardy as wild-reared quail (Roseberry, Ellsworth, and Klimstra 1987). How to halt the loss of this emblematic bird of the farmland of the Old South is a major challenge for game managers and bird students in Maryland and DC.

WALTER G. ELLISON

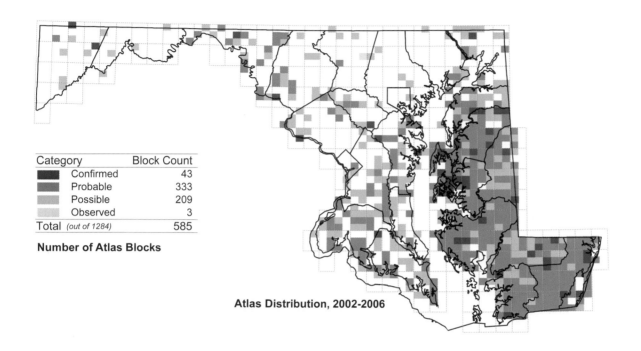

Category	Block Count
Confirmed	43
Probable	333
Possible	209
Observed	3
Total *(out of 1284)*	585

Number of Atlas Blocks

Atlas Distribution, 2002-2006

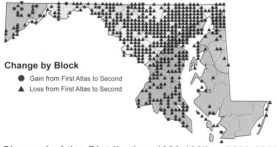

Change by Block
- ● Gain from First Atlas to Second
- ▲ Loss from First Atlas to Second

Change in Atlas Distribution, 1983-1987 to 2002-2006

Percent of Stops
- 50 - 100%
- 10 - 50%
- 0.1 - 10%
- < 0.1%

Relative Abundance from Miniroutes, 2003-2007

Atlas Region	1983-1987	2002-2006	Change No.	%
Allegheny Mountain	14	10	-4	-29%
Ridge and Valley	102	32	-70	-69%
Piedmont	284	55	-229	-81%
Upper Chesapeake	118	73	-45	-38%
Eastern Shore	311	288	-23	-7%
Western Shore	239	125	-114	-48%
Totals	1068	583	-485	-45%

Change in Total Blocks between Atlases by Region

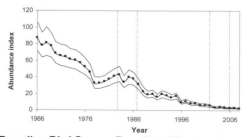

Breeding Bird Survey Results in Maryland

Ring-necked Pheasant
Phasianus colchicus

The hoarse crowing of the male Ring-necked Pheasant evokes misty, dewy mornings in pastoral landscapes redolent of the Old World. A well-regarded traditional upland game bird this pheasant has been introduced widely beyond its Asiatic homeland. It was introduced to England at least as early as the eleventh century, after the Norman conquest (Sharrock 1976), and first successfully introduced from English stock to North America in the mid- to late nineteenth century including a population in Washington County, Maryland, in the 1890s (Kirkwood 1895). Stocking for hunters by the DNR ceased in the early 1970s (S. A. Smith 1996b) and the species has declined greatly since then. A few populations have persisted, perhaps with the aid of private stocking. It is hard to tell if any truly self-sustaining population exists in Maryland.

Ring-necked Pheasants inhabit mixed open landscapes. Factors affecting the success of pheasant populations include protective cover allowing escape from predators, nesting sites in hay and fallow fields undisturbed by early mowing, and the availability of abundant food near cover. Habitats usually include some marshland with cattails or reeds for winter roosts in harsh weather, brushy forest edge or hedgerows near crop fields, and grassy fields for nesting (Giudice and Ratti 2001).

Where they are established, these pheasants are year-round residents. In some areas there is now a pattern of autumn, winter, and spring sightings with few summer observations, presumably related to stocking and attenuation of numbers as stocked birds die. Males begin spring crowing and display in March. Ring-necked Pheasants nest on the ground in dense grassy vegetation. Maryland egg dates range from 1 April to 18 July (S. A. Smith 1996b; MNRF).

Attempts to introduce Ring-necked Pheasants to Maryland date back at least to 1847 when the state legislature saw fit to proscribe hunting during the nesting season for pheasant (S. A. Smith 1996b). Populations did not become widely established in Maryland until well into the twentieth century, with most birds reported along the Mason-Dixon Line, presumably originating largely from birds released in Pennsylvania (Stewart and Robbins 1958). Pheasant populations reached their peak in Maryland in the 1960s and 1970s, and they were still fairly widely distributed in central Maryland from 1983 to 1987. They were found in 83% of the blocks in the Piedmont, and 56% of blocks in the Ridge and Valley, particularly in the Hagerstown Valley. They were scarce elsewhere and were unrecorded from four Eastern Shore counties, although there were 16 blocks with pheasants in Dorchester County presumably due to local releases.

During the 2002–2006 atlas Ring-necked Pheasants were not refound in a stunning 316 blocks, representing a 76% net loss, the largest such loss for a widespread bird between the two statewide atlas projects. They were reported from only 11% of blocks in the Ridge and Valley and 18% of blocks on the Piedmont. They still were found in 10 Dorchester County blocks; the clustering of blocks suggests local stocking. Most of the blocks with reports in central Maryland are along the Pennsylvania line. Ring-necked Pheasants are still being stocked by the Pennsylvania Game Commission.

Ring-necked Pheasants have been declining on BBS routes

continent-wide at 0.7% per year and at a much higher 2.2% per year in the east. In Maryland, pheasants have been declining at 7.3% per year with most of this decline since 1980. Recent changes in farming practices have been troublesome for pheasants. These include replacement of hay and fallow fields with crop fields, increased and earlier hay mowing, removal of weedy hedgerows, and increases in the use of herbicide and pesticide. The value of restocking in maintaining

a self-sustaining population is debatable. Stocked birds are naive, have low survival rates, and poor nesting success (P. Robertson 1993). The future of the Ring-necked Pheasant in Maryland is uncertain at best, although sportsmen will probably continue to stock them for hunting on larger properties.

WALTER G. ELLISON

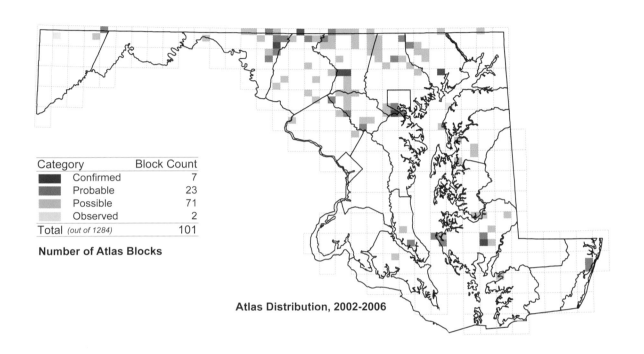

Category	Block Count
Confirmed	7
Probable	23
Possible	71
Observed	2
Total (out of 1284)	101

Number of Atlas Blocks

Atlas Distribution, 2002-2006

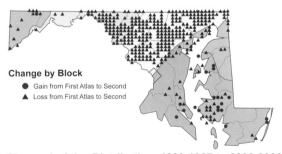

Change by Block
● Gain from First Atlas to Second
▲ Loss from First Atlas to Second

Change in Atlas Distribution, 1983-1987 to 2002-2006

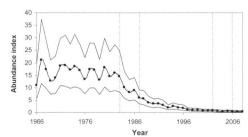

Percent of Stops
50 - 100%
10 - 50%
0.1 - 10%
< 0.1%

Relative Abundance from Miniroutes, 2003-2007

Atlas Region	1983-1987	2002-2006	Change No.	Change %
Allegheny Mountain	12	2	-10	-83%
Ridge and Valley	83	17	-66	-80%
Piedmont	264	57	-207	-78%
Upper Chesapeake	19	0	-19	-100%
Eastern Shore	22	18	-4	-18%
Western Shore	17	7	-10	-59%
Totals	417	101	-316	-76%

Change in Total Blocks between Atlases by Region

Breeding Bird Survey Results in Maryland

Ruffed Grouse
Bonasa umbellus

For most observers, their first encounter with the Ruffed Grouse was a heart-stopping thunderous explosion of whirring wings from underfoot followed by a glimpse of a brown rocket twisting between trees. This grouse has long been *the* upland game bird of woods, as much a part of autumn in the hills as falling multicolored leaves. Males produce one of the most unusual and distinctive sounds of spring and late autumn, the hollow accelerating thumping called drumming produced by the rapidly beating wings of a bird perched on a log, stump, or old stone wall. Ruffed Grouse once occurred more widely in Maryland eastward through the Piedmont, but the birds are now confined to western Maryland.

Although essentially a forest bird, the Ruffed Grouse requires a mosaic of young forest with abundant understory shrubs, small clearings, and more mature forest. North of Maryland, quaking and bigtooth aspen buds are important winter food (Rusch et al. 2000), but these trees are not major components in Maryland woodlands. In the southern Appalachians wintering grouse rely more on oak and beech mast, although the buds of birch, willow, alder, and cherry may also be important (Servello and Kirkpatrick 1987).

Ruffed Grouse are resident throughout the year wherever they occur and have only limited dispersal, generally 5 km (3.1 mi) or less (Rusch et al. 2000). Males begin drumming in March, and the peak of egg laying takes place from late April to mid-May. The nest is a well-concealed bowl nestled among leaves under a stump, log, brush pile, rock, or at the base of a shrub or sapling. Hens sit tightly, which makes nests very hard to find. Young leave the nest soon after hatching and are capable of short flights after 10 to 12 days (Johnsgard 1983). Most atlas confirmations are made by observers encountering broods of young accompanied by protective mothers. Maryland egg dates range from 8 April to 20 June (Therres 1996c; DNR unpubl. data; MNRF).

The Ruffed Grouse's original range in Maryland and DC encompassed all or most of the Piedmont and included parts of the Coastal Plain adjacent to the fall line, including DC (Coues and Prentiss 1862). By the second edition of Coues and Prentiss (1883) grouse were rare at best in DC and by the 1920s they were virtually absent from the Piedmont (Stewart and Robbins 1958). Stewart and Robbins described them as fairly common in the Ridge and Valley, although absent from the heavily farmed Hagerstown Valley, and common in the Allegheny Mountains. A similar range was documented by 1983–1987 atlas observers. They were not refound in a net total of 42 blocks including 48% fewer (29) in the Ridge and Valley during the 2002–2006 atlas project. They declined less in the Alleghenies but were still lost from a net of 14 blocks (19%). A single record was made in Frederick-SW, a block assigned to the Piedmont, but the report was from Catoctin Mountain, which belongs to the Ridge and Valley.

Although Ruffed Grouse are poorly surveyed from road-sides, it is worth noting that they have declined at a rate of 2.3% per year since 1966 on North American BBS routes. Other northeastern U.S. atlas maps either show scarcity (Ickes 1992a; Walsh et al. 1999) or declines (T. Post 2008) of grouse at lower elevations and in areas prone to habitat loss. It seems unlikely Ruffed Grouse will recolonize eastern Maryland in spite of a 1986 report from Harford County (Ringler 1987). Declines in western Maryland probably have

several causes. Changes in forest management that reduce clearing sizes and that lead to too uniformly mature forest may be a problem. Climatic warming may also be an issue as grouse may not be able to use snow roosts as often in winter and hot dry summers may reduce mast production. Although the future of this grouse seemed good in Maryland after the first statewide atlas (Therres 1996c), its future seems less secure in the early twenty-first century.

WALTER G. ELLISON

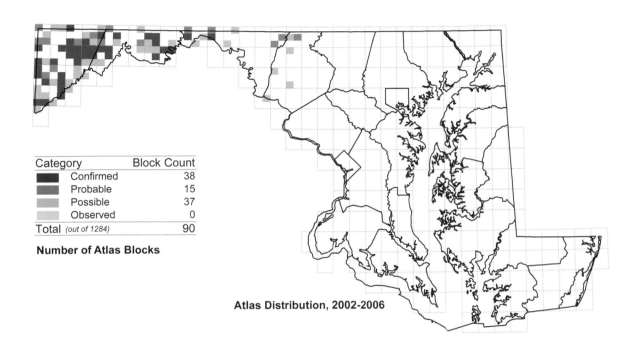

Category	Block Count
■ Confirmed	38
■ Probable	15
■ Possible	37
□ Observed	0
Total (out of 1284)	90

Number of Atlas Blocks

Atlas Distribution, 2002-2006

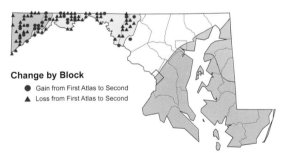

Change by Block
● Gain from First Atlas to Second
▲ Loss from First Atlas to Second

Change in Atlas Distribution, 1983-1987 to 2002-2006

			Change	
Atlas Region	1983-1987	2002-2006	No.	%
Allegheny Mountain	72	58	-14	-19%
Ridge and Valley	60	31	-29	-48%
Piedmont	0	1	+1	-
Upper Chesapeake	0	0	0	-
Eastern Shore	0	0	0	-
Western Shore	0	0	0	-
Totals	132	90	-42	-32%

Change in Total Blocks between Atlases by Region

Gary Smyle

Wild Turkey
Meleagris gallopavo

The Wild Turkey is an iconic bird, the great and iridescent game bird of the forests of North America. The domesticated turkey derived from the wild stock of pre-Columbian Mexico is one of the New World's major contributions to world agriculture and husbandry. A century ago the wary and fast-running Wild Turkey seemed on the way to extinction as a wild bird because of a combination of habitat loss and open season market hunting. But in the second half of the twentieth century the turkey responded to active management and to habitat regeneration and preservation and is once again widespread and often numerous in eastern North America.

At its population low ebb, the Wild Turkey was considered a denizen of extensive and largely uninhabited forest lands, but with carefully managed hunting seasons and modest-size woodlots, the Wild Turkey can thrive in farming and suburban landscapes with forested land. Turkeys require an abundance of mast-bearing trees such as oak, hickory, and beech; waste grain in fields and feedlots; and some understory brush for nesting. Turkeys will also feed on grasshoppers and other large insects in summer grasslands. Wild Turkeys are not multiple brooded, but they may renest often to hatch their large broods of highly mobile downy chicks. Nineteen egg dates for turkeys in this project ranged from 26

April (L. Schindler) to 19 July (J. Stasz), exceeding the dates reported in the first Maryland-DC atlas (Mathews 1996a). Turkey poults remain with adult hens into the winter, with young hens not departing until the following spring (Eaton 1992); young turkeys provided the bulk of atlas nesting confirmations, with 283 records from May to the third week of September.

By the late nineteenth century, the Wild Turkey was essentially limited to the oak-hickory forests of the more mountainous parts of the Ridge and Valley provinces of Pennsylvania (Wunz and Brauning 1992) and Maryland (Stewart and Robbins 1958), although a few were also found in the Allegheny Mountains. In the early twentieth century, attempts were made to reintroduce turkeys and bolster wild populations with farm-raised birds of wild ancestry, but despite the establishment of short-term populations (e.g., Worcester County; Stewart and Robbins 1958) these birds usually failed to thrive in the wild (Eaton 1992). Game managers eventually hit on the idea of translocating wild-caught turkeys among regions as a reintroduction and stock improvement method in the late 1940s (Eaton 1992). This practice, which began in 1956 in Pennsylvania (Wunz and Brauning 1992), and 1966 in Maryland (Mathews 1996a), led to the impressive 40-year recovery of the Wild Turkey.

In 1983–1987 the Wild Turkey was 20 years into its translocation-fueled recovery in Maryland. The atlas map from those years showed that this bird was still rare in 13 counties and widespread in only 5 counties. In 2002–2006 turkeys were recorded in a whopping 490 more blocks, an increase of over 200%. The greatest increases for Wild Turkeys were in the Western Shore and Eastern Shore regions, but the species increased in all regions, including its old Ridge and Valley stronghold. How much of this increase is attributable to ongoing translocation of turkeys and how much from population spread from release sites is unclear, but both certainly contributed to the turkey's success. Wild Turkeys are still thinly distributed in northern Carroll, southern Baltimore, Cecil, and north central Montgomery counties. Urban Wild Turkeys were found in a select few blocks in Baltimore City and DC, but they remain largely absent from urbanized blocks and the most open farm landscapes. However,

it should be noted that turkeys are furtive and hard to see in the peak atlas field season running from May through early August, so they may have been overlooked in several blocks.

Although Maryland BBS trends are positive, they are suspect because of the small number of Wild Turkeys detected (0.2 per survey route). Nonetheless, the change in this game bird's status is clear as it was not seen on BBS routes from 1966 to 1979 and was reported from 32 routes after 1980. The Wild Turkey appears to be well established in most of Maryland and should remain so, provided hunting seasons are carefully managed and sufficient forest land is preserved.

WALTER G. ELLISON

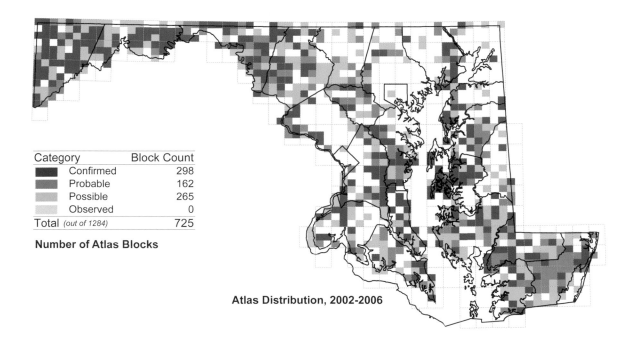

Category	Block Count
Confirmed	298
Probable	162
Possible	265
Observed	0
Total (out of 1284)	725

Number of Atlas Blocks

Atlas Distribution, 2002-2006

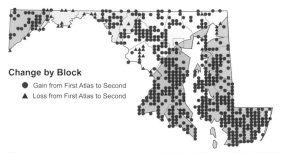

Change by Block
- ● Gain from First Atlas to Second
- ▲ Loss from First Atlas to Second

Change in Atlas Distribution, 1983-1987 to 2002-2006

Percent of Stops
- 50 - 100%
- 10 - 50%
- 0.1 - 10%
- < 0.1%

Relative Abundance from Miniroutes, 2003-2007

Atlas Region	1983-1987	2002-2006	Change No.	Change %
Allegheny Mountain	58	82	+24	+41%
Ridge and Valley	99	119	+20	+20%
Piedmont	42	129	+87	+207%
Upper Chesapeake	10	36	+26	+260%
Eastern Shore	14	212	+198	+1414%
Western Shore	9	144	+135	+1500%
Totals	232	722	+490	+211%

Change in Total Blocks between Atlases by Region

Breeding Bird Survey Results in Maryland

Mark Hoffman

Pied-billed Grebe
Podilymbus podiceps

The Pied-billed Grebe looks like a drab gray brown ball with a slender neck and round head. This little swimmer earned the vernacular name "helldiver" because it resembles a submarine when threatened: sinking out of sight, cruising underwater for minutes at a time, and swimming with only its nostrils breaking the surface. Its song is a bizarre war whoop that usually starts with chuckling, accelerates to a hooting laugh, and subsides into guttural muttering. These grebes are often noisy when they have neighbors, but can be largely silent when neighboring pairs are lacking. Although Pied-billed Grebes are fairly numerous transients and regularly winter in Maryland and DC, they have always been scarce and localized nesters here.

Pied-billed Grebes have two major breeding habitat requirements: a combination of luxuriant emergent and floating aquatic plants with adjacent open water. Maryland and DC have few wetlands that meet these criteria save on the Coastal Plain. Scattered pairs may nest on small marshy ponds, but such pairs are usually present for only a single nesting season. These grebes may occur in loose aggregations where there is extensive productive habitat.

Pied-billed Grebes are uncommon in winter, occurring largely on the Coastal Plain on tidewater. Migrations in Maryland and DC run from late February to late May and from mid-August to mid-December. Pied-billed Grebe nests are built on floating live or dead vegetation in water 30–80 cm (11.8–31.4 in) deep among obscuring emergent wetland plants (Muller and Storer 1999). Nests are often more subject to loss via drought or flooding than predation, although this varies with location (Muller and Storer 1999). Downy young grebes leave the nest soon after hatching and are distinctively striped with black and white about the head and neck. All nesting confirmations reported during this project were for downy young. Maryland egg dates range from 17 May to 5 June (Ringler 1996a; MNRF). An adult was seen on a nest on 8 July 2006 at Hart-Miller Island, Baltimore County (E. J. Scarpulla, pers. comm.). Small downy young have been reported to 11 August (Ringler 1996a; MNRF).

The Pied-billed Grebe has long been a rare to uncommon nester in Maryland. Kirkwood (1895) knew of no nest reports for Maryland. Stewart and Robbins (1958) listed them as uncommon on tidewater on the Western and Eastern Shore, and rare elsewhere on the Coastal Plain and Piedmont. The distributions mapped for 1983–1987 (Ringler 1996a) and 2002–2006 are broadly similar with some noteworthy differences in detail. They have become less widespread in Somerset County, with only a single report of possible breeding from Deal Island. On the other hand, they appear to have increased farther north with new records from Dorchester and Calvert counties, several reports from southern Cecil County near the C&D Canal, and a new, very large population on the impoundments of Hart-Miller Island. In 2006, counts of adult and juvenile grebes on Hart-Miller Island peaked at 175 on 15 July (E. J. Scarpulla, pers. comm.).

Breeding Bird Survey trends are unreliable for waterbirds with specialized habitats, but the existing survey data suggest a stable population and indicate a center of abundance in the prairie pothole region in the northern Great Plains. Maintenance of suitable wetlands with a mix of open water and tall emergent vegetation is necessary if a nesting population of Pied-billed Grebes is to remain in Maryland. The impoundments of Hart-Miller Island appear to be of considerable importance. Succession of wetland ponds to more heavily vegetated marshes with less open water on the lower Eastern Shore may have contributed to declines there. The increase of thick beds of invasive reeds may have accelerated this process.

WALTER G. ELLISON

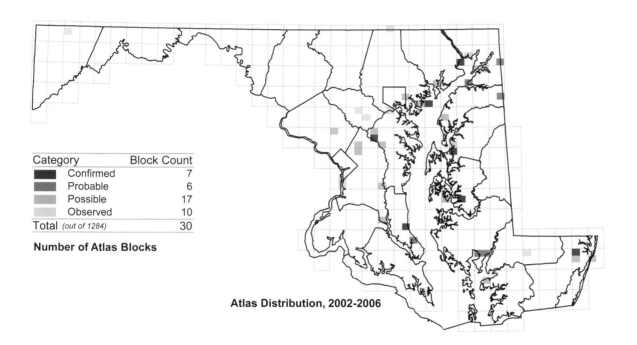

Category	Block Count
Confirmed	7
Probable	6
Possible	17
Observed	10
Total (out of 1284)	30

Number of Atlas Blocks

Atlas Distribution, 2002-2006

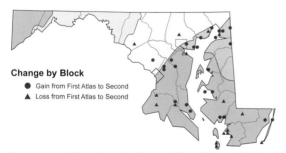

Change by Block
- ● Gain from First Atlas to Second
- ▲ Loss from First Atlas to Second

Change in Atlas Distribution, 1983-1987 to 2002-2006

Atlas Region	1983-1987	2002-2006	Change No.	%
Allegheny Mountain	0	0	0	-
Ridge and Valley	0	0	0	-
Piedmont	2	1	-1	-50%
Upper Chesapeake	2	10	+8	+400%
Eastern Shore	12	10	-2	-17%
Western Shore	6	9	+3	+50%
Totals	22	30	+8	+36%

Change in Total Blocks between Atlases by Region

Brown Pelican
Pelecanus occidentalis

With their prehistoric pterodactyl-like silhouette, Brown Pelicans have become a favorite bird of nearly everyone who visits Ocean City. The spectacular headfirst plunging dives that they make to obtain fish are a signature behavior. In summer, Brown Pelicans can frequently be seen along Maryland's oceanfront, throughout the coastal bays, and in the lower Chesapeake Bay. During late summer occasional observations are made in the upper portions of the Chesapeake Bay north of the Bay Bridge all the way up into Cecil County. Brown Pelicans have been observed in Maryland during every month of the year, and there are now annual stragglers that attempt to winter in the lower portions of the Chesapeake Bay, a bad decision that usually results in frostbitten gular pouches and feet. This is in stark contrast to the situation in the 1970s when the Brown Pelican population along the Atlantic and Gulf Coasts had been decimated by the impacts of endrin and DDT and it was listed as an endangered species (Shields 2002). The recovery of pelican populations after the banning of bioaccumulating organochlorine-based pesticides and other conservation actions has been a con-

servation success story that is well known to many schoolchildren.

Brown Pelicans are highly gregarious, gathering in colonies to nest on remote estuarine islands that are usually separated by considerable distances. Nests in Maryland have been found in small cedar trees, *Iva,* and groundsel shrubs, and ground nests have been discovered on the upper reaches of sand beaches. The largest pelican colony in Chesapeake Bay is located on one of the most remote pieces of land in the eastern United States, more than 16 km (10 mi) from the nearest public road. After spending the winter in the southeastern United States and the Caribbean, Brown Pelicans begin returning to Maryland in early to mid-March, the exact timing depending on the severity of the winter up to that point. Breeding phenology has changed since the 1983–1987 atlas and first eggs are now laid in early April, well before biologists are out doing colonial waterbird field surveys. The pelican nesting season is extended, and peak colony size is not reached until early June when chicks from the earliest nesting pairs are nearly large enough to fly. Chicks begin hatching in early May. For a detailed description of the history of Brown Pelican breeding in Maryland and more details on changes in breeding phenology in Maryland, see Brinker 1996a.

Brinker (1996a) documented the first nesting of Brown Pelicans in Maryland during the 1983–1987 atlas period. At that time Brown Pelicans were confirmed as breeding in 1 block and were recorded as observed in an additional 3 blocks, with all 4 blocks in the coastal bays of Worcester County. The number of blocks where Brown Pelicans were recorded as observed during the second atlas increased an astounding 2,000% over the first atlas to 60 blocks. Observations were mostly in the Eastern Shore region with observations recorded in only 4 blocks in the Western Shore region and 1 in the Upper Chesapeake region. During the second atlas the number of confirmed blocks increased 500% to 6 blocks. The original 1987 nesting site was last used in 1995 and that block was the only block lost between the two atlas periods. During the second atlas period, Brown Pelicans nested on Big Bay Marsh in Worcester County and on South Point Marsh (Virginia border block), Pry Island, Whitewood Cove Island (near Barren Island), and Middle Holland Island in the Chesapeake Bay. In the fall of 2003 Hurricane Isabel destroyed the nesting habitat on Whitewood Cove Island and that nesting site will never be occupied again. Erosion and habitat deterioration have significantly changed Pry Island and it is unlikely that pelicans will ever nest there again. During the intervening years between the two atlas periods, pelicans nested at Spring Island from 1998 through 2001 before moving to the current large colony on the remnants of Holland Island. As of the second atlas, Maryland remains the farthest north that Brown Pelicans nest along the Atlantic coast.

The nesting population of Brown Pelicans in Maryland

during 1987, the last year of the first atlas project, was 6 breeding pairs at a single colony site (Brinker 1996a). From 2002 through 2006, the number of breeding pairs of Brown Pelicans in Maryland varied from 51 pairs breeding at White-wood Cove Island during 2002 to 522 pairs nesting on Holland Island during 2006. The second atlas period was unusual for Brown Pelicans; it was marked by wide variation in breeding population size and frequent shifts in sites used by the nesting colonies (DNR, unpubl. data). BBS data are not useful for evaluating Brown Pelican population trends in Maryland and the Chesapeake Bay. A recent analysis of Chesapeake Bay and coastal Delmarva seabird populations

found that the Brown Pelican population had increased 350% from 386 breeding pairs in 1993 to 1,737 pairs in 2003, with the largest proportion of the breeding population located in Maryland (Brinker, Williams, and Watts 2007).

Brown Pelicans are now here to stay; they have discovered the rich fishery of the lower Chesapeake Bay and are content with our present climate. As long as stocks of menhaden, shad, and other forage fish preferred by Brown Pelicans remain healthy, vacationers along ocean beaches in the mid Atlantic states will be able to enjoy watching Brown Pelicans for many years to come.

DAVID F. BRINKER

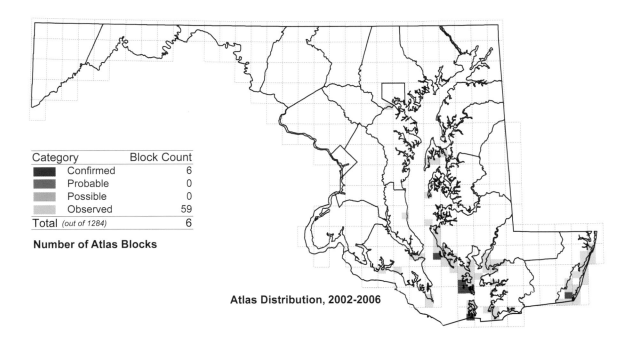

Category	Block Count
Confirmed	6
Probable	0
Possible	0
Observed	59
Total (out of 1284)	6

Number of Atlas Blocks

Atlas Distribution, 2002-2006

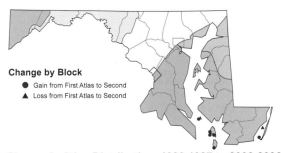

Change by Block
- ● Gain from First Atlas to Second
- ▲ Loss from First Atlas to Second

Change in Atlas Distribution, 1983-1987 to 2002-2006

Atlas Region	1983-1987	2002-2006	Change No.	%
Allegheny Mountain	0	0	0	-
Ridge and Valley	0	0	0	-
Piedmont	0	0	0	-
Upper Chesapeake	0	0	0	-
Eastern Shore	1	6	+5	+500%
Western Shore	0	0	0	-
Totals	1	6	+5	+500%

Change in Total Blocks between Atlases by Region

Mark Hoffman

Double-crested Cormorant
Phalacrocorax auritus

Double-crested Cormorants are so adept at catching fish that they have a long and mixed history with humans. In the Far East cormorants were captured and used to obtain fish for human consumption. In North America, for many years this bird's fishing prowess was rewarded with persecution that led to extirpation from the northeastern U.S. Atlantic coast in the late 1800s (Hatch and Weseloh 1999). As persecution lessened, Double-crested Cormorants gradually expanded back into the United States from Canada; they returned as a nesting species throughout much of New England during the 1920–1950 period. Cormorant populations were then seriously impacted by organochloride chemicals and many parts of their range again saw significant declines. With the end of persecution and the banning of detrimental pesticides, Double-crested Cormorants are now the most abundant and widespread of the six cormorant species in North America (Hatch and Weseloh 1999). The first nesting record for Double-crested Cormorant in Maryland occurred in 1991, three years after completion of the first atlas (Meritt 1996a).

In breeding condition Double-crested Cormorants are handsome birds with jet black feathers, bright green eyes, and striking orange gular skin, topped off with white feathers that form the "crests." On islands and other predator-free sites, cormorants build nests of sticks on the ground or in dead trees. They often build nests on artificial structures that are inaccessible to mammalian predators. Large colonies are known for the bad aroma from the abundant corrosive ammonium-based guano that the nesting colony produces. Although Double-crested Cormorants from farther north winter in Maryland, our breeding population begins returning from the south in mid-March. The earliest retuning individuals reoccupy breeding colonies and begin to sit on their nests by mid- to late April. Most egg laying by Double-crested Cormorants in Maryland occurs in May, but they have a long nesting season, and nests with fresh eggs can still be found in early to mid-July (D. Brinker, pers. obs.).

While Double-crested Cormorants did not breed in Maryland during the first atlas, they were recorded as observed in 54 blocks in the Upper Chesapeake and Eastern and Western Shore regions (Meritt 1996a). Following establishment of the first breeding colony of Double-crested Cormorants at Poplar Island in 1991, the population has expanded rapidly in Maryland. In the second atlas Double-crested Cormorants were recorded as observed in 247 blocks and confirmed in 16 blocks with the highest number of confirmations in the Eastern Shore region. Breeding was documented in all but the Allegheny Mountain and the Ridge and Valley regions. During the second atlas cormorants were recorded as observed in every region in Maryland.

From 2002 through 2006 the number of Double-crested Cormorants in Maryland increased from 1,246 to 1,890 breeding pairs (DNR unpubl. data). They were found at 14 different colony sites during the second atlas. Major cormorant colonies were located at the Poplar Island restoration project, Bodkin Island, Fort Carroll, and South Point Spoils Island (Worcester County). During 2006 approximately three hundred pairs attempted to nest on the understructure of the Chesapeake Bay Bridge, but their nests were removed to discourage continued expansion of nesting on artificial structures. Other artificial sites used in Maryland include the Route 301 bridge over the Potomac River and electrical transmission line towers crossing the Patuxent River near Chalk Point. There was also a small colony on an island in the Potomac River near Cabin John in Montgomery County. The colony that was forming on the Chesapeake Bay Bridge during the last year of the second atlas was located in two atlas blocks.

Unlike other species of colonial nesting waterbirds, Double-crested Cormorants are regularly recorded on BBS routes near bodies of water. BBS trends for Double-crested Cormorants, analyzed at Maryland, U.S., and North American levels, all show highly significant increasing trends ($P < 0.00$). These increases reflect the regional and continental increases being observed in cormorant populations. A recent analysis of Chesapeake Bay and coastal Delmarva seabird populations found that the number of Double-crested Cormorants had increased on the Delmarva from 686 breeding pairs in 1993 to 2,726 pairs in 2003, with most breeding occurring in the Maryland portion of the Chesapeake Bay (Brinker, Williams, and Watts 2007).

The nearly continent-wide explosive growth of the Double-crested Cormorant population over the past 30 years does not come without complications. During the winter cormorant predation at catfish farms in the southern United States, has become a significant nuisance. Double-crested Cormorants are also the center of fish management conflicts in the eastern Great Lakes and at some New England sites, especially those where Atlantic salmon are being reintroduced (Hatch and Weseloh 1999). In 2003 the USFWS finalized a public depredation order as part of a national Double-crested

Cormorant management plan to allow 24 states to manage cormorants that are affecting public resources as well as cormorants at aquaculture facilities and government-operated fish hatcheries. Cormorant control has been implemented in a number of northeastern states and Canada to reduce perceived conflicts with fisheries. The continuing debates are neither simple nor short. The real possibility exists that the increasing number of Double-crested Cormorants nesting in Maryland will eventually reach a level that will require control actions.

DAVID F. BRINKER

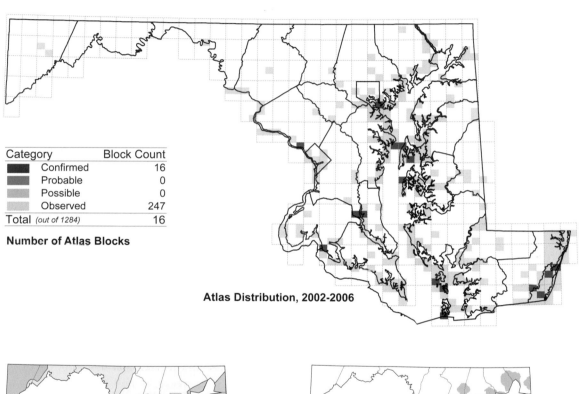

Category	Block Count
■ Confirmed	16
■ Probable	0
■ Possible	0
■ Observed	247
Total *(out of 1284)*	16

Number of Atlas Blocks

Atlas Distribution, 2002-2006

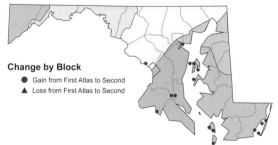

Change by Block
● Gain from First Atlas to Second
▲ Loss from First Atlas to Second

Change in Atlas Distribution, 1983-1987 to 2002-2006

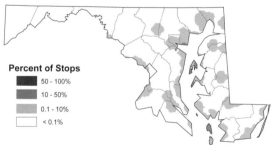

Percent of Stops
■ 50 - 100%
■ 10 - 50%
■ 0.1 - 10%
□ < 0.1%

Relative Abundance from Miniroutes, 2003-2007

Atlas Region	1983-1987	2002-2006	Change No.	%
Allegheny Mountain	0	0	0	-
Ridge and Valley	0	0	0	-
Piedmont	0	1	+1	-
Upper Chesapeake	0	1	+1	-
Eastern Shore	0	10	+10	-
Western Shore	0	4	+4	-
Totals	0	16	+16	-

Change in Total Blocks between Atlases by Region

American Bittern
Botaurus lentiginosus

The deep, resonant call of *pump-er-lunk* or *plum-puddin'* at dawn or dusk is often how the American Bittern reveals its presence. A master of camouflage, it is the most cryptically patterned and secretive member of the heron family. The tawny, streaked plumage blends perfectly with its marshy habitat and its freezing behavior, with bill pointed upward, completes its predator protection. Unlike its smaller cousin, the Least Bittern, this species usually prefers brackish rather than fresh marshes in Maryland (Stewart and Robbins 1958). Although both species are very difficult to survey, this bittern is the less abundant. It can be found throughout the year here; however, it is somewhat migratory since reports increase in spring from late March through May and in the fall from September through November (Robbins and Bystrak 1977). The American Bittern was included on the National Audubon Society Blue List from 1976 to 1986 (Tate 1986) and on the USFWS (1987) list of birds of management concern. Since 1987 DNR has listed this species as in need of conservation in Maryland.

Historically, the American Bittern was considered a fairly common breeder in Dorchester, Wicomico, and Somerset County tidal marshes, an uncommon breeder in remaining tidewater marshes, and a rare breeder in the Allegheny Mountains (Stewart and Robbins 1958). This bittern's breeding season in Maryland ranges from mid-April to mid-July (Davidson 1996a), although fledged young have been re-

ported as late as 28 August (Stewart and Robbins 1958). Platform nests of dead vegetation and sticks are built among cattails, tall grasses, and other tall vegetation. Clutches of four or five eggs are incubated about 24–28 days, and chicks fledge about 14 days later (Terres 1980). These carnivores prey on fish, crayfish, frogs, and small mammals (A. C. Martin et al. 1951).

This rare breeder was found in only 20 blocks during the 1983–1987 atlas (Davidson 1996a). The number of atlas blocks with this species declined 45% to only 11 blocks during the 2002–2006 atlas. During the first atlas, 1 block each was recorded in the Allegheny Mountain and Upper Chesapeake regions, 7 blocks were in the Western Shore region, and about half of the 20 blocks were located in the Eastern Shore region. The number of blocks in most regions was nearly unchanged between the two atlases; the Eastern Shore was an exception. The distributional difference between the atlases can be accounted for primarily in the lower Eastern Shore counties of Dorchester and Somerset, where the number of blocks dropped to just 2. In the Western Shore region, the number of blocks was nearly identical in both atlases, but the number of blocks in the riverine marshes of the Patuxent River dropped from 4 to none. There are too few BBS routes in Maryland to determine any population abundance trends.

Unfortunately, it is nearly impossible to determine the causes for the differences in distribution of the American Bittern between the first atlas in the 1980s and this atlas nearly 20 years later. The large difference in the lower Eastern Shore marshes may indicate an overall and perhaps subtle change in habitat or prey availability. Much of the brackish tidal marsh around Blackwater NWR was converted to open water during that period. However, that change would only account for 1 or 2 of the block differences. A more likely explanation to account for most of the differences in the lower Eastern Shore and the Patuxent River blocks may simply be differences in observer effort. For example, canoes were used to gain access to many otherwise inaccessible and extensive riverine marshes in these areas during the first atlas but not the second one. Very little canoe or boat work occurred in Dorchester County during the second atlas (Davidson, pers. obs.). In general, secretive marshbirds are quite difficult to survey and usually require specialized surveys using standard

taped call play-back methods over time to detect any meaningful population trends.

However, habitat conditions are known to be changing in the lower Chesapeake Bay marshes because of sea level rise and local land subsidence. As global climate change impacts these marshes through more frequent and intense flooding events, the American Bittern will likely disappear from its historical stronghold in these marshes. The creation and maintenance of appropriate emergent marshes elsewhere in Maryland will become even more important for maintaining this species as a viable component of our breeding avifauna.

LYNN M. DAVIDSON

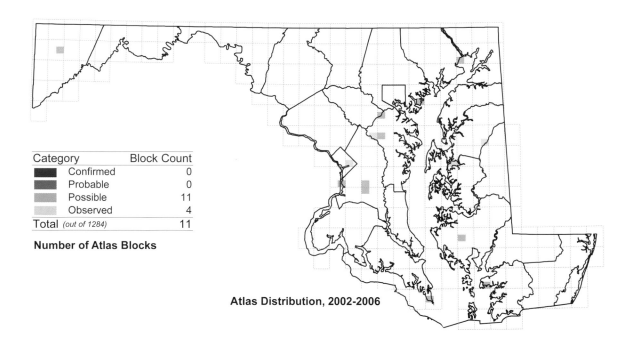

Category	Block Count
Confirmed	0
Probable	0
Possible	11
Observed	4
Total (out of 1284)	11

Number of Atlas Blocks

Atlas Distribution, 2002-2006

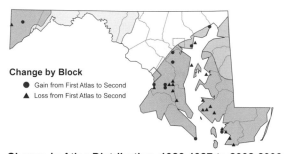

Change by Block
- ● Gain from First Atlas to Second
- ▲ Loss from First Atlas to Second

Change in Atlas Distribution, 1983-1987 to 2002-2006

Atlas Region	1983-1987	2002-2006	Change No.	%
Allegheny Mountain	1	1	0	0%
Ridge and Valley	0	0	0	-
Piedmont	0	0	0	-
Upper Chesapeake	1	2	+1	+100%
Eastern Shore	11	2	-9	-82%
Western Shore	7	6	-1	-14%
Totals	20	11	-9	-45%

Change in Total Blocks between Atlases by Region

Least Bittern
Ixobrychus exilis

About one foot tall, the Least Bittern is the smallest North American heron. Like its larger cousin, the American Bittern, it hunts stealthily in dense marsh vegetation and tends to freeze, reedlike, with bill pointed skyward when threatened. It reveals itself by the flash of buffy inner wing patches as it flies or by its low, soft *coo-coo-coo* call. Although it occurs in nearly any type of marsh, it prefers fresh marshes, especially those dominated by narrow-leaved cattails (Stewart and Robbins 1958). As wetlands have disappeared over the past two centuries, this species has declined to the point where it was added to the National Audubon Society Blue List in 1979 (Tate 1986) and the DNR listed it as in need of conservation in 1987.

The first local recorded specimen was collected in June 1843 from DC (Baird et al. 1858). Early accounts of its abundance in Maryland and DC ranged from rare (C. Richmond 1888) to rather uncommon (Coues and Prentiss 1883) to common (Kirkwood 1895). Its reclusive nature and the fact that under optimal habitat conditions it sometimes breeds in small, loose colonies likely account for these differences. Stewart and Robbins (1958) recorded its summer abundance as common in Chesapeake Bay marshes and occasional elsewhere. In April the Least Bittern arrives primarily from Mexican, Central American, and West Indian wintering grounds and departs beginning in early August and continuing into October (Robbins and Bystrak 1977).

Breeding territories are set up soon after their spring arrival. In New York these territories are a mean of 9.7 ha (23.9); the range is from 1.8 to 35.7 ha (4.4–88.2; Bogner and Baldassarre 2002). Males pick the nest sites, but females assist by weaving cup nests of dead vegetation around live plants, usually above standing water (Palmer 1962; Bent 1926). Clutches of 18 Maryland nests ranged from 1 to 5, with a mean of 3.2 eggs (MNRF), but the mean clutch size was higher in New York with 5.25 eggs at 64 nests (Bogner and Baldassarre 2002). Nests with eggs in Maryland have been documented from 10 May to 12 July (MNRF). Incubation lasts about 18 days (Weller 1961) and fledging occurs in 10–14 days, although the young cling to vegetation and stay within a mean of 29.4 m (32.1 yd) from the nest for nearly two more weeks, when they are then able to fly (Bogner and Baldassarre 2002). Nesting success rate averaged 48.2% in a 1999 and 2000 study in New York and some adults were double brooded (Bogner and Baldassarre 2002). They feed primarily on invertebrates, mostly insects and crustaceans, but also on amphibians, small fish, and small mammals (Bent 1926).

The Least Bittern was found in 77 blocks during the 1983–1987 atlas (Davidson 1996b) and 54 blocks during this atlas, a 30% decline in nearly 20 years. The species' overall distribution remains similar, however. This species occurs mostly in the Eastern Shore and Western Shore regions, with 80% of its 2002–2006 atlas blocks found there and only 20% of its blocks found in the Piedmont and Upper Chesapeake regions. During the 1983–1987 atlas, these percentages were 82% and 18% of its blocks, respectively.

The marshes along the Patuxent River and in southern Dorchester County were two concentration areas during the first atlas (Davidson 1996b). These areas also showed the greatest change between the first and second atlases. With a few notable exceptions, such as the disappearance of brackish marshes along much of the Blackwater River at Blackwater NWR, the distribution of marshes within these two areas has remained largely unchanged in the past 20 years. Some areas have no doubt been overtaken by the invasive plant phragmites, but the Least Bittern has been known to occur in marshes with phragmites. A more likely explanation involves the amount of observer effort between the two atlases. Because this species is secretive and occurs in a fairly inaccessible habitat, specialized survey efforts using boats or canoes would greatly increase detection. Such efforts occurred during the first atlas in these two areas of the state, but were more limited during the second atlas. Therefore, it is difficult if not impossible to interpret a comparison of the two atlas results. Unfortunately, the number of Mary-

land BBS routes recording this species is too few for accurate trend analysis. Within North America, BBS trend data show a 1.7% annual increase for this species from 1980 to 2007.

As with all wetland-dependent species, the Least Bittern has been threatened by the loss and degradation of its habitat. However, wetland protection laws enacted over the past few decades have greatly reduced this loss. As long as its habitat can be maintained and properly managed, this species is likely to remain a member of Maryland's marshland avifauna.

LYNN M. DAVIDSON

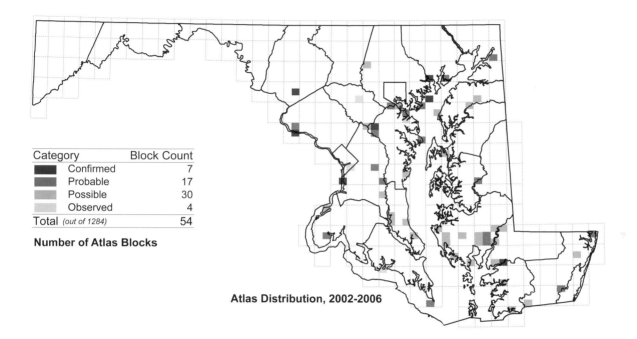

Category	Block Count
Confirmed	7
Probable	17
Possible	30
Observed	4
Total (out of 1284)	54

Number of Atlas Blocks

Atlas Distribution, 2002-2006

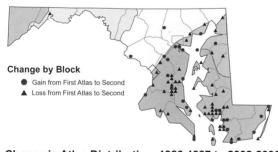

Change by Block

● Gain from First Atlas to Second
▲ Loss from First Atlas to Second

Change in Atlas Distribution, 1983-1987 to 2002-2006

Atlas Region	1983-1987	2002-2006	Change No.	Change %
Allegheny Mountain	0	0	0	-
Ridge and Valley	1	0	-1	-100%
Piedmont	4	4	0	0%
Upper Chesapeake	10	7	-3	-30%
Eastern Shore	36	24	-12	-33%
Western Shore	26	19	-7	-27%
Totals	77	54	-23	-30%

Change in Total Blocks between Atlases by Region

Great Blue Heron
Ardea herodias

Stately with measured movements, the Great Blue Heron is a usually silent, dignified presence on Maryland's and DC's ponds, lakes, rivers, creeks, coastlines, and marshes. Great Blue Herons are slow and stealthy hunters, watching vigilantly for prey as they stand at the water's edge or wade slowly, striking with lightning swiftness with their lethal spearlike bills. The diet of this heron consists mostly of fish and frogs, but it will also take crabs, crayfish, snakes, small mammals, and birds.

Great Blue Herons generally nest in colonies of between 20 and 50 nests (Butler 1992). In Maryland they nest primarily in single species colonies ranging from a few pairs to more than a thousand pairs (DNR unpubl. data). The largest Great Blue Heron colony observed in Maryland was 1,653 pairs on Pooles Island in 1999 (DNR unpubl. data). Colonies are usually far from human dwellings (Bendel and Therres 1999). Nests composed of sticks are usually built in trees and shrubs, unless the site is predator-free (such as on isolated islands). They are loosely constructed shallow platforms that become more substantial with reuse over several years. A few pairs of Great Blue Herons may nest singly in large trees near water or on ridges; 12% of a large sample of Ontario nests were not in a colony (Peck and James 1983). Because nest-building Great Blue Herons may carry sticks for long distances, often across block boundaries, the nest building atlas code (NB) cannot be applied to this bird without ambiguity. Confirmations of nesting must be made by using the

ON code (bird attending a nest of unknown contents), or a standard nest code (NE or NY) when contents of the nest are known. Similarly, the code for recently fledged young (FL) is inappropriate for young herons exploring the branches near their nests while they are still at their colony.

Great Blue Herons have a long nesting season in Maryland. Egg dates for the state range from 28 February to 14 June (Iliff, Ringler, and Stasz 1996). Dates for 80 atlas confirmations coded ON ran from 9 March to 10 July. Young have been reported in the nest in Maryland as late as 8 August (McKearnan 1996a).

The Great Blue Heron has been expanding its range inland in Maryland over the past 50 years. Eighteen colonies were limited to the fall line and below in Stewart and Robbins (1958). In 1977, 14 Great Blue Heron colonies were documented in Maryland, all on the Coastal Plain (Erwin and Korschgen 1979). The largest colony was 390 nesting pairs on Barren Island in Dorchester County.

In the 1983–1987 atlas 51 colonies and nesting pairs were found, with inland confirmations in Allegany and Frederick counties (McKearnan 1996a). Colonies ranged in size from 2 to 1,400 pairs (Gates et al. 1992). In the 2002–2006 atlas 95 nesting sites were found, including 3 in the Ridge and Valley in Washington and Frederick counties and 13 on the Piedmont. In 2003 the largest colony was on Pooles Island in Harford County where 1,146 pairs nested. That year a DNR (unpubl. data) survey documented 5,638 nesting pairs in 59 colonies. Large numbers of herons observed in western Maryland during 2002–2006 may indicate the presence of small, undiscovered colonies in those regions. Reasons for the increase in nesting Great Blue Herons, an 86% increase in colonies over the past two decades, are hard to specify but may include the building and stocking of artificial ponds, the banning of DDT in the early 1970s, and increased survey efforts for colonial waterbirds by DNR beginning in the late 1980s. DNR has surveyed and mapped colony locations every five years since 1987.

Breeding Bird Survey trends for Great Blue Heron have been strongly positive in Maryland. Between 1977 and 2003, this heron's breeding population in the Chesapeake Bay region has increased nearly fivefold and was estimated at 14,774 pairs in 2003 (B. Williams, Brinker, and Watts 2007). Recent increases have been attributed to the species' rapid expansion inland along tributaries of the bay. Trends in the phys-

iographic regions that include Maryland have also shown increases, although the trend for the Allegheny Plateau is indeterminate. Increases for the Piedmont and the Ridge and Valley have been 11% and 8% per year, respectively. The Great Blue Heron is a common sight in the wetlands and on the waterways of the Coastal Plain, and it is increasing and establishing a foothold as a nester on the Piedmont and in Central Maryland. As long as heron colonies are given a buffer zone from human disturbance and feeding sites continue clean and prey-rich, this majestic bird should remain widespread in Maryland and DC.

WALTER G. ELLISON

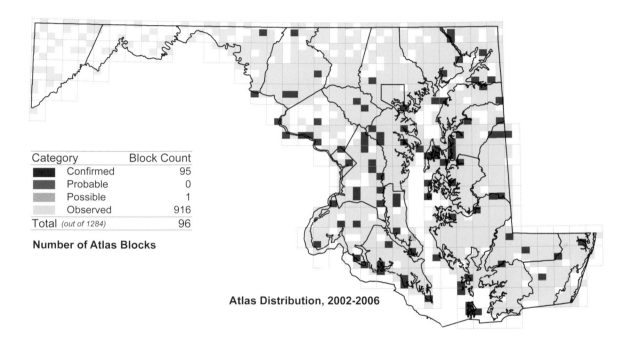

Category	Block Count
Confirmed	95
Probable	0
Possible	1
Observed	916
Total (out of 1284)	96

Number of Atlas Blocks

Atlas Distribution, 2002-2006

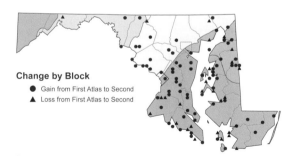

Change by Block
- Gain from First Atlas to Second
▲ Loss from First Atlas to Second

Change in Atlas Distribution, 1983-1987 to 2002-2006

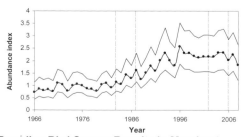

Percent of Stops
- 50 - 100%
- 10 - 50%
- 0.1 - 10%
- < 0.1%

Relative Abundance from Miniroutes, 2003-2007

Atlas Region	1983-1987	2002-2006	Change No.	%
Allegheny Mountain	0	0	0	-
Ridge and Valley	1	3	+2	+200%
Piedmont	3	16	+13	+433%
Upper Chesapeake	8	14	+6	+75%
Eastern Shore	22	29	+7	+32%
Western Shore	15	34	+19	+127%
Totals	49	96	+47	+96%

Change in Total Blocks between Atlases by Region

Breeding Bird Survey Results in Maryland

Great Egret
Ardea alba

The Great Egret is the largest of the three white egrets that nest in Maryland. Formally called Common Egret, this member of the heron family frequents marshes, swamps, shallow ponds, tidal flats, and the shorelines of tidal creeks and rivers. It is a colonial waterbird, usually nesting in colonies with other species of herons and egrets. It is known for its long plumes, called aigrettes, extending beyond the tail during the breeding season and used for courtship display (McCrimmon et al. 2001).

Populations of the Great Egret were decimated during the latter part of the nineteenth century and the early part of the twentieth century by extensive hunting of the species during the breeding season for its beautiful plumes (McCrimmon et al. 2001). The plumes were used in the millinery industry. Concern over the plight of this species and other migratory birds led to the formation of several conservation organizations and the passage of the federal Migratory Bird Treaty Act. Populations recovered following the passage of this important bird protection legislation.

By the mid-1950s the Great Egret was a fairly common breeding bird in the coastal areas of Worcester County and in the Pocomoke River swamp (Stewart and Robbins 1958). It was a rare breeder elsewhere in the Eastern Shore and Western Shore sections. Court (1936) found a few nests within a colony of Black-crowned Night-Herons in Charles County on 12 April 1931, which may be the first nesting record of Great Egret in Maryland. Vernon Stotts (in Stewart and Robbins 1958) reported nesting on Bodkin Island in Queen Anne's County. During the mid-1970s Osborn and Custer (1978) and Erwin and Korschgen (1979) documented 22 colonies of nesting Great Egrets primarily on the lower Eastern Shore. They also reported 1 colony in Talbot County and 2 in St. Mary's County.

The breeding season in Maryland begins when Great Egrets return to their nesting colonies in April. Colony sites, often used year after year, are generally located on offshore islands in the coastal bays or in Chesapeake Bay. Great Egrets routinely nest in mixed species colonies. Colony sizes range from a few pairs to more than a thousand pairs of nesting egrets and other heron species. Nests are constructed of sticks and twigs placed near the top of shrubs and trees (Palmer 1962; McCrimmon et al. 2001). The first eggs are laid in April (Robbins and Bystrak 1977) with clutch sizes averaging three eggs (McKearnan 1996b; McCrimmon et al. 2001). Incubation lasts 23–26 days and young start flying around seven weeks after hatching (Palmer 1962; McCrimmon et al. 2001).

The Great Egret was confirmed nesting in 14 blocks during the 1983–1986 atlas period (McKearnan 1996b). Since this colonial nesting waterbird can travel great distances from nest sites, all observations of Great Egrets other than of nests were considered observed records only. Most nesting occurred on the lower Eastern Shore in association with the tidal bays, both the Chesapeake Bay and coastal bays of Worcester County. Confirmed nesting occurred in 2 other blocks on the Coastal Plain, 1 each in Talbot and St. Mary's counties. A DNR-sponsored survey of colonial waterbirds during 1985–1988 documented Great Egrets nesting in 13–16 Maryland colonies each year with an average breeding population of 700 nesting pairs (Gates et al. 1992). During a comprehensive colonial waterbird survey in 1995, Brinker, Byrne, et al. (1996) documented 918 nesting pairs of Great Egrets in 20 colonies.

Results of the 2002–2006 atlas confirmed nesting in 17 atlas blocks. As in the first atlas, only records of birds at nests were accepted as proof of nesting. All other observations away from nest sites were considered observed records only. Confirmed nesting was concentrated on the lower Eastern Shore, with most occurring on the offshore islands in Dorchester and Somerset counties and the coastal bays of Worcester County. Great Egrets nested elsewhere in the Coastal Plain at 2 other locations on the Eastern Shore and 3 on the Western Shore. Nesting on the Western Shore was at Fort Carroll, Canoe Neck Creek, and Myrtle Point Park. D. Czaplak (Ringler 2004a) documented at least one pair of Great Egrets nesting in a large colony of Great Blue Herons on an island in the Potomac River in Washington County on 3 May 2003. In 2003 DNR (unpubl. data) documented 1,008 pairs of Great Egrets nesting in 24 colonies.

The distribution of nesting Great Egrets in Maryland has been consistent since the 1950s, with the one exception: expansion into Washington County. Trends in the Maryland breeding population appear to be increasing slightly. Within the Chesapeake Bay region, the number of Great Egret breeding pairs has increased by 153% since 1977 (B. Williams, Brinker, and Watts 2007). Given that many of the larger nesting colonies are on public lands and protection of colonial nesting waterbirds is afforded by both state and federal laws and regulations, the Great Egret breeding population will likely remain secure in Maryland.

GLENN D. THERRES

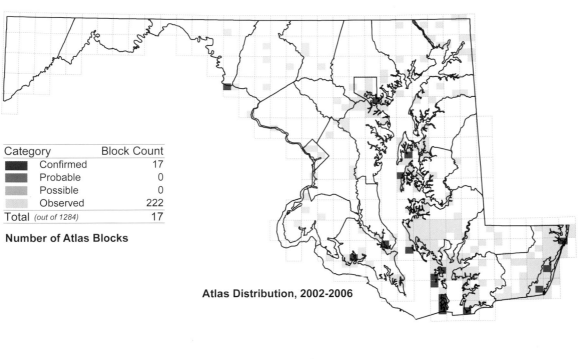

Category	Block Count
Confirmed	17
Probable	0
Possible	0
Observed	222
Total (out of 1284)	17

Number of Atlas Blocks

Atlas Distribution, 2002-2006

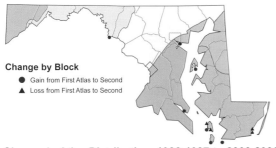

Change by Block
- ● Gain from First Atlas to Second
- ▲ Loss from First Atlas to Second

Change in Atlas Distribution, 1983-1987 to 2002-2006

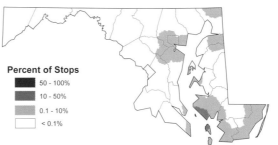

Percent of Stops
- 50 - 100%
- 10 - 50%
- 0.1 - 10%
- < 0.1%

Relative Abundance from Miniroutes, 2003-2007

Atlas Region	1983-1987	2002-2006	Change No.	Change %
Allegheny Mountain	0	0	0	-
Ridge and Valley	0	1	+1	-
Piedmont	0	0	0	-
Upper Chesapeake	0	1	+1	-
Eastern Shore	13	13	0	0%
Western Shore	1	2	+1	+100%
Totals	14	17	+3	+21%

Change in Total Blocks between Atlases by Region

Fran Saunders

Snowy Egret
Egretta thula

The Snowy Egret, a small white egret with black legs and yellow feet, is easily recognized. Like other members of the heron family, it inhabits a variety of wetlands and aquatic habitats, including marshes, swamps, tidal flats, ponds, creeks, and the shorelines of rivers and estuaries. Along the East Coast they seem to prefer saltwater and brackish habitats for nesting (Spendelow and Patton 1988). The Snowy Egret is a colonial nesting waterbird, nesting in mixed species colonies with other herons, egrets, and ibises. Its breeding distribution in North America is primarily along the coastal areas of the Atlantic and Pacific oceans and the Gulf of Mexico (Parsons and Master 2000) and in Central America. It also nests throughout much of South America.

During the late 1800s and early 1900s, the Snowy Egret was highly prized by hunters for its delicate, recurved back plumes, which were used to adorn women's hats (Parsons and Master 2000). Its population numbers plummeted dras-

tically as a result of this market hunting (J. Ogden 1978). Concern over the plight of this species and other colonial waterbirds led to protection of all migratory birds with the passage of the federal Migratory Bird Treaty Act. Snowy Egret populations recovered following the passage of this important bird protection legislation.

In Maryland the Snowy Egret was not known as a breeding species until Stewart and Robbins (1947) found two nesting colonies in 1946 in Worcester County. Prior to that, it occurred in Maryland during postbreeding dispersal in late summer (Hampe and Kolb 1947). By the mid-1950s it was considered a fairly common breeder in the coastal area of Worcester County and on Smith Island in Somerset County; it may also have nested elsewhere in tidewater areas of the lower Eastern Shore (Stewart and Robbins 1958). The largest documented colony was a hundred pairs on Mills Island in Worcester County in 1956. During the mid-1970s Snowy Egrets were documented nesting in five colonies in coastal Worcester County and seven colonies on offshore islands in Dorchester and Somerset counties (Osborn and Custer 1978; Erwin and Korschgen 1979). The total number of breeding Snowy Egrets averaged 1,224 pairs during 1975–1977.

The Snowy Egret was confirmed nesting in 11 blocks during the 1983–1987 breeding bird atlas (McKearnan 1996c). Only records of birds at nests were accepted as proof of nesting; all other observations away from nest sites were considered observed records only. All confirmed nesting occurred on the Eastern Shore, primarily in colonies on offshore islands in Dorchester, Somerset, Talbot, and Worcester counties. Two single nesting attempts were recorded on or near Tilghman Island in Talbot County. During 1985–1988 Gates et al. (1992) documented Snowy Egrets nesting in 11 colonies on the lower Eastern Shore of Maryland. The mean number of nesting Snowy Egrets was 1,251 pairs per year. The greatest number of nesting Snowy Egrets occurred in a mixed species colony on Smith Island in 1987 when 1,421 pairs were estimated in that colony.

The nesting season in Maryland begins in April when Snowy Egrets return to their colony sites. They almost always nest in mixed-species colonies (Parsons and Master 2000). In Maryland, colonies are usually located on offshore islands in the Chesapeake Bay or coastal bays of Worcester County. Nests constructed of loosely woven twigs and small sticks are placed in trees or shrubs (Parsons and Master 2000). Snowy Egrets will also nest in stands of common reed (Parsons 2003). Eggs are generally laid in May (Robbins and Bystrak 1977) with a mean clutch size of three eggs in Maryland (McKearnan 1996c). Incubation lasts 22–23 days (Parsons and Master 2000). Young disperse from the nest after seven weeks (Erwin, Haig, et al. 1996).

During the 2002–2006 atlas Snowy Egrets were confirmed nesting in 12 blocks. Only records of birds at nests were accepted as proof of nesting; all others were considered observed records. All nesting colonies, except one, were located on offshore islands on the Eastern Shore. Snowy Egrets nested in a mixed-species colony on Fort Carroll in the mouth of the Patapsco River on the Western Shore. In 2003 DNR (unpubl. data) documented Snowy Egrets nesting in 16 colonies and estimated a total of 1,624 breeding pairs in Maryland. The largest colony of Snowy Egrets was 813 pairs on Middle Holland Island in Dorchester County.

Trends of nesting Snowy Egrets in Maryland are stable (DNR, unpubl. data), but within the Chesapeake Bay region there has been a 58% decline since 1977 (B. Williams, Brinker, and Watts 2007). This decline is the result of an 80% decrease within Virginia coastal barrier island colonies between 1993 and 2003 (Watts and Byrd 2006). Causes of this decline include predation and loss of colony sites. Fortunately, similar declines did not occur in Maryland. Conservation of nesting and foraging habitat is key to sustaining Snowy Egrets. Protection of nesting colonies is afforded by both state and federal laws and regulations. Efforts to improve the Chesapeake Bay and its living resources will help to maintain the Snowy Egret as a breeding bird in Maryland.

GLENN D. THERRES

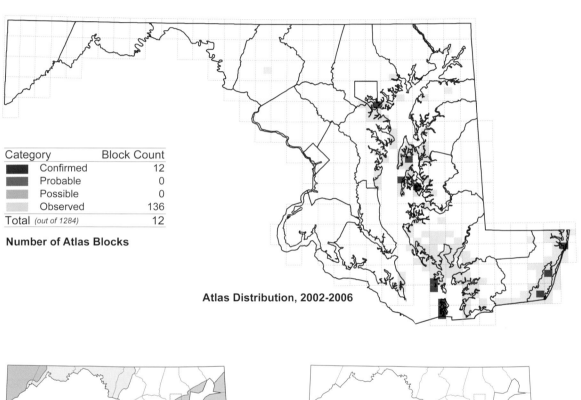

Category	Block Count
Confirmed	12
Probable	0
Possible	0
Observed	136
Total (out of 1284)	12

Number of Atlas Blocks

Atlas Distribution, 2002-2006

Change by Block
- ● Gain from First Atlas to Second
- ▲ Loss from First Atlas to Second

Change in Atlas Distribution, 1983-1987 to 2002-2006

Percent of Stops
- 50 - 100%
- 10 - 50%
- 0.1 - 10%
- < 0.1%

Relative Abundance from Miniroutes, 2003-2007

Atlas Region	1983-1987	2002-2006	Change No.	Change %
Allegheny Mountain	0	0	0	-
Ridge and Valley	0	0	0	-
Piedmont	0	0	0	-
Upper Chesapeake	0	1	+1	-
Eastern Shore	11	11	0	0%
Western Shore	0	0	0	-
Totals	11	12	+1	+9%

Change in Total Blocks between Atlases by Region

Mark Hoffman

Little Blue Heron
Egretta caerulea

Little Blue Heron is one of Maryland's smallest and least common waders. It is unique among herons in having two distinct, age-specific color morphs; first-year immature birds exhibit a white plumage while adults display a slate blue plumage (Rodgers and Smith 1995). This species is seldom seen far from coastal areas, and most breeding records are from islands in the Chesapeake Bay and coastal bays. Stewart and Robbins (1958) considered it a fairly common breeder in coastal Worcester County but rare and local elsewhere in tidewater areas of the Eastern Shore and Western Shore. It is generally more numerous and widespread during late summer, following postbreeding dispersal of adults and young (Iliff, Ringler, and Stasz 1996; McKearnan 1996d).

After wintering along the southeastern U.S. coast, Caribbean basin, and Central America (Rodgers and Smith 1995), Little Blue Herons begin arriving in Maryland in mid-March (Iliff, Ringler, and Stasz 1996). Nesting occurs exclusively in mixed heronries located on offshore islands. Nests consist of loosely woven stick platforms that are constructed 0.3–4.5 m (1–15 ft) above the ground in hardwood hammocks, shrub thickets, and phragmites stands. Clutch size averages 3–4 eggs with a mean first egg date of around 22 April. Incubation typically lasts 23 days and hatching usually begins in mid-May. Fledging begins in late June to early July. The Little Blue Heron forages primarily in shallow tidal and freshwater marshes, feeding diurnally on a variety of small fish, amphibians, and crustaceans (Rodgers and Smith 1995).

During the 1983–1987 atlas, breeding was confirmed in 5 blocks (McKearnan 1996d). Three blocks were in the Chesapeake Bay, with 2 on Smith Island and 1 on Poplar Island. The other 2 blocks were located in the coastal bays region. During 1987, the only year during the first atlas for which DNR statewide monitoring data are available, Little Blue Herons nested at a total of 8 sites supporting 314 breeding pairs, all of which occurred in large mixed heronries (Gates et al. 1992). Nearly 75% of the state's breeding pairs occurred on Smith Island, where 5 colonies contained 234 pairs. The largest colony was at Point Comfort on Smith Island with 192 pairs. Outside of Smith Island, Poplar Island, with 5 pairs, was the only other Chesapeake Bay colony.. The coastal bays region supported 2 colonies in 1987, 1 at South Point Spoils (31 pairs), an old dredge spoil island in northern Chincoteague Bay, and another at Heron Island (44 pairs) next to Ocean City in Assawoman Bay.

During the 2002–2006 atlas the number of confirmed breeding blocks increased from 5 to 8. DNR monitoring data (unpubl.) during 1985–2003 also show a small but significant ($P < 0.05$) increase in both the number of breeding pairs (from 314 to 354) and sites (from 8 to 14), although there is considerable year to year variation. Most blocks occurred in the Chesapeake Bay, with 3 blocks in the Smith Island area, 2 in the Holland Island archipelago of southwestern Dorchester County, and 1 in the Upper Chesapeake. The Smith Island area continued to have the greatest concentration; approximately half (48%–52%, 171–232 pairs) of the state's breeding pairs occurred there at 5–8 scattered sites. Other significant concentrations occurred in the Holland Island area with 1–3 colonies containing 52–129 pairs, and the coastal bays region with 2–3 colonies containing 40–154 pairs. Elsewhere, the only other breeding site was at Fort Carroll in Baltimore Harbor, where 9–15 Little Blue Heron pairs nested within a large mixed heronry. The only confirmed breeding outside of the Chesapeake Bay occurred in 2 blocks in the coastal bays region, where nesting continued to occur on South Point Spoils as well as Skimmer Island near the Ocean City inlet.

The small increase in the number of confirmed blocks during the second atlas, as well as a small but notable shift in breeding distribution in the Chesapeake Bay, can largely be attributed to the collapse of the Point Comfort mixed heronry on Smith Island. By the mid-1990s this large heronry disintegrated. (See the Glossy Ibis account for the cause.) The subsequent dispersion of nesting birds led to the establishment of several smaller heronries on Smith Island and the nearby Holland Island area. Also, by 1999 Poplar Island had nearly completely eroded away, eliminating the only colony in the central Chesapeake.

Although Little Blue Heron showed a small increase in breeding distribution and population size during the second atlas, it continues to be one of Maryland's least common wading birds. The majority of the state's breeding population occurs on a handful of low-lying offshore islands in the lower Chesapeake Bay and coastal bays. Like other waders, the greatest long-term threat to this species is the inundation of its coastal breeding and foraging habitat as a result of sea level rise. (See the Tricolored Heron account for management options for colonial waterbirds.)

JAMES M. MCCANN

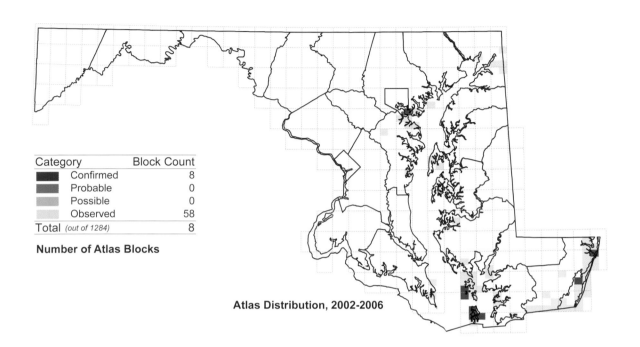

Category	Block Count
Confirmed	8
Probable	0
Possible	0
Observed	58
Total (out of 1284)	8

Number of Atlas Blocks

Atlas Distribution, 2002-2006

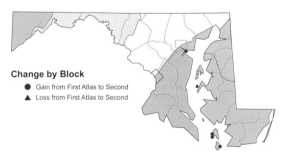

Change by Block

● Gain from First Atlas to Second
▲ Loss from First Atlas to Second

Change in Atlas Distribution, 1983-1987 to 2002-2006

Atlas Region	1983-1987	2002-2006	Change No.	%
Allegheny Mountain	0	0	0	-
Ridge and Valley	0	0	0	-
Piedmont	0	0	0	-
Upper Chesapeake	0	1	+1	-
Eastern Shore	5	7	+2	+40%
Western Shore	0	0	0	-
Totals	5	8	+3	+60%

Change in Total Blocks between Atlases by Region

Tricolored Heron
Egretta tricolor

Slender, elegant in form, and beautifully plumaged, Audubon was so impressed with the Tricolored Heron that he referred to it as the "lady of the waters." It is one of the most abundant herons in North America, exceeded only by Cattle Egret following its arrival during the 1950s (Frederick 1997). Although it has experienced declines in parts of its U.S. range, particularly in Florida and Louisiana, it remains a fairly common and widespread wader, with breeding colonies occurring throughout the Gulf Coast region, along the Atlantic seaboard as far north as Maine, and in a few isolated areas in the Midwest. Formerly known as the Louisiana Heron, it continues to be most abundant in its namesake state although coastal wetland loss has led to significant declines there. In Maryland, Stewart and Robbins (1958) recorded it as an uncommon, local breeder restricted to coastal Worcester County during the 1950s. By the 1970s it was a regular breeder in both the Chesapeake Bay and coastal bays region as it expanded its range northward along the Atlantic coast.

Like many waders that nest in Maryland, Tricolored Herons winter primarily in coastal areas in the southeastern United States, Caribbean basin, and Central America (Frederick 1997). Birds begin arriving in Maryland in mid- to late March (Iliff, Ringler, and Stasz 1996). They breed entirely in mixed heronries on offshore islands in hardwood hammocks, shrub thickets, and phragmites stands. Their loose twig platform nests and pale and unmarked greenish blue eggs are indistinguishable from those of Snowy Egret and Little Blue Heron, whose nests are often intermingled. Clutch size varies from 3 to 5 eggs. First eggs are usually laid around 22 April. Following a mean incubation period of 23 days, hatching typically begins in mid-May. The mean nestling period of 28 days is relatively short compared to other similar-size waders, with first fledging usually occurring in mid-June. Tricolored Herons forage singly or in small, loosely organized monospecific or mixed-species flocks. They hunt primarily in open salt marshes, feeding mostly on small estuarine fish. Tricolored Herons are very active diurnal hunters, displaying a variety of foraging techniques from slow stalking and quick chases to "canopy-feeding" (luring prey into the shadow cast by their outstretched wings) and "foot-raking" (using their feet to flush prey). For additional details on this species' life history and ecology, see McKearnan (1996e) and Frederick (1997).

Tricolored Heron remains a locally common breeder restricted to mixed heronries in the lower Chesapeake Bay and coastal bays region. The number of confirmed breeding blocks increased from 5 during the first atlas (McKearnan 1996e) to 7 during the second. During each atlas the majority of the blocks were located in the Chesapeake Bay with 2 blocks in the coastal bays region. The greatest concentrations of nesting birds continued to occur at Smith Island followed by the Holland Island area and coastal bays. DNR waterbird monitoring data show no significant change in the number of breeding pairs during 1985–2006. For those years when complete statewide monitoring data was collected, the breeding population varied from 705 pairs at 7 sites in 1987 to 582–605 pairs at 9–14 sites during 2002–2003. The increased number of sites during the second atlas was caused primarily by the breakup of the Point Comfort colony on Smith Island (see Glossy Ibis account). During the mid-1980s, this colony represented one of the state's largest mixed heronries with more than 3,700 nesting pairs, including 77% (548 pairs) of the state's Tricolored Heron breeding population. Following the breakup, 4–6 smaller colonies became established elsewhere on Smith Island and in the nearby Holland Island area.

Although Tricolored Heron populations are currently stable, this species nests exclusively on a small number of offshore islands in the lower bay and coastal Worcester County. As climate change continues to impact these islands and surrounding tidal marshes, this and many other coastal inhabitants face an uncertain future. If Maryland is to continue to

support this species and other colonial nesting waterbirds, key steps must be taken to offset additional island and salt marsh loss. Such interventions would include island creation and the bolstering of existing islands using dredge spoil materials and offshore sediments. The use of shoreline erosion control methods that maintain natural shoreline erosion and island building processes should be favored over those that harden shorelines and disrupt these processes (e.g., bulkheads, rock rip-rap). Without such measures, this and other species face an uncertain future in Maryland.

JAMES M. MCCANN

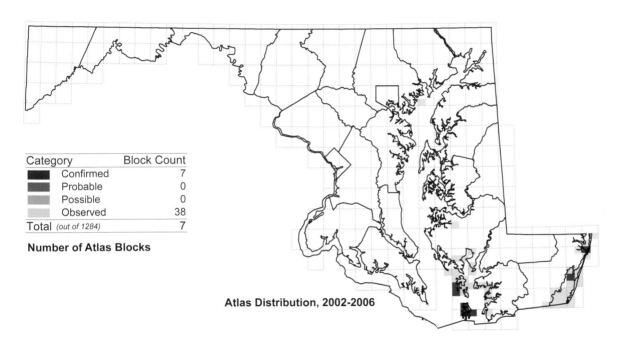

Category	Block Count
Confirmed	7
Probable	0
Possible	0
Observed	38
Total (out of 1284)	7

Number of Atlas Blocks

Atlas Distribution, 2002-2006

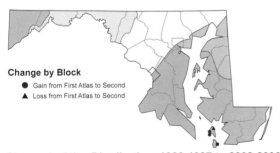

Change by Block
- Gain from First Atlas to Second
▲ Loss from First Atlas to Second

Change in Atlas Distribution, 1983-1987 to 2002-2006

Atlas Region	1983-1987	2002-2006	Change No.	%
Allegheny Mountain	0	0	0	-
Ridge and Valley	0	0	0	-
Piedmont	0	0	0	-
Upper Chesapeake	0	0	0	-
Eastern Shore	5	7	+2	+40%
Western Shore	0	0	0	-
Totals	5	7	+2	+40%

Change in Total Blocks between Atlases by Region

Cattle Egret
Bubulcus ibis

Robert Ampula

The Cattle Egret is atypical of wading birds in that it feeds primarily in open fields instead of wetlands and along shorelines. It feeds mainly on large-bodied insects, especially grasshoppers (Palmer 1962), and is often seen in pastures alongside grazing cattle, thus its name. The Cattle Egret is originally from Africa, where it fed among the large herds of ungulates on the African plains (Telfair 1994). Like other wading birds, the Cattle Egret nests in colonies, usually with other species of herons and egrets.

In the late 1800s Cattle Egrets crossed the Atlantic Ocean and were recorded on the northeast coast of South America (Palmer 1962; Telfair 1994). They expanded onto the Caribbean islands in the 1930s. The first North American record was in southern Florida in 1941, with breeding documented in 1953. By 1954 it had been observed on the East Coast as far north as Massachusetts. D. Davis (1960) believed the range expansion had stopped in the late 1950s and was confined to the Atlantic and Gulf coasts. The expansion may have briefly stalled, but Cattle Egrets now breed throughout most of North America (Telfair 1994).

The Cattle Egret was first recorded in Maryland on 25 April 1953 near Berlin in Worcester County (Stewart and Robbins 1958). The first nesting in Maryland was documented on 1 June 1957 when G. Miller (1959) found a nest with three eggs on Mills Island in Worcester County. By 1977 Cattle Egrets were known to be nesting in 9 colonies on the Eastern Shore, with an estimated population of 1,325 breeding pairs (Osborn and Custer 1978; Erwin and Korschgen 1979).

The Cattle Egret was confirmed nesting in 5 blocks during the 1983–1987 breeding bird atlas (McKearnan 1996f). Only records of birds at nests were accepted as proof of nesting; observations away from nest sites were considered observed records only. All confirmed nesting occurred on the Eastern Shore in colonies on offshore islands: two on Smith Island in Somerset County, two in Worcester County, and one in Talbot County. During 1985–1988 the Cattle Egret nesting population in Maryland ranged from 712 to 2,217 pairs (Gates et al. 1992). The greatest number of nesting Cattle Egrets occurred in a mixed species colony at Point Comfort on Smith Island in 1988 when 1,144 pairs were estimated in that colony.

The nesting season in Maryland begins in April, when Cattle Egrets return to their island colony sites. They nest in mixed-species colonies and usually arrive later than the other species (J. Burger 1978; Telfair 1994). In Maryland, colonies are located on offshore islands in the Chesapeake Bay or the coastal bays of Worcester County. The earliest egg date in Maryland is 9 April (Robbins and Bystrak 1977) and the latest is 30 August (McKearnan 1996f). Nests constructed of loosely woven twigs and small sticks are placed in trees or shrubs. In dense colonies where no nest sites are left in vegetation, they will nest on the ground (Telfair 1994). Cattle Egrets will also nest in stands of common reed, where they have been documented producing larger clutches with higher hatching rates (Parsons 2003). Clutch size may be as large as six eggs (Palmer 1962). Hatching is asynchronous (Telfair 1994). During the 2002–2006 atlas, nests with young ranged from 8 June to 11 July. Young begin to fly at about 25 days old and fledge at 30 days of age (Telfair 1994).

During the 2002–2006 atlas, Cattle Egrets were confirmed nesting in 9 blocks. Only records of birds at nests were accepted as proof of nesting; all others were considered observed records. The number of blocks with nonbreeding Cattle Egrets was considerably less than in 1983–1987. All nesting colonies except one were located on offshore islands on the Eastern Shore. This distribution is similar to that found during the first atlas. Cattle Egrets expanded their nesting distribution to a mixed-species colony on Fort Carroll in the mouth of the Patapsco River on the Western Shore. In 2003 DNR (unpubl. data) documented a population of 575 breeding pairs of Cattle Egrets in Maryland. This is a substantial decrease from the high of 2,217 breeding pairs in 1988 (Gates et al. 1992).

Trends of nesting Cattle Egrets in Maryland increased and then declined (DNR, unpubl. data). Within the Chesapeake Bay region there has been a 64% decline since 1977 (B. Williams, Brinker, and Watts 2007). Much of this decline is the result of a drastic population reduction in Virginia. Conservation of nesting colonies is key to sustaining Cattle Egrets and other colonial nesting waders in Maryland. Protection of nesting colonies is afforded by both state and federal laws and regulations. This species is expected to persist as a breeding bird in Maryland, though perhaps in reduced numbers.

GLENN D. THERRES

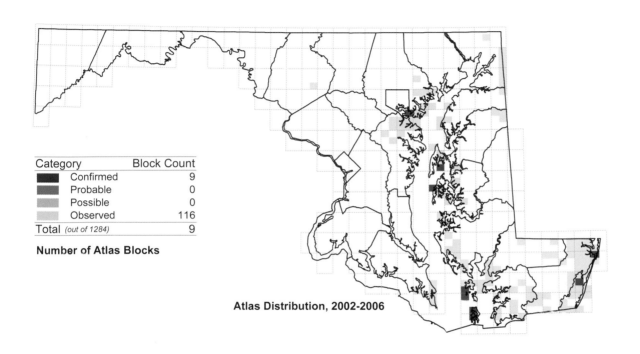

Category	Block Count
Confirmed	9
Probable	0
Possible	0
Observed	116
Total (out of 1284)	9

Number of Atlas Blocks

Atlas Distribution, 2002-2006

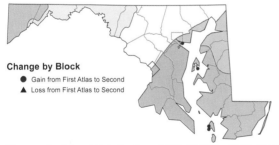

Change by Block
● Gain from First Atlas to Second
▲ Loss from First Atlas to Second

Change in Atlas Distribution, 1983-1987 to 2002-2006

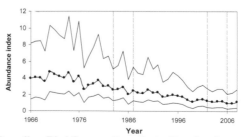

Percent of Stops
- 50 - 100%
- 10 - 50%
- 0.1 - 10%
- < 0.1%

Relative Abundance from Miniroutes, 2003-2007

			Change	
Atlas Region	1983-1987	2002-2006	No.	%
Allegheny Mountain	0	0	0	-
Ridge and Valley	0	0	0	-
Piedmont	0	0	0	-
Upper Chesapeake	0	1	+1	-
Eastern Shore	5	8	+3	+60%
Western Shore	0	0	0	-
Totals	5	9	+4	+80%

Change in Total Blocks between Atlases by Region

Breeding Bird Survey Results in Maryland

Green Heron
Butorides virescens

Short legged for a heron, small, compact, and dark, the Green Heron does not wade after prey but prefers to hunt along shorelines and from perches surrounded by shallow water. Green Herons are often seen flying high between feeding grounds and nest sites looking much like deep-chested, long-legged crows. This heron's sharp somewhat musical bark, *skiyow!* also announces its presence.

Green Herons are found almost anywhere there is shallow water with sufficient prey such as crayfish, frogs, and small fish. These herons frequent creeks (but not narrow wooded brooks), rivers, small ponds, lakeshores, and tidal shores, with the exception of oceanic beaches. Green Herons usually construct their loosely built stick nest in small to midsize trees or shrubs at heights at or below 3.1 m (10 ft), they occasionally nest on the ground (especially where predators are few) or in tall trees. They are less prone to nesting close to or above water than are other herons. In tidewater areas Green Herons sometimes nest among other herons in colonies or form small, loose nesting aggregations with their own kind (W. Davis and Kushlan 1994), but this heron is usually a solitary nester, almost entirely so inland. Green Herons are migratory and are regularly present in Maryland from April to October. The Green Heron raises a single brood in the northern part of its range. Maryland and DC dates for

nests containing eggs are 21 April to 26 July, with nestlings reported as late as 5 August (Rasberry 1996; MNRF).

The status and distribution of the Green Heron in Maryland and DC has changed little since the end of the nineteenth century. Kirkwood (1895) considered it common, and Eifrig (1904) called it not rare in Allegany and Garrett counties. Stewart and Robbins (1958) found the Green Heron common in tidewater areas and fairly common in the rest of the state and district. In spite of a slight loss of blocks between the 1983–1987 and 2002–2006 atlas periods, the status of this heron has remained essentially the same over the past two decades. Some of the losses and gains on the 2002–2006 map appear related to differences in observer effort and coverage between atlas periods.

North American BBS trends show a significant decline averaging 1.7% per year from 1966 to 2007. On Maryland BBS routes the trend has been negative but not statistically significant. Factors that have a negative impact on Green Herons include declines in water quality, which affects prey abundance, and the loss of nest sites as stands of trees and shrubs are removed along waterfronts for development. A factor that favors this heron, at least in the short term, is the construction of small ponds by landowners and developers for recreation and wetland mitigation. At present it appears that the Green Heron continues fairly common and widespread in Maryland and DC, but as a bird dependent on wetlands and undeveloped shores it bears watching.

WALTER G. ELLISON

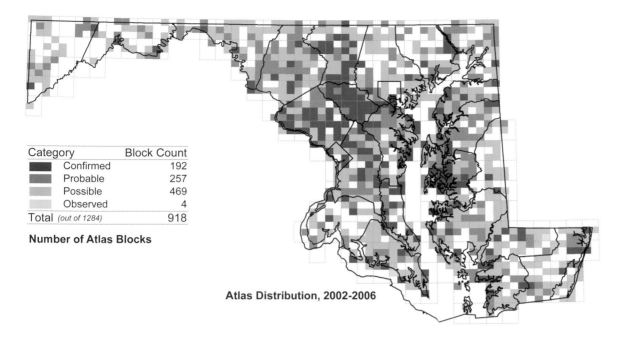

Category	Block Count
■ Confirmed	192
■ Probable	257
■ Possible	469
■ Observed	4
Total *(out of 1284)*	918

Number of Atlas Blocks

Atlas Distribution, 2002-2006

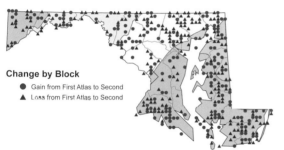

Change by Block
- ● Gain from First Atlas to Second
- ▲ Loss from First Atlas to Second

Change in Atlas Distribution, 1983-1987 to 2002-2006

Percent of Stops
- ■ 50 - 100%
- ■ 10 - 50%
- ■ 0.1 - 10%
- □ < 0.1%

Relative Abundance from Miniroutes, 2003-2007

Atlas Region	1983-1987	2002-2006	Change No.	Change %
Allegheny Mountain	40	37	-3	-8%
Ridge and Valley	102	101	-1	-1%
Piedmont	249	235	-14	-6%
Upper Chesapeake	85	87	+2	+2%
Eastern Shore	257	253	-4	-2%
Western Shore	201	202	+1	0%
Totals	934	915	-19	-2%

Change in Total Blocks between Atlases by Region

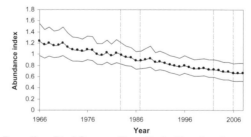

Breeding Bird Survey Results in Maryland

Eric Skrzypczak

Black-crowned Night-Heron
Nycticorax nycticorax

The Black-crowned Night-Heron is a stocky, medium-size wader with nocturnal and crepuscular feeding habits. It typically breeds in mixed heronries on islands. But monospecific colonies are sometimes formed, and nest sites can occur in or near a variety of wetland habitats in both coastal and inland areas, sometimes in close proximity to urban areas. Nest sites in Maryland have included shrub thickets in salt marshes, stands of phragmites, and deciduous riparian forests (McKearnan 1996g). It is one of the world's most widely distributed waders, with breeding populations on every continent except Antarctica and Australia (W. Davis 1993). It is widespread in the United States, nesting throughout most of the lower 48 states and in Hawaii. Stewart and Robbins (1958) documented 17 colonies in seven counties on the Eastern and Western Shore, as well as in Washington, DC. Most colonies were on islands or in tidal areas in the Chesapeake Bay and coastal bays. This species' ability to nest near urban areas and in inland settings is apparently not a recent phenomenon; colonies were reported more than a hundred years ago near Baltimore City and in Washington, DC (Kirkwood 1895; Cooke 1929).

Black-crowned Night-Herons winter primarily in coastal areas in the southeastern United States and California, the Caribbean basin, Mexico, and Central America (W. Davis 1993). Some individuals may overwinter in Maryland, mostly near tidal areas, especially during mild winters (W. Davis 1993; McKearnan 1996g; Iliff, Ringler, and Stasz 1996). Birds begin to arrive at Maryland breeding sites in March (DNR, unpubl. data). Nests consist of loose stick platforms roughly 0.6 m (2 ft) in diameter, placed in dense scrub, phragmites, and deciduous trees from near ground level to as high as 15.2 m (50 ft) above the ground (W. Davis 1993; McKearnan 1996g). First egg dates average around 14 April (DNR, unpubl. data). The mean incubation period is 25 days, with first chicks usually appearing about 9 May. The mean nestling period is relatively long compared to most waders, lasting an average of 46 days, with fledglings appearing in late June. It tends to forage singly, feeding in salt marsh tidal pools and guts, along ditches and streams, mudflats, inland marshes, and pond and lake shorelines (W. Davis 1993). Prey items include a variety of small fish, crustaceans, frogs, aquatic insect larvae, small mammals, and occasionally other wader eggs and chicks (W. Davis 1993).

During 1983–1987 breeding was confirmed in 16 blocks, 14 of which were located in the lower Chesapeake Bay and in coastal Worcester County (McKearnan 1996g). Most of these colonies were on islands in mixed heronries. Breeding was also recorded in 1 block each in Baltimore County and Washington, DC. In 1987 statewide surveys documented a total of 17 colonies containing 1,082 breeding pairs (Gates et al. 1992). The largest colony by far was at Point Comfort (633 pairs) on Smith Island followed by a colony near the Francis Scott Key Bridge in Baltimore County with 300 pairs. Together these two colonies supported 86% of the state's breeding population.

During the second atlas, there was little change in the number or general distribution of blocks. Atlas observers confirmed breeding in 18 blocks, the majority of which continued to be concentrated in the lower Chesapeake Bay and coastal bays. However, there were a greater number of blocks west of the Chesapeake Bay, with 7 widely scattered colonies in the following counties: St. Mary's, Baltimore, Harford, Montgomery, and Frederick. The vast majority of the unconfirmed observations were probably of foraging birds or postbreeding adults and juveniles. But it is quite possible that a few small colonies went undetected, particularly at inland sites on the Western Shore. Although Black-crowned Night-Herons were found in more colonies during the second atlas, DNR (unpubl. data) monitoring efforts show a significant decline during 1985–2006. The decline resulted, in part, from the collapse of two colonies at Point Comfort and near Francis Scott Key Bridge during the 1990s. Numbers also dropped in the coastal bays from 74–219 pairs during 1985–1987 to 50 pairs in 2003. Lower numbers also resulted from the loss of the Heron Island colony near Ocean City, which at one point supported as much as 70% of the coastal bays' breeding population. Larger, regional shifts in breeding distribution and population size may have also contributed to declines.

Black-crowned Night-Heron is one of the most widely distributed and adaptable of Maryland's colonial waterbirds. Although sparsely distributed away from coastal areas, it has become more widespread in inland areas west of the Chesapeake Bay. Nonetheless, recent declines are of concern. The heart of its Maryland breeding range and the majority of the state's breeding population continues to lie in the vulnerable offshore islands of the lower Chesapeake Bay and coastal bays. As with most waterbirds and other coastal nesting birds, sea level rise caused by climate change poses a serious threat to these island habitats. Global actions aimed at reducing greenhouse gas emissions along with efforts to maintain suitable nesting and foraging habitat are urgently needed.

JAMES M. MCCANN

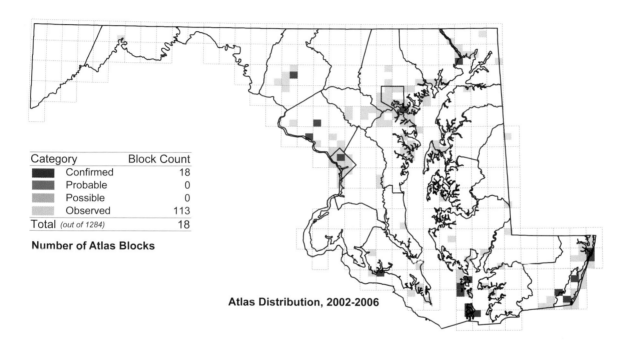

Category	Block Count
Confirmed	18
Probable	0
Possible	0
Observed	113
Total (out of 1284)	18

Number of Atlas Blocks

Atlas Distribution, 2002-2006

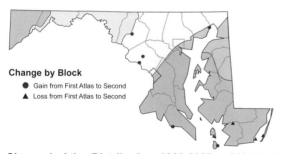

Change by Block

● Gain from First Atlas to Second
▲ Loss from First Atlas to Second

Change in Atlas Distribution, 1983-1987 to 2002-2006

Atlas Region	1983-1987	2002-2006	Change No.	%
Allegheny Mountain	0	0	0	-
Ridge and Valley	0	0	0	-
Piedmont	1	4	+3	+300%
Upper Chesapeake	1	2	+1	+100%
Eastern Shore	14	11	-3	-21%
Western Shore	0	1	+1	-
Totals	16	18	+2	+13%

Change in Total Blocks between Atlases by Region

Yellow-crowned Night-Heron
Nyctanassa violacea

Yellow-crowned Night-Herons are solitary hunters, specializing on crustaceans. Although often breeding in large mixed heronries, they do so less than most colonial waterbirds, frequently nesting singly or in small monospecific colonies (Watts 1995). They are most numerous in coastal areas but commonly breed in swamps and riparian forests and near lake and pond shorelines. From the 1920s through the 1960s, this species exhibited a significant northward range expansion along the East Coast as far north as Massachusetts and into the upper Midwest. Many of these areas had been previously occupied during the mid- to late 1800s but populations declined sharply by 1900 for reasons that are unclear (Watts 1995). The first DC record was a specimen collected by W. Palmer in 1901 on Smithsonian Institution property (Cooke 1929). Not until 1921 was the first breeding record for Maryland-DC confirmed, when E. Court collected a clutch of five fresh eggs near Seneca in Montgomery County (McKearnan 1996h). The first Maryland specimen records were documented in 1927 by B. Overington who shot two birds in Laurel, Prince George's County (Ball 1932). It was regarded as a rare summer resident by Hampe and Kolb (1947). Stewart and Robbins (1958) considered it a rare, local breeder with small, very widely distributed colonies on the Eastern Shore, Western Shore, and Piedmont.

Most Yellow-crowned Night-Herons winter in subtropical and tropical areas in south Florida, Central America, and the Caribbean, where their crustacean prey remain active and plentiful (Watts 1995). Birds usually begin arriving at Maryland breeding sites in late March. Most large colonies occur in mixed heronries on offshore islands. Smaller colonies may occur in deciduous floodplain forests along streams and rivers. They construct loosely woven, stick platform nests about 50.8 cm (20 in) in diameter, placed in shrub thickets and small to large deciduous trees from 0.9 to 18.2 m (3–60 ft) above the ground (McKearnan 1996h; Watts 1995). Mean clutch size is 2.8 eggs; the mean first egg date is 21 April (DNR, unpubl. data); mean first chick dates fall around 15 May. Chicks first begin fledging around the third week in June (DNR, unpubl. data). Birds usually forage alone along water margins in tidal marshes, mudflats, tidal pools, ponds, streams, and rivers (Watts 1995). In Maryland's coastal areas, this species feeds primarily on fiddler crabs while, in inland areas, it specializes on crayfish (McKearnan 1996h). Compared to other waders, it exhibits highly sedentary foraging techniques, often remaining stationary for long periods of time before striking at passing prey or using a slow, deliberate "walk and stalk" approach (Watts 1995).

Fran Saunders

During 1983–1987 breeding was confirmed in 7 blocks in two distinctly different habitats and regions (McKearnan 1996h). In 1987 statewide surveys documented 11 colonies in 4 blocks in the lower Chesapeake Bay (Gates et al. 1992). Occurring mostly in mixed heronries, these colonies contained a total of 234 pairs. Nine of these colonies occurred on Smith Island, which together represented 95% of the state's breeding population. Two small colonies also occurred on nearby Holland Island. Outside of the lower bay, the only confirmed breeding occurred in 3 blocks in Baltimore County, each containing a single nesting pair. Most of the 32 blocks with unconfirmed records were probably of strays, immature birds, or postbreeding adults and fledglings. However, some observations may have been of undetected pairs nesting alone or in small colonies.

During the second atlas, Yellow-crowned Night-Heron was considerably more widespread, with 20 confirmed blocks. Fourteen of these blocks were on the Western Shore in Washington, DC; Baltimore City; and Baltimore, Harford, and Montgomery counties. The greatest concentration of blocks occurred in Baltimore City and adjacent Baltimore County. Most, perhaps all, Western Shore records were of solitary nesting pairs or small colonies of fewer than 10 pairs. These breeding sites typically occurred in narrow corridors of deciduous riparian forest along streams and rivers, sometimes in very close proximity to busy roads and development. The lower Chesapeake Bay continued to support most (>90%) of the state's breeding population. In 2003 DNR (unpubl. data) surveys recorded 14 colonies in 6 blocks with a total of 257 pairs. Although birds were more widely distributed during the second atlas, statewide waterbird monitoring efforts by DNR indicate no significant change in population size from 1985 to 2006.

Yellow-crowned Night-Heron continues to be one of Maryland's least common waders. The vast majority of the state's breeding population remains restricted to offshore islands in the lower Chesapeake Bay. These low-lying salt marsh islands are highly vulnerable to sea level rise resulting from climate change. This precarious status may be offset to some degree by the species' ability to breed at inland sites, often in close proximity to highly developed areas. To what extent this crustacean specialist can persist along degraded streams and rivers, as well as in areas where the native crayfish fauna has been reduced or displaced by non-native species, is uncertain and worthy of study.

JAMES M. MCCANN

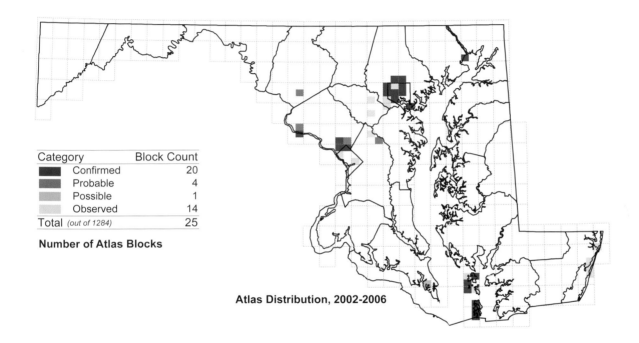

Category	Block Count
Confirmed	20
Probable	4
Possible	1
Observed	14
Total (out of 1284)	25

Number of Atlas Blocks

Atlas Distribution, 2002-2006

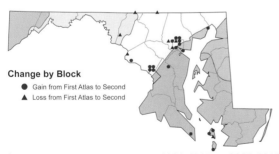

Change by Block
● Gain from First Atlas to Second
▲ Loss from First Atlas to Second

Change in Atlas Distribution, 1983-1987 to 2002-2006

Atlas Region	1983-1987	2002-2006	Change No.	Change %
Allegheny Mountain	0	0	0	-
Ridge and Valley	2	0	-2	-100%
Piedmont	9	14	+5	+56%
Upper Chesapeake	0	2	+2	-
Eastern Shore	4	6	+2	+50%
Western Shore	0	3	+3	-
Totals	15	25	+10	+67%

Change in Total Blocks between Atlases by Region

Monroe Harden

Glossy Ibis
Plegadis falcinellus

Glossy Ibis is a gregarious, medium-size wading bird with a distinctive, long, decurved bill. In proper lighting, the plumage reveals its namesake, a brilliant metallic bronze sheen tinged in green. Seldom seen far from coastal areas, it usually nests in large mixed-species wading bird colonies in salt marsh hammocks and coastal islands. It is Maryland's only nesting ibis and the most widespread ibis in the world, occurring on every continent except Antarctica (W. Davis and Kricher 2000). Sometimes ranging widely after breeding, it probably colonized North America during the early 1800s. By 1900 it was a rare and local breeder in Florida. It slowly expanded its range northward along the Atlantic coast and westward into the Gulf states (W. Davis and Kricher 2000). Glossy Ibis was a highly rare breeder in Maryland some 50 years ago, at the time of Stewart and Robbins' (1958) seminal publication. The first Maryland nesting record was in 1956 (R. Stewart 1957). In 1962 it was recorded nesting in the Maryland portion of the Chesapeake Bay (Weske and Fessenden 1963). Since then, this species has become a well-established breeding bird in Maryland's coastal areas. Today, it regularly breeds as far north as Maine and throughout the U.S. Gulf Coast (W. Davis and Kricher 2000).

After wintering along the southeastern U.S. coast, Glossy Ibises arrive in Maryland in March to early April. Since the bird's first appearance in Maryland, all nest sites have occurred in mixed heronries on islands, mostly natural but a few man-made (e.g., dredge spoil; Fort Carroll in the Baltimore harbor), in the coastal bays and Chesapeake Bay. Nests consist of loose stick platforms, often lined with leaves or grasses. They are constructed from near ground level to 7 m (23 ft) high in a variety of trees and shrub substrates, as well as in phragmites stands. Egg laying begins in late April, hatching starts by mid- to late May, and fledging begins in mid- to late June. The Glossy Ibis is a tactile forager, feeding mostly in shallow tidal marshes and mudflats on a variety of mollusks (snails, clams, mussels), insects, and other aquatic invertebrates (W. Davis and Kricher 2000). It commonly feeds in loose, single species flocks that range from several birds to scores of individuals.

During the 1983–1987 atlas, Glossy Ibis was a confirmed breeder in 3 blocks (McKearnan 1996i). Two blocks occurred on Smith Island in the lower Chesapeake Bay. The third block occurred on South Point Spoils, an old dredge spoil island in northern Chincoteague Bay. During 1986–87, the only years during the first atlas for which complete statewide monitoring data are available, the number of breeding pairs varied from 836 in 1986 to 1,733 in 1987, with 7 active colonies during each year (Gates et al. 1992). The state's largest colony then was Point Comfort, a hardwood hammock island within the Smith Island area. It supported 71% (592 pairs) to 77% (1,339 pairs) of all breeding pairs in Maryland in 1986 and 1987, respectively. The second largest colony was South Point Spoils with 220 (1986) to 350 pairs (1987). Together, these two colonies supported 97% of all nesting Glossy Ibises during that two-year period.

During the 2002–2006 atlas, confirmed breeding was documented in 10 blocks. Despite this block increase, the total number of breeding pairs has remained relatively stable since the first atlas. The number of breeding pairs varied from 978 to 1,467 pairs during 2002 and 2003 (DNR, unpubl. data), comparable to the breeding population sizes reported during the first atlas. However, the number of colonies increased substantially, from 7 colonies during 1986–1987 to 11 colonies in 2002 and 20 colonies in 2003. Much of this increase can be attributed to the abandonment of the Point Comfort mixed heronry during the mid-1990s (DNR, unpubl. data), followed by the redistribution of nesting birds in subsequent years to other sites on Smith Island and the nearby Holland Island archipelago. In an unusual example of the effects of non-native species, the birds' abandonment of Point Comfort has been attributed to the introduction of feral goats to that particular part of Smith Island. Abandonment appears to have been a direct result of goat predation on chicks and eggs of Glossy Ibis, as well as other low-nesting waders. By 2003 this species and other waders had relocated to 13 other mixed heronry

sites in the scrub thickets and hardwood hammocks of Smith Island and Holland Island. In the coastal bays it continued to nest in large numbers at South Point Spoils with 63% (925 pairs) of the state's breeding pairs occurring here in 2003, along with several other smaller nearby colonies.

The greatest long-term threat to Glossy Ibis in Maryland is the inundation of its coastal breeding and foraging habitat from sea level rise. (See the Tricolored Heron account for management options.) It remains to be seen whether this seemingly adaptive, far-wandering species will continue to grace our coastal marshes and islands well into the future.

JAMES M. MCCANN

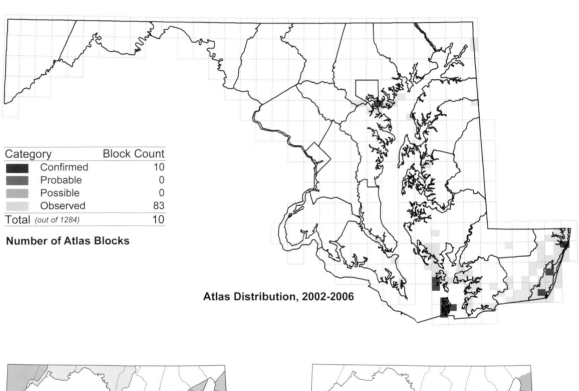

Category	Block Count
■ Confirmed	10
■ Probable	0
■ Possible	0
■ Observed	83
Total *(out of 1284)*	10

Number of Atlas Blocks

Atlas Distribution, 2002-2006

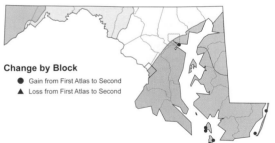

Change by Block
- ● Gain from First Atlas to Second
- ▲ Loss from First Atlas to Second

Change in Atlas Distribution, 1983-1987 to 2002-2006

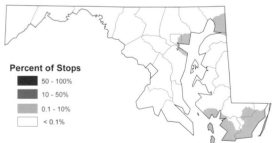

Percent of Stops
- ■ 50 - 100%
- ■ 10 - 50%
- ■ 0.1 - 10%
- □ < 0.1%

Relative Abundance from Miniroutes, 2003-2007

			Change	
Atlas Region	1983-1987	2002-2006	No.	%
Allegheny Mountain	0	0	0	-
Ridge and Valley	0	0	0	-
Piedmont	0	0	0	-
Upper Chesapeake	0	1	+1	-
Eastern Shore	3	9	+6	+200%
Western Shore	0	0	0	-
Totals	3	10	+7	+233%

Change in Total Blocks between Atlases by Region

Black Vulture
Coragyps atratus

The undertaker dressed in funereal black with white-gloved hands, the Black Vulture soars very high over great distances seeking large carrion meals. Black Vultures use their broad wings and short fanned tails to rise on thermals high above the Turkey Vultures they often rely upon to locate food for them. With their long sharp bills and greater weight, Black Vultures outcompete Turkey Vultures and crows at carcasses. Among Maryland scavengers only Bald Eagles can keep Black Vultures from taking their cut first. The Black Vulture is also impatient, sometimes attacking the living if they are sufficiently defenseless (Buckley 1999). These vultures live in mixed landscapes of open fields and woodlots; they require the open land to generate the thermals that carry them long distances in search of food; and they need trees in which to roost and for nesting cover. Black Vultures generally have huge home ranges, averaging nearly 16,000 ha (39,536 ac) in a study in Pennsylvania and Maryland (Coleman and Fraser 1989). Any flying or feeding bird could be far from where it has its nest; because of this a very high 38% of 2002–2006 atlas records were given the observed status code.

Like Turkey Vultures, Black Vultures use an array of hidden and sheltered sites for nesting, including dense thorny thickets, hollows in trees and logs, rocky recesses in cliffs and talus, and abandoned man-made structures (J. Jackson 1983). These vultures do not build a nest as such, laying their eggs directly on the substrate of the nest site. Black Vultures ap-

pear to be more prone to using man-made structures than Turkey Vultures. They use not only old barns and derelict houses but also duck blinds and abandoned upper floors of buildings still in use. Although the birds are single brooded, the nesting season is long, partly because of renesting; the early date for eggs in Maryland is 14 March (Stewart and Robbins 1958; MNRF). The late date comes from this project, 30 June 2006 near Allen's Fresh, Charles County (G. Jett). New records for nestling dates were also established during this project: 13 April 2006 at Wye Island, Queen Anne's County (D. Poet) and 13 August 2005 west of Hughesville, Charles County (K. Pardieck).

The Black Vulture expanded its breeding range northward into Maryland during the early to mid-twentieth century. The first Black Vulture report for Maryland was from 1895 (Kirkwood 1895); the first nest record was from St. Mary's County in 1922 (Court 1924). There were still only 10 records for the DC area by the late 1920s (Cooke 1929). Stewart and Robbins (1958) reported that Black Vultures nested west along the Potomac to southeastern Washington County; north to central Howard and southwestern Baltimore counties; along the Susquehanna; and sparingly on the Eastern Shore south to the Pocomoke Valley. During the 1983–1987 Maryland-DC atlas project, Black Vultures had spread into the northern Piedmont, west sparingly to Allegany County (two nesting reports), and over much of the Upper Chesapeake and Eastern Shore, although they were still scarce in Dorchester and Wicomico counties (Gregoire and Therres 1996a).

The Black Vulture's range expansion has continued over the past two decades, both in Maryland and in states to the north and east. From 2002 to 2006 there were eight breeding season records from Garrett County, where they were unreported in atlas blocks from 1983 to 1987 (Gregoire and Therres 1996a). Otherwise Black Vultures spread into holes in the 1983–1987 distribution including more records from Allegany County, from the northern Piedmont, and increases on the lower Eastern Shore.

BBS trends are statistically unreliable owing to the small number of Black Vultures seen on the early morning routes run before they become active, but this vulture has increased rapidly on Maryland routes. Black Vultures are also increasing significantly on BBS routes elsewhere throughout their U.S. range. Black Vultures currently appear to benefit from human activities as they nest in abandoned man-made struc-

tures, roost on telecommunication and power line towers, haunt chicken farms, and feed on road-killed carrion, particularly deer carcasses. These vultures can be a nuisance around communal roosts near houses (e.g., they sometimes damage roofs), and can be a threat to young livestock at times, but these problems are solved by limited management tactics that do not threaten the wider population. It appears likely the Black Vulture will continue widespread in Maryland for decades to come, perhaps even increasing in the Allegheny Mountains.

WALTER G. ELLISON

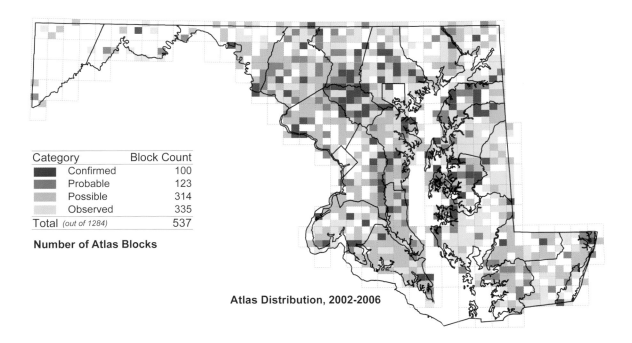

Category	Block Count
Confirmed	100
Probable	123
Possible	314
Observed	335
Total (out of 1284)	537

Number of Atlas Blocks

Atlas Distribution, 2002-2006

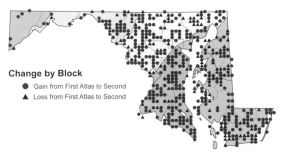

Change by Block
- Gain from First Atlas to Second
- ▲ Loss from First Atlas to Second

Change in Atlas Distribution, 1983-1987 to 2002-2006

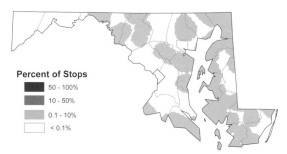

Percent of Stops
- 50 - 100%
- 10 - 50%
- 0.1 - 10%
- < 0.1%

Relative Abundance from Miniroutes, 2003-2007

Atlas Region	1983-1987	2002-2006	Change No.	Change %
Allegheny Mountain	0	7	+7	-
Ridge and Valley	31	51	+20	+65%
Piedmont	92	156	+64	+70%
Upper Chesapeake	34	42	+8	+24%
Eastern Shore	78	127	+49	+63%
Western Shore	54	151	+97	+180%
Totals	289	534	+245	+85%

Change in Total Blocks between Atlases by Region

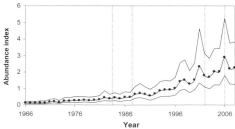

Breeding Bird Survey Results in Maryland

Robert Ampula

Turkey Vulture
Cathartes aura

The Turkey Vulture is by far the most widely distributed and frequently encountered vulture in North America, including in Maryland and DC. This long-winged, long-tailed dark brown vulture is one of our most conspicuous birds as it soars over forests, fields, and roadsides in search of deer carcasses and the remains of opossums, rabbits, and other mammals. These vultures are superb gliders, making use of both wind updrafts and thermals, rocking back and forth on wings held in a shallow to steep V depending on wind speed and the strength of rising thermals. Turkey Vultures cruise over large areas, carefully searching the landscape for any sign of carrion, aided by a keen sense of smell not demonstrated in the Black Vulture or in crows.

Like the Black Vulture, the Turkey Vulture requires a mixed landscape combining open land to help them spot carrion and woods for roosting and for nest cover. Turkey Vultures appear to be more frequent in heavily wooded country than Black Vultures, perhaps in part because of their ability to use light air to keep themselves aloft.

Turkey Vultures do not build a nest as such, laying their eggs on whatever substrate is found at a nest site. These nest sites are highly variable, but they are usually secluded and

dimly lit. Among the sites used by Turkey Vultures in descending order of preference are crevices in cliff faces and talus; hollows in standing trees, stumps, and logs; dense, vine-covered thickets and brush piles (often in conjunction with an old mammal den); abandoned man-made structures such as duck blinds, barns, derelict houses, and old sheds, even abandoned cars (pers. obs.); and the abandoned nests of other large birds (Kirk and Mossman 1998). Egg dates for Maryland and DC are 3 April to 10 June (Stewart and Robbins 1958; MNRF). Nestling dates range from 4 May 2005 near Glenelg, Howard County (P. Smith, this project) to 29 August 1942 (Stewart and Robbins 1958). Turkey Vultures have larger home ranges than Black Vultures. Radio-tracked birds in southern Pennsylvania and northern Maryland covered an average of 27,000 ha (66,713 ac), and two nesting birds used a smaller home range of about 7,000 ha (17,297 ac) (Coleman and Fraser 1989). Wide-ranging foraging adults cannot be given the possible (X) atlas code and must instead be given an observed (O) code; 38% of 2002–2006 atlas records were coded O versus 28% in 1983–1987.

Turkey Vultures were reported in a similar distribution of atlas blocks in the 2000s as they were in the 1980s (Gregoire and Therres 1996b). It is plausible that they may have increased between atlas projects, but this is obscured by the greater caution in the use of the observed and possible codes by observers between 1983–1987 and 2002–2006. The breeding confirmation rate increased from 8% to 11%, and Turkey Vultures were found in many more blocks (11%) when observed records are taken into account. Turkey Vultures were absent from the urban industrial region in southeastern Baltimore City and southeastern Baltimore County, but they were otherwise very widespread, even occurring on Smith Island and much of Assateague Island because they are more likely than Black Vultures to cross large bodies of water (Kirk and Mossman 1998).

Turkey Vultures have increased on Maryland BBS routes at a rate of 2.7% per year over the past four decades, although the rate of increase has been slow over the last two and a half decades. This may reflect an increase in road-

killed carrion coupled with a growing deer herd and warmer winters that promote winter survival at roosts by this largely resident bird. Turkey Vultures are scarce at best in industrial urban settings and are absent from completely deforested landscapes devoted to row crop farming (Mossman 1991). As long as a mixed landscape of farms and woodlots with abundant carrion exists, Turkey Vultures will remain numerous in Maryland and DC, but they may decline if housing becomes more common in the vicinity of vulture roosts, the landscape becomes ever more urban, and farmers remove hedgerows and woodlots in order to increase cropland.

WALTER G. ELLISON

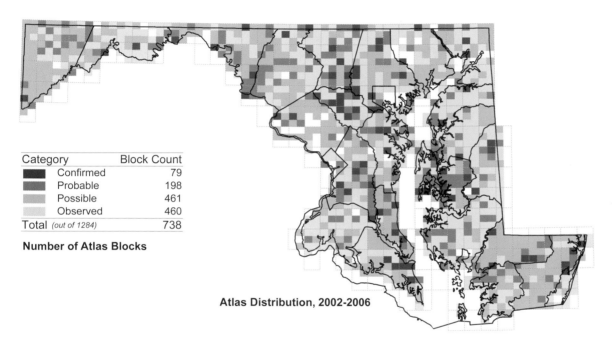

Category	Block Count
■ Confirmed	79
■ Probable	198
■ Possible	461
■ Observed	460
Total (out of 1284)	738

Number of Atlas Blocks

Atlas Distribution, 2002-2006

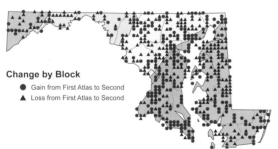

Change by Block
● Gain from First Atlas to Second
▲ Loss from First Atlas to Second

Change in Atlas Distribution, 1983-1987 to 2002-2006

Percent of Stops
■ 50 - 100%
■ 10 - 50%
■ 0.1 - 10%
□ < 0.1%

Relative Abundance from Miniroutes, 2003-2007

Atlas Region	1983-1987	2002-2006	Change No.	%
Allegheny Mountain	78	75	-3	-4%
Ridge and Valley	115	96	-19	-17%
Piedmont	209	180	-29	-14%
Upper Chesapeake	69	47	-22	-32%
Eastern Shore	199	189	-10	-5%
Western Shore	109	148	+39	+36%
Totals	779	735	-44	-6%

Change in Total Blocks between Atlases by Region

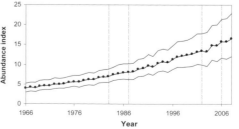

Breeding Bird Survey Results in Maryland

Melissa Boyle

Osprey
Pandion haliaetus

The Osprey, also called the fish hawk, has parlayed the trade of fishing into a vast cosmopolitan range, but there are few places on earth that rival Chesapeake Bay for population density of this well-adapted catcher of the finned and scaled. Ospreys prefer large bodies of water such as bays, estuaries, broad rivers, lakes, and ponds. A large fish population (not necessarily diverse) is also a prerequisite for Ospreys to set up housekeeping, although if a suitable nest site is within a 16-km (10-mi) commute to those fish a pair will likely nest there. Prior to European settlement of North America, Osprey nests were placed in such sites as tall snags, live trees with dead crown limbs, upturned root clusters on shorelines, and on the ground on islands inaccessible to terrestrial predators. Ospreys began accepting man-made sites almost as soon as they became available, particularly favoring duck blinds, active or defunct, on Chesapeake Bay (Reese 1996b; S. Cardano, pers. comm.). When the Coast Guard replaced

floating markers with fixed channel markers in the 1970s, Ospreys immediately took to them, often nesting far from shore (S. Cardano, pers. comm.). Other artificial sites used by Ospreys include tall, high-voltage transmission towers, communication towers, silos, and wooden platforms specially prepared for Ospreys, indeed almost anything that even vaguely resembles a tall dead tree.

Ospreys in Maryland and DC are migratory, wintering for the most part in the western Amazon basin and to a lesser extent along tropical coastlines. Ospreys arrive as early as late February; most are back by mid-March. They leave Maryland from late August through October, with most local nesters gone by the end of September (Stewart and Robbins 1958; Iliff, Ringler, and Stasz 1996). Ospreys are reliable harbingers of spring on Maryland's tidal waters, seemingly timed to the departure of wintering Canada Geese. Nesting dates for this atlas project ranged from nest building on 17 March to a report of fledged young on 21 August. Maryland egg dates range from 26 March to 20 July (Stewart and Robbins 1958, Reese 1996b).

Ospreys were common in tidewater Maryland from the nineteenth century to the 1950s (Kirkwood 1895; Stewart and Robbins 1958). During the 1960s it became clear that something was seriously amiss with the Osprey, as nesting populations fell to about 50% of previous levels around Chesapeake Bay (Poole et al. 2002). Eventually the declines were linked to poisoning by the pest control agent DDT and its derivative DDE, largely through their effect on calcium metabolism, which adversely affected eggshell thickness (Anderson and Hickey 1972; Wiemeyer et al. 1975). The banning of DDT in the United States and Canada in 1972 lessened this threat (DDT is still acquired by Ospreys on the winter range). In the 1970s and 1980s artificial nest sites proliferated wittingly, in the form of Osprey nest platforms, and unwittingly, in the form of towers and channel markers.

The Osprey had only begun to recover in 1983–1987 (Reese 1996b), but the population continued to increase over the next two decades. Nesting Ospreys were located in 112 more blocks in 2002–2006 than in the preceding statewide atlas, and they have gained tentative footholds in the Piedmont, Ridge and Valley, and Allegheny Mountains. Gains are evident along most of the state's river systems on both sides of Chesapeake Bay, and Ospreys have increased in both Baltimore City and Washington, DC. Nesting was documented along the Potomac in Washington County (although Ospreys were missing from their lone 1980s Montgomery County block), and Ospreys bred on two mountain reservoirs in Garrett County. The latter birds may have originated from reintroduction programs in other states (the closest being in West Virginia).

BBS trends agree with the trends on the maps seen here. Ospreys have increased by more than 200% both nationally and in Maryland. Maryland had the good fortune of having a very large Osprey population when DDT was severely affecting Atlantic coast populations. The worst effects were seen from New Jersey northward (Poole et al. 2002). Indeed Maryland was the source for birds used in Pennsylvania's reintroduction program in the 1980s (Ryman 1992). Whether the current Maryland Osprey population will exceed historical levels is debatable, but many pairs now nesting on artificial structures far from water and far offshore unquestionably

occupy places inaccessible to the species in the past. Unless human harassment, some as yet unidentified contaminant as pernicious as DDT, or degradation of Chesapeake Bay and overfishing harm the fish hawk, this spectacular bird will remain a symbol of Maryland's tidal waters for decades to come.

WALTER G. ELLISON

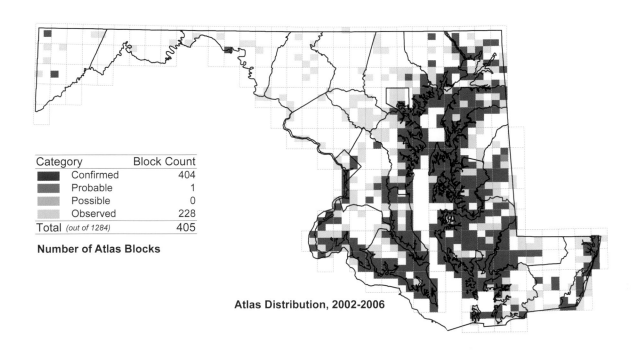

Category	Block Count
Confirmed	404
Probable	1
Possible	0
Observed	228
Total (out of 1284)	405

Number of Atlas Blocks

Atlas Distribution, 2002-2006

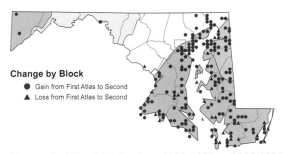

Change by Block
● Gain from First Atlas to Second
▲ Loss from First Atlas to Second

Change in Atlas Distribution, 1983-1987 to 2002-2006

Percent of Stops
- 50 - 100%
- 10 - 50%
- 0.1 - 10%
- < 0.1%

Relative Abundance from Miniroutes, 2003-2007

Atlas Region	1983-1987	2002-2006	Change No.	Change %
Allegheny Mountain	0	2	+2	-
Ridge and Valley	0	1	+1	-
Piedmont	3	14	+11	+367%
Upper Chesapeake	26	64	+38	+146%
Eastern Shore	169	190	+21	+12%
Western Shore	92	131	+39	+42%
Totals	290	402	+112	+39%

Change in Total Blocks between Atlases by Region

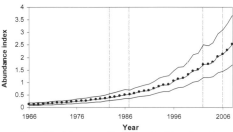

Breeding Bird Survey Results in Maryland

Bald Eagle
Haliaeetus leucocephalus

The Bald Eagle serves not only as the national symbol of the United States of America but also as a symbol of the endangered species efforts in this country. It demonstrated that if adequate conservation attention is given to an endangered species, then recovery is possible. The Bald Eagle was listed as an endangered species in the lower 48 states by the federal government in 1967. Its population decline was the result of several causes, the most significant of which was effects of organochlorine pesticides, such as DDT, on reproduction (Stickel et al. 1966). In 1972 the U.S. Environmental Protection Agency banned the use of these pesticides in the United States and the recovery of this species began. The breeding population in the Chesapeake Bay area, including Maryland, doubled every eight years (Watts, Therres, and Byrd 2008). The Bald Eagle was removed from the federal threatened and endangered species list in August 2007.

In Maryland the Bald Eagle nesting season starts with pair formation and nest building as early as October. Breeding pairs construct a nest of large sticks in the upper crotch of a tall tree. Some Bald Eagles have adapted to nesting on transmission towers, with the first occurring in Maryland in 2002 (G. Therres, pers. obs.). Egg laying generally occurs in Maryland during the month of February, though pairs nesting on the lower Eastern Shore or in extreme southern Maryland have laid eggs in late January (Therres 2005). Some late nesting eagles will lay their eggs in March. The clutch is usually one to three eggs and incubation starts after the last egg is laid. In Maryland, eggs hatch usually by mid-April. Nesting success in Maryland and the Chesapeake Bay area is one of the highest in North America with greater than 70% success per nesting attempt (Therres 2005; Watts, Therres, and Byrd

2008). Nestlings fledge in May or June. Survival rates for immature Bald Eagles are fairly high in Maryland (Buehler et al. 1991).

The Chesapeake Bay area has always supported nesting Bald Eagles. Tyrrell (1936), who conducted a survey of nesting Bald Eagles in Maryland and Virginia for the National Audubon Society, estimated the number of nesting pairs at 600–800. Since Maryland supports about half of the Chesapeake Bay nesting population, there may have been 300–400 nesting pairs of Bald Eagles in the state at that time. By the 1960s Bald Eagle populations in the lower 48 states were extremely depressed. Abbott (1963) documented only 150 pairs in the Chesapeake Bay area in 1962. Maryland's nesting Bald Eagle population was at an all-time low.

In 1977 aerial surveys in Maryland documented only 44 pairs nesting (Therres 2005). As the environment rid itself of organochlorine pesticides, the Bald Eagle population increased. Maryland had 390 nesting pairs by 2004 (Therres 2005) and an estimated 400 pairs or more in 2006 (DNR, unpubl. data).

The distribution of nesting Bald Eagles in Maryland is heavily associated with the Chesapeake Bay and its tidal tributaries. Confirmed nesting occurred in 316 atlas blocks in 2002–2006, with a large concentration of these on the Eastern Shore and Charles County. These concentration areas, associated with tidal freshwater, support the highest densities of breeding eagles in the Chesapeake Bay area (Watts, Markham, and Byrd 2006). Bald Eagles were confirmed nesting in every Maryland county during the second atlas period.

As in the 1983–1987 atlas, nesting records for Bald Eagles were treated differently from those of other species. Owing to the high mobility of nesting eagles and the presence of migrant adults from other parts of eastern North America during the breeding season for Maryland's eagles (Watts,

Therres, and Byrd 2007), only confirmations at nest sites (i.e., ON, NE, NY) were accepted as evidence of breeding in the block. All other records were treated as observed.

There was a significant gain in blocks with confirmed Bald Eagles compared to the 1983–1987 atlas period (Therres 1996a). There were nearly 200 more blocks with nesting Bald Eagles during the 2002–2006 atlas. Fewer than 20 blocks lost nesting Bald Eagles from the 1983–1987 atlas period. The distribution of nesting eagles expanded onto the Piedmont, Ridge and Valley, and Allegheny Mountain physiographic provinces. Historically, nesting Bald Eagles were documented along the nontidal portions of the Potomac River as far west

as Hancock in Washington County (Kirkwood 1895). No nesting was known in Allegany or Garrett counties until this atlas.

The increase in the Bald Eagle breeding population and expansion of its breeding distribution is attributed to the recovery efforts for this former endangered species. The ban on DDT and other organochlorine pesticides, in combination with the protection measures provided by state and federal agencies, allowed the breeding Bald Eagle population to recover to a level similar to that of the 1930s.

GLENN D. THERRES

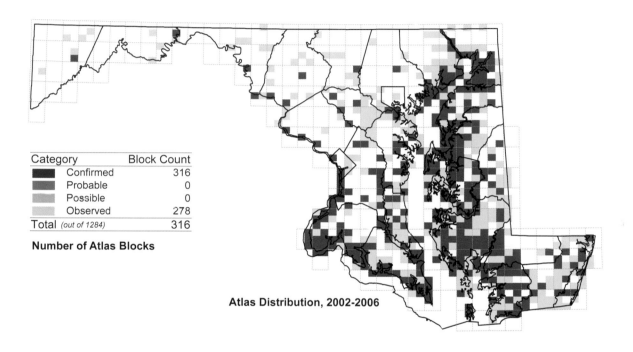

Category	Block Count
■ Confirmed	316
■ Probable	0
■ Possible	0
■ Observed	278
Total *(out of 1284)*	316

Number of Atlas Blocks

Atlas Distribution, 2002-2006

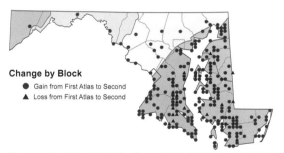

Change by Block
- ● Gain from First Atlas to Second
- ▲ Loss from First Atlas to Second

Change in Atlas Distribution, 1983-1987 to 2002-2006

Percent of Stops
- ■ 50 - 100%
- ■ 10 - 50%
- ■ 0.1 - 10%
- □ < 0.1%

Relative Abundance from Miniroutes, 2003-2007

Atlas Region	1983-1987	2002-2006	Change No.	Change %
Allegheny Mountain	0	1	+1	-
Ridge and Valley	0	4	+4	-
Piedmont	3	18	+15	+500%
Upper Chesapeake	20	46	+26	+130%
Eastern Shore	57	148	+91	+160%
Western Shore	37	96	+59	+159%
Totals	117	313	+196	+168%

Change in Total Blocks between Atlases by Region

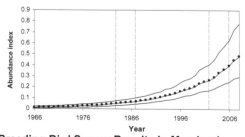

Breeding Bird Survey Results in Maryland

Northern Harrier
Circus cyaneus

The Northern Harrier is the lone North American representative of the harriers (genus *Circus*), a fairly small but well-defined group of birds of prey that nest on the ground in dense vegetation. They hunt by coursing low over open land, scanning and listening for small rodents and birds. This is a moderate-size hawk with a prominent contrasting white rump. Males and females have the most distinct plumages among Maryland and DC's birds of prey. Adult males are pale gray above and white below with black wingtips; adult females are medium brown with brown streaks on a creamy ground color below. Long known in North America as the Marsh Hawk, it was linked to the other harriers by name in 1973 (AOU 1973). Its Old World English name is Hen Harrier (which some now consider a separate species), a name too close to the inappropriate vernacular "chicken hawk." Another bird is called the Marsh Harrier in the Old World, hence the modifier Northern for this most northerly nesting of harriers.

Nesting Northern Harriers are found in essentially two regions in Maryland. There is a small population in the Allegheny Mountains, but most Maryland harriers nest in wetlands on the lower Eastern Shore. These are birds of open areas with few or no trees. Nesting habitat usually includes some tall rank vegetation such as cattails, cordgrasses, or brushy shrubs to provide nest sites. Specific Maryland haunts include salt marshes with tall cordgrass and marsh elder; freshwater marshes; wet hayfields; sedge meadows and the margins of bog ponds; fallow fields; and reclaimed strip mines.

The Northern Harrier is found year-round in Maryland, but rather few winter west of the fall line and only as long as fields remain snow free; they are regular on the Coastal Plain and numerous in lower Eastern Shore wetlands during the nonbreeding seasons (Hatfield et al. 1994). Nesting harriers begin to return in late February and March and depart from mid-August to October. Maryland nests with eggs have been found from 28 April to 23 June, with nestlings reported as late as 1 July (Stewart and Robbins 1958; MNRF).

Kirkwood (1895) considered the Northern Harrier common in tidewater Maryland but made no mention of the Allegheny Mountain population, nor did Eifrig (1904) report them in western Maryland. Stewart and Robbins (1958) called them fairly common on the lower Eastern Shore and in the Alleghenies, and found them uncommon in other tidewater areas. By the 1980s statewide atlas these raptors had greatly declined in the Alleghenies and in the Upper Chesapeake region but remained fairly well established on the eastern shore of lower Chesapeake Bay and in the coastal bays of Worcester County (S. A. Smith 1996a).

Since the 1980s the general distribution has changed little. In 2002–2006 there were a few more reports from the Allegheny Mountains, but they were found in fewer blocks on the lower Eastern Shore. Most of the Allegheny reports were clustered near the Allegheny Front, and many appear to have been associated with reclaimed strip mines. They were not found in most blocks in the Blackwater River drainage although they remain widespread in southern Dorchester and western Somerset counties. The Worcester County nesting population appears to have moved away from the mainland and islands in the coastal bays onto Assateague Island.

Continental BBS trends for the Northern Harrier suggest the species had been slowly declining since 1966, but this trend has apparently stabilized since 1980. Northeastern North American harrier populations are low and their distribution is patchy. Several eastern states have listed them officially or unofficially as being of conservation concern.

New Jersey lists them as endangered (Walsh et al. 1999). Many Maryland wetlands are becoming dominated by reed beds that are less productive than native marsh vegetation, and this may be leading to declines in this harrier's prey. At present these elegant and graceful birds of prey appear to be slowly declining in Maryland. It seems likely that declines in wetland quality and productivity may be contributing to this deterioration.

WALTER G. ELLISON

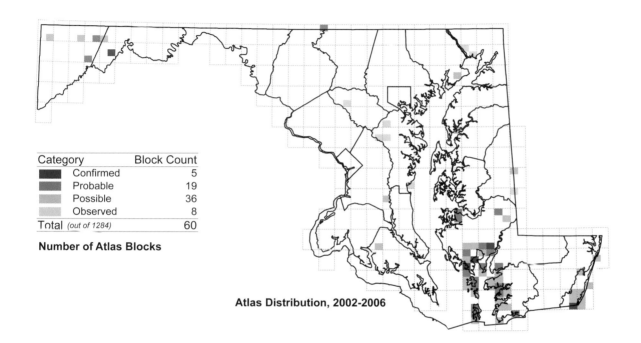

Category	Block Count
Confirmed	5
Probable	19
Possible	36
Observed	8
Total *(out of 1284)*	60

Number of Atlas Blocks

Atlas Distribution, 2002-2006

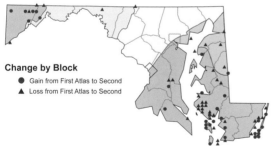

Change by Block
- ● Gain from First Atlas to Second
- ▲ Loss from First Atlas to Second

Change in Atlas Distribution, 1983-1987 to 2002-2006

Atlas Region	1983-1987	2002-2006	Change No.	Change %
Allegheny Mountain	4	5	+1	+25%
Ridge and Valley	1	1	0	0%
Piedmont	2	1	-1	-50%
Upper Chesapeake	5	1	-4	-80%
Eastern Shore	53	48	-5	-9%
Western Shore	3	2	-1	-33%
Totals	68	58	-10	-15%

Change in Total Blocks between Atlases by Region

Sharp-shinned Hawk
Accipiter striatus

A small to tiny blue gray raptor, the Sharp-shinned Hawk is primarily a hunter of small birds. Estimates of its diet are all in excess of 90% birds (Bildstein and Meyer 2000). They often hunt at bird feeders and are sometimes unpopular with feeder watchers. This is this hawk's way of life; feeders concentrate its food in the same way they offer concentrated food to the hawk's prey. Bird feeders bring the life and death drama of nature into people's yards. Hawks are part of nature's interactive machinery of winnowing and survival and are not nearly as threatening to songbird populations as is wholesale loss of habitat. These hawks reach the southeastern edge of their largely northern breeding range in Maryland, with most nesting in the westernmost counties (Bildstein and Meyer 2000).

Most descriptions of this hawk's habitat emphasize a preference for extensively wooded landscapes (Bildstein

and Meyer 2000), unlike the Cooper's Hawk, which thrives in a mixed landscape of fields with scattered woodlots. The Sharp-shin occurs in both mixed and coniferous woodlands but shows a strong preference for nesting in evergreens, often nesting in conifer plantations. Forest maturity does not appear to be as important to Sharp-shinned Hawks as it apparently is for the Northern Goshawk.

Sharp-shinned Hawks, which are found year-round in Maryland, occur statewide in winter, with most wintering in central and eastern Maryland. This is marginally the latest nesting of Maryland and DC's three accipiters, with birds returning to nesting territories from mid-April to early May. The broad, shallow stick platform nest is placed on horizontal branches near the trunk in thick cover just below the crown, generally from 7.6 to 12 m (25–40 ft) above the ground (Bildstein and Meyer 2000). Males bring prey to feed their mates before the eggs hatch, so some confirmations of birds presumably carrying prey to young (FY) may have referred to this behavior instead. Almost a third of records for this project were nesting confirmations, but only two were of active nests. Maryland egg dates range from 15 May to 30 June (Titus and Brinker 1996; MNRF), and nestlings have been observed to 11 July (Stewart and Robbins 1958).

Sharp-shinned Hawks were apparently more widely distributed east of the western Maryland mountains in the nineteenth century and just after the beginning of the twentieth century. C. Richmond (1888) found them rare as nesters around DC; Kirkwood (1895) called them "not very numerous in summer." But Eifrig (1904) wrote it was "the most common of the hawks." The basic distributional pattern described by Stewart and Robbins (1958) was similar, although numbers were probably lower to the east as they were found to be fairly common in the Alleghenies, uncommon in the Ridge and Valley, and rare on the Piedmont. In 1983–1987 they were almost entirely limited to Garrett County, with a single breeding confirmation in Green Ridge State Forest in eastern Allegany County (Titus and Brinker 1996). Records east of the Allegheny Mountains were few between 1987 and 2002, although an adult was observed carrying prey at Morgan Run, Carroll County on 27 June 1998 (Southworth 1999).

From 2002 to 2006 the Sharp-shinned Hawk appears to have increased notably east of the Allegheny Front, with records from 8 additional blocks in the Ridge and Valley, including a nest with young from Catoctin Mountain near Harmony, Frederick County, on 1 July 2002 (D. Smith). More remarkably there were three records from the Piedmont with two confirmations—a nest with young west of Ijamsville, Frederick County, on 4 July 2006 (D. Smith) and an adult seen carrying prey northwest of Norrisville, Harford County, on 27 July 2005 (P. Kreiss; P. Webb). These records seem a bit less surprising in light of atlas records from all of Pennsylvania's southern tier counties during the first Pennsylvania atlas project, including confirmations east to Adams

County (Goodrich 1992). A scatter of breeding season observations occurred south to Taylors Island, Dorchester County. Although not indicating breeding per se, these observations signal a fairly large nonbreeding summer population, sufficient to fuel further range expansion.

Sharp-shinned Hawks, which are wary and stealthy, occur at low population densities, making them very difficult to survey from roadsides. BBS trends are unreliable, although they suggest these hawks may have stable or slowly increasing numbers. Declines at traditional migration concentration points in the 1980s and 1990s caused some alarm until it was demonstrated that they were increasing on northern Christmas Bird Counts, presumably being short-stopped by abundant prey at bird feeders (Duncan 1996; Viverette et al. 1996). For the Sharp-shinned Hawk to continue to occur in Maryland, large tracts of woodland with good numbers of evergreens need to be retained.

WALTER G. ELLISON

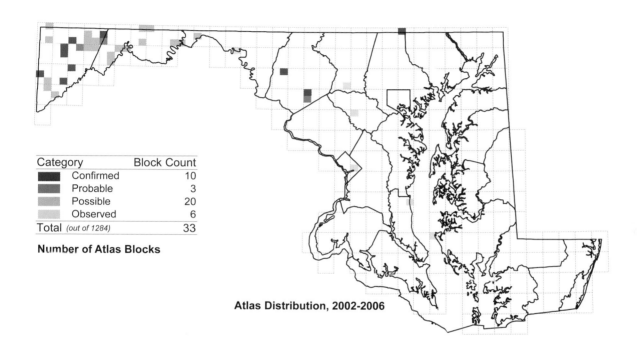

Category	Block Count
Confirmed	10
Probable	3
Possible	20
Observed	6
Total (out of 1284)	33

Number of Atlas Blocks

Atlas Distribution, 2002-2006

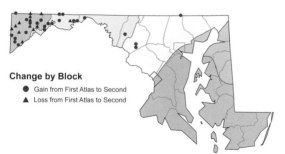

Change by Block

● Gain from First Atlas to Second
▲ Loss from First Atlas to Second

Change in Atlas Distribution, 1983-1987 to 2002-2006

Atlas Region	1983-1987	2002-2006	Change No.	%
Allegheny Mountain	20	20	0	0%
Ridge and Valley	2	10	+8	+400%
Piedmont	0	3	+3	-
Upper Chesapeake	0	0	0	-
Eastern Shore	0	0	0	-
Western Shore	0	0	0	-
Totals	22	33	+11	+50%

Change in Total Blocks between Atlases by Region

Mark Hoffman

Cooper's Hawk
Accipiter cooperii

The Cooper's Hawk is a medium-size, gray-backed hawk built for maneuverability. Its long legs aid it in capturing cornered prey. The Cooper's Hawk's plumage and general shape are similar to the smaller Sharp-shinned Hawk and it takes considerable practice for bird observers to consistently identify the two species. A few atlas reports could possibly refer to the other species, but the Cooper's Hawk greatly outnumbers its smaller relative during the summer in Maryland, save perhaps in Garrett County.

The Cooper's Hawk was once known by people in the country as the chicken hawk. H. Bailey (1913) said of it, "Of our harmful species, this is probably the worst of all." This persistent, opportunistic, and ferocious hawk would sometimes take a chicken in the presence of someone feeding a flock. This boldness is now seen by people watching birds scatter from their feeder as the local Cooper's Hawk arrives for a brief, businesslike visit. Fortunately the misplaced moral judgment that was once visited upon this hawk and other raptors has abated. Bounties are no longer offered on the bird, and all species of birds of prey have been given legal protection.

The Cooper's Hawk is a woodland and forest hawk, although it hunts in the open more often than the state's other two accipiter species and does well in a mixed landscape of mature woodlots and open areas. In recent decades, Cooper's Hawks have become ever more likely to nest in older suburban and urban communities as trees have matured. Cooper's Hawks nest high in tall trees, usually in trunk forks just below the canopy (Titus and Mosher 1981). These hawks generally form a stick and bark–lined cup atop a preexisting structure such as the used nest of a squirrel, a crow, or another hawk. They will often nest in mature pine plantations. This hawk is one of the latest nesting raptors in Maryland and DC, probably awaiting the arrival of migratory songbirds in order to feed their ravenous nestlings. Egg dates range from 21 April to 5 June in Maryland and DC (Stewart and Robbins 1958; MNRF). An adult was reported on its nest on 15 April 2004 at College Park, Prince George's County, during this project (J. Saba). Nestlings have been reported in Maryland as late as 23 July (Stewart and Robbins 1958; MNRF), including a 2002 report near Coopstown, Harford County (D. Kirkwood), during this atlas.

Though Cooper's Hawk was always widespread in Maryland and DC, it was uncommon to rare. At the close of the nineteenth century, this accipiter was less numerous than Sharp-shinned Hawk, which is now far less common than Cooper's in the nesting season (Kirkwood 1895; Eifrig 1904), largely because of relentless shooting of resident hawks and unrestricted hunting at migration concentration points. Stewart and Robbins (1958) called the Cooper's Hawk uncommon throughout Maryland, save for the Western Shore where it was fairly common.

In the 1983–1987 atlas project, this raptor was found throughout Maryland, but it was found in a substantial portion of blocks only on the Western Shore, and in Allegany and Garrett counties, where Kimberly Titus and colleagues made a special effort to find active nests. Cooper's Hawk was considered scarce elsewhere. It was not found in Queen Anne's County, DC, or Baltimore City and was observed in only 1 block each in Baltimore and Cecil counties (Titus 1996). This species has increased greatly, 226% over the last 20 years, an increase of nearly 400 blocks. The greatest increases were on the Piedmont and in the Upper Chesapeake region. The only area showing a net loss was Allegany County.

Cooper's Hawk BBS trend statistics are suspect because these stealthy and somewhat shy hawks are seldom seen on BBS routes (only 0.04 per route in Maryland on 24 routes), nonetheless far more are now seen on BBS routes than in the past, implying a true increase. The spectacular increase can be attributed to reduced persecution by humans; habituation

to humans and their dwellings; and restrictions on pesticide application, especially DDT, which apparently severely curtailed its nesting success (Snyder et al. 1973). So successful has been the recovery of this hawk that it is now a conservation threat to other bird species, such as the declining American Kestrel (Farmer et al. 2006). Maintaining the Cooper's Hawk in Maryland and DC requires the retention of fairly large mature stands of trees for nesting and roosting.

WALTER G. ELLISON

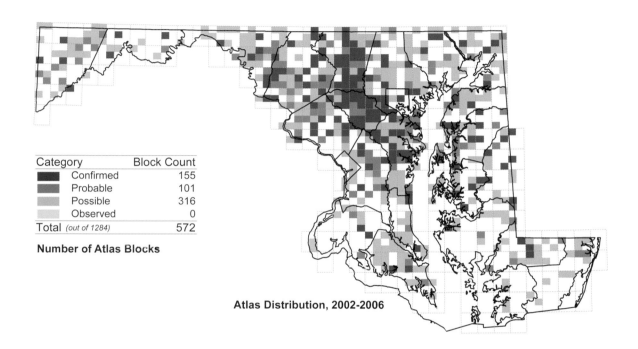

Category	Block Count
Confirmed	155
Probable	101
Possible	316
Observed	0
Total (out of 1284)	572

Number of Atlas Blocks

Atlas Distribution, 2002-2006

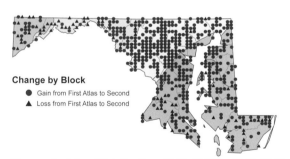

Change by Block
● Gain from First Atlas to Second
▲ Loss from First Atlas to Second

Change in Atlas Distribution, 1983-1987 to 2002-2006

Percent of Stops
- 50 - 100%
- 10 - 50%
- 0.1 - 10%
- < 0.1%

Relative Abundance from Miniroutes, 2003-2007

Atlas Region	1983-1987	2002-2006	Change No.	%
Allegheny Mountain	37	38	+1	+3%
Ridge and Valley	40	64	+24	+60%
Piedmont	18	207	+189	+1050%
Upper Chesapeake	2	47	+45	+2250%
Eastern Shore	22	88	+66	+300%
Western Shore	56	127	+71	+127%
Totals	175	571	+396	+226%

Change in Total Blocks between Atlases by Region

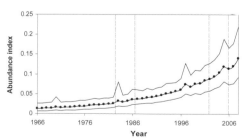

Breeding Bird Survey Results in Maryland

Northern Goshawk
Accipiter gentilis

Known for aggressive defense of nests when they are attending chicks, Northern Goshawks should be a relatively easy breeding bird for atlas participants to locate. But because of their low population density and somewhat secretive nature, goshawks are perhaps the most difficult eastern raptor to detect. Goshawks have large territories and inhabit landscapes that are primarily forested. In the eastern United States they prefer to nest in maturing mixed forests of hardwoods and conifers that provide nest trees with diameters at breast height (1.2 m [4 ft] above the ground) of more than 30.4–38.1 cm (12–15 in; Speiser and Bosakowski 1987). Breeding goshawks are found throughout New England and the northern United States, across boreal Canada, and south in the mountainous regions of the west to Mexico (Squires and Reynolds 1993). In the central Appalachian Mountains, goshawks are permanent residents (D. Brinker, unpubl. data) that have been documented breeding as far south as North Carolina (Hader 1975).

Northern Goshawks are presumed to have bred on the Appalachian Plateau in Maryland prior to European settlement and up to the period of exploitive timber harvesting that took place in Garrett County from the 1880s through the early 1900s (D. Boone 1996a). The first documented breeding in Maryland was near Jennings in 1901 (Behr 1914). This was the last record of goshawks breeding in Maryland for nearly 80 years, when a nest was discovered in 1980 (D. Boone 1984). From the 1960s through about 2000, goshawk populations were increasing in the northeastern United States as forests recovered from earlier harvesting (J. Bull 1974; Crocoll 2008), and during the 1990s goshawk breeding spread to states south of Pennsylvania (Buckelew and Hall 1994; D. Brinker, unpubl. data).

In February, Northern Goshawks begin pair bonding and courtship activities in forested landscapes of the central Appalachian Mountains (D. Brinker, unpubl. data). During courtship in March and early April, goshawks are readily detected within a few hundred meters of their future nest site (Dewey et al. 2003). Once females begin incubating, goshawks become quiet and secretive until incubation is completed, normally in May. When chicks are a few days old, goshawks become highly aggressive near nest sites. By mid-June to early July, they are leaving nest sites and are no longer highly aggressive toward nest site intruders.

The 1983–1987 atlas project reported Northern Goshawks in 4 blocks, consisting of 3 possible and a single confirmed record (D. Boone 1996a). The confirmed record was at the same site as the 1980 nest (D. Boone 1984). In 1988, just after the first atlas was completed, a second nest was located in Maryland by a bird bander in southern Garrett County (DNR, unpubl. data). As goshawk populations continued to grow in Maryland, a third nesting was recorded in 1996 when 2 chicks were banded at the same location in southern Garrett County as in the 1988 record (DNR, unpubl. data). No other Northern Goshawk nests were recorded in Maryland until 2001, the year before data collection began for the second atlas, when a nest was located in Savage River State Forest (DNR, unpubl. data).

The map on the facing page does not illustrate the nesting records for the Northern Goshawk in order to protect nesting pairs from unwarranted disturbance. From 2002 through 2006, 4 territories produced 7 nesting attempts resulting in 5 confirmed blocks in Garrett County (DNR, unpubl. data). This represents a significant increase from the first atlas and parallels trends from the mid-1980s to 2000 in Pennsylvania (D. Gross, pers. comm.). The core nesting areas of all 4 goshawk territories were located on public land or lands dedicated to conservation purposes. All recent goshawk nests in Maryland have been in association with habitats that contain significant conifer cover. Based on chick age at banding, first egg dates for Maryland breeding goshawks range from 29 March through 10 April (D. Brinker, unpubl. data). Unfledged chick dates range from 4 May through late June (D. Brinker, unpubl. data). Five of the recent nesting attempts were successful; one failed; results of the other nesting efforts are unknown. Brood sizes in the five nests that produced near-fledging young were two or three chicks. Chicks begin fledging during June and depend on their parents for food for at least eight weeks thereafter, although eventually they range widely (Squires and Reynolds 1993).

As of 2006 Northern Goshawks are firmly established as breeding in Maryland. Since 1990 goshawks found in Garrett and Allegany counties have been listed as endangered. There are habitats in Allegany County nearly identical to occupied nesting habitat used in the Ridge and Valley portion of nearby Pennsylvania where there are accepted atlas records

(Brauning 1992). As long as Garrett and Allegany counties retain significant areas of forested landscape that include native conifers such as hemlock and white pine, along with areas of larger trees suitable for nesting, Northern Goshawks should continue to nest in Maryland. If populations to the north decline, the continued presence of breeding Northern Goshawks in Maryland could be in jeopardy. Sustainable conservation of the matrix of forests in Garrett and Allegany counties is essential to the continued presence of nesting goshawks in Maryland.

DAVID F. BRINKER

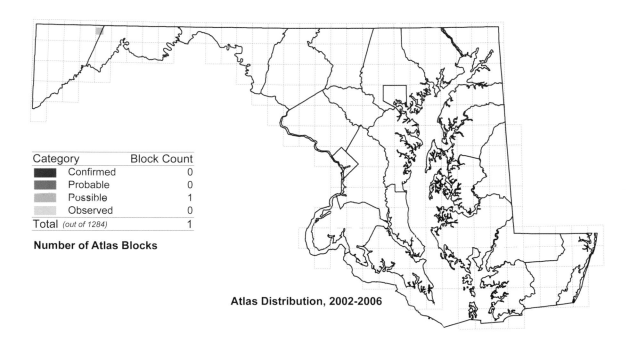

Category	Block Count
■ Confirmed	0
■ Probable	0
■ Possible	1
■ Observed	0
Total (out of 1284)	1

Number of Atlas Blocks

Atlas Distribution, 2002-2006

Change by Block
● Gain from First Atlas to Second
▲ Loss from First Atlas to Second

Change in Atlas Distribution, 1983-1987 to 2002-2006

Atlas Region	1983-1987	2002-2006	Change No.	%
Allegheny Mountain	4	1	-3	-75%
Ridge and Valley	0	0	0	-
Piedmont	0	0	0	-
Upper Chesapeake	0	0	0	-
Eastern Shore	0	0	0	-
Western Shore	0	0	0	-
Totals	4	1	-3	-75%

Change in Total Blocks between Atlases by Region

George M. Jett

Red-shouldered Hawk
Buteo lineatus

The Red-shouldered Hawk is a handsome, rust and black and white forest buteo. It may be seen quietly awaiting prey as it sits on a medium-height perch beneath the forest canopy or along the edge of wetlands and meadows. These hawks, which prey on amphibians, reptiles, and small mammals, prefer the wetter parts of woods. Red-shouldered Hawks can be overlooked, but they may be conspicuous in late winter and early spring when they do their aerial breeding displays, escort migrants passing over their territories, and are most vocal, giving their loud strident *keee-ya!* calls repeated in series (commonly imitated in convincing style by Blue Jays).

Red-shouldered Hawks inhabit mature timber usually near water. Home ranges are routinely more than 100 ha (247 ac; Dykstra et al. 2008), with the majority forested; so this hawk is limited to some extent by forest area. Red-shouldered Hawks are most often found in floodplain forest, and near forested ponds and small streams. They are resident year-round in central and eastern Maryland; in western Maryland most are migratory (E. Martin 1996). Nesting begins in late February. Maryland egg dates range from 4 March to 31 May (MNRF). Nestling dates for this project, which exceeded prior records, were from 30 March 2004 south of Elmer, Montgomery County (D. DeRycke; a 22 March report may have involved a male feeding an incubating female), to 13 July 2005 near Scotland, St. Mary's County (P. Craig).

In the late nineteenth and early twentieth centuries, the Red-shouldered Hawk was common in eastern Maryland (Kirkwood 1895) and uncommon in western Maryland (Eifrig 1904). In the 1950s this raptor was called local on the Eastern and Western Shore, fairly common in the Upper Chesapeake and Piedmont, and uncommon in western Maryland (Stewart and Robbins 1958). By the 1983–1987 atlas project they had become scarce in the Upper Chesapeake and more widespread on the Western Shore; they remained very local on the Eastern Shore (E. Martin 1996). Red-shouldered Hawks have expanded their atlas breeding range by 26% over the past two decades. By far the largest increase was in the Ridge and Valley region, where this hawk has been traditionally uncommon, with an increase of 35 blocks (78%). Increases occurred in all physiographic regions except the Allegheny Mountains, where this hawk essentially maintained its same distribution. Red-shouldered Hawk retains a spotty distribution in the Hagerstown Valley, in the Upper Chesapeake, and on the Eastern Shore (largely along the Delaware line and in the Pocomoke drainage).

Trends from BBS routes are not reliable for Red-shouldered Hawk largely because very few are seen per route

(0.37/route in Maryland; 0.53 in all of North America). Nonetheless, Red-shouldered Hawks have increased significantly in Maryland (2.9%/year) and North America (3.1%/year). Overall this raptor appears to be maintaining a stable or slightly increasing population in eastern North America (Dykstra et al. 2008). E. Martin (2004) reported a 75% decline in nests in a population he monitored in and near the Patuxent Wildlife Research Center, Prince Georges County, from 1975 to 2002. He speculated that the concomitant increase on BBS routes was due to Red-shouldered Hawks becoming more wide ranging and conspicuous from roadsides, as their population has become smaller and more food stressed. The increases in this hawk's geographic range in Maryland and DC has several possible causes, including a rising population forcing individuals to spread into new locations; habituation to human activity and habitations; and possibly a compensatory increase as the rapid decline of Broad-winged Hawk in eastern Maryland has removed a competitor. The Red-shouldered Hawk appears to be holding its own in Maryland better than in states immediately to the north (Bednarz 1992; Walsh et al. 1999), but it still requires preservation of mature forest and wetlands, and protection from excessive human disturbance near nests.

WALTER G. ELLISON

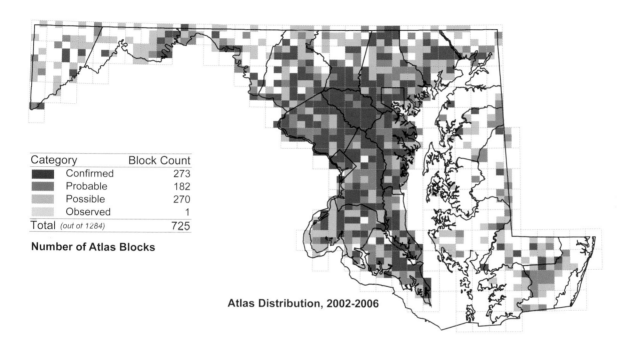

Category	Block Count
Confirmed	273
Probable	182
Possible	270
Observed	1
Total (out of 1284)	725

Number of Atlas Blocks

Atlas Distribution, 2002-2006

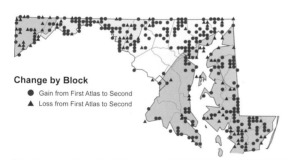

Change by Block
- ● Gain from First Atlas to Second
- ▲ Loss from First Atlas to Second

Change in Atlas Distribution, 1983-1987 to 2002-2006

Percent of Stops
- 50 - 100%
- 10 - 50%
- 0.1 - 10%
- < 0.1%

Relative Abundance from Miniroutes, 2003-2007

Atlas Region	1983-1987	2002-2006	Change No.	Change %
Allegheny Mountain	38	36	-2	-5%
Ridge and Valley	45	80	+35	+78%
Piedmont	199	259	+60	+30%
Upper Chesapeake	31	39	+8	+26%
Eastern Shore	63	84	+21	+33%
Western Shore	199	224	+25	+13%
Totals	575	722	+147	+26%

Change in Total Blocks between Atlases by Region

Breeding Bird Survey Results in Maryland

Broad-winged Hawk
Buteo platypterus

The Broad-winged Hawk is a small perch-hunting buteo of mature moist deciduous and mixed forest. These hawks are not particularly conspicuous because they are not very active, spending much of their time quietly perched awaiting prey beneath the forest canopy or sitting along forest margins. Widely distributed in Maryland, they are fairly numerous in Allegany and Garrett counties but sparsely distributed and declining from central Washington County eastward. They are most easily found when they are soaring and occasionally displaying over the forest or when their distinctive thin, piping, whistled calls can be heard. Broad-winged Hawks are remarkably approachable, often not flushing until people are very near. These hawks are most famous for their spectacular seasonal migrations when they form large soaring flocks (called kettles by hawk watchers); these kettles are largest along inland ridges during mid-September.

Broad-winged Hawks live in extensive second-growth to mature deciduous and mixed forests. Although they are forest birds, they tend to nest near water, openings, and edges; presumably to be near abundant prey such as frogs, snakes, and chipmunks (Titus and Mosher 1981). Nonetheless, this hawk is rarely found in small woodlots (Robbins, Dawson, and Dowell 1989), and it is rare in largely open or heavily built-up landscapes.

Broad-winged Hawks use thermals and light ridge top updrafts to remove most of the energetic costs of steady flapping during their migrations. By soaring high and making long glides, they use the swirling flocks of other broad-wings as cues for finding the next patch of rapidly rising air on their long journey to wintering grounds in Central and South America (Bildstein 1999). They are present in Maryland from early April to mid-October. The nest is a fairly substantial stick structure, usually built in a relatively low main trunk fork of a medium to large tree, 7.5 m (25 ft) or higher. They tend to select nest trees from among the most abundant large species; in Maryland white oak and northern red oak are frequent choices (Titus and Mosher 1987). Maryland and DC egg dates range from 23 April to 16 June (Gregoire and Brinker 1996; MNRF). Nestlings have been reported as late as 26 July 2002 near Dameron, St. Mary's County (P. Craig, this project).

The early status of the Broad-winged Hawk in Maryland and DC is clouded somewhat by apparent frequent confusion with Red-shouldered Hawk (Griscom 1923). Many early authors including Kirkwood (1895) and Eifrig (1904) called this long-distance migrant a year-round resident in Maryland. These hawks apparently were not common in the early twentieth century; they seem to have increased through the century. Stewart and Robbins (1958) called them common in western Maryland eastward through the Ridge and Valley, fairly common on the Piedmont and Western Shore, and uncommon elsewhere on the Coastal Plain.

There were breeding season records from all but two counties (Talbot and Caroline) from 1983 to 1987, but the Broad-winged Hawk's range was spotty from the western edge of the Hagerstown Valley eastward, and it was rare on the Delmarva Peninsula, with only one nesting confirmation (Gregoire and Brinker 1996). It was almost uniformly distributed from western Washington County westward, and it also had fairly solid distribution on South and Catoctin mountains, in Howard County, and in southern Prince George's County. Broad-winged Hawks were lost from a net total of 172 blocks from the 1980s to the 2002–2006 atlas. The greatest losses were from Hancock, Washington County, eastward, with a loss of 97 blocks from the Piedmont alone. They actually increased by 6 blocks in the Allegheny Mountains and remained widespread in Allegany County.

The number of Broad-winged Hawks sighted on BBS routes is so low that they show no significant trend, although their numbers appear at least stable. Steep declines in eastern Maryland appear to be related to forest fragmentation in farming and urbanizing areas. Three other hawks (Cooper's, Red-shouldered, and Red-tailed) appear to be relatively well adjusted to suburban landscapes and small woodlots and all have increased in number. This trend likely places intense competitive pressure on the Broad-winged Hawk in eastern Maryland. This hawk bears watching because of its sudden decline in eastern Maryland. The relatively few large woodlands still found east of the western mountains need to be preserved, and its status should be monitored in western Maryland where it is still widespread.

WALTER G. ELLISON

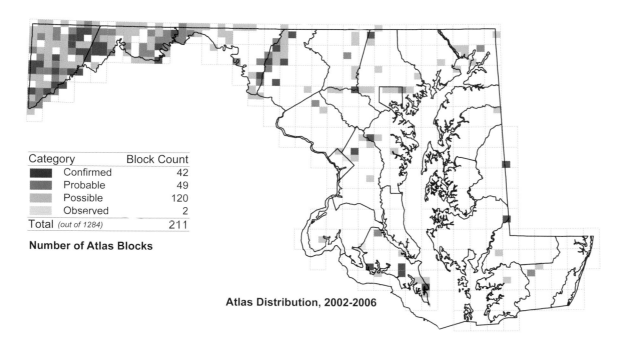

Category	Block Count
■ Confirmed	42
■ Probable	49
■ Possible	120
■ Observed	2
Total (out of 1284)	211

Number of Atlas Blocks

Atlas Distribution, 2002-2006

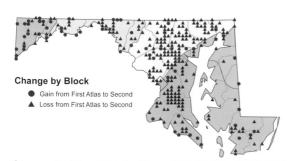

Change by Block
● Gain from First Atlas to Second
▲ Loss from First Atlas to Second

Change in Atlas Distribution, 1983-1987 to 2002-2006

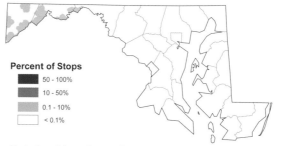

Percent of Stops
■ 50 - 100%
■ 10 - 50%
■ 0.1 - 10%
□ < 0.1%

Relative Abundance from Miniroutes, 2003-2007

Atlas Region	1983-1987	2002-2006	Change No.	Change %
Allegheny Mountain	72	78	+6	+8%
Ridge and Valley	92	75	-17	-18%
Piedmont	128	31	-97	-76%
Upper Chesapeake	8	3	-5	-63%
Eastern Shore	17	5	-12	-71%
Western Shore	66	19	-47	-71%
Totals	383	211	-172	-45%

Change in Total Blocks between Atlases by Region

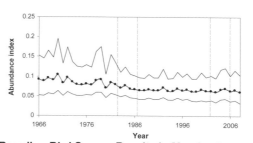

Breeding Bird Survey Results in Maryland

Red-tailed Hawk
Buteo jamaicensis

Red-tailed Hawks can be seen year-round on perches, awaiting unwary prey along byways and around the perimeters of woods and fields throughout Maryland and DC. The Red-tail is by far the most conspicuous and widespread hawk in the state and is seen soaring high in the sky as often as it is seen perch-hunting. This is a robust hawk with a brown back speckled with splashes of white, with whitish underparts usually crossed by a band of variable blackish streaks dividing the lower breast from the belly; adults have the emblematic brick red upper tail surface. The most common call of this raptor is a hoarse, hissing, drawn-out scream, often dubbed into television shows and movies as the voice of all manner of hawks and eagles, sometimes even Red-tailed Hawks.

Red-tailed Hawks need large trees to support their big stick nests and an abundance of hunting perches such as snags, electric poles, power line towers, and fence posts. They also need some open ground or widely spaced trees for hunting because these long-winged buteos cannot hunt efficiently among close-ranked trees. Red-tailed Hawks, which have become increasingly tolerant of humans and their habitations, frequently nest in treed areas of suburbs and cities, sometimes even nesting on buildings.

Red-tailed Hawks may be seen throughout the year in Maryland although they are less numerous in western Maryland in winter. Local pairs may be seen perched together from January onward, and nests are started or refurbished beginning in mid-February. Red-tailed Hawk nests are built in main trunk crotches of large trees, usually in the highest available fork with substantial branches, often at 15 m (50 ft) or more above the ground (Titus and Mosher 1981). Egg laying commences by early March; egg dates for Maryland and DC Red-tailed Hawks range from 8 March to 28 June (Gregoire 1996; MNRF), and nestlings have been reported from 13 April 2005 near St. Mary's City, St. Mary's County (E. Willoughby, this project) to 16 July (MNRF).

At the end of the nineteenth century, the Red-tailed Hawk was widespread in Maryland but subject to considerable shooting and trapping in the countryside because it was viewed as a menace to poultry and game; and it provided good target practice when it perched in the open. Kirkwood (1895) noted that this hawk was less numerous in summer than winter, and Eifrig (1904) remarked on pole-trapping by western Maryland farmers. This buteo continued to decline into the middle of the twentieth century, especially in agricultural districts (Pough 1951; Hess et al. 2000). It has been increasing at least since the 1960s. It was found statewide during the 1983–1987 atlas and expanded into more atlas blocks over the two intervening decades as it was located in 223 more blocks from 2002 to 2006. Red-tailed Hawks increased almost uniformly except for a 46% gain in Garrett County where it may have benefited from some forest clearing. The largest absolute number of blocks added in any region was the 56 on the Eastern Shore. Notable increases were also seen in southern Baltimore County and in DC and its suburbs. These raptors were not found in some of the extensive marshes on the lower Eastern Shore, parts of Assateague Island, or in some heavily forested upland blocks in the Allegheny Mountains.

BBS trends in North America and Maryland show significant increases. In Maryland the rate of increase was greatest from 1966 to 1979. Benefiting from a significant reduction in human persecution in the mid-twentieth century, Red-tailed Hawks increased in farmland, a good habitat with its traditional abundance of small rodents and groves of mature trees. As shooting declined, Red-tailed Hawks became habituated to humans and are now spreading into more urbanized habitats. Conservation of the Red-tailed Hawk requires preserving stands of mature timber and maintaining open land with an abundance of rodent prey.

WALTER G. ELLISON

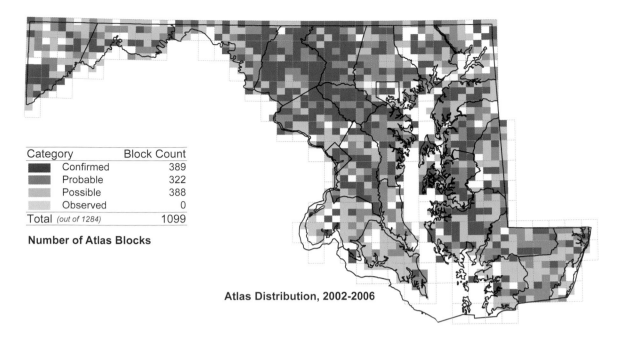

Category	Block Count
Confirmed	389
Probable	322
Possible	388
Observed	0
Total (out of 1284)	1099

Number of Atlas Blocks

Atlas Distribution, 2002-2006

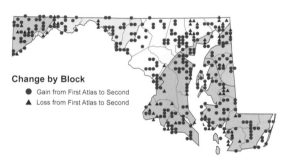

Change by Block

● Gain from First Atlas to Second
▲ Loss from First Atlas to Second

Change in Atlas Distribution, 1983-1987 to 2002-2006

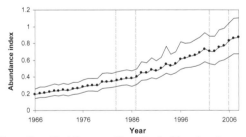

Percent of Stops

■	50 - 100%
■	10 - 50%
■	0.1 - 10%
□	< 0.1%

Relative Abundance from Miniroutes, 2003-2007

Atlas Region	1983-1987	2002-2006	Change No.	Change %
Allegheny Mountain	56	82	+26	+46%
Ridge and Valley	103	133	+30	+29%
Piedmont	246	291	+45	+18%
Upper Chesapeake	78	99	+21	+27%
Eastern Shore	222	278	+56	+25%
Western Shore	168	213	+45	+27%
Totals	873	1096	+223	+26%

Change in Total Blocks between Atlases by Region

Breeding Bird Survey Results in Maryland

American Kestrel
Falco sparverius

The Mourning Dove–size American Kestrel is frequently seen perched on roadside wires looking down intently as it elegantly dips its tail. Or it might be seen staring just as intently from the air as it hovers into a headwind over an open field. This is our smallest, most numerous, and most well-distributed falcon. It is a bird of open places that have elevated perches, sufficient cover for prey, and cavities in which to nest. Kestrels were once fairly common throughout Maryland and DC, but although they are still found statewide, they have declined noticeably in recent decades.

American Kestrels generally live in open country that has woodlots or scattered trees, but they can live in treeless landscapes if suitable nesting cavities are available. Prey captured from perches and hovering flight include small mammals, large insects, and small flocking birds. Nest sites are often old woodpecker holes in snags and dead limbs or crevices in man-made structures including building eaves and silos. But kestrels are very flexible and will even use burrows in soil or crevices in rock ledges (Wise-Gervais 2005). They also readily use nest boxes. Bird and Palmer (1988) published detailed plans for building kestrel boxes. Kestrels are tolerant of human activity, often nesting close to human habitation and active industrial sites.

American Kestrels are present throughout the year in Maryland, although in winter they are rare in the Allegheny Mountains, becoming gradually more numerous to the east until they are fairly common on the Coastal Plain. It is uncertain what percentage of wintering birds are actually resident rather than northern migrants replacing summer birds wintering elsewhere. Nesting activities commence in mid- to late March and continue until late summer; fledglings sometimes depend upon parents into August. Kestrels are single brooded but will renest on the loss of a first clutch, thus extending the nesting season. Maryland and DC egg dates range from 23 March to 4 August, the latter is likely the result of a very late renesting attempt (D. R. Smith 1996; MNRF).

Both Kirkwood (1895) and Eifrig (1904) called the American Kestrel numerous, but considered it less so in summer. Eifrig noted that kestrels were more common at higher elevations in western Maryland. Stewart and Robbins (1958) called them fairly common throughout Maryland and DC. From 1983 to 1987 they were recorded widely in Maryland with a fairly solid distribution from western Washington County eastward through the Piedmont to the northern Western Shore and Upper Chesapeake (D. R. Smith 1996). They had a spottier distribution in Charles and St. Mary's counties, the Eastern Shore, and Allegany and Garrett counties. The American Kestrel was the second most widely distributed diurnal bird of prey behind Red-tailed Hawk in the 1980s.

American Kestrels were lost from a net total of 228 blocks, representing a range-shrinkage of one-third, becoming the fourth most widespread diurnal raptor, after the Red-tailed, Cooper's, and Red-shouldered hawk. They have lost the least ground in the Allegheny Mountains and the Ridge and Valley, actually registering a small gain in the former. The greatest losses have been from the Piedmont and Coastal Plain, especially in areas experiencing rapid suburbanization. Nonetheless, they continue to nest in Baltimore City and Washington, DC. They have also declined in some intensively farmed areas on the Delmarva Peninsula.

The American Kestrel has been declining slowly but significantly on North American BBS routes. Declines have been especially severe in the Northeast. This includes Maryland, where they have declined at 2.7% per year. Several factors appear to be combining to promote these declines. Perhaps the most important is the loss of open farmland to development and to reforestation, the latter mostly north of Maryland. Modern crop farming practices also destroy potential nest sites through removal of dead trees, changes in farm buildings, reduction in prey abundance, and secondary poisoning of kestrels with pesticides and herbicides (Smallwood and Bird 2002). Cooper's Hawks, which are rapidly increasing, prey on kestrels; they may have contributed to kestrel declines in suburban settings (Farmer et al. 2006). American Kestrels are also sometimes infected with West Nile virus

(Joyner et al. 2006), although its effect on kestrel populations is not known. The maintenance of open land with sufficient prey is necessary to preserve this little falcon in Maryland. An active nest box program might prove helpful in maintaining populations in some areas. This is a resilient bird that is generally tolerant of humans; it should be possible to halt its decline and maintain it in Maryland's and DC's bird fauna.

WALTER G. ELLISON

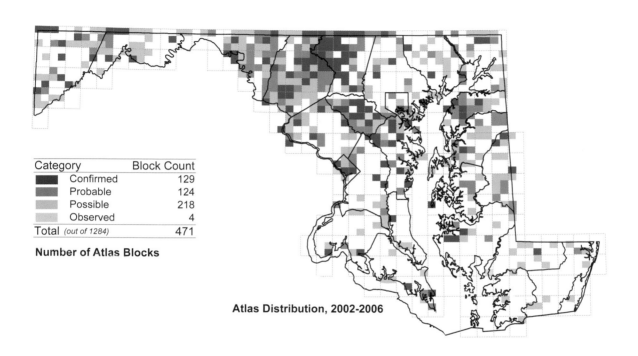

Category	Block Count
Confirmed	129
Probable	124
Possible	218
Observed	4
Total (out of 1284)	471

Number of Atlas Blocks

Atlas Distribution, 2002-2006

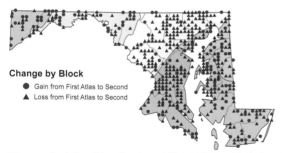

Change by Block

● Gain from First Atlas to Second
▲ Loss from First Atlas to Second

Change in Atlas Distribution, 1983-1987 to 2002-2006

Percent of Stops

■	50 - 100%
■	10 - 50%
■	0.1 - 10%
□	< 0.1%

Relative Abundance from Miniroutes, 2003-2007

			Change	
Atlas Region	1983-1987	2002-2006	No.	%
Allegheny Mountain	38	43	+5	+13%
Ridge and Valley	90	87	-3	-3%
Piedmont	267	166	-101	-38%
Upper Chesapeake	78	55	-23	-29%
Eastern Shore	93	55	-38	-41%
Western Shore	133	65	-68	-51%
Totals	699	471	-228	-33%

Change in Total Blocks between Atlases by Region

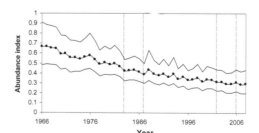

Breeding Bird Survey Results in Maryland

Craig Koppie, U.S. Fish & Wildlife Service

Peregrine Falcon
Falco peregrinus

The Peregrine Falcon disappeared as a nesting bird in the eastern United States by the mid-1960s (Hickey 1969). At that time, the last known active nest site in Maryland had been recorded in 1952 (Stewart and Robbins 1958). Reproductive failure from contamination by organochlorine pesticides, such as DDT, led to the demise of this species (Peakall 1976). It was listed as an endangered species by the USFWS in 1970.

Traditional nest sites of Peregrine Falcon were cliff faces and rock outcrops. Historically, this fast and agile bird of prey nested on cliffs along the Potomac River and elsewhere in the mountains of western Maryland. An aerie at Maryland Heights, opposite Harpers Ferry and known to be occupied in the late 1800s (Kirkwood 1895), was used as late as 1947 (Stewart and Robbins 1958).

Efforts to reintroduce the Peregrine Falcon to the eastern United States began in the 1970s as a collaborative effort between the Peregrine Fund at Cornell University, the USFWS, and state wildlife agencies, including the Maryland DNR. Between 1975 and 1984, 19 releases of young captive-reared Peregrine Falcons were conducted in Maryland through a technique called hacking (Therres, Dawson, and Barber 1993). The reintroduction efforts in Maryland did not occur on traditional cliff faces because of concerns over predation by Great Horned Owls. Instead, hacking was conducted from specially constructed towers on the Coastal Plain to avoid the owls and take advantage of the greater abundance of prey.

The first nesting attempts of the reintroduced population in Maryland occurred in 1983 when two pairs of Peregrine Falcons nested (Therres, Dawson, and Barber 1993). By the end of the 1983–1987 atlas, six nesting pairs were confirmed (Therres 1996b). All occurred on the Coastal Plain: three on towers, two on bridges, and one on a tall building in Baltimore City. By 1992 eight pairs attempted nesting in Maryland, seven successfully (Therres, Dawson, and Barber 1993).

During the 2002–2006 atlas, the number of confirmed nesting pairs increased to 18. All were still located on the Coastal Plain. Pairs on the lower Eastern Shore nested on specially constructed towers. Those on the Western Shore nested on bridges except for a pair that nested on a tall building in Baltimore City and one that nested in a lighthouse near Hart-Miller Island. Bridges with confirmed nesting included the Chesapeake Bay Bridge, Key Bridge, Governor Nice Bridge (Route 301 bridge across the Potomac River), Solomons Island Bridge, and Woodrow Wilson Bridge. An unconfirmed pair was found in Montgomery County that in 2007 nested on the American Legion Bridge (DNR, unpubl. data).

The expansion of the breeding population of Peregrine Falcons in Maryland can be attributed, in part, to the high nesting success rate and productivity of young. Therres (1996d) documented 81% nesting success of Maryland's Coastal Plain nesting falcons with nearly three young produced per successful nesting attempt. Owing to successful reintroduction efforts in the eastern United States, including in Maryland, the Peregrine Falcon was removed from the federal endangered species list in 1999.

Unfortunately, the breeding population of this species has not yet reoccupied traditional cliff sites in Maryland. One such site in western Virginia was reoccupied beginning in 2005 thanks to reintroduction efforts there (Virginia De-

partment of Game and Inland Fisheries, unpubl. data). Four years of reintroduction efforts at Maryland Heights by the National Park Service from 2001 through 2004 (B. Hepp, unpubl. data) have yet to result in a nesting pair of Peregrine Falcons on natural substrate in Maryland. Hopefully, nesting peregrines will return to Maryland's mountains in the near future and continue to expand elsewhere in the state.

GLENN D. THERRES

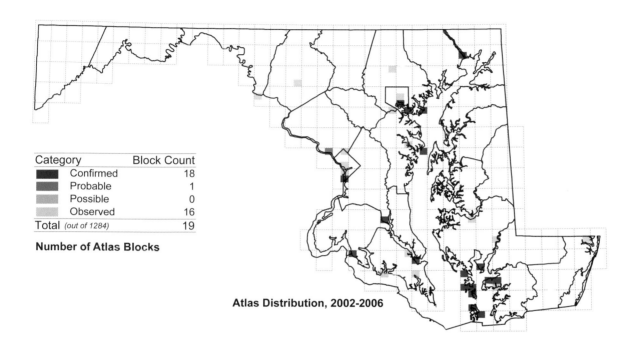

Category	Block Count
Confirmed	18
Probable	1
Possible	0
Observed	16
Total (out of 1284)	19

Number of Atlas Blocks

Atlas Distribution, 2002-2006

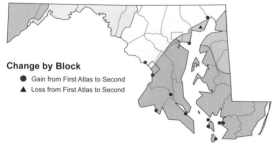

Change by Block
- ● Gain from First Atlas to Second
- ▲ Loss from First Atlas to Second

Change in Atlas Distribution, 1983-1987 to 2002-2006

Atlas Region	1983-1987	2002-2006	Change No.	%
Allegheny Mountain	0	0	0	-
Ridge and Valley	0	0	0	-
Piedmont	0	2	+2	-
Upper Chesapeake	3	3	0	0%
Eastern Shore	3	8	+5	+167%
Western Shore	1	5	+4	+400%
Totals	7	18	+11	+157%

Change in Total Blocks between Atlases by Region

Black Rail
Laterallus jamaicensis

The sparrow-size Black Rail is one of the rarest and most enigmatic breeding birds in Maryland. Because of the bird's secretive, nocturnal habits and somewhat inaccessible tidal marsh habitat, detecting and studying it are difficult. Rarely seen, its distinctive, incessant *ki-ki-krr* reveals its presence during the breeding season. The first breeding-season record in our area was of a specimen near Washington, DC, in 1879 (Cooke 1929); the first Maryland nest was not found until 1931, even though the Black Rail was thought to be a fairly common but local breeder on the lower Eastern Shore (Stewart and Robbins 1958).

Generally, the Black Rail breeds in shallow-water wetlands, including wet, grassy fields and meadows, as well as a variety of fresh, brackish, and salt marshes (Ripley 1977). In Maryland this species is almost exclusively found in "high marsh" areas of salt and brackish marshes. This vegetation in these communities is dominated by saltmeadow cordgrass, smooth cordgrass, and seashore saltgrass, with scattered patches of taller vegetation such as needlerush, cattail, or Olney's bulrush (H. Wierenga, pers. obs.). Two exceptions in Maryland are the wet, grassy fields of Carroll Island in Baltimore County in 1990 (DNR, unpubl. data) and the grassy sewage treatment fields of Easton in Talbot County in the 1990s and 2000s.

Most Black Rails are believed to migrate south for the winter, but a few are known to occur in Maryland during the winter (Bystrak 1974; Reese 1975; DNR, unpubl. data; H. Wierenga, pers. comm.). Adults are thought to begin arriving on breeding grounds in early to mid-April (Robbins and Bystrak 1977). The nest is a tightly woven cup placed deep in dense vegetation and covered with a thin canopy (Ripley 1977). Egg dates from the eight known Maryland nests range from 20 May (Stewart and Robbins 1958) to 8 August (W. Burt, unpubl. data). Maryland clutch sizes range from 5 to 8 eggs (W. Burt, unpubl. data, R. Edwards, unpubl. data). Incubation of two Maryland clutches lasted no more than 13 days (W. Burt, unpubl. data; R. Edwards, unpubl. data). The tiny, black chicks are precocial and it is not known how long they remain dependent on their parents.

This rare species was detected in 18 blocks during the 1983–1987 atlas (Wierenga 1996a) and only 12 blocks during the 2002–2006 atlas, a 33% decline. The Breeding Bird Survey does not detect this species; therefore, no regional or national trends are available for comparison. The overall distribution remained similar for both atlas projects: its core range is the extensive marshes of southern Dorchester and western Somerset counties, with scattered, isolated records elsewhere, including the western shore of the Upper Chesapeake Bay. The first atlas documented 3 blocks along the coastal salt marshes of Worcester County, where the Black Rail was not detected during the 2002–2006 atlas.

Because many marsh species are so difficult to survey, DNR conducted a targeted study of several marsh birds in 1990 and 1991. Black Rails were detected in 40 atlas blocks, nearly all on the lower Eastern Shore (Brinker, Therres, et al. 2002). The Black Rail portion of this study was duplicated in 2007 with a survey of the three lower Eastern Shore counties (Dorchester, Somerset, and Worcester). The species was detected in only 8 atlas blocks at that time, indicating that Black Rails experienced a decline of more than 80% between 1991 and 2007 (DNR, unpubl. data).

In 1987 the Department of Natural Resources listed the Black Rail as in need of conservation; its status was changed to endangered in 2007. Based on decades-long, informal monitoring of the Elliott Island population by a few birders, this species has probably been in gradual decline for many years. Factors that affect its survival include food availability, weather patterns, hydrology, predation levels, and habitat management practices. Of particular concern is the role that prescribed fire and illegal burning of lower Eastern Shore marshes may have on overwintering individuals and on early spring food availability and habitat. The more recent precipitous decline may be the result of factors directly related to local sea level rise brought on by marsh subsidence and compounded by global warming, primarily an increase in the duration, frequency, and depth of flooding events and a concurrent increase in predation. Flooding and predation were noted as the primary causes of nest failure in Florida, whereas hydrology and water levels potentially had the greatest impact on nest-site selection and nesting success (Legare and Eddleman 2001). More than 60 years ago, Grinnell and

Miller (1944) noted that the "most important hazards to existence [of Black Rails] on salt marshes appear to be extra high tides." Those hazards include both nest loss and increased mortality from predation (Evens and Page 1986).

As the effects of marsh subsidence and sea level rise continue to increase, the Black Rail populations in tidal marshes may become extirpated. However, this species is known to inhabit inland shallow wetlands and may be able to persist in Maryland if sufficient alternate habitat is created and maintained for shallow-water wetland species.

LYNN M. DAVIDSON

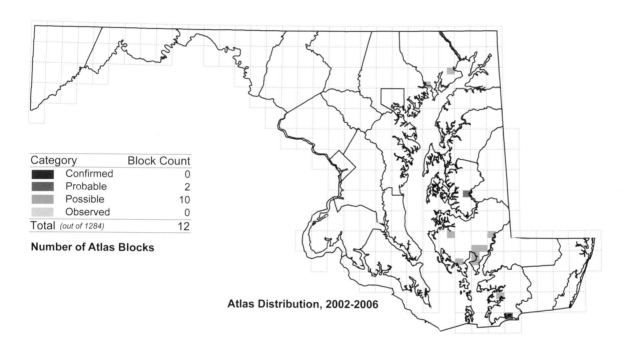

Category	Block Count
Confirmed	0
Probable	2
Possible	10
Observed	0
Total (out of 1284)	12

Number of Atlas Blocks

Atlas Distribution, 2002-2006

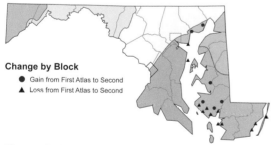

Change by Block
- ● Gain from First Atlas to Second
- ▲ Loss from First Atlas to Second

Change in Atlas Distribution, 1983-1987 to 2002-2006

Atlas Region	1983-1987	2002-2006	Change No.	%
Allegheny Mountain	0	0	0	-
Ridge and Valley	0	0	0	-
Piedmont	0	0	0	-
Upper Chesapeake	1	2	+1	+100%
Eastern Shore	16	10	-6	-38%
Western Shore	1	0	-1	-100%
Totals	18	12	-6	-33%

Change in Total Blocks between Atlases by Region

Clapper Rail
Rallus longirostris

The Clapper Rail is a denizen of salt and brackish marshes. Its ascending *kek kek kek* call can be heard from the marsh vegetation day or night. Its breeding distribution in North America is concentrated in the salt marshes along the coasts of the Atlantic and Pacific oceans and the Gulf of Mexico (Eddleman and Conway 1994). The Clapper Rail is the salt marsh counterpart to the King Rail, which occurs in freshwater and brackish wetlands. There is a zone of overlap between these two species in brackish marshes; they are known to occasionally hybridize in those areas (Meanley 1969, 1985).

Optimum breeding habitat along the East Coast is tidal salt marshes dominated by smooth cordgrass (Meanley 1985; Eddleman and Conway 1994). Clapper Rail nests are usually built in the taller cordgrass along the edges of tidal creeks and guts or adjacent to shorter cordgrass (R. Stewart 1951); however, they will nest in salt marshes dominated by other marsh grasses as well (Meanley 1985). The nesting season in Maryland is generally from May through August. The peak of egg laying and incubation is late May to June. Known egg dates in Maryland range from 20 May to 20 July (Stewart and Robbins 1958). Clutch sizes range from 5 to 12 (Meanley 1985), with a mean of 7.6 from 36 nests in Maryland (Blom 1996a). The young are precocial, leaving the nest shortly after hatching. The downy young follow the adults for 9–10 weeks (Meanley 1985) learning how to feed and avoid predators and learning other essential survival behavior.

Historically in Maryland, Clapper Rail breeding populations were found in the tidal salt marshes of Worcester County and the Chesapeake Bay salt marshes on the lower Eastern Shore. Kirkwood (1895) described the Clapper Rail as fairly common in Maryland salt marshes in the late 1800s.

Stewart and Robbins (1958) reported this species as fairly common in coastal Worcester County and uncommon and local in the outer fringe marshes of the Chesapeake Bay, including southern St. Mary's County and on the Eastern Shore north to Parson Island in Queen Anne's County.

The breeding distribution of the Clapper Rail during the 1983–1987 atlas (Blom 1996a) was similar to that described by Stewart and Robbins (1958). Seventy atlas blocks had Clapper Rail, with the greatest concentration of those in the coastal marshes of Worcester County and the salt marshes in Somerset and Dorchester counties. Clapper Rails were also documented in interior tidal marshes in Dorchester County. The northern limit of Clapper Rail breeding records was in the southwestern portion of Talbot County around Tilghman Island. Two atlas blocks on the Western Shore had Clapper Rails during 1983–1987: one at Patuxent Naval Air Station near the mouth of the Patuxent River and one at Sandy Point State Park in Anne Arundel County (Blom 1996a).

During 1990–1992, Brinker, Therres, et al. (2002) conducted an extensive nocturnal survey of tidal marshes in Maryland for breeding marsh birds using tape playback methodology. They documented Clapper Rail concentrations in the coastal bay marshes and the salt marshes of Somerset and Dorchester counties, including interior tidal marshes. Clapper Rail was also documented on the Western Shore in southern St. Mary's County and at Black Marsh in Baltimore County.

Clapper Rails documented during the 2002–2006 atlas are distributed in a manner comparable to those reported by Blom (1996a) and Brinker, Therres, et al. (2002), with a few exceptions. Of the 84 blocks with Clapper Rails in the second atlas project, the vast majority were concentrated along the coastal bays in Worcester County and the salt marshes of Somerset, Wicomico, and Dorchester counties. The northernmost block with Clapper Rail on the Eastern Shore was on Eastern Neck Island, Kent County. Clapper Rails were recorded in only 2 blocks on the Western Shore, both in southern St. Mary's County near the mouth of the Potomac River.

There was an overall gain in blocks with Clapper Rail in the 2002–2006 atlas compared to the first atlas. Results of the more recent atlas suggest that Clapper Rails may have disappeared from the interior tidal marshes of Dorchester County, an area they were not known to occupy historically. Wilson et al. (2007) postulated that the expansion of breeding Clapper Rails into this area may have been the result of habitat changes caused by sea level rise. Salt intrusion may have altered marsh vegetation to that more suitable for Clap-

per Rails. The interior tidal marshes of Dorchester County are now being converted to more open water areas as a result of continued sea level rise. As a result, suitable marsh habitat for Clapper Rails may now be disappearing in that area.

The Clapper Rail breeding population is currently secure in Maryland. Tidal wetlands generally receive high levels of protection from human alteration, which helps conserve habitat. The effects of climate change and sea level rise will play a more significant role for breeding Clapper Rails in the future.

GLENN D. THERRES

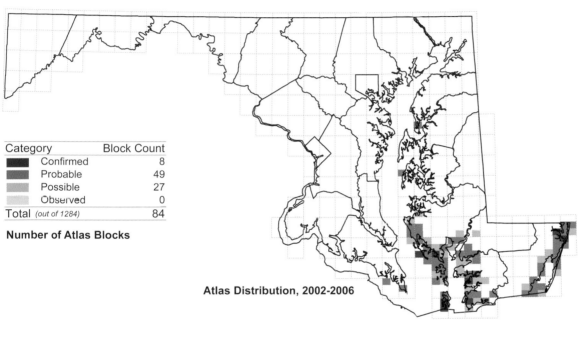

Category	Block Count
Confirmed	8
Probable	49
Possible	27
Observed	0
Total (out of 1284)	84

Number of Atlas Blocks

Atlas Distribution, 2002-2006

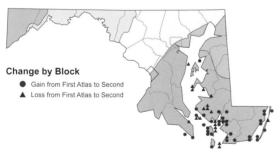

Change by Block
- ● Gain from First Atlas to Second
- ▲ Loss from First Atlas to Second

Change in Atlas Distribution, 1983-1987 to 2002-2006

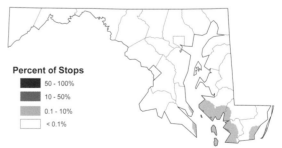

Percent of Stops
■	50 - 100%
▨	10 - 50%
▨	0.1 - 10%
□	< 0.1%

Relative Abundance from Miniroutes, 2003-2007

Atlas Region	1983-1987	2002-2006	Change No.	%
Allegheny Mountain	0	0	0	-
Ridge and Valley	0	0	0	-
Piedmont	0	0	0	-
Upper Chesapeake	0	0	0	-
Eastern Shore	67	81	+14	+21%
Western Shore	2	2	0	0%
Totals	69	83	+14	+20%

Change in Total Blocks between Atlases by Region

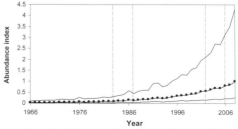

Breeding Bird Survey Results in Maryland

King Rail
Rallus elegans

Stan Arnold

The King Rail is a secretive marsh bird that calls the brackish and freshwater marshes of the Chesapeake Bay home. It is similar in appearance to the Clapper Rail, and some consider the two to be conspecific (B. Taylor and van Perlo 1998). Even the calls of the two species are similar, though the mating call of the King Rail is usually a series of fewer than 10 *kek kek kek* notes evenly spaced whereas the Clapper Rail's call is a series of 10 or more notes accelerating and then slowing (J. Dunn and Blom 1983). During the breeding season, both species can occur in brackish marshes though the Clapper Rail is more frequently found in salt marsh habitat.

Historically, the King Rail nested primarily in the large tidal brackish and freshwater river marshes in Maryland. Stewart and Robbins (1958) described this species as fairly common in the tidewater areas of the Eastern Shore, Western Shore, and Upper Chesapeake regions and locally uncommon in the Piedmont and interior Eastern Shore regions. Kirkwood (1895) reported nesting in the marshes of the Patapsco River near Baltimore and at Tolchester in Kent County. Meanley (1975) reported breeding in the marshes of the Blackwater, Choptank, Nanticoke, and Patuxent rivers, with the greatest numbers found in the marshes of Dorchester County (Meanley 1969).

The nesting season in Maryland starts during the first warm days in April with calling males establishing territories and attracting mates (Meanley 1975). Most King Rails nest in May and June. Clutch sizes generally range from 10 to 12 eggs (Meanley 1969). After 21 or 22 days of incubation, the black downy young emerge from the egg and are mobile within a day. The brood remains with the adults for more than a month.

No precocial young were observed during the 2002–2006 atlas, as there were no confirmations within the 37 blocks in which King Rails were documented. The lack of confirmations does not mean that this species did not nest in Maryland. Finding nests or young in nesting habitat requires extensive field efforts and a little luck.

During the 1983–1987 atlas King Rails were concentrated in the marshes of Dorchester County, Tanyard wetlands, and along the Patuxent River (Blom 1996b). No occurrences of King Rails were recorded at the head of the Chesapeake Bay during the first atlas; limited access to the army base at Aberdeen Proving Ground may have precluded documenting King Rails during that time period. During the 2002–2006 atlas this obligate marsh bird was detected in only half the number of blocks in which it had been found in the first atlas. King Rails were still concentrated in the marshes of Dorchester County, Tanyard wetlands along the Choptank

River, marshes along the Wicomico River on the Eastern Shore, and some were found at the head of the Chesapeake Bay (primarily Aberdeen Proving Ground). King Rails were also recorded in scattered wetlands on the Western Shore and at Lilypons in Frederick County.

From 1990 to 1992 an extensive nocturnal survey of tidal marshes for breeding marsh birds was conducted using tape playback methodology (Brinker, Therres, et al. 2002). King Rails were found to be locally abundant only in the marshes along the Choptank River, especially the Tanyard wetlands. They were found at only a few other sites, mostly in Dorchester County. None were recorded on the Western Shore except at Black Marsh. Aberdeen Proving Ground was not included in the nocturnal surveys.

Gains in King Rail records between the first and second atlas were mostly in the freshwater wetlands at the head of the Chesapeake Bay and in the brackish marshes at the mouth of the Nanticoke River. A few new blocks with King Rails were recorded on the Western Shore, including Point Lookout in St. Mary's County. Losses in blocks with King Rails were concentrated along the Patuxent River, Dorchester County, and Somerset County. These were traditional strongholds for this species in Maryland. Because of the secretive nature of King Rails and the difficulty in surveying wetland habitats, it is possible that some blocks shown as losses may have had undetected birds. However, it is unlikely that the 50 percent reduction in blocks with this species can be blamed entirely on undetected rails.

King Rail populations in Maryland have declined since the 1950s. The species is declining throughout its North American range as well (Meanley 1992; T. Cooper 2008). BBS trends suggest a significant population decline, which is believed to be primarily the result of habitat loss. Changes in wetland structure and vegetative composition may be responsible for

the reduction in distribution and abundance of the King Rail in Maryland (Brinker, Therres, et al. 2002). Clapper Rails may have displaced King Rails from their traditional wetlands because of habitat changes. Saltwater intrusion of freshwater habitats from sea level rise may have altered marsh vegetation to conditions less suitable for King Rails in portions of the Chesapeake Bay area (Wilson et al. 2007). Conservation of brackish and freshwater wetlands will be essential to the survival of this species in Maryland.

GLENN D. THERRES

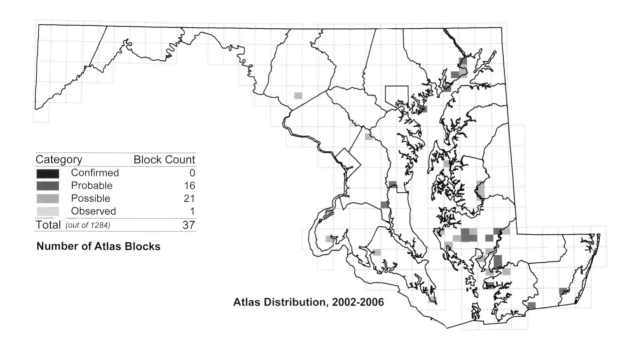

Category	Block Count
■ Confirmed	0
■ Probable	16
■ Possible	21
■ Observed	1
Total (out of 1284)	37

Number of Atlas Blocks

Atlas Distribution, 2002-2006

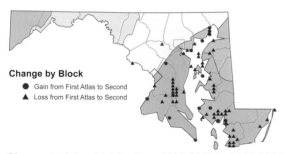

Change by Block
- ● Gain from First Atlas to Second
- ▲ Loss from First Atlas to Second

Change in Atlas Distribution, 1983-1987 to 2002-2006

Atlas Region	1983-1987	2002-2006	Change No.	Change %
Allegheny Mountain	0	0	0	-
Ridge and Valley	0	0	0	-
Piedmont	3	1	-2	-67%
Upper Chesapeake	5	5	0	0%
Eastern Shore	47	23	-24	-51%
Western Shore	18	7	-11	-61%
Totals	73	36	-37	-51%

Change in Total Blocks between Atlases by Region

Virginia Rail
Rallus limicola

The Virginia Rail is the most abundant and widespread of the breeding rails in Maryland. It is primarily a nocturnal bird, though it is most active and vocal at dawn or dusk. Maryland is near the southern edge of its eastern breeding range in North America (Conway and Eddleman 1994). The Virginia Rail occurs in freshwater, brackish, and salt marshes and in other wetlands with emergent vegetation. Virginia Rails need standing water, moist soil, or mud flats interspersed within these wetlands (Conway and Eddleman 1994).

Because of the extensive tidal marshes of the Chesapeake Bay, Virginia Rail was historically a common breeder and migrant in the tidewater areas of the Eastern Shore and Upper Chesapeake and fairly common in the tidewater areas of the Western Shore (Stewart and Robbins 1958). They were also fairly common locally in the Allegheny Mountain section of Maryland, but rare elsewhere in the state.

The breeding season in Maryland is generally from late April through August (Stewart and Robbins 1958). After pair formation, the female selects a nest site and the pair constructs the nest (Kaufmann 1989). Virginia Rail nests are well concealed within a variety of emergent vegetation types (Conway and Eddleman 1994). Clutch sizes in North America range from 4 to 13 (Kaufmann 1989); clutch sizes from 6 Maryland nests ranged from 4 to 9 eggs (Wierenga 1996b). Incubation lasts for 19 days and the young are mobile shortly after hatching (Conway and Eddleman 1994). Young stay with their parents as a family group for 3–4 weeks until they become independent (Kaufmann 1989).

During the 1983–1987 atlas, breeding Virginia Rails were found in most of the extensive tidal marshes on the Coastal Plain, with large concentrations in Dorchester and Somerset counties and along the tidal portions of the Patuxent and Choptank rivers (Wierenga 1996b). Smaller concentrations occurred along the tidal portions of the Chester River on the Eastern Shore and along tidal rivers of Baltimore County. Scattered records occurred elsewhere on the Coastal Plain. The western counties of Frederick, Washington, Allegany, and Garrett had one record each during the 1983–1987 atlas.

Brinker, Therres, et al. (2002) conducted intensive nocturnal surveys of tidal marshes in Maryland for rails and other marsh birds using tape-playback methodology in 1990–1992 and found similar distributions for Virginia Rail to those of the 1983–1987 atlas. They documented the highest abundance of Virginia Rails in Dorchester and Somerset counties.

During the 2002–2006 atlas, Virginia Rail was documented in 110 blocks, mostly on the Coastal Plain. The largest concentration of this breeding species occurred in the tidal

Bill Sherman

marshes of Dorchester, Somerset, and Wicomico counties. A smaller concentration occurred in the Tanyard wetlands of the Choptank River. Scattered breeding records occurred elsewhere on the Coastal Plain. Every inland county, except Washington County, had 1 or 2 blocks with Virginia Rail. It may no longer be locally common in the Allegheny Mountain section.

Overall, there were 15 fewer blocks with breeding Virginia Rail during the 2002–2006 atlas compared to the 1983–1987 atlas. The greatest number lost was along the Patuxent River on the Western Shore. This was surprising, and the cause of the apparently reduced breeding population of these rails is unclear. Haramis and Kearns (2007a) described the emergent marshes of the Patuxent River as being as luxuriant and productive today as at any time in the past. They documented high numbers of migrating Soras and Virginia Rails using the freshwater marshes during the 1990s. Impacts of resident Canada Geese through alteration of marsh vegetation in this river system may have contributed to the absence of Virginia Rail. Since that problem has been addressed (Haramis and Kearns 2007b), Virginia Rails may move into regrowing marshes as soon as there is adequate cover (Conway and Eddleman 1994). Limited access to the marshes by field-workers, especially at night, may also have contributed to fewer Virginia Rails being documented. Because of the secretive nature of this nocturnal bird, losses from other blocks may be a function of varying survey effort.

Population trends for Virginia Rail currently appear to be stable or increasing in North America according to the BBS. But detection of secretive marsh birds is difficult, and analysis of population trends is not as reliable as for some other species. The Virginia Rail population is believed to have de-clined in the mid-1900s because of habitat loss, especially in the Midwest (Conway and Eddleman 1994). As for any obligate marsh bird, the conservation of emergent wetlands is essential for the persistence of the species in Maryland and elsewhere.

GLENN D. THERRES

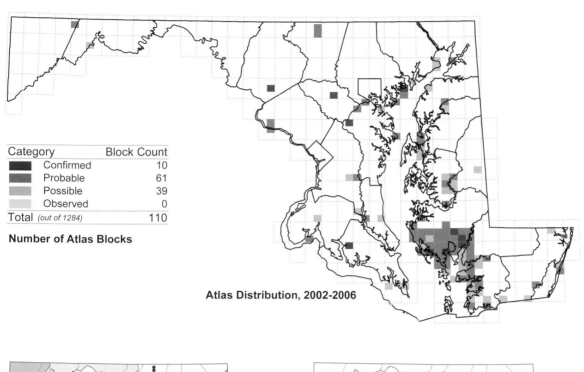

Category	Block Count
Confirmed	10
Probable	61
Possible	39
Observed	0
Total (out of 1284)	110

Number of Atlas Blocks

Atlas Distribution, 2002-2006

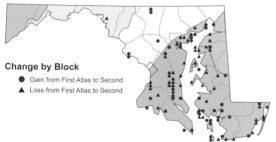

Change by Block
- ● Gain from First Atlas to Second
- ▲ Loss from First Atlas to Second

Change in Atlas Distribution, 1983-1987 to 2002-2006

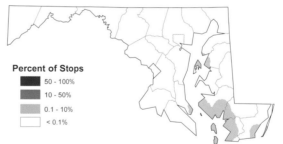

Percent of Stops
- 50 - 100%
- 10 - 50%
- 0.1 - 10%
- < 0.1%

Relative Abundance from Miniroutes, 2003-2007

Atlas Region	1983-1987	2002-2006	Change No.	Change %
Allegheny Mountain	1	1	0	0%
Ridge and Valley	2	1	-1	-50%
Piedmont	1	6	+5	+500%
Upper Chesapeake	13	10	-3	-23%
Eastern Shore	82	77	-5	-6%
Western Shore	26	15	-11	-42%
Totals	125	110	-15	-12%

Change in Total Blocks between Atlases by Region

George M. Jett

Sora
Porzana carolina

Found throughout much of North America, the Sora is a small rail easily identified by its yellow bill and black mask. It is less secretive than other rails and is found in Maryland more frequently as a migrant than as a breeder. Its migratory behavior and patterns have been studied extensively in wild rice wetlands along the Patuxent River, an area thought to be a critical migratory stopover habitat for this species (Kearns et al. 1998; Haramis and Kearns 2000, 2007a). Despite relatively low population levels, Soras are still hunted in Maryland.

The Sora lives in a wide variety of marshes, including bogs, ponds with dense emergent vegetation, and tidal fresh and brackish marshes; but it occurs primarily in freshwater wetlands and breeds rarely in salt marsh habitats (Ripley 1977). Returning as early as late March, most migrants arrive from mid-April to mid-May (Robbins and Bystrak 1977). Nesting probably begins in early May, with nests constructed of dead cattails and other vegetation woven into a basket up to 15.2 cm (6 in) above standing water (Bent 1926; Ripley 1977). Only 2 records of Maryland nests exist: a Harford County nest had 7 eggs on 25 May 1899, and one at PWRC had 3 eggs and 1 young on 3 June 1965 (MNRF). Fall migrants return from early August to late October; small numbers overwinter (Robbins and Bystrak 1977).

A rare breeder in Maryland, the Sora reaches the southeastern limit of its North American breeding range here. The 1983–1987 atlas recorded this species in just 6 blocks (David-son 1996c); the number of blocks increased to 15 during the 2002–2006 atlas project. This increase of 9 blocks may not be significant because the normal annual variation in breeding populations is unknown and because the differences in observer effort between the atlas projects specifically related to surveys for nocturnal marsh birds is also unknown. Surveys designed specifically for nocturnal marsh birds were undertaken by DNR in 1990 and 1991 (Brinker, Therres, et al. 2002). Soras were found in greater abundance, especially on the lower Eastern Shore, than during the 1983–1987 atlas. DNR (unpubl. data) also conducted specific marsh bird surveys during 2005, which likely contributed to the increase in detection of Soras during the second atlas project.

This species' population may be stable, if not increasing, over the 20-year period between the two atlases. Sora was reported from some of the same wetland areas in both atlases, such as Lilypons (Frederick County), Jug Bay (Anne Arundel / Prince Georges counties), and Deal Island WMA (Somerset County). Absent from the Allegheny Mountain and Upper Chesapeake regions during the 1983–1987 atlas (Davidson 1996c), Sora was found in every physiographic region during the current atlas. BBS trend data show population declines in North America of about 4% per year from 1966 to 1979 and nearly stable populations from 1980 to 2007. The BBS trend map for 1966 to 2003 shows that northeastern populations from the eastern Great Lakes to the Atlantic have been increasing more than 1.5% per year. Although no BBS trend data exist specifically for Maryland, this regional trend may be influencing Maryland populations.

The greatest threat to the Sora is the loss and degradation of its breeding habitat, especially freshwater emergent wetlands. This habitat is vulnerable to ditching, draining, fill-ing, and other hydrologic alterations, as well as vegetation manipulation, such as excavating cattails from stormwater management ponds. Additionally, this species requires abundant invertebrate prey in the summer, including snails, crustaceans, beetles, and spiders, as well as the seeds of aquatic plants. Factors that affect prey abundance, such as pesticide contamination, pollution, and invasive species, are likely to impact Sora populations as well. Conservation measures for maintaining the Sora as a Maryland breeder must emphasize the protection and appropriate management of wetlands for rails.

LYNN M. DAVIDSON

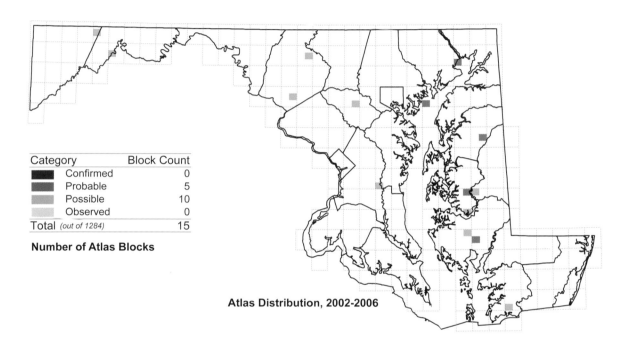

Category	Block Count
Confirmed	0
Probable	5
Possible	10
Observed	0
Total (out of 1284)	15

Number of Atlas Blocks

Atlas Distribution, 2002-2006

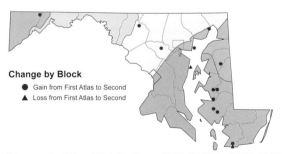

Change by Block
● Gain from First Atlas to Second
▲ Loss from First Atlas to Second

Change in Atlas Distribution, 1983-1987 to 2002-2006

Atlas Region	1983-1987	2002-2006	Change No.	%
Allegheny Mountain	0	1	+1	-
Ridge and Valley	1	1	0	0%
Piedmont	1	3	+2	+200%
Upper Chesapeake	0	3	+3	-
Eastern Shore	1	6	+5	+500%
Western Shore	2	1	-1	-50%
Totals	5	15	+10	+200%

Change in Total Blocks between Atlases by Region

Common Moorhen
Gallinula chloropus

The Common Moorhen is a member of the rail family (Rallidae) but behaves more like a duck and is much more active during daylight hours than its rail cousins. It resembles the American Coot, but has a red bill and frontal shield and a prominent white stripe along its sides. Its habitat is freshwater marshes, ponds, and other emergent wetlands where open water is interspersed with submerged or floating vegetation (Bannor and Kiviat 2002). This species was formerly called Common Gallinule or Florida Gallinule in North America. It occurs on every continent around the world except Australia and Antarctica (B. Taylor and van Perlo 1998). Its breeding range in North America is primarily in the eastern United States, with scattered populations in the West (Bannor and Kiviat 2002). The Common Moorhen's North American breeding range is thought to have expanded northward during the twentieth century.

The first nest in Maryland was found in Dorchester County on 10 May 1916 (R. Jackson 1941). Historically, the Common Moorhen was considered a common breeder only in the marshes of the Gunpowder River at the head of the Chesapeake Bay (Meanley 1975). Stewart and Robbins (1958) described its breeding status elsewhere in Maryland as uncommon and local in tidewater areas of southern Dorchester County and possibly breeding sparingly in other tidewater areas. Meanley (1975) reported breeding Common Moorhens were fairly common near Dames Quarter in Somerset County by the 1970s.

The nesting season in Maryland is from early May through July. Four types of nest structures are built by the breeding pair: trail nests, egg nests, brood nests, and elevated platforms (Helm et al. 1987). Nesting structures, which are composed mainly of twigs, stems, and leaves of whatever vegetation is nearby, are located close to open water. Known egg dates in Maryland range from 10 May (Stewart and Robbins 1958) to 4 July (Meanley 1975). The precocial young are mobile within a day after hatching (Greij 1994). Young moorhens are brooded on brood platforms or in the egg nest for about 14 days (Frederickson 1971; Greij 1994) when not foraging with the adults in the open water. Both adults tend to the young.

During the 1983–1987 breeding bird atlas project, Common Moorhens were recorded in 42 blocks (N. Stewart 1996a). They were locally distributed throughout the Coastal Plain with small concentrations in the Patuxent, Choptank, and Blackwater river marshes and near Deal Island in Somerset County. The few records in the Baltimore area were generally in disturbed habitats. Common Moorhens were recorded from a few locations west of the Coastal Plain, including 2 sites in Howard County and 1 site each in Montgomery, Frederick, and Allegany counties.

Brinker, Therres, et al. (2002) conducted extensive breeding marsh bird surveys of tidal marshes in Maryland from 1990 through 1992 and documented Common Moorhens at 22 survey locations. Concentrations were found in the Tanyard wetlands along the Choptank River and in the Fishing Bay marshes of Dorchester County. They were also found at 1 site along the Patuxent River and 1 on the Sassafras River. No Common Moorhens were documented in the Gunpowder River wetlands.

Results from the 2002–2006 atlas documented Common Moorhens in 37 blocks, mostly on the Coastal Plain. The majority of records were in the marshes of Dorchester and Somerset counties and along the Choptank River, especially in the Tanyard wetlands. Scattered records occurred along the Patuxent River, upper Eastern Shore, and in the Baltimore area. Nesting was confirmed on Hart-Miller Island. Unfortunately, it seems this species is no longer a common breeding bird in the Gunpowder River marshes. Common Moorhens were seen in suitable habitat west of the Coastal Plain at two human-created wetlands, namely Lilypons in Frederick County and McKee-Beshers WMA in Montgomery County.

Though the number of blocks with Common Moorhens and the bird's breeding distribution did not change drastically between the two atlas periods, losses and gains did occur. There were a greater number of losses than gains in the Western Shore section, especially along the Patuxent River and in the Baltimore area. Some of this may have resulted from differences in survey effort. There was an overall increase in the number of blocks in Dorchester County with breeding Common Moorhens. This is likely because of the extensive emergent wetlands there and the county's emphasis on waterfowl habitat management.

Though the Common Moorhen is protected by the DNR as a wildlife species in need of conservation, its breeding status in the state seems secure. Across its North American range, there are no significant BBS population trends. Wetland conservation is essential for sustaining this species, and in Maryland, wetlands generally receive high levels of protection. Given the Common Moorhen's affinity for impoundments managed for waterfowl and for other shallow water conditions, the species is likely to remain an uncommon breeder across Maryland's Coastal Plain and elsewhere.

GLENN D. THERRES

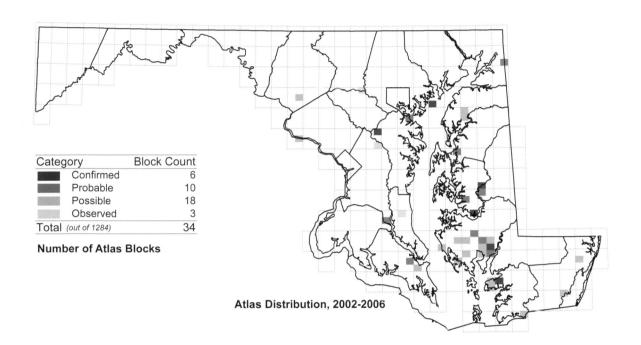

Category	Block Count
Confirmed	6
Probable	10
Possible	18
Observed	3
Total (out of 1284)	34

Number of Atlas Blocks

Atlas Distribution, 2002-2006

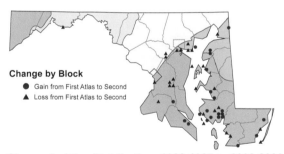

Change by Block
- ● Gain from First Atlas to Second
- ▲ Loss from First Atlas to Second

Change in Atlas Distribution, 1983-1987 to 2002-2006

			Change	
Atlas Region	1983-1987	2002-2006	No.	%
Allegheny Mountain	0	0	0	-
Ridge and Valley	1	0	-1	-100%
Piedmont	4	2	-2	-50%
Upper Chesapeake	4	4	0	0%
Eastern Shore	19	23	+4	+21%
Western Shore	11	5	-6	-55%
Totals	39	34	-5	-13%

Change in Total Blocks between Atlases by Region

Eric Skrzypczak

American Coot
Fulica americana

American Coot is an easily recognizable waterbird with its slate gray body, which appears black at a distance, and contrasting white bill. It is a member of the rail family, but acts more like a diving duck than a typical rail. During the winter, it occupies habitats used by dabbling ducks, including ponds, lakes, wetlands, tidal creeks, and rivers.

Historically, American Coot was common in migration and during winter in Maryland, frequenting all the tidewater sections of Maryland and also inland reservoirs, ponds, and lakes (Stewart and Robbins 1958). It occurred in tremendous numbers on the Susquehanna flats at the head of the Chesapeake Bay, feeding on the large beds of submerged aquatic vegetation.

Until 1970, American Coot was not known to nest in Maryland. Stewart and Robbins (1958) described it only as a casual summer vagrant in Queen Anne's and Prince George's counties and in the District of Columbia. The first breeding evidence in Maryland was found on 16 August 1970, when Armistead (1970) documented downy young with adults at Deal Island WMA in Somerset County. He later documented adults with downy young at Fairmont WMA (Robbins 1973), also in Somerset County.

The breeding range for American Coot is throughout the United States, Mexico, and much of Canada, with the highest breeding densities in the U.S. prairie potholes region (Brisbin and Mowbray 2002). Breeding in the eastern United States is somewhat discontinuous, with isolated breeding along the Atlantic coast. The northern edge of the species'

eastern breeding range is Prince Edward Island, Canada (Erskine 1992).

During the 1983–1987 atlas, American Coot was confirmed nesting in only 1 block in Somerset County and observed in 5 other blocks on the Coastal Plain (Meritt 1996i). Deal Island WMA was the only location that supported breeding American Coot during that atlas period. The other observations were scattered reports from impoundments and sewage ponds, and these occurrences were not thought to indicate breeding.

The breeding season in Maryland appears to begin in late May and continue through August, based on the limited nesting records. Multiple nests are built before one is selected for egg laying (Brisbin and Mowbray 2002). The nest is a woven basket of vegetation with a hollowed inner cavity of sufficient size to hold the eggs (Bent 1926). The nest is generally a floating structure attached to upright stalks (Brisbin and Mowbray 2002). There are currently no Maryland records of clutch sizes. The young are precocial and leave the nest within a day or two. They are attended by their parents for approximately one month before they form brood flocks.

American Coot was confirmed nesting in 2 blocks during the 2002–2006 atlas, considered possible in 2, and observed in 5 other blocks. Danny Poet found a pair of adults swimming with eight downy young on 2 July 2004 near Grasonville in Queen Anne's County. The adults were observed feeding the young regularly. The birds were seen in a shallow pond created and managed for waterfowl. The pond had a combination of open water interspersed with emergent wetland plants. A year later, breeding was confirmed at Hart-Miller Island in Baltimore County. Gene Scarpulla and Don Burggraf observed a pair of American Coots with four half-grown young on 6 August 2005. The young were seen pecking at the adults' bills, presumably feeding. The young, no longer downy, were about one-third the size of adult coots. Earlier that season (4 June 2005), the adult pair had been observed in the same pond (E. Scarpulla, pers. obs.). The other possible breeding observations of American Coot during the breeding season occurred in Cecil and Montgomery counties.

The number of blocks with American Coot were similar in the 1983–1987 and 2002–2006 atlas efforts. During the second atlas, no breeding coots were documented at Deal Island or elsewhere in Somerset County, where they were first discovered nesting in Maryland and had been found during the 1983–1987 atlas (Meritt 1996i). Because of the rarity

of this nesting species and the extensive wetlands in Somerset County, it is possible that American Coot has not disappeared as a breeder there but simply went undetected. During the 2002–2006 atlas, its breeding distribution expanded to 2 other man-made wetlands on the Coastal Plain.

Given its affinity for shallow water impoundments created by humans, the American Coot is expected to stay a breeding bird in Maryland. It is likely to remain a rare breeder, scattered across the Coastal Plain and elsewhere, utilizing impoundments managed for waterfowl and other shallow water conditions.

GLENN D. THERRES

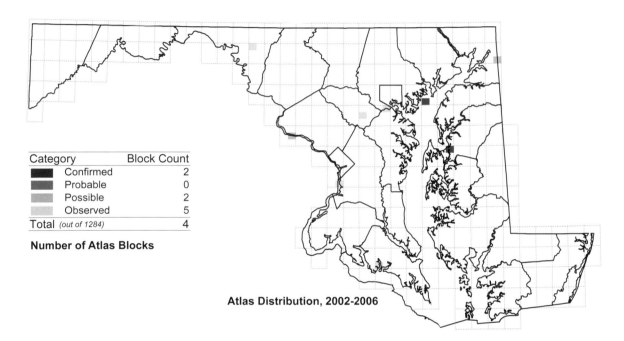

Category	Block Count
■ Confirmed	2
■ Probable	0
■ Possible	2
■ Observed	5
Total (out of 1284)	4

Number of Atlas Blocks

Atlas Distribution, 2002-2006

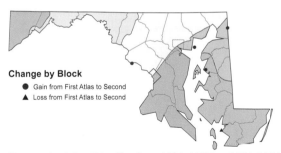

Change by Block
- ● Gain from First Atlas to Second
- ▲ Loss from First Atlas to Second

Change in Atlas Distribution, 1983-1987 to 2002-2006

Atlas Region	1983-1987	2002-2006	Change No.	%
Allegheny Mountain	0	0	0	-
Ridge and Valley	0	0	0	-
Piedmont	0	1	+1	-
Upper Chesapeake	0	2	+2	-
Eastern Shore	1	1	0	0%
Western Shore	0	0	0	-
Totals	1	4	+3	+300%

Change in Total Blocks between Atlases by Region

Piping Plover
Charadrius melodus

The Piping Plover is a small, ghostly pale bird with a black band across its chest; it lives on beaches and adjacent to poorly vegetated dunes. In the Missouri River drainage and the Canadian Prairie, this plover also frequents the sandy shores of alkali lakes and river bars. It is this bird's misfortune to depend upon a habitat greatly valued by humans as a playground. As a result of this, the Piping Plover is listed as threatened by the USFWS and is considered endangered in Maryland, where its nesting is now limited to Assateague Island.

Although the Piping Plover has long been known as an uncommon breeding bird on Maryland's Atlantic barrier beaches (Stewart and Robbins 1958), careful censuses of the nesting population were not made until the 1980s (Hoffman 1996b). Piping Plovers formerly bred on the Ocean City barrier beach but have not done so since the 1940s (Hoffman 1996b). By the early 1980s Atlantic Piping Plover populations had fallen to between 634 and 662 pairs (Haig and Oring 1985). The species was federally listed as threatened in 1986 (Paxton 1998).

The Maryland Piping Plover breeding population reached a low of 14 pairs in 1990 (Kumer 2004); it has since increased substantially with aid from intensive management and research by the National Park Service and the DNR. In 1994 Piping Plovers experienced a notably productive nesting season on the East Coast, followed by an increase to 61 pairs in

Maryland in 1996 (Kumer 2004). The Maryland nesting population has been fairly stable over the past decade, ranging from 56 to 66 pairs (USFWS 2004; USFWS, unpubl. data).

In the 1983–1987 Maryland-DC atlas the Piping Plover nested in 4 blocks (Hoffman 1996b); that total rose by 1 in 2002–2006 with nesting reported in 2 new blocks and probable breeding in 1 of the 1983–1987 blocks. The map shows 2 areas on Assateague that currently support nesting plovers: north Assateague with 4 blocks (3 confirmed, 1 probable), and south Assateague, not far north of the Virginia line (2 confirmed). Presumably this distribution reflects the Piping Plover's preference for washouts or sparsely vegetated dunes for nest sites, away from most summer high tides, above the beaches used for feeding (Haig 1992). These conditions are not found on all of Assateague. After chicks hatch, they follow their parents and feed on the lower beach. Chicks are often at risk from beach traffic, both pedestrian and vehicular, and pairs that nest on popular beaches often have low fledging success. This plover is most successful on those rare remote beaches with few vehicles or beach walkers. Assateague Island offers more of these conditions than most barrier beaches on the East Coast because much of it is undeveloped and roadless parkland. Maryland has a comparatively large Piping Plover population in spite of the limited amount of Atlantic coastline in the state.

The Atlantic Coast Piping Plover population has risen to about 1,743 pairs as of 2006 (USFWS, unpubl. data) and is approaching the recovery goal of 2,000 pairs set for the species. But this has been accomplished through intensive management of popular coastal resort beaches, including efforts to exclude predators, humans and their pets, and vehicles from nesting areas during the summer. If the Piping Plover were to be removed from federal listing under the Endangered Species Act and if states followed suit, removing the birds from local listings, with the cessation of vigilant management, the bird could quickly lose all its gains. Global climate change also could pose a threat from heavy beach erosion caused by more frequent, violent coastal storms and from rising sea levels on the outer coast.

WALTER G ELLISON

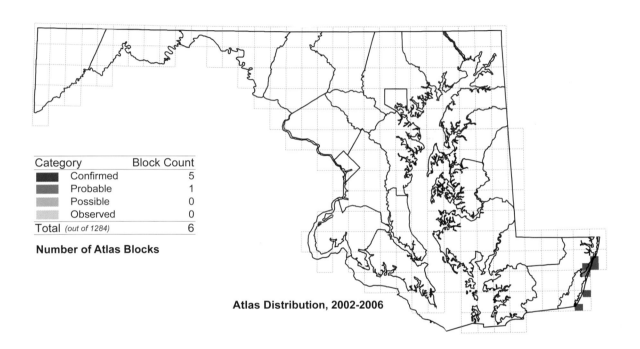

Category	Block Count
Confirmed	5
Probable	1
Possible	0
Observed	0
Total (out of 1284)	6

Number of Atlas Blocks

Atlas Distribution, 2002-2006

Change by Block
- ● Gain from First Atlas to Second
- ▲ Loss from First Atlas to Second

Change in Atlas Distribution, 1983-1987 to 2002-2006

Atlas Region	1983-1987	2002-2006	Change No.	%
Allegheny Mountain	0	0	0	-
Ridge and Valley	0	0	0	-
Piedmont	0	0	0	-
Upper Chesapeake	0	0	0	-
Eastern Shore	4	6	+2	+50%
Western Shore	0	0	0	-
Totals	4	6	+2	+50%

Change in Total Blocks between Atlases by Region

Charles Lentz

Killdeer
Charadrius vociferus

The Killdeer is the most widespread and best known shorebird in Maryland and DC. This plover is named for its keen piercing calls easily rendered as *kill-dee!* This basic utterance is often shortened to a repetitive *dee-dee-dee* or even a single *dee* depending on the bird's level of agitation. Like many other plovers, the Killdeer is boldly patterned, with sharp countershading between dark brown upperparts and white underparts, two black bands across its breast, and a prominent white wing stripe and burnt orange rump visible in flight and during distraction displays. Killdeer are famous for their distraction displays to deflect predators from their nests and offspring, ranging from feigning incubation posture to the spreading of a limp apparently broken wing accompanied by much plaintive trilling. The Killdeer had by far the most DD (distraction display) nesting confirmations (136) among the birds recorded on this project.

Although most Killdeer live within easy flight distance of water, they are less tied to it than most waterbirds. Killdeer require short or sparse vegetation and open treeless spaces, although these can sometimes be rather small openings in an otherwise wooded landscape. Typical haunts for this plover include crop fields before crops grow too high, heavily grazed pastures and stock enclosures, large lawns with gravel roadways, wash flats among dunes, and even suburban malls with nearby lawns or fields (where they nest on graveled rooftops). Killdeer overwinter in eastern Maryland,

becoming progressively scarcer to the west with increasing elevation. Migrants and nesting pairs arrive on Maryland and DC nesting territories from mid-February onward. Egg laying begins in March, and from renests and second broods can continue well into the summer; 65 egg dates for this atlas ranged from 21 March to 29 July. The earliest egg date for Maryland is 8 March (Iliff, Ringler, and Stasz 1996). Downy young Killdeer leave the nest scrape soon after hatching; they do not fly for more than three weeks (B. Jackson and Jackson 2000). Fledglings (FL) was by far the most common confirmation code used for Killdeer in this project (n = 251). Fledged young have been found in Maryland from 8 April (Fletcher and Farrell 1996) to 10 August (this project). An extraordinary report was made of four downy young in Salisbury on 8 November 2004 (J. Juriga; D. and C. Broderick, pers. comm.). November nesting by Killdeer has also been reported in Mississippi, well south of Maryland (B. Jackson and Jackson 2000).

During 2002–2006, Killdeer were reported from most of their haunts during the 1983–1987 atlas (Fletcher and Farrell 1996), with a loss of only 16 blocks from the previous total. But the changes between projects merit comment. Increases in block occupancy were reported from St. Mary's County, presumably because of improved coverage, on Assateague Island, and in lower Eastern Shore blocks dominated by open wetlands. Block losses are most noticeable in the Baltimore to Washington suburban corridor, especially in Baltimore and Harford counties. Other clusters of empty blocks were along the Potomac River in Allegany County and in central and eastern Kent County. Some of these losses may be attributable to changes in the thoroughness of block coverage, but it is possible land use changes may have reduced the amount and quality of nesting habitat. Another factor that might have come into play on suburban lawns and athletic fields is the burgeoning population of grazing Canada Geese, which are large and can be aggressive toward other waterbirds. Killdeer went through a fairly short-lived decline during the period of heavy market hunting of shorebirds from 1905 to 1918 (H. Bailey 1913; B. Jackson and Jackson 2000), and this shorebird is also prone to short-term declines after cold snowy winters on their southeastern U.S. wintering grounds (Robbins, Bystrak, and Geissler 1986).

Current BBS trends for Killdeer suggest it has a relatively stable population in Maryland. However, trends for the Northern Piedmont region that includes Maryland have been downward. The Killdeer currently does not evince the strong declines shown by many birds of agricultural grasslands, but some of the factors affecting other grassland birds may be influencing populations of this adaptable upland plover. In particular, further study seems warranted to examine the possible reasons for the Killdeer's decline on the Piedmont.

WALTER G. ELLISON

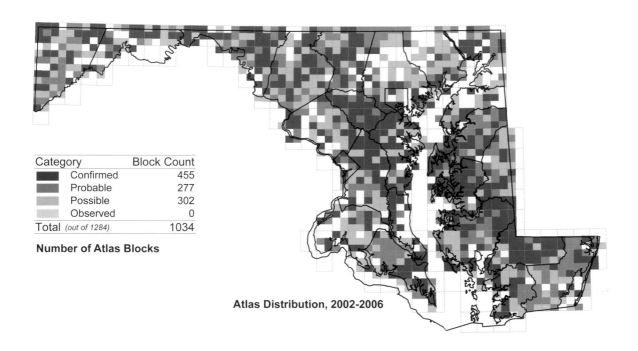

Category	Block Count
Confirmed	455
Probable	277
Possible	302
Observed	0
Total (out of 1284)	1034

Number of Atlas Blocks

Atlas Distribution, 2002-2006

Change by Block
- ● Gain from First Atlas to Second
- ▲ Loss from First Atlas to Second

Change in Atlas Distribution, 1983-1987 to 2002-2006

Percent of Stops
- 50 - 100%
- 10 - 50%
- 0.1 - 10%
- < 0.1%

Relative Abundance from Miniroutes, 2003-2007

Atlas Region	1983-1987	2002-2006	Change No.	%
Allegheny Mountain	84	76	-8	-10%
Ridge and Valley	128	116	-12	-9%
Piedmont	286	253	-33	-12%
Upper Chesapeake	100	91	-9	-9%
Eastern Shore	258	293	+35	+14%
Western Shore	192	203	+11	+6%
Totals	1048	1032	-16	-2%

Change in Total Blocks between Atlases by Region

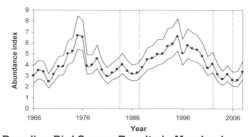

Breeding Bird Survey Results in Maryland

Mark Hoffman

American Oystercatcher
Haematopus palliatus

American Oystercatchers, striking, large, vociferous, black and white shorebirds with conspicuous long orange red bills, usually nest in coastal areas where human intrusion is minimal. They are often observed foraging for mole crabs and other invertebrates along the beach at Ocean City. When nesting they are particularly fond of small predator-free estuarine islands in the coastal bays of Worcester County and along the salt marshes and islands of the lower Chesapeake Bay (Traut et al. 2006). They also nest in the dune systems along the ocean beaches of Assateague Island and formerly nested in similar habitats on Fenwick Island. In Maryland oystercatchers feed on ribbed and blue mussels, marine worms, mole crabs, and other marine bivalves more than they feed on oysters. Although formerly much less abundant up and down the Atlantic coast (Nol and Humphrey 1994) and considered summer residents and migrants in Maryland (Stewart and Robbins 1958), they are now uncommon permanent residents here (Brinker 1996b).

One of the earliest breeding shorebirds in Maryland, American Oystercatchers begin courtship and pair bonding activities during March on small islands in the Chesapeake and coastal bays. Once considered primarily a beach nester, they have adapted to nesting on salt marsh islands (Lauro and Burger 1989), especially on small sand beaches where they can be partially hidden by areas of low salt marsh elder and groundsel shrubs along the edge of the beach (Nol and Humphrey 1994). In Maryland oystercatchers have been using salt marsh islands since at least the mid-1980s (Brinker, pers. obs.) They also nest in the open on the beaches of Assateague and rarely on wrack lines deposited by storm tides on salt marsh islands. Egg laying begins in early to mid-April, with most first clutches established before May. Oystercatcher chicks begin hatching in mid-May and fledge from late June into August.

The 1983–1987 atlas found American Oystercatchers confirmed as breeding in 16 blocks; 10 of these were in the coastal bays of Worcester County and 6 within the Chesapeake Bay (Brinker 1996b). At 84%, the confirmation rate for oystercatchers during the first atlas was relatively high because of a concerted effort to locate nests conducted in conjunction with the Maryland Colonial Waterbird Project. A survey and census of oystercatchers was conducted by the DNR during 2003, in part to provide a comparable effort for the second atlas (Traut et al. 2006). Results from the 2002–2006 atlas document an increase in the number of blocks, with limited expansion of the breeding range northward within the Chesapeake Bay. Only 2 blocks occupied during the first atlas were not reconfirmed during the second atlas and an additional 30 blocks were added, for an increase of nearly 160%. Over the past 20 years, expansion of American Oystercatchers has filled in all blocks with suitable breeding habitat in Worcester County and most of the blocks with suitable habitat in Dorchester and Somerset counties. Since the first atlas, oystercatchers have expanded their breeding range northward into Talbot County where a pair has nested successfully within the Poplar Island restoration project.

Increases in the breeding population of American Oystercatchers have been documented throughout the northeastern United States, and atlas results very similar to those found here in Maryland were documented on Long Island during the second New York atlas project, where a 51% increase was observed (Wasilco 2008a). To the south of Maryland, population trends are reversed, with declines reported from Virginia to Florida (Davis, Simmons, et al. 2001).

The unoccupied habitat that oystercatchers have exploited over the past 20 years in the Chesapeake Bay was recognized during the first atlas project (Brinker 1996b) and the increase in their breeding population in Maryland was not surprising. The comprehensive survey and census of American Oystercatchers conducted by the DNR during 2003 found a total population of 108 pairs, with 57 in the Chesapeake Bay, 39 in the coastal bays, and 12 on Assateague Island (Traut et al. 2006). In the coastal bays, populations may be near carrying capacity as some small islands support multiple nesting pairs (D. Brinker, pers. obs.), and there have been instances of multiple females pairing with a single male, laying two clutches in a single nest scrape (D. Brinker, pers. obs.)

Maryland's breeding population of American Oystercatchers appears to be healthy and secure. If estimates of the breeding population from the mid-1980s are close to correct, then Maryland has seen a 30 to 50 percent increase in oystercatcher numbers since the first atlas. Most American Oystercatchers in Maryland nest in remote locations where interaction with human populations is minimal. Oystercatchers even seem to have adapted to human disturbance, with a number of pairs nesting in close proximity to devel-

opment and recreation areas in Ocean City. Potential foreseeable threats to American Oystercatchers in Maryland are increases in nest predation that would result from increases in predator populations and habitat changes that may accompany sea level rise. At least for the near future, American Oystercatcher populations in Maryland should remain healthy and secure.

DAVID F. BRINKER

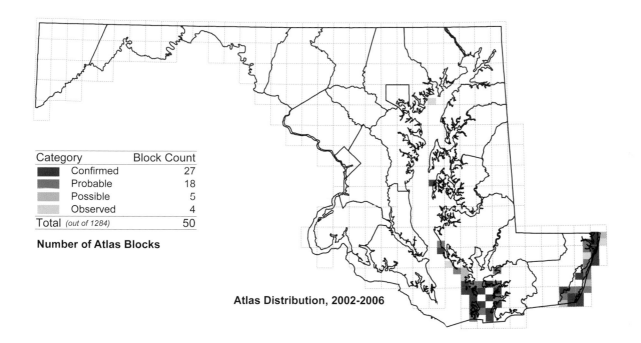

Category	Block Count
Confirmed	27
Probable	18
Possible	5
Observed	4
Total (out of 1284)	50

Number of Atlas Blocks

Atlas Distribution, 2002-2006

Change by Block
- ● Gain from First Atlas to Second
- ▲ Loss from First Atlas to Second

Change in Atlas Distribution, 1983-1987 to 2002-2006

Atlas Region	1983-1987	2002-2006	Change No.	%
Allegheny Mountain	0	0	0	-
Ridge and Valley	0	0	0	-
Piedmont	0	0	0	-
Upper Chesapeake	0	0	0	-
Eastern Shore	19	49	+30	+158%
Western Shore	0	0	0	-
Totals	19	49	+30	+158%

Change in Total Blocks between Atlases by Region

Black-necked Stilt

Himantopus mexicanus

One of fewer than 10 species of shorebirds that breed in Maryland, the distinctively tall and slender Black-necked Stilt has black and white plumage offset by very long pinkish red legs. These characteristics, along with its loud alarm call, place it, along with Willet (*Tringa semipalmata*) and American Oystercatcher (*Haematopus palliatus*), among the large, vociferous shorebirds that breed in Maryland's coastal marshes. In eastern North America, this species occurs primarily in coastal areas, and Maryland is near the northeastern limit of its breeding range. The first documented breeding in Maryland was during the 1983–1987 atlas project (Davidson 1996d).

Kirkwood (1895) provided the only early suggestion that this species belonged to Maryland's avifauna when he wrote that they were once common on the East Coast from Florida north to Maine, although they had become rare by the time he was writing. In 1810 Wilson stated the Black-necked Stilt regularly nested in Cape May County, N.J., and a nest was reported from Egg Island, N.J. (Turnbull 1869). A small population has bred in Delaware coastal marshes since 1964. A rare migrant to Maryland's coastal wetlands, this species was first substantially documented in Maryland in 1967 in Dorchester County (Armistead and Russell 1967). It has been found in this state nearly every year since, with spring records from 23 March to 2 June (Ringler 1977, 1986) and fall migration concentrated in late August. Breeding season observations have been reported in *Maryland Birdlife* from four counties:

Baltimore, Dorchester, Somerset, and Worcester. Up to six birds were found in 1983 at Deal Island WMA in Somerset County (Ringler 1983), and adults were seen with immatures there from 21 July to 6 August 1985 (Ringler 1985). Breeding in Maryland was unequivocally confirmed in 1987 when a nest was discovered in May and downy young located in June (Armistead 1987). Nesting may have also occurred at Blackwater NWR.

Black-necked Stilts in the eastern United States inhabit brackish ponds in salt marshes and fresh to brackish ponds behind coastal beaches (Terres 1980). In Maryland this species favors extensive brackish marshes in areas with relatively stable hydrology, such as inside the Deal Island WMA impoundment and within shallow-water perched or permanent ponds (L. Davidson, pers. obs.). In California clutches averaging 4 to 6 eggs are laid in sparse scrapes, incubation averages 22 to 26 days, and the precocial young can fly about 28 to 32 days after hatching (R. Hamilton 1975). In 2006 Armistead found a nest with eggs on 13 May along Elliott Island Road in Dorchester County. After the atlas period, participants in an MOS conference field trip saw fledged young at Deal Island WMA on 28 July 2007. Stilts extended their range far to the north in 2008, using the man-made impoundments on both major island restoration projects in the Upper Chesapeake Bay with nesting birds at Poplar Island, Talbot County, 24 June to 2 August (J. Reese) and downy young at Hart-Miller Island in July (E. J. Scarpulla; M. Adams and Hafner 2009).

The Black-necked Stilt is among the rarest of Maryland's breeding birds. In the 1983–1987 atlas it was first documented as a breeder in the state and was found in only 3 blocks (Davidson 1996d). During the 2002–2006 atlas, it expanded to 7 blocks, of which 1 was confirmed and 6 were probable; it was recorded as observed in 2 additional blocks. The current atlas recorded the species in the same lower Eastern Shore marshes as nearly 20 years ago, with concentrations at Fishing Bay WMA, Blackwater NWR, and Deal Island WMA. Additional records came from a marsh in Talbot County and one in southern Somerset County.

Reasons for the relatively recent appearance of the Black-necked Stilt in Maryland marshes and subsequent successful breeding are unknown. Maryland's birds may have arrived from the small population breeding in Delaware or from farther south. The southeastern breeding population in southern Georgia, Florida, and along the Gulf Coast has had an in-

creasing trend on Breeding Bird Survey routes of more than 1.5% per year from 1966 to 1993. This species may be taking advantage of increasing populations and temperature shifts in more northerly latitudes owing to global warming to expand its breeding population farther north. Within a few decades, however, additional effects of global warming, especially sea level rise, could inundate the lower Chesapeake Bay tidal marshes to the point where this species may no longer find breeding habitat in Maryland except in managed impoundments.

LYNN M. DAVIDSON

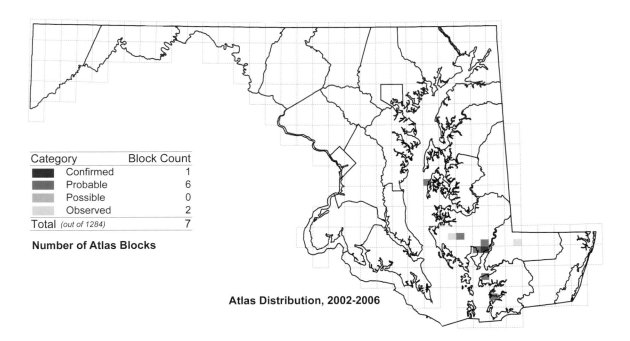

Category	Block Count
Confirmed	1
Probable	6
Possible	0
Observed	2
Total (out of 1284)	7

Number of Atlas Blocks

Atlas Distribution, 2002-2006

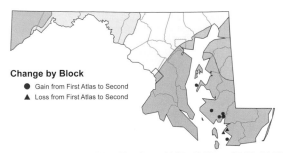

Change by Block
- ● Gain from First Atlas to Second
- ▲ Loss from First Atlas to Second

Change in Atlas Distribution, 1983-1987 to 2002-2006

			Change	
Atlas Region	1983-1987	2002-2006	No.	%
Allegheny Mountain	0	0	0	-
Ridge and Valley	0	0	0	-
Piedmont	0	0	0	-
Upper Chesapeake	0	0	0	-
Eastern Shore	3	7	+4	+133%
Western Shore	0	0	0	-
Totals	3	7	+4	+133%

Change in Total Blocks between Atlases by Region

Gary Van Velsir

Spotted Sandpiper
Actitis macularius

The small, medium brown Spotted Sandpiper, with its distinctive bobbing hindquarters, its stiff-winged fluttering flight skimming low over water, and its plaintive sweet *peet-weet* alarm call is one of North America's most widely distributed and recognizable shorebirds. Although common during migration in Maryland and DC, they are scarce but broadly distributed breeding birds close to the southern edge of their broad nesting range. As with many other sandpipers, this bird has an unusual breeding biology. Females are emancipated from almost all incubation and care of the downy young, leaving them free to mate with different males in succession. This mating system seems to be the upshot of ephemeral nesting habitat prone to flooding and high rates of nest predation (Oring, Lank, and Maxson 1983).

Although Spotted Sandpipers may occur almost anywhere there is water with sufficient shoreline for feeding, their nesting habitat requirements are less flexible. The nest is always on the ground in sparsely vegetated flat places generally some distance from water. Nesting sites include sandbars and gravel bars in rivers; stony or pebbly lake and pond shores; crop fields near farm ponds and streams; gravel pits with ponds; dredge spoil near water; and sparsely vegetated islands. Maryland and DC do not lack for streams and ponds, but many of these do not have nest sites or are prone to regular human disturbance.

The nonbreeding range of the Spotted Sandpiper is at least as broad as its breeding range; they winter from the southern tier of states south to southern South America (Oring, Gray,

and Reed 1997). Spring migrants reappear in Maryland and DC in April, and in the autumn the last are usually seen in October. Northbound migrants may be present until early June and southbound birds are on the move by early July, so this sandpiper has a safe date window of only three weeks during June. This short window may have excluded a few true nesting birds, but the resulting map is reassuringly conservative. Given the low number of these shorebirds breeding here, it seems likely that few Maryland and DC birds are polyandrous, and many females probably help their mates raise their offspring. Maryland and DC egg dates range from 10 May to 15 July (Stewart and Robbins 1958; Ringler 1996b; MNRF). Downy young have been seen as late as 21 July 2006 on the lower Patapsco River northwest of Linthicum, Anne Arundel County (S. Arnold, this project) and 21 July 1991 at Hart-Miller Island.

Early authorities may have confused migrants with nesting birds because many referred to Spotted Sandpiper as a common summer resident and nester (Coues and Prentiss 1883; Kirkwood 1895; Eifrig 1904). C. Richmond's (1888) assessment of them as rather uncommon breeders around DC appears to be closer to the current status of the species. Stewart and Robbins (1958) noted that they were uncommon in most of Maryland but considered them fairly common nesters in tidewater Maryland. This last statement was partly based on the observations of R. Jackson (1941) in Dorchester County. Much of the tidal habitat used then may have been lost to rising sea levels.

In 1983–1987 Spotted Sandpipers were reported from all physiographic regions, but they were very thinly distributed, with breeding codes assigned to only 53 blocks (Ringler 1996b). They were also scarce in tidewater areas, although such records did represent nearly 38% of all blocks with breeding reports (Ringler 1996b). From 2002 to 2006 the overall distribution was very similar, but there were far more reports of nonbreeding birds given the observed code. Reports in tidewater blocks fell to a little more than 27% of the blocks with records. Oddly, few were reported along the Potomac River, which appears to have much suitable nesting habitat.

The BBS does not sample stream sides and ponds very well; Maryland has insufficient data to produce a trend estimate. Overall, this bird's population has apparently been fairly stable since 1966 but has shown declines since 1980.

They appear to be declining rather rapidly in eastern North America because of habitat loss and disturbance. Keys to retaining the Spotted Sandpiper in Maryland and DC are clean water both inland and on Chesapeake Bay, and maintenance of undisturbed river bars, islands, and shorelines. The many small farm ponds in the state might aid this shorebird and it is uncertain if they were adequately surveyed during this project because many ponds are far from roads and on private property.

WALTER G. ELLISON

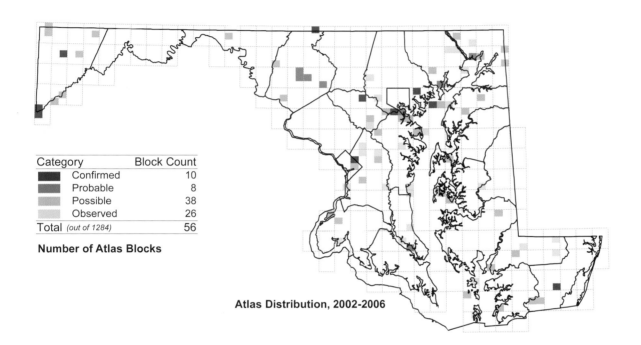

Category	Block Count
■ Confirmed	10
■ Probable	8
■ Possible	38
■ Observed	26
Total *(out of 1284)*	56

Number of Atlas Blocks

Atlas Distribution, 2002-2006

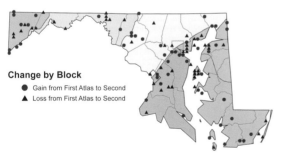

Change by Block
- ● Gain from First Atlas to Second
- ▲ Loss from First Atlas to Second

Change in Atlas Distribution, 1983-1987 to 2002-2006

Atlas Region	1983-1987	2002-2006	Change No.	%
Allegheny Mountain	5	9	+4	+80%
Ridge and Valley	9	2	-7	-78%
Piedmont	7	13	+6	+86%
Upper Chesapeake	12	7	-5	-42%
Eastern Shore	10	11	+1	+10%
Western Shore	10	14	+4	+40%
Totals	53	56	+3	+6%

Change in Total Blocks between Atlases by Region

Willet
Tringa semipalmata

The Willet is one of the three large, vociferous shore-birds that breed in Maryland's coastal marshes. Although it is mostly gray above, it flashes a bold black and white wing pattern as it flies, and its raucous alarm call, from which its name is derived, can be heard long after the threat to its nest has retreated. It defends nesting territories that include several types of habitat components, primarily salt marsh meadows or "high marsh" dominated by saltmeadow cordgrass, sometimes interspersed with patches of taller vegetation, and containing areas of open water, such as tidal creeks or shallow, permanent ponds. Other habitats include sandy areas with beach grass or with other types of vegetation above the normal tidal zone that include a mixture of grasses and shrubs, such as wax myrtle and marsh elder (Bent 1929; Stewart and Robbins 1958; Howe 1982).

The breeding season spans from April when birds arrive from wintering areas in Latin America and southern Atlantic states (AOU 1998) into August when first the adults and then the juveniles depart for the winter (Robbins and Bystrak 1977). Birds of the inland-breeding and slightly larger western subspecies of Willet are sometimes found from late summer into winter in coastal areas (Tomkins 1955, 1965). Nests are depressions lined with vegetation; they are usually well hidden within marsh grasses but infrequently may be on bare sand (Bent 1929). Eggs are usually laid 1.5 to 2 weeks after arrival on breeding grounds (Howe 1982), and egg dates for 67 Maryland nests range from 23 April to 12 July, with a peak from 1 to 12 June (MNRF). Thirteen records of nests with eggs from 2002–2006 ranged from 14 May to 24 June, both extreme dates reported by M. Hoffman. Incubation of an average clutch of 3 or 4 eggs lasts about 25 days, and the precocial young leave the nest within a day (Bent 1929; Howe 1982). Downy young have been reported in Maryland from 5 June to 13 July (MNRF). During this atlas fledged young were seen to 10 August (S. Arnold). The female abandons the brood within 2 to 3 weeks after hatching and the male follows suit in another 2 weeks (Ehrlich et al. 1988).

According to Wilson in 1832, the Willet bred "in great numbers" in Maryland and nearby coastal states (Bent 1929). But not until 1904 was the first Maryland nest documented (Hampe and Kolb 1947). In the Northeast, this game bird was hunted and eggs were harvested for food to the point where it appeared "destined to disappear" from this part of its range (Bent 1929). Although the Willet is still considered a game species and is regulated as such by the USFWS under the Migratory Bird Treaty Act, it is no longer hunted and its population in the Northeast seems to have recovered. By mid-twentieth century it was considered locally common in tidewater areas of the lower three Chesapeake Bay counties (Dorchester, Wicomico, and Somerset) and uncommon in coastal Worcester County (Stewart and Robbins 1958).

During the 1983–1987 breeding bird atlas, the Willet was found in 8 counties (O'Brien 1996a), with most of its atlas blocks in the same lower four counties as reported by Stewart and Robbins (1958) and a few scattered records in Kent, Queen Anne's, Talbot, and St. Mary's counties. The number of blocks dropped 14%, from 100 in the first atlas to 85 in the second. During the 2002–2006 atlas, no blocks were documented on the Western Shore, and 1 block in Talbot County was the only area outside its core range of the lower four counties. Even within its core area there was a net loss of 10 blocks. The BBS trend data show a fairly stable population for eastern North America from 1966 to 1989 and for the United States from 1980 to 2007, but the data are too few to estimate an accurate Maryland trend.

Salt marshes in the lower Chesapeake Bay and coastal bays of Worcester County are highly threatened by many factors, including fragmentation and hydrologic changes from ditching, inundation and erosion from subsidence and from sea level rise induced by global climate change, and invasion and

conversion of short grass meadows by phragmites. Invasion by phragmites has been reported to negatively impact this species (Benoit and Askins 1999). Of particular interest and concern is the impact that prescribed fire and illegal burning of some lower Eastern Shore marshes may have on early spring food availability and suitability of nesting habitat. As sea levels continue to rise and the duration, frequency, and depth of flooding events increase, the Willet may no longer be able to successfully reproduce without the restoration and proper management of high marsh meadows.

LYNN M. DAVIDSON

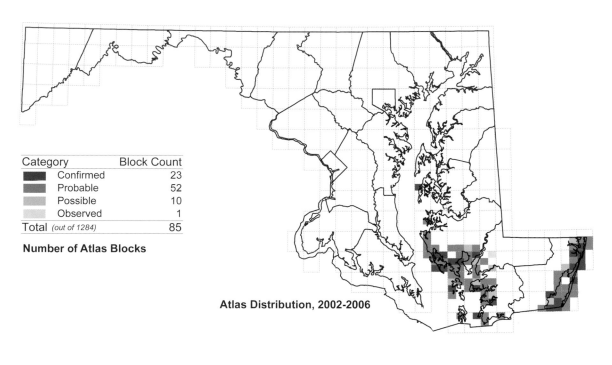

Category	Block Count
Confirmed	23
Probable	52
Possible	10
Observed	1
Total (out of 1284)	85

Number of Atlas Blocks

Atlas Distribution, 2002-2006

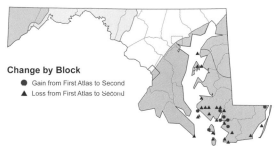

Change by Block
- ● Gain from First Atlas to Second
- ▲ Loss from First Atlas to Second

Change in Atlas Distribution, 1983-1987 to 2002-2006

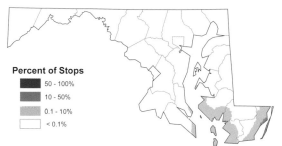

Percent of Stops
- 50 - 100%
- 10 - 50%
- 0.1 - 10%
- < 0.1%

Relative Abundance from Miniroutes, 2003-2007

Atlas Region	1983-1987	2002-2006	Change No.	%
Allegheny Mountain	0	0	0	-
Ridge and Valley	0	0	0	-
Piedmont	0	0	0	-
Upper Chesapeake	0	0	0	-
Eastern Shore	98	85	-13	-13%
Western Shore	1	0	-1	-100%
Totals	99	85	-14	-14%

Change in Total Blocks between Atlases by Region

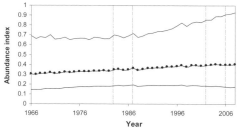

Breeding Bird Survey Results in Maryland

Upland Sandpiper
Bartramia longicauda

One of the most memorable experiences in birding is hearing the rising whistle of a flight-displaying Upland Sandpiper, then catching sight of him high in the air circling with shallow rapid wing-beats. This tall, slender, long-tailed grassland shorebird is elegant and graceful. It uses elevated perches such as fence posts, hay bales, and power poles to survey the open landscapes it inhabits. It has become increasingly rare as farmland disappears and as farming has moved away from hay and dairy farming to crops and silage feeding for cattle. Populations of this species in eastern North America are in a precarious state. They were found in only a single block during this atlas project and are listed by the DNR as endangered in Maryland.

Upland Sandpipers live in extensively open lands dominated by grass and forbs with few shrubs. They require a mix of tall and short grass for nesting and feeding, respectively. They appear to be best served by a combination of well-cropped pastures and large hayfields, and they are usually found in gently rolling or flat lands.

As with many Northern Hemisphere shorebirds, the Upland Sandpiper is a long-distance migrant with a two-month breeding season that leaves time to raise only a single brood. They winter in South America as far south as the Argentine Pampas, well south of the equator. They return to Maryland in April and May and begin leaving by July. Some presumably failed breeders from the northern part of the range may appear in Maryland by early July. The nest is a loosely arranged grass-lined depression in tall grass with good overhead cover

(Houston and Bowen 2001). Downy young leave the nest soon after hatching. Meanley (1943) reported nesting dates for a Baltimore County population. He found eggs from 10 May to 10 June and observed downy flightless young from 25 May to 21 June. The lone confirmation for this atlas project cited dependent flying young on 29 June 2005 south of Oakland, Garrett County (F. Pope).

Maryland is at the southeastern edge of the Upland Sandpiper's nesting range. They have also nested on a few occasions in northern and central Delaware (Hess et al. 2000) and have nested in northern Virginia south of Frederick County (S. A. Smith 1996c). In the past they were far more widespread in Maryland, at least until the late 1950s. Stewart and Robbins (1958) considered them fairly common but localized, with nesting in the Worthington Valley, Baltimore County; western Montgomery County, the Frederick Valley, the Hagerstown Valley, and in Garrett County. By the 1983–1987 atlas project they had become restricted to southwestern Garrett County, where they were found in 5 blocks with 3 nesting confirmations (S. A. Smith 1996c). The last breeding season report from the Hagerstown Valley was in 1965 (Robbins and Boone 1985), and the last nesting season reports from Montgomery and Frederick counties were from 1983 (S. A. Smith 1996c). There were only two breeding season reports for 2002–2006, the breeding birds near Oakland and a bird prudently listed as observed at Alpha Ridge Landfill, Howard County, 19 June 2004.

The 40-year North American BBS trend for the Upland Sandpiper is stable, but there has been a modest 0.9% per year decline continent-wide since 1980. The decline has been much steeper in eastern North America at 2.5% per year. Changes in farming in Maryland have not been kind to this prairie shorebird. There has been a long-term loss of hayfield

acreage; pastureland has also declined; hay-cutting has become more frequent; and crop fields have replaced meadows.
The combined effect of these changes has reduced available
cover and foraging habitat for this species. At present the future seems bleak for the Upland Sandpiper in Maryland and
the rest of northeastern North America.

WALTER G. ELLISON

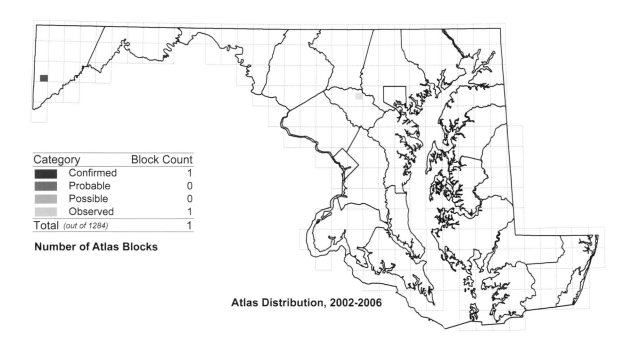

Category	Block Count
Confirmed	1
Probable	0
Possible	0
Observed	1
Total (out of 1284)	1

Number of Atlas Blocks

Atlas Distribution, 2002-2006

Change by Block
- ● Gain from First Atlas to Second
- ▲ Loss from First Atlas to Second

Change in Atlas Distribution, 1983-1987 to 2002-2006

Atlas Region	1983-1987	2002-2006	Change No.	%
Allegheny Mountain	5	1	-4	-80%
Ridge and Valley	1	0	-1	-100%
Piedmont	1	0	-1	-100%
Upper Chesapeake	0	0	0	-
Eastern Shore	0	0	0	-
Western Shore	0	0	0	-
Totals	7	1	-6	-86%

Change in Total Blocks between Atlases by Region

George M. Jett

American Woodcock

Scolopax minor

The American Woodcock has one of the most elaborate and spectacular displays of Maryland's nesting birds. Most birders have their first experience with this portly upland sandpiper during a chilly early spring dusk when they hear a mysterious buzzy *peeent* emanating from the gloom over a damp meadow. Suddenly a rotund twittering form shoots upward against the darkening sky and continues to flutter aloft, twittering all the time, until it starts intermittent chirping. Silence for several breathless seconds and then the *peeent* starts again from the same spot. Woodcocks live in wet woodlands and thickets adjacent to grassy openings. The bird has a bizarre appearance with its short legs set far back, large eyes set high and well back on the head, and a long bill. Because they flush at the last second and fly evasively, woodcocks are popular game birds. The American Woodcock is widespread in Maryland but it is generally scarce and local and it has been declining for several decades.

American Woodcock habitat combines wet to moist woods with understory shrubs, thickets of young trees and shrubs on forest edges, and grassy open ground for males to perform their displays. Woodcocks roost by day and feed on earthworms and other soil invertebrates at night in muddy ground (Keppie and Whiting 1994). Nests are built on the ground in dense stands of shrubs or saplings.

American Woodcocks are resident in small numbers in most winters on the Coastal Plain with a few westward through the eastern Ridge and Valley region. Prolonged freezes drive woodcocks south of Maryland and DC. This is one of Maryland's earliest spring migrants, usually arriving in February. Males display as soon as they return; migrating males headed north of Maryland will also display. It is unclear if any migratory males mate with Maryland females, but this species is promiscuous, with males mating with multiple partners and uninvolved with parental care (McAuley, Longcore, and Sepik 1993). Migration does not end until early April, therefore safe dates for excluding migrants extend to mid-April—five days later in 2002–2006 than in 1983–1987. These dates probably excluded some genuine nesters because displays begin to decline after the end of March. Almost all males cease display by mid-May. The early nesting season and the need for dawn and dusk coverage also lowered the number of block records. Woodcock eggs have been found in Maryland and DC from 25 February to 25 May (Mathews 1996c; MNRF).

Numbers of American Woodcock appear to have varied across Maryland at the end of the nineteenth century. Kirkwood (1895) called them fairly common despite June and July hunting, and Eifrig (1904) called them common in Allegany and Garrett counties. Stewart and Robbins (1958) found them fairly common but localized in most of Maryland, but uncommon on the Piedmont. Woodcocks were found in nearly a third of the blocks in the 1983–1987 atlas, with the highest levels of occurrence on the Western Shore and in the Upper Chesapeake region and good numbers of blocks along Chesapeake Bay and in the Patuxent River valley (Mathews 1996c). Notably few occupied blocks were found in the heavily farmed Frederick and Hagerstown valleys.

American Woodcocks suffered a 48% net reduction in occupied blocks in the 2002–2006 atlas. The Western Shore still has the highest percentage of its blocks with this sandpiper (23%), with many blocks along the Patuxent River, but it declined by 46% there. Large losses were also noted on the Eastern Shore, in the Upper Chesapeake, and on the Piedmont. Woodcock records actually increased substantially in the Allegheny Mountains, possibly because of improved coverage.

American Woodcocks display early in spring and are poorly sampled by the largely daytime BBS routes. The U.S. Fish and Wildlife Service has conducted volunteer singing ground counts of woodcocks since 1968. Woodcocks are far less numerous on Maryland song count surveys than in states along the Canadian border. Woodcocks have shown a significant 1.2% per year long-term decline from 1968 to 2008 in the eastern seaboard states, and they have declined on the few Maryland routes at a rate of 4% per year (T. Coo-

per et al. 2008). Over the past 10 years, the decline appears to have bottomed out along the eastern seaboard, at least to the north of Maryland (T. Cooper et al. 2008). These declines appear to be largely attributable to loss of habitat, especially successional shrublands, to reforestation, and to development. Disturbance and predation may also be contributing factors in urbanizing regions. Forests should be managed for uneven age with occasional openings for display grounds and to regenerate nesting thickets. It would be a great shame to lose the annual sky-dance of this unique sandpiper.

WALTER G. ELLISON

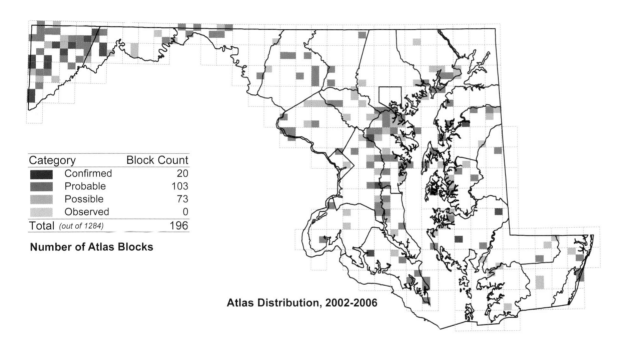

Category	Block Count
Confirmed	20
Probable	103
Possible	73
Observed	0
Total (out of 1284)	196

Number of Atlas Blocks

Atlas Distribution, 2002-2006

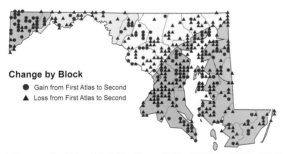

Change by Block
- Gain from First Atlas to Second
▲ Loss from First Atlas to Second

Change in Atlas Distribution, 1983-1987 to 2002-2006

Atlas Region	1983-1987	2002-2006	Change No.	%
Allegheny Mountain	24	39	+15	+63%
Ridge and Valley	39	20	-19	-49%
Piedmont	79	28	-51	-65%
Upper Chesapeake	44	17	-27	-61%
Eastern Shore	84	33	-51	-61%
Western Shore	108	58	-50	-46%
Totals	378	195	-183	-48%

Change in Total Blocks between Atlases by Region

Laughing Gull
Leucophaeus atricilla

Although the Laughing Gull is the common gull of the Delmarva and mid Atlantic coast during summer, it is an infrequent to rare breeding bird in Maryland. A notable characteristic of Laughing Gulls is the large breeding colonies (up to 25,000 pairs) that they often form (Burger 1996). This handsome dark-headed gull is commonly seen foraging for worms and invertebrates in flooded agricultural fields across the Delmarva, and they frequently approach beachgoers to scavenge stray scraps of food. Laughing Gulls are also often observed kiting insects high over farm fields as one travels the Eastern Shore. As with other gull species, Laughing Gull populations were seriously reduced by demands of the millinery trade and by egging prior to the passage of the Migratory Bird Treaty (Wasilco 2008b). Populations have since recovered along the Atlantic Coast, and in the United States these birds now breed from Maine to Florida and the Gulf of Mexico (Burger 1996). The Laughing Gull is a very common breeding species with large populations on either side of Maryland in nearby New Jersey and Virginia. During the late fall, Laughing Gulls depart Maryland for the warmer climates of the Gulf of Mexico, the Caribbean, and northern South America (Burger 1996).

In mid-April Laughing Gulls return to the Delmarva Peninsula and estuarine waters along Maryland's coastal beaches. When Laughing Gulls nest in Maryland, they choose small predator-free coastal and estuarine salt marsh islands, where they construct a nest of grass and other fibrous vegetative material that in most cases is built high enough to escape all but extreme storm tides. With the expansion of breeding Herring Gulls into Maryland, Laughing Gulls have modified their nest placement behavior. In Maryland, Laughing

Gull nests are now often located in dense patches of marsh elder in higher portions of the salt marsh to provide better cover from the larger gull species. The Laughing Gull is the latest-nesting gull in Maryland; egg laying begins in mid- to late May, with most clutches established by the first week of June. Clutch size varies from 1 to 3 eggs, with 2 or 3 the most commonly observed clutch size. Chicks begin hatching in mid-June and fledge during late July or early August.

During the first atlas Laughing Gulls were confirmed as breeding in 3 blocks, all in the coastal bays of Worcester County (Brinker 1996c). During both atlases Laughing Gulls were widely recorded as observed in blocks throughout the Eastern and Western Shore regions with fewer occurrences in the Upper Chesapeake region. The number of blocks where Laughing Gulls were recorded as observed during the second atlas increased 14% over the first atlas. From 2002 through 2006, Laughing Gulls were recorded as confirmed in 2 blocks, with 1 of these in the Chesapeake Bay. But none of the confirmations for Laughing Gulls were the result of locating breeding colonies of significant size (greater than 100 pairs). During the second atlas Laughing Gull breeding was confirmed at Reedy Island and South Goose Pond Islands in Worcester County and Western Island in Somerset County.

The nesting population of Laughing Gulls in Maryland ranged from 72 to 792 breeding pairs spread among 6 colony sites with only 3 colony sites active in any one year (DNR, unpubl. data) during the first atlas period. During the 1980s, relatively large colonies of Laughing Gulls used Bridge Island and Gray's Cove Island in Worcester County. These 2 colony sites have not been used by breeding colonies of Laughing Gulls since the late 1980s (DNR, unpubl. data). From 2002 through 2006, the number of breeding pairs of Laughing Gulls in Maryland ranged from 2 to 20, generally occupying only 1 site during any breeding season (DNR, unpubl. data). An analysis of Chesapeake Bay and Coastal Delmarva seabird populations found that the Laughing Gull population has remained stable on the Delmarva at approximately 45,000 breeding pairs between 1993 and 2003, with most breeding occurring on coastal barrier islands in Virginia (Brinker, Williams, and Watts 2007).

Laughing Gulls are known for their high annual turnover of colony sites (Burger 1996). Between 1985 and 2006 Laughing Gull nesting colonies have formed at 15 different sites during 16 of these 22 breeding seasons (DNR, unpubl. data). In only 4 of those years did the colony size exceed 200 breeding pairs, and during 10 seasons fewer than 25 pairs bred in Maryland. In Maryland, Laughing Gulls are now restricted to small salt marsh islands barely above high tides, and nest-

ing often fails as a result of storm-driven tides. The potential impact on Laughing Gulls from continued sea level rise is serious (Erwin, Sanders, et al. 2006; Brinker, Williams, and Watts 2007). The future for Laughing Gulls as a breeding species in Maryland will continue to be characterized by infrequent formation of colonies numbering a few hundred pairs, occasional solitary nesting by isolated individuals, and small colonies of fewer than 50 pairs. In the long term, loss of salt marshes to sea level rise could further reduce the potential for Laughing Gulls to continue nesting in the Free State.

DAVID F. BRINKER

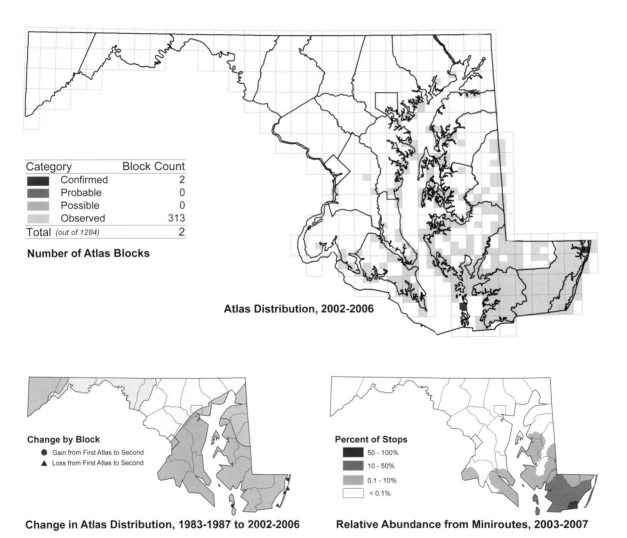

Category	Block Count
Confirmed	2
Probable	0
Possible	0
Observed	313
Total (out of 1284)	2

Number of Atlas Blocks

Atlas Distribution, 2002-2006

Change by Block
● Gain from First Atlas to Second
▲ Loss from First Atlas to Second

Change in Atlas Distribution, 1983-1987 to 2002-2006

Percent of Stops
50 - 100%
10 - 50%
0.1 - 10%
< 0.1%

Relative Abundance from Miniroutes, 2003-2007

Atlas Region	1983-1987	2002-2006	Change No.	%
Allegheny Mountain	0	0	0	-
Ridge and Valley	0	0	0	-
Piedmont	0	0	0	-
Upper Chesapeake	0	0	0	-
Eastern Shore	3	2	-1	-33%
Western Shore	0	0	0	-
Totals	3	2	-1	-33%

Change in Total Blocks between Atlases by Region

George M. Jett

Herring Gull
Larus argentatus

Now the most abundant breeding gull in Maryland, Herring Gulls were nearly extirpated along the Atlantic coast as a result of persecution by gunners and egg collectors in the 1800s (Pierotti and Good 1994; M. Richmond 2008). During the mid-1900s Laughing Gulls were the most abundant nesting gull in Maryland (Stewart and Robbins 1958) and Herring Gulls were only common winter residents. Herring Gulls are highly adaptable and are undeterred by human development. Once the downward pressure from persecution and collecting was removed by passage of the Migratory Bird Treaty, Herring Gulls began to reoccupy coastal areas of the Northeast and to prosper. In 1955 the first observation of a nesting Herring Gull in Maryland was recorded from Sharps Island in Talbot County (Kleen 1956). Since then breeding Herring Gulls have expanded rapidly in Maryland. By the first atlas the number of breeding Herring Gulls in Maryland exceeded the number of breeding Laughing Gulls in the state (Brinker 1996d). Herring Gulls, which continued to expand southward to the Carolinas, are now a secure nesting species throughout the northeastern and middle Atlantic states (Pierotti and Good 1994).

In early March, Herring Gulls begin gathering at their breeding colonies throughout the Chesapeake and coastal bays. They prefer nesting on salt marsh islands, especially in denser areas of vegetation, particularly under low salt marsh elder and groundsel shrubs. They will also nest in open salt marsh cordgrass, on exposed salt pans, in dredge material containment facilities, on channel markers, on old barges (Poplar Island), and a variety of other places. In 2008 Herring Gulls were recorded nesting on a building roof near the Patapsco River in Baltimore City (D. Brinker, pers. obs.). Egg

laying begins in mid-April, with most clutches established by the middle of May. Herring Gull chicks begin hatching in mid-May and fledge from late June into August.

During the first atlas Herring Gulls were confirmed as breeding in 13 blocks; 8 of these were in the coastal bays of Worcester County and 5 within the Chesapeake Bay (Brinker 1996d). Herring Gulls were recorded as observed in slightly more blocks during the first atlas than the second, but the distribution was similar in both atlases. The number of confirmed blocks increased to 21 in the 2002–2006 atlas, but there was no significant change in statewide distribution. Colonial waterbird breeding colonies are not static, and the social group referred to as the breeding colony will readily move to alternate breeding sites should a colony's habitat conditions deteriorate. For example, the gull colony that was breeding on the slag pile at Bethlehem Steel's Sparrows Point foundry moved to Fort Carroll between the two atlas periods. The large gull colony occupying the Easter Point dredged material facility on Smith Island dispersed and settled into a number of widespread smaller colonies when new spoils were deposited at Easter Point during the 1990s.

The nesting population of Herring Gulls in Maryland appears to have stabilized, possibly having reached carrying capacity for the Chesapeake and coastal bays. The breeding population of Herring Gulls in Maryland during the late 1980s was approximately 4,000–4,500 pairs (DNR, unpubl. data). There has been a decline since then with population estimates of approximately 2,500–3,000 breeding pairs during the second atlas period (DNR, unpubl. data). Although BBS data do not monitor Herring Gull population changes well, there is an indication of Herring Gull declines in BBS data. BBS trends for Herring Gulls show slight declines that are significant at the continental level (–5.4, $P = 0.01$) and a nonsignificant decline (–2.1, $P = 0.16$) for the United States. When analyzed at the state level BBS data shows a nonsignificant ($P = 0.07$) increase of 11.2% in Maryland. The changes in Herring Gull breeding population are not limited to Maryland; a recent analysis of Chesapeake Bay and coastal Delmarva populations found a 32% decline in the number of breeding Herring Gulls from the peak census of 10,931 breeding pairs in 1993 to 7,484 pairs in 2003 with breeding spread among a greater number of colonies in 2003 than 1993 (Brinker, Williams, and Watts 2007).

Over the next 20 years it will be interesting to see if Herring Gull breeding populations stabilize or continue slowly declining. Loss of breeding habitat for colonial nesting waterbirds to shoreline and island erosion continues. Most Herring Gull nests are situated barely above high tide level

and continued sea level rise is perhaps the most serious environmental change to which Herring Gulls will need to adjust over the next 100 years (Erwin, Sanders, et al. 2006; Brinker, Williams, and Watts 2007). Herring Gull is a good example of an adaptable species that recovered well and expanded in response to the protection afforded by the Migratory Bird Treaty. But Herring Gulls may have reached their peak population levels in the late 1900s, especially if sea level rise is on the high end of the predictions and results in major loss of salt marsh islands along the Atlantic coast over the next hundred years.

DAVID F. BRINKER

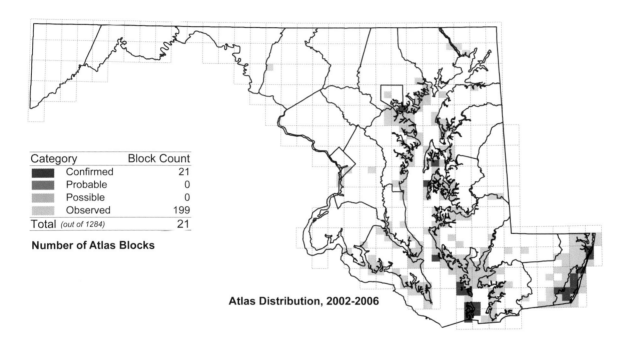

Category	Block Count
■ Confirmed	21
■ Probable	0
■ Possible	0
■ Observed	199
Total *(out of 1284)*	21

Number of Atlas Blocks

Atlas Distribution, 2002-2006

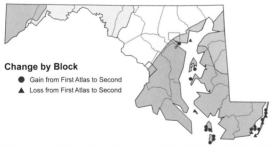

Change by Block

● Gain from First Atlas to Second
▲ Loss from First Atlas to Second

Change in Atlas Distribution, 1983-1987 to 2002-2006

Atlas Region	1983-1987	2002-2006	Change No.	%
Allegheny Mountain	0	0	0	-
Ridge and Valley	0	0	0	-
Piedmont	0	0	0	-
Upper Chesapeake	1	1	0	0%
Eastern Shore	12	20	+8	+67%
Western Shore	0	0	0	-
Totals	13	21	+8	+62%

Change in Total Blocks between Atlases by Region

Great Black-backed Gull
Larus marinus

One of the largest gulls in the world, the Great Black-backed is also one of the most aggressive coastal nonraptorial avian predators. These birds take a wide variety of prey, including large marine invertebrates, fish and mammals, as well as adults, eggs, and chicks of other seabirds and waterfowl (Good 1998). They also scavenge on carrion and garbage. Great Black-backed Gulls were temporarily extirpated along the East Coast as a result of persecution by gunners and egg collectors in the late 1800s (Good 1998). Now they are thriving, while learning to coexist with human populations throughout the northeastern United States. After human persecution ended in the early 1900s, Great Black-backed Gulls expanded south from the Canadian Maritimes. The first recorded nesting in Maryland was documented in Worcester County in 1972 (D. Boone 1975). Breeding Great Black-backed Gulls have also expanded as far west in the Great Lakes as Wisconsin, where their first nesting occurred at Spider Island in 1994 (Tessen 1994). Since first breeding in Maryland, Great Black-backed Gulls have continued to expand, even evicting Herring Gulls from some smaller traditional colony sites and converting them to sites occupied exclusively by nesting Great Black-backed Gulls.

The earliest breeding gull in Maryland, Great Black-backed Gulls begin gathering during March at their breeding colonies throughout the Chesapeake and coastal bays. They prefer nesting on salt marsh islands, especially in denser areas of vegetation under low salt marsh elder and groundsel shrubs. They will also nest in open salt marsh cordgrass, on exposed salt pans, in dredge material containment facilities, on channel markers, on old barges (Poplar Island), and a variety of other places. Egg laying begins in early to mid-April, with most clutches established by the middle of May. Great Black-backed Gull chicks begin hatching in mid-May and fledge from late June into August.

During the first atlas Great Black-backed Gulls were confirmed as breeding in 9 blocks; 5 of these were in the coastal bays of Worcester County and 4 within the Chesapeake Bay (Brinker 1996e). Great Black-backed Gulls were reported as observed in 72 blocks during the first atlas and 121 in the second, an increase of 68%. The distribution of blocks they were observed in was similar in both atlas efforts. The number of confirmed blocks increased 156% to 23 in the 2002–2006 atlas, and only 2 blocks that had Great Black-backed Gull colonies in the first atlas did not have active colonies during the second atlas period. The newly occupied blocks were the result of breeding population expansion in Worcester and Somerset counties. Between the two atlas periods, the Great Black-backed Gull colony on Hart-Miller Island moved to Fort Carroll near the Key Bridge. A colony made up exclusively of Great Black-backed Gulls now occupies the islands that support the towers for the eastbound lanes of the Chesapeake Bay bridge. Great Black-backed Gulls have driven the Common Tern colony from Bodkin Island and now exclusively occupy the western end of the island. Some small islands in Worcester and Somerset counties also support Great Black-backed nesting colonies.

The nesting population of Great Black-backed Gulls in the Maryland portion of the Chesapeake and coastal bays has increased from 100–150 breeding pairs during the late 1980s to approximately 560 pairs spread among 29 colonies during 2003 (DNR, unpubl. data). BBS data do not monitor

Great Black-backed Gull population changes well nor show any significant trends. The increase in Great Black-backed Gull breeding population is not limited to Maryland. A recent analysis of Chesapeake Bay and Coastal Delmarva populations found a 195% increase from a census of 600 breeding pairs in 1993 to 1,770 pairs in 2003 with about a third of the breeding population in Maryland (Brinker, Williams, and Watts 2007).

Since the first atlas Great Black-backed Gulls have continued their population expansion in Maryland. Over the next 20 years it will be interesting to see if Great Black-backed Gull breeding populations stabilize or continue increasing.

Great Black-backed Gull population expansion may have played a role in the small decline of Maryland's Herring Gull population that has occurred over the past 20 years. Great Black-backed Gulls are a significant, aggressive addition to the Chesapeake and coastal bays' avifauna that will have impacts on the food web and other bird species. Localized reduction of Great Black-backed Gulls to reduce impacts to other species has occurred in New England (Good 1998). If they continue to increase in Maryland, managers may have to implement localized control in the Free State as well.

DAVID F. BRINKER

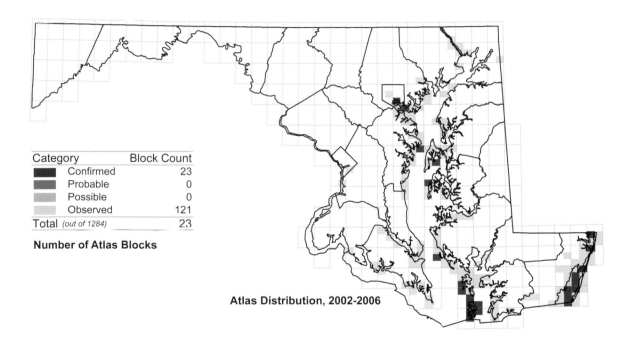

Category	Block Count
■ Confirmed	23
■ Probable	0
■ Possible	0
■ Observed	121
Total (out of 1284)	23

Number of Atlas Blocks

Atlas Distribution, 2002-2006

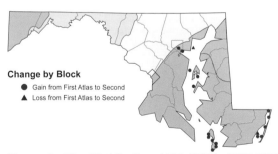

Change by Block

● Gain from First Atlas to Second
▲ Loss from First Atlas to Second

Change in Atlas Distribution, 1983-1987 to 2002-2006

Atlas Region	1983-1987	2002-2006	Change No.	%
Allegheny Mountain	0	0	0	-
Ridge and Valley	0	0	0	-
Piedmont	0	0	0	-
Upper Chesapeake	1	2	+1	+100%
Eastern Shore	8	20	+12	+150%
Western Shore	0	1	+1	-
Totals	9	23	+14	+156%

Change in Total Blocks between Atlases by Region

George M. Jett

Least Tern
Sternula antillarum

Least Terns are aptly named for they are the smallest tern in North America. The Least Tern and the nearly identical and closely related Little Tern (*Sternula albifrons*) of Europe are two of the smallest species of terns in the world. This diminutive tern makes up in tenacity what it lacks in size. Least Terns are fierce nest defenders that repeatedly strafe intruders, human or otherwise, that enter into their nesting colonies, often delivering a sharp peck to the top of the intruder's scalp. Least Terns frequent shallow waters where they forage for small shallow-bodied fresh, estuarine, and marine fish. Although widespread throughout the United States, they are very patchily distributed and never abundant away from nesting colonies. Least Terns are found nesting on open beaches and similar habitats on both coasts and on sand or gravel islands in the Mississippi and Missouri river systems (Thompson et al. 1997).

Least Terns begin returning to Maryland in mid-April and spend more than a month preparing for breeding, which begins in mid-May. Historically Least Terns nested on barren sand beaches throughout the Chesapeake Bay and coastal Worcester County (Stewart and Robbins 1958). With the exception of Assateague Island, natural sites for nesting by Least Terns are exceedingly rare in Maryland. More than half the breeding population of Least Terns in Maryland nest on building roofs, particularly the flat pea gravel–covered roofs of schools and large stores. With increases in mammalian predator populations that frequent the marsh side beaches formerly used by Least Terns as nesting sites, these birds have adapted to man-made mammalian predator–free islands within a sea of blacktop and people. Roof-top Least Tern colonies are usually located within 1.6 km (1 mi) of water, al-though some Maryland colonies have been found almost 3.2 km (2 mi) from the nearest suitable foraging habitat. Least Terns are relatively late nesters in Maryland; egg laying begins in late May, with most clutches established during the first week of June. Chicks begin hatching in mid-June, grow quickly, and fledge during early July. A second wave of nesting involving young first-time breeders and failed pairs from the first nesting wave often occurs from late June to early July, especially in larger colonies (McKearnan 1996j).

During the first atlas period, Least Terns were confirmed as breeding in 23 blocks, rather evenly spread throughout the Upper Chesapeake, Western Shore, and Eastern Shore regions (McKearnan 1996j). During the second atlas the number of blocks where Least Terns were recorded as observed increased 87% over the first atlas, but the second atlas finds Least Terns recorded as observed in blocks throughout the Upper Chesapeake and Eastern Shore regions with far fewer occurrences in the Western Shore region. No confirmations were recorded during the 2002–2006 atlas of Least Tern breeding in the Western Shore region and the number of confirmed blocks dropped 22% to 18. Least Tern colonies on roofs of buildings are difficult to locate. The divergence between increased observations and a decline in the number of confirmed blocks could indicate that some roof-top nesting occurrences were missed by atlas volunteers.

The nesting population of Least Terns in Maryland ranged from 403 to 723 breeding pairs spread among 10–16 colonies (DNR, unpubl. data) during the first atlas period. From 2002 through 2006, the number of breeding pairs of Least Terns in Maryland ranged from 563 to 819, forming breeding colonies at 8 to 14 sites each breeding season (DNR, unpubl. data). Although colony sites were redistributed, the total breeding population in Maryland did not change substantially between the two atlas projects. An analysis of Chesapeake Bay and Coastal Delmarva seabird populations found that the Least Tern population declined only slightly (2.5%)

between 1993 and 2003 and consisted of approximately 1,500 breeding pairs, with breeding about equally divided between Maryland and Virginia (Brinker, Williams, and Watts 2007).

Although the total breeding population of Least Terns in Maryland has remained at approximately the same size over the past 20 years, this tern's breeding distribution has shifted. Least Terns remain in a precarious position in those places where they are dependent on buildings for a significant proportion of their annual reproductive effort. If habitat conditions on the northern portion of Assateague Island change significantly, Maryland could lose its remaining stronghold for Least Terns that nest in a natural situation. In consider-

ation of continued nesting habitat loss and movement of breeding colonies to building roofs, the status of Least Terns in Maryland was changed from in need of conservation to threatened between the two atlas projects. More attention should be paid to locating colonies of roof-nesting Least Terns so that building managers or owners may be informed of the conservation implications of their rooftops and advised of steps necessary to coexist with this rare little tern. For the foreseeable future, Least Terns will remain dependent on the habitat provided by rooftops for their continued existence in Maryland.

DAVID F. BRINKER

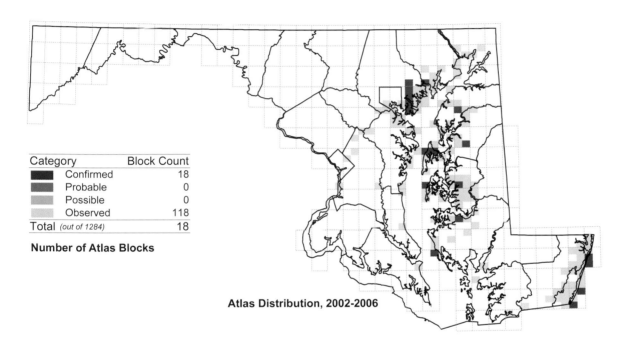

Category	Block Count
■ Confirmed	18
■ Probable	0
■ Possible	0
■ Observed	118
Total (out of 1284)	18

Number of Atlas Blocks

Atlas Distribution, 2002-2006

Change by Block

● Gain from First Atlas to Second
▲ Loss from First Atlas to Second

Change in Atlas Distribution, 1983-1987 to 2002-2006

Atlas Region	1983-1987	2002-2006	Change No.	%
Allegheny Mountain	0	0	0	-
Ridge and Valley	0	0	0	-
Piedmont	0	2	+2	-
Upper Chesapeake	3	5	+2	+67%
Eastern Shore	13	11	-2	-15%
Western Shore	7	0	-7	-100%
Totals	23	18	-5	-22%

Change in Total Blocks between Atlases by Region

George M. Jett

Gull-billed Tern
Gelochelidon nilotica

Gull-billed Terns have a decidedly un-ternlike diet. Their varied diet does not focus on fish as the diets of other tern species do. They are fond of insects such as grasshoppers; small crabs, including ghost crabs; and other prey that they snatch while on the wing. Only infrequently do they take fish. They have even been observed eating small chicks of other tern species (Parnell et al. 1995). Gull-billed Terns are widely distributed, breeding on every continent except Antarctica. Along the East Coast, Gull-billed Terns are known to have bred as far north as western Long Island (Mitra 2008a). The largest concentration of Gull-billed Terns in the United States is along the Texas Coast (Molina and Erwin 2006). Despite this wide distribution, Gull-billed Terns are rare breeders north of Virginia. In Maryland they breed in colonies with other terns, primarily Common Terns.

Gull-billed Terns return to the mid Atlantic and northeast as early as very late April, but they are usually not detected in Maryland until May. Gull-billed Terns nest in small colonies usually in association with Common Terns and Black Skimmers. All three species prefer nesting on barren or sparsely vegetated islands and beaches, preferably in protected locations above spring and summer high tides. The nests of Gull-billed Terns, which are different from those of Common Terns, are characterized by the shells and debris used to decorate the nest rims. The combination of decorated nests and the noticeably larger size of Gull-billed Tern eggs allows keen observers to readily separate Gull-billed and Common Tern nests. Egg laying begins in mid-May. Chicks begin hatching in mid-June and fledge during July. For a de-

tailed description of the history of Gull-billed Tern breeding in Maryland and details of breeding phenology and biology, see Brinker 1996f.

During the first atlas, Gull-billed Terns were confirmed as breeding in 2 blocks, both in the coastal bays of Worcester County (Brinker 1996f). During both atlases Gull-billed Terns were recorded as observed in several blocks in addition to those where breeding was confirmed. Gull-billed Terns were confirmed as breeding in 1 block during the second atlas, right at the Maryland-Virginia state line in Chincoteague Bay. This confirmation came from two nests discovered among a colony of Forster's Terns and Black Skimmers at the Cedar Islands on 6 June 2002 (D. Brinker, pers. obs.). These were the only nesting pairs of Gull-billed Terns observed during annual colonial nesting waterbird survey efforts by the DNR during the 2002–2006 atlas.

From 1985 through 2006 the nesting population of Gull-billed Terns in Maryland ranged from 1 to 33 breeding pairs (DNR, unpubl. data). During any single year there have never been more than two Gull-billed Tern colonies in Maryland. Except for the colony of 33 pairs that bred at Big Bay Marsh during 1986, the maximum colony observed in Maryland was 5 pairs and most breeding attempts in Maryland have consisted of single pairs nesting within a Common Tern and Black Skimmer colony. From 1985 to 2007 Gull-billed Terns have bred in Maryland during 8 summers, with 2002 being the last year of known breeding in the state. An analysis of Chesapeake Bay and coastal Delmarva Gull-billed Tern populations documented a 47% decline in breeding pairs from the peak census of 607 breeding pairs in 1993 to 322 in 2003, with virtually all breeding occurring on coastal barrier islands in Virginia (Brinker, Williams, and Watts 2007).

Continued presence of Gull-billed Terns as a breeding species in Maryland depends on a healthy breeding population on the coastal barrier islands of the nearby Virginia coast. The infrequent breeding of Gull-billed Terns in Maryland is primarily the result of limited nest habitat availability. In Virginia, predation by red fox, habitat fragmentation, and habitat loss are relevant factors in the observed decline of Gull-billed Terns (Erwin, Truit, and Jimenez 2001; Molina and Erwin 2006). Recently implemented predator management on some Virginia coastal islands may help stabilize or increase Gull-billed Tern breeding populations there. Conservation measures that increase Gull-billed Tern populations in neighboring states should increase the chance of breeding by small colonies of Gull-billed Terns in Maryland. Breed-

ing habitat for open sand nesters is severely limited in Maryland. Conservation measures to increase the availability of open sand nesting habitat on coastal bay islands in Worcester County, an area that attracts a large colony of Common or Royal Terns, may encourage Gull-billed Terns to regularly breed in Maryland for a few years. Given the implications of habitat change accompanying the expected rise in sea level over the next hundred years (Erwin, Sanders, et al. 2006) and the present precarious state of open sand nesting habitat in Maryland, the future for Gull-billed Terns in Maryland does not look bright.

DAVID F. BRINKER

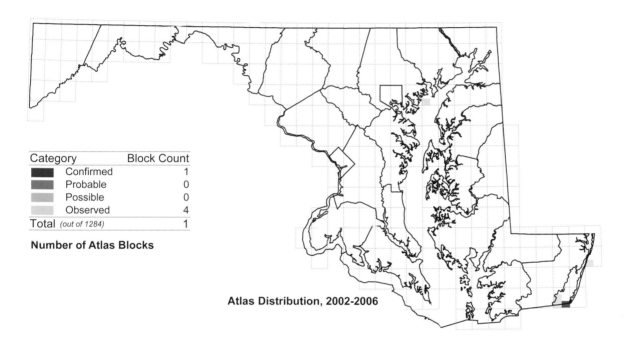

Category	Block Count
■ Confirmed	1
■ Probable	0
■ Possible	0
■ Observed	4
Total *(out of 1284)*	1

Number of Atlas Blocks

Atlas Distribution, 2002-2006

Change by Block

● Gain from First Atlas to Second
▲ Loss from First Atlas to Second

Change in Atlas Distribution, 1983-1987 to 2002-2006

Atlas Region	1983-1987	2002-2006	Change No.	Change %
Allegheny Mountain	0	0	0	-
Ridge and Valley	0	0	0	-
Piedmont	0	0	0	-
Upper Chesapeake	0	0	0	-
Eastern Shore	2	1	-1	-50%
Western Shore	0	0	0	-
Totals		2	1	-1	-50%

Change in Total Blocks between Atlases by Region

Fran Saunders

Common Tern
Sterna hirundo

The Common Tern, widespread along the East Coast, St. Lawrence Seaway, and Great Lakes, and throughout the Canadian prairies, prefers to nest on predator-free grassy islands, beaches, and other barren or open habitats. Breeding Common Terns are found throughout the Chesapeake and coastal bays, where they nest on small areas of sand resembling beaches, or occasionally on wrack lines just above high tide on isolated salt marsh islands. The nest structure of most Common Terns is a simple scrape or depression in the sand that keeps eggs from rolling away. Many pairs will add small amounts of dried grass to form a weak nest rim, but they never build a structure as substantial as that of a Forster's Tern nest. Common Terns forage for small fish over open water and tend to be more pelagic than Forster's Terns. Their population was nearly extirpated on the Atlantic coast by the millinery trade in the late 1800s (Nisbet 2002). Common Tern populations recovered quickly, reoccupying most of their range by the 1930s (Nisbet 2002) and spreading south as far as South Carolina by the 1960s (McNair and Post 1993).

Most Common Terns winter on the coasts of Central and South America, with only a small number of individuals remaining in the United States on the northern coast of the Gulf of Mexico from Texas through Florida (Nisbet 2002). They begin returning to Maryland from late March to early April. Egg laying begins in mid-May, with most clutches established by the first week of June. Clutch size varies from 1 to 4 eggs, with 3 the most commonly observed clutch size. Chicks begin hatching in early June, grow quickly, and fledge during July.

During the first atlas, Common Terns were confirmed as breeding in 14 blocks, with 1 in the Upper Chesapeake region, none in the Western Shore region, 7 in the Chesapeake Bay, and 6 in the coastal bays of the Eastern Shore region (Brinker 1996h). During the second atlas the number of blocks where Common Terns were recorded as observed was essentially the same although there were fewer observations in the Western Shore region than during the first atlas. The number of confirmed blocks in the second atlas increased by 1 block, to 15. Despite this suggestion of a stable distribution and population, 7 blocks, 50% of those confirmed during the first atlas, were not occupied during the second atlas. Common Tern breeding colonies switch sites in response to yearly changes in available nest site habitat and loss of islands to shoreline erosion. During the second atlas the most consistently used sites supporting the largest nesting colonies were the Poplar Island restoration project in the Chesapeake Bay and Skimmer and Reedy islands in the coastal bays. Increased use of Skimmer Island by nesting Herring and Great Black-backed Gulls resulted in abandonment of the site by breeding terns and skimmers before the end of the second atlas.

The nesting population of Common Terns in Maryland ranged from 2,125 to 2,435 breeding pairs spread among 13 to 17 colonies each year during the first atlas period (DNR, unpubl. data). From 2002 through 2006, the number of breeding pairs of Common Terns in Maryland showed a substantial decline: 649 to 1,345 pairs breeding at 5 to 10 sites each breeding season out of the 14 different locations used during the five-year censusing period (DNR, unpubl. data). An analysis of Chesapeake Bay and Coastal Delmarva seabird populations found that the Common Tern population had declined 60% from 8,130 breeding pairs in 1993 to 3,236 pairs in 2003 with the largest proportion of the breeding population located in Virginia (Brinker, Williams, and Watts 2007).

Since 1987 Common Terns have noticeably declined in Maryland. Shoreline erosion in the Chesapeake Bay has claimed a number of islands that were formerly used as nesting sites by colonies of Common Terns, and quality nesting sites are now in short supply. As island loss was occurring, artificial nesting islands created as part of the Poplar Island environmental restoration project attracted most of the Common Tern nesting pairs in the Maryland portion of the Chesapeake Bay. While Poplar Island's artificial nesting islands were safe from storm tides and erosion, Great Horned Owls from nearby Coaches Island located the nesting colonies and owl predation resulted in several years of no reproduction by Common Terns (DNR, unpubl. data). At the same time, in the coastal bays a combination of island erosion and increasing predation pressure from gull popu-

lations adversely impacted Common Tern nesting there, in particular excluding Common Terns from Skimmer Island. Sea level rise and its impact on nesting habitat will be a serious challenge for Common Terns (Erwin, Sanders, et al. 2006; Brinker, Williams, and Watts 2007). Apparently Common Terns are declining regionally and their future in Maryland appears to be a continuing decline. Conservation solutions to stabilize or reverse the decline of Common Terns are necessarily regional in nature and may be challenging to identify and implement.

DAVID F. BRINKER

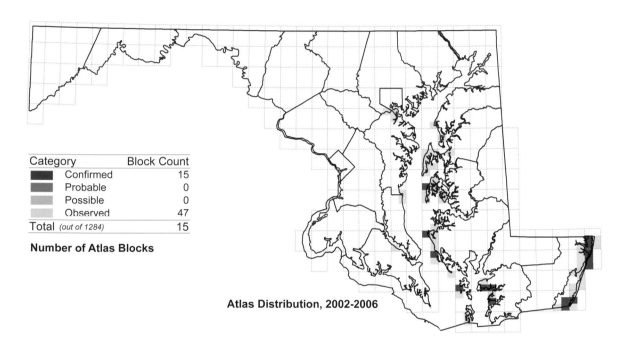

Category	Block Count
■ Confirmed	15
■ Probable	0
■ Possible	0
Observed	47
Total (out of 1284)	15

Number of Atlas Blocks

Atlas Distribution, 2002-2006

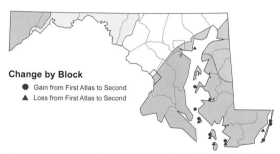

Change by Block

● Gain from First Atlas to Second
▲ Loss from First Atlas to Second

Change in Atlas Distribution, 1983-1987 to 2002-2006

Atlas Region	1983-1987	2002-2006	Change No.	Change %
Allegheny Mountain	0	0	0	-
Ridge and Valley	0	0	0	-
Piedmont	0	0	0	-
Upper Chesapeake	1	0	-1	-100%
Eastern Shore	13	15	+2	+15%
Western Shore	0	0	0	-
Totals	14	15	+1	+7%

Change in Total Blocks between Atlases by Region

Mark Hoffman

Forster's Tern
Sterna forsteri

Forster's Tern is a marsh-nesting tern that resembles a close relative, the Common Tern. Differentiating between these two tern species is perhaps the most difficult field identification issue of the colonial nesting waterbird guild inhabiting Maryland. Distributed across North America, Forster's Terns nest in freshwater and estuarine marshes, never on beaches. Breeding Forster's Terns are found throughout the Chesapeake and coastal bays where they nest on mats of floating or semifloating vegetation on small predator-free salt marsh islands. In the prairie pothole wetlands of the Great Plains, they nest on floating platforms constructed out of the dried remnants of cattails (McNicholl et al. 2001). Maryland nests are built on lines of wrack (mostly dried eel grass) washed up onto salt marsh islands by storm tides or on platforms of dried grass the terns construct in flooded centers of islands. Unlike Common Terns, which simply scrape a depression in the sand, Maryland Forster's Terns always build a nest of dried vegetation. The extensive salt marshes of the mid Atlantic coast are an important breeding area for Forster's Tern and they have expanded their breeding range north to Long Island in the past 25 years (Wasilco 2008c).

After wintering in the southeastern United States, Forster's Terns begin returning to Maryland in March. Egg laying begins in mid-May, with most clutches established by the first week of June. Chicks begin hatching in early June, grow quickly, and fledge during July.

During the first atlas, Forster's Terns were confirmed as breeding in 14 blocks, with 4 in the Chesapeake Bay and 10 in the coastal bays of the Eastern Shore region (Brinker 1996i). The number of blocks where Forster's Terns were recorded as observed during the second atlas increased 31% with the overall distribution similar during both atlases. The number of confirmed blocks in the second atlas increased 36%, to 19 blocks, but 7 blocks, 50% of those confirmed during the first atlas, were not occupied during the second atlas. Forster's Tern breeding colonies, like those of the Common Tern, routinely switch sites in response to annual changes in available nest site habitat and loss of islands to shoreline erosion. Winter storms change the distribution of wrack lines each winter, necessitating yearly relocation of breeding colonies to the set of most secure islands. During the second atlas the most consistently used sites that supported the largest nesting colonies were Manbone and North Holland islands in the Chesapeake Bay and Cedar, Horse, and Reedy islands in the coastal bays.

The nesting population of Forster's Terns in Maryland ranged from 1,337 to 1,916 breeding pairs spread among 14 to 18 colonies each year during the first atlas period (DNR, unpubl. data). From 2002 through 2006, the number of breeding pairs of Forster's Terns in Maryland showed a slight decline, to approximately 1,250 pairs breeding at approximately 20 sites each breeding season out of the 36 different locations used during the five-year period (DNR, unpubl. data). Although observed on a limited number of routes, BBS trends for Forster's Terns show a significant increase ($P < 0.01$) within Maryland. BBS trends at the U.S. and North American levels show no significant trends. A recent analysis of Chesapeake Bay and Coastal Delmarva seabird populations found that the Forster's Tern population had declined 5.6%, from 3,692 breeding pairs in 1993 to 3,484 pairs in 2003, with a larger proportion of the breeding population located in Virginia (Brinker, Williams, and Watts 2007).

Although the total breeding population of Forster's Terns in Maryland has remained approximately the same size since 1987, conditions for this species have become more precarious and the breeding population could soon start to decline more seriously. Forster's Terns nest on small islands very near the level of mean high tide and they depend on winter storm tides to create secure nesting habitat. Spring through early summer storm and tidal flooding is a regular cause of breeding failure. Because the availability and condition of Forster's Tern nesting habitat is associated with tidal variation and loss of small islands, the consequences of continued

sea level rise could be serious (Erwin, Sanders, et al. 2006; Brinker, Williams, and Watts 2007). Forster's Terns build substantial nests that do have some protection from tidal events when they float on a gradually rising tide. That nest design characteristic, when combined with their marsh nesting behavior, may be an important reason that their populations have been relatively stable. In the long term, the significant reduction of tidal marsh area that is expected to accompany sea level rise would significantly reduce the area of foraging and nesting habitat for Forster's Tern in Maryland.

DAVID F. BRINKER

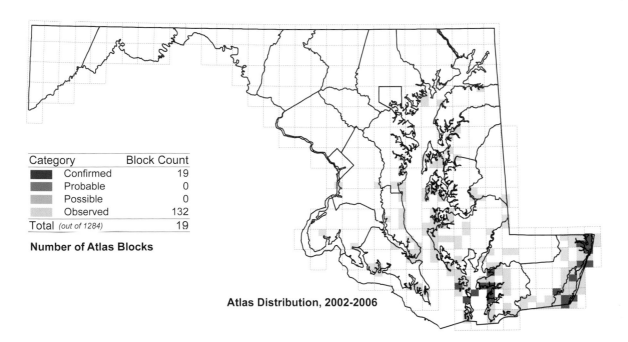

Category	Block Count
Confirmed	19
Probable	0
Possible	0
Observed	132
Total (out of 1284)	19

Number of Atlas Blocks

Atlas Distribution, 2002-2006

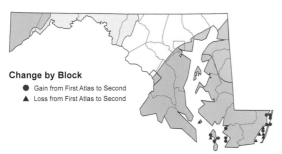

Change by Block

● Gain from First Atlas to Second
▲ Loss from First Atlas to Second

Change in Atlas Distribution, 1983-1987 to 2002-2006

Atlas Region	1983-1987	2002-2006	Change No.	%
Allegheny Mountain	0	0	0	-
Ridge and Valley	0	0	0	-
Piedmont	0	0	0	-
Upper Chesapeake	0	0	0	-
Eastern Shore	14	19	+5	+36%
Western Shore	0	0	0	-
Totals	14	19	+5	+36%

Change in Total Blocks between Atlases by Region

Royal Tern
Thalasseus maximus

The Royal Tern is Maryland's largest breeding tern species. When in peak breeding condition it has a bright orange bill, distinctive black cap and crest, and is much deserving of its name. Their loud rolling contact call, *Keer-reet* is often heard as they forage in the shallow estuarine bays and surf near Ocean City and Assateague Island. In North America the Royal Tern nests from central New Jersey (T. Pover, pers. comm.) south to the Gulf of Mexico and the Caribbean, and along the Pacific coast in the Gulf of California and nearby southern California (P. Buckley and Buckley 2002). Royal Terns are highly social and nest in densely packed colonies on usually remote, predator-free islands. Colonies are widely spaced. Royal Terns forage long distances, often as much as 65 km (40 mi) or more from nesting colonies (P. Buckley and Buckley 2002). They feed primarily on small fish and crabs, and farther south, shrimp. They once nested on barrier island beaches that are now occupied by humans and their recreational accoutrements or infested with mammalian predators. This has forced much of the mid Atlantic coast population to switch to alternate nesting sites such as bare dredge material islands (Emslie et al. 2009). Colony sites are characteristically open areas of sand usually safely above the reach of most spring and summer storm tides. Vegetation succession often renders nesting sites unusable. When nesting habitat becomes limited, long-established colonies will sometimes select marginal nesting sites rather than abandon traditional breeding areas.

Royal Terns winter from the southeastern United States south to the coasts of Central and South America (P. Buckley and Buckley 2002). They begin returning to Maryland in late March and are here in full force by late April. Egg laying begins in mid- to late May. Most Royal Terns lay a single egg; fewer than 5% of nests contain two eggs (D. Brinker, unpubl. data). Chicks begin hatching in early June, grow quickly, and fledge during July.

During the first atlas, although Royal Terns did not breed in Maryland, they were recorded as observed in 55 blocks (Brinker 1996g). These were certainly foraging Royal Terns from a breeding colony in the Tangier Island area just south of the Maryland-Virginia state line. Royal Terns previously nested in the state, with a high of 1,160 nesting pairs in 1976 (Therres, Weske, and Byrd 1978). During the second atlas the number of blocks where Royal Terns were recorded as observed increased 42%, and they were recorded as confirmed in 2 blocks in the coastal bays of Worcester County. During the second atlas Royal Terns nested at Skimmer Island from 2002 through 2005 and at Reedy Island during 2006, as well as just south of the Maryland-Virginia state line in the Chesapeake Bay. Increased use of Skimmer Island by nesting Herring and Great Black-backed Gulls resulted in abandonment of Skimmer Island by breeding Royal Terns after the 2005 nesting season. The breeding population of Royal Terns in Maryland ranged from 685 in 2002 to 149 breeding pairs in 2006 (DNR, unpubl. data). The loss in breeding pairs that was associated with the move to Reedy Island coincides with the first significant nesting by Royal Terns in New Jersey (T. Pover, pers. comm.). A recent analysis of Chesapeake Bay and Coastal Delmarva seabird populations found that the Royal Tern population had declined 49% from 6,586 breeding pairs in 1993 to 3,332 pairs in 2003 with most of the breeding population located in Virginia (Brinker, Williams, and Watts 2007). A review of mid Atlantic Royal Tern breeding populations showed a peak in 1984 followed by a decline of approximately 35% to a relatively stable level since the late 1990s (Emslie et al. 2009).

Royal Terns expanded into Maryland and successfully used Skimmer Island from 1991 through 2005. Increasing predation pressure from expansion of gulls nesting on Skimmer Island eventually pushed the Royal Tern breeding colony to a very marginal site on Reedy Island (D Brinker, pers. obs.). The available sand at Reedy Island is limited and this probably resulted in the substantial reduction in breeding population between 2005 and 2006. Royal Tern habitat in Maryland is rare and severely restricted. The open sands of Skimmer Island were a boon to terns and skimmers from 1990 until the gull colony became overwhelming in 2005. Recovery of Royal Tern breeding populations in Maryland to levels similar to the past 20 years will require disruption and removal of the gull colony from Skimmer Island. If this conservation objective can be achieved, it should result in the return of breeding Royal Terns to Skimmer island and provide badly needed nesting habitat for Common Terns and Black Skimmers. Since most Royal Terns north of South Carolina now nest on artificial islands that require human management, the conservation solutions necessary to stabilize or even increase Royal Tern populations are necessarily regional in nature and challenging to implement.

DAVID F. BRINKER

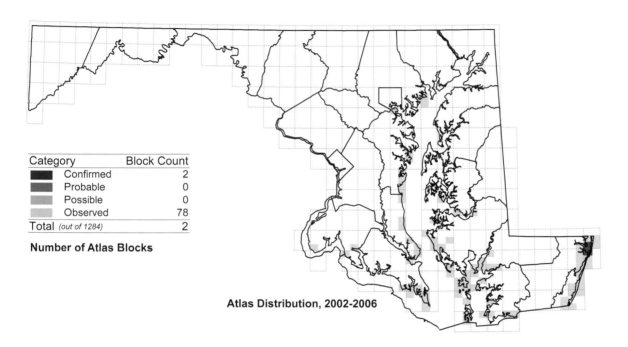

Category	Block Count
■ Confirmed	2
■ Probable	0
■ Possible	0
■ Observed	78
Total (out of 1284)	2

Number of Atlas Blocks

Atlas Distribution, 2002-2006

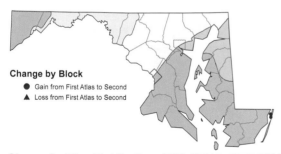

Change by Block
● Gain from First Atlas to Second
▲ Loss from First Atlas to Second

Change in Atlas Distribution, 1983-1987 to 2002-2006

Atlas Region	1983-1987	2002-2006	Change No.	%
Allegheny Mountain	0	0	0	-
Ridge and Valley	0	0	0	-
Piedmont	0	0	0	-
Upper Chesapeake	0	0	0	-
Eastern Shore	0	2	+2	-
Western Shore	0	0	0	-
Totals	0	2	+2	-

Change in Total Blocks between Atlases by Region

Black Skimmer
Rhynchops niger

Black Skimmers possess a unique set of adaptations among the breeding birds of the state of Maryland. They are the only Maryland breeding bird species that has a slit pupil similar to those of felines, an adaptation that facilitates their nocturnal foraging life while protecting them from the excessive sunlight of their favored nesting beaches (Zusi and Bridge 1981). Worldwide, skimmers are the only group of birds with a knife-thin lower mandible that is approximately one-third longer than the upper mandible. Along with other skeletal features of the skull and neck, the combination of this unusual bill cross section and quick reflexes are essential to the Black Skimmer's method of catching fish near the water's surface (Zusi 1962). The aspect ratio of their wings increases ground effect, helping to maximize the efficiency of their flight while they are foraging just above the water's surface (Withers and Timko 1977). Black Skimmers are nearly perfectly suited to their nocturnal existence at the boundary between air and water within Maryland's estuaries.

In mid-April Black Skimmers return to the narrow coastal zone between the saline and estuarine waters along Maryland's coastal beaches. They prefer nesting on barren to sparsely vegetated islands and beaches just out of reach of high tide. Their nesting habitat depends on exceptionally high tides during the nonbreeding season to maintain the preferred open sand substrate that skimmers use for nesting. Although they have occasionally been recorded using unusual substrates such as wrack in salt marshes (Frohling 1965) and gravel on flat roofs (Fisk 1978), these unusual nest site selection behaviors have not taken hold in Maryland. Egg laying begins in mid-May, with most clutches established by the first week of June. Chicks begin hatching in mid-June and fledge during late July or early August. For a detailed description of the history of Black Skimmer breeding in Maryland and more details of breeding phenology and biology, see Brinker 1996j.

During the first atlas, Black Skimmers were confirmed as breeding in 5 blocks, 4 of which were in the coastal bays of Worcester County (Brinker 1996j). The fifth block, in the Chesapeake Bay, was a colony near Barren Island in Dorchester County. During both atlases Black Skimmers were recorded as observed in a similar number of blocks, but none were farther north in the Chesapeake Bay than the vicinity of Poplar Island in Talbot County. Although confirmed in a similar number of blocks (6) during the second atlas, Black Skimmers no longer breed within the Maryland portion of the Chesapeake Bay. The number and distribution of confirmed atlas blocks is misleading and does not reflect the significant change in the Maryland breeding population

Monroe Harden

of Black Skimmers that occurred between the first and second atlases.

The nesting population of Black Skimmers in Maryland ranged from 150 to 328 breeding pairs spread in 5 to 6 colony sites (DNR, unpubl. data) during the first atlas period. At that time Black Skimmers were already listed by the DNR as a threatened species. The plight of Black Skimmers has worsened since then, and their breeding population has significantly declined in the coastal bays. During 2002 a total of 143 breeding pairs of Black Skimmers were spread among 4 colonies. Between 2002 and 2006 the breeding population of skimmers plummeted to a total of two breeding pairs at a single colony (DNR, unpubl. data). This decline is not limited to Maryland; a recent analysis of Chesapeake Bay and coastal Delmarva Black Skimmer populations found a 43% decline in breeding pairs from the peak census of 3,359 breeding pairs in 1993 to 1,924 in 2003 with most breeding occurring on coastal barrier islands in Virginia (Brinker, Williams, and Watts 2007).

Deterioration of nesting habitats for Black Skimmers in Maryland is responsible for their continued decline. In 2007 Black Skimmers were listed as endangered in Maryland. Since 1985, erosion of beach habitats essential to nesting skimmers reduced the suitable nesting areas for Black Skimmers in Maryland. The largest single piece of nesting habitat, Skimmer Island just north of the U.S. Route 50 bridge into Ocean City, has evolved into a substantial gull nesting colony and predation pressure from nesting gulls has forced Black Skimmers to vacate Skimmer Island for small beach remnants on salt marsh islands in the northern half of the coastal bays. Black Skimmers in Maryland are now restricted to small sand beaches barely above high tides, and nesting often fails as a result of storm-driven tides. The potential impacts on Black Skimmers from continued sea level rise are serious (Erwin, Sanders et al. 2006; Brinker et al. 2007). To reverse the decline of Black Skimmers in Maryland will require reducing gull competition and predation as well as increasing the availability of safe beach nesting habitat in the coastal bays of Worcester County. The future for Black Skimmers in Maryland will be difficult and only time will tell if conservation actions can improve their plight.

DAVID F. BRINKER

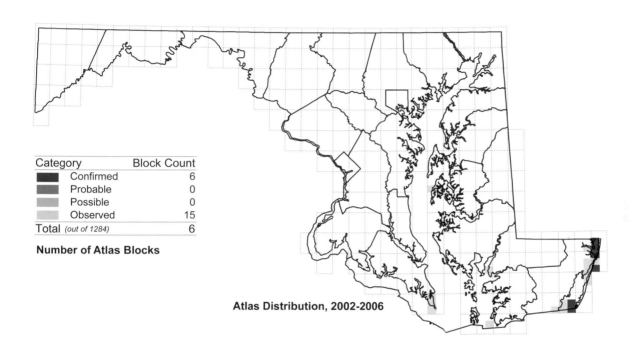

Category	Block Count
Confirmed	6
Probable	0
Possible	0
Observed	15
Total (out of 1284)	6

Number of Atlas Blocks

Atlas Distribution, 2002-2006

Change by Block
- ● Gain from First Atlas to Second
- ▲ Loss from First Atlas to Second

Change in Atlas Distribution, 1983-1987 to 2002-2006

			Change	
Atlas Region	1983-1987	2002-2006	No.	%
Allegheny Mountain	0	0	0	-
Ridge and Valley	0	0	0	-
Piedmont	0	0	0	-
Upper Chesapeake	0	0	0	-
Eastern Shore	5	6	+1	+20%
Western Shore	0	0	0	-
Totals	5	6	+1	+20%

Change in Total Blocks between Atlases by Region

Rock Pigeon
Columba livia

Fast and highly maneuverable on the wing and renowned for its homing abilities, the Rock Pigeon is a portly waddling pedestrian on city streets where it has become an avian symbol of urbanity. Although their excretions on nest ledges and public monuments and their grain-feeding habits on farms have led some to call them pests, the effect of pigeons on native birds is more benign than their Old World compatriots the European Starling and the House Sparrow. The Rock Pigeon has been used since ancient times as a food source, for carrying messages, and for the pleasure their mastery of the air gives owners. Their association with humans is so intimate that most early ornithological references in North America did not consider them wild birds. Thus information on their history and biology on this continent was almost nonexistent until the past few decades. They were introduced to North America during the seventeenth century. There is reference to a shipment to Virginia in 1621 (Long 1981). Rock Pigeons are very widespread in Maryland, but not ubiquitous.

The Rock Pigeon is a bird of open to lightly forested landscapes. The major factor limiting its occurrence is the availability of nest sites. They nest on ledges. Natural sites were cliffs in mountainous country and sea cliffs, but these have long been supplemented with man-made structures in cities and farms, such as buildings, bridges, and towers. Pigeons occasionally still nest on rock ledges in Maryland, for example, on the Sideling Hill I-68 cut in Washington County (pers. obs.). Although these pigeons may range far from nest sites,

a lack of accessible building ledges and bridges can prevent them from living in an area.

The Rock Pigeon is a gregarious and year-round resident in Maryland and DC. It also nests virtually year-round although there appears to be a distinct breeding peak from March to July. Young remain in the nest from 25 to 32 days (Johnston 1992) and can be quite loud when fed, giving loud squeals. Adults do not visibly carry food to nestlings. As do all pigeons and doves, they feed their young with a secretion called crop milk supplemented with seeds. Egg dates for Maryland and DC range from 2 January to 27 October (Robbins and Bystrak 1977).

Little published information on the Rock Pigeon's status in Maryland and DC was available through 1958. They were not counted systematically until 1965 with the advent of the BBS (Robbins, Bystrak, and Geissler 1986), and they were not counted on Christmas Bird Counts until 1974 (Monroe 1974). The 1983–1987 bird atlas was one of the earliest accounts of this common bird's occurrence in the state and the District. They were found in 79% of the blocks surveyed in the 1980s, with gaps and thin distribution in southern Maryland on both the Eastern and Western Shore and in well-forested parts of Allegany and Garrett counties (Blom 1996c).

In 2002–2006 pigeons were found in fewer blocks and were reported from only 72% of surveyed blocks. Losses were especially large in southern Maryland. Some losses on the Piedmont may relate to variation in coverage by fieldworkers. The gaps in forested parts of western Maryland continue. Note the trace of blocks across Green Ridge State Forest in eastern Allegany County indicating the presence of highway bridges on I-68. Another indication that Rock Pigeons may have declined is a drop in the breeding confirmation rate from 45% in the first atlas to 31% in the second.

Rock Pigeon shows a significant decline since 1980 on North American BBS routes, but its long-term status has been essentially stable. The trend in Maryland since 1980 has been steeply declining, at an annual rate of 5.4%. Declines may be attributable to cleaner farming practices, the general decline of farming, declines in pigeon-keeping (reducing a major source for sustaining wild populations), and perhaps some concerted pest control efforts. In spite of declines, urban and farm populations of Rock Pigeons in Maryland and DC are likely to remain.

WALTER G. ELLISON

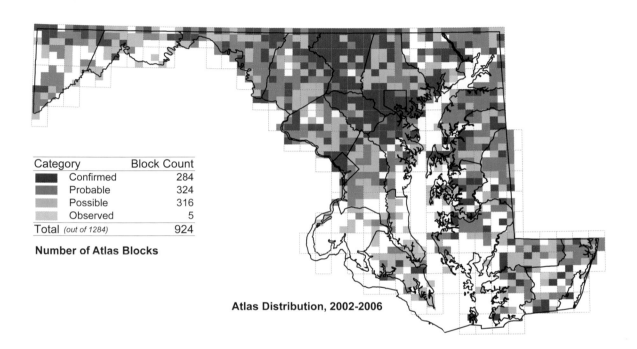

Category	Block Count
Confirmed	284
Probable	324
Possible	316
Observed	5
Total *(out of 1284)*	924

Number of Atlas Blocks

Atlas Distribution, 2002-2006

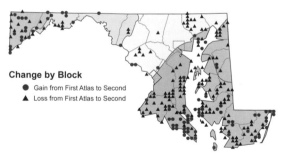

Change by Block
- ● Gain from First Atlas to Second
- ▲ Loss from First Atlas to Second

Change in Atlas Distribution, 1983-1987 to 2002-2006

Percent of Stops
- 50 - 100%
- 10 - 50%
- 0.1 - 10%
- < 0.1%

Relative Abundance from Miniroutes, 2003-2007

Atlas Region	1983-1987	2002-2006	Change No.	%
Allegheny Mountain	59	62	+3	+5%
Ridge and Valley	124	124	0	0%
Piedmont	308	291	-17	-6%
Upper Chesapeake	110	103	-7	-6%
Eastern Shore	227	195	-32	-14%
Western Shore	171	149	-22	-13%
Totals	999	924	-75	-8%

Change in Total Blocks between Atlases by Region

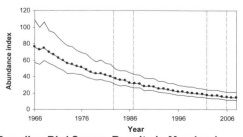

Breeding Bird Survey Results in Maryland

Gary Smyle

Mourning Dove
Zenaida macroura

Pedestrian and ubiquitous, stolid and persevering, Mourning Doves thrive among humans in agricultural, suburban, and urban landscapes. In the air, Mourning Doves are dashingly fast and maneuverable on whistling wings. The graceful looping glides of displaying males are conspicuous from February to October, as are the mournful sighs of territorial doves that characterize early mornings over virtually all of Maryland and DC. Mourning Doves nest where there is thick cover and abundant support for their flimsy stick platform nests, often utilizing young conifers and vine tangles (Mirarchi and Baskett 1994). As long as there are suitable nest sites and cover in which to retreat from predators, combined with seeds and open ground for feeding, doves can survive and often thrive. All that they require for nesting in an otherwise open landscape are ornamental conifers or a thick hedgerow. Doves are rare only in extensive open marsh land with few trees and in extensive mature forest.

Mourning Doves have almost certainly increased greatly in North America since the colonial period. During the twentieth century, doves have become more common and increasingly sedentary, wintering even in the coldest parts of their range. These changes took place as the climate warmed and bird feeding became a more common practice. As re-

cently as the 1950s, the Mourning Dove was rare in winter in the Allegheny Mountains (Stewart and Robbins 1958). Since the end of the previous statewide atlas in 1987 (Blom 1996d), the breeding range of the Mourning Dove in Maryland and DC has changed very little. This dove remains one of the five most widespread nesting birds in the Old Line State. Mourning Doves added a handful of blocks to their already broad range in mountainous parts of western Maryland and in the southern Dorchester County wetlands.

Half of all blocks had records of confirmed breeding by Mourning Doves. Nests with eggs have been found in Maryland from 12 February to 27 October (MNRF). Recently fledged Mourning Doves have shorter tails than their parents and distinctive pale scaling on their upperparts. Nearly 60% of dove confirmations were for fledged young (11 April to 2 October). The window of opportunity for confirming nesting is long, as doves attempt up to three broods in a nesting season, ceasing nesting activity only in the coldest months of the year.

In spite of this dove's geographic expansion and population increase over much of the twentieth century (Mirarchi and Baskett 1994), BBS trends for Mourning Dove have been stable over the past four decades. Trends within the physiographic regions spanning Maryland have also been stable, save for a modest decline on the Upper Coastal Plain (Maryland to North Carolina). It is possible that changes in agricultural practices do not agree with Mourning Doves; such practices include destruction of hedgerows, harvesting techniques that leave less waste grain in fields, and the treatment of seeds and fields with pesticides. On the other hand, doves are popular autumn game birds. Many farmers plant crops (e.g., sunflowers) to attract them to fields from September to November. Mourning Doves should continue to be ubiquitous and familiar to Marylanders for many decades to come.

WALTER G. ELLISON

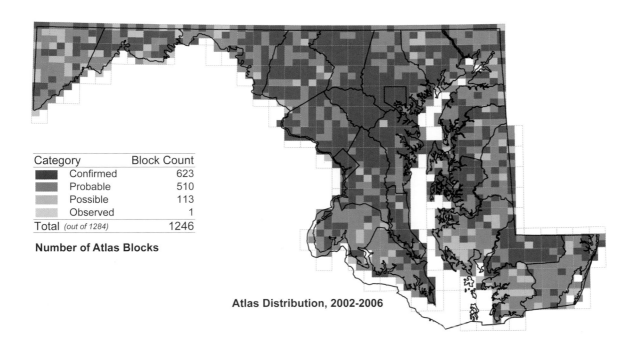

Category		Block Count
■	Confirmed	623
■	Probable	510
■	Possible	113
■	Observed	1
Total (out of 1284)		1246

Number of Atlas Blocks

Atlas Distribution, 2002-2006

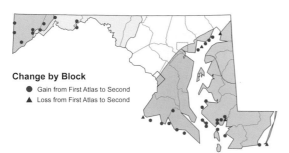

Change by Block
● Gain from First Atlas to Second
▲ Loss from First Atlas to Second

Change in Atlas Distribution, 1983-1987 to 2002-2006

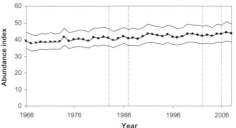

Percent of Stops
■ 50 - 100%
■ 10 - 50%
■ 0.1 - 10%
□ < 0.1%

Relative Abundance from Miniroutes, 2003-2007

Atlas Region	1983-1987	2002-2006	Change No.	Change %
Allegheny Mountain	87	93	+6	+7%
Ridge and Valley	144	148	+4	+3%
Piedmont	316	316	0	0%
Upper Chesapeake	117	119	+2	+2%
Eastern Shore	305	315	+10	+3%
Western Shore	244	249	+5	+2%
Totals	1213	1240	+27	+2%

Change in Total Blocks between Atlases by Region

Breeding Bird Survey Results in Maryland

Yellow-billed Cuckoo

Coccyzus americanus

A slender, shy, slow-moving bird, the Yellow-billed Cuckoo is nomadic; it is at once elusive and hard-to-know, yet familiar enough to be known as the Rain Crow, the foreteller of humid and rainy weather in the countryside. These cuckoos are best detected by their calls, an accelerating wooden and guttural clatter and a sad-sounding deliberately delivered series of hollow *coos*. The sounds are ventriloquial, making it hard to locate the singer. Yellow-billed Cuckoos feed on large insects, especially species prone to large population fluctuations and localized abundance such as tent caterpillars, fall webworms, gypsy moth caterpillars, and 17-year cicadas. Because of this association, these cuckoos are prone to wandering within and among breeding seasons and may vary greatly in numbers through a summer or between years.

Although they are arboreal, Yellow-billed Cuckoos are not forest interior birds. They prefer edges, small trees and shrubs, and canopy gaps in forest; abundant vines that hide the birds and their nests are often a feature of this cuckoo's haunts. Yellow-billed Cuckoos tend to dwell near a stream or pond.

Yellow-billed Cuckoos appear to search for abundant prey and settle to nest in such areas. Because these conditions may arise at any time from this cuckoo's arrival in the last few days of April to the dog days of August, there is broad overlap among migrants, nesters, and within-season nomads making use of the possible (X) atlas code dicey. Safe dates for this cuckoo were limited to 45 days. Observers used care with the Yellow-billed Cuckoos they located, and nearly 70% of records were either probable or confirmed breeding birds. Yellow-billed Cuckoos can start nesting and bring off fledglings in as little as 17 days (Hughes 1999), reflecting their need for flexibility when they come upon abundant prey. Yellow-billed Cuckoo nests are generally built fairly low, often 6.1 m (20 ft) or lower in small trees and shrubs. Maryland egg dates for Yellow-billed Cuckoo nearly cover the annual time of stay in Maryland ranging from 13 May to 28 August (Stewart and Robbins 1958).

The Yellow-billed Cuckoo is found throughout Maryland during the nesting season, from the scrubby wooded dunes of Assateague Island to the cove hardwoods of the Allegheny Mountains. There was essentially no change in this cuckoo's distribution between the 1980s (Hilton 1996a) and 2000s, but it appears to have increased slightly in the Alleghenies and in the pine savannahs of southern Dorchester County. Yellow-billed Cuckoos were even recorded in several blocks in Baltimore City and DC, nesting in city parks.

In spite of maintaining its presence in essentially the same number of blocks between statewide atlas projects, the Yellow-billed Cuckoo has declined significantly over the past two decades on BBS routes, wiping out earlier increases. From 1966 to 2006 these cuckoos have declined by 1.7% per year throughout North America and by 2% per year in Maryland. These declines may reflect habitat loss to housing development as farmland and woodlots are converted to open lawn with only scattered ornamental plantings. Another possible source of trouble would be heavy use of pesticides to kill insects, especially *Bacillus thuringiensis* applied during caterpillar outbreaks because it kills all caterpillars indiscriminately. Eastern Yellow-billed Cuckoos are less tied to floodplains than western Yellow-billed Cuckoos and have generally proven adaptable if there are small trees and shrubs in which to nest and insect outbreaks to exploit. If these resources continue to exist we should continue to have these cuckoos foretelling rain in the future.

WALTER G. ELLISON

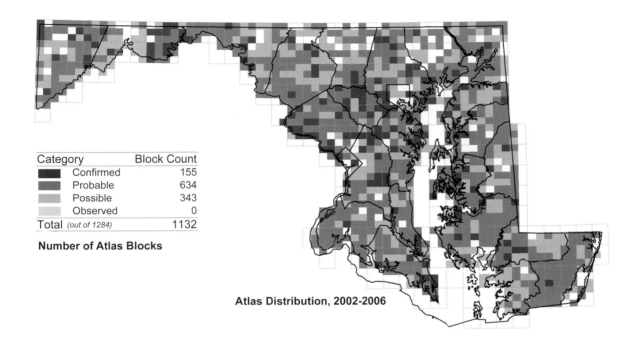

Category	Block Count
■ Confirmed	155
Probable	634
Possible	343
Observed	0
Total *(out of 1284)*	1132

Number of Atlas Blocks

Atlas Distribution, 2002-2006

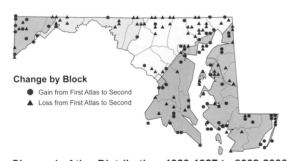

Change by Block
- ● Gain from First Atlas to Second
- ▲ Loss from First Atlas to Second

Change in Atlas Distribution, 1983-1987 to 2002-2006

Percent of Stops
- ■ 50 - 100%
- 10 - 50%
- 0.1 - 10%
- □ < 0.1%

Relative Abundance from Miniroutes, 2003-2007

Atlas Region	1983-1987	2002-2006	Change No.	%
Allegheny Mountain	79	83	+4	+5%
Ridge and Valley	144	135	-9	-6%
Piedmont	300	284	-16	-5%
Upper Chesapeake	113	112	-1	-1%
Eastern Shore	267	279	+12	+4%
Western Shore	227	233	+6	+3%
Totals	1130	1126	-4	0%

Change in Total Blocks between Atlases by Region

Breeding Bird Survey Results in Maryland

Black-billed Cuckoo
Coccyzus erythropthalmus

The Black-billed Cuckoo is similar in most respects to the Yellow-billed Cuckoo including general appearance, vocalizations, diet, nesting behavior, and habitat preference. It is more subtly plumaged in gray brown, gray, creamy white, and buff than the more contrastingly patterned Yellow-billed. The two species' voices are distinct but sufficiently overlapping to cause confusion. The Black-billed's repetitive hollow and musical *coo-coo-coo* is the species' most distinctive vocalization. Its other calls include a guttural, accelerating *croo-croo-croo-croo-croop-croop* similar to the Yellow-billed's clatter although less wooden and with less distinct notes. Its slow *coo* series is less far-carrying and less slurred. The Black-billed Cuckoo has a more northerly breeding range than the Yellow-billed. It is close to the southeastern edge of its range in Maryland and is far less common than its relative is in Maryland, save perhaps in Garrett County.

Black-billed Cuckoos live in successional habitats including forest clearings and edges, overgrown brushy farmland, shrub swamps, stream banks, and lake shores. They seem to prefer a landscape that has a higher percentage of forest than Yellow-billed Cuckoo, and they are less prone to nest in small woodlots or in urban parks and wooded suburbs (Ickes 1992b; Walsh et al. 1999). They are usually associated with local or regional outbreaks of hairy caterpillars and cicadas. They appear to spend some time nomadically seeking such outbreaks after spring migration (Nolan and Thompson 1975; Sealy 1978).

These cuckoos winter in South America, arriving in Maryland and DC in late April and early May, with apparently migratory birds still moving into June. Fall migration commences in July and continues until late September. The safe dates for Black-billed Cuckoo were set at 30 days; in spite of this precaution some birds reported as possible breeders may have been wandering birds looking for insect outbreaks. Cuckoos have a lengthy nesting season reflecting the seasonality of insect outbreaks from the late April emergence of eastern tent caterpillars to the early August hatching of fall webworms. The loosely built stick nests are usually placed fairly low, often about 2 m (6 ft), in shrubs and small trees (Spencer 1943). Maryland nest records are few but they capture much of the long nesting season with eggs reported from 18 May to 19 July with nestlings to 26 July (Stewart and Robbins 1958; MNRF). During this project fledglings were detected as early as 26 May 2006 near Oldtown, Allegany County (N. Martin), and as late as 9 August 2005 near Kingstown, Queen Anne's County (J. Gruber).

The Black-billed Cuckoo was considered rare during the nesting season in eastern Maryland during the late nineteenth century (C. Richmond 1888; Kirkwood 1895). Eifrig (1904) called it not common as a migrant but noted that it bred at higher elevations in Allegany and Garrett counties. Half a century later Stewart and Robbins (1958) called them fairly common in the Allegheny Mountains, uncommon eastward to the Upper Chesapeake, and rare on the lower Coastal Plain. This pattern has generally held from the 1980s

(Stasz 1996a) to the 2002–2006 statewide atlas project although it has been demonstrated that they are widespread in well-forested parts of the Ridge and Valley in Allegany and western Washington counties. There was a net loss from 33% of blocks between atlas projects, although the general outlines of the range remained similar and they declined less dramatically in the Alleghenies.

Black-billed Cuckoos have been declining on BBS routes at 1.8% per year since 1966 and at an accelerated 3.1% per year since 1980. In Maryland they are far too scarce on routes to detect a clear trend. Potential causes for the long-term declining trend include habitat loss, both on breeding and wintering grounds; declines in prey abundance; and perhaps global climate change. Methods for attacking caterpillar outbreaks have become sufficiently effective that they may also be affecting cuckoos' ability to find enough prey to nest successfully in many years. Both cuckoos deserve further research because of their unusual life history traits and because of their apparently declining populations.

WALTER G. ELLISON

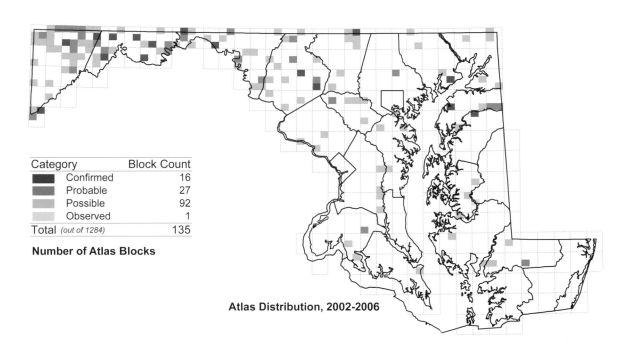

Category	Block Count
Confirmed	16
Probable	27
Possible	92
Observed	1
Total (out of 1284)	135

Number of Atlas Blocks

Atlas Distribution, 2002-2006

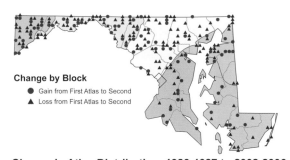

Change by Block
- ● Gain from First Atlas to Second
- ▲ Loss from First Atlas to Second

Change in Atlas Distribution, 1983-1987 to 2002-2006

Percent of Stops
- 50 - 100%
- 10 - 50%
- 0.1 - 10%
- < 0.1%

Relative Abundance from Miniroutes, 2003-2007

Atlas Region	1983-1987	2002-2006	Change No.	%
Allegheny Mountain	48	42	-6	-13%
Ridge and Valley	58	36	-22	-38%
Piedmont	49	27	-22	-45%
Upper Chesapeake	17	8	-9	-53%
Eastern Shore	16	9	-7	-44%
Western Shore	11	12	+1	+9%
Totals	199	134	-65	-33%

Change in Total Blocks between Atlases by Region

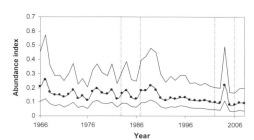

Breeding Bird Survey Results in Maryland

Barn Owl
Tyto alba

Superstitious humans might find the Barn Owl's unearthly shrieking, reptilian hissing, and gnomish heart-shaped visage unsettling. That it haunts lonely uninhabited structures only adds to its reputation. This owl's great skill in preying on small rodents in almost profound darkness, its penchant for adopting a wide array of dark crevices for nesting, and its strong flight have aided it in occupying the broadest world range of any owl. It has even colonized remote oceanic islands. Barn Owls are scarce but fairly widespread in the eastern two-thirds of Maryland. They have a long history in the region, with documentation extending back to nesting during the 1860s in the Smithsonian castle tower in DC (C. Richmond 1888; Bendire 1895).

Barn Owls inhabit open grasslands with abundant small rodents and accessible nesting cavities that can range from tree hollows to caves to all manner of man-made structures. If there is sufficient open, rodent-infested land, these owls can nest in urban settings. They are rare or absent in areas dominated by row crops or lacking available nest sites. In tidewater parts of Maryland and neighboring states, Barn Owls live in extensive brackish and salt marshes, nesting in tree cavities or in structures such as duck blinds (Reese 1972; Bendel and Therres 1990; Hess et al. 2000).

Barn Owls are present throughout the year in Maryland and DC although some birds may move seasonally as documented in nearby New Jersey (Duffy and Kerlinger 1992). Nesting seems to be tied to prey abundance and may take place at any time of year (Marti et al. 2005). Almost any relatively dark secluded recess may be used as a nest site. As with other owls no nest is actually built; eggs are laid on the sheltered substrate. Incubation of eggs begins with the first laid; young, therefore, vary greatly in size. Some late-hatched nestlings are lost when food is in short supply (Marti et al. 2005). Eggs have been found in Maryland and DC from 2 January to 17 October (Stewart and Robbins 1958; MNRF) and nestlings have been seen as late as 1 January (E. A. T. Blom in Bendel and Therres 1990). The peak of nesting activity is from late March to mid-July (Reese 1972; Bendel and Therres 1990).

Kirkwood (1895) did not consider Barn Owls common in Maryland but noted their occurrence in DC and in tidewater areas near Baltimore. Stewart and Robbins (1958) called them fairly common but local on the Eastern Shore, uncommon elsewhere west to the Allegheny Front, and rare in Garrett County. They were widely recorded in 1983–1987 westward to southwestern Garrett County (2 blocks) with the largest percentage of the 183 blocks from open farmland in the eastern Ridge and Valley and Piedmont, and on the Eastern Shore where most reports were from tidewater and wetlands (Jeschke and Therres 1996).

Gary Van Velsir

Barn Owls were lost from a net total of 131 blocks between atlas projects, a decline of 72%. This was the largest loss between projects for any bird occurring in a hundred or more blocks in 1983–1987. The basic apportionment of blocks among regions was similar between atlas projects with 54% of the 52 current blocks being from the Piedmont and Eastern Shore. There was a disproportionate loss from the Ridge and Valley, especially in the Hagerstown Valley, and a less drastic loss on the Western Shore. The latter may have arisen from an active nest box program on the Patuxent River and in southern Charles County (M. Callahan, pers. comm.).

The Barn Owl's strongly nocturnal habits make BBS trends unreliable at best. Christmas Bird Count trends from other eastern states have been declines (Walsh et al. 1999; Hess et al. 2000; McGowan 2008a). The large loss of block records between atlas projects suggests a large decline in Maryland and DC as well. Given the difficulties in finding Barn Owls, a targeted survey should be undertaken that includes contacting farmers and other landowners for information about owls. Nest box programs could also be established more broadly. But to retain this owl in Maryland, the preservation and maintenance of open land with sizable rodent populations will also be necessary (Colvin 1985). That such an opportunistic and flexible predator is declining on Maryland farmland does not bode well for other open country wildlife.

WALTER G. ELLISON

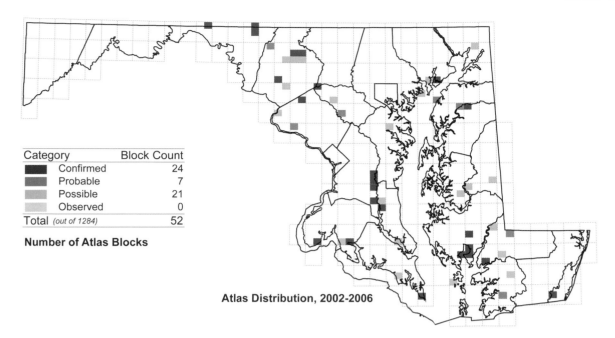

Category	Block Count
Confirmed	24
Probable	7
Possible	21
Observed	0
Total *(out of 1284)*	52

Number of Atlas Blocks

Atlas Distribution, 2002-2006

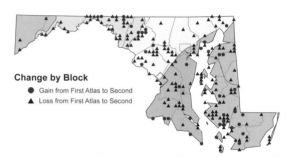

Change by Block

● Gain from First Atlas to Second

▲ Loss from First Atlas to Second

Change in Atlas Distribution, 1983-1987 to 2002-2006

			Change	
Atlas Region	1983-1987	2002-2006	No.	%
Allegheny Mountain	3	0	-3	-100%
Ridge and Valley	34	5	-29	-85%
Piedmont	47	13	-34	-72%
Upper Chesapeake	18	6	-12	-67%
Eastern Shore	51	15	-36	-71%
Western Shore	30	13	-17	-57%
Totals	183	52	-131	-72%

Change in Total Blocks between Atlases by Region

Robert Ampula

Eastern Screech-Owl
Megascops asio

A tiny bundle of ferocity, the Eastern Screech-Owl may be one of Maryland and DC's most common and widespread birds of prey. But its strongly nocturnal habits, soft whinnies and tremolo whistles, and small size render it tricky to find in atlas blocks. Although these owls were widespread, with records from all counties and DC, they were recorded in only a little over half of the blocks covered in 2002–2006. This owl is polymorphic with a common gray morph, a frequent rufous morph, and uncommon brown intermediates.

Eastern Screech-Owls are often found near water. They live in small woodlots and near the edges of larger tracts of forest where cavities are available for nesting. These include either natural or man-made nest boxes, especially those put out for Wood Ducks. Maryland and DC haunts include riparian corridors, forest edges near farmland, urban parks, wooded suburbs, swamps, and the edges of marshland.

Screech-owls are resident year-round; young birds disperse after they achieve independence and can move up to 16 km (10 mi) from their birthplace (Belthoff and Ritchison 1989). Safe dates outside of dispersal times for young birds were set from 1 April to 15 August; unfortunately for atlas surveyors this was outside of the documented peak of ter-

ritorial calling (D. G. Smith et al. 1987). Territorial establishment begins in mid-February and nests are under way by early March. Maryland and DC egg dates range from 8 March to 9 May (Jeschke 1996a; MNRF). During this project nestlings were found as late as 19 July 2005 near Coopstown, Harford County (C. Robbins), suggesting egg dates later than 9 May. The most frequent confirmation method for this project, 70% of records, was the observation of dependent young.

The Eastern Screech-Owl was generally referred to as common in historical references, although Eifrig (1904) termed it relatively less common than in other states, alluding to bounties paid out for all hawks and owls. Stewart and Robbins (1958) called it uncommon in all sections, with pockets where the birds were fairly common. During the 1983–1987 atlas, this owl was found throughout Maryland and in the District of Columbia (Jeschke 1996a).

Most owls, including Eastern Screech-Owl, present difficulties for volunteer surveyors because effective surveying requires nighttime fieldwork. In Maryland and DC use of playback of screech-owl territorial calls has been encouraged in order to obtain a reasonable atlas distribution. In the final year of fieldwork in this project (2006), MP3 audio files were offered to observers, and block busters were paid to do nocturnal surveying in addition to diurnal fieldwork. Maps from both atlas projects show a patchy distribution; gaps appear to be as much attributable to variation in survey effort as lack of resident birds. For instance there was an obvious improvement in coverage in Worcester County in 2002–2006.

BBS data are unreliable for recording population trends for Eastern Screech-Owl, as they are for other night birds. Indirect evidence suggests these owls may have declined between atlas projects. Determining whether this owl has actually declined and if such declines are chronic or cyclic requires more systematic monitoring. As long as hollow and dead trees are left standing in woodlots and suitable nest boxes are provided, this adaptable owl should be able to hold its own in Maryland and DC.

WALTER G. ELLISON

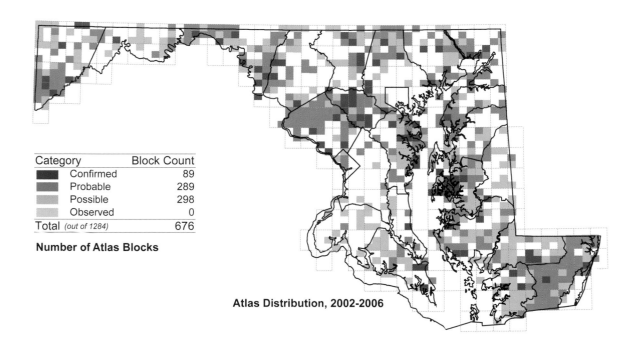

Category	Block Count
Confirmed	89
Probable	289
Possible	298
Observed	0
Total (out of 1284)	676

Number of Atlas Blocks

Atlas Distribution, 2002-2006

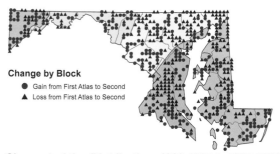

Change by Block

● Gain from First Atlas to Second
▲ Loss from First Atlas to Second

Change in Atlas Distribution, 1983-1987 to 2002-2006

				Change	
Atlas Region		1983-1987	2002-2006	No.	%
	Allegheny Mountain	40	54	+14	+35%
	Ridge and Valley	82	83	+1	+1%
	Piedmont	187	166	-21	-11%
	Upper Chesapeake	67	48	-19	-28%
	Eastern Shore	187	225	+38	+20%
	Western Shore	115	97	-18	-16%
Totals		678	673	-5	-1%

Change in Total Blocks between Atlases by Region

Great Horned Owl
Bubo virginianus

Ferocious, powerful, adaptable, and tough, the Great Horned Owl is the most widespread of Maryland and DC's three common owl species; indeed it is often a deadly competitor for hunting range and resources against the Barred Owl and Eastern Screech-Owl. This aggressive and almost fearless predator elicits mixed emotions from people. As with any strong and efficient predator capable of taking a wide range of prey, the Great Horned Owl is impressive, imposing, and awe inspiring. But some people view it as dangerous and destructive because it readily assaults poultry, game, endangered wildlife, and even the occasional house pet. These owls generally require mature trees in which to nest and roost, but Great Horned Owls that are hunting prefer to work from perches on the edge of woodland or in the midst of open landscapes. Great Horned Owls nest state-wide (Stewart and Robbins 1958; Jeschke 1996b). They are least numerous in treeless landscapes and in extensive closed canopy forest; they are most common in areas with modest-size woodlots and farmland.

The peak of territorial calling by Great Horned Owls is from late October to late December (atlas safe dates were set to begin late in this period). Although Great Horned Owls are persistent hooters, they generally do not respond vocally to broadcast recordings of their calls, and they often approach recordings silently if they do so at all; it is most effective to survey for them by ear. Nesting takes place from mid-January to mid-May (rarely early June); egg dates for Maryland nests range from 8 January to 15 May (MNRF). Most young fledge before May. Young Great Horned Owls will stay in their parents' home range until early autumn, begging and occasionally receiving food (Houston et al. 1998). For this atlas, 96 of 137 confirmations were for young out of the nest.

Effective surveying for nesting Great Horned Owls requires atlas workers to survey at night and at times of year when most other birds are not nesting, As such this owl's range map shows an uneven distribution that is largely attributable to these censusing issues. Although there was a net increase of 72 blocks from the 1983–1987 to the 2002–2006 Maryland-DC atlas project this apparent increase should be interpreted with caution. For instance, Great Horned Owls were not found in 208 blocks where they had been found two decades ago, and most of the apparent losses are concentrated in regions that had less owl surveying in the 2000s versus the1980s. In other areas it is evident that coverage for night birds increased between atlas projects, for example, in Worcester County. The Great Horned Owl has definitely held its own over the past 20 years. It may have increased marginally in some areas, perhaps because of reduced shooting and trapping since the 1960s when legal protections for birds of prey increased. Also, the public's attitude toward owls and other raptors has improved somewhat.

Great Horned Owls and other night birds are not suited for daytime roadside surveys, so BBS trends must be viewed cautiously. Nonetheless it is of some interest that there has been a long-term strong increase by Great Horned Owls on Maryland BBS routes. It is possible that the Great Horned Owl has declined in some parts of Maryland because of exposure to West Nile virus. Raptors are not as prone to severe illness and morbidity as some other birds are (e.g., American Crow), but West Nile virus can produce severe enough

damage to suppress survival, and captive raptors have experienced die-offs from West Nile infection (Ludwig et al. 2002; Nemeth et al. 2006; Joyner et al. 2006). The effects of West Nile virus are not evident in 2002–2006 atlas results. The Great Horned Owl has proven resilient. It survived an extended period of persecution by sportsmen and farmers and is still widely distributed throughout Maryland and DC. Great Horned Owls should continue to broadcast their deep ominous hoots across the state and district for decades to come.

WALTER G. ELLISON

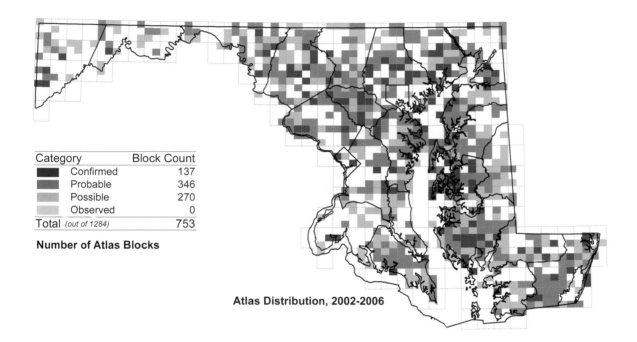

Category	Block Count
■ Confirmed	137
■ Probable	346
■ Possible	270
■ Observed	0
Total (out of 1284)	753

Number of Atlas Blocks

Atlas Distribution, 2002-2006

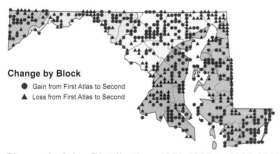

Change by Block
- ● Gain from First Atlas to Second
- ▲ Loss from First Atlas to Second

Change in Atlas Distribution, 1983-1987 to 2002-2006

Atlas Region	1983-1987	2002-2006	Change No.	%
Allegheny Mountain	24	30	+6	+25%
Ridge and Valley	62	69	+7	+11%
Piedmont	176	202	+26	+15%
Upper Chesapeake	74	65	-9	-12%
Eastern Shore	205	239	+34	+17%
Western Shore	138	146	+8	+6%
Totals	679	751	+72	+11%

Change in Total Blocks between Atlases by Region

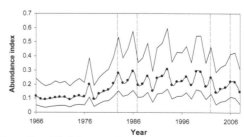

Breeding Bird Survey Results in Maryland

Barred Owl
Strix varia

Of Maryland's two widespread "hoot owls" the Barred Owl is the one with a rounded, tuftless head and large dark brown eyes. Its distinctive barking hoot "who cooks for you, who cooks for you-all" and its array of caterwauls and cackles can be heard in mature wet or swampy woods nearly throughout Maryland.

Barred Owls inhabit mature deciduous and mixed forest, often, but not always, extensive. They appear to prefer wet woods and often live in cool damp ravines, swamps, and floodplains. In open and urbanized landscapes they will sometimes occur in mature timber on riparian corridors and in parks. Mature and large trees are necessary to provide roosting and nest sites.

The Barred Owl is resident year-round in Maryland and DC; young birds can disperse considerable distances after becoming independent (Mazur and James 2000). These owls are most vocal, and therefore easiest to detect, from late January to mid-March; there is a resurgence of spontaneous calling from September through October when young birds are dispersing. Fortunately for atlas surveyors, adults are strongly territorial year-round and can be induced to call

or approach when they hear recordings or imitations of their hoots. Barred Owls usually nest in cavities in large trees and snags but will also reuse old stick platform nests. Fledglings are dependent on their parents for food until August, and their thin, rising, breathy begging calls are distinctive but hard to hear. Four-fifths of confirmations during this project were for fledged young. Although these owls are single brooded, the nesting season is long. Maryland and DC egg dates extend from 25 February to 26 May with unfledged nestlings to 25 June (Stewart and Robbins 1958; MNRF).

Kirkwood (1895) called the Barred Owl "numerous down the necks" (eastern Baltimore and southern Harford counties) and fairly common elsewhere. Eifrig (1904) found it less common than Great Horned Owl in Allegany and Garrett counties. Stewart and Robbins (1958) called it common on the lower Coastal Plain and fairly common elsewhere, echoing Kirkwood's description.

Variation in observer effort affects atlas maps for any night bird because these birds require special efforts at odd hours and outside the peak nesting season of most other birds. Given this caveat, atlas maps for both Maryland and DC projects give a good impression of the Barred Owl's statewide range and the larger gaps within them probably reflect actual areas of scarcity if not absence. From 1983 to 1987 this owl was found statewide, with biologically relevant gaps in heavily farmed parts of the eastern Ridge and Valley,

Piedmont, and Upper Chesapeake, and in the pinelands and marshes of the lower Eastern Shore (Jeschke 1996c). From 2002 to 2006 this owl was found in 115 more blocks, an increase of nearly 20%. Some of these gains may be related to improved coverage, as appears likely in Garrett County where all owls increased, but gains in the northern Piedmont and eastern Ridge and Valley may reflect genuine distributional increases.

Breeding Bird Survey routes do not sample night birds effectively; nonetheless, Barred Owls have consistently increased on BBS routes almost throughout their range over

the past four decades. Reasons for the apparent increase in the nesting range for this owl are probably related to reforestation in some areas and the maturation of timber along riparian corridors and in parks. Over the same period the Pileated Woodpecker, a major provider of potential nest sites for this owl, has increased. Another plausible explanation for the apparent increase is that these owls have become more tolerant of living in close proximity to humans. At present the future of the Barred Owl in Maryland and DC seems secure.

WALTER G. ELLISON

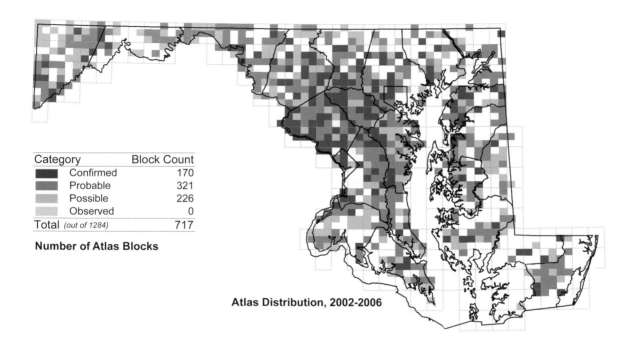

Category	Block Count
Confirmed	170
Probable	321
Possible	226
Observed	0
Total (out of 1284)	717

Number of Atlas Blocks

Atlas Distribution, 2002-2006

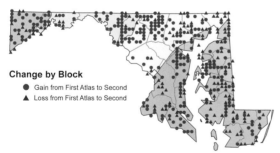

Change by Block
- ● Gain from First Atlas to Second
- ▲ Loss from First Atlas to Second

Change in Atlas Distribution, 1983-1987 to 2002-2006

Atlas Region	1983-1987	2002-2006	Change No.	Change %
Allegheny Mountain	45	67	+22	+49%
Ridge and Valley	67	93	+26	+39%
Piedmont	185	221	+36	+19%
Upper Chesapeake	59	61	+2	+3%
Eastern Shore	106	103	-3	-3%
Western Shore	139	171	+32	+23%
Totals	601	716	+115	+19%

Change in Total Blocks between Atlases by Region

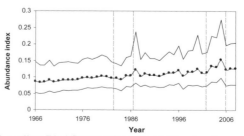

Breeding Bird Survey Results in Maryland

Mark Hoffman

Long-eared Owl
Asio otus

Although the Long-eared Owl is widespread in the Northern Hemisphere including a broad North American range, it is thinly distributed and something of a mystery to many bird observers. Low numbers and a spotty distribution may arise from competition with larger owls (Sharrock 1976). It nests and roosts in dense stands of trees, often conifers, but it hunts in the open. These owls are similar in general appearance to the larger Great Horned Owl but are more slender with longer ear-tufts and with longer wings that are employed in buoyant coursing flight similar to that of the closely related Short-eared Owl. The territorial call, a low *hoo* given at fairly long intervals in an almost interminable series, is distinctive but soft and easily missed. In contrast to the Short-eared Owl, the Long-eared is almost entirely nocturnal. It is most often seen by birders at winter roost sites that are often communal. When communal roosts break up, some birds may remain to nest at the roost site (Santner 1992a). There are few Maryland and DC nesting records,

only two over the past 60 years, all of which are from the western Coastal Plain and westward.

Long-eared Owls are sometimes characterized as birds of extensive mature forest, but they actually inhabit mixed open and closed landscapes. They feed primarily on small rodents, often voles or mice caught in grassland or in open shrubland. They nest in old platform nests built by other animals, often used crows' nests, in dense stands of trees. In eastern North America they commonly use mature pine and spruce plantations. They sometimes nest on ledges or on the ground (Marks et al. 1994), but Long-eared Owls are much more arboreal than the otherwise similar Short-eared Owl.

Although they are often called year-round residents in their nesting range, Long-eared Owl populations show some seasonal movement and a propensity for opportunistic nesting where prey is abundant into the spring months (Marks et al. 1994). The broad overlap of early spring nesting with late departure by roosting owls makes surveying for breeders difficult (Santner 1992a). Five Maryland and DC egg dates range from 3 April to 1 May, and six nests with nestlings have been reported from 14 April (indicating a March laying date) to 1 June (Stewart and Robbins 1958; MNRF).

Some early bird references refer to greater numbers in

the nineteenth and early twentieth centuries, but the Long-eared Owl has apparently never been more than uncommon in Maryland and DC. C. Richmond (1888) called it rare in DC, and Kirkwood (1895) called it "resident, but as far as I know not common." Stewart and Robbins (1958) found it a rare permanent resident and cited nesting only from the upper Western Shore and in the Piedmont. There were no reports of nesting in Maryland from 1950 to 2000, including the 1983–1987 atlas project. Iliff, Hafner, and Armistead (2000) found a nest with four nestlings in a pine grove in southeastern Garrett County on 16 May 2000. They reported occupancy of the site by territorial and young owls from 25 March to 1 July. In 2002, early in this atlas project, another nest with young was found by Iliff in the same grove on 20 and 22 May (Iliff, pers. comm.). No other breeding report of this owl was made in the remaining four years of the atlas project.

There is no BBS trend for Long-eared Owl because of its nocturnal habits and its early nesting season, which takes place before BBS routes are conducted. This owl poses serious problems for general purpose, volunteer-powered surveys such as an atlas project. To consistently find this bird requires highly motivated and dedicated observers working in the field at night from late February to May. The 2000 nest was started in late March while nonbreeding birds were still at the site (Iliff, Hafner, and Armistead 2000), an indication of how tricky it can be to establish that nesting birds are present. Birds found roosting in March, April, and May in Maryland and DC could be nesting. They should be treated circumspectly by anyone lucky enough to find them to reduce harassment by understandably overenthusiastic birders wanting to observe this seldom-seen owl. Thorough dedicated surveys should be conducted for Long-eared Owls to obtain a better understanding of their status.

WALTER G. ELLISON

Note: The single 2002–2006 atlas record for nesting Long-eared Owl came from Garrett County. The exact location is being kept confidential to protect this rare owl from disturbance.

Short-eared Owl
Asio flammeus

The Short-eared Owl is the only locally nesting owl regularly seen hunting in the open during daylight hours, although its primary feeding hours are dawn and dusk. It is most often seen coursing low over open grassy land, either wet or dry, with deep snappy wing-beats. It could be briefly taken for a Northern Harrier, a raptor that patrols its habitat by day; these two birds of prey can sometimes be seen in aerial skirmishes in shared haunts. This owl's biology is largely tied to the population cycles of its prey, small mammals such as *Microtus* voles. Although some populations appear to be fairly sedentary where prey is generally plentiful, most are nomadic and opportunistic, wintering and nesting where prey populations are at or close to maximum levels (Glue 1993). This owl has a worldwide range over five continents and many oceanic islands. It is a very rare breeder in Maryland with few proven nesting records, although it occurs regularly in winter in modest numbers. The Short-eared Owl is officially listed as endangered in Maryland by the DNR.

The Short-eared Owl's basic habitat requirements are met by extensively open land with tall grass, abundant prey, and low disturbance from predators and humans (Sharrock 1976). Nesting habitats in Maryland have included large hayfields, the drier upland parts of tidal marshes, and grassland on reclaimed strip mines. Areas that have sizable winter roosts would be appropriate places to anticipate nesting attempts.

This owl's breeding season begins in March and extends into August (Holt and Leasure 1993). Some wintering birds may display before moving on, but the spectacular aerial courtship display is a good indication of where nesting pairs may be established. The nest is a shallow bowl nestled into tall grassy vegetation, often tucked next to a rock or shrub (R. Clark 1975). Female Short-eared Owls sit tight on the nest, so the usual way they are confirmed is by males' mobbing of predators or observers or by regular prey drops to his mate, who rises to catch the proffered food (Sharrock 1976).

This owl was unknown as a breeding bird in Maryland or DC to Kirkwood (1895) and most other early bird authorities although some reports existed for the summer months (Stewart and Robbins 1958). The first definite nesting attributed to Short-eared Owl in Maryland was of a "marsh owl" nest with eggs in Dorchester County in June 1923 (Stewart and Robbins 1958). Another nesting was reported from Dorchester County in 1958 (Robbins and Boone 1985). It was confirmed breeding based on mobbing, distraction displays, and prey exchanges near Poolesville, Montgomery County during the first Montgomery County breeding bird

George M. Jett

atlas (Klimkiewicz 1972). There were no nesting reports for the next 28 years, including during the 1983–1987 atlas project. A nest with young was found on 17 June 2000 in a reclaimed strip mine grassland in southeastern Garrett County by Iliff, Stasz, and Hafner (Ringler 2001). Two more nesting season reports were made at the location in 2001 and a bird was seen on Dans Mountain in Allegany County in August 2001 (Ringler 2002a, 2002b). The only 2002–2006 atlas report was from a large meadow on Dans Mountain on 4 May 2002, a single sighting with no further nesting evidence (S. Sires, pers. comm.).

There are no BBS trend data for Short-eared Owl in eastern North America. However, several sources suggest a long-term decline in North America (Wiggins 2004; Schneider 2008). Loss of large undisturbed grasslands is a major cause of these declines, but prey also appears to be less numerous and consistent on modern agricultural lands. Reclaimed strip mines have recently been used for nesting in Pennsylvania (Master 1992a) as well as in Maryland, but these will eventually be lost to succession unless managed as grassland. Extensive natural or near-natural grasslands and open shrublands must exist for the Short-eared Owl to continue among Maryland's nesting birds.

WALTER G. ELLISON

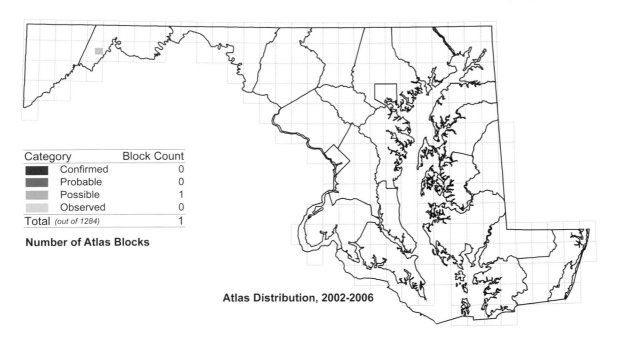

Category	Block Count
■ Confirmed	0
■ Probable	0
■ Possible	1
■ Observed	0
Total (out of 1284)	1

Number of Atlas Blocks

Atlas Distribution, 2002-2006

Change by Block

● Gain from First Atlas to Second
▲ Loss from First Atlas to Second

Change in Atlas Distribution, 1983-1987 to 2002-2006

			Change	
Atlas Region	1983-1987	2002-2006	No.	%
Allegheny Mountain	0	1	+1	-
Ridge and Valley	0	0	0	-
Piedmont	0	0	0	-
Upper Chesapeake	0	0	0	-
Eastern Shore	0	0	0	-
Western Shore	0	0	0	-
Totals	0	1	+1	-

Change in Total Blocks between Atlases by Region

Northern Saw-whet Owl
Aegolius acadicus

Northern Saw-whet Owls, the smallest species of owl that breeds in Maryland, are among the most charming bird species to inhabit the state, their large yellow eyes capturing the attention of virtually any observer fortunate enough to locate one. While relatively rare in Maryland, they are common but seldom-observed birds in forested areas of the northern tier of the United States and across boreal Canada (Godfrey 1986; Cannings 1993). As nearby as Pennsylvania they are a widely dispersed uncommon breeding owl (http://bird. atlasing.org/Atlas/PA/; D. Gross, pers. comm.). Saw-whet owls breed in West Virginia, Virginia, and North Carolina in the higher elevations of the central and southern Appalachian Mountains (Simpson and Range 1974; Buckelew and Hall 1994; Trollinger and Reay 2001). Northern populations are highly migratory and during the winter saw-whet owls can be found in nonbreeding habitats as far south as northern Florida (Lesser and Stickley 1967). They regularly winter in the Piedmont and Coastal Plain provinces of Maryland.

In early March, Northern Saw-whet Owls begin singing on breeding territories in the conifer wetlands of Garrett County (D. Brinker and K. Dodge, unpubl. data). These small owls frequent forests with dense understory structure that offers security from predation by larger owls. They are most often found in conifer swamps, alder wetlands, areas of dense eastern hemlock, or hardwood forest habitats with a thick understory of rhododendron and/or mountain laurel as well as in red spruce forests at higher elevations south of Maryland. Northern Saw-whet Owls are secondary cavity nesters most often found using old flicker or Pileated Woodpecker cavities or natural cavities formed in broken tree limbs (Cannings 1993). They will readily use nest boxes that are available in the appropriate habitats.

When the 1983–1987 atlas was conducted, there had been no confirmed nests of Northern Saw-whet Owls in Maryland (Jeschke and Brinker 1996). The first confirmed nest was located on 23 April 1993 in a nest box placed specifically for saw-whet owls in the southern portion of Cranesville Swamp (Brinker and Dodge 1993). From 1993 through 2008 nine additional nests were recorded in Garrett County (D. Brinker and K. Dodge, unpubl. data). Maryland egg dates range from 11 April through 20 May (D. Brinker and K. Dodge, unpubl. data), with most observations of eggs occurring during late April when nest boxes are traditionally checked for the first time each season. Chick dates range from 12 May through 22 June, with most dates in mid- to late May (D. Brinker and K. Dodge, unpubl. data). In these same nest boxes clutch and brood sizes have ranged from two to five. Chicks begin

fledging during late May and are dependent on their parents for food for at least a month thereafter. An adult male and chick were mist netted together in an occupied breeding territory on 30 June 1992 near Cunningham Swamp (Brinker and Dodge 1993).

Saw-whet owls are infrequent breeders in Garrett County and although undocumented by either atlas effort, they may be very rare irregular breeders at higher elevations as far east as Washington and western Frederick counties. During the first atlas Northern Saw-whet Owls were found in 3 atlas blocks but were not confirmed as breeding in any block (Jeschke and Brinker 1996). The increase in natural history information available for saw-whet owls since the first atlas project may have helped increase their detection in the second atlas. The 2002–2006 atlas found evidence of breeding Northern Saw-whet Owls in a total of 9 blocks, with 2 confirmed records and 1 additional block where the species was observed. Both confirmed records were from nest boxes in Cranesville Swamp. One atlas record (1 of only 2 probable records) was of an adult Northern Saw-whet Owl in a woodpecker cavity in a lone white pine snag located in a clear-cut in Garrett State Forest. This owl was first noticed in the cavity on 7 April 2006; it was observed in the cavity entrance for several weeks thereafter (D. Brinker, unpubl. data). For Northern Saw-whet Owls the increased number and distribution of confirmed atlas blocks is misleading and probably does not reflect a significant change in the Maryland breeding population of this species between the first and second atlases. Since the first atlas there has been one additional observation of a juvenile Northern Saw-whet Owl on Catoctin Mountain, a fledged individual photographed by Jim McGibney (pers. comm.).

Northern Saw-whet Owls have now been adequately documented as regular, but rare, breeding birds in western Maryland. Atlas efforts will probably never provide enough data on saw-whet owl populations to assess population trends, although when sufficient nocturnal effort is part of a project, an atlas project can provide insight into the extent of the owls' normal breeding range. As long as western Maryland retains adequate forest cover with associated snags, healthy populations of primary cavity excavators, and healthy con-

ifer swamps, Northern Saw-whet Owls may remain a rare breeding species in Maryland. But with the potential loss of hemlock cover to hemlock wooly adelgids and the possibility that global climate change will significantly warm the higher elevations of Garrett County, Maryland may eventually lose one of its most charming bird species, the Northern Saw-whet Owl. Only time will tell.

DAVID F. BRINKER

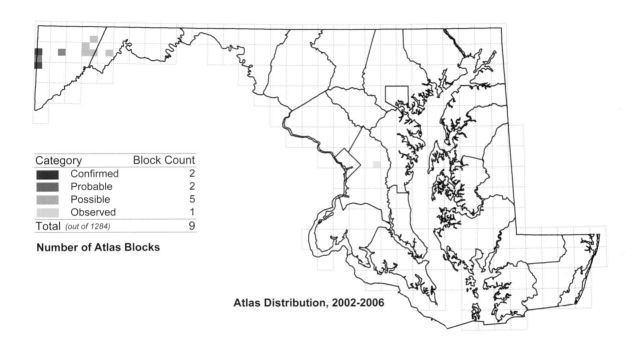

Category	Block Count
Confirmed	2
Probable	2
Possible	5
Observed	1
Total (out of 1284)	9

Number of Atlas Blocks

Atlas Distribution, 2002-2006

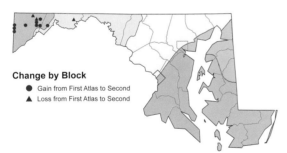

Change by Block
- ● Gain from First Atlas to Second
- ▲ Loss from First Atlas to Second

Change in Atlas Distribution, 1983-1987 to 2002-2006

Atlas Region		1983-1987	2002-2006	Change No.	%
	Allegheny Mountain	2	9	+7	+350%
	Ridge and Valley	1	0	-1	-100%
	Piedmont	0	0	0	-
	Upper Chesapeake	0	0	0	-
	Eastern Shore	0	0	0	-
	Western Shore	0	0	0	-
Totals		3	9	+6	+200%

Change in Total Blocks between Atlases by Region

Common Nighthawk
Chordeiles minor

For some older birders the Common Nighthawk calls to mind warm summer evenings in towns with brick storefront Main Streets as the sky was crisscrossed by pirouetting, saber-winged, batlike birds. On occasion the nighthawks would utter a breathy *peeert* or rise high and make a swooping dive accompanied by a windy rush of air through the wingtips. Sadly the "bullbat" is no longer widespread or a regular summer bird over most of Maryland and DC. It has become increasingly scarce and localized over the past 50 years.

Common Nighthawks nest in flat, open, unvegetated places and feed on the wing on hatches of flying insects, often near water in eastern North America. Traditional nest sites include sand dunes, beaches, sandbars, burns, overgrazed pastureland, rock ledges, and boulders. The last two nest sites suggest why nighthawks started nesting on flat graveled or pea-stone and asphalt rooftops in towns and cities during the mid-nineteenth century (A. Gross 1940). The earliest reference to roof-nesting by nighthawks in Maryland was from Baltimore (Kirkwood 1895). Whether nighthawks abandoned natural nesting sites for rooftops for a time is not clear, but current populations occur in both habitats. Many blocks with displaying birds in 2002–2006 were away from large towns with flat roofing.

Common Nighthawks winter in southern South America (Poulin et al. 1996). They arrive in Maryland during late April and early May and depart by late September or early October. Autumn migrants often occur in large flocks sometimes numbering into the hundreds that feed on swarms of flying ants. About one-third of nests from Maryland and DC listed by Sauer (1996; MNRF) were found on beaches in St. Mary's County and just over half on rooftops. Walbeck (1989) suggested that Common Nighthawks might prefer relatively dark rooftops in Frostburg, Allegany County. However, when roofs are black, with rubberized or pure asphalt surfaces, they often become dangerously hot in summer and provide no camouflaging background for sitting nighthawks or chicks (Marzilli 1989). Maryland and DC egg dates range from 23 May to 13 July with nestlings to 31 July (Walbeck 1989).

C. Richmond (1888) called Common Nighthawks uncommon nesters in DC, although a quarter century earlier they were considered common by Coues and Prentiss (1862). Kirkwood (1895) called them common but local. Stewart and Robbins (1958) found them uncommon and local throughout Maryland and DC. The distribution mapped in 1983–1987 (Sauer 1996) at first glance appears similar but is complicated by the small number of probable and confirmed breeding reports, just one-third of reports. If a proportion of possible breeders may have been wanderers or long-range foraging birds, the true nesting range was likely smaller than the 139 blocks mapped. Nighthawks were found in 62% fewer blocks from 2002 to 2006. Records of probable nesting fell to about 25% of all reports and there were no confirmed nestings. There were only three reports of probable nesting above the fall line, with two from Frostburg. A number of reports from the lower Eastern Shore appear to be in traditional habitats, including the sand dunes of Assateague Island (7 blocks) and the pinelands of western Dorchester County (5 blocks).

Although the BBS technique is ill suited for sampling

Common Nighthawk populations, BBS routes show a 1.7% per year continental decline and a steeper 5% per year fall on eastern regional routes. Eastern nighthawk populations may have risen during the nineteenth-century peak of agricultural clearing and declined steadily since then. There is a strong contrast in the number of nighthawks on BBS routes between eastern and western North America (Price et al. 1995). Most of the decline of these aerial acrobats may be laid to habitat losses and perhaps to the shortcomings of

roofs as nesting habitat. Rooftops do not appear to be preferred in populations with an apparent choice of nesting sites (Brigham1989). Fewer graveled roof surfaces may also reduce nesting success on rooftops (Marzilli 1989). Nesting habitat for nighthawks needs to be preserved and improved (e.g., by providing gravel nest boxes on roofs; Marzilli 1989) for this species to remain among Maryland's regular nesting birds.

WALTER G. ELLISON

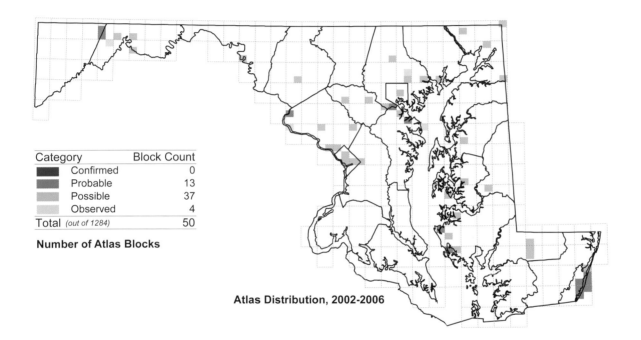

Category	Block Count
Confirmed	0
Probable	13
Possible	37
Observed	4
Total (out of 1284)	50

Number of Atlas Blocks

Atlas Distribution, 2002-2006

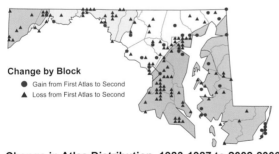

Change by Block
- ● Gain from First Atlas to Second
- ▲ Loss from First Atlas to Second

Change in Atlas Distribution, 1983-1987 to 2002-2006

Atlas Region	1983-1987	2002-2006	Change No.	%
Allegheny Mountain	12	3	-9	-75%
Ridge and Valley	24	3	-21	-88%
Piedmont	29	15	-14	-48%
Upper Chesapeake	7	3	-4	-57%
Eastern Shore	20	19	-1	-5%
Western Shore	41	7	-34	-83%
Totals	133	50	-83	-62%

Change in Total Blocks between Atlases by Region

George M. Jett

Chuck-will's-widow
Caprimulgus carolinensis

The "hollerin' boys" of warm humid moonlit nights down the necks in Chesapeake Bay country go on and on about the late Chuck Will's spouse. The Chuck-will's-widow is a large but otherwise fairly typical nightjar: strictly nocturnal, a persistent caller at dawn and dusk and on nights around the full moon, and an aerial and ground feeder on large insects and the occasional small vertebrate. In essence a native of the southeastern United States, it has slowly spread northward over the past 80 years and is now a regular resident in the piney woods of Maryland's Coastal Plain.

The habitat of the Chuck-will's-widow includes dry woodland and open land for foraging flights. They prefer open pine and pine-oak woods, but they will also occupy oak-hickory woods. They appear to be outnumbered by the related Whip-poor-will in extensively wooded country but are more numerous in mixed settings that include open fields and pastureland (R. Cooper 1981).

Chuck-will's-widows are migratory, wintering in Mexico and Central America, Colombia, and the Greater Antilles (Straight and Cooper 2000). They arrive in Maryland in late April and are gone by late September (Iliff, Ringler, and Stasz 1996). They sing regularly until late June, vocalize only sporadically in July, and rarely in August. Their nests are shallow depressions on the ground in leaf litter. The clutch is small, usually two eggs (Straight and Cooper 2000). In order to have any chance of flushing an incubating or brooding adult, observers must walk off-trail, even then birds will not flush unless nearly trodden upon. Adults near nests will fan their tails to display the white-edged outer feathers and croak agitatedly (Ricciardi 1995). Maryland has few nest records for Chuck-will's-widow, nonetheless egg dates range from 9 May to 8 July (Stewart and Robbins 1958; MNRF). Downy young have been seen on 2 June 1994 (Ricciardi 1995), and on 18 June (Reese 1996c; MNRF).

The Chuck-will's-widow's range expansion, which has been largely near the coast, has been ongoing since at least 1921 when they were found nesting in Maryland and Cape May, New Jersey (Court 1921; Fables 1955). Early Maryland reports date back to the 1890s (Kirkwood 1895; M. Richmond in Stewart and Robbins 1958). They appear to have increased most steadily on the Coastal Plain in Maryland and Delaware in the 1930s (Reese 1996c; Hess et al. 2000). Stewart and Robbins (1958) delineated a range north to northeastern Anne Arundel County, west to eastern Charles County on the Western Shore, and north to Kent Island on the Eastern Shore, with most on tidewater save for the lower Eastern Shore and St. Mary's County.

The Chuck-will's-widow's nesting range in Maryland was generally similar from 1983 to 1987 (Reese 1996c), although they had expanded to fill much of the Coastal Plain including much of southern Prince George's County and southwestern Kent County. There were a smattering of records just west of the fall line on the Piedmont and an exceptional report in central Washington County just north of a pioneering population in West Virginia (Hall 1983). In the 2002–2006 atlas they appear to have retrenched somewhat, with no records away from the Coastal Plain and with losses in suburban Anne Arundel and Prince George's counties. Nonetheless they were recorded from about the same number of blocks and there was no significant change in the number of probable and confirmed records.

The overall trend on BBS routes in North America has been downward at 1.7% per year since 1966. Chuck-will's-widows are heard on only the earliest 5 or so stops per 50-stop route, so sample sizes are very small. The declines probably reflect changes in land use, including loss of open land adjacent to woodland because of reforestation and urbanization, changes to pure crop agriculture from hay and pasture, and removal of small woodlots and hedgerows. The north-eastern range expansion reached Long Island, New York, in the 1970s and they are still present there in small numbers (Mitra 2008b). They may also have nested for a time on the outer coast of Massachusetts (Veit and Petersen 1993). At present the Chuck-will's-widow appears to be holding its own in Maryland, but it bears watching given the broader decline suggested by the BBS data.

WALTER G. ELLISON

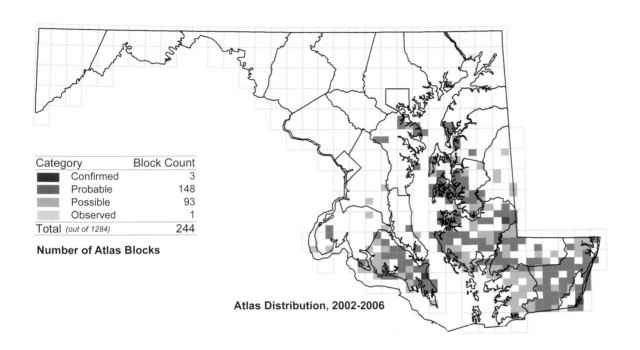

Category	Block Count
Confirmed	3
Probable	148
Possible	93
Observed	1
Total (out of 1284)	244

Number of Atlas Blocks

Atlas Distribution, 2002-2006

Change by Block
- ● Gain from First Atlas to Second
- ▲ Loss from First Atlas to Second

Change in Atlas Distribution, 1983-1987 to 2002-2006

Atlas Region	1983-1987	2002 2006	Change No.	%
Allegheny Mountain	0	0	0	-
Ridge and Valley	1	0	-1	-100%
Piedmont	4	0	-4	-100%
Upper Chesapeake	8	3	-5	-63%
Eastern Shore	162	181	+19	+12%
Western Shore	63	58	-5	-8%
Totals	238	242	+4	+2%

Change in Total Blocks between Atlases by Region

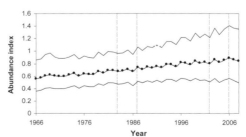

Breeding Bird Survey Results in Maryland

Eastern Whip-poor-will
Caprimulgus vociferus

The Whip-poor-will chants its name almost incessantly until dawn breaks into day or until dusk falls utterly into night, unless the moon is bright and rides high. Poets and storytellers have called this voice of the night many things, from nostalgic to demonic. By day this beautifully cryptic bird, arrayed in a complex dead leaf pattern, roosts on the ground or along low branches. It feeds on night-flying insects, especially medium to large moths. This nightjar is a widespread but very localized and declining summer resident in Maryland and DC.

The Whip-poor-will's habitat is open woodland, but the type of woodland and its extent and age appear to vary across the bird's broad North American range. Whip-poor-wills appear to avoid extensive mature, closed canopy forest and also seem to eschew heavy undergrowth, although a well-developed understory is not unusual in occupied woods. They require some openings for feeding, but they prefer larger woodlots with relatively small gaps to open lands with small woodlots (R. Cooper 1981). Some habitats that have large populations are pine-oak barrens, dry oak-hickory woodland, and second-growth forest adjacent to brushy overgrown farmland. Whip-poor-wills may be at a competitive disadvantage versus Chuck-will's-widow in loblolly pine woodlands and in more open landscapes with small woodlots. In southern Maryland the Whip-poor-will is found in forested upland areas and the Chuck-will's-widow in pine lots in farmland and near tidewater (Reese 1996d).

Whip-poor-wills winter in the Gulf States, eastern Mexico, and northern Central America. They return to Maryland in mid-April and stay through September. Their singing is most persistent through the night for the five days before and after the full moon and they are most reliably censused when 50% or more of the moon's face is illuminated (Wilson and Watts 2006). Nesting is apparently timed to the phase of the moon with egg laying 10 days before the full moon (Mills 1986). Whip-poor-wills nest on the ground on unmodified leaf litter usually near a small tree or shrub that shades the site from the afternoon sun (Cink 2002). Maryland egg dates range from 24 April to 7 July with nestlings to 17 July (Reese 1996d; MNRF).

Kirkwood (1895) called Whip-poor-wills locally common; Eifrig (1904) found them widespread and fairly numerous in Garrett and Allegany counties. Stewart and Robbins (1958) described them as common in the Ridge and Valley, Eastern Shore, and Western Shore, and fairly common in the Alleghenies, the Piedmont, and the Upper Chesapeake. From 1983 to 1987 they were found in all counties, with one report from DC, but they were not uniformly distributed (Reese 1996d). There were large gaps in heavily farmed regions such as the Hagerstown Valley, in urbanized areas, and in the tidewater wetlands of the lower Eastern Shore. They were most uniformly distributed in places with extensive low to middle elevation woodland such as the Pocomoke basin.

Whip-poor-wills were not refound in 243 blocks from 2002 to 2006, a decline of 57%. They were missing from western Montgomery County, and many Western Shore blocks, places where they had been widespread in the 1980s. There were still clusters of occupied blocks in the central Pocomoke drainage, in eastern Dorchester County and western Charles counties, in and near the Patuxent Wildlife Research Center, in eastern Allegany County, and along the Allegheny Front.

Although BBS routes do not sample nocturnal birds well,

it is noteworthy that Whip-poor-wills have declined at a rate of 2.1% per year in North America and at 3.6% per year in Maryland. Many observers can recall when they last heard Whip-poor-wills near their homes. The second New York State atlas also recorded nearly a three-fifths loss of blocks between atlas projects (Medler 2008). The causes of these declines are multifaceted but can be essentially traced to habitat loss and degradation. Forest fragmentation, maturation of second-growth forest into closed canopy forest, and urbanization have all contributed to habitat losses. Regular

applications of pesticide may have severely reduced moth populations, reducing suitable prey. Whip-poor-wills are also often killed by vehicles as they forage along roads. The publicly owned forest blocks that hold many of the state's Whip-poor-wills need to be managed for uneven aged stands and clearings, and larger blocks of forest need to be planted in other areas if this haunting night singer is to remain among Maryland's avifauna.

WALTER G. ELLISON

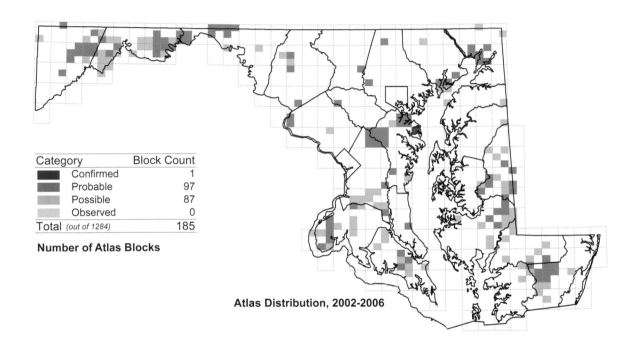

Category	Block Count
Confirmed	1
Probable	97
Possible	87
Observed	0
Total (out of 1284)	185

Number of Atlas Blocks

Atlas Distribution, 2002-2006

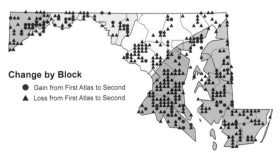

Change by Block
- ● Gain from First Atlas to Second
- ▲ Loss from First Atlas to Second

Change in Atlas Distribution, 1983-1987 to 2002-2006

Atlas Region	1983-1987	2002-2006	Change No.	%
Allegheny Mountain	34	20	-14	-41%
Ridge and Valley	49	37	-12	-24%
Piedmont	72	16	-56	-78%
Upper Chesapeake	33	13	-20	-61%
Eastern Shore	117	47	-70	-60%
Western Shore	122	51	-71	-58%
Totals	427	184	-243	-57%

Change in Total Blocks between Atlases by Region

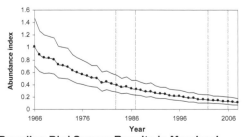

Breeding Bird Survey Results in Maryland

Mark Hoffman

Chimney Swift
Chaetura pelagica

Swifts are indeed among the fastest of birds on the wing. The towns and cities of Maryland host chittering parties of Chimney Swifts, dark and scimitar-winged, skittering across the heavens on long summer evenings. Chimney Swifts are essentially the daytime replacement for bats, living on the wing save at the nest, and snapping insects out of the air. Chimney Swifts have almost entirely abandoned their native nest sites, hollow snags and cave mouths, for human-made sites: particularly chimneys, but also the interior walls of dimly lit barns and other buildings not occupied by people. The nearly exclusive use of chimneys for nesting can be a problem for swifts if fires are lit during cool summer evenings or a chimney sweep visits during the nesting season (May to August). Another serious problem has been the decline of nest sites as homeowners switch to heating methods that do not require a working chimney. In many new houses the trend is toward decorative chimneys or no chimney at all, and working chimneys are often designed in ways that prevent nesting by swifts. Modern barns and silos are also not as accessible to birds as they once were. It is debatable whether the Chimney Swift increased greatly after it made the move into chimneys, since it is hard to know whether there were actually more chimneys in settled North America than there were hollow trees in pre-Columbian forests.

Chimney Swifts were reported in slightly fewer blocks (–2%) in 2002–2006 than in 1983–1987 (Zucker 1996). Whether this is evidence of a population decline is not certain. Parts of Maryland showing net losses and gains were areas that appeared to show worse and better coverage, respectively, than in the 1983–1987 atlas. Another confounding factor for interpreting block losses between projects was a presumably short-term decline in East Coast Chimney Swift populations connected to the large number of migrating swifts killed when Hurricane Wilma swept birds into northeastern North America and out to sea on 24–26 October 2005 (Dinsmore and Farnsworth 2006). In the summer of 2006, Chimney Swifts were scarce during atlas fieldwork (W. Ellison, pers. obs.); observers visiting lightly covered blocks to wrap up atlas work may have missed the species in some blocks that had small swift populations prior to 2006. Gaps in the Chimney Swift's breeding range were noticeable in heavily forested blocks in western Maryland, reflecting the species' shift away from natural nest sites, and on the Coastal Plain, most notably the wetlands of the Lower Shore and the uninhabited dunes of much of Assateague Island.

Chimney Swifts have been declining on BBS routes for several decades. This is also the case in Maryland. The current atlas distribution does not appear to reflect these losses; coverage artifacts may be involved in some apparent losses on the map; and the rates of confirmed, probable, and possible breeding are similar between this project and the 1983–1987 project. Confirming swifts is difficult because their nests are usually inaccessible to field-workers, and it is hard to identify fledglings because they are essentially independent of adults once they leave the immediate vicinity of the nest (Cink and

Collins 2002). Most confirmations were of adults seen entering nest sites. As such, 40% of atlas records referred to no more than possible nesting, and some possible records may reflect wandering nonbreeding birds or birds feeding in blocks neighboring the one in which they nested. It appears that swifts are suffering from long-term loss of nest sites as fewer homes are being built with more than decorative chimneys. Chimney Swifts will likely continue to decline and a future atlas project may reflect these declines more clearly.

WALTER G. ELLISON

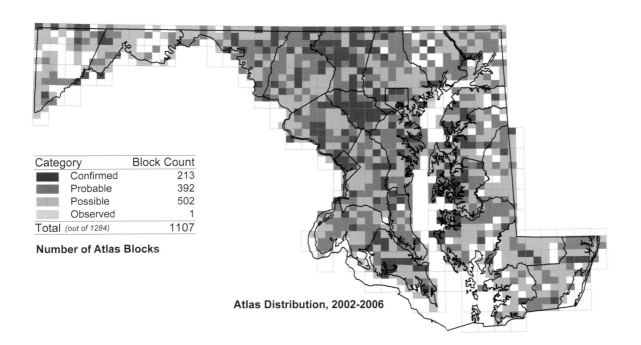

Category	Block Count
Confirmed	213
Probable	392
Possible	502
Observed	1
Total (out of 1284)	1107

Number of Atlas Blocks

Atlas Distribution, 2002-2006

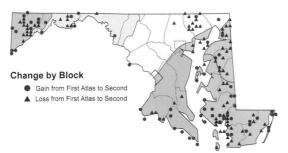

Change by Block
- ● Gain from First Atlas to Second
- ▲ Loss from First Atlas to Second

Change in Atlas Distribution, 1983-1987 to 2002-2006

Percent of Stops
- 50 - 100%
- 10 - 50%
- 0.1 - 10%
- < 0.1%

Relative Abundance from Miniroutes, 2003-2007

Atlas Region	1983-1987	2002-2006	Change No.	%
Allegheny Mountain	70	68	-2	-3%
Ridge and Valley	133	132	-1	-1%
Piedmont	311	303	-8	-3%
Upper Chesapeake	108	93	-15	-14%
Eastern Shore	267	267	0	0%
Western Shore	233	238	+5	+2%
Totals	1122	1101	-21	-2%

Change in Total Blocks between Atlases by Region

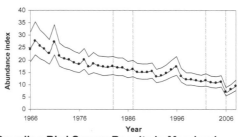

Breeding Bird Survey Results in Maryland

Ruby-throated Hummingbird
Archilochus colubris

Miraculously tiny and maneuverable, hummingbirds feed primarily on nectar, an uncommon diet for vertebrates in temperate climes. The Ruby-throated Hummingbird is the only nesting hummingbird in the eastern third of North America. Hummingbirds that nest at temperate latitudes, including the Ruby-throated, are strongly migratory as befits birds that depend on the nectar of flowers. Examples of flowers frequently visited by hummingbirds in Maryland are trumpet creeper, cardinal flower, bee-balm, jewelweed, wild columbine, and mimosa.

The Ruby-throated Hummingbird occurs throughout Maryland, although nowhere in very large numbers. This hummingbird requires trees for nest sites and sunny places with abundant nectar-rich flowers for food. The highest population densities for this hummingbird in Maryland have been recorded in floodplain forest (Stewart and Robbins 1958). Apparently hummingbirds have become increasingly common in suburban neighborhoods as homeowners have enticed them into their yards by putting out nectar-filled feeders and planting flowers that are attractive to hummingbirds.

The current geographic scope of the Ruby-throated Hummingbird's nesting range in Maryland and DC is essentially the same as it was in 1983–1987 (Ricciardi 1996a), but it is now more extensively reported in blocks dominated by urban, suburban, and agricultural land. Hummingbirds have also become more uniformly distributed in Maryland's western mountains. Ruby-throated Hummingbirds were reported in 114 more blocks in the present atlas project than in the first statewide atlas, an increase of slightly more than 10%. Hummingbirds are easy to overlook because of their tiny size and rapid flight; their weak vocalizations are hard to hear at a distance. One of the best ways to find hummingbirds in an atlas block is to spend some time watching blossoming food plants (e.g., mimosa; Ricciardi 1996a) or hummingbird feeders (if such observation is not too obvious and disturbing to homeowners). It is likely that many of the small gaps seen in the 2002–2006 range arose because hummingbirds were hard to find rather than absent from many of those blocks. An exception to this was the absence of hummingbirds from the islands and marshes on the lower Eastern Shore in Dorchester and Somerset counties.

It is difficult to prove breeding by Ruby-throated Hummingbird; nearly two-fifths of the reports in this atlas project referred to possible breeding. Nonetheless, more than a quarter of records were for confirmed nesting. Maryland egg dates range from 14 May to 20 August (MNRF). Although the nest is very small and well camouflaged, resembling a lichen-covered knot on a tree limb, 83 of 292 breeding reports were of active nests. Another 150 records were for recently fledged young.

Few Ruby-throated Hummingbirds are detected on BBS routes (0.4/route) so trend information derived from these surveys is unreliable. The long-term trend in Maryland has been downward, but not significantly so. In general the Ruby-throated Hummingbird appears to be holding its own in Maryland and DC and may even have increased over the past 20 years because of an increase in hummingbird feeders and nectar-rich flowers planted in gardens.

WALTER G. ELLISON

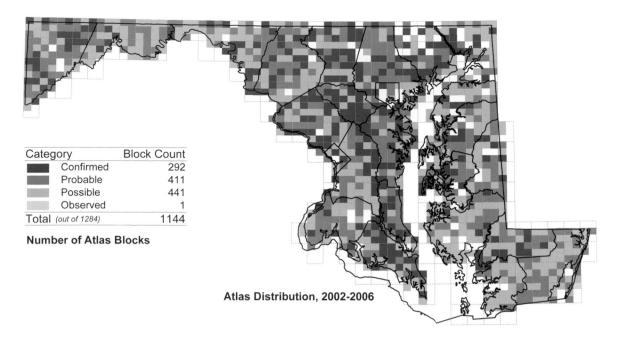

Category	Block Count
■ Confirmed	292
■ Probable	411
■ Possible	441
■ Observed	1
Total (out of 1284)	1144

Number of Atlas Blocks

Atlas Distribution, 2002-2006

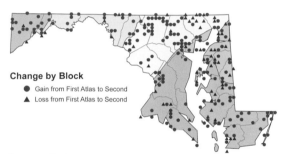

Change by Block

● Gain from First Atlas to Second
▲ Loss from First Atlas to Second

Change in Atlas Distribution, 1983-1987 to 2002-2006

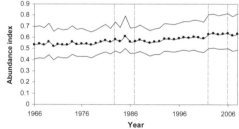

Percent of Stops

■ 50 - 100%
■ 10 - 50%
■ 0.1 - 10%
□ < 0.1%

Relative Abundance from Miniroutes, 2003-2007

Atlas Region	1983-1987	2002-2006	Change No.	%
Allegheny Mountain	76	89	+13	+17%
Ridge and Valley	127	142	+15	+12%
Piedmont	251	287	+36	+14%
Upper Chesapeake	95	99	+4	+4%
Eastern Shore	258	288	+30	+12%
Western Shore	219	235	+16	+7%
Totals	1026	1140	+114	+11%

Change in Total Blocks between Atlases by Region

Breeding Bird Survey Results in Maryland

Bill Sherman

Belted Kingfisher
Megaceryle alcyon

The Belted Kingfisher cuts an ungainly figure when perched, with its fly-away ragged crest, outsize head and bill, and tiny feet clinging invisibly to waterside perches. An airborne kingfisher is transformed into an aerial fishing-spear capable of diving from a high hover or from low-level flight above the water. Kingfishers will also attack prey by diving from a perch. The shivering clatter of kingfishers may be heard on watercourses statewide, although their preferences tend to broad shallow creeks, rivers, bays, ponds, and lakes rather than narrow, forested brooks. They are driven out of their haunts during the colder months when ice spreads over the water, although they remain if there is open water with available prey. In summer, kingfishers need nest sites at earthen banks for their long nesting tunnels, as well as water in which to fish. These banks need not be adjacent to water. Kingfishers will use suitable artificial banks in gravel pits and road-cuts up to 1.6 km (1 mi) away from their fishing grounds (Hamas 1974; W. Ellison, pers. obs.), but most pre-

fer their holes to be close to their fishing grounds. Maryland egg dates range from 11 April to 5 June (MNRF). Kingfishers feed on good-size aquatic animals besides fish, including frogs and crayfish. This bird is primarily a visual hunter so it prefers clear, pollution-free waters.

Belted Kingfishers are spottily distributed throughout Maryland and DC, tending toward major stream valleys (Dupree 1996a), notably in the western Ridge and Valley, the Allegheny Plateau, and the Lower Eastern Shore. Kingfishers are often scarce along heavily wooded streams and are rare along the small brooks that outnumber larger creeks in mountainous country. A clear break in the evenness of this kingfisher's nesting range is visible below the fall line as the species is even more localized on the Coastal Plain. Kingfishers are notably thin in eastern Queen Anne's County and southern Dorchester County and are not much more thickly distributed in Somerset, Wicomico, and Worcester counties. The flat landscapes of the Eastern Shore have few nest sites and many of these are far from water in gravel pits requiring long travel distances for parent kingfishers. This puts a premium on high prey abundance and accessibility.

The Belted Kingfisher has been declining on North American BBS routes over the last four decades. In Maryland, kingfisher numbers have shown no clear trend. Trends for the physiographic regions that cross Maryland include a long-term significant decline on the Upper Coastal Plain from Maryland to North Carolina. This decline is reflected in the net loss of 26 blocks on the Eastern Shore since the 1983–1987 Maryland-DC atlas, with most of this loss on the Lower Shore. These losses encompass the entire 3% loss of blocks between statewide atlas projects. Although kingfish-

ers have been lost from many blocks on the Western Shore, Piedmont, and Western Maryland, these losses have been balanced by gains in other blocks. The continued presence of kingfishers in Maryland and DC depends on the availability of suitable banks for nest tunnels and clean waters with an abundance of prey untainted by pollutants.

WALTER G. ELLISON

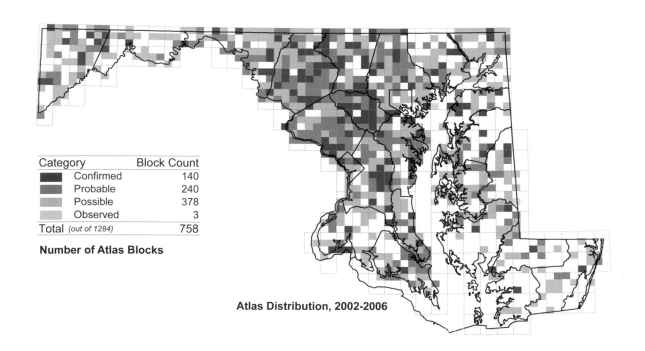

Category	Block Count
Confirmed	140
Probable	240
Possible	378
Observed	3
Total *(out of 1284)*	758

Number of Atlas Blocks

Atlas Distribution, 2002-2006

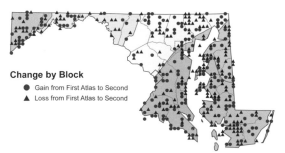

Change by Block
- ● Gain from First Atlas to Second
- ▲ Loss from First Atlas to Second

Change in Atlas Distribution, 1983-1987 to 2002-2006

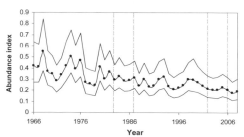

Percent of Stops
- 50 - 100%
- 10 - 50%
- 0.1 - 10%
- < 0.1%

Relative Abundance from Miniroutes, 2003-2007

			Change	
Atlas Region	1983-1987	2002-2006	No.	%
Allegheny Mountain	47	56	+9	+19%
Ridge and Valley	119	106	-13	-11%
Piedmont	260	261	+1	0%
Upper Chesapeake	69	56	-13	-19%
Eastern Shore	132	106	-26	-20%
Western Shore	151	170	+19	+13%
Totals	778	755	-23	-3%

Change in Total Blocks between Atlases by Region

Breeding Bird Survey Results in Maryland

Red-headed Woodpecker
Melanerpes erythrocephalus

*This very erratic species; common one year . . . and the next
entirely absent . . . , is resident, migratory or anything else,
apparently at its own sweet will.*

Frank C. Kirkwood

The Red-headed Woodpecker remains hard to pin down.
It is a colorful and distinctive woodpecker with a completely
cherry red head, white underparts and lower back, black upper back, and largely black wings with bright white secondaries. Its food habits are varied. In summer and fall it eats fruit;
it also catches insects and spiders by pouncing on them from
a perch or by sallying like a flycatcher. It feeds on acorns and
beechnuts during winter, caching what it does not consume
on the spot. These woodpeckers are very spottily distributed
in Maryland and DC, with the majority in the Piedmont and
eastern Ridge and Valley.

Red-headed Woodpecker habitat combines grassy open
ground, scattered large trees, and standing snags for nest and
roost holes (Conner 1976; Conner and Adkisson 1977). In
Maryland its haunts include open parklike woods in agricultural settings such as shady pastureland, flooded timber adjacent to open mature woods along reservoir and beaver pond
margins, open loblolly pine–oak woodland with numerous
snags often at the upland edge of tidal wetlands, and golf
courses. Good mast years promote winter survival (K. G.
Smith 1986), and nesting success may be aided by outbreaks
of the 17-year cicada. Although starlings may usurp Red-
headed Woodpecker nest holes, the woodpecker's relatively
late nesting season lessens the starling's impact on nesting
success (D. Ingold 1994).

The Red-headed Woodpecker's migratory behavior is erratic and somewhat nomadic as birds in autumn seek areas
with good acorn and beechnut crops. Most Maryland nesters
winter outside of the state although small pockets of wintering birds occur on the Piedmont and Coastal Plain in most
years (Hatfield et al. 1994). Nesting birds arrive on territory
from mid-April to early May. They dig cavities in long-dead
snags lacking all or most of their bark. They will sometimes
also nest in telephone poles that lack creosote treatment
(a rare practice of late), and old wooden fence posts (Bent
1939). Maryland and DC egg dates range from 25 April to
23 June, with nestlings to 10 September, suggesting later egg
dates (Wilmot 1996b; MNRF).

From the early to middle 1800s the Red-headed Woodpecker was common in the pastoral landscape of those times,
but it declined through the latter half of the century and it
continues to decline today (K. G. Smith, Withgott, and Rode-

Gary Van Velsir

wald 2000; Peterjohn 2001). C. Richmond (1888) described
it as common in DC, but informed Kirkwood (1895) it was
"not very common and [was] local" there by 1895. Eifrig
(1904) noted it was "abundant" in "deadenings" in Garrett
County but that few nested in Allegany County. Stewart and
Robbins (1958) found it local throughout, fairly common in
the Allegheny Mountains, uncommon to rare elsewhere,
not nesting on the lower Coastal Plain save for older records
from Dorchester County.

This woodpecker's status was similar in 1983–1987 although the major regions of occurrence shifted (Wilmot
1996b). It was largely restricted to southwestern Garrett
County near Oakland and had become widespread in the
Hagerstown Valley and in the western and northern Piedmont. It was found spottily along the Patuxent and lower
Potomac and was widely distributed on the lower Eastern
Shore (Wilmot 1996b). In 2002–2006 Red-headed Woodpeckers were found in 27 fewer blocks, with modest net losses
from all regions save the Upper Chesapeake, where it fell
from uncommon to very rare.

Red-headed Woodpeckers have declined steeply on North
American BBS routes with the highest rate since 1980. But
it has increased on Maryland BBS routes although this conclusion is rendered tenuous by the low numbers seen on
routes. Although these woodpeckers appear to have held
their own on the southeastern Piedmont and Coastal Plain,
they have declined steeply in the north. Their distribution
declined by 76% between the two New York State atlas proj-

ects (McGowan 2008b). Maryland is the northernmost East Coast state where these woodpeckers have a reasonably stable distribution and numbers. The farming landscapes of the eastern Ridge and Valley and northern Piedmont are particularly amenable to this species at present. However, removal of dead trees and conversion of pastoral lands to crop fields could easily change the situation as could die-offs of oak and beech. This woodpecker could easily become rare in the near future if its habitat deteriorates.

WALTER G. ELLISON

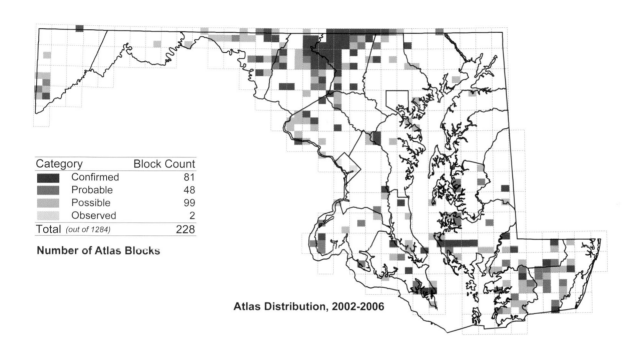

Category	Block Count
Confirmed	81
Probable	48
Possible	99
Observed	2
Total (out of 1284)	228

Number of Atlas Blocks

Atlas Distribution, 2002-2006

Change by Block
- ● Gain from First Atlas to Second
- ▲ Loss from First Atlas to Second

Change in Atlas Distribution, 1983-1987 to 2002-2006

Percent of Stops
- 50 - 100%
- 10 - 50%
- 0.1 - 10%
- < 0.1%

Relative Abundance from Miniroutes, 2003-2007

Atlas Region	1983-1987	2002-2006	Change No.	%
Allegheny Mountain	12	9	-3	-25%
Ridge and Valley	38	33	-5	-13%
Piedmont	89	83	-6	-7%
Upper Chesapeake	11	2	-9	-82%
Eastern Shore	68	64	-4	-6%
Western Shore	36	36	0	0%
Totals	254	227	-27	-11%

Change in Total Blocks between Atlases by Region

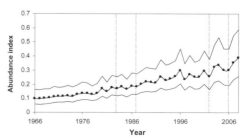

Breeding Bird Survey Results in Maryland

Red-bellied Woodpecker
Melanerpes carolinus

A lively, acrobatic, and loquacious bird, the Red-bellied Woodpecker is found virtually throughout Maryland and DC and is one of the 20 most prevalent breeding birds in the state. This was not always so; this woodpecker has spread fairly gradually from east to west in Maryland for more than a century. Red-bellied Woodpeckers need mature trees for nesting and feeding but do not require many of them, often occurring in small woodlots and narrow strips of forest on floodplains. In general these woodpeckers prefer their woods on the wet side, often with dense understory. Red-bellied Woodpeckers can be vocal, even noisy; atlas field-workers often first notice them via their far-carrying calls including a musical husky *vweeer!* a less emphatic *churr,* and an array of squirrel-like clucking and squawking.

The Red-bellied Woodpecker has a long nesting season, particularly for a woodpecker. Dates for confirmed nesting for 2002–2006 range from an occupied nest hole on 10 April to fledglings observed on 15 September. Maryland egg dates range from 18 April (Van Ness 1996a) to 14 June 2002 at Rocky Gap State Park, Allegany County (J. Kerns-McClellend, this project). Red-bellied Woodpeckers are at least occasionally double brooded in the northern part of their nesting range (Sullivan 1992; Rines 1998). Fledged young were by far the most common method of proving breeding in this project; begging young Red-bellied Woodpeckers utter a breathy, un-woodpecker-like high-pitched wheeze, which is distinctive, once learned.

Over the past century the Red-bellied Woodpecker has expanded its nesting range farther than all but a handful of eastern birds, although anecdotal evidence suggests the bird may have reconquered some of that range (J. Jackson and Davis 1998). Nineteenth-century references termed the Red-bellied Woodpecker rare in the District of Columbia (Coues and Prentiss 1883) and uncommon in eastern Maryland (Kirkwood 1895). By the mid-1950s Red-bellied Woodpeckers were fairly common to common east of Catoctin Mountain, but rare to the west (Stewart and Robbins 1958). Results from the first statewide Maryland-DC atlas showed this woodpecker had become ubiquitous in the Ridge and Valley region and had become established, largely at lower elevations, in Garrett County (Van Ness 1996a). Since the first atlas Red-bellied Woodpeckers have consolidated their range in the Alleghenies, with reports from 49 more blocks (an increase of more than 150%). The species has now completed its expansion across the Free State.

The only gaps in the Red-bellied Woodpecker's range in Maryland are now in a few blocks in the extensive marshy savannahs of southern Dorchester and western Somerset counties, parts of Assateague Island, and 15 high-elevation blocks in western Maryland. Possible factors in the Red-bellied Woodpecker's range expansion are listed in J. Jackson and Davis (1998). Among the most influential appear to be maturing forests on farmlands in the northeast, the advent of widespread bird feeding, and climatic warming. Although adult Red-bellied Woodpeckers are sedentary, young birds often disperse long distances, and such birds are recorded as migrants at migration hot spots such as Cape May, New Jersey (Walsh et al. 1999); these long-range dispersals likely aid the species' rapid spread into new locations.

Red-bellied Woodpeckers have increased steadily on BBS routes in Maryland over the past 40 years, with an overall increase of 30%. The greatest rates of increase among the physiographic regions in Maryland and neighboring states have been on the Allegheny Plateau, Ridge and Valley, and Northern Piedmont. The Red-bellied Woodpecker appears firmly established in Maryland and has few places left to conquer here.

WALTER G. ELLISON

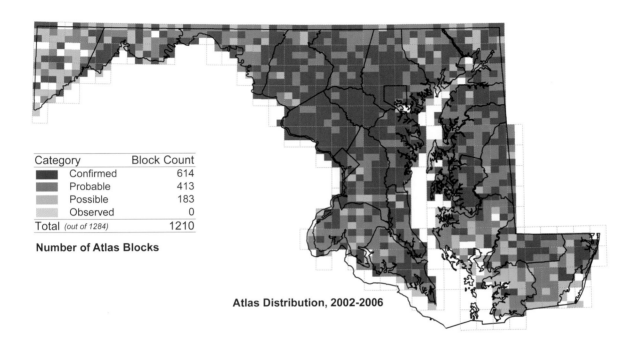

Category	Block Count
Confirmed	614
Probable	413
Possible	183
Observed	0
Total (out of 1284)	1210

Number of Atlas Blocks

Atlas Distribution, 2002-2006

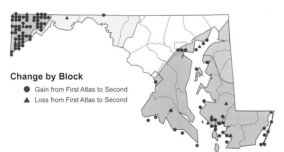

Change by Block
● Gain from First Atlas to Second
▲ Loss from First Atlas to Second

Change in Atlas Distribution, 1983-1987 to 2002-2006

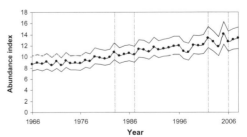

Percent of Stops
■ 50 - 100%
■ 10 - 50%
■ 0.1 - 10%
□ < 0.1%

Relative Abundance from Miniroutes, 2003-2007

Atlas Region	1983-1987	2002-2006	Change No.	%
Allegheny Mountain	30	79	+49	+163%
Ridge and Valley	147	147	0	0%
Piedmont	316	316	0	0%
Upper Chesapeake	115	119	+4	+3%
Eastern Shore	275	292	+17	+6%
Western Shore	243	251	+8	+3%
Totals	1126	1204	+78	+7%

Change in Total Blocks between Atlases by Region

Breeding Bird Survey Results in Maryland

Yellow-bellied Sapsucker

Sphyrapicus varius

George M. Jett

The Yellow-bellied Sapsucker is a characteristic bird of the northern hardwood forest. Its staccato stuttering drumming is a distinctive morning sound in maple, birch, beech, and hemlock woods in steep hilly country of northeastern North America. In winter and during migration, sapsuckers are often quiet and unobtrusive, but in the center of their nesting range they are usually conspicuous, with nasal mewing cries and regular loud drumming on resonant soundingboards including television antennas and metal roof flashing. These woodpeckers derive their unusual name from their habit of drilling rows of wells to the cambium layer in trees in order to lap up, or less elegantly "suck," the sap. At times they may girdle trees and kill them, thus gaining a bad reputation among foresters. This sapsucker reaches the southeastern edge of its nesting range in Maryland, where it has been recorded in varying numbers in the state's western mountains since at least the late nineteenth century. At present it is a rare summer resident restricted to Garrett County.

Yellow-bellied Sapsuckers are forest birds preferring fairly large woodlots with little forest fragmentation, although forest age seems less important. They are often seen near forest edges and tend to nest in early successional trees such as aspen and birch. They prefer mesic to wet mixed forest. Sapsuckers almost never occupy pure conifer woods because they require deciduous trees for making their sap wells.

These sapsuckers are uncommon to fairly common winter residents in eastern Maryland, particularly on the Coastal Plain (Hatfield et al. 1994), and they are fairly common migrants throughout Maryland and DC. They summer at high elevations, 730 m (2,400 ft) and higher in the Allegheny Mountains, arriving there in April and departing by late October (Stewart and Robbins 1958). A constant creaky, grating muttering emanates from nest cavities with nestlings; this muttering increases in volume when adults arrive with food. North country atlas workers call such nest trees "talking trees." Noisy nestlings and fledglings contribute to the highest confirmation rate among woodpeckers in states where they are numerous (e.g., McGowan 2008c). There have been relatively few documented nesting records of Yellow-bellied Sapsuckers in Maryland, beginning with a family group seen at Deer Park, Garrett County, on 6 July 1895 (Tylor in Kirkwood 1895). The only Maryland egg date is 5 June 1925, but young have been found in nests on dates ranging from 29 May 1949 to 16 July 2003 (Stewart and Robbins 1958; F. Pope, pers. comm.; MNRF). Ontario egg dates range from 15 May to 12 July (Peck and James 1983).

Yellow-bellied Sapsuckers appear to have been fairly common and widespread at upper elevations in the Allegheny Mountains through at least the first half of the twentieth century (Stewart and Robbins 1958). In spite of this assessment, only a single Garrett County breeding season report was made from 1958 to the beginning of the first statewide Maryland atlas project in 1978 (Hilton 1996b). Sapsuckers were recorded in 3 blocks from 1983 to 1987 with reports of possible and probable nesting in northeastern Garrett County. A bird seen at a feeder in northeastern Allegany County was treated as a nonbreeder although there is suitable habitat in Green Ridge State Forest.

There was another long hiatus in summer reports from the late 1980s through 1999. In 2000 and 2001, Yellow-bellied Sapsuckers were confirmed breeding at an undisclosed location near Oakland, Garrett County (Ringler 2001, 2002b). Breeding continued to be rumored at this location during the 2002–2006 atlas project, but the only official report was of adults feeding a fledgling on 11 August 2006 north of Avilton in the Two Mile Run drainage of northeastern Garrett County (S. Arnold, pers. comm.).

Yellow-bellied Sapsuckers have been increasing on North American BBS routes at 0.8% per year since 1966 and at a more robust 3.6% per year on U.S. routes. They declined significantly on eastern North American routes from 1966 to 1979 but have increased strongly since 1980. Recent increases may simply represent recovery from losses incurred during harsh winters in the eastern United States from 1976 to 1978 (Robbins, Bystrak, and Geissler 1986), but they might also be linked to maturation and regeneration of forests. Bird ob-

servers in the western Maryland mountains should listen for this woodpecker's distinctive drum (Ellison 1992a). If the increasing trend north of Maryland continues, Yellow-bellied Sapsuckers may become more numerous and widespread nesters in Maryland's western high country.

WALTER G. ELLISON

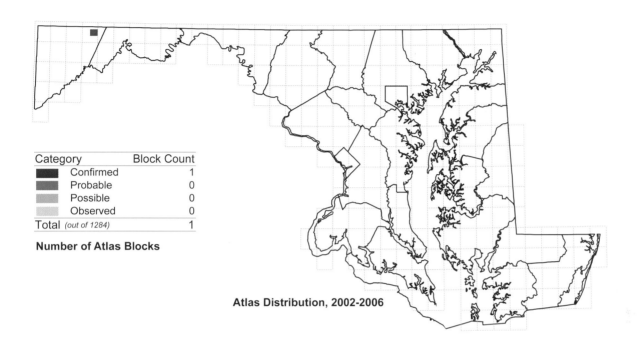

Category	Block Count
Confirmed	1
Probable	0
Possible	0
Observed	0
Total (out of 1284)	1

Number of Atlas Blocks

Atlas Distribution, 2002-2006

Change by Block
- ● Gain from First Atlas to Second
- ▲ Loss from First Atlas to Second

Change in Atlas Distribution, 1983-1987 to 2002-2006

Atlas Region	1983-1987	2002-2006	Change No.	%
Allegheny Mountain	2	1	-1	-50%
Ridge and Valley	0	0	0	-
Piedmont	0	0	0	-
Upper Chesapeake	0	0	0	-
Eastern Shore	0	0	0	-
Western Shore	0	0	0	-
Totals	2	1	-1	-50%

Change in Total Blocks between Atlases by Region

Eric Skrzypczak

Downy Woodpecker
Picoides pubescens

Compact and small, the Downy Woodpecker is Maryland and DC's most widespread woodpecker. Although its small size and rather unobtrusive voice make it easy to overlook particularly compared to the larger and more loquacious Northern Flicker and Red-bellied Woodpecker, the Downy Woodpecker is familiar to anyone who follows up on insistent light tapping on tree limbs or maintains a feeding operation (especially one that offers suet). This woodpecker's small size allows it to exploit small trees for food and nest sites and even nonwoody plants such as corn and goldenrod (attacking galls) for feeding. Downy Woodpeckers tend to nest in limbs, dead branches, and snags smaller than those used by other woodpeckers. And even small woodlots often have a pair of Downy Woodpeckers.

The Downy Woodpecker is one of two widespread white-backed woodpecker species with white-spangled black wings. The Hairy Woodpecker differs in small details

of plumage, but is best identified by its larger size, longer bill (nearly equaling its head length), and sharper and louder calls. The territorial drumming of the Downy Woodpecker is reasonably distinguishable from the drumming of its larger relative and the Northern Flicker. Downy Woodpeckers drum from late December through June with a peak in March and April, producing a fairly slow evenly spaced roll with audibly distinct taps and short time intervals between distinct drum rolls (Ellison 1992a). The Red-bellied Woodpecker has a similar slow-paced drum, but it often includes one or two hesitant introductory taps, drums fewer times in a session, and intersperses drums with its distinctive calls (W. Ellison, pers. obs.). Maryland egg dates extend from 5 April to 5 June (MNRF).

Ricciardi (1996b) reviewed the historical status of Downy Woodpecker in Maryland and DC and found its distribution and abundance had not changed over the decades. The same conclusion may be drawn from the 2002–2006 atlas map; this woodpecker continues to occur almost uniformly through the state and district. Downy Woodpeckers were found in 37 more blocks in 2002–2006 than in 1983–1987, but it appears that these additional blocks were mostly the result of improved coverage. New blocks are either in areas of known better coverage such as at Aberdeen Proving Ground in Harford County or in blocks that were surrounded by occupied blocks in 1983–1987. Downy Woodpeckers are reluctant to cross large bodies of water, so they were not found on the marshy islands of lower Chesapeake Bay and were scarce on Assateague Island.

Downy Woodpecker numbers sampled by BBS routes in Maryland apparently have been generally stable over the past 40 years, although prone to occasional downward fluctuations caused by cold winters. But Downy Woodpeckers have been declining in most of the physiographic regions that include Maryland, except for the Northern Piedmont. Maryland appears to offer good habitat for Downy Woodpecker as it ranks second among states in the average number seen per BBS route. As long as woodlands with dead limbs and snags are common in Maryland and DC the Downy Woodpecker should continue to thrive.

WALTER G. ELLISON

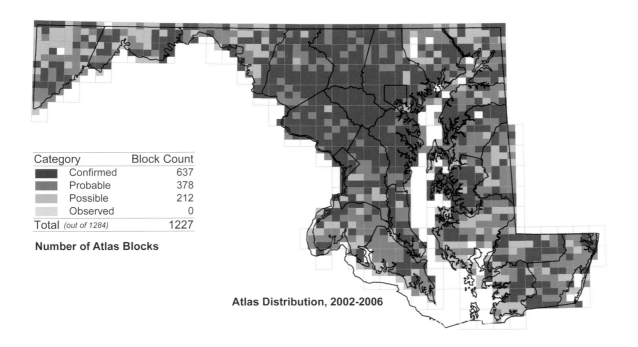

Category	Block Count
Confirmed	637
Probable	378
Possible	212
Observed	0
Total (out of 1284)	1227

Number of Atlas Blocks

Atlas Distribution, 2002-2006

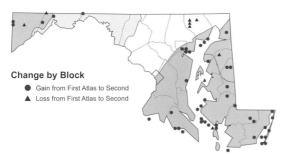

Change by Block
- ● Gain from First Atlas to Second
- ▲ Loss from First Atlas to Second

Change in Atlas Distribution, 1983-1987 to 2002-2006

Percent of Stops
- 50 - 100%
- 10 - 50%
- 0.1 - 10%
- < 0.1%

Relative Abundance from Miniroutes, 2003-2007

Atlas Region	1983-1987	2002-2006	Change No.	%
Allegheny Mountain	90	91	+1	+1%
Ridge and Valley	146	148	+2	+1%
Piedmont	316	312	-4	-1%
Upper Chesapeake	111	120	+9	+8%
Eastern Shore	279	302	+23	+8%
Western Shore	242	248	+6	+2%
Totals	1184	1221	+37	+3%

Change in Total Blocks between Atlases by Region

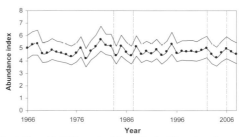

Breeding Bird Survey Results in Maryland

Hairy Woodpecker
Picoides villosus

The Hairy Woodpecker, a big brother of the familiar Downy Woodpecker, is one of those in-between species: neither so common as to be immediately familiar to everyone and passed by without a second thought, nor rare enough to be actively sought out and celebrated as a prize. It is not especially common, but in proper habitat it is likely to be present in small numbers. Because of its similarity to the much more common Downy Woodpecker, it can easily escape notice. This may be the reason for the reported rarity of the species before the widespread availability of high-quality optical equipment. Its main distinguishing characteristic is its size, almost half again as long as its very similar relative (about 24 cm [9.4 in] compared to 17 cm [6.6 in] for the Downy). When size is difficult to judge, the relative proportion of the bill separates them (about 29 mm [1.1 in] long for the Hairy, compared to 16 mm [0.6 in] short for the stubby bill of the Downy). Like many of our woodpeckers, the plumage is strongly patterned in black and white: mostly black above with white-spotted wings and a white streak down the middle of the back, and white below. The black and white pattern is set off by a red patch on the head of the male. In the adult male, the red patch is toward the back of the head; in the juvenile male, the crown is red-streaked. A close look reveals that the outer tail feathers of the Hairy are pure white, compared to the sparsely black-barred white outer tail feathers of the Downy. The call is a sharp *peek,* and sometimes a series of harsher notes run together in a sort of whinny. The single call is a little shriller and louder than the call of the Downy, but as with plumage, the call is similar enough to make distinguishing between the two difficult. In common with most woodpeckers, a drum roll on a resonant branch serves in place of a song. Both sexes engage in drumming.

The Hairy Woodpecker is more restricted to forest than the Downy, although the forest need not be dense. It occasionally ventures into more open suburban areas. It forages mostly on trunks and larger branches, and seldom uses the smaller branchlets and twigs favored by the Downy Woodpecker. Food is mostly invertebrates such as beetles, caterpillars, spiders, and millipedes. The grubs of wood-boring beetles are an important food, especially in winter when other meals are scarce. Vegetable food (about 20%) includes fruits and seeds. Like the Downy, this species is not averse to feasting on suet at feeders. Nesting is usually in a deciduous tree, commonly within the interior of larger wooded areas. The nest is located in an excavated hole, usually in a live hardwood tree, 1 to 20 m (3–85 ft) above the ground. Holes are often lost to thieves such as starlings or squirrels.

This species is a permanent resident in Maryland. Some birds in the northern part of the species range withdraw southward during the winter, and the Maryland resident population of Hairy Woodpeckers may be augmented in the winter by the addition of northern birds.

From the swamp forests of the lower Eastern Shore to the ridges of the Allegheny Mountains, the entire state of Maryland supports Hairy Woodpeckers, subject only to the existence of proper habitat. About 85% of all the blocks in the state reported Hairies. Blocks lacking Hairy Woodpeckers are mostly open farmland, urbanized, marshland, or a combination of these environments. The species has become more widespread since the completion of the previous atlas (Ricciardi 1996c), with increases from 10% to 15% in total blocks for all regions except the Ridge and Valley.

T. DENNIS COSKREN

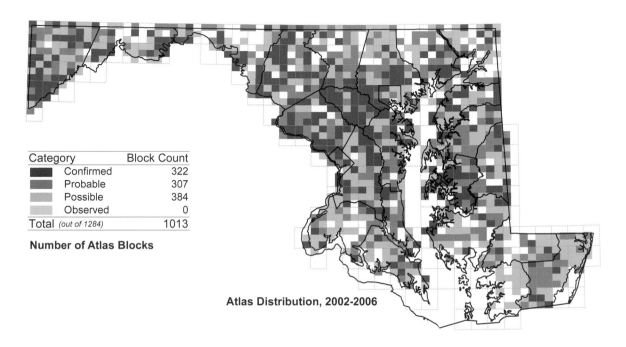

Category	Block Count
■ Confirmed	322
■ Probable	307
■ Possible	384
■ Observed	0
Total (out of 1284)	**1013**

Number of Atlas Blocks

Atlas Distribution, 2002-2006

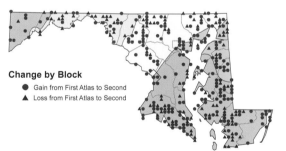

Change by Block
- ● Gain from First Atlas to Second
- ▲ Loss from First Atlas to Second

Change in Atlas Distribution, 1983-1987 to 2002-2006

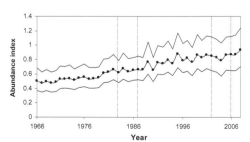

Percent of Stops
- ■ 50 - 100%
- ■ 10 - 50%
- ■ 0.1 - 10%
- □ < 0.1%

Relative Abundance from Miniroutes, 2003-2007

Atlas Region	1983-1987	2002-2006	Change No.	Change %
Allegheny Mountain	81	89	+8	+10%
Ridge and Valley	118	110	-8	-7%
Piedmont	250	274	+24	+10%
Upper Chesapeake	89	100	+11	+12%
Eastern Shore	202	233	+31	+15%
Western Shore	179	203	+24	+13%
Totals	**919**	**1009**	**+90**	**+10%**

Change in Total Blocks between Atlases by Region

Breeding Bird Survey Results in Maryland

Northern Flicker
Colaptes auratus

The Northern Flicker is Maryland and DC's second largest woodpecker, but it is an atypical woodpecker. Flickers spend much time on the ground in areas of sparse or short vegetation instead of feeding by boring into wood or under bark as do most woodpeckers. Flickers often frequent lawns; a large part of their warm-season diet is ground-dwelling insects, especially ants (Moore 1995). In the colder months they feed on fruit. They frequently perch on branches and treetops like a songbird instead of clinging to a tree trunk like most woodpeckers. Flickers are strikingly marked, flashing bright yellow under the wings and tail in flight while displaying a large white rump that contrasts with their pale brown backs. They are vocal, with loud calls, and are often seen in the open, which makes them familiar to anyone who notices birds.

Northern Flickers live in more open haunts than most other woodpeckers, preferring open woodlands with well-spaced trees, floodplains, flooded timber, and lightly wooded parklike places in farmland or in more urban settings. Northern Flickers usually excavate nest holes in snags in the open or near forest edges; as such they are prone to competition from European Starlings (Brackbill 1957; D. Ingold 1994).

Maryland egg dates range from 12 April to 26 June, and nestlings have been found in nest cavities until 8 August (Van Ness 1996b; MNRF). Maryland flickers are probably not double brooded; later nests may be attributable to renesting after cavity usurpation by starlings (Master 1992b). Sometimes a Northern Flicker will use a nest box or nest in a burrow in the ground (Moore 1995); in May 2006 a flicker in Talbot County excavated a cavity into loose soil in a garden and laid seven eggs (L. Roslund, pers. comm.).

Early historical bird lists reviewed in Van Ness (1996b) termed the Northern Flicker either abundant or common. By the late 1950s they had become uncommon east of the Allegheny Mountains (Stewart and Robbins 1958). In the 1983–1987 atlas project, flickers were still distributed statewide and were among the 30 most widespread bird species in Maryland and DC (Van Ness 1996b). In 2002–2006 the Northern Flicker was still among the top 30 most prevalent bird species and occurred in only 9 fewer blocks. Some hard-to-explain gaps occurred in the flicker's range, in particular on the lower Western Shore along the lower Patuxent River and just north of Point Lookout. Northern Flickers were not found on lower Chesapeake Bay islands (e.g., Smith Island) or in the marshes of Elliott Island. On the other hand, these adaptable woodpeckers occurred essentially throughout the District of Columbia and Baltimore City.

The population decline that appeared evident for the Northern Flicker in Maryland in 1958 was demonstrated on

BBS routes over the next four decades. Flickers have declined at a rate of 1.8% per year in North America (all subspecies), and in Maryland at a similar 2% per year. The trend has been less steeply declining since the 1980s. Much of this negative population trend has been attributed to competition with starlings over nest cavities both freshly excavated and preexisting (Robbins, Bystrak, and Geissler 1986). Other possible contributors may be loss of snags to intensive management of farmland and yards, and pesticide applications on farms and on suburban lawns that kill a large number of the flicker's prey. The Northern Flicker should be monitored carefully; in spite of population declines it continues to occupy the same nesting range it has had for decades but this may not remain the case indefinitely.

WALTER G. ELLISON

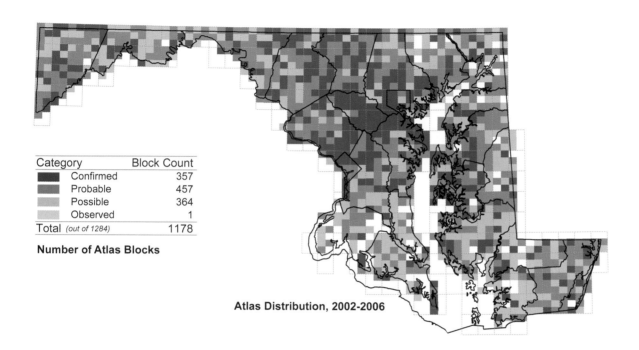

Category	Block Count
Confirmed	357
Probable	457
Possible	364
Observed	1
Total (out of 1284)	1178

Number of Atlas Blocks

Atlas Distribution, 2002-2006

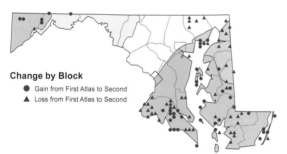

Change by Block
- ● Gain from First Atlas to Second
- ▲ Loss from First Atlas to Second

Change in Atlas Distribution, 1983-1987 to 2002-2006

Percent of Stops
■	50 - 100%
■	10 - 50%
■	0.1 - 10%
□	< 0.1%

Relative Abundance from Miniroutes, 2003-2007

			Change	
Atlas Region	1983-1987	2002-2006	No.	%
Allegheny Mountain	88	93	+5	+6%
Ridge and Valley	146	146	0	0%
Piedmont	316	310	-6	-2%
Upper Chesapeake	114	110	-4	-4%
Eastern Shore	296	302	+6	+2%
Western Shore	221	211	-10	-5%
Totals	1181	1172	-9	-1%

Change in Total Blocks between Atlases by Region

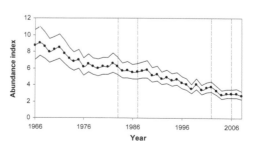

Breeding Bird Survey Results in Maryland

Evelyn Ralston

Pileated Woodpecker
Dryocopus pileatus

The Pileated Woodpecker is traditionally considered the big woodpecker of the big woods; long believed to be a creature of extensive old growth forest, Maryland and DC's largest woodpecker has been gradually shedding this identity over the past half century. This crow-size woodpecker appears largely black when seen on a tree save for streaks of white on the head and neck and a flaming red, peaked cap on its crown, or pileum, from which it derives its odd name. In flight, Pileated Woodpeckers show a white patch at the base of the primaries, and the underside of the wing is mostly white. These shy woodpeckers have loud laughing calls and a loud rolling territorial drum. The loud measured thwacks made by a feeding Pileated are also distinctive. Even when no woodpecker is around, deep, oblong pits up to a foot long in trees and snags give mute evidence of the Pileated's tenancy in woodlands.

The habitat requirements of the Pileated Woodpecker can be met by large trees, even when these are scattered among younger trees, large snags, stumps, and old downed logs (E. Bull and Jackson 1995). Pileated Woodpeckers feed as much near the ground as higher up in trees. Assiduous in their pursuit of carpenter ants and wood-boring beetles, a feeding Pileated may even ignore the quiet presence of a human observer. Over the past five decades, Pileated Woodpeckers have become more frequent denizens of small wood-

lots with mature trees and parks in urbanized landscapes, but they remain most numerous in extensive mature forest. This woodpecker is single brooded, starting the excavation of its large slightly ovoid 8.5-cm-wide (3.3-in) nest hole in mid- to late March. Pileated Woodpecker eggs have been recorded in Maryland and DC from 18 April to 2 June (Van Ness 1996c; MNRF), and nestlings have been found in cavities as late as 15 July 2002 near Colesville, Montgomery County (G. Mackiernan).

The Pileated Woodpecker population has been slowly increasing since reaching a low ebb in the late nineteenth century. Although Kirkwood (1895) called this woodpecker fairly common, he also found it very local and absent from much of the state. Eifrig (1904) called the bird rare in western Maryland except at higher elevations. The range of the Pileated Woodpecker remained restricted in the late 1950s when Stewart and Robbins (1958) placed it in Garrett and Allegany counties; on Catoctin and South mountains; along the Potomac in Montgomery County; in the middle Patuxent Drainage; in southern Charles County (especially Zekiah Swamp); in southern Dorchester County; and in the Nanticoke and Pocomoke drainages. Over the next quarter of a century, these woodpeckers spread throughout the Ridge and Valley region and the southern Piedmont remaining scarce to the north. They colonized the Upper Chesapeake region and adjacent Eastern Shore, although they remained scarce there (Van Ness 1996c). In the current atlas the range expansion has continued as Pileated Woodpeckers gained an additional 251 atlas blocks, an increase of 34%. The largest additions were made in the Upper Chesapeake and northern Piedmont, but the "log cock" increased or was stable in all regions. Pileated Woodpeckers are still unrecorded in most blocks around Hagerstown, in central Carroll County, and much of southern Cecil and central Kent counties.

BBS trends for the Pileated Woodpecker have been increases, including 1.8% per year from 1966 to 2007 in North America, and a robust 3.2% per year in Maryland. The main reason for the Pileated Woodpecker's recovery over the past hundred years has been maturation of second-growth woodlands and some reforestation. Pileated Woodpeckers have also lost some of their shyness around humans. Once shot by market hunters (as alluded to by Kirkwood [1895] and Eifrig [1904]), these woodpeckers are now protected by law. As their reclusiveness abated, Pileateds took to smaller woodlots, nearer human habitation. In order to ensure the continued success of Pileated Woodpecker in Maryland and DC, forest managers should manage for uneven-aged stands and retain large trees, snags, and deadwood. The establishment of invasive non-native wood-boring beetles may cause managers to remove older trees and dead timber in forest lands and this could lead to declines for this impressive woodpecker.

WALTER G. ELLISON

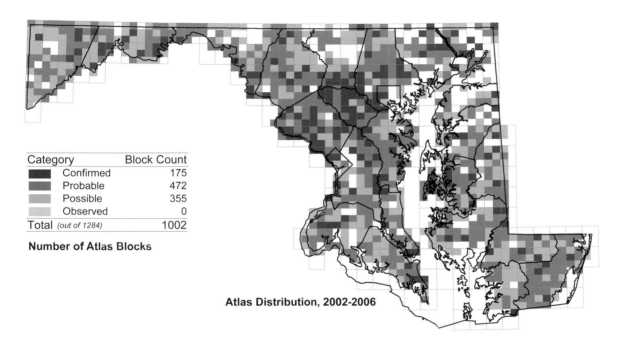

Category	Block Count
Confirmed	175
Probable	472
Possible	355
Observed	0
Total (out of 1284)	1002

Number of Atlas Blocks

Atlas Distribution, 2002-2006

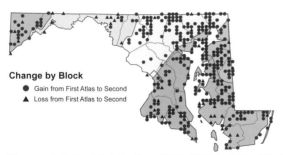

Change by Block
- ● Gain from First Atlas to Second
- ▲ Loss from First Atlas to Second

Change in Atlas Distribution, 1983-1987 to 2002-2006

Percent of Stops
- 50 - 100%
- 10 - 50%
- 0.1 - 10%
- < 0.1%

Relative Abundance from Miniroutes, 2003-2007

Atlas Region	1983-1987	2002-2006	Change No.	Change %
Allegheny Mountain	70	85	+15	+21%
Ridge and Valley	135	132	-3	-2%
Piedmont	181	264	+83	+46%
Upper Chesapeake	25	68	+43	+172%
Eastern Shore	175	236	+61	+35%
Western Shore	159	211	+52	+33%
Totals	745	996	+251	+34%

Change in Total Blocks between Atlases by Region

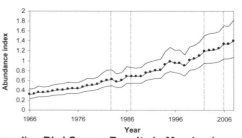

Breeding Bird Survey Results in Maryland

Eric Skrzypczak

Eastern Wood-Pewee
Contopus virens

The sweet plaintive song of the Eastern Wood-Pewee is often the avian voice of a hot and humid summer's day. This small, rather drab flycatcher sings its name, *pee-ur-weee?* often followed by an answer, a sighing *pee-urr*. Pewees are honest-to-goodness flycatchers, unlike some other members of their family, capturing many of their prey by sallying from an open perch, often producing an audible "snap" as they capture medium-size flying insects. This pewee inhabits open forests and woodland, often with little understory, and canopy breaks at edges and tree-fall gaps (McCarty 1996). Eastern Wood-Pewees are not limited by foliage type, being at home in open pine forests as well as mixed and deciduous forests, but they are scarce in dark, closed-canopy conifer forests such as mature hemlock stands. Because this pewee lives around edges and clearings, it can often be found in wooded subdivisions and small woodlots. Eastern Wood-Pewees do not appear to be severely limited by woodlot size or habitat fragmentation (Robbins, Dawson, and Dowell 1989).

Eastern Wood-Pewees are relatively late migrants, with the vast majority arriving in early May. They are usually not widespread until 10 May or later; pewees remain in Maryland until early autumn with most gone by the end of September. This pewee's nest is compact, decorated with lichen, and inconspicuously saddled well out on a medium to large tree limb, usually at 6.1 m (20 ft) or higher. Maryland nests have held eggs from 21 May to 15 August, with reports of nestlings to 13 September (Stewart and Robbins 1958; MNRF). The length of this pewee's Maryland nesting season may indicate occasional second broods or perhaps late nestings by migrants that arrived in June.

Kirkwood (1895) and Eifrig (1904) called the Eastern Wood-Pewee common in Maryland, and Stewart and Robbins (1958) called it fairly common throughout Maryland and DC. This pewee retained essentially the same status during the 1983–1987 atlas project when it was one of the 20 most widespread bird species—not found in just 64 blocks out of 1,256 (Vaughn 1996a). Little has changed in 20 years, as this flycatcher retained its rank among the state's most widespread birds and was missed from 63 blocks out of the 2002–2006 total of 1,284 surveyed blocks. The only blocks where it is likely the bird actually did not occur are a few heavily urbanized blocks and several blocks dominated by salt marsh on the lower Eastern Shore.

In spite of its continued ubiquity, the Eastern Wood-Pewee has been declining on BBS routes. North American routes have shown a 1.7% per year decline, and Maryland routes have shown a less steep 1% per year downward trend. Among the physiographic regions shared by Maryland and neighboring states, the largest declines have been on the Allegheny Plateau and the Ridge and Valley; pewees have also declined on the Upper Coastal Plain but have been stable in the Northern Piedmont. What factors might contribute to this decline are hard to specify given that the Eastern Wood-Pewee remains very widespread. The range wide nature of the decline based on BBS data and trend maps (USGS data) implies that the factors either are acting throughout the breeding range or originate in winter quarters. This may be related to wholesale deforestation. This pewee's winter range habitat is similar to its breeding haunts as it inhabits edges and successional forests in Brazil (Fitzpatrick 1980; Stotz et al. 1992). At present the Eastern Wood-Pewee continues prevalent throughout Maryland and DC although it bears watching as its population continues to slowly decline.

WALTER G. ELLISON

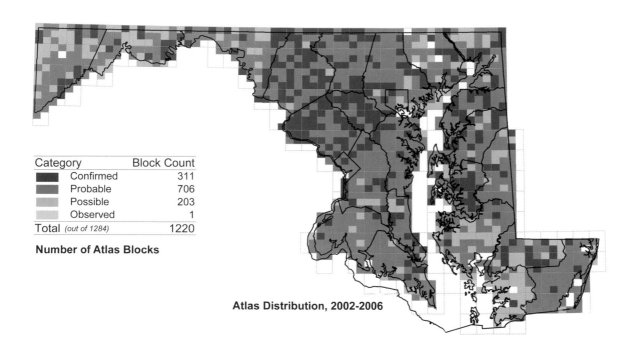

Category	Block Count
Confirmed	311
Probable	706
Possible	203
Observed	1
Total *(out of 1284)*	1220

Number of Atlas Blocks

Atlas Distribution, 2002-2006

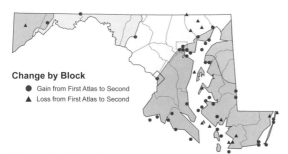

Change by Block
- ● Gain from First Atlas to Second
- ▲ Loss from First Atlas to Second

Change in Atlas Distribution, 1983-1987 to 2002-2006

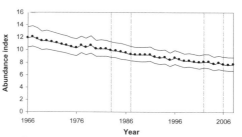

Percent of Stops
- 50 - 100%
- 10 - 50%
- 0.1 - 10%
- < 0.1%

Relative Abundance from Miniroutes, 2003-2007

Atlas Region	1983-1987	2002-2006	Change No.	%
Allegheny Mountain	92	92	0	0%
Ridge and Valley	147	148	+1	+1%
Piedmont	312	312	0	0%
Upper Chesapeake	115	121	+6	+5%
Eastern Shore	288	297	+9	+3%
Western Shore	236	245	+9	+4%
Totals	1190	1215	+25	+2%

Change in Total Blocks between Atlases by Region

Breeding Bird Survey Results in Maryland

George M. Jett

Acadian Flycatcher
Empidonax virescens

The Acadian Flycatcher is by far the most common and widespread member of the infamously difficult flycatcher genus *Empidonax* in Maryland and DC. Plumage is very similar among *Empidonax* species, and some migratory individuals prove impossible to identify for even the most acute observers. But nesting members of the genus have distinctive calls and songs and their nesting habitats are generally distinct. The Acadian Flycatcher is a relatively large, notably greenish *Empidonax* with an explosive high-pitched sneeze of a song, *spe-cheee!* As with most flycatchers, the Acadian is a rather deliberate feeder; it sits still for long periods then darts out and hovers below leaves, deftly picking prey off them. As with many other flycatchers it is also aggressive. Acadian Flycatchers inhabit mature moist forest, often in stream valleys; these include beech–tulip poplar forest, floodplain forest, hemlock ravines, and deciduous swamps.

Acadian Flycatchers winter in Central America and northwestern South America and reside in Maryland and DC from the last days of April to mid-September. They usually raise just one brood, but it is likely that a few females that raise an early-fledging brood attempt a second nesting as they do in the Midwest (Whitehead and Taylor 2002). The Acadian Flycatcher's nest is a distinctive, flimsy hammock of plant fibers, with a beard of catkins hanging below it, slung across a branch-tip fork at medium height, usually from 3 to 6 m (10–20 ft). Maryland egg dates range from 19 May to 3 August, with nestlings reported to 21 August (Blom 1996e; MNRF).

Kirkwood (1895) considered the Acadian Flycatcher common in eastern Maryland, but Eifrig (1904) called it very rare in western Maryland. Stewart and Robbins (1958) described this flycatcher's status as common on the Coastal Plain, fairly common westward through Allegany County, and very local and uncommon in the Allegheny Mountains. The 1983–1987 atlas project revealed a great increase in the Alleghenies, with records from 78% of blocks in the region (Blom 1996e). Acadian Flycatchers were also reported from nearly 80% of the blocks in the rest of the state and DC in the 1980s. The results for 2002–2006 show little change in this distribution. Acadian Flycatchers are absent or very sparsely reported in areas with little or no suitable habitat, including much of the Hagerstown Valley, most of Assateague Island, the wet pine woodlands and marshes of Dorchester and Somerset counties, and some upper elevation blocks in Garrett County. Nonetheless the Acadian Flycatcher is one of the most prevalent forest birds in major urban centers and in the heavily farmed Upper Chesapeake and Eastern Shore regions.

Trends from the BBS for Acadian Flycatcher currently show it is stable over the long term surveywide; early increases from 1966 to 1979 have slowed. In Maryland, Acadian Flycatcher trends have been stable over the past four decades. Although the Acadian Flycatcher has similar numbers and the same nesting range in Maryland and DC between the 1980s and 2000s, it is still prone to cowbird brood parasitism, as well as habitat loss and fragmentation, although it can be found in some very small forest patches (Robbins, Dawson, and Dowell 1989; Robinson et al. 1995). Preservation of riparian forest in the critical area on Chesapeake Bay tributaries may be helpful in preserving this flycatcher's habitat. Forest habitat should be preserved, preferably in large blocks in many areas in Maryland, to help retain the Acadian Flycatcher as a familiar member of our avifauna.

WALTER G. ELLISON

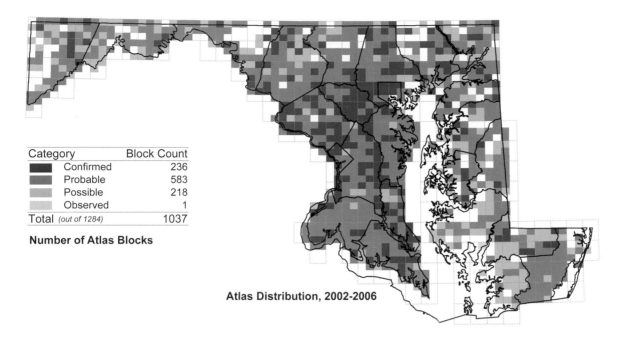

Category	Block Count
■ Confirmed	236
■ Probable	583
■ Possible	218
■ Observed	1
Total *(out of 1284)*	1037

Number of Atlas Blocks

Atlas Distribution, 2002-2006

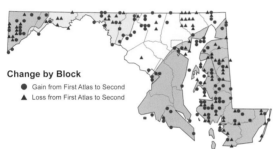

Change by Block

● Gain from First Atlas to Second
▲ Loss from First Atlas to Second

Change in Atlas Distribution, 1983-1987 to 2002-2006

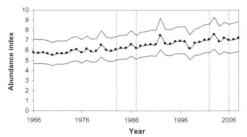

Percent of Stops

■ 50 - 100%
■ 10 - 50%
■ 0.1 - 10%
□ < 0.1%

Relative Abundance from Miniroutes, 2003-2007

Atlas Region	1983-1987	2002-2006	Change No.	%
Allegheny Mountain	73	76	+3	+4%
Ridge and Valley	122	124	+2	+2%
Piedmont	289	289	0	0%
Upper Chesapeake	102	97	-5	-5%
Eastern Shore	191	197	+6	+3%
Western Shore	232	248	+16	+7%
Totals	1009	1031	+22	+2%

Change in Total Blocks between Atlases by Region

Breeding Bird Survey Results in Maryland

Alder Flycatcher
Empidonax alnorum

This "little brown job" is notoriously difficult to identify by sight. Like its congeners, it is a small flycatcher with plain brown upperparts, dark wings with well-marked whitish wing-bars (buffier in first-year birds), and a white throat and belly shading to a slightly dusky breast. It has an eye-ring that is less distinct than most other *Empidonax,* a characteristic it shares with the Willow Flycatcher. Subtle differences in the exact shade of brown (averages slightly more greenish) and minor differences in average lengths of bill and primaries (slightly longer for both) are not helpful in the field. Fortunately, the typical song is distinctive: a burry three-syllable *rray-BEE-oh,* with a distinct accent on the second syllable. The call note of the Alder may also be helpful: "a flat *pip* reminiscent of [the] single note of Olive-sided [Flycatcher], unlike the *whit* of Willow and other *Empidonax*" (Sibley 2000).

The usual habitat of the Alder Flycatcher, open northern bogs with thickets of alder, is reflected in its name, but it will also accept dry brushy pastures. Like most North American flycatchers, the Alder Flycatcher is a sally feeder, pursuing flying insects from an exposed perch in a shrub, to which it habitually returns. Other prey taken in minor amounts include spiders and nonflying insects; berries may also be eaten occasionally.

The somewhat loosely built nest is constructed from coarse grass, twigs, bark strips, mosses, and cattail down. Given the habitat, it is not surprising that it is placed fairly low (below 2 m [6 ft]) in a shrub (e.g., alder, willow, spirea, hawthorn) (H. Harrison 1975).

The range of the species is generally northern, extending down the Appalachians to North Carolina. Like other boreal species, it is to be expected only in the Allegheny Mountain region of Garrett County. During the present atlas project, it was found only there. But summering individuals, apparently without mates, have been found in several Piedmont counties on occasion. It is possible that Alder Flycatchers may rarely breed elsewhere in the state.

The Alder Flycatcher winters farther to the south than other *Empidonax* flycatchers, its winter range extending from Colombia to northern Argentina. The exact limits of the range are uncertain due to the difficulty in identifying non-singing birds. Migrants may arrive in Maryland as early as the end of April, but most appear in mid-May. It seems to leave a little earlier than the other flycatchers, and the last stragglers have usually departed before the end of September.

This species has increased since the completion of the first atlas in 1987 (W. Murphy 1996). The number of blocks with possible to confirmed breeders rose from 28 to 38, an increase of 36%. This is in accord with data from Maryland Breeding Bird Surveys, which show a consistent upward trend (Willow/Alder combined; only data from Allegheny Mountain region considered). This upward trend in Maryland does not reflect the status elsewhere, which shows a slight decrease in population.

T. DENNIS COSKREN

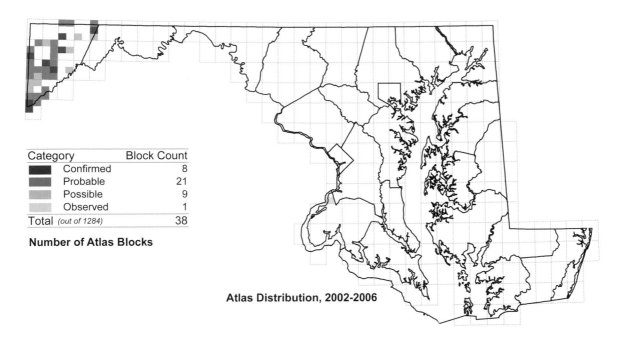

Category	Block Count
■ Confirmed	8
■ Probable	21
■ Possible	9
■ Observed	1
Total (out of 1284)	38

Number of Atlas Blocks

Atlas Distribution, 2002-2006

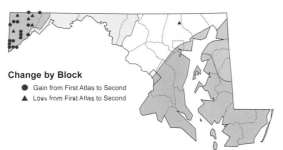

Change by Block

● Gain from First Atlas to Second
▲ Loss from First Atlas to Second

Change in Atlas Distribution, 1983-1987 to 2002-2006

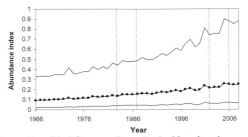

Percent of Stops

■ 50 - 100%
■ 10 - 50%
■ 0.1 - 10%
□ < 0.1%

Relative Abundance from Miniroutes, 2003-2007

Atlas Region	1983-1987	2002-2006	Change No.	%
Allegheny Mountain	27	38	+11	+41%
Ridge and Valley	0	0	0	-
Piedmont	1	0	-1	-100%
Upper Chesapeake	0	0	0	-
Eastern Shore	0	0	0	-
Western Shore	0	0	0	-
Totals	28	38	+10	+36%

Change in Total Blocks between Atlases by Region

Breeding Bird Survey Results in Maryland

Willow Flycatcher
Empidonax traillii

Itz-bew! the male Willow Flycatcher sings, usually from the highest perch in his territory, sometimes even from a power line. Its song is distinctly different from the song of the Alder Flycatcher, from which it cannot reliably be distinguished by sight, even when in the hand. Willow Flycatcher and Alder Flycatcher were long considered "song-forms" of *Empidonax traillii* and were not recognized as distinct species until 1973 (AOU 1973). Observations of either species during the non-breeding season still must be reported as Traill's Flycatcher.

Willow Flycatchers are associated with patches of woody vegetation in wetlands, old fields, or dry gullies, generally surrounded by farmland or other open habitats. They tend to avoid early successional habitats that are available only for the short-term, seeming to prefer habitats dominated by shrubs, rather than young trees, or where succession has been arrested, such as occurs in a power line right-of-way or atop old landfills. The shrubs may be dogwoods, willows, elderberry, or hawthorn (Walkinshaw 1966); Willow Flycatchers also use multiflora rose.

Willow Flycatchers likely winter in Mexico and Central America, and possibly as far south as northern South America (AOU 1998; Sedgwick 2000); the extent to which their winter distribution overlaps with that of Alder Flycatcher is unknown. Willow Flycatchers have been reported in Maryland as early as 1 May, but most individuals arrive in the sec-

ond half of May. For the atlas, the species was not considered safe for records of possible nesting until 10 June. Departure dates from Maryland are hard to establish because of the identification issue but occur mostly during August and September, with adults departing before young birds because their postbreeding molt occurs after they reach the wintering grounds, unlike most other passerines.

Willow Flycatcher nests, built entirely by the female, are placed 0.6 to 2.8 m (1.9–9.1 ft) above the ground (mean 1.3 m [4.2 ft]; Walkinshaw 1966) either in an upright crotch or on a horizontal branch of a shrub or small tree and are well concealed by the foliage. Nests are often at the outer edge of a shrub or thicket (Sedgwick 2000). In this atlas project, most confirmations of breeding (n = 69) were from observations of nest building (20%), adults carrying food for young (39%), or fledglings (26%). The earliest active nest was observed on 21 May, and nest building was observed from 27 May to 17 June, with eggs reported from 4 June through 28 July (Van Ness 1996d; MNRF). Fledglings were observed during late June through mid-August, with one report from early September.

Willow Flycatchers were not known to occur in Maryland before the 1950s (Van Ness 1996d); nesting was first documented in 1961 in Bladensburg, Prince George's County, and in nearby DC (Bridge and Riedel 1962). Their distribution expanded thereafter, with territorial birds also observed in several Piedmont counties by 1966 (Robbins 1966). They were documented in 328 atlas blocks during 1983–1987 (Van Ness 1996d) and in 350 blocks in 2002–2006, an increase of 7%. The Willow Flycatcher's stronghold in Maryland is the Piedmont, where it was found in 215 blocks in 2002–2006, 11% more than in 1983–1987. It also appears to be more common on the Western Shore, where it was found in 26 blocks during 2002–2006, 86% more than in 1983–1987, including several blocks in and around DC. The distribution remained essentially the same in the Ridge and Valley region and on the Allegheny Plateau, where Willow Flycatcher co-occurs with Alder Flycatcher. The species is largely absent from southern Maryland (Calvert, Charles, and St. Mary's counties) and from the Eastern Shore, except for a few locations on Assateague Island and one near the Chester River.

Long-term population trend estimates from BBS data are difficult to interpret because Willow and Alder Flycatcher have generally been combined. Rangewide, trends are negative, but not significant. Trend estimates for Maryland are positive, increasing at a rate of 2.4% per year since 1966, and at 13% per year since 1999. Losses between atlas projects generally are concentrated in areas undergoing suburban expansion, suggesting that habitat loss is the greatest threat for Willow Flycatchers in Maryland. Efforts should be made to preserve or to manage the shrub habitats that Willow Flycatchers need. The species can adapt to living in settled areas if habitat, surrounded by a buffer of open space, remains.

DEANNA K. DAWSON

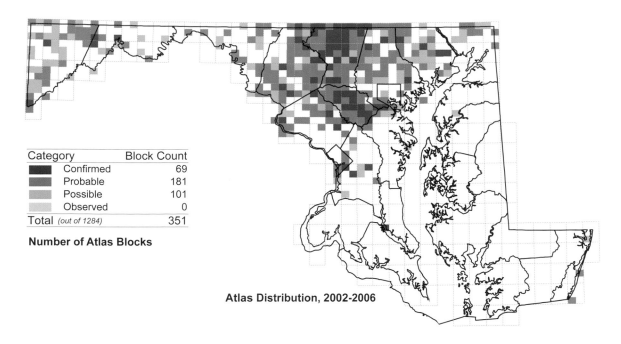

Category	Block Count
Confirmed	69
Probable	181
Possible	101
Observed	0
Total *(out of 1284)*	351

Number of Atlas Blocks

Atlas Distribution, 2002-2006

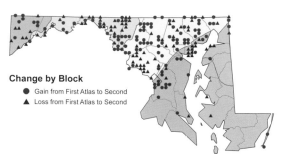

Change by Block

● Gain from First Atlas to Second
▲ Loss from First Atlas to Second

Change in Atlas Distribution, 1983-1987 to 2002-2006

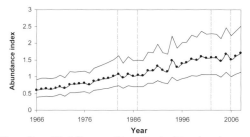

Percent of Stops

■	50 - 100%
■	10 - 50%
■	0.1 - 10%
□	< 0.1%

Relative Abundance from Miniroutes, 2003-2007

Atlas Region	1983-1987	2002-2006	Change No.	Change %
Allegheny Mountain	49	50	+1	+2%
Ridge and Valley	53	49	-4	-8%
Piedmont	193	215	+22	+11%
Upper Chesapeake	16	7	-9	-56%
Eastern Shore	3	3	0	0%
Western Shore	14	26	+12	+86%
Totals	328	350	+22	+7%

Change in Total Blocks between Atlases by Region

Breeding Bird Survey Results in Maryland

Least Flycatcher
Empidonax minimus

Least Flycatchers are aggressive, active, vocal, and tend to settle in clusters of small territories in open woodlands. This smallest flycatcher east of the Mississippi River is olive gray above and whitish below, with a dingy grayish vest and sharp white eye-ring. They are best located by their repetitive and almost incessant songs, *chi-bik, chi-bik, chi-bik;* they are highly conversational within their territory clusters, giving off a testy dry *wip* that is mixed with a purring growl when fights take place.

Least Flycatchers are birds of the mixed and deciduous woodlands of the north, and of mountain and hill country in the southeastern United States. They are most often found in young open forest or near openings and along streams in more mature forests in areas where there is little foliage from 6.1 to 12.2 m (20–40 ft) above the ground (R. Holmes and Sherry 2001). The species' tendency to settle in clustered territories that often leave other seemingly suitable areas unoccupied is distinctive and may be related to females preferring to settle where there are many males holding territories (Tarof et al. 2005).

Least Flycatchers spend the winter in Mexico and Central America, returning to Maryland and DC in late April and early May. They stay until mid- to late September. These diminutive birds build their well-woven nests in branch and trunk forks at various heights, although nest height in most studies is below 6.1 m (20 ft). Maryland egg dates range from 19 May to 17 June (Stewart and Robbins 1958; MNRF); during this project a female was seen on a nest as late as 20 June 2003 near Table Rock on Backbone Mountain, Garrett County (J. Stasz). An active nest was reported 15 August 1949 near Friendsville, Garrett County (A. Wright in Stewart and Robbins 1958), presumably a second brood or late renest. Least Flycatchers are only rarely double brooded (Briskie and Sealy 1987) because adults molt on the winter range and begin fall migration in August.

Breeding Least Flycatchers have generally been limited to the mountainous parts of western Maryland. Kirkwood (1895) found birds on territory on Dans Mountain, Allegany County; Eifrig (1904) noted their presence in the summer but found them much rarer than in migration. Stewart and Robbins (1958) called these flycatchers fairly common in the Allegheny Mountains, uncommon eastward to western Washington County, with rare nesting records east to the fall line. This distribution had changed little by the 1983–1987 atlas project; Least Flycatchers were widespread west of the Allegheny Front, thinly distributed in the western Ridge and Valley to just east of Hancock, Washington County, with one record of probable breeding from Baltimore County

Mark Hoffman

just north of Pikesville (Blom 1996f). During 2002–2006 they were still widely distributed in Garrett County but had become rare in the Ridge and Valley with reports from only 4 blocks; the easternmost was on South Mountain northeast of Hagerstown. There were also three safe-date reports from Howard, Baltimore, and Cecil counties, all of which were prudently coded as observed rather than possible because they might have been late spring migrants. Many more reports were listed as possible breeders in this atlas than in 1983–1987 (55% versus 45%); there were only two confirmed nesting records versus seven in the 1980s.

Least Flycatchers have declined on BBS routes in North America at a rate of 1.3% per year. This decline has been most pronounced in the East where the downward trend has been 1.9% per year. Maryland trend data are unreliable because of few routes and few birds (0.5 per route). The decline may be related to patterns of forest succession and loss (R. Holmes and Sherry 2001) or the effects of deer browsing on the understory (DeCalesta 1994). Although they are still widespread in the Alleghenies, there is good reason for concern over the future of the Least Flycatcher as a Maryland nesting bird because they have declined at 1.9% per year in the Allegheny Plateau region, including in neighboring Pennsylvania and West Virginia. Managed forests should be of uneven age to support populations of this feisty little flycatcher.

WALTER G. ELLISON

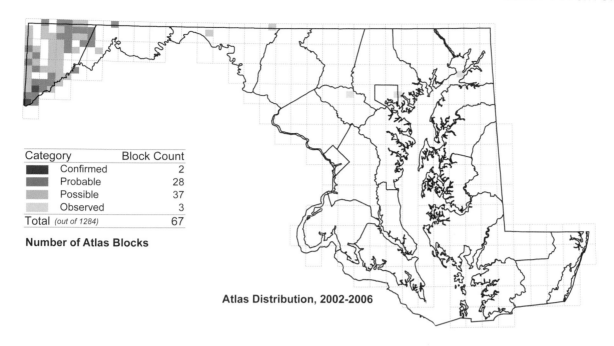

Category	Block Count
■ Confirmed	2
■ Probable	28
■ Possible	37
■ Observed	3
Total *(out of 1284)*	67

Number of Atlas Blocks

Atlas Distribution, 2002-2006

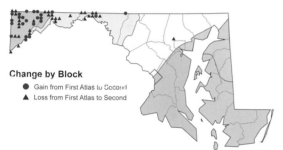

Change by Block

● Gain from First Atlas to Second
▲ Loss from First Atlas to Second

Change in Atlas Distribution, 1983-1987 to 2002-2006

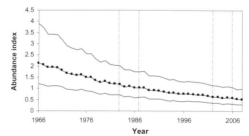

Percent of Stops

■ 50 - 100%
■ 10 - 50%
■ 0.1 - 10%
□ < 0.1%

Relative Abundance from Miniroutes, 2003-2007

			Change	
Atlas Region	1983-1987	2002-2006	No.	%
Allegheny Mountain	59	63	+4	+7%
Ridge and Valley	15	4	-11	-73%
Piedmont	1	0	-1	-100%
Upper Chesapeake	0	0	0	-
Eastern Shore	0	0	0	-
Western Shore	0	0	0	-
Totals	75	67	-8	-11%

Change in Total Blocks between Atlases by Region

Breeding Bird Survey Results in Maryland

Bill Sherman

Eastern Phoebe
Sayornis phoebe

The Eastern Phoebe is a small nondescript, but charming, gray green bird with a habitual flirting downward dip of the tail. The bird is named for its repetitive husky two-note *fee-bee!* song. This phoebe inhabits wooded areas near water, usually close to some sort of clearing. Phoebes feed primarily by hawking insects from the air, thus the need for open flight lanes, but they also capture prey from surfaces more than 20% of the time, often by pouncing from a perch, which requires open ground free of leaf litter (Weeks 1994). This flycatcher is a common sight near highway bridges and large culverts over creeks and small rivers in much of the Old Line State and throughout the District of Columbia.

Eastern Phoebes place their substantial mud and moss nests on artificial and natural shelves, such as bridge girders, bolts in culverts, houses and outbuildings, docks, overhanging cut-banks, and mossy forested rock ledges. Phoebes frequently reuse the bases of old nests by building a new superstructure on them (Weeks 1994). This is one of the best species for employing the atlas UN (used nest) breeding code. Phoebes do not appear to be as wedded to artificial nest sites as Chimney Swifts or Barn Swallows. Nonetheless such nest sites are predominant among Eastern Phoebes, and given the likely scarcity of natural nest sites in pre-Columbian North America, phoebes are probably more common today than before European settlement.

Eastern Phoebes were found throughout Maryland and DC during this atlas project, but they were far less prevalent and numerous on the Eastern Shore, a situation that was also true in the first statewide atlas project (Cheevers 1996a). A combination of more open landscapes with fewer trees along watercourses, fewer large bridges and culverts, and salt and brackish waters (providing fewer large aquatic insects) may help explain the lack of phoebes from parts of the Eastern Shore. In general the Eastern Phoebe was found in far more atlas blocks in 2002–2006 than in 1983–1987, with a net increase of 41 blocks. Among physiographic regions the greatest net increases in blocks were in the Allegheny Mountains and on the Western Shore. These increases were offset somewhat by a net loss of 12 blocks on the lower Eastern Shore.

Early in the history of the BBS, the Eastern Phoebe showed a decline. Over the entire run of the survey, however, the trend has been an increase of slightly under 1% per year. On Maryland BBS routes phoebe populations have fluctuated, but they also have recovered somewhat from initial declines. Some of the early decline registered on the BBS was attributable to precipitous declines after two very hard winters in the Eastern Phoebe's southeastern United States wintering range in the late 1970s (Robbins, Bystrak, and Geissler 1986). Trends within the broader physiographic strata spanning Maryland have been stable or positive. The long-term prospects for this phoebe appear to be good given its use of human artifacts for nest sites, but loss of streamside forests and changes in bridge and building design have the potential to deprive this engaging bird of habitat and nest sites.

WALTER G. ELLISON

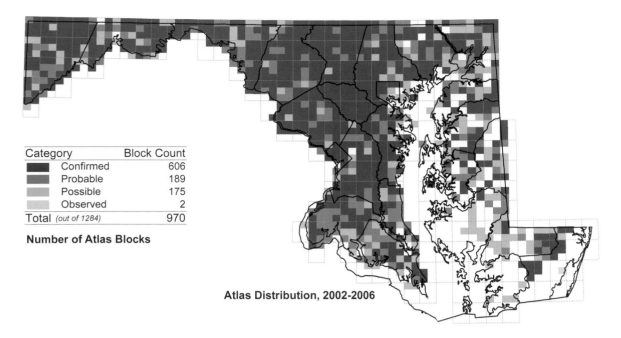

Category	Block Count
Confirmed	606
Probable	189
Possible	175
Observed	2
Total (out of 1284)	970

Number of Atlas Blocks

Atlas Distribution, 2002-2006

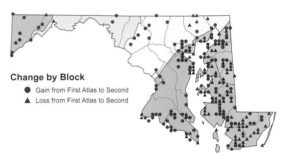

Change by Block
- ● Gain from First Atlas to Second
- ▲ Loss from First Atlas to Second

Change in Atlas Distribution, 1983-1987 to 2002-2006

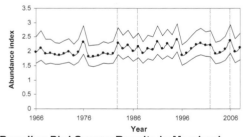

Percent of Stops
- 50 - 100%
- 10 - 50%
- 0.1 - 10%
- < 0.1%

Relative Abundance from Miniroutes, 2003-2007

Atlas Region	1983-1987	2002-2006	Change No.	%
Allegheny Mountain	83	92	+9	+11%
Ridge and Valley	146	147	+1	+1%
Piedmont	298	311	+13	+4%
Upper Chesapeake	79	84	+5	+6%
Eastern Shore	124	112	-12	-10%
Western Shore	194	219	+25	+13%
Totals	924	965	+41	+4%

Change in Total Blocks between Atlases by Region

Breeding Bird Survey Results in Maryland

Great Crested Flycatcher
Myiarchus crinitus

Large and colorful, but surprisingly hard to see because of its canopy-dwelling habits and generally slow movements when foraging, the Great Crested Flycatcher is often over-looked when it is not singing or calling. The song is powerful and far-carrying, an aggressive rising *weeeep!* usually combined with a burst of another three to five hoarse *weeps*. It also gives an array of burry short calls similar in tone. This is the only cavity-nesting flycatcher east of the Mississippi River. It will nest in a variety of natural cavities as well as in artificial sites. The latter include nest boxes placed on the edge of or in the interior of the bird's woodland habitat and such unusual sites as mailboxes (Ricciardi 1996d; MNRF) and cannon barrels in city parks or battlefields (Ickes 1992c; Hess et al. 2000). These elegant flycatchers are found virtually throughout Maryland and DC although they are seldom more than fairly common.

Great Crested Flycatchers are birds of open woodlands and forest edges. They generally avoid pure conifer stands but often occur in mixed pine-oak woodlands. They adapt well to stands of scattered trees that mimic open, parklike forests, such as those found in city parks and orchards. They are often most common near water, especially in strips of woods along streams and in swampy woods.

The Great Crested Flycatcher migrates to the neotropics for the winter, residing from southeastern Mexico to northwestern South America (Lanyon 1997). They return to Maryland and DC in late April and generally remain into September. Great Crested Flycatchers are most detectable in May and June, when they are quite vocal. By mid-July they become almost silent and are hard to find; the window of opportunity for finding them in an atlas block is fairly narrow. Nonetheless, during the second atlas they were located in almost 1,200 atlas blocks with a respectable 30% nesting confirmation rate. Maryland and DC egg dates have been recorded from 13 May (Stewart and Robbins 1958) to 16 July 2003 at Sandy Point State Park in Anne Arundel County (D. Walbeck, this project). Nestlings have been found as late as 4 August (Ricciardi 1996d; MNRF).

The status of the Great Crested Flycatcher appears to have changed remarkably little over the past one hundred years. Kirkwood (1895) called it common; Eifrig (1904) termed it only locally common in western Maryland. Stewart and Robbins (1958) described a variable status among Maryland's physiographic regions, calling it common on the lower Coastal Plain, fairly common from the Upper Chesapeake

George M. Jett

westward through the Ridge and Valley, and uncommon in the Allegheny Mountains. Lower numbers at high elevations have also been reported in other eastern states (Ickes 1992c; McGowan 2008d).

In Maryland and DC the Great Crested Flycatcher was one of the three most broadly distributed tyrant flycatchers in the 1983–1987 (Ricciardi 1996d) and 2002–2006 atlas projects. There were a few small holes in the 1980s range; some of these appear to have been caused by gaps in coverage. Absences on lower Chesapeake Bay islands, Assateague Island, and in some heavily farmed and urban blocks were likely genuine. They were found in a net total of 40 more blocks from 2002 to 2006. Some increases reflect better coverage, but other gains were on lower Eastern Shore islands, in the Frederick Valley, and in Washington, DC, and Baltimore City. On the other hand, they clearly lost ground in the Allegheny Mountains in higher elevation blocks. They have also declined significantly on Allegheny Plateau region BBS routes since 1980.

Great Crested Flycatcher numbers have been stable on North American BBS routes from 1966 to 2007. They have also been stable on Maryland routes since 1966 with a significant increase of 1.1% per year beginning in 1980. At present there is no indication that the brawny *weep* of this flycatcher will cease to be a feature of Maryland woodlands into the foreseeable future.

WALTER G. ELLISON

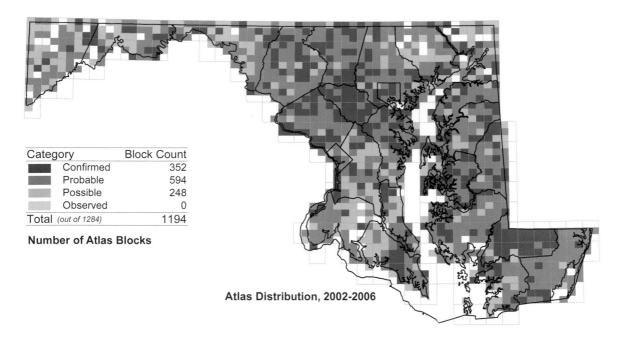

Category	Block Count
■ Confirmed	352
■ Probable	594
■ Possible	248
■ Observed	0
Total (out of 1284)	1194

Number of Atlas Blocks

Atlas Distribution, 2002-2006

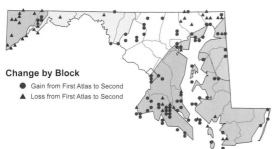

Change by Block
- ● Gain from First Atlas to Second
- ▲ Loss from First Atlas to Second

Change in Atlas Distribution, 1983-1987 to 2002-2006

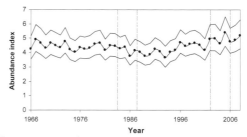

Percent of Stops
- ■ 50 - 100%
- ■ 10 - 50%
- ■ 0.1 - 10%
- □ < 0.1%

Relative Abundance from Miniroutes, 2003-2007

			Change	
Atlas Region	1983-1987	2002-2006	No.	%
Allegheny Mountain	89	76	-13	-15%
Ridge and Valley	143	143	0	0%
Piedmont	299	305	+6	+2%
Upper Chesapeake	111	119	+8	+7%
Eastern Shore	293	307	+14	+5%
Western Shore	213	238	+25	+12%
Totals	1148	1188	+40	+3%

Change in Total Blocks between Atlases by Region

Breeding Bird Survey Results in Maryland

Eastern Kingbird
Tyrannus tyrannus

The Eastern Kingbird derives its name from its reputation for aggressive and pugnacious behavior in defense of its nesting territory. These kingbirds are particularly hard on any large bird that presents an apparent threat to precious eggs and young. They hound crows, raptors, vultures, and herons, even landing on their backs like miniature bull-riders to tug at feathers. Eastern Kingbirds are most numerous on the Great Plains (Price et al. 1995) as befits an essentially open country bird. On the eastern seaboard in precolonial times these flycatchers were largely limited to floodplains, the edges of wetlands (especially coastal marshes), the margins of lakes and ponds, and the prairies that were regularly burned by Native Americans. With the advent of European-style farming practices many more habitats resembling the open lands and edges preferred by kingbirds proliferated, including orchards, hedgerows, and farmyards with stately ornamental trees.

Eastern Kingbirds, which migrate to the Amazon basin and become flocking fruit-eaters in floodplain forests (M. Murphy 1996), are generally present in Maryland and DC from late April to mid-September. These kingbirds raise only one brood per year, but they are persistent renesters. Maryland nesting ranges from nest-building in early May to the last few dependent fledglings in late August. Eastern Kingbirds are not furtive around their nests and seldom hide them as assiduously as other songbirds, presumably relying on their ferocious mobbing behavior to protect the nest. Over 48% of

Eastern Kingbird reports from 2002 to 2006 were breeding confirmations, including 181 records of active nests (10 May to 29 July), and 209 records of dependent young (1 June to 17 August).

The Eastern Kingbird is one of the 30 most widespread birds in Maryland and DC; it is widely but rather thinly distributed across the state and is not as numerically dominant as some other widespread species. The only relatively large holes in the distribution are in the heavily forested mountains of Garrett and Allegany counties, and perhaps less explicably, inland on the lower Eastern Shore. There was no change in the Eastern Kingbird's range between the 1980s and the 2000s; it was found in essentially the same number of blocks and showed similar rates of possible, probable, and confirmed nesting. This kingbird's status has not changed since the 1950s when it was considered fairly common statewide (Stewart and Robbins 1958).

Maryland BBS trends for Eastern Kingbird have revealed a slow steady decline of 2.3% per year, although it has been less steep over the past two and a half decades; 1.5% per year versus 6.4% per year from 1966 to 1979. In spite of the long-term decline, this kingbird retained a similar nesting range from 1983 to 1987 in Maryland and DC (Wilkinson 1996). Reasons for the decline may include hedgerow removal in farmland, some reforestation in western Maryland, use of pesticides on farmland and lawns, and perhaps declines in the large-bodied flying insects that are the summer staple for adults and young. At present the Eastern Kingbird remains widespread, if less numerous, in the Old Line State and is likely to be here when another atlas project takes place.

WALTER G. ELLISON

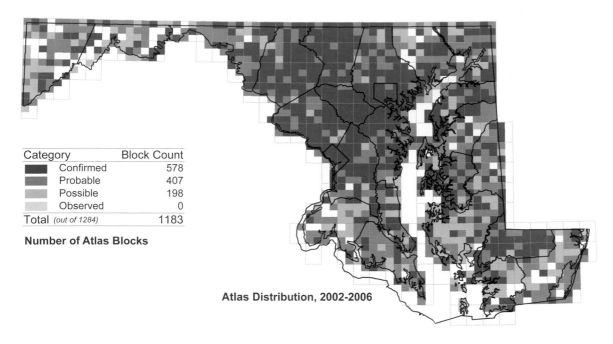

Category	Block Count
■ Confirmed	578
■ Probable	407
■ Possible	198
■ Observed	0
Total (out of 1284)	1183

Number of Atlas Blocks

Atlas Distribution, 2002-2006

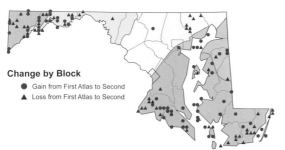

Change by Block
● Gain from First Atlas to Second
▲ Loss from First Atlas to Second

Change in Atlas Distribution, 1983-1987 to 2002-2006

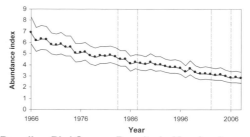

Percent of Stops
■ 50 - 100%
■ 10 - 50%
■ 0.1 - 10%
□ < 0.1%

Relative Abundance from Miniroutes, 2003-2007

Atlas Region	1983-1987	2002-2006	Change No.	%
Allegheny Mountain	76	73	-3	-4%
Ridge and Valley	133	134	+1	+1%
Piedmont	315	309	-6	-2%
Upper Chesapeake	112	117	+5	+4%
Eastern Shore	307	309	+2	+1%
Western Shore	232	236	+4	+2%
Totals	1175	1178	+3	+0%

Change in Total Blocks between Atlases by Region

Breeding Bird Survey Results in Maryland

Loggerhead Shrike
Lanius ludovicianus

Similar in size and appearance to the Northern Mockingbird, the Loggerhead Shrike has black in the wings and tail, a black mask, and a thick, hook-tipped bill for ripping prey. Sometimes called the "butcherbird," the Loggerhead Shrike is a predatory songbird that hunts large insects, such as grasshoppers, crickets, and beetles, as well as small birds and mammals, and impales them on thorny shrubs or barbed wire. This species does not possess the powerful talons of true raptors, and it has learned to use thorns to aid in feeding, as well as in food storage. This unique songbird has been on Maryland's endangered species list since 1987 and is now nearly extirpated from the state.

Like many other grassland species in Maryland, the Loggerhead Shrike may have been present in low numbers prior to European settlement; it is believed to have expanded its population north and east through the region as extensive forested areas were cleared for family farms in the eighteenth and nineteenth centuries. It was first documented in Maryland during the breeding season around 1880 (Kirkwood 1895), and nesting was finally confirmed in 1910 (Stewart and Robbins 1958). The Loggerhead Shrike was considered an uncommon permanent resident in the early twentieth century (Cooke 1929), while Stewart and Robbins (1958) considered it an uncommon breeder in Prince George's County and rare and local elsewhere in Maryland outside the Allegheny Mountain region, where it was absent.

Loggerhead Shrikes require open habitat, such as pastures, meadows, and fallow fields, scattered with appropriate thorny shrubs and trees, such as hawthorns and locusts. This species has adapted to use barbed wire in place of natural thorns, and it frequently hunts from roadside fences and telephone wires. Since it hunts from perches (rather than employing the hovering technique of the American Kestrel), perch abundance, height, and the height of ground vegetation affect prey detectability and probably reproductive success (Luukkonen and Fraser 1987). Nesting data from Maryland are scant, but nests in Virginia are frequently placed in eastern red cedars and hawthorns covered with vines, about 1.7 to 5.5 m (5.5–18 ft) from the ground (Luukkonen and Fraser 1987). Loggerhead Shrikes are typically double brooded, and egg dates in DC and Maryland range from 19 April (MNRF) to 11 July (Davidson 1996f), and nestling dates range from 9 May to 19 July (Davidson 1996f). Clutch sizes of 6 nests range from 4 to 6 eggs and broods range from 1 to 6 young (Davidson 1996f; MNRF). During the 2002–2006 atlas, an adult was carrying food for young on 13 June (D. Weesner).

Maryland's Loggerhead Shrike population declined 77% in the past 20 years, as measured by the drop in recorded atlas blocks from 13 in the 1983–1987 atlas (Davidson 1996f) to only 3 blocks in the 2002–2006 atlas project. This decline is mirrored by regional and national declines as determined through BBS trend data. From 1966 to 1979 the North American decline was 4.6% per year; this trend has lessened to a loss of 2.9% per year since 1980. No BBS trend exists specifically for Maryland because of insufficient data.

In both the previous atlas project and the current one, the primary remaining habitat areas for the Loggerhead Shrike in Maryland are the short-grass fields of dairy farms and horse pastures in the valleys of central and eastern Washington County and the Monocacy Valley in Frederick County. This species was found in 3 blocks within the Western Shore region during the 1983–1987 atlas (Davidson 1996f), but it is now absent from this region and its range has become very restricted as it teeters on the brink of extirpation as a member of Maryland's breeding avifauna.

Among the many factors that have led to this species'

precipitous decline is the loss of dairy and horse farms with scrubby hedgerows of hawthorns and other thorny species. Barbed wire fences largely have been replaced with electric fences, further reducing this bird's ability to store prey. Those hedgerows and barbed wire fences that remain are frequently found along roadsides. Shrikes tend to fly close to the ground, and collisions with automobiles are a major cause of mortality, as documented in both Virginia (Blumton et al.1989) and Maryland (Davidson 1996f). Finally, pesticide contamination, which has been documented in several states, may also be a contributing factor since this species is near the top of the food chain. However, a direct link to reduced reproductive success or increased mortality has not been determined (Luukkonen and Fraser 1987; Blumton et al. 1989). Targeted management assistance at farms with remaining breeding pairs may provide short-term relief for this species' survival in Maryland, but the scale of its decline throughout North America speaks to a larger need, including perhaps federal attention.

LYNN M. DAVIDSON

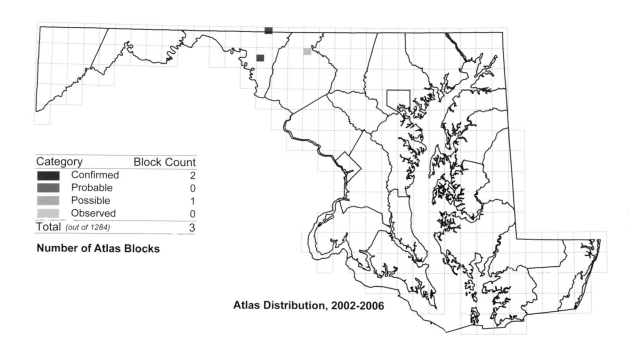

Category	Block Count
Confirmed	2
Probable	0
Possible	1
Observed	0
Total (out of 1284)	3

Number of Atlas Blocks

Atlas Distribution, 2002-2006

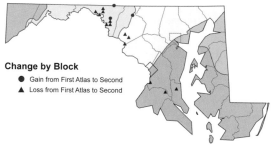

Change by Block
- Gain from First Atlas to Second
▲ Loss from First Atlas to Second

Change in Atlas Distribution, 1983-1987 to 2002-2006

Atlas Region	1983-1987	2002-2006	Change No.	%
Allegheny Mountain	0	0	0	-
Ridge and Valley	7	2	-5	-71%
Piedmont	3	1	-2	-67%
Upper Chesapeake	0	0	0	-
Eastern Shore	0	0	0	-
Western Shore	3	0	-3	-100%
Totals	13	3	-10	-77%

Change in Total Blocks between Atlases by Region

White-eyed Vireo
Vireo griseus

The White-eyed Vireo has an air of impertinence about it with its jaunty switching tail, pale-eyed glare, and fussy mechanical song that warns *Check! Ya got another tick!* apparently enjoying an observer's need to brave ticks and chiggers to make the little scrub vireo's acquaintance. White-eyed Vireos require dense shrubs and small trees, usually with a generous admixture of thorny plants such as greenbrier, roses, brambles, and hawthorns. These vireos will accept sites where a canopy closes over shrubs as long as it is near an opening and more often than not there are some trees exceeding the average height of the thicket.

White-eyed Vireos, which are migratory, are present in Maryland from mid-April to early October (a few), wintering farther south in the United States and in Mesoamerica south to northwestern Honduras. The White-eyed Vireo nests lower than other Maryland vireos, often building nests at 1 m (3.3 ft) or lower (Hopp et al. 1995). This vireo builds the well-woven and prettily decorated hanging pouch nest characteristic of vireos, although its nest is more conical than those of other vireos. Over half of Maryland atlas breeding confirmations for this vireo were of recently fledged juveniles. Records of nests with eggs for Maryland and DC range from 22 April to 26 July, with nestlings reported to 19 August (Cheevers 1996b; MNRF).

Although Kirkwood (1895) called White-eyed Vireo common in eastern Maryland, Eifrig (1904) listed no records on his list of western Maryland birds. By the late 1950s Stewart and Robbins (1958) described the White-eyed Vireo as common to fairly common on the Coastal Plain and along the Piedmont portion of the Potomac valley, uncommon west to Allegany County, and rare in Garrett County. By the time of the 1983–1987 Maryland-DC atlas project (Cheevers 1996b), these vireos had increased on the Piedmont (solid distribution), in the Ridge and Valley (75% of blocks), and in the Alleghenies (56% of blocks). In the 2002–2006 atlas project, White-eyed Vireos were not refound in 73 blocks, a 7% decline. Most of these losses (57%) were from western Maryland with a 48% loss in the Allegheny Mountains and an 18% loss in the Ridge and Valley, with the greatest losses there in Allegany County and the northern Hagerstown Valley. The fluctuation in western Maryland over the past 50 years may indicate an increase in successional habitat in the 1960s and 1970s followed by a decline after the 1980s because of reforestation and loss to development.

White-eyed Vireo populations as measured by BBS surveys appear to be stable in North America, although the BBS trend map shows a patchwork of increases and decreases across the range indicative of varying trends in the status of scrub and successional habitats in different parts of this vireo's distribution. The Maryland BBS trend is significantly negative (1.1% per year) probably reflecting the twin pressures of reforestation and development on White-eyed Vireo habitat. Preservation or creation of scrub and thickets is necessary for this vireo to maintain its numbers in Maryland and DC. Managers of public lands, especially state forestry lands, should consider prescribing timber-cutting policies that favor creation of suitable transitional scrublands for this vireo as well as Yellow-breasted Chat and other denizens of shrubby places.

WALTER G. ELLISON

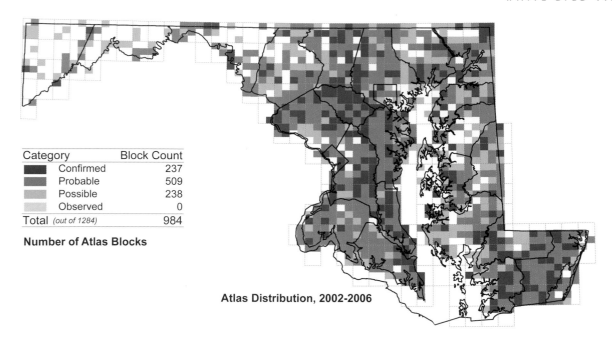

Category	Block Count
Confirmed	237
Probable	509
Possible	238
Observed	0
Total (out of 1284)	984

Number of Atlas Blocks

Atlas Distribution, 2002-2006

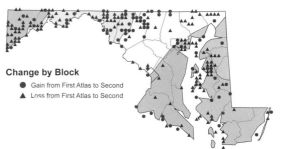

Change by Block

● Gain from First Atlas to Second
▲ Loss from First Atlas to Second

Change in Atlas Distribution, 1983-1987 to 2002-2006

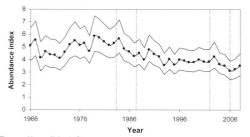

Percent of Stops

■	50 - 100%
■	10 - 50%
■	0.1 - 10%
□	< 0.1%

Relative Abundance from Miniroutes, 2003-2007

Atlas Region	1983-1987	2002-2006	Change No.	Change %
Allegheny Mountain	52	27	-25	-48%
Ridge and Valley	92	75	-17	-18%
Piedmont	282	282	0	0%
Upper Chesapeake	114	105	-9	-8%
Eastern Shore	275	253	-22	-8%
Western Shore	237	237	0	0%
Totals	1052	979	-73	-7%

Change in Total Blocks between Atlases by Region

Breeding Bird Survey Results in Maryland

Yellow-throated Vireo
Vireo flavifrons

A green and yellow bird that is hard to see amid green leaves dappled with yellow sunlight, the Yellow-throated Vireo is most often encountered via its drawling husky song that sounds like a hoarse Blue-headed Vireo or a very slow Scarlet Tanager. This is a bird of the treetops and midstory of tall deciduous trees, but not a bird of forest depths, preferring river banks, edges of clearings, and roadsides. This vireo is not as common as the ubiquitous Red-eyed Vireo, but is widespread in rather small numbers over much of Maryland and DC although there are large gaps in its nesting range.

Although the Yellow-throated Vireo is rarely found in the interior of deep woods, it nonetheless appears to occur most commonly in a landscape dominated by woodland. This may have to do with its preference for mature tall trees or because such landscapes have relatively few cowbirds; this vireo is a very frequent host to cowbird eggs and nestlings (Peck and James 1987). It is also seldom encountered where evergreens dominate and is scarce at higher elevations (Master 1992c). Yellow-throated Vireos capture much of their insect prey by gleaning from twigs and branches, helping them coexist with the leaf-gleaning Warbling and Red-eyed vireos (James 1976).

Yellow-throated Vireos, which winter in southern Mexico, Central America, and northwestern South America (Rodewald and James 1996), reside in Maryland and DC from mid-April to late September. The nest is a basket liberally decorated with bark and lichen hung over the twigs of an outer branch fork. In general they nest high in trees; most nests are found from 6 to 15 m (20–50 ft; Rodewald and James 1996) above the ground. Nests are hard to find and adults can be difficult to observe, so only about 10% of atlas records refer to confirmed nesting. The 40% of records assigned possible nesting status emphasizes the generally low population densities of this vireo. Maryland egg dates range from 9 May to 21 July (Stewart and Robbins 1958; MNRF).

Apparently Yellow-throated Vireos have never been very common in Maryland and DC. C. Richmond (1888) called them rather common around DC; Kirkwood (1895) wrote that they were not very common; and Eifrig's (1904) assessment for Allegany and Garrett counties was not common. Stewart and Robbins' (1958) more detailed discussion of this vireo's status termed them fairly common in eastern Maryland including the Potomac Valley in the Piedmont, uncommon on the rest of the Piedmont and in the Ridge and Valley, and rare in the Allegheny Mountains. By the 1983–1987 atlas project they had apparently declined in the Upper Chesapeake and the northern Eastern Shore remaining widespread in the Pocomoke and Wicomico drainages (Van Ness 1996e). They increased on the Piedmont in the central and western Ridge and Valley, and in the Alleghenies with centers of occurrence there on the Potomac River and in northwestern Garrett County. Their status was similar in the second atlas, but there was a loss of 25% (net loss of 13 blocks) from the Upper Chesapeake and an increase of 11% (net gain of 20 blocks) on the Piedmont, especially in northwestern Carroll County.

The 40-year BBS trend for the Yellow-throated Vireo has been a modest increase of 1.2% per year. But nowhere are they very numerous, averaging just 0.8 birds per route. In Maryland they have been more or less stable with a nonsignificant declining tendency. Although the Yellow-throated Vireo's atlas range map and BBS trends speak of stability, its numbers have never been very great and its habitat requirements make it vulnerable to habitat loss. This remains a bird that could decline to rarity in Maryland and DC if it is not properly studied and monitored.

WALTER G. ELLISON

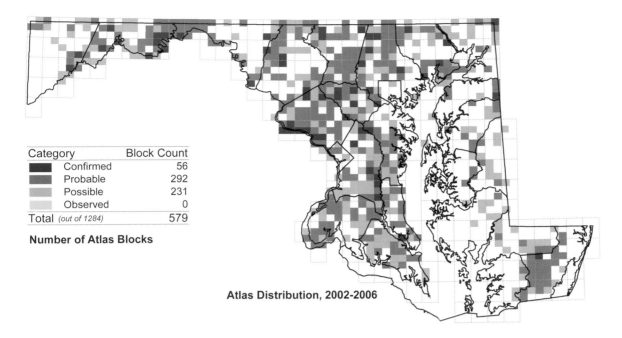

Category	Block Count
Confirmed	56
Probable	292
Possible	231
Observed	0
Total (out of 1284)	579

Number of Atlas Blocks

Atlas Distribution, 2002-2006

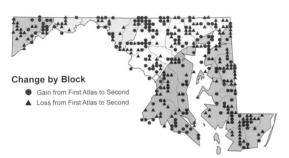

Change by Block

● Gain from First Atlas to Second
▲ Loss from First Atlas to Second

Change in Atlas Distribution, 1983-1987 to 2002-2006

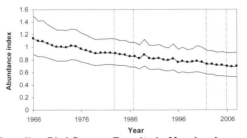

Percent of Stops

50 - 100%
10 - 50%
0.1 - 10%
< 0.1%

Relative Abundance from Miniroutes, 2003-2007

Atlas Region	1983-1987	2002-2006	Change No.	Change %
Allegheny Mountain	30	28	-2	-7%
Ridge and Valley	89	90	+1	+1%
Piedmont	182	202	+20	+11%
Upper Chesapeake	51	38	-13	-25%
Eastern Shore	77	72	-5	-6%
Western Shore	142	146	+4	+3%
Totals	571	576	+5	+1%

Change in Total Blocks between Atlases by Region

Breeding Bird Survey Results in Maryland

David Ziolkowski

Blue-headed Vireo
Vireo solitarius

The Blue-headed Vireo is usually heard before it is seen. It sounds rather like a Red-eyed Vireo carefully trying to reproduce its song on a slide whistle. Although the Blue-headed, with its prominent wing-bars, bluish gray hood, and obvious white spectacles, looks nothing like the Red-eyed, its song of paired couplets can be mistaken for the latter vireo causing it to be overlooked at times. This vireo was once combined with two similar species from western North America under the name Solitary Vireo but has been demonstrated to be sufficiently distinct to be a separate species (N. Johnson et al. 1988; AOU 1997). It is a bird of northern coniferous and mixed forest, usually mature with a well-developed but not dense understory (James 1998). In Maryland it is predominantly found in the mountains of the western third of the state from the eastern edge of the Ridge and Valley region westward.

The Blue-headed Vireo's habitat differs from the rather similar Yellow-throated Vireo's in having more understory, more conifers, and generally smaller trees (James 1979). It often overlaps its territories with Red-eyed Vireo but usually does not strongly compete with it because it gleans much of its prey from twigs, branches, and trunks leaving deciduous foliage for the leaf-gleaning and hovering Red-eyed (S. Robinson and Holmes 1982). This use of woody surfaces that are characterized by lower prey densities than foliage may be correlated with this species' large territories and low population densities, a quarter of those for Red-eyed Vireos in Maryland (Stewart and Robbins 1958).

Blue-headed Vireos, which are migratory, winter from the southeastern United States north to coastal North Carolina and southward through eastern Mexico to northern Central America. This is the earliest of the vireos to return and the last to depart Maryland and DC, with spring arrivals usually in early April (rarely in late March) and late departures generally in late October, with some remaining into January. These vireos usually build their well-decorated basketlike nests rather low, often between 3 and 6 m (10–20 ft) off the ground. Maryland egg dates range from 30 May to 22 July (Farrell 1996a; MNRF). Egg dates from West Virginia and Virginia range from 25 April to 25 July (J. DeCecco in James 1998). Given these dates, an early nestling date from this project of 22 May 2004 from the southwestern part of Green Ridge State Forest is not suspect (J. Green).

The first observer to report breeding season Blue-headed Vireos in Maryland was Preble (1900) who found them fairly common in northeastern Garrett County and on Dans Mountain, Allegany County. J. H. Sommer found nests in Garrett County in 1918, 1919, and 1925 (Stewart and Robbins 1958). Stewart and Robbins (1958) found them fairly common in the Allegheny Mountain region above 610 m (2,000 ft). Two subspecies occur in Maryland, the transient *solitarius* and the local nesters ascribed to *alticola* of the southern Appalachian Mountains (Phillips 1991).

Blue-headed Vireos continued widespread at upper elevations in Garrett County in 1983–1987, with absences from lower elevation blocks in southeastern and northwestern parts of the county (Farrell 1996a). They expanded into the higher mountains of the western Ridge and Valley in eastern Allegany County and westernmost Washington County. Several birds were found within nonmigratory safe dates on South Mountain and at scattered locations on the Piedmont including birds apparently holding territory in southern Howard County and in the Hanover Watershed in southern York County, Pennsylvania.

The slow range expansion of the Blue-headed Vireo continued over the two decades between the two atlas projects. It was found in 28 more blocks, a 39% range expansion, in the 2002–2006 atlas. It increased in southeastern Garrett County and in southwestern Allegany County, remained well established in eastern Allegany County, and bred on South Mountain. This vireo also nested for the first time on the Piedmont at an elevation of 200 m (660 ft) in the Soldiers Delight Natural Environment Area in southwestern

Baltimore County, where a nest-building pair was seen on 9 June 2002 (K. E. Costley). A bird had been found singing on territory at Soldiers Delight in the summer of 2001 (Ringler 2002b).

Blue-headed Vireos have increased on North American BBS routes at an impressive 4.5% per year, 3.6% per year since 1980. This included an increase of 4.1% per year on 94 Allegheny Plateau BBS routes. Although this vireo tends to be absent from small woodlots (Robbins, Dawson, and Dowell 1989) it has been increasing, unlike many other forest interior birds. The most likely explanation for the increase is regeneration of mature forest at upper elevations in eastern North America. This is also a relatively short-distance migrant that may be benefiting from ameliorating winters in the southeastern United States. At present the Blue-headed Vireo is doing well in Maryland, but development and forest clearing in western Maryland could reduce its numbers in the future.

WALTER G. ELLISON

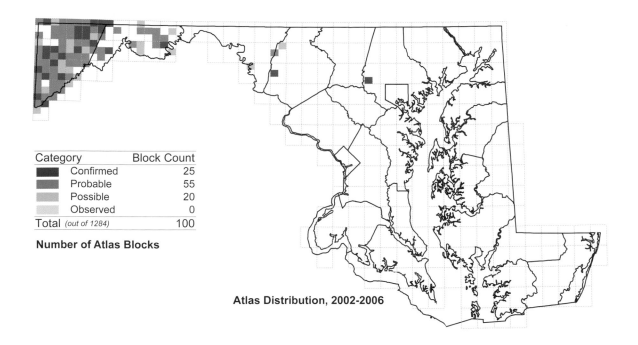

Category	Block Count
Confirmed	25
Probable	55
Possible	20
Observed	0
Total (out of 1284)	100

Number of Atlas Blocks

Atlas Distribution, 2002-2006

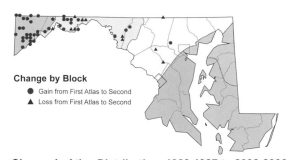

Change by Block
- ● Gain from First Atlas to Second
- ▲ Loss from First Atlas to Second

Change in Atlas Distribution, 1983-1987 to 2002-2006

Percent of Stops
- 50 - 100%
- 10 - 50%
- 0.1 - 10%
- < 0.1%

Relative Abundance from Miniroutes, 2003-2007

Atlas Region	1983-1987	2002-2006	Change No.	%
Allegheny Mountain	56	80	+24	+43%
Ridge and Valley	13	19	+6	+46%
Piedmont	3	1	-2	-67%
Upper Chesapeake	0	0	0	-
Eastern Shore	0	0	0	-
Western Shore	0	0	0	-
Totals	72	100	+28	+39%

Change in Total Blocks between Atlases by Region

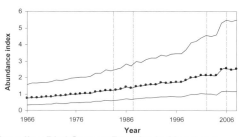

Breeding Bird Survey Results in Maryland

Mark Hoffman

Warbling Vireo
Vireo gilvus

Drab, gray, small, slow moving, and a bird of the canopy of tall trees, the Warbling Vireo can be difficult to see and identify. This vireo's incessant song, which is pleasantly musical and as meandering as the streams along which the bird often lives, draws attention to it. This bird has a limited distribution in Maryland and DC; it is scarce at upper elevations and in well-wooded regions, and approaches the southern edge of its range on the Coastal Plain. The breeding range is largely bounded to the east by the fall line.

Warbling Vireos live in tall trees in or adjacent to large open areas such as tall open floodplain forests, edges of deciduous and mixed forest, and stands of shade trees in urban parks and along town or city streets. One study of vireo habitat reported that Warbling Vireo territories had an average canopy coverage of just 34% (James 1976). These unobtrusive but vociferous birds winter in western Mexico and Central America; they reside in Maryland from late April to late September. Their nests, neatly woven vireo baskets that are well hidden by surrounding leaves, are slung on outer branch forks from the midstory to the canopy, often between 6 and 9 m (about 20–30 ft; Peck and James 1987). The reported mean height of Maryland nests is closer to 12 m (39 ft; S. Clarkson 1996c; MNRF). Maryland egg dates range from 16 May to 22

June (S. Clarkson 1996c; MNRF), and nestlings have been reported as late as 31 July in 2003 and 2005 near Hollofield, Baltimore County (K. Heffernan, this project). The latter dates imply some clutches are completed in early July.

Historical accounts suggest that the Warbling Vireo has always had a localized distribution in Maryland and DC. Kirkwood (1895) cited evidence of breeding at DC, Hagerstown, Cumberland, and perhaps Talbot County, but said they did not breed in Baltimore County. Eifrig (1904) called it not common in western Maryland. The broad outline of the nesting range illustrated in Stewart and Robbins (1958) is similar to those delineated in the two atlas projects, but the species is not as widespread on the Eastern Shore and is less numerous in the Upper Chesapeake—it is now uncommon there as opposed to fairly common. Other differences include a small nesting population now present in the Allegheny Mountains (nesting confirmed at Broadford Lake) and more records on the Western Shore, particularly in Prince George's County. They have increased their distribution, notably on the Piedmont; they were found in 50% of the region's blocks in the 1980s (S. Clarkson 1996c) and were recorded in 61% of regional blocks from 2002 to 2006.

The Warbling Vireo has been increasing at a modest 0.9% per year on North American BBS routes since 1966, with a faster rate of increase in the east (1.2% per year). They are also increasing in Maryland, but the change does not reach statistical significance because of low numbers seen on Maryland BBS routes. The long-term increase may reflect reductions in spraying of shade trees with pesticides or the maturation of shade trees planted in neighborhoods established in the 1950s through the 1970s. This vireo is still spottily distributed in most of Maryland and is not numerous near the southeastern edge of its range. Its status will probably continue to fluctuate in Maryland and DC.

WALTER G. ELLISON

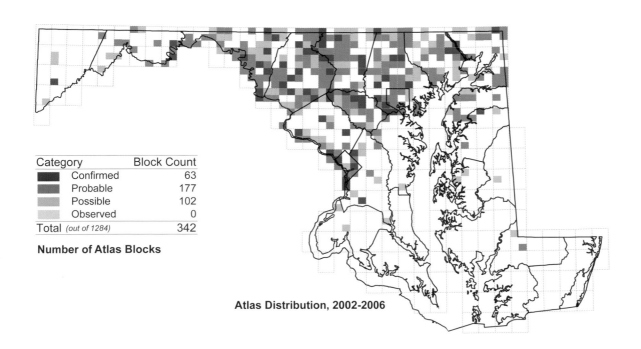

Category	Block Count
Confirmed	63
Probable	177
Possible	102
Observed	0
Total (out of 1284)	342

Number of Atlas Blocks

Atlas Distribution, 2002-2006

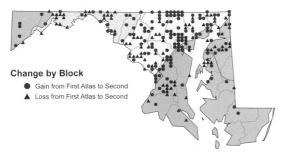

Change by Block
- ● Gain from First Atlas to Second
- ▲ Loss from First Atlas to Second

Change in Atlas Distribution, 1983-1987 to 2002-2006

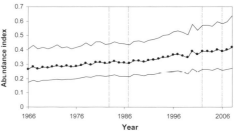

Percent of Stops
- 50 - 100%
- 10 - 50%
- 0.1 - 10%
- < 0.1%

Relative Abundance from Miniroutes, 2003-2007

Atlas Region	1983-1987	2002-2006	Change No.	%
Allegheny Mountain	3	7	+4	+133%
Ridge and Valley	86	79	-7	-8%
Piedmont	158	192	+34	+22%
Upper Chesapeake	23	28	+5	+22%
Eastern Shore	6	7	+1	+17%
Western Shore	21	27	+6	+29%
Totals	297	340	+43	+14%

Change in Total Blocks between Atlases by Region

Breeding Bird Survey Results in Maryland

George M. Jett

Red-eyed Vireo
Vireo olivaceus

An almost inescapable part of the ambiance of woodlands in eastern North America is the sing-song repetitive caroling of the Red-eyed Vireo. This vireo seems to be holding a one-sided conversation in the forest canopy, *Here I am—way up—in the treetop—and you can't see me* repeated ad infinitum, or so it seems. Red-eyed Vireos are most likely to occur in large blocks of forest, but they may also be found in small wood-lots and seem more prone than many other forest birds to live in them (Robbins, Dawson, and Dowell 1989). Red-eyed Vireos evince a broad habitat tolerance, although they are most numerous in mature moist deciduous and mixed decid-uous and conifer forests (Stewart and Robbins 1958). They will also occupy medium stature successional woodlands and well-treed suburban neighborhoods and city parks.

North American Red-eyed Vireos are migratory, wintering in the western Amazon basin and residing in Maryland from late April to early October. In spite of their verbosity, Red-eyed Vireos are sufficiently unobtrusive to make confirming nesting difficult; nearly two-fifths of atlas records were for probable nesting and only about a third were for confirmed breeding. Most nesting confirmations were via young fledg-lings (173 records) or of adults carrying food (156 records). This vireo's nest, a distinctive little gray basket hung from low- to medium-height forks near branch tips, is surprisingly hard to find. Nests with eggs have been reported in Maryland from 9 May to 11 August (Russell 1996d; MNRF).

The Red-eyed Vireo is one of the most widespread for-est birds in Maryland and DC. It was found during the sec-ond atlas project in essentially the same number of blocks as in 1983–1987 (Russell 1996d) with similar rates of possible, probable, and confirmed breeding. This vireo was termed common by authors back at the turn of the twentieth cen-tury (Kirkwood 1895; Eifrig 1904). Between the first statewide atlas and the current project these vireos declined slightly in the Hagerstown area and were lost from several shoreline blocks on the Eastern Shore. The largest holes in this vireo's nesting range were in areas that lacked sufficient deciduous woodland, such as the suburban and heavily farmed Hagers-town Valley, and the pine savannas, marshes, and dunes of the lower Eastern Shore.

In spite of some sensitivity to woodlot size and to frag-mentation of woodlands, the Red-eyed Vireo was recorded in more blocks than most other forest birds such as Ovenbird and Wood Thrush. It seems likely that Red-eyed Vireos have lower nesting success in small woodlots than in more exten-sive forests. The Red-eyed Vireo is a common host of the Brown-headed Cowbird, nearly a quarter of a sample of 162 nests from Maryland contained cowbird eggs (Russell 1996d; MNRF), and cowbird parasitism is greatest in small woodlots and in fragmented woodlands (S. Robinson 1994).

On a continental scale, the Red-eyed Vireo has been increasing on BBS routes, and it has increased long-term in neighboring Pennsylvania and Virginia. However, the Red-eyed Vireo has shown declines on Maryland BBS routes, with the trend beginning in the 1980s after a period of increase from 1966 to 1979. Maryland has had less reforestation than most other eastern states and has experienced rapid suburbanization from central Washington County eastward, conditions that would contribute to the slow decline of the forest-dwelling Red-eyed Vireo. If trends continue in this direction the Red-eyed Vireo may show more obvious losses in atlas blocks in a future atlas project but will likely remain reasonably common where it still occurs.

WALTER G. ELLISON

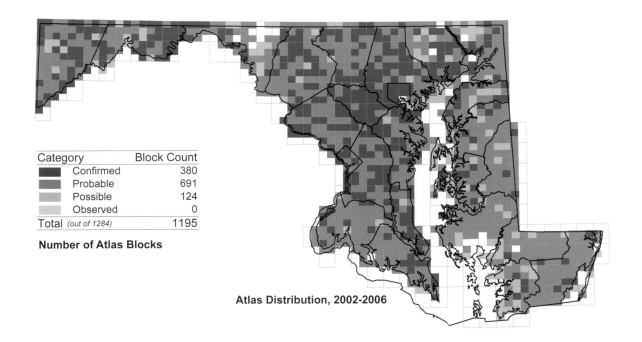

Category	Block Count
Confirmed	380
Probable	691
Possible	124
Observed	0
Total (out of 1284)	1195

Number of Atlas Blocks

Atlas Distribution, 2002-2006

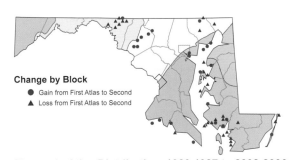

Change by Block
- ● Gain from First Atlas to Second
- ▲ Loss from First Atlas to Second

Change in Atlas Distribution, 1983-1987 to 2002-2006

Percent of Stops
- 50 - 100%
- 10 - 50%
- 0.1 - 10%
- < 0.1%

Relative Abundance from Miniroutes, 2003-2007

Atlas Region	1983-1987	2002-2006	Change No.	%
Allegheny Mountain	93	93	0	0%
Ridge and Valley	146	140	-6	-4%
Piedmont	309	313	+4	+1%
Upper Chesapeake	116	121	+5	+4%
Eastern Shore	287	275	-12	-4%
Western Shore	244	247	+3	+1%
Totals	1195	1189	-6	-1%

Change in Total Blocks between Atlases by Region

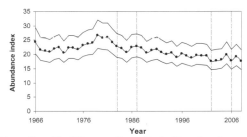

Breeding Bird Survey Results in Maryland

Eric Skrzypczak

Blue Jay
Cyanocitta cristata

This colorful, sometimes boisterous, moderate-size bird evokes mixed reactions from people familiar with it. Blue Jays are sometimes characterized as "bullies" and "loudmouths" because of their domineering behavior at bird feeders, their taste for the eggs and nestlings of smaller birds, and their enthusiastic strident outbursts when predators are detected. Blue Jays are also among the planters of our forests, as many a forgotten jay-cached acorn, hickory nut, and beechnut has matured over the millennia.

The Blue Jay has been common in Maryland for more than a hundred years (Kirkwood 1895; Stewart and Robbins 1958; Russell 1996a) and remains so. The Blue Jay was the 18th most widespread bird in Maryland during this atlas project, with reports from 1,212 blocks. Birds of forest and woodland, Blue Jays are less common in large tracts of mature closed forest than they are in more broken woodland. This is especially evident on the Allegheny Plateau of Garrett County (Price et al. 1995). They are scarce in extensively open landscapes. The only parts of the state where Blue Jays were absent, or perhaps overlooked, were the marshlands and open pine savanna of southern Dorchester County, the marshy islands of the lower Chesapeake Bay, and the dunes of Assateague Island.

Although Blue Jays are widespread and fairly numerous, they present some difficulties for field-workers. There are migratory and resident populations of Blue Jays. Many migratory jays continue their spring return flight well into June thus broadly overlapping the nesting season of Maryland's resident jays and preventing assignment of "safe dates" until many jays have completed nesting. Blue Jays are also very stealthy around nest sites and are not territorial, save for the nest site itself. Jay nests containing eggs have been found in Maryland and DC from 8 April to 30 July (MNRF) and nestlings have been recorded from 20 April 2002 north of Ellicott City, Howard County (R. Todd, this project) to 22 August (MNRF). Many observers are not aware that jays are nesting in their areas until fledglings make their appearance. Fortunately those fledglings give insistent whining calls and often visit feeders with their parents. The Blue Jay had a relatively robust 46% nesting confirmation rate and was recorded as confirmed or probable in 81% of the blocks in which it was found.

Blue Jays have shown a chronic slow decline on North

American and Maryland BBS routes of similar magnitude (1.1% per year continental; 1.4% per year in Maryland). Reasons for this decline include loss of forest to suburbanization, increases in major jay predators (especially Cooper's Hawk), and, possibly, declines of mast-bearing tree species—oak and beech have been subject to pathogen outbreaks in recent decades. Blue Jays also are prone to die when infected with West Nile virus (Komar et al. 2003). They declined sharply after the arrival of West Nile virus to North America in 1999 but appear to be showing signs of recovery (LaDeau et al. 2007). In general Blue Jays appear to tolerate most human modification of the landscape and are resilient to challenges such as introduced diseases, implying that they are likely to remain fairly common and widespread in Maryland and the District of Columbia.

WALTER G. ELLISON

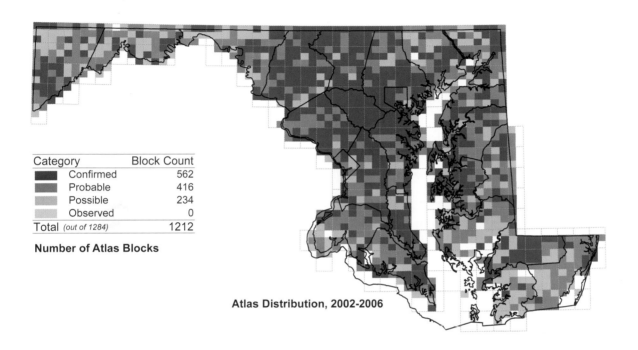

Category	Block Count
Confirmed	562
Probable	416
Possible	234
Observed	0
Total (out of 1284)	1212

Number of Atlas Blocks

Atlas Distribution, 2002-2006

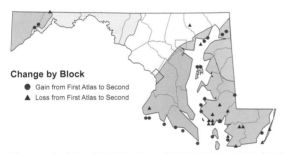

Change by Block
- ● Gain from First Atlas to Second
- ▲ Loss from First Atlas to Second

Change in Atlas Distribution, 1983-1987 to 2002-2006

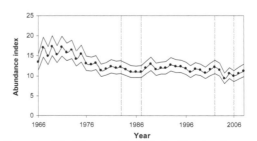

Percent of Stops
- 50 - 100%
- 10 - 50%
- 0.1 - 10%
- < 0.1%

Relative Abundance from Miniroutes, 2003-2007

Atlas Region	1983-1987	2002-2006	Change No.	%
Allegheny Mountain	92	92	0	0%
Ridge and Valley	147	148	+1	+1%
Piedmont	316	315	-1	0%
Upper Chesapeake	116	119	+3	+3%
Eastern Shore	287	284	-3	-1%
Western Shore	243	248	+5	+2%
Totals	1201	1206	+5	+0%

Change in Total Blocks between Atlases by Region

Breeding Bird Survey Results in Maryland

American Crow
Corvus brachyrhynchos

The crow needs no description.
    Roger Tory Peterson, *A Field Guide to the Birds,* 1934

On most continents the large black corvids have captured the imagination, dislike, and grudging respect of humans. The American Crow, as our exemplar, produces equally mixed feelings. An integral part of most landscapes, it is common, widespread, opportunistic, and equally at home in urban and rural landscapes. Our most common crow does best in a mixed open and wooded landscape. Open areas provide abundant and easily found food, and trees offer their branches for roosting, for nesting, and for refuge from most predators. Much to the chagrin of farmers, agriculture renders a heavily wooded landscape more congenial to these large, black, noisy, wary birds that can have a considerable impact on crops. The caws of crows around dumpsters, fields, and from lawns often serve as our alarm clock, and the almost hysterical calls of mobbing crows can alert us to the presence of otherwise unseen predators, mammals and birds alike. In winter the roosts of crows are often spectacular, if unsettling, in their numbers and their raw, raucous sound at dusk and dawn. There are few more ubiquitous or easily recognized birds in Maryland and the District of Columbia.

The American Crow is the most widespread of Maryland's three large black corvids. The Fish Crow occurs in the eastern two-thirds of the state and the Common Raven is largely limited to the western third. The American Crow was reported in 1,237 blocks, more than 96% of the blocks covered in this atlas, and ranked sixth among Maryland and the District of Columbia's most widespread bird species. These totals are comparable to those reported in the 1983–1987 atlas (Russell 1996b). The only gaps in this crow's atlas map are in the marsh and pinelands of southern Dorchester County, Bloodsworth and Smith islands, and a few blocks on southern Assateague Island (all areas occupied by the smaller Fish Crow).

Although the large nests of American Crows are generally carefully hidden (often in tall conifers), and adults and nestlings are usually quite silent, the nesting confirmation rate for this species was a relatively high 57%. Fledgling crows are often noisy and conspicuous, although atlas workers needed to exercise some caution because of the similarity of the calls of young American Crows to the high-pitched calls of the Fish Crow. Adult crows can often be observed carrying sticks in March, and used nests may be found by searching groves frequented by adult American Crows after (and before) nesting. Crows' nests with eggs have been found in Maryland and DC from 13 March to 21 May with nestlings observed to 16 June (MNRF).

American Crows appear to have increased slightly in southern Dorchester County but have been lost from Smith Island. Long-term BBS trends posit a modest increase in numbers for this crow over the past four decades. However, American Crows are highly susceptible to infection with West Nile virus, recently arrived (in 1999) from the Old World. Experimentally infected American Crows have shown 100% mortality from this mosquito-borne disease (Komar et al. 2003). Recent, well-documented declines in American Crow popu-

lations have been precipitous, including in BBS data (LaDeau et al. 2007). LaDeau and her colleagues (2007) reported that this crow had yet to recover from these declines by 2005. In spite of this disease-related decline, American Crows remain as well-distributed in Maryland and the District of Columbia as they were in the 1980s. Because its population was so high when the disease arrived and its habitat remains widespread and intact, American Crows may eventually develop some immunity to West Nile virus and recover fully over time.

WALTER G. ELLISON

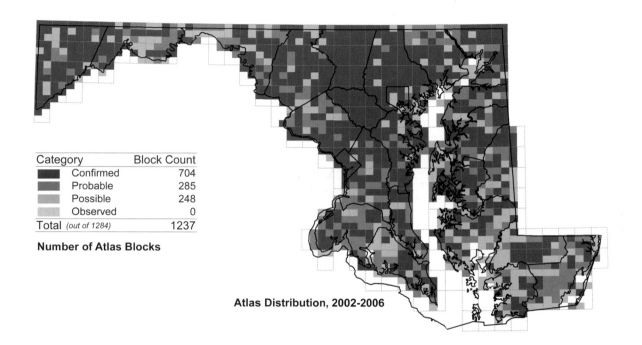

Category	Block Count
Confirmed	704
Probable	285
Possible	248
Observed	0
Total (out of 1284)	1237

Number of Atlas Blocks

Atlas Distribution, 2002-2006

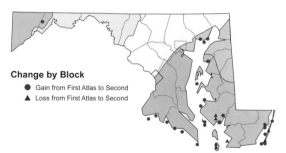

Change by Block

● Gain from First Atlas to Second
▲ Loss from First Atlas to Second

Change in Atlas Distribution, 1983-1987 to 2002-2006

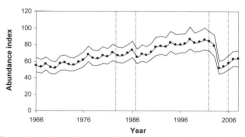

Percent of Stops

■	50 - 100%
■	10 - 50%
■	0.1 - 10%
□	< 0.1%

Relative Abundance from Miniroutes, 2003-2007

Atlas Region	1983-1987	2002-2006	Change No.	Change %
Allegheny Mountain	92	93	+1	+1%
Ridge and Valley	148	148	0	0%
Piedmont	316	316	0	0%
Upper Chesapeake	116	119	+3	+3%
Eastern Shore	297	304	+7	+2%
Western Shore	245	251	+6	+2%
Totals	1214	1231	+17	+1%

Change in Total Blocks between Atlases by Region

Breeding Bird Survey Results in Maryland

Fish Crow
Corvus ossifragus

If it were less vociferous, the Fish Crow would be virtually invisible to most bird observers. In spite of very subtle, but consistent, structural and plumage distinctions, the Fish Crow is best known from the American Crow by its less raspy, higher-pitched calls. The Fish Crow is endemic to the eastern United States, ranging up the Atlantic seaboard and major coastal rivers and up the Mississippi River from the Gulf of Mexico, including the lower reaches of its major tributaries. In the northeastern part of its range, this small crow has been expanding inland over the past three decades. Although widespread in the eastern two-thirds of Maryland, the Fish Crow is more local in distribution and, in spite of West Nile virus' impact on American Crow, much less numerous than its larger relative. Fish Crows prefer more open landscapes than American Crows, but they still require stands of trees to provide nest sites. Fish Crows also, as the name suggests, are seldom found very far from coastlines, rivers, and creeks. In spite of these habitat tendencies there is considerable overlap in habitat use by the two crow species with only average differences in behavior and food preferences separating them in sympatry.

Over the past hundred years, the Fish Crow has been increasing in Maryland. It has changed from a bird of Coastal Plain rivers, bays, and coastlines to a bird widespread on Piedmont and Ridge and Valley creeks, rivers, and ponds from its avenue of invasion, the Potomac River (Stewart and Robbins 1958; Russell 1996c). This increase has continued slowly over the past 20 years, with Fish Crows now reported from 76 more blocks than in 1983–1987. They have penetrated farther inland along the Potomac from their 1987 western outpost of Cumberland to Westernport in far western Allegany County. The Fish Crow has also expanded into new blocks on the northern and western Piedmont, and inland on the Eastern Shore. The Fish Crow's relatively low breeding confirmation rate (25%) and its correspondingly high number of possible breeding records (49%) reflect not only its scarcity compared to the American Crow but also its somewhat colonial habits and wide-ranging feeding forays. Egg dates in Maryland and DC range from 30 March to 8 June, with nestlings noted to 26 July (MNRF). During this atlas project Jan Reese reported an adult sitting on a nest at St. Michael's, Talbot County, on 16 March 2002.

Continental BBS trends for the Fish Crow have been

stable over the last 40 years, incorporating a reversal from a strong increase over the BBS's first 14 years, and a nonsignificant declining trend from 1980 to 2005. In Maryland the Fish Crow still evinces a strong increasing trend of 3.3% per year, although this trend has slowed of late. Reasons for the rapid increase of the Fish Crow and its spread inland along the eastern seaboard are not clear, but the rise of large urban landfills and extensive suburban sprawl, with its malls, golf courses, and well-mowed parks with nest trees, could not have done any harm. West Nile virus, which is particularly virulent in corvids (Komar et al. 2003), may have contributed

to recent declining trends in the Fish Crow. However, Fish Crows appear to suffer less morbidity from West Nile virus than American Crows and Blue Jays (Komar et al. 2003). LaDeau and her colleagues (2007) could not establish that West Nile virus had a significant effect on Fish Crow population trends in spite of implicating the disease in a strong downturn in American Crow populations. It is possible that Fish Crows derived a temporary benefit via less competition and less nest predation when their larger relative declined over the past half-decade.

WALTER G. ELLISON

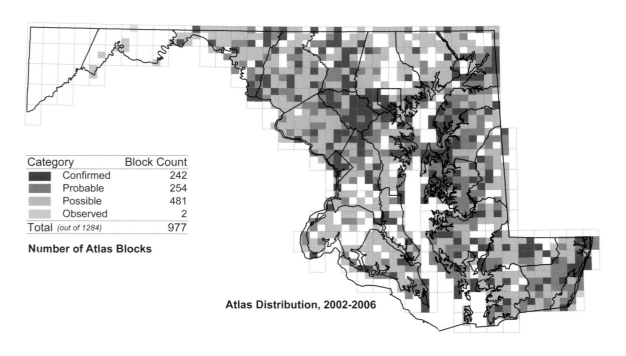

Category	Block Count
Confirmed	242
Probable	254
Possible	481
Observed	2
Total (out of 1284)	977

Number of Atlas Blocks

Atlas Distribution, 2002-2006

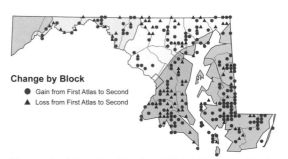

Change by Block
● Gain from First Atlas to Second
▲ Loss from First Atlas to Second

Change in Atlas Distribution, 1983-1987 to 2002-2006

Percent of Stops
- 50 - 100%
- 10 - 50%
- 0.1 - 10%
- < 0.1%

Relative Abundance from Miniroutes, 2003-2007

Atlas Region	1983-1987	2002-2006	Change No.	%
Allegheny Mountain	0	1	+1	-
Ridge and Valley	83	90	+7	+8%
Piedmont	258	264	+6	+2%
Upper Chesapeake	98	112	+14	+14%
Eastern Shore	256	307	+51	+20%
Western Shore	202	199	-3	-1%
Totals	897	973	+76	+8%

Change in Total Blocks between Atlases by Region

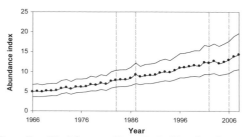

Breeding Bird Survey Results in Maryland

Eric Skrzypczak

Common Raven
Corvus corax

The Common Raven is a legendary bird among the human cultures of the Northern Hemisphere. It may well be the most intelligent of birds, as well as the world's largest songbird (by taxonomic definition). This great black bird, known as a creature of wild, uninhabited mountainous landscapes, has a reputation for shunning human company. In truth, ravens benefit from human influences including hunting, haying, open landfill operations, and road-killed animals, and they will use artificial nest sites in the form of towers, quarries, and highway road cuts. In the early twentieth century ravens were very scarce in eastern North America south of the Canadian border. They were confined to remote mountains and northern coasts, largely because of persecution by farmers and hunters. As this persecution lessened in the waning years of the twentieth century, ravens began to habituate to humans in a reforesting Appalachian landscape. They have recovered and are now slowly expanding their breeding range toward the Coastal Plain, which appears to present a limit to further colonization for this bird of hilly to mountainous lands.

Like crows, ravens are big birds with powerful deep throaty harsh voices and solid black coloration. They are bigger than crows (comparable to Red-tailed Hawk) and have a heavy-billed and large-headed profile, and a goiterlike tuft of long pointed throat hackles. They also have more tapered wings, held flat or subtly bent down at the wrist in gliding flight, more effortless soaring flight, often accompanied by almost clownish acrobatics, and a long, slightly wedge-shaped tail. Ravens begin nesting earlier than most regional songbirds. Maryland egg dates have been reported from 10 to 23 March (MNRF). Adults were recorded sitting on nests during this atlas from 7 March 2005 at Rocky Gap, Allegany County (D. Brinker) to 29 March 2002 at Grantsville, Garrett County (F. Pope). Nestlings have been reported until 7 May (MNRF).

It is unclear from literature references whether ravens might have died out briefly in Maryland, as they did in a few other eastern states such as Massachusetts (Griscom and Snyder 1955) and New Jersey (Walsh et al. 1999), or persisted in small numbers, as they did in Pennsylvania (Mulvihill 1992a). What is known is that several turn-of-the-century records reported nesting in Garrett and Allegany counties (Eifrig 1904, 1905) and that reports ceased until 1936 when ravens were reported from Garrett County (Brooks 1936). In the 1920s and 1930s ravens occasionally strayed to the Maryland

Piedmont, so it is plausible they might have been nesting in Garrett and Allegany counties at the time (Kirkwood 1930; Stewart and Robbins 1958). The Common Raven has been resettling the eastern United States since the 1950s and 1960s (Oatman 1985; J. Peterson 1988); this increase greatly accelerated in the 1970s (Robbins, Bystrak, and Geissler 1986).

The outline of the 2002–2006 atlas map for the Common Raven is identifiable from the map produced in the previous Maryland-DC atlas (Hilton 1996c), but the distribution has filled in greatly since then and ravens have occupied a few new Piedmont nest sites. Piedmont ravens have proven resourceful, nesting on a cell phone tower in southern Fred-

erick County (J. Stasz and M. Hafner, unpubl. data) and in a rock quarry in western Baltimore County in 2005 and 2006 (K. E. Costley, pers. comm.). Ravens may maintain their slow increase on the Piedmont if they continue to accept artificial nest sites. A spread onto the Coastal Plain, with its hot humid summer climate and relatively high abundance of competitors, such as Bald Eagle, Black Vulture, and American Crow, may prove very difficult for this large black scavenger. It is also plausible that West Nile virus, well known as virulent in corvids (Komar et al. 2003), may slow the Common Raven's increase in Maryland at least in the foreseeable future.

WALTER G. ELLISON

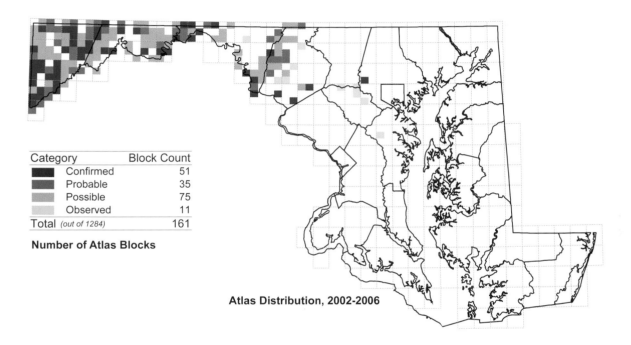

Category	Block Count
Confirmed	51
Probable	35
Possible	75
Observed	11
Total (out of 1284)	161

Number of Atlas Blocks

Atlas Distribution, 2002-2006

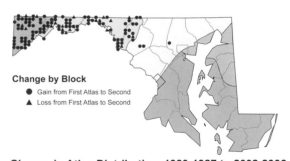

Change by Block
- ● Gain from First Atlas to Second
- ▲ Loss from First Atlas to Second

Change in Atlas Distribution, 1983-1987 to 2002-2006

Percent of Stops
■	50 - 100%
■	10 - 50%
■	0.1 - 10%
□	< 0.1%

Relative Abundance from Miniroutes, 2003-2007

Atlas Region	1983-1987	2002-2006	Change No.	Change %
Allegheny Mountain	41	80	+39	+95%
Ridge and Valley	44	76	+32	+73%
Piedmont	2	5	+3	+150%
Upper Chesapeake	0	0	0	-
Eastern Shore	0	0	0	-
Western Shore	0	0	0	-
Totals	87	161	+74	+85%

Change in Total Blocks between Atlases by Region

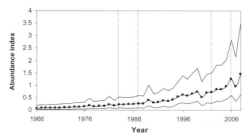

Breeding Bird Survey Results in Maryland

Horned Lark
Eremophila alpestris

Among Maryland and DC's birds, the Horned Lark is the most closely tied to the fortunes of large-scale crop farming. This lark's range is dictated to a large extent by the presence of the bare soil of crop fields just after planting and after summer harrowing and harvests. This is a typical lark in being a bird of open country with wide open skies where it performs its towering song flights. In spite of living in the open, these are well-camouflaged birds, hard to see except when in motion among dirt clods and crop furrows, against a matching pale brown soil background. Two subspecies of this lark occur in Maryland and DC: *praticola,* often referred to by the English name Prairie Horned Lark, is the breeding subspecies; the Northern Horned Lark, *alpestris,* which arrives in late autumn and spends the winter augmenting the regional population, departs in late winter just before the vernal equinox.

Horned Larks occur in areas with minimal to no vegetation and no plants taller than a few centimeters. Typical haunts include reclaimed strip mines (Whitmore 1980), fallow stubble fields, bare harrowed fields, airport runway margins, sod farms, and coastal dunes. Prairie Horned Larks occur year-round in Maryland, but it is unclear whether local nesters migrate out of parts of Maryland or are largely sedentary. Males begin defending territories in February and

nest building starts late that month. The nest is a grass-lined bowl on the ground, usually built against some kind of cover such as a furrow, dirt clod, or tuft of vegetation (Pickwell 1931). These birds usually attempt a second brood where possible; many fields in Maryland, however, mature to tall crops by late May or June. Perhaps lark pairs will move among suitable fields between nesting attempts, for example, to cut-over wheat fields from maturing corn in June. Maryland egg dates range from 3 March to 6 July, and nestlings have been reported to 30 August (Fletcher 1996a; MNRF).

Prairie Horned Larks were unknown in the eastern United States and southeastern Canada before the late nineteenth century (Hurley and Franks 1976). The earliest report from Maryland and DC was of two collected near Washington by Palmer in February 1881 (H. Smith and Palmer 1888). The first documented breeding record was from Garrett County in 1904 (Eifrig 1920, 1923). The first central Maryland nesting record was from Laurel in 1922 (Swales 1922). The first nesting on the Maryland Eastern Shore is unknown but apparently they first bred in Delaware in the early 1930s (Hess et al. 2000). Stewart and Robbins (1958) documented a state-wide distribution, calling them fairly common east to the fall line and uncommon on the Coastal Plain save for coastal Worcester County where they were fairly common. By the 1980s they had become well established on the Eastern Shore and in Western Shore farmland although their range was far from uniform in the latter region (Fletcher 1996a).

The situation for Horned Larks in 2002–2006 was mixed. They showed a modest net increase of 24 blocks overall, but the atlas map reveals some areas with major losses and others with notable gains. These larks have been lost from 73% of 1980s blocks in the Allegheny Mountains and from 27% of the blocks in which they were found on the Western Shore. They increased on the Piedmont, adding 70% more blocks, mostly to the northwest, and in the Ridge and Valley, adding 32% more blocks, largely in the Hagerstown Valley. Larger crop fields are featured in the latter areas, whereas crop fields have declined in the former.

Horned Larks have declined at a rate of 2% per year on BBS routes in North America over the last four decades, with a higher rate since 1980. The BBS trend in Maryland has been more or less stable. These larks have declined over much of the recently occupied eastern part of their North American range but continue to show increases on the southeastern Coastal Plain. Horned Larks tend to benefit from agricultural practices that harm many other grassland birds, especially removal of hedgerows and conversion to crop agriculture from hay fields. These birds almost certainly will lose ground to increasing residential development on the Piedmont and Eastern Shore in the near future.

WALTER G. ELLISON

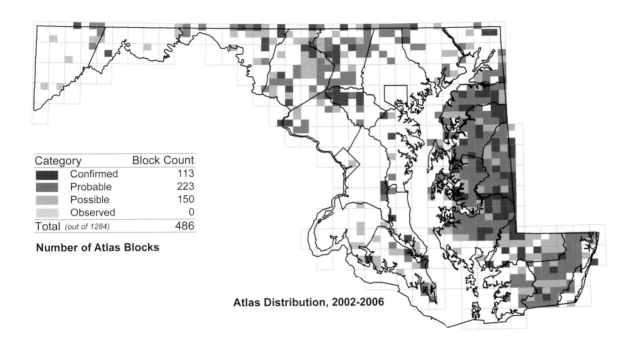

Category	Block Count
Confirmed	113
Probable	223
Possible	150
Observed	0
Total (out of 1284)	486

Number of Atlas Blocks

Atlas Distribution, 2002-2006

Change by Block
● Gain from First Atlas to Second
▲ Loss from First Atlas to Second

Change in Atlas Distribution, 1983-1987 to 2002-2006

Percent of Stops
50 - 100%
10 - 50%
0.1 - 10%
< 0.1%

Relative Abundance from Miniroutes, 2003-2007

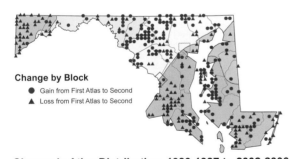

		Change	
Atlas Region	1983-1987	2002-2006	No.	%
Allegheny Mountain	45	12	-33	-73%
Ridge and Valley	25	33	+8	+32%
Piedmont	64	109	+45	+70%
Upper Chesapeake	80	81	+1	+1%
Eastern Shore	192	210	+18	+9%
Western Shore	55	40	-15	-27%
Totals	461	485	+24	+5%

Change in Total Blocks between Atlases by Region

Breeding Bird Survey Results in Maryland

Purple Martin
Progne subis

The Purple Martin is Maryland and DC's largest swallow, dashing in level flight and capable of effortless soaring. The pleasant, syrupy chattering of martins is one of the most pleasing summer sounds in Maryland farm country. Martins are intimately associated with humans; all those in Maryland and DC nest in nesting structures set out for them. Martin houses generally take the form of large multiple-compartment boxes, although collections of artificial gourds are becoming popular. The latter mimic the arrays of hollowed gourds Native Americans set out to attract nesting martins, which helped disperse crows and raptors from their villages and fields (Sprunt 1942). The last published report in Maryland of these swallows nesting away from a martin house was in 1970 (Klimkiewicz 1996a). They also have nested in crevices and holes in buildings in DC, such as the post office building, before European Starlings displaced them from such sites (C. Richmond 1888).

Beyond nest boxes, habitat requirements for Purple Martins include an abundance of open ground around the colony for their low swooping approach flights. They also need expanses of water or open land for feeding. Colonies are very often situated near a body of water or a wetland that serves as a source of abundant flying insect prey.

Purple Martins winter in South America, largely in south-ern and eastern Brazil (C. Brown 1997). A few may return to Maryland in March but most arrive in April, often making their first appearances around colony sites. They depart after their young fledge in July and August, forming large premigratory roosts at favored localities. All are gone by the end of September in most years. Although nest building commences in April, egg laying does not begin until several weeks afterward. Maryland and DC egg dates range from 9 May to 10 August (Klimkiewicz 1996a). Nestling dates from this project exceeded prior reported limits, ranging from 25 May 2002 southeast of Scaggsville, Howard County (J. Friedhoffer), to 28 August 2004 southwest of Wittman, Talbot County (L. Roslund).

Early references called the Purple Martin common, including in Allegany and Garrett counties (Eifrig 1904). Stewart and Robbins (1958) reported that they were common on the Coastal Plain, fairly common on the Piedmont, and uncommon to the west. Their status was similar from 1983 to 1987, including 6 blocks with nesting birds in Garrett County, and 3 in Allegany County. Although they were nearly ubiquitous on the Coastal Plain, there were gaps in eastern Queen Anne's County and in the Pocomoke drainage. These swallows had apparently bounced back in western Maryland in the 1980s from declines recorded after the June 1972 rains of Hurricane Agnes; prolonged rains can make prey scarce causing adult and nestling martins to starve to death (Robbins, Bystrak, and Geissler 1986; Klimkiewicz 1996a).

The distribution documented from 2002 to 2006 is similar to the first atlas, with a modest loss of 58 blocks. There were clear and steep losses in the western mountains, and the birds were lost from 40 blocks on the Piedmont. Although there were losses on the Western Shore, martins increased east of Chesapeake Bay, including in Queen Anne's County and in the Pocomoke basin. Prolonged rains in the spring of 2003 may have contributed to declines in central and western Maryland.

Continental trends on BBS routes, which had been stable, have recently trended toward declines, including a modest declining trend of 0.7% per year on eastern BBS routes since 1980. Maryland BBS routes have been stable, but recent trends in two physiographic strata of western and central

Maryland, the Ridge and Valley and Northern Piedmont, have shown declines. Purple Martins compete for nest sites with European Starlings and House Sparrows, which can occupy boxes before martins arrive in spring and drive martin pairs from cavities. Nest boxes require vigilant management to protect against these competitors. Boxes should be cleaned after nesting and closed or taken down until martins return in the spring. A changing global climate may not favor these swallows, as it may lead to wetter late springs and increases in early season tropical storms.

WALTER G. ELLISON

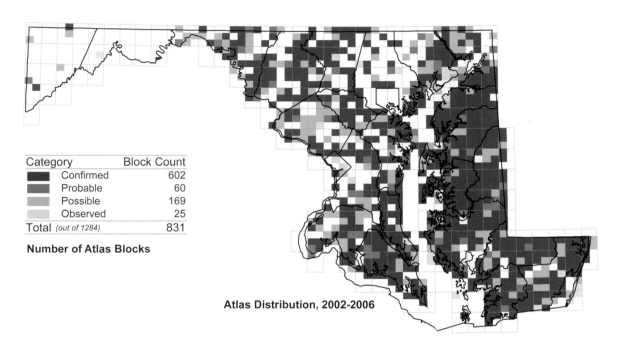

Category	Block Count
Confirmed	602
Probable	60
Possible	169
Observed	25
Total (out of 1284)	831

Number of Atlas Blocks

Atlas Distribution, 2002-2006

Change by Block
- ● Gain from First Atlas to Second
- ▲ Loss from First Atlas to Second

Change in Atlas Distribution, 1983-1987 to 2002-2006

Percent of Stops
- 50 - 100%
- 10 - 50%
- 0.1 - 10%
- < 0.1%

Relative Abundance from Miniroutes, 2003-2007

Atlas Region	1983-1987	2002-2006	Change No.	%
Allegheny Mountain	21	6	-15	-71%
Ridge and Valley	77	63	-14	-18%
Piedmont	224	184	-40	-18%
Upper Chesapeake	89	101	+12	+13%
Eastern Shore	263	288	+25	+10%
Western Shore	211	185	-26	-12%
Totals	885	827	-58	-7%

Change in Total Blocks between Atlases by Region

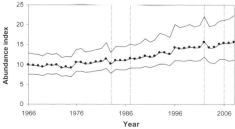

Breeding Bird Survey Results in Maryland

Scott Berglund

Tree Swallow
Tachycineta bicolor

Tree Swallows effortlessly pirouetting over placid waters or vacant fields, flashing their china white underparts and steely blue green upperparts, are an early harbinger of spring in much of Maryland. These swallows feed in the open, almost tirelessly swooping after flying prey. They nest in tree cavities, usually snags, and in nest boxes. Egg dates in Maryland range from 24 April to 3 July (MNRF). Tree Swallows are seldom found far from water, usually ponds or broad, slow-flowing streams.

Tree Swallows are much more abundant in the northern states and Canada than in the mid Atlantic region. Evidence indicates that they have increased and expanded their breeding range southward over the past century or more (Robertson et al. 1992; Walsh et al. 1999). The natural cavities traditionally used by Tree Swallows were most often in trees killed by flooding, and many swallows continue to nest in such trees. The twentieth-century penchant for damming streams to create reservoirs and control floods probably benefited Tree Swallows as long as trees still stood on the shore. The return of the beaver to its old haunts, after its extirpation in the nineteenth century, has also been beneficial for Tree Swallows. Yet another trend favoring this swallow has been an increase in nest boxes provided by bird lovers, partic-

ularly bluebird boxes. Many landowners have built ponds on their property in combination with erecting nest boxes; this also has likely promoted the increase of this pleasant bird with its sweet chirping calls and song.

Fifty years ago, Tree Swallows in Maryland were essentially limited to the Allegheny Mountains and tidewater on Chesapeake Bay and the Atlantic Ocean (Stewart and Robbins 1958). The 1983–1987 Maryland-DC atlas project revealed a considerable range expansion into much of the Ridge and Valley, Piedmont, and Western Shore, as well as substantial spread inland on the Eastern Shore and an increase in the Allegheny Mountains (S. Clarkson 1996a). In 2002–2006 observers documented a net increase of 561 blocks between the two statewide atlas projects, more than a 140% increase in two decades. Increases were documented in all physiographic regions in Maryland, but the greatest increases were in the northern part of the Piedmont and inland on the lower Eastern Shore. The percentage of confirmed nesting rose from 36% in 1983–1987 to 62% in 2002–2006. This may be explained in part by increased nesting in bluebird boxes, but it also reflects the increase in swallow population.

There are still some gaps and thin spots in the Tree Swallow's range in Maryland and DC. These include much of southern Maryland where it is largely limited to the Chesapeake shore and the Patuxent and Potomac rivers; inland Worcester County except for a few records on the Pocomoke River; and blocks along the ridge of Savage Mountain on the Allegheny Front.

The Tree Swallow has increased on Maryland BBS routes at a rate in excess of 10% per year, and it has increased in all of the physiographic regions that include Maryland. Increases in the United States have been more modest although significant, but Tree Swallows have been declining significantly on the continental scale over the last quarter century. There are still frontiers to conquer for the Tree Swallow in Maryland and DC and it is plausible to think that they will eventually expand to them. As long as people put out suitable nest boxes, Tree Swallows will occupy a generous portion of them. Even without further expansion, the Tree Swallow is likely to hold onto most of its gains in the Old Line State into the foreseeable future.

WALTER G. ELLISON

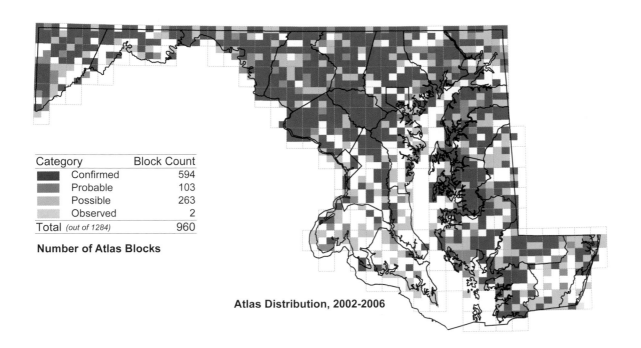

Category	Block Count
Confirmed	594
Probable	103
Possible	263
Observed	2
Total (out of 1284)	960

Number of Atlas Blocks

Atlas Distribution, 2002-2006

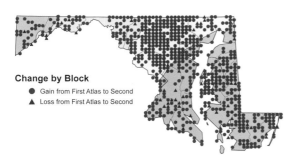

Change by Block
● Gain from First Atlas to Second
▲ Loss from First Atlas to Second

Change in Atlas Distribution, 1983-1987 to 2002-2006

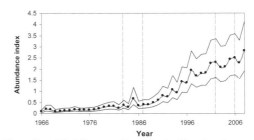

Percent of Stops
■ 50 - 100%
■ 10 - 50%
■ 0.1 - 10%
□ < 0.1%

Relative Abundance from Miniroutes, 2003-2007

			Change	
Atlas Region	1983-1987	2002-2006	No.	%
Allegheny Mountain	47	71	+24	+51%
Ridge and Valley	70	123	+53	+76%
Piedmont	66	278	+212	+321%
Upper Chesapeake	42	93	+51	+121%
Eastern Shore	111	252	+141	+127%
Western Shore	60	140	+80	+133%
Totals	396	957	+561	+142%

Change in Total Blocks between Atlases by Region

Breeding Bird Survey Results in Maryland

Mark Hoffman

Northern Rough-winged Swallow
Stelgidopteryx serripennis

North America's counterpart to the Southern Rough-winged Swallow (*Stelgidopteryx ruficollis*) of Central and South America, the Northern Rough-winged Swallow is distinguished by its dull, nondescript plumage. Brown above and grayish white below with a buffy brown throat, this swallow was named for the serrated edges of its outer primary feathers. It is similar in appearance to the Bank Swallow and shares that swallow's behavior of nesting in burrows within earthen cliffs and banks. However, unlike the colonial Bank Swallow, this species is a solitary nester and does not excavate its own burrows. It nests in the abandoned burrows of other species or in artificial surrogates, such as culverts, drainpipes, quarries, niches under bridges, and crevices in shoreline structures, cliffs, and buildings (MNRF). Like most swallows, the Northern Rough-winged feeds almost exclusively on flying insects, such as ants, beetles, and flies (Beal 1918).

This neotropical migrant was thought to be somewhat rare or uncommon in the mid- to late 1800s (Coues and Prentiss 1862, 1883; Kirkwood 1895). A century later it was considered fairly common along the Potomac River and uncommon and local elsewhere in the state (Stewart and Robbins 1958). Migrants return from Mexico and Central America as early as late March, but primarily in April, when nest construction begins. The mean depth of nests measured by J. Hill (1988) from 44 drainpipes was 82 cm (32.4 in), and the mean height of 50 nests was 3 m (10 ft) (MNRF). The earliest evidence of nest building from this atlas came from C. Stirrat on 14 April 2002 and the latest from K. Costley on 28 June

2003. Egg dates range from 19 April to 24 June; clutches of 39 nests range from 3 to 7 with a mean of 5.3 eggs (Davidson 1996e; MNRF). Incubation takes about 11 days, and nestlings fledge about 20 days later (Dingle 1942). Fledglings were found as early as 25 May in 2004 (L. Eastman) and as late as 3 August in 2002 (W. Bell). This species is considered to be single brooded (Lunk 1962). Fall migration begins as early as late June, and the species is mostly gone from Maryland by November (Robbins and Bystrak 1977).

The Northern Rough-winged Swallow was the 84th most widely reported species during the 1983–1987 atlas and 83rd during the 2002–2006 atlas; its distribution and status have remained fairly stable over the past 20 years. It was reported from 111 more blocks during the second atlas, an increase of 25%. BBS trend data show an increase in Maryland of 3.5% per year from 1980 to 2007. The North American BBS trend increased by 2.5% per year from 1966 to 1979; it decreased by 1.1% per year from 1980 to 2007. Within Maryland, the regions with the greatest increases in atlas block numbers were the Western Shore (58% increase), the Upper Chesapeake (50% increase), and the Piedmont (36% increase). The only region that decreased was the Eastern Shore with 11 fewer blocks (20% decrease).

As shown in the first atlas, this species tends to concentrate along the larger rivers, especially the Potomac, Patuxent, and Susquehanna, along the bayside cliffs of Calvert County and the Upper Chesapeake Bay, and at the larger reservoirs in Baltimore, Allegany, and Garrett counties. It is locally distributed throughout much of the state, except for the lower Eastern Shore, where shoreline banks may not be high enough or may be too sandy for stable burrows. It remains widespread throughout much of the Piedmont and the Ridge and Valley regions.

Like Barn and Cliff swallows, this species' ability to adapt its nesting preferences to include man-made structures has likely enabled the gradual expansion of its population. The relative stability of its distribution and increasing population trends over the past 20 years bode well for the future existence of this species in Maryland and for the overall health of Maryland's environment. As long as Maryland's waterways remain healthy enough to produce sufficient populations of flying insects, this species should continue to do well.

LYNN M. DAVIDSON

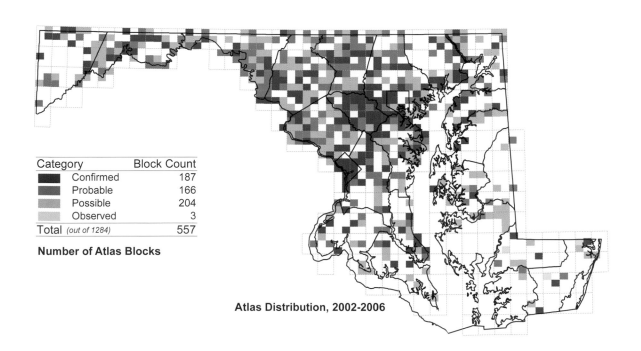

Category	Block Count
Confirmed	187
Probable	166
Possible	204
Observed	3
Total (out of 1284)	557

Number of Atlas Blocks

Atlas Distribution, 2002-2006

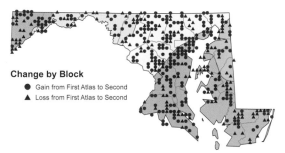

Change by Block
- ● Gain from First Atlas to Second
- ▲ Loss from First Atlas to Second

Change in Atlas Distribution, 1983-1987 to 2002-2006

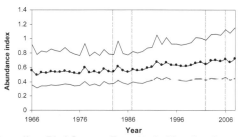

Percent of Stops
- 50 - 100%
- 10 - 50%
- 0.1 - 10%
- < 0.1%

Relative Abundance from Miniroutes, 2003-2007

Atlas Region	1983-1987	2002-2006	Change No.	%
Allegheny Mountain	28	32	+4	+14%
Ridge and Valley	96	100	+4	+4%
Piedmont	160	217	+57	+36%
Upper Chesapeake	30	45	+15	+50%
Eastern Shore	56	45	-11	-20%
Western Shore	73	115	+42	+58%
Totals	443	554	+111	+25%

Change in Total Blocks between Atlases by Region

Breeding Bird Survey Results in Maryland

Bank Swallow
Riparia riparia

Of our colonial swallows the Bank Swallow tends to have the largest colonies. These are bustling communities of dark-mouthed tunnels in sand and gravel banks over water and in man-made sand and gravel pits. Dependence on such habitat enforces a localized and clumped distribution on this swallow, even where it is common. This is the smallest and most maneuverable of Maryland and DC swallows. They are highly vocal, uttering an almost constant whispery husky chatter as they fly about. They are similar to Northern Rough-winged Swallows, which also often nest in sand banks. Thus observers should use some care in identifying swallows seen nesting in burrows. They are near the southern edge of their broad boreal and cold temperate North American nesting range in Maryland and DC and are currently almost restricted to the Coastal Plain. This bird also breeds across northern Eurasia where its English name is Sand Martin (Cramp et al. 1988).

In addition to the obvious limitation imposed by nest sites, Bank Swallows also require good feeding conditions near their colonies. These are provided by large bodies of water that produce good flying insect hatches and open areas for this bird's low, swooping feeding flights (Bryant and Turner 1982).

Bank Swallows are migratory, with most North American birds wintering in southern South America. They reside in Maryland from mid-April to mid-September. As with Maryland's other highly gregarious colonial small swallow, the Cliff Swallow, they arrive late in spring and depart early in autumn. Large numbers are on the move by mid-July. Safe

dates are accordingly brief. The ephemeral nature of their nest sites forces Bank Swallows to regularly abandon colony sites, making their distribution changeable. Nonetheless, occasional traditional sites with extensive potential nesting habitat may be tenanted for decades. Maryland and DC egg dates range from 5 May to 23 June and nestlings have been reported as late as 19 July (Saunders and Saunders 1996; MNRF).

Early bird lists called the Bank Swallow a common summer resident (C. Richmond 1888; Kirkwood 1895; Eifrig 1904) although it is not clear how much of this perceived abundance was attributable to May and July migrants. Stewart and Robbins (1958) largely limited this swallow's range to tidewater, plus two colonies on the Potomac River in Washington County and a few colonies on the eastern Piedmont. The 1983–1987 atlas distribution was similar, with a few Piedmont and western Potomac valley colonies (west to Allegany County) and most colonies on the Coastal Plain. The Coastal Plain colonies, however, were mostly along major tidal river valleys, particularly the Patuxent, rather than along the shores of Chesapeake Bay (Saunders and Saunders 1996). Only 54 colonies were mapped, making it the state's least widely distributed swallow.

During the second atlas the Bank Swallow was confirmed in only 33 blocks (32 from blocks covered in 1983–1987) a loss of more than 40%. There were fewer blocks with colonies on the Patuxent, especially in the lower valley. There were no colonies in the upper Potomac valley, and only 1 block on the Piedmont proper in Montgomery County. There are still large colonies on the high bluffs along the lower Sassafras River. Colonies are also found in inland sand and gravel mines in Kent, Queen Anne's, Caroline, and Dorchester counties. A colony on the Pocomoke River on the Wicomico and Worcester county line is well south of the historic southern limit on the Eastern Shore.

Trends on BBS routes for Bank Swallow are unreliable because of large year-to-year fluctuations along routes, in part because of the loss and gain of colonies near routes. Bank Swallows are apparently declining on North American and eastern BBS routes, with an annual rate of decline of 4.6% on eastern routes since 1980. Factors contributing to declines include natural habitat loss to erosion and bank revegetation; erosion is a particular problem along Chesapeake Bay. Bank

stabilization with rip-rap and walls destroys nest sites, as does abandoning sand and gravel mines or excavating banks in intensively worked pits. Management practices could include creation of artificial sand banks rather than reliance on sand mining entrepreneurs, and existing bluffs and cut banks could be actively maintained. The current situation does not bode well for this little swallow, but this species is opportunistic and would respond well to the creation of new nest sites.

WALTER G. ELLISON

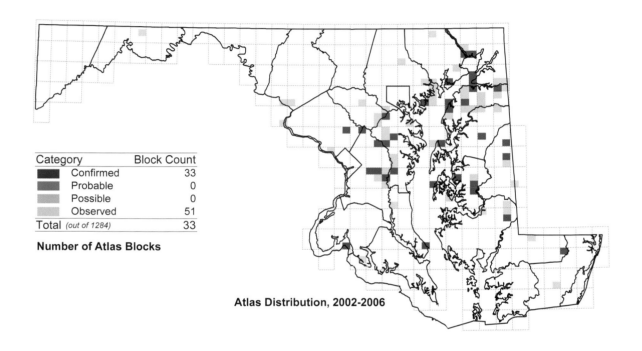

Category	Block Count
Confirmed	33
Probable	0
Possible	0
Observed	51
Total (out of 1284)	33

Number of Atlas Blocks

Atlas Distribution, 2002-2006

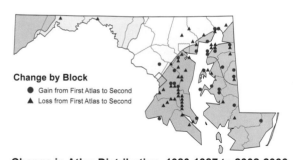

Change by Block
- ● Gain from First Atlas to Second
- ▲ Loss from First Atlas to Second

Change in Atlas Distribution, 1983-1987 to 2002-2006

Percent of Stops
- 50 - 100%
- 10 - 50%
- 0.1 - 10%
- < 0.1%

Relative Abundance from Miniroutes, 2003-2007

Atlas Region	1983-1987	2002-2006	Change No.	%
Allegheny Mountain	0	0	0	-
Ridge and Valley	2	0	-2	-100%
Piedmont	6	4	-2	-33%
Upper Chesapeake	11	11	0	0%
Eastern Shore	4	7	+3	+75%
Western Shore	31	10	-21	-68%
Totals	54	32	-22	-41%

Change in Total Blocks between Atlases by Region

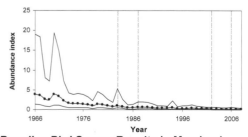

Breeding Bird Survey Results in Maryland

Cliff Swallow
Petrochelidon pyrrhonota

The Cliff Swallow is one of the two most colonial of Maryland's swallows. They feed in flocks and gather mud for their distinctive nests in flocks, often following one another when seeking these important resources (C. Brown and Brown 1995). The Cliff Swallow is rather stocky for a swallow, with broad-based wings and a short square tail. Its rusty brown rump and whitish headlight above the bill are distinctive. These swallows are most common in western North America, where they can still be found nesting on canyon walls and mountain cliffs as well as on man-made structures. They were absent or rare in eastern North America into the early nineteenth century but quickly became widespread and fairly numerous there after they started nesting on building walls under protecting eaves (A. Gross 1942). Cliff Swallows are still prospecting new territory at the southeastern edge of their range including in Maryland, where they are numerous in the western mountains and more local to the east.

Cliff Swallow habitat includes open areas for feeding flights in pursuit of swarms of flying insects, usually near bodies of water, where insect hatches occur and mud can be collected for nest building (Emlen 1954; C. Brown and Brown 1995). Maryland Cliff Swallows nest exclusively on man-made structures including barns (usually on the outside), other buildings with eaves, concrete bridges, culverts, and dams.

Cliff Swallows are long-distance migrants that winter in South America. They return to nesting areas rather late in the spring and are generally present in Maryland from late April to mid-September. Their nests are remarkable structures that resemble chimneys or retorts plastered together from small clay or silt-mud pellets. The nests are always clustered together in tight colonies ranging from fewer than 10 to more than 100, even thousands in the west (C. Brown and Brown 1995). Maryland egg dates range from 22 May to 9 July, with nestlings to 24 August (Dowell 1996a; MNRF).

The dichotomy in abundance between Cliff Swallows in westernmost Maryland and those in eastern Maryland has long been the case. Kirkwood (1895) called the Cliff Swallow localized and not common in eastern Maryland, and Eifrig (1904) said it was common in Allegany and Garrett counties. Stewart and Robbins (1958) found them common to fairly common eastward to western Washington County and rare on the Piedmont. They were found nesting on the upper Coastal Plain in Prince George's County just before the first statewide atlas (R. Patterson 1981). From 1983 to 1987 the range was similar but had increases along major Piedmont rivers, with birds nesting on dams and bridges, and some spread onto the Coastal Plain on the Western Shore (Dowell 1996a).

In 2002–2006 the Cliff Swallow remained numerous in Garrett and western Allegany counties, declined in western Washington County, and lost some blocks on the Piedmont and the Western Shore. Nesting was documented for the first time from the Eastern Shore on bridges in the upper Nanticoke drainage in eastern Dorchester County in 2002. The species had nested on the Delmarva Peninsula in New Castle County, Delaware in 1993 (Hess et al. 2000), but had nested no farther south since then.

The clustered and sometimes ephemeral nature of colonial species such as Cliff Swallow makes trend estimates unreliable. The North American trend estimate, which was weakly positive from 1966 to the present, has been stable since 1980. Maryland trend estimates are volatile and are derived from only small numbers of birds and routes. This swallow has long maintained essentially the same distribution in Maryland although colony locations shift across the decades. House Sparrows usurping nests have probably contributed to abandonment of some farm colonies; some colonies on buildings are deliberately destroyed for the sake of tidiness;

and changes in barn maintenance and architecture may also have discouraged nesting at times. Whether regular painting has prevented nests from adhering to modern barns is uncertain although anecdotal claims of this exist (A. Gross 1942). But some colonies may have relocated to concrete structures in preference to wooden ones.

WALTER G. ELLISON

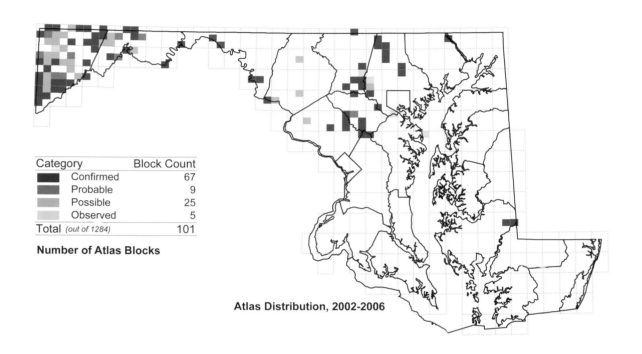

Category	Block Count
Confirmed	67
Probable	9
Possible	25
Observed	5
Total (out of 1284)	101

Number of Atlas Blocks

Atlas Distribution, 2002-2006

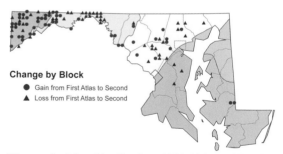

Change by Block
- Gain from First Atlas to Second
▲ Loss from First Atlas to Second

Change in Atlas Distribution, 1983-1987 to 2002-2006

Percent of Stops
- 50 - 100%
- 10 - 50%
- 0.1 - 10%
- < 0.1%

Relative Abundance from Miniroutes, 2003-2007

Atlas Region	1983-1987	2002-2006	Change No.	Change %
Allegheny Mountain	39	52	+13	+33%
Ridge and Valley	22	18	-4	-18%
Piedmont	34	28	-6	-18%
Upper Chesapeake	0	0	0	-
Eastern Shore	0	2	+2	-
Western Shore	4	1	-3	-75%
Totals	99	101	+2	+2%

Change in Total Blocks between Atlases by Region

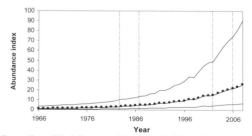

Breeding Bird Survey Results in Maryland

Barn Swallow
Hirundo rustica

A true harbinger of spring and a familiar companion of dooryards in landscapes both rural and urban, the Barn Swallow was the fourth most widely distributed bird in this atlas. Although on rare occasion a stray pair of Barn Swallows might nest in the mouth of a natural cave, over 99% of these agile aerial insect hunters now nest on man-made structures (C. Brown and Brown 1999). Such sites include barns (as the bird's name implies), house porches, outbuildings, highway bridges (usually over water), and large culverts. Nesting dates for Maryland and DC Barn Swallows range for eggs from 21 April 2004 near Bay View, Cecil County (B. Olsen) to 4 August (MNRF) and for nestlings from 14 May 2003 at Catonsville, Baltimore County (P. Kreiss), to 28 August 2003 at Worton, Kent County (pers. obs.; Ringler 2004b).

Barn Swallows have been common and widespread in Maryland and DC as far back as published records exist (Coues and Prentiss 1862; Kirkwood 1895), The atlas maps show little change in the distribution of this swallow since the first atlas project. This is true despite a 40-year declining trend in the BBS for Maryland and North America. This trend has been most pronounced over the 25 years since the previous atlas. That this decline is not reflected in the mapped range in Maryland and DC indicates how numerous this bird was two decades ago.

Barn Swallows were missed in just 43 of 1,284 atlas blocks receiving coverage (many dominated by water) showing no obvious pattern to the resulting pinhole gaps in the range. A lack of Barn Swallows in some of the extensive marshes and pinelands of southern Dorchester County repeats similar absences in 1983–1987 (S. Clarkson 1996b). This swallow was absent from a few western Maryland blocks, probably because they were extensively wooded and lacked livestock farming. Other blocks without Barn Swallows seem to represent oversights, perhaps because of late seasonal coverage. The Barn Swallow is an early autumn migrant and therefore has an early cutoff in atlas safe dates.

Although Barn Swallows are now dependent upon man-made structures and benefit from the open landscapes created by humans, some of the species' decline may be traced to human activities. Many farms are maintained in a much more sanitary fashion than in the past, including efforts to discourage nesting by wild birds. Livestock farming, with its large barns and outbuildings, is being replaced by crop agriculture in some parts of Maryland, particularly the Eastern Shore. Because of their close association with humans, it is likely that Barn Swallows will continue to be widespread and fairly common in Maryland and DC, but how abundant they might be is another matter.

WALTER G. ELLISON

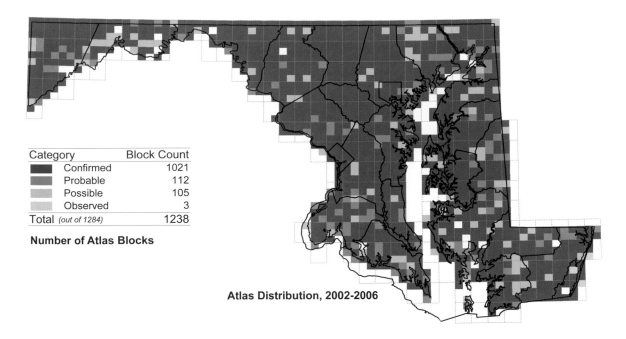

Category		Block Count
	Confirmed	1021
	Probable	112
	Possible	105
	Observed	3
Total *(out of 1284)*		1238

Number of Atlas Blocks

Atlas Distribution, 2002-2006

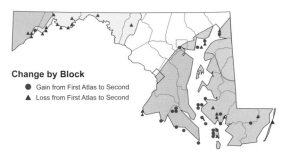

Change by Block

● Gain from First Atlas to Second
▲ Loss from First Atlas to Second

Change in Atlas Distribution, 1983-1987 to 2002-2006

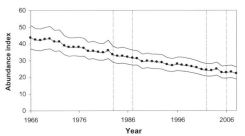

Percent of Stops

	50 - 100%
	10 - 50%
	0.1 - 10%
	< 0.1%

Relative Abundance from Miniroutes, 2003-2007

			Change	
Atlas Region	1983-1987	2002-2006	No.	%
Allegheny Mountain	93	90	-3	-3%
Ridge and Valley	147	137	-10	-7%
Piedmont	316	316	0	0%
Upper Chesapeake	119	119	0	0%
Eastern Shore	315	321	+6	+2%
Western Shore	242	248	+6	+2%
Totals	1232	1231	-1	0%

Change in Total Blocks between Atlases by Region

Breeding Bird Survey Results in Maryland

Carolina Chickadee
Poecile carolinensis

The Carolina Chickadee is endemic to the southeastern United States, where it replaces the Black-capped Chickadee of the upper elevations in the southern Appalachian Mountains and the north. These two chickadees are extremely similar, exhibiting very slight average differences in plumage, size, structure, and somewhat more obvious differences in voice. In areas where their breeding ranges meet they overlap but narrowly, defend their territories against the other species, hybridize, and learn each other's songs. As such, atlas range maps must be viewed with some caution. Nonetheless, the stability of the general outlines of the two chickadee's ranges between the two statewide Maryland-DC atlases speaks to the care exerted by atlas field-workers in the regions where the chickadees have traditionally been known to overlap.

Carolina Chickadees are forest birds that are able to breed and persist in very small woodlots in urban parks and in well-planted suburbs. Egg dates for this familiar bird range from 12 April (MNRF) to 17 July 2002 northeast of Leonardtown, St. Mary's County (P. Craig), and nestlings have been detected from 7 April 2002 at Salisbury, Wicomico County (S. Dyke) to 22 July 2002 northwest of Clover Hill, Frederick County (N. Parker). Carolina Chickadees are more prone than Black-capped Chickadees to use nest boxes instead of natural cavities (Mostrom et al. 2002). These chickadees will also excavate new cavities and use natural cavities in rotting dead limbs, stumps, snags, fence posts, and old dock pilings for nests. Consequently the Carolina Chickadee is almost universally distributed in Maryland east of the high country occupied by the Black-capped Chickadee. The only parts of eastern Maryland that appear to lack this approachable and spunky little bird are some areas in the extensive marshlands in southeastern Dorchester County, the marshy islands of lower Chesapeake Bay, and some of the dune scrub and marshes of southern Assateague Island. The latter two examples point up this chickadee's unwillingness to cross long stretches of open water.

In spite of the current population being depressed by the West Nile virus, the Carolina Chickadee was found in slightly more blocks in the current atlas than in the first, a net total gain of 24 (2% more than in 1983–1987). Six of these blocks were in western Washington County in the overlap zone with Black-capped Chickadee. In eastern Maryland much of the increase was on the lower Eastern Shore (net gain of 13). Carolina Chickadees were found in 2 Allegany County blocks in 1983–1987 (Ringler 1996d) but were not reported there in 2002–2006, in spite of apparent increases across the Potomac River in Morgan County, West Virginia (Buckelew and Hall 1994).

In the BBS, Carolina Chickadees have shown a survey-wide modest declining trend of 0.7% per year. Maryland has shown a similar downward trend of 0.8% per year. This chickadee shows a particularly strong decline in the upper

Coastal Plain (from Maryland to North Carolina) in the BBS suggesting that the species may be declining most steeply on the Coastal Plain in Maryland in spite of a net increase in the number of occupied atlas blocks there. Despite these declines the Carolina Chickadee appears to be hardy and adaptable enough to remain widespread and conspicuous in Maryland and DC well into the foreseeable future. Detailed study of hybridization and behavior would be of value in the zone of overlap between the two chickadees in Maryland as it would complement studies in nearby West Virginia, Virginia, and Pennsylvania (Reudink et al. 2007; Sattler and Braun 2000).

WALTER G. ELLISON

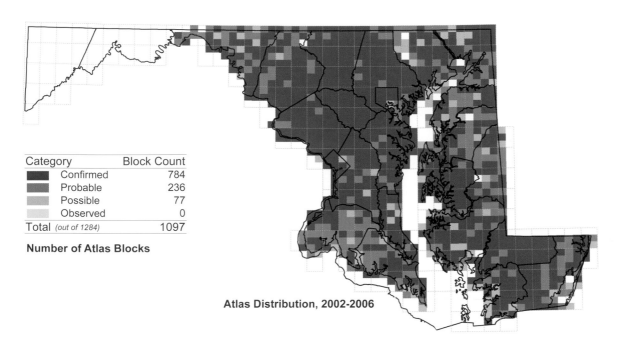

Category	Block Count
Confirmed	784
Probable	236
Possible	77
Observed	0
Total (out of 1284)	1097

Number of Atlas Blocks

Atlas Distribution, 2002-2006

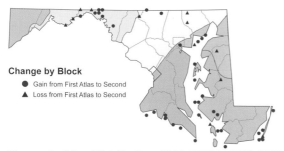

Change by Block
● Gain from First Atlas to Second
▲ Loss from First Atlas to Second

Change in Atlas Distribution, 1983-1987 to 2002-2006

Percent of Stops
- 50 - 100%
- 10 - 50%
- 0.1 - 10%
- < 0.1%

Relative Abundance from Miniroutes, 2003-2007

Atlas Region	1983-1987	2002-2006	Change No.	%
Allegheny Mountain	0	0	0	-
Ridge and Valley	90	93	+3	+3%
Piedmont	314	313	-1	0%
Upper Chesapeake	118	120	+2	+2%
Eastern Shore	300	313	+13	+4%
Western Shore	245	252	+7	+3%
Totals	1067	1091	+24	+2%

Change in Total Blocks between Atlases by Region

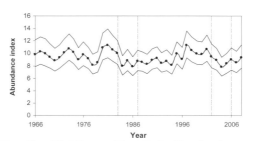

Breeding Bird Survey Results in Maryland

Gary Smyle

Black-capped Chickadee

Poecile atricapillus

The Black-capped Chickadee is very similar to its southern counterpart the Carolina Chickadee, but it is slightly larger and appears to possess other traits that suit it for colder winters. Black-capped Chickadees tend to form larger winter flocks that allow individuals to spend more time feeding and less time watching for predators (S. M. Smith 1993; E. Dunn and Tessagila-Hymes 1999; Mostrom et al. 2002). This chickadee replaces the Carolina Chickadee in the western Ridge and Valley region and on the Allegheny Plateau. The two chickadee species overlap in central Washington County and in the northern part of the Catoctin and South Mountain ridges where block elevations exceed 450 m (1,500 ft).

Black-capped Chickadees, which occupy a wide array of wooded habitats, are cavity nesters that apparently have a strong preference for natural sites as opposed to bird houses. Nesting dates for this feisty but personable bird in Maryland range from 11 May 2003 north of Friendsville, Garrett County (S. Hinebaugh), to 2 June (MNRF) for nests with eggs, and 27 May (MNRF) to 30 June 2005 at Deep Creek Lake State Park (C. Skipper) for nestlings.

The Black-capped Chickadee's distribution is quite similar in outline to its range in the 1983–1987 atlas (Ringler 1996c), but it has lost a total of 10 blocks. Most of these were in areas of overlap with Carolina Chickadee 20 years ago. Black-capped Chickadees were not found from 2002–2006 on the central Catoctin and South Mountain ridges and were also reported from fewer blocks in central and western Washington County. Some of the 17 overlap blocks may not actually contain both chickadee species because the chickadees can learn each other's songs and the possibility of hybrids is another complication. Losses can be interpreted more straightforwardly because such blocks lack birds giving Black-capped Chickadee songs, implying there were no birds to serve as song tutors for resident Carolina Chickadees.

Why Black-capped Chickadees appear to have declined in the area of overlap with Carolina Chickadee is uncertain. Recent studies of chickadee hybrid zones in neighboring states indicate that the hybridization zone between the two chickadees is widening and moving northward and uphill in favor of Carolina Chickadee (Sattler and Braun 2000; Reudink et al. 2007). Male Carolina Chickadees may have a mating advantage over male Black-cappeds (Reudink et al. 2007). It is also possible that climatic warming might favor the Carolina Chickadee.

Population trends for the Black-capped Chickadee from the BBS have been increasing on the continental scale. This species has also increased in the Allegheny Plateau region in Pennsylvania, Maryland, and West Virginia although it has been stable on its few Maryland survey routes (n = 12). The only place where Black-capped Chickadees seem to be declining is in the narrow hybrid zone with the Carolina Chickadee. In southeastern Pennsylvania the breadth of the hybrid zone is presently estimated to be about 35 km (20 mi) (Reudink et al. 2007), and this appears to be roughly correct for Maryland as well given the recent northward and westward shifts in the apparent zone of overlap in the state.

WALTER G. ELLISON

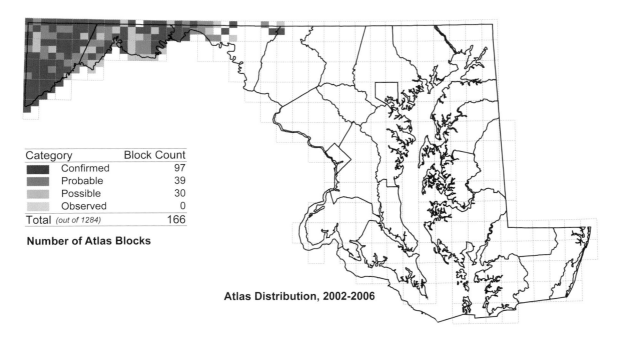

Category	Block Count
■ Confirmed	97
■ Probable	39
■ Possible	30
■ Observed	0
Total (out of 1284)	166

Number of Atlas Blocks

Atlas Distribution, 2002-2006

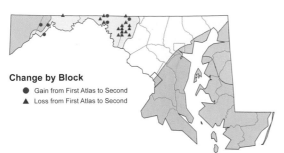

Change by Block
- ● Gain from First Atlas to Second
- ▲ Loss from First Atlas to Second

Change in Atlas Distribution, 1983-1987 to 2002-2006

Percent of Stops
- ■ 50 - 100%
- ■ 10 - 50%
- ■ 0.1 - 10%
- □ < 0.1%

Relative Abundance from Miniroutes, 2003-2007

Atlas Region	1983-1987	2002-2006	Change No.	%
Allegheny Mountain	92	93	+1	+1%
Ridge and Valley	84	73	-11	-13%
Piedmont	0	0	0	-
Upper Chesapeake	0	0	0	-
Eastern Shore	0	0	0	-
Western Shore	0	0	0	-
Totals	176	166	-10	-6%

Change in Total Blocks between Atlases by Region

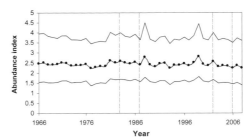

Breeding Bird Survey Results in Maryland

Tufted Titmouse
Baeolophus bicolor

The rich sweet *peter-peter-peter* of the Tufted Titmouse is a prominent part of the late winter and early spring morning songbird chorus. Larger and plainer than the closely related chickadees, the lead gray titmouse sports a distinctive short pointed crest, black forehead, and large, easily seen dark eyes. Found throughout Maryland and DC, it is a familiar bird feeder patron wherever there is a sufficient supply of mature trees.

Tufted Titmice are found in their highest population densities in wet and mesic deciduous woodlands (Stewart and Robbins 1958); they also occur in mixed deciduous and conifer woods and in Coastal Plain pinelands. Unlike chickadees these titmice rely entirely on preexisting natural cavities and on nest boxes for nesting. Most woodlands are suitable for them as long as mature trees with dead limbs, snags, or nest boxes near woods are available. They are often common in well-wooded parks and suburban neighborhoods. Their frequently acrobatic feeding activities are similar to those of chickadees although they are more adept at breaking into hard nuts and they forage on the ground more often especially where there is deep leaf litter.

The Tufted Titmouse is resident year-round wherever it nests in Maryland and DC. Nest building begins in late March and peaks in April. They are generally quiet and se-

cretive around their nest holes; just 9% of confirmations during this project involved active nests. On the other hand, fledged young are often noisy and follow their parents around begging for food for up to six weeks after leaving the nest (Brackbill 1970). Young birds sometimes remain in their parents' company until the next breeding season (Brackbill 1970). Fledged dependent young were cited as the basis for breeding confirmations in 64% of the records in this project. Another 19% of records were of adults carrying food; some of the earliest of these reports may have referred to courtship feeding of mates by males. Maryland and DC egg dates range from 14 April to 7 July (Solem 1996a; MNRF).

Early Maryland and DC bird references all refer to Tufted Titmouse as common (e.g., Kirkwood 1895). Stewart and Robbins (1958) described diminishing abundance westward in Maryland, with them common west to the Blue Ridge at South Mountain, fairly common west through Allegany County, and uncommon in the Allegheny Mountains. In 1983–1987 these titmice were mapped throughout Maryland and DC and the species was one of the 20 most widespread nesting birds in the state (Solem 1996a). They were missed in a few high-elevation blocks in Garrett County, the most open wetland savannas of southern Dorchester and western Somerset counties, the lower Bay islands, and from most of Assateague Island. They are also not found on barrier islands in New Jersey (Walsh et al. 1999), reflecting their reluctance to cross wide expanses of water. They were found in essentially the same range of blocks in 2002–2006, with small increases in Garrett County and in lower Eastern Shore wetlands.

Tufted Titmice have increased slowly on BBS routes since 1966 with a surveywide increase of 0.9% per year. They increased at a slightly lower rate of 0.6% per year in Maryland. To the north of Maryland this species has been inexorably expanding its range from a limit in southern Pennsylvania and New Jersey in the 1930s to southern Ontario, northern New York, and northern New England in the 1980s (Nichols 1985; Meade 1988; Walsh et al. 1999). Presumably this expansion came in response to warming climate and increased winter bird feeding (Beddall 1963; Robbins, Bystrak, and Geissler 1986). Titmice have also apparently increased in Garrett County during the same period. BBS routes in the three-state Allegheny Plateau region have shown a 3.8% per year increase since 1966. If woodlands with snags and nearby nest boxes continue to exist in Maryland and DC the future of this adaptable bird should remain secure.

WALTER G. ELLISON

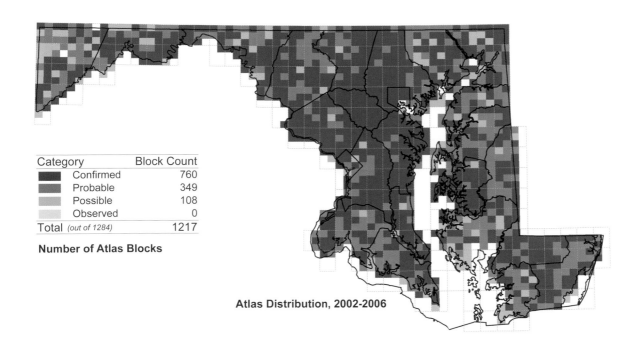

Category	Block Count
Confirmed	760
Probable	349
Possible	108
Observed	0
Total (out of 1284)	1217

Number of Atlas Blocks

Atlas Distribution, 2002-2006

Change by Block
- ● Gain from First Atlas to Second
- ▲ Loss from First Atlas to Second

Change in Atlas Distribution, 1983-1987 to 2002-2006

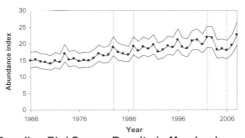

Percent of Stops
■	50 - 100%
■	10 - 50%
■	0.1 - 10%
□	< 0.1%

Relative Abundance from Miniroutes, 2003-2007

			Change	
Atlas Region	1983-1987	2002-2006	No.	%
Allegheny Mountain	86	91	+5	+6%
Ridge and Valley	148	148	0	0%
Piedmont	316	316	0	0%
Upper Chesapeake	116	119	+3	+3%
Eastern Shore	276	286	+10	+4%
Western Shore	245	251	+6	+2%
Totals	1187	1211	+24	+2%

Change in Total Blocks between Atlases by Region

Breeding Bird Survey Results in Maryland

Mark Hoffman

Red-breasted Nuthatch
Sitta canadensis

The Red-breasted Nuthatch is a rather fussy-sounding, busy, inquisitive, and approachable little bird of the coniferous trees of the North Woods. It feeds on insects primarily gleaned from the bark and needles of evergreen trees, and on the seeds from cones, acrobatically scampering around on the trunk and limbs after the fashion of all nuthatches. In Maryland this nuthatch inhabits stands of various conifers on the Allegheny Plateau. These include red spruce around mountain bogs and glades, hemlock and white pine on north-facing ridges and in ravines, and forestry plantations of Norway spruce. Red-breasted Nuthatches have long been suspected of nesting in the mountains of western Maryland based on a handful of records in the summer months including an apparent family group seen by Preble (1900) in 1899 in the Bittinger area. During the first Maryland-DC atlas project, a fledgling was found in a Norway spruce plantation in the Hanover watershed of the Manchester quadrangle just north of the Pennsylvania state line in York County. In late April 1990, a pair of Red-breasted Nuthatches was seen at Finzel Swamp in northeastern Garrett County excavating an apparent nest cavity (Blom 1996g).

The Red-breasted Nuthatch has been increasing over the past 40 years in Canada according to BBS data, and it has been expanding at the southeastern edge of its northern and Appalachian range including notable expansion into New Jersey (Walsh et al. 1999) and southward through central and eastern Pennsylvania (Santner 1992b). During the first statewide atlas, Red-breasted Nuthatches were reported from only 8 blocks (6 of them with breeding codes): 4 in Garrett County, 1 in Allegany County, 2 from northern Carroll County (along the Pennsylvania line), and 1 in Prince George's County (Blom 1996g). In the current atlas (2002–2006), this nuthatch was found in far more blocks, 29 overall, marking a 263% increase over the previous atlas. The three confirmations of nesting were fledglings in the York County, Pennsylvania, Norway spruces of the Hanover Watershed, and adults with food for young and fledglings observed at New Germany State Park and the Savage River State Forest in northeastern Garrett County. There were 11 reports of probable nesting in Garrett County, spanning the county's upper elevations from the Pennsylvania line southwest along Backbone Mountain.

If food resources are good in the boreal forest during the autumn few Red-breasted Nuthatches venture south, and those that do seldom go far south of the Canadian border and northern states. However, good cone crops occur only at two- to four-year intervals, and in the down years wintering Red-breasted Nuthatches irrupt into regions far south of the boreal forest (Ghalambor and Martin 1999). This irruptive behavior allows these nuthatches to explore potentially suitable habitat south of the main breeding range and contributes to the recent range expansion in the eastern United States. Planting of coniferous tree farms and plantations peaked in the latter Great Depression and the 1940s and these stands have matured into good Red-breasted Nuthatch habitat including on the Piedmont of Pennsylvania and northern Carroll County. With the climate warming and mature conifer stands ripe for felling, it remains to be seen if Red-breasted Nuthatches will remain in their small, if expanding, Maryland outpost.

WALTER G. ELLISON

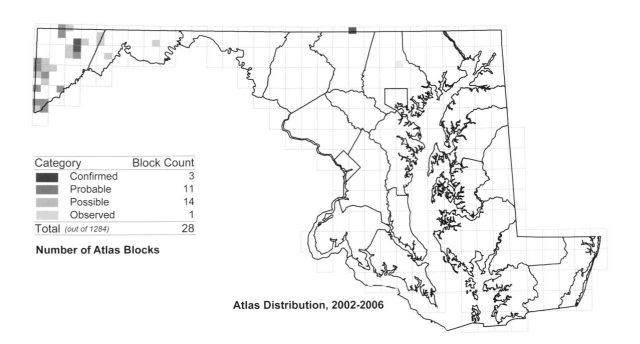

Category	Block Count
■ Confirmed	3
Probable	11
Possible	14
Observed	1
Total (out of 1284)	28

Number of Atlas Blocks

Atlas Distribution, 2002-2006

Change by Block
- ● Gain from First Atlas to Second
- ▲ Loss from First Atlas to Second

Change in Atlas Distribution, 1983-1987 to 2002-2006

			Change	
Atlas Region	1983-1987	2002-2006	No.	%
Allegheny Mountain	3	26	+23	+767%
Ridge and Valley	0	1	+1	-
Piedmont	2	1	-1	-50%
Upper Chesapeake	0	0	0	-
Eastern Shore	0	0	0	-
Western Shore	1	0	-1	-100%
Totals	6	28	+22	+367%

Change in Total Blocks between Atlases by Region

Eric Skrzypczak

White-breasted Nuthatch

Sitta carolinensis

An entertaining bird familiar to feeder watchers, the White-breasted Nuthatch is best known as the chunky long-billed bird that advances head-downward on the boles of trees using only its strong feet with no assistance from its stubby tail. Whereas the smaller Red-breasted and Brown-headed nuthatches are conifer specialists that often depend on cones for winter food, the White-breasted is a specialist of largely broad-leaved mature forest and woodland depending upon acorns, hickory, and beechnuts for its winter diet. White-breasted Nuthatches are widely distributed in rather small numbers; they are often inconspicuous by May as females incubate eggs and males cut back on singing. Maryland and DC nests with eggs have been recorded from 7 April to 20 May (MNRF). This nuthatch is essentially sedentary once it establishes a home range and attracts a mate. Strong southward autumnal movements may occur in some years in response to lean masting. Presumably the majority of the birds in these flights are birds of the year. It is plausible that

these occasional dispersals may be the source of colonists to mature hardwood forests on the Eastern Shore and on the southern Western Shore where the species is expanding its range.

Over the past 50 years the White-breasted Nuthatch has been increasing on the Coastal Plain. Stewart and Robbins' (1958) nesting range map for this nuthatch showed it to be uniformly distributed from the fall line westward, with two regional populations on the Coastal Plain in the greensand soil district of east central Prince George's County and along the Pocomoke River and its tributaries on the lower Eastern Shore. A few isolated nesting records were documented elsewhere on the Coastal Plain. The nesting range during the first Maryland-DC atlas showed increases on the southern Western Shore and on the upper and central Eastern Shore although the bird remained scarce and less than uniformly distributed (Solem 1996b). In fact it was almost absent from the Piedmont in Cecil County, where it had been mapped as widespread in 1958. Over the past two decades the White-breasted Nuthatch has continued to increase in southern Maryland and the upper and central Eastern Shore where it is now widely, if thinly, distributed. This nuthatch remains rare to absent in much of Dorchester County and western Wicomico and Somerset counties.

Trends from the BBS have all been positive or stable for

the White-breasted Nuthatch over the past 40 years from the continental scale down to Maryland, where the species has shown an annual increase of 6.8% per year. This increasing trend has been stronger in Maryland over the past quarter century. The designation of riparian critical areas established to protect the Chesapeake Bay watershed may have helped preserve and expand the mature floodplain hardwood timber (including an increasing number of snags and dead limbs) that forms this nuthatch's preferred habitat on the Eastern Shore and in southern Maryland. As long as there is mature hardwood timber to provide nest and roost sites, this nuthatch should have a secure future in Maryland.

WALTER G. ELLISON

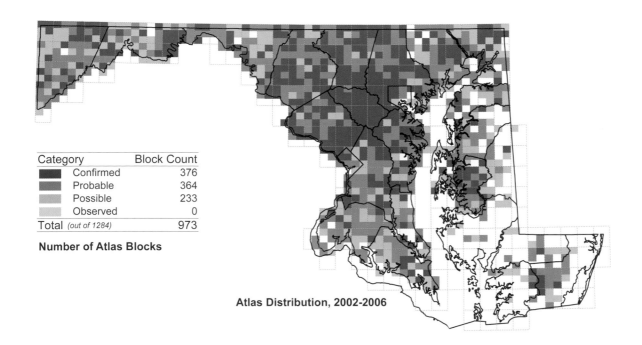

Category	Block Count
Confirmed	376
Probable	364
Possible	233
Observed	0
Total (out of 1284)	973

Number of Atlas Blocks

Atlas Distribution, 2002-2006

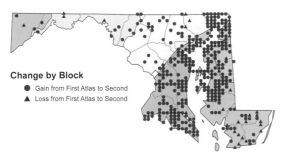

Change by Block
- ● Gain from First Atlas to Second
- ▲ Loss from First Atlas to Second

Change in Atlas Distribution, 1983-1987 to 2002-2006

Percent of Stops
- 50 - 100%
- 10 - 50%
- 0.1 - 10%
- < 0.1%

Relative Abundance from Miniroutes, 2003-2007

Atlas Region	1983-1987	2002-2006	Change No.	%
Allegheny Mountain	91	92	+1	+1%
Ridge and Valley	139	144	+5	+4%
Piedmont	255	312	+57	+22%
Upper Chesapeake	23	80	+57	+248%
Eastern Shore	62	120	+58	+94%
Western Shore	104	220	+116	+112%
Totals	674	968	+294	+44%

Change in Total Blocks between Atlases by Region

Breeding Bird Survey Results in Maryland

Eric Skrzypczak

Brown-headed Nuthatch

Sitta pusilla

The Brown-headed Nuthatch is most common in Maryland in loblolly pine and willow oak woodland along the edge of tidal wetlands, often nesting in the frequent saltwater-killed snags in this ecotonal habitat. Over the past few decades this nuthatch has spread inland and now occurs sparingly in the frequently logged upland pine woods of the lower Eastern Shore. This is a rotund little bird, no bigger than the pine cones that form its major winter food source. Busy and highly sociable birds, Brown-headed Nuthatches form tightly knit flocks that pore over cones, under bark, and in the terminal needle clusters of pines, usually well up in the crowns of taller trees. It is a bird that justifiably registers as cute with even the most jaundiced of observers because of its tiny size, relative tameness, squealing rubber-ducky calls, and the conversational nature of its other fussy twittering and squeaky vocalizations. This is one of only two birds that regularly breeds cooperatively in Maryland (the other being American Crow). Young birds often remain in the parents' home range into subsequent nesting seasons, helping with territorial defense, excavating roost and nest holes, and feeding nestlings. Eggs have been recorded in Maryland from 15 April to 7 May, with nestlings to 9 July (MNRF).

The extent of Brown-headed Nuthatch's tidewater distribution appears to have fluctuated over the past 50 years. Stewart and Robbins (1958) illustrated the bird as ranging northward to southern Calvert County and to the south shore of the Chester River but essentially absent inland on the lower Eastern Shore. The first Maryland-DC atlas map showed records of this nuthatch only as far north as the northwestern part of the Point Lookout Peninsula in St. Mary's County on the Western Shore and to Claiborne, Talbot County, on the Eastern Shore (Willoughby 1996a). In the current atlas project this situation appears to have reversed. The species was widely recorded in coastal St. Mary's County with 2 records north to southern Calvert County, reports from 4 blocks in western Queen Anne's County, and nesting confirmations in 2 blocks on Eastern Neck Island, Kent County, north of the Chester River. Eastern Neck Island is the northernmost locality in this nuthatch's southeastern United States distribution. The first Maryland atlas map showed an apparent large increase inland on the lower Eastern Shore. This inland expansion has continued and consolidated over the past two decades. The Brown-headed Nuthatch's atlas range has increased by 71 blocks since the 1983–1987 atlas, a 76% increase in the number of blocks with records.

Increases in Maryland are in opposition to the long-term trend seen in the Brown-headed Nuthatch's entire range where an annual decrease of 1.6% per year has been recorded in BBS data. Hess et al. (2000) also reported that this nuthatch has been declining in southern Delaware. Withgott and Smith (1998) noted that the Brown-headed Nuthatch has a reputation for not recolonizing parts of its range once it has been extirpated, for example in southeastern Missouri. This seems in keeping with its reputation as a highly sedentary bird, although there are a handful of records of vagrants far from its normal haunts (AOU 1998). In the light of these observations it is noteworthy that Brown-headed Nuthatches appear to have reclaimed substantial portions of their tidewater range over the past 20 years in Maryland.

If suitable habitat remains available or regenerates, and distances are modest, Brown-headed Nuthatches may recover lost ground. Protected forested shoreline buffers, set aside as part of the critical area protecting the Chesapeake Bay watershed, may have had a role in this species' reclamation of historic strongholds.

WALTER G. ELLISON

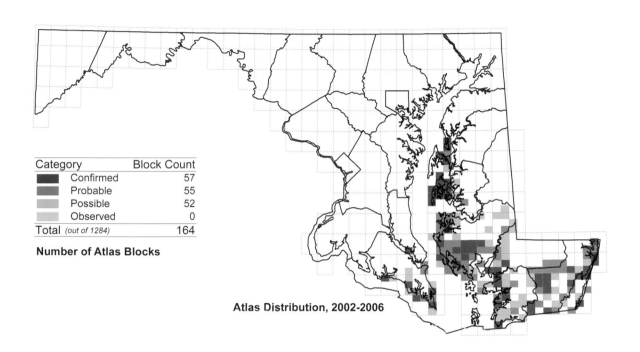

Category	Block Count
Confirmed	57
Probable	55
Possible	52
Observed	0
Total (out of 1284)	164

Number of Atlas Blocks

Atlas Distribution, 2002-2006

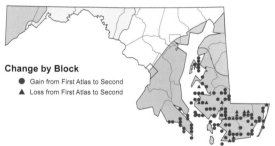

Change by Block
- ● Gain from First Atlas to Second
- ▲ Loss from First Atlas to Second

Change in Atlas Distribution, 1983-1987 to 2002-2006

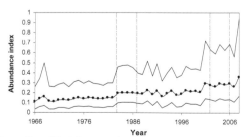

Percent of Stops
- 50 - 100%
- 10 - 50%
- 0.1 - 10%
- < 0.1%

Relative Abundance from Miniroutes, 2003-2007

Atlas Region	1983-1987	2002-2006	Change No.	%
Allegheny Mountain	0	0	0	-
Ridge and Valley	0	0	0	-
Piedmont	0	0	0	-
Upper Chesapeake	0	0	0	-
Eastern Shore	90	147	+57	+63%
Western Shore	3	17	+14	+467%
Totals	93	164	+71	+76%

Change in Total Blocks between Atlases by Region

Breeding Bird Survey Results in Maryland

Eric Skrzypczak

Brown Creeper
Certhia americana

The Brown Creeper has been described as an animated bark flake (Tyler 1948), an apt description. This tiny brown bird streaked with buff spends most of its time hitching up bole and branch often disappearing on the far sides as if playing hide-and-seek. Its voice is thin and high-pitched. The call is similar to the contact notes of Golden-crowned Kinglet, chickadees, and titmice, and its song is a short but complex, creaky ditty sung most often in early spring, almost ceasing after mid-May. These characteristics render this creeper inconspicuous, and it is often overlooked even by experienced observers. The Brown Creeper is at the southeastern edge of its range here and has been a documented nesting bird in Maryland for only 50 years.

Brown Creepers live in mature forests, often mixed or coniferous, with frequent snags that have retained at least some of their bark in adherent strips and plates. They build their nests behind the loose bark and forage on mature trees with rough bark. They often live in wet and streamside forest on floodplains, in mountain ravines, and drowned lands including swamps, reservoirs, and beaver ponds. These creepers

also appear to avoid small forest fragments, preferring extensive woodland (Askins et al. 1987; Robbins, Dawson, and Dowell 1989).

Brown Creepers are found throughout the year in Maryland and DC. There is nonetheless clear seasonal movement: they are much more numerous and widespread in eastern Maryland in winter and migratory peaks are evident in spring and autumn. Creeper nests are wedged behind loose bark, usually on dead trees, rarely on living trees with shaggy loose bark (Santner 1992c), and the birds rarely nest in cavities (Hejl, Newlon et al. 2002). Nests are well concealed and parents are secretive once eggs are laid; they are seldom found after the bustle of nest building. No active nests were listed among this project's nesting confirmations. A nest was found being built on the early date of 14 March 2006 at Halethorpe, Baltimore County, during atlas work (S. Arnold). There are no Maryland egg dates, but nests with young have been reported from 28 May to 15 June (MNRF); fledged young were seen on 13 May 2006 by Bryan Sykes (this atlas). Egg dates for Ontario cover a range from 23 April to 13 July (Peck and James 1997).

The few summer records of Brown Creeper for Maryland through the first half of the twentieth century include a female collected at Bittinger, Garrett County, in June 1899 (Preble 1900), and three reports from the Piedmont and Western Shore from 1944 to 1953 (Stewart and Robbins 1958). Nesting was first documented on 14 June 1958 when adults and fledglings were seen on Meadow Mountain, Garrett County (Fletcher and Fletcher 1959). Adults were found carrying food for young at PWRC in 1964, and Maryland's first nest was found there in 1967 (Van Velzen 1967). Additional summer records arose from the Pocomoke River valley at Shad Landing State Park in 1965 and 1966 (Van Velzen 1967).

During the first statewide atlas (1983–1987), Brown Creepers proved to have an unexpectedly broad breeding range, encompassing 118 blocks in 19 counties and DC. Robbins (1996a) observed a clear association with river valleys including the Patuxent, Pocomoke, Monocacy, and Potomac. Creepers were confirmed nesting in DC and were widely distributed in the mountains of Garrett County. In 2002–2006 they were reported from 15 counties and DC but only in 63 blocks, a 46% decline. The largest losses were in the Pocomoke drainage, but declines were also large from Washington County eastward. Interestingly, Brown Creepers increased by 10 blocks in Garrett County, suggesting a different state of affairs there than in the rest of Maryland.

Brown Creepers are observed in such small numbers on BBS routes that they show no reliable trend, although there has been a tendency toward lower numbers on routes in the United States since 1980. This creeper's range has shrunken in central and eastern Maryland, but it appears to be stable or increasing in the Allegheny Mountains. Why this should be so is unclear. The same pattern also occurred in New York State between atlas projects, with increases at upper elevations and losses at lower elevations (McGowan 2008e). Possible reasons for these changes may include differing forest cover trends, with increase and maturation at higher elevations and habitat fragmentation combined with snag removal in many existing stands at lower elevations. These changes might also be related to climatic warming as lower elevations become less suitable for creepers as summer temperatures increase.

WALTER G. ELLISON

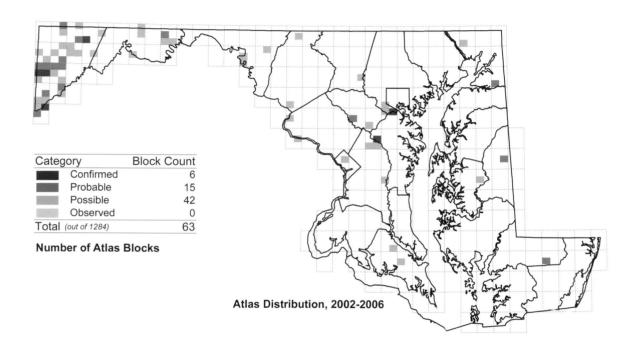

Category	Block Count
■ Confirmed	6
■ Probable	15
■ Possible	42
■ Observed	0
Total (out of 1284)	63

Number of Atlas Blocks

Atlas Distribution, 2002-2006

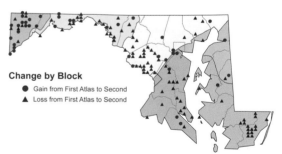

Change by Block
- ● Gain from First Atlas to Second
- ▲ Loss from First Atlas to Second

Change in Atlas Distribution, 1983-1987 to 2002-2006

Atlas Region	1983-1987	2002-2006	Change No.	%
Allegheny Mountain	22	32	+10	+45%
Ridge and Valley	23	10	-13	-57%
Piedmont	30	9	-21	-70%
Upper Chesapeake	2	2	0	0%
Eastern Shore	19	2	-17	-89%
Western Shore	21	8	-13	-62%
Totals	117	63	-54	-46%

Change in Total Blocks between Atlases by Region

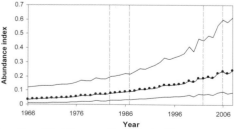

Breeding Bird Survey Results in Maryland

Carolina Wren
Thryothorus ludovicianus

The Carolina Wren can burst into its joyous ringing song even on the coldest mornings in winter. This is Maryland and DC's largest wren, a jaunty, active cinnamon and rust bird. It is unusual among our wrens in being essentially sedentary (Haggerty and Morton 1995) and almost entirely monogamous (Haggerty, Morton, and Fleischer 2001). Their year-round residency can prove disastrous if harsh winter conditions such as deep snow, heavy icing, and frigid temperature last long. This leads to large population fluctuations especially at high elevations and in the northern parts of their breeding range (Robbins, Bystrak, and Geissler 1986). This wren, which is almost ubiquitous in Maryland and DC, was the seventh most widely recorded bird during this atlas project.

Carolina Wrens occur almost anywhere there is thick shrub cover and sufficiently large hollows and cavities to accommodate its bulky domed nest of dead leaves and stems. They are often found near a permanent water source. At higher elevations they are generally restricted to stream valleys and the vicinity of human settlements where there are shelters and feeders. Examples of Maryland and DC haunts of these wrens include floodplain forest, swamps, brushy hedgerows, suburban and urban neighborhoods with abun-

dant ornamental plantings, and upland forest with thick shrubs in the understory.

Carolina Wrens sing throughout the year, although the frequency of vocalization is greatest in spring (Haggerty and Morton 1995). Nesting begins during March as the male starts a number of dummy nests, one of which is ultimately chosen and lined by his mate (Laskey 1948; Haggerty and Morton 1995). Nest sites, which vary widely, include natural cavities; bird houses; shelves inside sheds, derelict houses, and barns; planters on porches; even the pockets of coats hung up in outbuildings (Bent 1948). Carolina Wrens usually have two broods, sometimes three. Maryland egg dates range from 3 March to 13 September (MNRF), with nestlings reported to 26 September (Stewart and Robbins 1958; MNRF).

Both Kirkwood (1895) and Eifrig (1904) called the Carolina Wren common, although Eifrig restricted this status to the "low parts" of western Maryland. Stewart and Robbins (1958) wrote that they were common on the lower Coastal Plain, fairly common from the upper Coastal Plain westward to the Allegheny Front, and rare in the Allegheny Mountains. During the 1983–1987 atlas they were found in 96% of the blocks east of the Allegheny Front but in just 30% of the blocks in the Alleghenies (Joyce 1996a). They were also absent from some Eastern Shore tidewater wetland blocks. But in spite of their sedentary nature, they were found on several Chesapeake Bay islands. In 2002–2006 they had increased substantially in the Allegheny Mountains and were found in

80% of the region's blocks. They remain rare or absent in the highest elevation blocks in the region.

These lively wrens have increased on BBS routes, including Maryland routes where they increased 1.6% per year from 1966 to 2007. They declined steeply after the exceptionally harsh winters of 1976–1978 (Robbins, Bystrak, and Geissler 1986) and have been recovering robustly, with only temporary setbacks, since then. The winter climate has ameliorated greatly since the end of the Little Ice Age and Carolina Wrens have been haltingly but inexorably expand-

ing their range northward to a current limit in southeastern Canada and northern New England (Beddall 1963; Haggerty and Morton 1995). The upslope expansion in Garrett County is a part of this process. They have increased at an impressive rate of 7.2% per year on 74 Allegheny Plateau BBS routes since 1966. Unless winter climate takes a sharp turn for the worse, this feisty bird will remain a prominent feature of Maryland and DC dooryards into the foreseeable future.

WALTER G. ELLISON

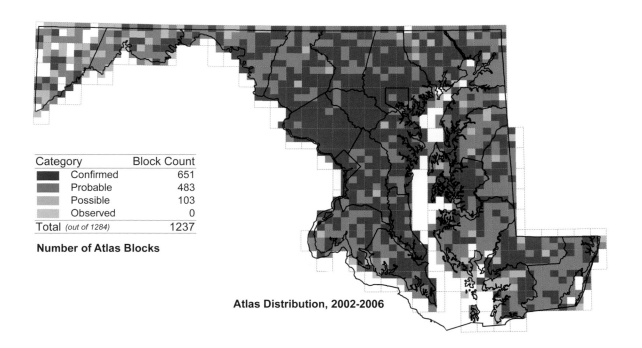

Category		Block Count
■	Confirmed	651
■	Probable	483
■	Possible	103
■	Observed	0
Total (out of 1284)		**1237**

Number of Atlas Blocks

Atlas Distribution, 2002-2006

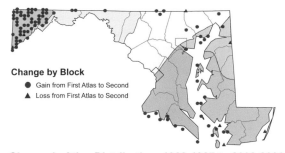

Change by Block
- ● Gain from First Atlas to Second
- ▲ Loss from First Atlas to Second

Change in Atlas Distribution, 1983-1987 to 2002-2006

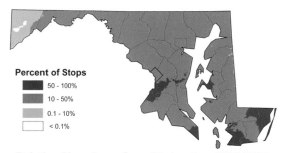

Percent of Stops
■	50 - 100%
■	10 - 50%
■	0.1 - 10%
□	< 0.1%

Relative Abundance from Miniroutes, 2003-2007

Atlas Region	1983-1987	2002-2006	Change No.	%
Allegheny Mountain	28	74	+46	+164%
Ridge and Valley	143	148	+5	+3%
Piedmont	310	315	+5	+2%
Upper Chesapeake	115	120	+5	+4%
Eastern Shore	313	321	+8	+3%
Western Shore	244	253	+9	+4%
Totals	**1153**	**1231**	**+78**	**+7%**

Change in Total Blocks between Atlases by Region

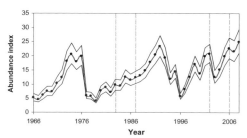

Breeding Bird Survey Results in Maryland

Gary Van Velsir

House Wren
Troglodytes aedon

The House Wren makes up for appearing in a plain brown wrapper by being fearless, curious, peppery, and vociferous. The male's torrential cascading and bubbling song is often sung throughout the day in spring and remains frequent to the end of summer. This bird often lives in close association with people, frequenting yards and gardens near houses. This wren is found almost everywhere above the fall line and, although more spottily distributed, is widespread on the Coastal Plain.

The House Wren is not a forest bird although it is usually associated with some woody vegetation. It may occupy open woodland near an opening and all manner of woodland edges including small openings in forest such as beaver ponds and selectively logged areas. It is also a bird of parks, backyards, overgrown fields, and vacant lots, wherever natural or artificial cavities for nesting exist. On the Eastern Shore they are often associated with the margins of marshes where there is standing dead timber, otherwise they are generally restricted to lower Shore towns where nest boxes are provided.

House Wrens are migratory in Maryland and DC although a few overwinter on the lower Eastern Shore (Stewart and Robbins 1958). They return in early to mid-April and most have departed by the end of October. Males usually build an array of stick nests in cavities in their territories; if a female accepts one, she will remodel it to hold her eggs. Males will remove the contents of other cavities in their territories, including nests of wasps, mammals, and other birds, killing young and piercing eggs (L. Johnson 1998). Some males attract and nest with more than one female (L. Johnson 1998), and females have been documented to raise the offspring of more than one male (Soukup and Thompson 1997). House Wrens usually attempt two broods unless the nesting season is too brief. The ease of locating an occupied wren nest box is reflected in the relatively high (49%) breeding confirmation rate for this project. Maryland and DC egg dates range from 18 April to 16 August with nestlings reported to 2 September (Dupree 1996b; MNRF).

Early Maryland and DC bird authorities all proclaimed the House Wren a common breeding bird, remarking on the wide array of nest sites claimed by them (C. Richmond 1888;

Kirkwood 1895). Stewart and Robbins (1958) also called them common nesters in all regions. The 1983–1987 atlas revealed a localized distribution on the Coastal Plain, with these wrens increasingly associated with tidal wetlands, shorelines, and towns to the south (Dupree 1996b). The 2002–2006 distribution was similar, with ubiquitous distribution above the fall line, and spotty distribution on the Coastal Plain. They have increased their range somewhat on the lower Western Shore and Eastern Shore, presumably in response to an increase in suburban tract housing inland.

North American BBS route trends are relatively flat long term, although declines have occurred in some regions, for example, New England. Maryland BBS trends have been stable. The House Wren appears to continue to be well established and secure in Maryland and DC. L. Johnson (1998) reviewed a long-term southward range expansion by these wrens in eastern North America. This expansion may have contributed to the loss of Appalachian populations of the Bewick's Wren (including in Maryland; Blom 1996h), which appears to be at a competitive disadvantage versus the highly aggressive House Wren (Hall 1983; Kennedy and White 1996).

WALTER G. ELLISON

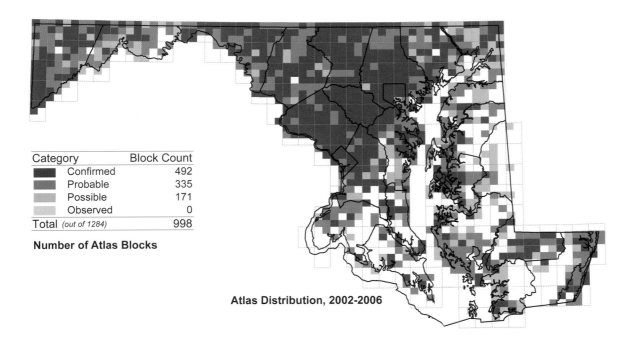

Category	Block Count
Confirmed	492
Probable	335
Possible	171
Observed	0
Total (out of 1284)	998

Number of Atlas Blocks

Atlas Distribution, 2002-2006

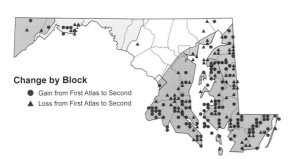

Change by Block
● Gain from First Atlas to Second
▲ Loss from First Atlas to Second

Change in Atlas Distribution, 1983-1987 to 2002-2006

Percent of Stops
- 50 - 100%
- 10 - 50%
- 0.1 - 10%
- < 0.1%

Relative Abundance from Miniroutes, 2003-2007

Atlas Region	1983-1987	2002-2006	Change No.	%
Allegheny Mountain	91	91	0	0%
Ridge and Valley	139	140	+1	+1%
Piedmont	315	313	-2	-1%
Upper Chesapeake	84	73	-11	-13%
Eastern Shore	181	212	+31	+17%
Western Shore	154	167	+13	+8%
Totals	964	996	+32	+3%

Change in Total Blocks between Atlases by Region

Breeding Bird Survey Results in Maryland

Winter Wren
Troglodytes hiemalis

The Winter Wren is usually heard before it is seen. The sound might be the fussy, querulous *cut-cut* call, a testy sputtering chitter, or, if one is fortunate, the torrential cascading song composed of tinkling trills and high-pitched whistles that is seldom given on migration. The song is experienced to best effect in chilly half light at dawn or dusk in a dark forested ravine. When seen, this dark brown bird gives the impression of a tiny, hyperactive mouse or shrew as it scrambles over logs, peeks from tangled brush, or bounces nervously on its haunches atop a rock. This is a scarce nesting bird of the mountains of westernmost Maryland with only a few definitely proven nesting records.

Nesting Winter Wrens prefer to live near water in places with a cool microclimate to support good numbers of short-needled conifers and with old enough growth for lots of dead wood to be strewn on the ground. They live in ravines, the conifers surrounding bogs, and on cool steep slopes usually with ledges or talus.

Winter Wrens are present year-round in Maryland and DC. In winter they are more numerous in eastern Maryland, especially on the lower Eastern Shore (Stewart and Robbins 1958). Summering birds presumably return to nesting territories in late March and early April and depart some time in September. The superstructure of the globular moss nest is built by the male and lined by the female. As with other wrens, males build several nests, only a few of which are used for nesting by his mate or mates. Nest sites vary but are always some sort of cavity. The majority of nests in most studies are found placed in upturned root clusters and under stream banks (Hejl, Holmes, and Kroodsma 2002). The few active and dummy nests found in Maryland have all been lodged in crevices in sandstone ledges (Wennerstrom 1996). Dates for confirmed breeding in Maryland are few. An incomplete clutch of eggs was found by Edward Thompson on 7 June 1990 near Loch Lynn, Garrett County. Thompson later found two other inactive nests on 19 July 1990 at the same location (Wennerstrom 1996). Fledged young were seen in Cranesville Swamp on 9 June 2002, the only confirmed nesting during this atlas project (F. Pope).

It appears that the Winter Wren was a regular nesting species in Garrett County old growth forests during the nineteenth century although the evidence for this is anecdotal (Behr 1914). They declined as mature forests were removed, but they were still present in 1914 in the upper Casselman River valley (Eifrig 1915). When Eifrig (1920) returned in 1918 he could no longer find them. Stewart and Robbins (1958) considered them extirpated in Garrett County although Brooks (1936) had reported one during summer in Cranes ville Swamp. A few summer reports were made from 1958 to the mid-1980s (Robbins and Boone 1985). Some of these were early atlas reports; a total of 15 blocks limited to Garrett County had records from 1983 to 1987 (Wennerstrom 1996). None of these reports were of confirmed breeding although in nine instances probable nesting behavior was observed. They were subsequently confirmed in 1990 (Wennerstrom 1996).

Winter Wrens were found in a net total of 7 more blocks

in the second atlas. They were found along the main ridges of the Alleghenies from southwest to northeast from Backbone Mountain northward over Meadow and Savage mountains. They were also found in the historic Casselman drainage and confirmed in Cranesville Swamp.

North American BBS trends for the Winter Wren have been stable or slightly increasing. Numbers have fluctuated strongly over the past 40 years; in particular there were steep declines after harsh eastern North American winters in 1976–1978 (Robbins, Bystrak, and Geissler 1986). Since 1980 the trend in eastern North America has been a fairly strong increase of 1.7% per year. This wren seems to have increased in Maryland over the past several decades and become better established in Garrett County between statewide atlas projects. This may be attributable to maturation of forests in the Allegheny Mountains, notably on state-managed forest lands. It is plausible that warming winters in the Winter Wren's southeastern United States winter range may also have aided a recent steady increase in distribution in Maryland and other northeastern states (e.g., Bonney 1988; Walsh et al. 1999; McGowan 2008f).

WALTER G. ELLISON

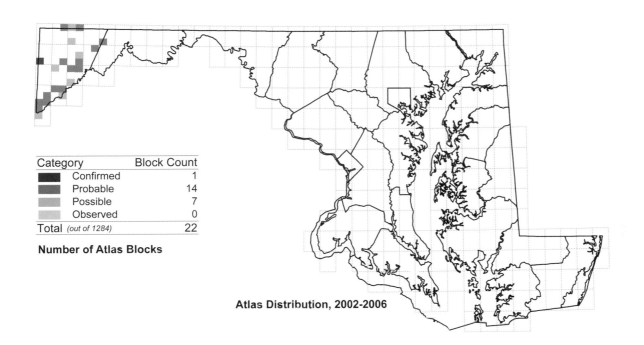

Category	Block Count
■ Confirmed	1
■ Probable	14
■ Possible	7
■ Observed	0
Total (out of 1284)	22

Number of Atlas Blocks

Atlas Distribution, 2002-2006

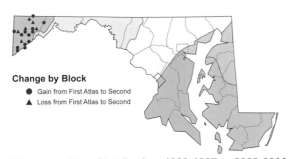

Change by Block
● Gain from First Atlas to Second
▲ Loss from First Atlas to Second

Change in Atlas Distribution, 1983-1987 to 2002-2006

Atlas Region	1983-1987	2002-2006	Change No.	%
Allegheny Mountain	15	22	+7	+47%
Ridge and Valley	0	0	0	-
Piedmont	0	0	0	-
Upper Chesapeake	0	0	0	-
Eastern Shore	0	0	0	-
Western Shore	0	0	0	-
Totals	15	22	+7	+47%

Change in Total Blocks between Atlases by Region

Sedge Wren
Cistothorus platensis

Tiny and unobtrusive, the Sedge Wren may go unnoticed even when it is fairly numerous. The song is high pitched, with the scratchy quality of a katydid, a staccato *chip-chip-chitter*. It is easily lost to listeners at a distance in the open breezy places it is sung. They were never truly common in Maryland, with two main centers of distribution on the southern Coastal Plain and in the Allegheny Mountains. They are now very rare and are officially listed as endangered by Maryland's DNR.

Once called Short-billed Marsh Wren, it was rechristened to reflect its true habitat proclivities. It is not a bird of regularly flooded land with tall emergent vegetation, as is its close relative the Marsh Wren. It occupies relatively dry ephemerally flooded or very shallow wetlands dominated by sedges, relatively fine (sometimes tall) grasses, usually with a scatter of small shrubs. In Maryland and DC it has nested in the upper parts of tidal marshes where marsh elder and other shrubs invade the grass and sedge, in wet switchgrass meadows, in the wetter parts of upland hay fields where reed canary grass and orchard grass grow, and in damp sedge meadows associated with the glades (bogs) of the Allegheny Mountains.

Sedge Wrens are migratory; they also make long-range movements during the nesting season. They may move long distances between nesting attempts; whether these moves happen only after failed nestings or also after successful breeding is not documented (Bedell 1996; Herkert, Kroodsma, and Gibbs 2001). They tend to nest earliest in the northwestern part of their breeding range in the northern Great Plains and western Prairies and arrive to nest later in the southern and eastern parts of their known range (Kroodsma et al. 1999). They are generally uncommon to rare in Atlantic seaboard states and often show low site fidelity there (Leberman 1992b; McGowan 2008g). In Maryland and DC they occur from April to November; they winter in small numbers on the lower Eastern Shore.

The nest is a woven ball of grass with a side entrance integrated into grasses or shrubs less than a meter (3 ft) above the ground. The male generally builds several nests, then the female lines the one of her choice. Where they are numerous, male Sedge Wrens will often have more than one mate (Burns 1982). They have not been confirmed nesting in Maryland in several decades, including during the two statewide atlas projects. The first proven nesting in Maryland was an adult with dependent young seen on 13 September 1896 in the Dulaney Valley, Baltimore County, by Frank Kirkwood (Stewart and Robbins 1958; MNRF). Wetmore (1935) found a nest with eggs at Point Lookout, St. Mary's County on 25 June 1935, and dummy nests were built by males in

Mark Hoffman

DC during the same year (Ulke 1935; Ball and Wallace 1936). A nest with eggs was also found at Elliott Island, Dorchester County, on 12 July 1958 (A. and B. Meanley; MNRF; Blom 1996i).

Among the early Maryland and DC bird compendia only Kirkwood (1895) listed the Sedge Wren. It continued to be a rarity worthy of publication until the late 1940s when Stewart and Robbins (1947) discovered good numbers of nesting birds in lower Eastern Shore marshland. Stewart and Robbins (1958) considered them common in southern Dorchester County and western Wicomico and Somerset counties, uncommon in other tidewater areas and in the glades of Garrett County. The species declined in intervening years and was found in just 12 atlas blocks during 1983–1987. Nine of these were possible nesters, some of which were probably transient males seeking a nesting territory. Three probable records were listed from Montgomery County, on the Patuxent River in Prince George's County, and from the traditional site of Elliott Island (Blom 1996i).

During 1988–2001 there were 25 reports of Sedge Wrens that may have been on breeding territory. More than two-thirds of these were from tidewater locations. There were no nesting records and little evidence of birds persistently holding territories. Six records, all listed as probable nesters, were obtained during the current atlas project (2002–2006). Five records were of long-staying males holding territory. In at least two cases, at Kingstown, Queen Anne's County (J. Gruber et al.), and at Mathias Point, Charles County (G. Jett), more than one male was present. A male was on territory east of McHenry, Garrett County from late May to late June 2002; Walter Ellison and others saw it carrying nest material on 9 June. The two Coastal Plain reports were not from the species' former tidal wetland stronghold on the lower Eastern Shore.

North American BBS trends suggest that the Sedge Wren may be increasing or stable in the core of its range in the northern Midwest. It is hard to develop management options for such a rare and sporadic nesting bird. Nonetheless, developing and maintaining Conservation Reserve grassland on wet parts of fields and preserving sedge meadows in montane bogs and grassy areas in the drier parts of tidal marshes may encourage birds to continue nesting in Maryland.

WALTER G. ELLISON

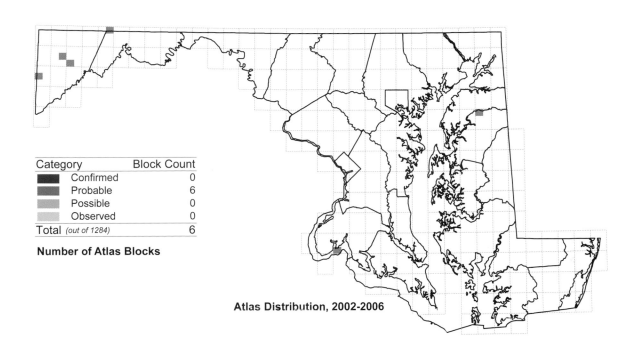

Category	Block Count
Confirmed	0
Probable	6
Possible	0
Observed	0
Total (out of 1284)	6

Number of Atlas Blocks

Atlas Distribution, 2002-2006

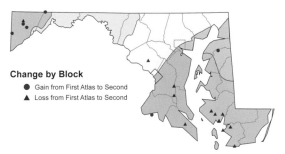

Change by Block
- ● Gain from First Atlas to Second
- ▲ Loss from First Atlas to Second

Change in Atlas Distribution, 1983-1987 to 2002-2006

Atlas Region	1983-1987	2002-2006	Change No.	%
Allegheny Mountain	1	4	+3	+300%
Ridge and Valley	0	0	0	-
Piedmont	1	0	-1	-100%
Upper Chesapeake	0	1	+1	-
Eastern Shore	7	0	-7	-100%
Western Shore	3	1	-2	-67%
Totals	12	6	-6	-50%

Change in Total Blocks between Atlases by Region

Bill Sherman

Marsh Wren
Cistothorus palustris

The clattering chattering songs of Marsh Wrens are among the most characteristic sounds of extensive marshlands in tidewater Maryland and DC. Often very common in their preferred haunts, these diminutive, skulking wrens are familiar only to people who brave the noisome and numerous biting insects of marshes. The rather unmusical but complex and varied song is important for males in attracting multiple mates to nest in their territories. Males have 40 to 50 different songs in their repertoires (Kroodsma and Verner 1987), and the frequency of males with extra mates often exceeds 25% (Kroodsma and Verner 1997). As with other wrens, males build several cock, or dummy, nests on their territories; only a few will be selected by their mates to lay eggs in (Kroodsma and Verner 1997).

Marsh Wrens nest in a variety of open wetlands with standing water and emergent grasslike plants. In Maryland these include needlerush marsh, waterside stands of big and smooth cordgrasses, high salt marsh dominated by salt-meadow cordgrass and marsh elder, three-square marshes, narrow-leaf cattail marsh, and regularly flooded reed beds. They are seldom seen away from wetlands, even during migration. Population densities may be very high; 257 territorial males per hundred hectares (104 males/100 ac; Springer and Stewart 1948) were recorded in needlerush marsh in Somerset County.

Although some winter in Maryland, many local Marsh Wrens winter to the south along the southeastern Atlantic coast. Wintering Marsh Wrens are most numerous here in the marshes of the lower Eastern Shore, although even these birds may withdraw in harsh winters. Returning migrants arrive fairly late from mid-April to early May. Fall departure is hard to establish but most are probably gone by early November. The nest is a globular woven structure with a side entrance made from the grassy leaves of emergent marsh plants. Because males build several dummy nests each season, nests must be watched carefully or inspected to establish breeding. Multiple broods, high nest predation rates, and late nestings by secondary and higher order mates extend the nesting season into late summer (Kale 1965). Maryland and DC egg dates have been recorded from 3 May (exceptionally early; most egg laying starts in late May) to 17 August (Joyce 1996b; MNRF).

The Marsh Wren was called common in tidewater marshes up to the fall line in the late nineteenth century (C. Richmond 1888; Kirkwood 1895). Stewart and Robbins (1958) called it abundant in appropriate habitat on the Coastal Plain. During the 1983–1987 atlas it was recorded from approximately the same distribution although it was only spottily distributed along the lower Potomac River, was local in Cecil and Kent counties in the Upper Chesapeake region and in the coastal bays of Worcester County (Joyce 1996b). A few were reported on the Piedmont but were not confirmed breeding there. There was a net loss of 40 blocks during the 2002–2006 atlas although the general outline of the range remains the same. These wrens were found in many fewer blocks in the Upper Chesapeake and were not refound on the western Piedmont.

Although Marsh Wrens have been increasing surveywide on North American BBS routes, they have declined in the East, 3.2% per year since 1980. Reasons for eastern declines include loss of relatively small wetlands to drainage and development, invasive plants supplanting more prey-rich native species, and loss of habitat to erosion and flooding, especially because of sea level rise along the coast. The preservation of extensive marshland is important for retaining the Marsh Wren among Maryland and DC's nesting birds. Although many wetlands are publicly owned and legally protected, they are still prone to degradation and should be protected from invasive plants, pollution, and loss because of climate change.

WALTER G. ELLISON

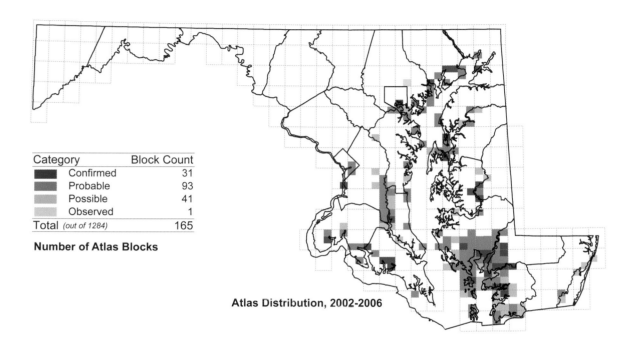

Category	Block Count
Confirmed	31
Probable	93
Possible	41
Observed	1
Total (out of 1284)	165

Number of Atlas Blocks

Atlas Distribution, 2002-2006

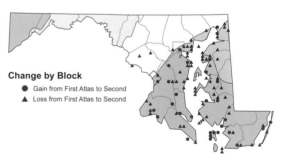

Change by Block
- Gain from First Atlas to Second
▲ Loss from First Atlas to Second

Change in Atlas Distribution, 1983-1987 to 2002-2006

Percent of Stops
- 50 - 100%
- 10 - 50%
- 0.1 - 10%
- < 0.1%

Relative Abundance from Miniroutes, 2003-2007

Atlas Region	1983-1987	2002-2006	Change No.	%
Allegheny Mountain	0	0	0	-
Ridge and Valley	0	0	0	-
Piedmont	6	1	-5	-83%
Upper Chesapeake	31	18	-13	-42%
Eastern Shore	113	99	-14	-12%
Western Shore	53	45	-8	-15%
Totals	203	163	-40	-20%

Change in Total Blocks between Atlases by Region

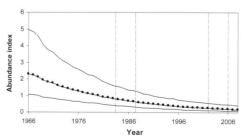

Breeding Bird Survey Results in Maryland

Mark Hoffman

Golden-crowned Kinglet

Regulus satrapa

A hyperactive tiny ball of fluff, the Golden-crowned Kinglet breeds in mature evergreen forest and winters widely in a broad array of woodlands, often close to water. This kinglet's voice is high-pitched, with thin piping calls similar to creepers, chickadees, and titmice, and its song begins with several high rising notes and descends into a chickadee-like chatter. Their small size and high-pitched voice combine to make them easily overlooked. They are local breeding birds in Maryland, with most found at high elevations in Garrett County, although they could potentially occur anywhere there are mature and dense conifer plantations (Mulvihill 1992b) as they do in northern Carroll County along the Mason-Dixon Line.

Golden-crowned Kinglets inhabit relatively dense stands of evergreens during their nesting season. Although they will sometimes occupy red pine plantations, hemlock groves, and fir stands north and south of Maryland, they prefer mature stands of spruce, both native red and planted non-native Norway. They have increased over the past four decades at relatively low elevations by invading Depression-era conservation plantings of spruce (Andrle 1971; Mulvihill 1992b).

These kinglets occur in winter throughout Maryland but are generally most numerous on the Coastal Plain and least frequently seen in the Alleghenies (Hatfield et al. 1994). Breeding birds arrive in nesting areas in April and depart by late September. The pouchlike nest is slung on outer branch forks of conifers, often quite high, and is constructed from moss, lichen, and bark strips with a feather lining (Galati and Galati 1985). Clutch sizes can be as many as nine, and Golden-crowned Kinglets often raise two broods (Galati 1991), presumably because of a short lifespan and frequent high winter mortality. Galati (1991) found eggs in Minnesota from 12 May to 1 July. Maryland nesting dates have been few. Dan Boone observed nest-building on 13 May 1982 at Rock Lodge, Garrett County, and Jon Boone recorded eggs on 1 July and nestlings on 29 July 1982 at New Germany State Park, Garrett County (J. Boone 1982). During this project, Stan Arnold reported fledglings as late as 12 August 2006 on Fourmile Ridge, Garrett County.

Golden-crowned Kinglets have been recorded during the nesting season in the Maryland Allegheny Mountains since at least the late nineteenth century, although they became very rare after almost all the virgin spruce had been logged (Eifrig 1904; Behr 1914). Nesting was first documented, via fledglings in Garrett County, in 1945 (Stewart and Robbins 1958). The next confirmation did not come until 1982, with the two nests cited above (J. Boone 1982). On 28 June 1981 Golden-crowned Kinglets were discovered in spruce plantings at the Hanover Watershed in Carroll County (Robbins and Boone 1985). Kinglets were found in 15 blocks from 1983 to 1987 including 2 in the Hanover Watershed in Carroll County and York County, Pennsylvania, and 13 blocks spanning Garrett County from northeast to southwest (Blom 1996j). The basic range recorded from 2002 to 2006 is the same. Kinglets still occur in 1 block in the Hanover Watershed, and they consolidated and expanded their range in the Garrett County high country, adding 17 blocks.

Golden-crowned Kinglets are declining continent-wide on BBS routes because of declines in the west. In eastern North America the BBS trend has been stable to increasing. Populations fluctuate with the harshness of winters (Robbins, Bystrak, and Geissler 1986). These tiny birds are still found in the same places they were found in the 1980s. Why they have yet to be found in more conifer plantings in the

Ridge and Valley region is unclear, but observers should continue checking such locations. As long as spruce plantations are maintained and regenerating native spruce is protected, Golden-crowned Kinglets should remain nesters in the Free State.

WALTER G. ELLISON

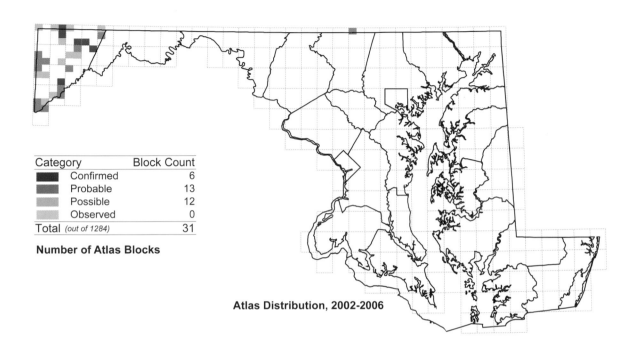

Category	Block Count
Confirmed	6
Probable	13
Possible	12
Observed	0
Total (out of 1284)	31

Number of Atlas Blocks

Atlas Distribution, 2002-2006

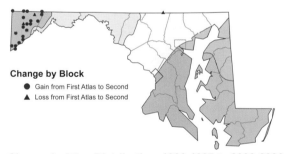

Change by Block
- ● Gain from First Atlas to Second
- ▲ Loss from First Atlas to Second

Change in Atlas Distribution, 1983-1987 to 2002-2006

Atlas Region	1983-1987	2002-2006	Change No.	Change %
Allegheny Mountain	13	30	+17	+131%
Ridge and Valley	0	0	0	-
Piedmont	2	1	-1	-50%
Upper Chesapeake	0	0	0	-
Eastern Shore	0	0	0	-
Western Shore	0	0	0	-
Totals	15	31	+16	+107%

Change in Total Blocks between Atlases by Region

Eric Skrzypczak

Blue-gray Gnatcatcher
Polioptila caerulea

A tiny, bluish pixie with an insignificant, but persistent, high wheezy voice, the Blue-gray Gnatcatcher might be easy to overlook were it not an energetic and sometimes acrobatic hunter of tiny prey. It regularly flirts and fans its long black and white tail and often chases flushed insects and spiders with a dipping pirouette. It is a bird of open woodlands and forest edges. Maryland was at the northern edge of its range until the 1940s, but a northward range expansion has carried the limit north to the Canadian border and slightly beyond (Ellison 1992b). It is currently numerous virtually throughout Maryland and DC.

Blue-gray Gnatcatchers inhabit broadleaved woodlands, sometimes with a smattering of conifers. They usually choose open woodlands, clearings, forest edges, or young successional woods. These include dry oak-hickory woods, open oak-pine woods, floodplain forest, swamps, and old fields, pastures, and timber cuts with an abundance of small trees. They are scarce or absent where woodlots are few and small (Robbins, Dawson, and Dowell 1989) preferring the

majority of the landscape wooded, but they are not birds of extensive mature closed-canopy forest. Given sufficient mature trees and undergrowth they will nest in wooded suburban neighborhoods and city parks.

These gnatcatchers are migratory, with an extensive winter range from the Coastal Plain of the southeastern United States south through Cuba and Mexico to northern Central America. Blue-gray Gnatcatchers are early migrants capable of feeding in early spring tree flowers and from twigs and branches before leaves unfurl. They arrive in Maryland and DC as early as the last week of March, with most arriving in early April. They also depart fairly early, with peak migration in late August and early September; most are gone by early October. This gnatcatcher's nest is a neat, tightly woven cup of fine plant fibers and down bound with webbing and decorated with lichen. Usually saddled far out on a limb, the nest may be built at various heights, but the average is from 9 to 12 m (30–40 ft; Ellison 1992b) off the ground. Both members of a pair work on the nest, and they are often vocal as they build, making confirmation relatively easy. Almost 20% of nesting confirmations in this project were of nest-building pairs, but most confirmations (44%) were for fledged young. Maryland and DC egg dates span 11 April to 10 June (Blom 1996k; MNRF), with nestlings reported to 13 July in Caroline County (J. Reese, pers. comm.).

Although the Blue-gray Gnatcatcher appears to have regularly nested north to eastern Pennsylvania in the early to mid-nineteenth century (Leberman 1992c), it declined in the late nineteenth century. By the end of the century it no longer nested in eastern Pennsylvania (Leberman 1992c), was a localized nester in eastern Maryland (Kirkwood 1895), had declined at Washington, DC (Coues and Prentiss 1883), and was scarce in Allegany County (Eifrig 1904). After a population rebound in the 1930s and 1940s, gnatcatchers were once again nesting in eastern Pennsylvania by the 1950s. Major spring migration overshooting occurred in a handful of years from 1947 to the early 1980s accompanied by extralimital nestings and range expansion (Ellison 1993). Stewart and Robbins (1958) found gnatcatchers common on the Coastal Plain, in the Ridge and Valley, and along the Potomac in the Piedmont, and uncommon to rare elsewhere.

The Blue-gray Gnatcatcher increased its distribution greatly on the Piedmont and in the Allegheny Mountains by the 1983–1987 Maryland-DC atlas project (Blom 1996k). It was found in 75% of the blocks west of the Allegheny Front and nearly 90% of the blocks on the Piedmont. It was still lacking or thinly distributed at the state's highest elevations and in heavily farmed areas, including the Hagerstown and Frederick valleys and east of Chesapeake Bay. It was also largely absent from lower Eastern Shore wetlands and offshore islands. Gnatcatchers were found in a net total of 75 more blocks in 2002–2006, showing gains in urban and agricultural areas and modest losses in the Alleghenies.

Blue-gray Gnatcatcher numbers on North American BBS

routes have been stable. They have been stable or increasing slowly in Maryland, but the 1.2% per year increase since 1980 here is not statistically significant. This gnatcatcher is relatively adaptable and resilient, as its stability during a period of strong suburban development in many parts of Maryland shows. Still it requires fairly large and heterogeneous wood-

land areas to persist. Apparent block losses in rapidly developing parts of Cecil and Harford counties suggest this bird is vulnerable to rapid loss of woodland. Nonetheless it is likely there will be nesting gnatcatchers in Maryland and DC for many years to come.

WALTER G. ELLISON

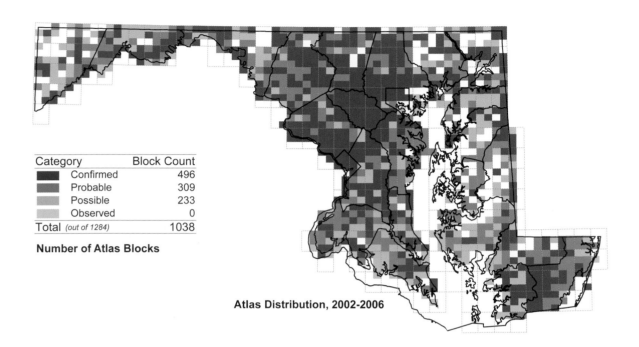

Category	Block Count
Confirmed	496
Probable	309
Possible	233
Observed	0
Total *(out of 1284)*	1038

Number of Atlas Blocks

Atlas Distribution, 2002-2006

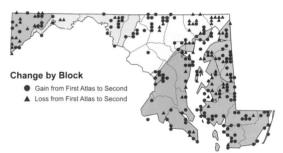

Change by Block

● Gain from First Atlas to Second
▲ Loss from First Atlas to Second

Change in Atlas Distribution, 1983-1987 to 2002-2006

Percent of Stops

■ 50 - 100%
■ 10 - 50%
■ 0.1 - 10%
□ < 0.1%

Relative Abundance from Miniroutes, 2003-2007

Atlas Region	1983-1987	2002-2006	Change No.	Change %
Allegheny Mountain	68	64	-4	-6%
Ridge and Valley	129	137	+8	+6%
Piedmont	282	289	+7	+2%
Upper Chesapeake	88	98	+10	+11%
Eastern Shore	182	212	+30	+16%
Western Shore	209	233	+24	+11%
Totals	958	1033	+75	+8%

Change in Total Blocks between Atlases by Region

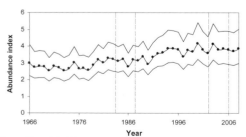

Breeding Bird Survey Results in Maryland

George M. Jett

Eastern Bluebird
Sialia sialis

One of the most beloved birds in the country, the Eastern Bluebird is common in open country across the state. Its azure and rust plumage and soft sweet warble are familiar to all who have even a minimal knowledge of or interest in birds. The male is deep blue above, with a chestnut red throat and breast and a white belly. The female is a more washed-out grayish blue (some are almost gray on the back), the breast is much paler (almost cinnamon), and the throat is white with a grayish malar stripe. Juvenile birds (May through August) are brownish gray above with some blue in the wings and tail, and a heavily speckled breast. The song of the male is a warbled *cheer-cheerily cheer-up.*

Bluebirds are found in open habitats mixing short grass and tall foraging perches: forest edges, orchards, overgrown brush fields, open groves, and cutover woodlands. Their annual diet is about two-thirds animal and one-third vegetable. The young are fed almost entirely from the animal portion of the menu. Orthoptera (grasshoppers, katydids, and crickets) and beetles are the dominant sources of protein; caterpillars are next in importance. Prey are taken from the air, in trees and bushes, and on the ground. Sallying from a perch, hovering, and gleaning are all used. The vegetable portion is mostly wild fruits: dogwood, Virginia creeper, blueberry, raspberry, holly, and poison ivy are all accepted (Bent 1949).

In December and January vegetable food makes up a clear majority of the fare. Bluebirds are hole nesters, and originally the species used natural tree cavities and abandoned woodpecker holes. Today, however, most Eastern Bluebirds use artificial nest boxes. The nest itself is loosely built by the female out of grasses and weed stalks. Each pair normally raises two broods, and sometimes three (H. Harrison 1975). Young from an early nesting will sometimes assist with a later brood (Zeleny 1976). The early and late egg dates are 12 March and 5 September (MNRF), and the normal clutch size is from 3 to 5 eggs. Nest-site competition is a severe problem for the species; European Starling, House Sparrow, House Wren, Tree Swallow, and even southern flying squirrels may try to evict bluebirds. The latter two are not generally a problem; the sparrow, wren, and especially the starling are more likely to succeed. Nest box holes are sized to shut out the starling, but the present near-exclusion of bluebirds from natural sites may be blamed on starlings. The bluebird itself may be the aggressor, appropriating holes from chickadees or woodpeckers. Cowbirds may parasitize the nest, but are not a major problem.

Throughout Maryland, from the mountains to the sea, bluebirds are common summer residents of the rural countryside. In the eastern parts of the state, many bluebirds can be found year-round, but they are vulnerable to severe winter weather, and the colder climate toward the west pushes more birds southward. The species remains common in winter in the Eastern and Western Shore regions, but the abundance decreases westward until it is a winter rarity in the Allegheny Mountain region (Stewart and Robbins

1958). Those that remain in areas of harsher climate are in more sheltered stream valleys and coastal areas. Particularly cold, snowy, and especially icy winters decimate the wintering population, as the birds are cut off from their supply of berries. The effect of the winter of 2002–2003 is clear in the Breeding Bird Survey data: following that winter, the number of birds recorded on the Maryland BBS routes was halved. Spring migrants arrive early, usually from the second week of February through mid-April. Fall migration normally extends from mid-September through mid-November, with a few stragglers continuing into December.

Bluebirds have done well in Maryland since the first atlas (Dupree 1996c), continuing a long-term trend that began around 1980, as shown in the BBS data. The number of blocks with bluebirds increased in every region of the state, with an increase of 31 blocks (44%) in the Upper Chesapeake region, where the previous atlas had noted low numbers. Statewide, the increase was 147 blocks (14%), leaving only 117 blocks (out of a total 1,284 blocks) without bluebirds.

T. DENNIS COSKREN

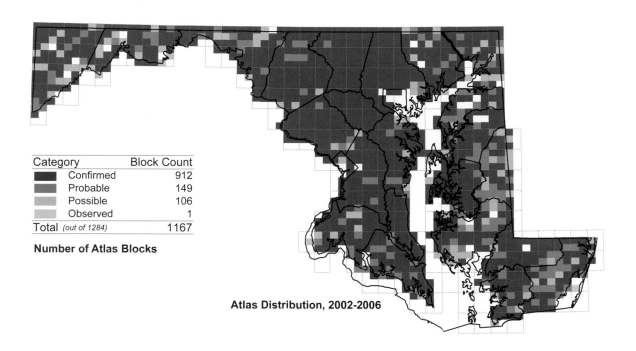

Category	Block Count
Confirmed	912
Probable	149
Possible	106
Observed	1
Total (out of 1284)	1167

Number of Atlas Blocks

Atlas Distribution, 2002-2006

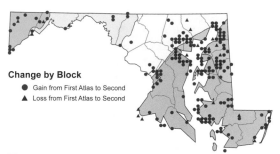

Change by Block

● Gain from First Atlas to Second
▲ Loss from First Atlas to Second

Change In Atlas Distribution, 1983-1987 to 2002-2006

Percent of Stops

■ 50 - 100%
■ 10 - 50%
■ 0.1 - 10%
□ < 0.1%

Relative Abundance from Miniroutes, 2003-2007

Atlas Region	1983-1987	2002-2006	Change No.	Change %
Allegheny Mountain	78	88	+10	+13%
Ridge and Valley	139	144	+5	+4%
Piedmont	285	309	+24	+8%
Upper Chesapeake	70	101	+31	+44%
Eastern Shore	233	283	+50	+21%
Western Shore	211	238	+27	+13%
Totals	1016	1163	+147	+14%

Change in Total Blocks between Atlases by Region

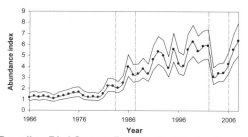

Breeding Bird Survey Results in Maryland

Veery
Catharus fuscescens

As with all spotted brown woodland thrushes of the Americas, the Veery is an accomplished songster. Its song, unlike the measured chanting of the Wood and Hermit thrushes, is a hurried cascade of mixed harmonic notes. Louis Halle (1947) called it "a soft, tremulous, utterly ethereal sound, swirling downward and ending, swirling downward and ending again. Heard in the gloom of twilight across the marshes, it gives the impression that this is no bird at all but some spirit not to be discovered." Eastern Veeries are more warmly colored overall than other forest thrushes, with uniform cinnamon brown upperparts, gray flanks, and a dusting of blurry brown spots across the upper breast. This is a bird of forests, preferring areas with an abundance of rank shrubs and tall ferns. Maryland is at the southeastern edge of this thrush's breeding range. It is widely, but locally, distributed from the western mountains to the fall line.

Veeries prefer moist to wet woods with dense understory shrubs and saplings (Bertin 1977). They usually occur where the forest canopy is thin or absent. Typical sites include brook banks, ravines, floodplains, shrub swamps, older regenerating clear-cuts (two or more years after cutting), and blow downs. The birds occupy both pure deciduous and mixed woods. Despite their preference for breaks in mature forest they generally require a large portion of the landscape to be forested in order to thrive and are rare in areas with small isolated woodlots (Robbins, Dawson, and Dowell 1989; Herkert 1995).

The Veery is a long-distance migrant to the neotropics, wintering in central and southeastern Brazil (Remsen 2001). It returns to Maryland and DC from the last week of April to mid-May and commences its departure in late August, with almost all gone by late September. Most, if not all, pairs appear to raise only a single brood. The nest is built either on the ground or slightly raised above ground level on low rotting stumps, fern clusters, and brush piles usually at 0.6 m (2 ft) or less. The nest is always well hidden and hard to find; no active nests were reported during this atlas project. Maryland and DC egg dates range from 5 May to 26 June with nestlings reported to 11 July (Czaplak 1996; MNRF).

Kirkwood (1895) considered the Veery to be strictly a seasonal transient through eastern Maryland, but he cited an 1895 nesting record at Deer Park, Garrett County. Eifrig (1904) wrote that it was a common nester near Frostburg but that he had yet to encounter it at Cumberland. Veeries began to nest on the Piedmont during the 1940s with the first nesting at DC's Rock Creek Park (Halle 1943). They were found at several additional Piedmont locations during the ensuing decade with five sites in Harford, Baltimore, and Montgomery counties mapped by Stewart and Robbins (1958). Stewart and Robbins also indicated that they bred on the northern part of South Mountain in the eastern Ridge and Valley region; they were very rare in the western Ridge and Valley, but were common and widespread in Garrett County.

During the 1983–1987 atlas project the Veery's Maryland and DC nesting range had expanded further (Czaplak 1996). They remained widespread in the Alleghenies with records from 70% of that region's blocks. They were also in most of the blocks in the northern part of Maryland's Blue Ridge on South and Catoctin mountains. They had increased just as dramatically on the Piedmont where they were well distributed along the fall line and westward through Howard, Montgomery, Baltimore, and Carroll counties. There were even a few breeding season reports from southern Cecil and northern Kent counties on the Coastal Plain. The 2002–2006 atlas distribution is similar although there have been modest net losses in the Allegheny Mountains and the eastern Piedmont (especially in Harford County, where coverage may have been an issue). They were found as possible breeders in more blocks in the western Ridge and Valley, but they remain scarce there.

This thrush has declined on North American BBS routes at a rate of 1.5% per year with a pronounced downturn from 1980 onward. The trend on 22 Maryland BBS routes appears to be stable or slightly negative. They have declined significantly on 112 routes in the Allegheny Plateau region. Although there was rather little range contraction between atlas projects, the declining trend for the Veery is worrisome. Forest fragmentation may have contributed to declines. Loss of wintering habitat in Brazil is also a probable contributor to declines. The Veery's future here is not as secure as it might be.

WALTER G. ELLISON

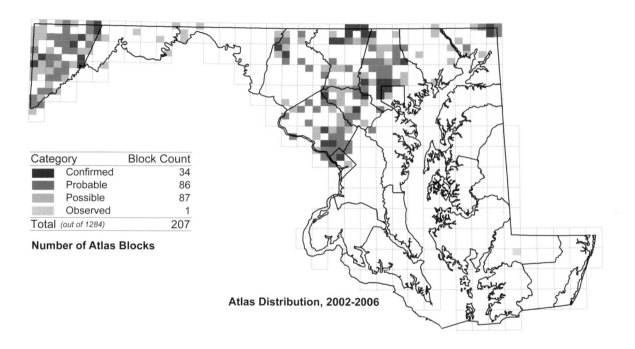

Category	Block Count
■ Confirmed	34
■ Probable	86
■ Possible	87
■ Observed	1
Total *(out of 1284)*	207

Number of Atlas Blocks

Atlas Distribution, 2002-2006

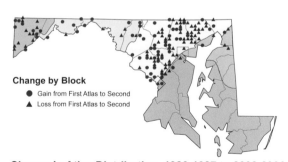

Change by Block
- ● Gain from First Atlas to Second
- ▲ Loss from First Atlas to Second

Change in Atlas Distribution, 1983-1987 to 2002-2006

Percent of Stops
- ■ 50 - 100%
- ■ 10 - 50%
- ■ 0.1 - 10%
- □ < 0.1%

Relative Abundance from Miniroutes, 2003-2007

Atlas Region	1983-1987	2002-2006	Change No.	%
Allegheny Mountain	65	60	-5	-8%
Ridge and Valley	13	21	+8	+62%
Piedmont	132	121	-11	-8%
Upper Chesapeake	3	1	-2	-67%
Eastern Shore	0	0	0	-
Western Shore	5	4	-1	-20%
Totals	218	207	-11	-5%

Change in Total Blocks between Atlases by Region

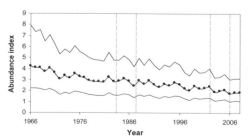

Breeding Bird Survey Results in Maryland

Hermit Thrush
Catharus guttatus

The Hermit Thrush is an unobtrusive brown bird with a black-spotted whitish breast. It is best known for its incomparable song, which is integral to hushed evergreen and mixed forests at dawn and dusk and sufficiently evocative to inspire beautiful prose from the most prosaic of naturalists, poetry from poets, and music from musicians. The Hermit Thrush is found in summer across northern North America, southward through the western mountain ranges, and along the Pacific Northwest coast. It nests in the southeastern United States only at high elevations in the Appalachians. In Maryland this thrush is restricted to Garrett County west of the Allegheny Front.

The breeding season haunts of the Hermit Thrush are varied, ranging from wet to dry, mature to young, and forest interior to edge. It generally prefers areas with canopy openings, and the edges of clearings, wetlands, and overgrown fields rather than deep forest, and there are almost always at least a few conifers among the trees in its territories. Maryland habitats include the conifers surrounding bogs, conifer plantations, hemlock stands, and mixed northern hardwoods with pine or hemlock. In coastal states to the north of Maryland, Hermit Thrushes also occupy pine-oak barrens (Walsh et al. 1999).

Hermit Thrushes winter in Maryland, being fairly common on the Coastal Plain and diminishing westward to the Alleghenies where they are rare (Stewart and Robbins 1958). Most eastern Hermit Thrushes winter farther south, mostly along the Atlantic and Gulf Coastal Plains (Root 1988). Nesting Hermit Thrushes return to the Maryland mountains in April and are nesting by mid-May. Nests are usually placed on the ground nestled into leaf litter under ferns, shrubs, or saplings. Hermit Thrushes may also nest in a shrub or small tree at less than 3 m (10 ft), a more frequent behavior in western North America (Jones and Donovan 1996). The few Maryland records of nests with eggs appear to delimit the peak of early nesting, these range from 25 May to 12 July (MNRF). Pennsylvania egg dates range from 12 May to 10 August (McWilliams and Brauning 2000). There are no nestling dates for Maryland, but adults have been seen carrying food for young to 26 August (F. Pope, pers. comm.) indicating that some nests are active as late as in Pennsylvania.

Early historic references to Hermit Thrush are sparse. Eifrig (1904) called it a summer resident and considered it common at upper elevations, but he was unclear about nesting status and exact localities. Stewart and Robbins (1958) found this thrush to be uncommon at elevations above 760 m (2,500 ft) at scattered places in Garrett County from Finzel southwest to Herrington Manor. Atlas results from 1983 to 1987 extended the Hermit Thrush's range southwest to Backbone Mountain, with reports from 35 blocks, and probable or confirmed (1 nest record) breeding in 14 blocks (Blom 1996l). In 2002–2006 the range had increased substantially in Garrett County. Hermit Thrushes were located in 21 more blocks, a 60% increase, and 70% of the reports referred to probable or confirmed nesting (9 records). Hermit Thrushes were missed in blocks around Oakland and in the northwestern part of the county where the elevation is lower.

Hermit Thrushes have increased on BBS routes in North America, especially in the eastern region. These thrushes are heard on too few routes (7) in Maryland to obtain statistically significant trend information, but there has been a large increase in the absolute numbers of Hermit Thrushes on these routes. The Hermit Thrush is the only spotted thrush

(*Catharus* and *Hylocichla*) that winters regularly in temperate North America, which insulates it somewhat from habitat loss in the tropics; however, its winter habitat in the southeastern United States is also subject to development and degradation. As a bird that winters in forest thickets near edges, feeding largely on fruit, the Hermit Thrush is less affected by habitat disruption than forest interior and wetland birds. Hermit Thrushes are also prone to periodic population declines after harsh winters in the southeast (Robbins, Bystrak, and Geissler 1986).

Given that there are nesting populations on South Mountain in Adams County, Pennsylvania (Mulvihill 1992c), just north of Maryland's portion of the Blue Ridge, and on the mountains of Bedford County, Pennsylvania, north of Allegany County (Mulvihill 1992c), observers should be alert for further range extensions into Maryland. Summer records east of Garrett County in Maryland are few; one was collected in Howard County in July 1890 (Kirkwood 1895), and a singing male summered in Trappe, Talbot County, in 2004 (W. Bell, pers. comm.). The Hermit Thrush should be secure in the publicly owned upper elevation enclaves that are its summer home in the Old Line State, but climatic warming may degrade its habitat over the next several decades.

WALTER G. ELLISON

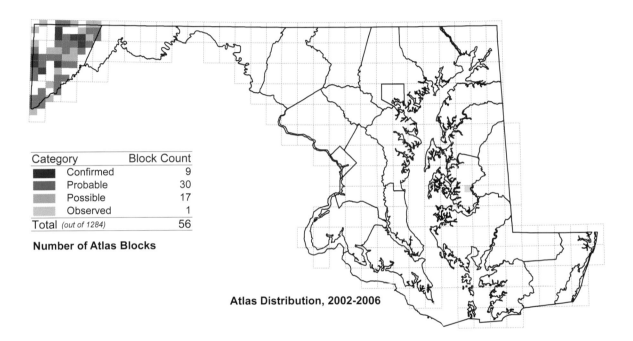

Category	Block Count
Confirmed	9
Probable	30
Possible	17
Observed	1
Total (out of 1284)	56

Number of Atlas Blocks

Atlas Distribution, 2002-2006

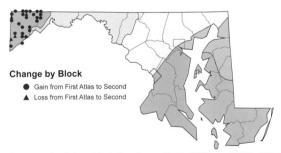

Change by Block
● Gain from First Atlas to Second
▲ Loss from First Atlas to Second

Change in Atlas Distribution, 1983-1987 to 2002-2006

	Atlas Region	1983-1987	2002-2006	Change No.	Change %
	Allegheny Mountain	35	56	+21	+60%
	Ridge and Valley	0	0	0	-
	Piedmont	0	0	0	-
	Upper Chesapeake	0	0	0	-
	Eastern Shore	0	0	0	-
	Western Shore	0	0	0	-
Totals		35	56	+21	+60%

Change in Total Blocks between Atlases by Region

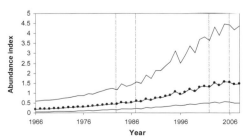

Breeding Bird Survey Results in Maryland

Bill Sherman

Wood Thrush
Hylocichla mustelina

The voice of the Wood Thrush is one of the most prominent and beautiful sounds of Maryland woodlands. The song is elegant and complex, ranging from sepulchral chuckles to intense pure fluty whistles to seemingly electronic high-pitched mixed harmonics. It is a virtuoso performance. This thrush is almost as hefty as a robin but shorter-tailed, with a warm brown back and china white underparts with crisp black spotting.

Wood Thrushes are birds of the forest floor and under-story saplings and shrubs. They feed on the ground on invertebrate prey and move up to the understory in late summer and fall for fruit. These thrushes reach their highest population densities in extensive mature moist predominantly deciduous forest with an ample understory, but occupied woods cover a broad range from dry oak and hickory with scattered small trees and shrubs to mixed forest dominated by mature hemlock. Although nesting success is often poor in small woodlots (Robinson et al. 1995), they will settle in urban parks and wooded suburban developments.

Wood Thrushes, which winter in Mexico and Central America, are generally seen in Maryland and DC from mid-April to mid-October. The nest, built in branch and trunk forks of small trees and large shrubs, is a substantial cup. It is similar to the American Robin's in incorporating mud in the walls, and it is festooned with pale dead leaves (and sometimes pieces of plastic). Most nests are built below 6.1 m (20 ft) from the ground, but occasional nests may be placed higher. Eggs have been found in Maryland and DC from 1 May to 6 August (Solem 1996c; MNRF). During this project Danny Poet saw a female on her nest as early as 4 May 2002 on Tilghman Neck, Queen Anne's County. Wood Thrushes are often double brooded, unlike many other neotropical migrant birds; they do not depart for winter quarters until late August or September.

The overall status and distribution of the Wood Thrush has changed little in more than a century. The species was called common throughout Maryland and DC by all early authors (e.g., Kirkwood 1895; Stewart and Robbins 1958). It ranked among the top 30 most widely distributed birds in Maryland and DC in the 1983–1987 atlas project (Solem 1996c) and retained that status, with the small difference of just 3 blocks, in the second atlas. Although it was found in 3 more blocks, the net gain arose from widely scattered additions balancing losses concentrated in western Dorchester and Somerset counties, which were also marginal areas in 1983–1987. Wood Thrushes were absent from most of Assateague Island; scarcity on barrier beaches is seen in other atlas projects (Hess et al. 2000; Walsh et al. 1999). Some urbanized blocks in Baltimore also lacked these thrushes, but they were found in every block in Washington, DC.

The Wood Thrush has declined on BBS routes virtually throughout its range, with greatest rates of loss in areas where it is most numerous. It has declined at a comparable rate in Maryland (1.7% per year). The causes of these declines are varied. Because Wood Thrushes winter in lowland wet forest on the Atlantic and Pacific slopes of Mexico and Central America, conversion of forest to farmland has likely had a significant effect on wintering thrushes. The on-going fragmentation of mature woodland on the breeding grounds also reduces nesting success for these birds, but double brooding may help lessen the impact (Roth and Johnson 1993). Although still widespread and relatively common, the Wood Thrush is a species of concern because of long-term declines over the past three decades. Large blocks of forest should be maintained to increase nesting success, and forest preservation and management should be encouraged in Central America and Mexico.

WALTER G. ELLISON

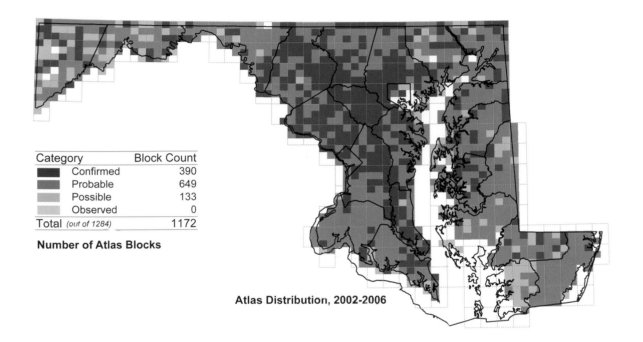

Category	Block Count
Confirmed	390
Probable	649
Possible	133
Observed	0
Total (out of 1284)	1172

Number of Atlas Blocks

Atlas Distribution, 2002-2006

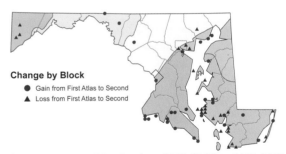

Change by Block
- ● Gain from First Atlas to Second
- ▲ Loss from First Atlas to Second

Change in Atlas Distribution, 1983-1987 to 2002-2006

Percent of Stops
- 50 - 100%
- 10 - 50%
- 0.1 - 10%
- < 0.1%

Relative Abundance from Miniroutes, 2003-2007

Atlas Region	1983-1987	2002-2006	Change No.	%
Allegheny Mountain	93	90	-3	-3%
Ridge and Valley	145	147	+2	+1%
Piedmont	315	315	0	0%
Upper Chesapeake	113	114	+1	+1%
Eastern Shore	263	254	-9	-3%
Western Shore	235	247	+12	+5%
Totals	1164	1167	+3	+0%

Change in Total Blocks between Atlases by Region

Breeding Bird Survey Results in Maryland

American Robin
Turdus migratorius

Roger Tory Peterson (1934) once called the American Robin "the one bird everybody knows." This may not be strictly true, as crows are more prevalent in popular culture as symbols of grim foreboding and are marginally more conspicuous and readily recognizable, but the robin is certainly the best-known small songbird in America. The American Robin is associated with the advent of spring, unafraid of humans during the summer breeding season, reasonably colorful with its brick red underparts, and portly but jaunty as it hops across lawns interspersed with the occasional strutting run. Robins also have a sweet cheery rollicking song that often starts long before sunup and starts again near dusk continuing long after sundown. This large thrush is often found in close association with humans, occupying any scrap of unsprayed green grass in an urbanized landscape where it can hunt earthworms and large soft-bodied insects; robins are virtually ubiquitous in Maryland and DC. occurring almost anywhere there are trees for nests and some open ground for feeding.

The American Robin was among the top 10 most widespread birds in Maryland and DC in both the 1983–1987 atlas project (Chestem 1996a) and the current one. Robins were recorded in slightly more blocks in 2002–2006 than in 1983–1987. There were more block records on Aberdeen Proving Ground because of improved coverage, and robins occurred in a few more blocks in the lower shore marshes and pinelands of Dorchester and Somerset counties. Robins are largely absent from the marshy islands in southern Chesapeake Bay and are scarce on Assateague Island. Sandy soils and mud saturated with saltwater lack earthworms (Sallabanks and James 1999).

American Robins are easy to confirm as nesters; 86% of atlas reports were for confirmed breeding, second only to European Starling among Maryland and DC's nesting birds. By far the most confirmations referred to fledged young (664 records, 19 April to 25 August) or adults seen carrying food to young birds (551 records). Robins have an extended nesting season; egg dates for Maryland range from 26 March to 23 August (Robbins and Bystrak 1977; MNRF). Some pairs may even attempt three broods over a spring and summer. Most of the bulky mud and grass nests made by robins are placed in small to medium-size trees, including ornamentals in yards; some pairs may use artificial structures such as building ledges.

Breeding Bird Survey trends for American Robin have

been increasing at a modest but steady rate, both for North America as a whole and Maryland. In Maryland physiographic regions there has been a modest declining trend in the Allegheny Mountains from Pennsylvania through West Virginia and an increase on the Upper Coastal Plain from Maryland south through North Carolina. The latter reflects a long-term increase and range expansion in the southeastern United States over the past century (Sallabanks and James 1999). Robin populations have been growing over the past 40 years. After exposure to DDT spraying (Mehner and Wallace

1959) during the 1950s and 1960s, they experienced several conspicuous localized declines that were clearly linked to the pesticide (Wurster et al. 1965). These incidents contributed to the outcry that led to the banning of organochlorine pesticides in the early 1970s. At present American Robins remain widespread and common breeders in Maryland and DC and they should continue to be familiar to Marylanders for a long time to come.

WALTER G. ELLISON

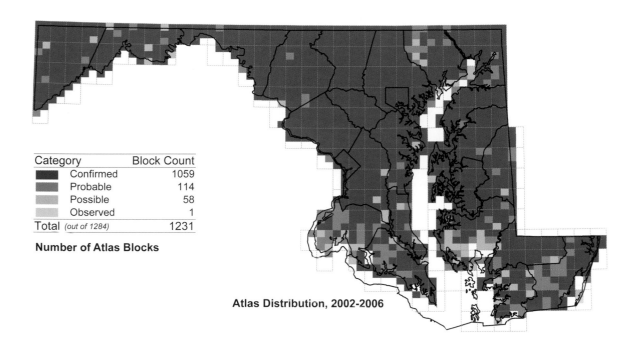

Category		Block Count
■	Confirmed	1059
■	Probable	114
■	Possible	58
■	Observed	1
Total (out of 1284)		1231

Number of Atlas Blocks

Atlas Distribution, 2002-2006

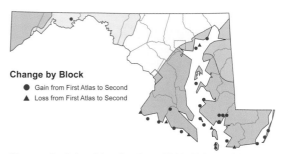

Change by Block
- ● Gain from First Atlas to Second
- ▲ Loss from First Atlas to Second

Change in Atlas Distribution, 1983-1987 to 2002-2006

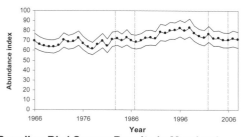

Percent of Stops
- ■ 50 - 100%
- ■ 10 - 50%
- ■ 0.1 - 10%
- □ < 0.1%

Relative Abundance from Miniroutes, 2003-2007

Atlas Region	1983-1987	2002-2006	Change No.	%
Allegheny Mountain	93	93	0	0%
Ridge and Valley	147	148	+1	+1%
Piedmont	316	316	0	0%
Upper Chesapeake	118	119	+1	+1%
Eastern Shore	296	305	+9	+3%
Western Shore	242	244	+2	+1%
Totals	1212	1225	+13	+1%

Change in Total Blocks between Atlases by Region

Breeding Bird Survey Results in Maryland

Gray Catbird
Dumetella carolinensis

The Gray Catbird is well named. It is predominantly a smooth, lead gray; the bird's neutral color is relieved by a charcoal black cap and a touch of burgundy under the base of its long black tail. A person's usual encounter with a catbird is as a disembodied catlike mew (hence its group name) or via its complex, disjointed song, which manages to be squeaky and musical by turns. Although catbirds are generally furtive, they are common and inquisitive enough to be fairly easy to observe among their favored haunts of thick, often thorny, shrubs; vine tangles; and dense ornamental plantings. Any reasonably thick plant cover might harbor a catbird wheezing, squeaking, and chortling, seemingly to itself.

Gray Catbird was the 10th most frequently recorded species in the 1983–1987 atlas (Hitchner 1996a). The catbird occurred in 14 fewer blocks in 2002–2006, and it fell to the 20th position among the state's most widespread nesting birds. Most of this relatively small decline was in southern Maryland and the lower Eastern Shore as the catbird suffered a net loss of 13 blocks on the Coastal Plain. The few gaps that existed in the catbird's nesting range in the first atlas were also most evident on the lower Coastal Plain. The miniroute map in the previous atlas (Hitchner 1996a) and the abundance map for Gray Catbird in Price et al. (1995) based on

BBS data demonstrate that the catbird has long had relatively low populations on the Coastal Plain from Maryland southward. Perhaps the warm and humid climate of the Upper Coastal Plain is not optimal for Gray Catbird; the species may also face stronger competition from the closely related Brown Thrasher and mockingbird in this region. Although the catbird has been stable or increasing on BBS routes in Maryland, trends for the Upper Coastal Plain physiographic region have been declining for the past 40 years. Much as with the House Wren, the Gray Catbird seems to be more common near tidewater on the lower Eastern Shore, occupying the thick shrub line along shores.

In spite of declines on the Coastal Plain, long-term prospects for the Gray Catbird appear reasonably good in Maryland and DC. Catbirds do not appear to be overly choosy about habitat in most of the Free State as long as there is thick growth from ground level up to about 3 m (10 ft) and an abundance of insects and high-energy fruit on which to feed. Catbirds are fairly tolerant of suburban development provided there are suitable hedges and ornamentals available for cover, fruit, and nest sites. Egg dates for Maryland and DC range from 1 May to 19 August with nestlings to 27 August (MNRF). This species bears watching as shrub habitat is lost to development, reforestation, and understory browsing by deer. A study of the reasons for the catbird's low numbers and decline on the Coastal Plain is a worthwhile topic for future research.

WALTER G. ELLISON

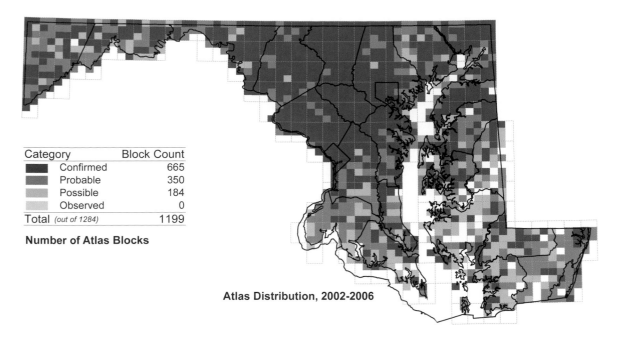

Category	Block Count
Confirmed	665
Probable	350
Possible	184
Observed	0
Total (out of 1284)	1199

Number of Atlas Blocks

Atlas Distribution, 2002-2006

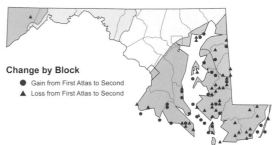

Change by Block
- ● Gain from First Atlas to Second
- ▲ Loss from First Atlas to Second

Change in Atlas Distribution, 1983-1987 to 2002-2006

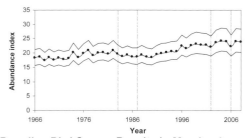

Percent of Stops
- 50 - 100%
- 10 - 50%
- 0.1 - 10%
- < 0.1%

Relative Abundance from Miniroutes, 2003-2007

Atlas Region	1983-1987	2002-2006	Change No.	%
Allegheny Mountain	93	92	-1	-1%
Ridge and Valley	148	148	0	0%
Piedmont	316	316	0	0%
Upper Chesapeake	117	114	-3	-3%
Eastern Shore	304	292	-12	-4%
Western Shore	231	233	+2	+1%
Totals	1209	1195	-14	-1%

Change in Total Blocks between Atlases by Region

Breeding Bird Survey Results in Maryland

Northern Mockingbird
Mimus polyglottos

This bird is among the few known to almost everyone. The Northern Mockingbird is a year-round resident that is common in suburbs and towns. It is bold, conspicuous, and highly vocal. Its scientific name, translated as "many-tongued mimic," is a tribute to its ability to imitate other birds. Although the mockingbird's plumage is not bright it is difficult to mistake for any other species. It is moderately large, about the size of an American Robin, but more slender. Male and female share the same plumage. The back and head are gray, shading to grayish white from the belly to the throat. The wings and tail are mostly grayish black. The outer tail feathers are conspicuously white, and the wings flash a large white patch in flight. The juvenal plumage is similar but the underparts are spotted and streaked. The long tail is frequently cocked. Intruders into its territory may be attacked fiercely; victims include trespassing mockingbirds, of course, but also dogs, cats, crows, snakes, and even people. The song of the mockingbird, performed by both sexes, is its premier attribute. It consists of most anything and everything. Local bird songs are its mainstay, but the repertoire has included "the crowing of chanticleer, the cackling of the hen, the barking of the house dog, the squeaking of the unoiled wheelbar-

row, the postman's whistle." One bird is said to have changed its tune 87 times in 7 minutes, and 57 of the imitations were recognizable (Forbush, quoted in Bent 1948). During the breeding season the bird frequently sings during the night, especially when the moon is bright or where artificial light is present. The effect is pleasant enough early in the evening; but for those attempting to sleep at three in the morning with a nonstop mockingbird a short distance away, the nuisance value may be considerable.

The Northern Mockingbird is found in open areas that retain some cover. It is most common and conspicuous in urban and suburban backyards. In rural areas, it is found in agricultural land that still has brushy edges, in abandoned fields, and around forest edges. A characteristic ground feeder, the bird runs or hops in pursuit of its prey, but it also hawks for flying insects at low heights. When it first alights, it often raises its extended wings, displaying its white wing patch. It is catholic in its food choices, about half of which are animal and half vegetable. The spread of multiflora rose has been advantageous to the mockingbird as the brambly thickets provide both food and cover. A small proportion of the diet may be commercial fruit (e.g., grape, raspberry). The nest is usually in a shrub or bush, from 1 to 3 m (3–10 ft) above the ground, although it has been recorded as high as 19 m (60 ft). Sites near houses seem to be preferred. The male lays material in several locations as part of the courtship ritual but the female makes the final choice (H. Harrison 1975). Both sexes assist in nest construction; the male has most of the responsibility for the foundation, while the female does the lining

(D. Brewer 2001). The nest is sturdy and bulky. The outer part is made up of twigs (often thorny) and an inner layer is composed of finer material, such as plant stems and moss. Grass, fine rootlets, and hair constitute the lining.

This species is a permanent resident in Maryland, and is common throughout most of the state. The Allegheny Mountain region has the sparsest population. Historically, the mockingbird was a bird of the southern United States, and Maryland was on the northern edge of its range. At the beginning of the twentieth century, it was resident in the southern parts of the state, regularly nesting only as far north as Anne Arundel and Kent counties (Kirkwood 1895).

During the twentieth century, the species has extended its range far to the north, having occupied almost all of the eastern United States and adjacent Canada. By the 1950s mockingbirds were present in all of Maryland but were rare in the western reaches of the state (Stewart and Robbins 1958). The first atlas showed them common everywhere except in the Allegheny Mountain region, where it was mostly absent (Robbins 1996b). The number of reporting blocks in that region has grown from 22 to 40, an increase of 82%. Elsewhere the number of blocks has increased only slightly, but this is because there are so few blocks left without mockingbirds.

T. DENNIS COSKREN

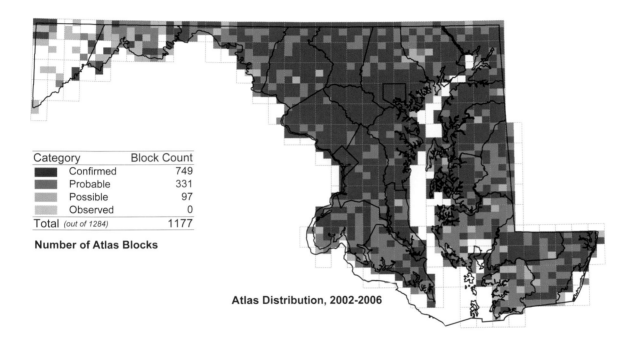

Category	Block Count
Confirmed	749
Probable	331
Possible	97
Observed	0
Total (out of 1284)	1177

Number of Atlas Blocks

Atlas Distribution, 2002-2006

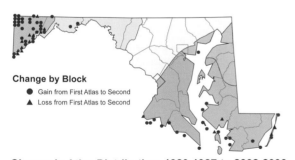

Change by Block
- ● Gain from First Atlas to Second
- ▲ Loss from First Atlas to Second

Change in Atlas Distribution, 1983-1987 to 2002-2006

Percent of Stops
- 50 - 100%
- 10 - 50%
- 0.1 - 10%
- < 0.1%

Relative Abundance from Miniroutes, 2003-2007

Atlas Region	1983-1987	2002-2006	Change No.	%
Allegheny Mountain	22	40	+18	+82%
Ridge and Valley	141	143	+2	+1%
Piedmont	316	316	0	0%
Upper Chesapeake	118	119	+1	+1%
Eastern Shore	296	304	+8	+3%
Western Shore	243	250	+7	+3%
Totals	1136	1172	+36	+3%

Change in Total Blocks between Atlases by Region

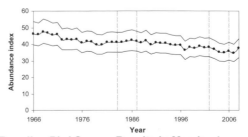

Breeding Bird Survey Results in Maryland

Brown Thrasher
Toxostoma rufum

Although the Brown Thrasher is comparably widespread to the related mockingbird and catbird, its secretive habits and relatively brief peak singing season render it less familiar and detectable. It was located in fewer blocks and had a lower breeding confirmation rate than the other Maryland and DC mimic thrushes. This is a lanky, short-winged, long-tailed bird with long legs and a strong slightly decurved bill used for thrashing in leaf litter and digging in loose soil in search of invertebrates. Thrashers are also partial to fruit, especially in the colder months. These birds are most noticeable in early spring when the males sing persistently from high song perches. The song is varied, generally melodic, and delivered almost entirely in paired coupled phrases, many of which mimic other birds. The calls of this thrasher include a loud smacking *chuck* and a mewing *quirr?*

Brown Thrashers live in dry shrubby places with interspersed grassy openings, often with scattered tall song perches. Some typical haunts in Maryland and DC include hedgerows, regenerating upland fields, pastures, and timber cuts, and thick ornamental plantings among buildings. Winter quarters are in the southeastern United States including Maryland's Coastal Plain where they are uncommon. Nesting thrashers return in late March and remain until October. Nests are close to the ground, sometimes on it, usually placed in shrubs or saplings under 3 m (10 ft). This species normally attempts only a single brood, although they are persistent renesters and a few successful nesters will try a second brood (Cavitt and Haas 2000). Maryland and DC egg dates range from 18 April to 30 July (Hitchner 1996b; MNRF).

Brown Thrashers have been widespread and common in Maryland and DC since at least the late nineteenth century (Kirkwood 1895; Eifrig 1904); they were considered fairly common throughout in the 1950s (Stewart and Robbins 1958). They remained widespread during the 1983–1987 atlas project, being scarce only on the lower Eastern Shore, particularly in the wetlands of southern Dorchester and Somerset counties. This changed little during the second atlas; indeed thrashers were found in 30 more blocks, 28 of them on the lower Western and Eastern Shore.

The Brown Thrasher is declining over most of its range; the long-term BBS trend has been a decline of 1.2% per year since 1966. In Maryland the long-term trend has also been downward, 1.9% per year, but the trend since 1980 has been essentially stable at a statistically nonsignificant 0.1% loss per year. This bird bears careful watching as its thicket habitat is caught between residential and commercial development and forest succession. For now this fox red bird with its baleful yellow eyes is holding its own, but recent range losses in states to the north of Maryland give ample reason for concern (R. Renfrew, pers. comm.; K. Corwin, pers. comm.).

WALTER G. ELLISON

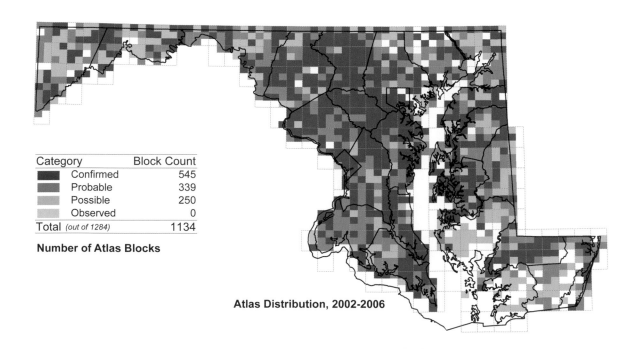

Category	Block Count
Confirmed	545
Probable	339
Possible	250
Observed	0
Total (out of 1284)	1134

Number of Atlas Blocks

Atlas Distribution, 2002-2006

Change by Block
- ● Gain from First Atlas to Second
- ▲ Loss from First Atlas to Second

Change in Atlas Distribution, 1983-1987 to 2002-2006

Percent of Stops
- 50 - 100%
- 10 - 50%
- 0.1 - 10%
- < 0.1%

Relative Abundance from Miniroutes, 2003-2007

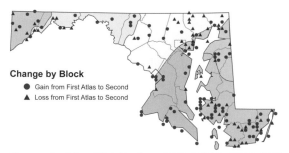

Atlas Region	1983-1987	2002-2006	Change No.	%
Allegheny Mountain	84	86	+2	+2%
Ridge and Valley	140	141	+1	+1%
Piedmont	300	296	-4	-1%
Upper Chesapeake	106	109	+3	+3%
Eastern Shore	240	252	+12	+5%
Western Shore	229	245	+16	+7%
Totals	1099	1129	+30	+3%

Change in Total Blocks between Atlases by Region

Breeding Bird Survey Results in Maryland

Eric Skrzypczak

European Starling
Sturnus vulgaris

The European Starling has become an all-too-familiar immigrant in all parts of populated North America. Starlings are highly gregarious at all seasons, adopting only a semblance of a dispersed population during nesting when adults defend discrete nest sites. This dark, rather chunky, short-tailed bird with pointed wings and ice-pick bill requires open landscapes with at least some bare ground and short grass for feeding, cavities for nests and roosting, and fresh water. When these needs are met, starlings prove both hardy and fecund, occurring most abundantly in cities, suburbs, towns, and farmland; they are scarce in relatively undisturbed native habitats. Starlings have a long nesting season, often attempting two or more broods. Maryland and DC nests with eggs have been recorded from 22 March (MNRF) to 24 July 2004 northwest of Trappe, Talbot County (J. Reese, this project). Troops of dusty brown, harshly murmuring juvenile starlings are commonplace in open places from late May onward.

North American starlings trace their lineage back to two modest-size flocks released in New York City's Central Park in 1890 and 1891. The starling arrived in Maryland as early as 1906 and was nesting widely by 1920. Starlings had reached the Allegheny Plateau in westernmost Maryland by 1928 (Stewart and Robbins 1958; Yingling 1996). There has been no noticeable change in the starling's breeding range in Maryland and DC since the first statewide atlas project. The tiny holes in the starling's distribution continue to be a few heavily wooded blocks in Garrett and Allegany counties, some of the wetland-dominated blocks of southern Dorchester and western Somerset counties, and 2 blocks on Assateague Island with its dunes and shrubby vegetation.

Continental Breeding Bird Survey trends for the starling have shown steady declines for the past four decades. These trends were also evident in Maryland and the decline has accelerated over the past quarter century. As the atlas results reveal, however, these declines have not affected the breeding range of the starling, nor have they affected the ease of confirming nesting. Atlas workers were able to prove breeding by starlings in 91% of the blocks where they were found, the highest rate of confirmation for any bird during this atlas.

WALTER G. ELLISON

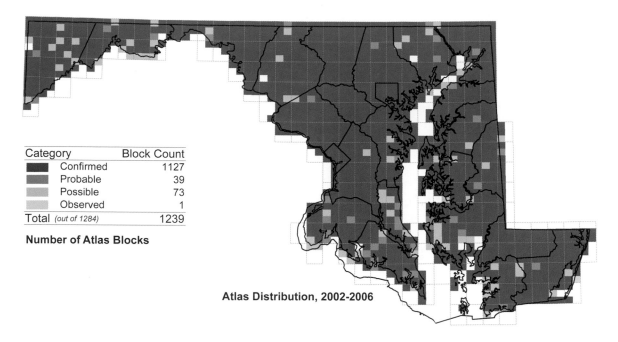

Category	Block Count
Confirmed	1127
Probable	39
Possible	73
Observed	1
Total (out of 1284)	1239

Number of Atlas Blocks

Atlas Distribution, 2002-2006

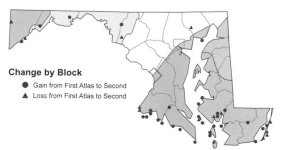

Change by Block

● Gain from First Atlas to Second
▲ Loss from First Atlas to Second

Change in Atlas Distribution, 1983-1987 to 2002-2006

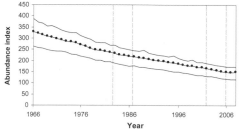

Percent of Stops

■	50 - 100%
■	10 - 50%
■	0.1 - 10%
□	< 0.1%

Relative Abundance from Miniroutes, 2003-2007

Atlas Region	1983-1987	2002-2006	Change No.	%
Allegheny Mountain	91	91	0	0%
Ridge and Valley	144	145	+1	+1%
Piedmont	316	315	-1	0%
Upper Chesapeake	119	120	+1	+1%
Eastern Shore	308	315	+7	+2%
Western Shore	240	248	+8	+3%
Totals	1218	1234	+16	+1%

Change in Total Blocks between Atlases by Region

Breeding Bird Survey Results in Maryland

Cedar Waxwing
Bombycilla cedrorum

The Cedar Waxwing is distinctly crested, sleek, brown, pale yellow, and gray with a bright yellow band at the tail tip. The red waxy growths on the ends of the secondaries of most adults that give birds in this family their names are hard to see in the field. Much of this bird's life history is dictated by its diet, which is dominated by fruit. Fruit occurs in abundance, but only in patches that cannot easily be exploited by a single bird or pair. Waxwings occur in wandering flocks when not nesting, do not defend nesting territories, and lack a song for attracting mates and holding territory. Waxwings eat fruit year-round, switching among the array of plants that fruit throughout the year. They will eat insects in the summer and are often seen flycatching over water. They eat flower petals and nectar in April and May when fruit is less common. The nesting season is late, beginning in early June and running into autumn, and is correlated with the ripening of summer fruit (Witmer 1996).

Most fruiting trees and shrubs grow best in sunlight and Cedar Waxwings gravitate to forest openings and edges. Ornamental and agricultural fruit plantings are also attractive to waxwings. Typical haunts include open woods, floodplains, beaver ponds, orchards, parks, and well-planted residential areas.

The wandering life of following fruit crops means Cedar Waxwings have no fixed address. Their site fidelity is very low (Witmer et al. 1997). They sometimes winter as far south as Panama, but they are often seen year-round in areas with fruit that persists through the winter including Maryland and DC. In the nonbreeding period waxwings occur in flocks, sometimes quite large, and they are still gregarious in the nesting season in smaller flocks. Nest building commences in late May and nesting continues until midautumn. The nest is large and substantial and a bit unkempt looking, with much stringy decorative material. The range of nest heights varies, but most are built below 6.1 m (20 ft) from the ground. Maryland egg dates range from 3 June to 9 September. Nestlings have been found as late as 18 October (M. Clarkson 1996; MNRF). October nests have also been reported in New York State (W. Hamilton 1933). The presence of juveniles in winter flocks (W. Ellison, pers. obs.) probably reflects late nesting and perhaps suspension of post-juvenile molt with the onset of cold weather.

The historical status of the Cedar Waxwing prior to the 1950s is uncertain. Kirkwood (1895) called it a resident with no further comment on numerical status. Eifrig (1904) considered it very abundant in western Maryland, which seems exaggerated relative to later assessments. Stewart and Robbins (1958) considered them common in the nesting season only in the Allegheny Mountains. They were uncommon east to the Piedmont and Upper Chesapeake and rare on the lower Coastal Plain. In 1983–1987 they increased greatly, becoming widespread in the Ridge and Valley, more numerous but not uniformly distributed in the Piedmont, Upper Chesapeake, and Western Shore, and scattered across the Eastern Shore, especially in Talbot County. They were confirmed nesting in all counties (M. Clarkson 1996). In 2002–2006 waxwings were found in 328 more blocks and were more likely to be given probable and confirmed status codes (69% versus 48% in 1983–1987). The greatest increases in blocks were on the Piedmont, the Upper Chesapeake, and the Eastern Shore. Waxwings remained scarce in southeastern Dorchester, western Wicomico, and Somerset counties.

North American BBS trends for the Cedar Waxwing have been modestly positive since 1966 at 0.5% per year. Maryland trends have shown a greater rate of increase (6% per year), with most of this increase since 1980. Increases have probably been aided by extensive planting of ornamental fruit trees in urban and suburban areas. Waxwings have also benefited from farmland reverting to shrubland in some areas, and they may have responded positively in the short term to the introduction and spread of invasive fruiting plants such as tatarian honeysuckle. At present the Cedar Waxwing is increasing and has become almost ubiquitous in Maryland and DC.

WALTER G. ELLISON

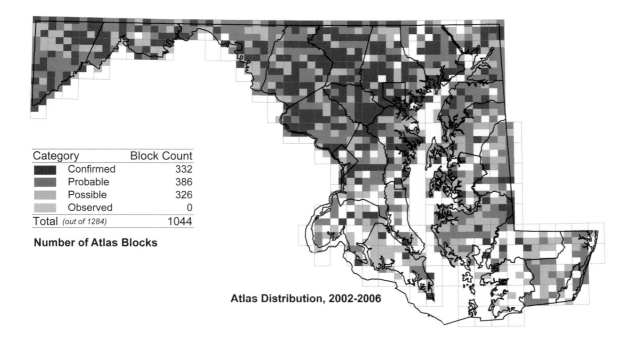

Category	Block Count
Confirmed	332
Probable	386
Possible	326
Observed	0
Total (out of 1284)	1044

Number of Atlas Blocks

Atlas Distribution, 2002-2006

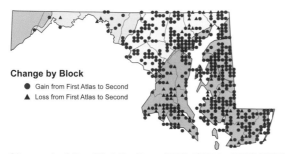

Change by Block
- ● Gain from First Atlas to Second
- ▲ Loss from First Atlas to Second

Change in Atlas Distribution, 1983-1987 to 2002-2006

Percent of Stops
■	50 - 100%
■	10 - 50%
■	0.1 - 10%
□	< 0.1%

Relative Abundance from Miniroutes, 2003-2007

Atlas Region	1983-1987	2002-2006	Change No.	%
Allegheny Mountain	92	93	+1	+1%
Ridge and Valley	134	145	+11	+8%
Piedmont	191	295	+104	+54%
Upper Chesapeake	45	108	+63	+140%
Eastern Shore	104	218	+114	+110%
Western Shore	147	182	+35	+24%
Totals	713	1041	+328	+46%

Change in Total Blocks between Atlases by Region

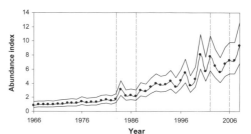

Breeding Bird Survey Results in Maryland

Blue-winged Warbler

Vermivora cyanoptera

The Blue-winged Warbler is inconspicuous despite its bright yellow head and underparts, staying within thick foliage most of the time. It is a bird of brushy places with abundant tall saplings. Males mount to upper branches of the saplings in their haunts to sing their peculiar high-pitched Bronx-cheer song: *eeee-bzzzzzzt* occasionally alternating with a chittering trill-buzz. Blue-winged Warblers will hybridize with the closely related Golden-winged Warbler and they may contribute to the latter's local decline where they co-occur (F. Gill 1980). The precolonial range of this warbler appears to have been west of the Appalachians in the Mississippi-Ohio drainage basin; they have spread northeastward since the early nineteenth century (F. Gill 1980). This small, slender warbler reaches the eastern edge of its range in Maryland, nesting from the fall line westward.

Blue-winged Warbler breeding habitat is generally grassy with abundant tall herbs and shrubs and young trees. The moisture regime ranges from fairly dry to swampy. On average, Blue-winged Warbler territories are more heavily treed with older saplings than is ideal for Golden-winged Warbler, but the former occurs over a broader range of habitat, from old field to open young woodland, than the latter and overlaps extensively (Confer and Knapp 1981).

Blue-winged Warblers winter in Mexico and Central America and return to Maryland and DC during the last third of April. The last autumn migrants generally leave by early September. They nest on the ground amid herb and shrub stems, often next to a small tree (F. Gill et al. 2001). Their nesting season is brief; they are single brooded; and the young leave the nest after a combined incubation and nestling period of only three weeks. Maryland egg dates range from 17 May to 18 June; nestlings have been reported to 12 July (D. Holmes 1996a; MNRF). Eggs have been found as early as 4 May in neighboring West Virginia (F. Gill et al. 2001).

The Blue-winged Warbler was scarce and local in eastern Maryland in the late nineteenth century (Kirkwood 1895) and apparently absent from western Maryland (Eifrig 1904). In the 1950s they were still localized; Stewart and Robbins (1958) found them fairly common on the northeastern Piedmont in Cecil and Harford counties and in the Blue Ridge section of the Ridge and Valley but otherwise rare and sporadic in eastern Maryland. The breeding distribution had expanded greatly by the 1980s with Blue-winged Warblers found in 133 blocks in 15 counties from 1983 to 1987 (D. Holmes 1996a); the major concentration of blocks was on a northeast to southwest axis from Cecil to Montgomery and southeastern Frederick counties. They were in many blocks in central Howard County where much land had been taken out of farming pending creation of the new town of Columbia. They were also on Catoctin and South mountains in the Blue Ridge and scattered westward including 4 blocks in Garrett County. There were also a few Coastal Plain records with two unusual records of territorial males in the Pocomoke drainage.

The range recorded during the 2002–2006 atlas was similar, but Blue-winged Warblers were lost from a net total of 39 blocks. The largest losses were on the Piedmont, primarily just west of the fall line and in southwestern Montgomery County. There were only two marginal reports from the Coastal Plain.

Trends on BBS routes for Blue-winged Warbler have been declining at a rate of 1.8% per year since 1980, although there are regions of decline and stability across its range presumably reflecting the changing status of old field habitats. In Maryland the trend appears to be negative but cannot be considered significantly so because of the small number of

BBS routes with records. This warbler's nesting range continues to expand at its northern limit as shown in New York (Confer 2008) and Vermont (R. Renfrew, pers. comm..). It has declined recently farther south, presumably because of loss of shrubby fields to forest maturation and suburban development. The greatest losses between Maryland atlas projects have been in areas prone to rapid suburbanization. Successional habitats need to be preserved and produced via forest management for this warbler to remain widespread in Maryland.

WALTER G. ELLISON

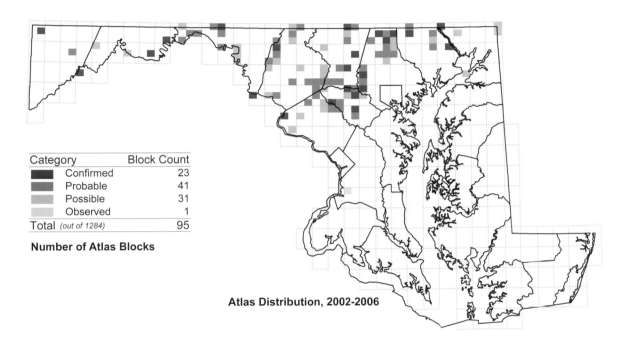

Category	Block Count
Confirmed	23
Probable	41
Possible	31
Observed	1
Total *(out of 1284)*	95

Number of Atlas Blocks

Atlas Distribution, 2002-2006

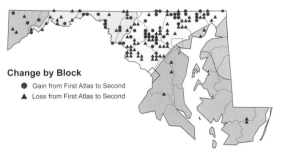

Change by Block
- ● Gain from First Atlas to Second
- ▲ Loss from First Atlas to Second

Change in Atlas Distribution, 1983-1987 to 2002-2006

Percent of Stops
- 50 - 100%
- 10 - 50%
- 0.1 - 10%
- < 0.1%

Relative Abundance from Miniroutes, 2003-2007

Atlas Region	1983-1987	2002-2006	Change No.	%
Allegheny Mountain	3	3	0	0%
Ridge and Valley	31	27	-4	-13%
Piedmont	92	64	-28	-30%
Upper Chesapeake	2	0	-2	-100%
Eastern Shore	2	0	-2	-100%
Western Shore	3	0	-3	-100%
Totals	133	94	-39	-29%

Change in Total Blocks between Atlases by Region

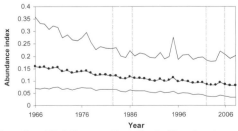

Breeding Bird Survey Results in Maryland

Golden-winged Warbler
Vermivora chrysoptera

With males elegantly attired in pearl gray, black, and white with patches of bright yellow on the crown and coverts, the Golden-winged Warbler is as uncommon as it is attractive. This bird of wet brushy places is recognized by its thin insectlike song composed of four or more flat buzzes. It is a largely upland bird of northeastern North America and the Appalachians at the southeastern edge of its range in western Maryland. This warbler is declining throughout its range and is on Partners in Flight's watch list (Rich et al. 2004).

Golden-winged Warblers nest in successional habitats; in particular they tend to settle in a short-lived stage between shrubland and forest with emergent deciduous saplings. Some typical haunts include alder swamps, overgrown pastures with low shrubs and saplings, power line rights of way, regenerating timber cuts, and shrubby bog margins. The closely related Blue-winged Warbler has overlapping habitat preferences and often mates with Golden-wingeds producing fertile hybrids. It appears that Blue-wingeds have a competitive advantage over Golden-wingeds and their arrival in areas hosting the latter usually coincides with decline and disappearance of their predecessors (F. Gill 1980; Confer and Knapp 1981). However, in a few places these two warblers coexist in slightly different habitats (Confer 1998).

Golden-winged Warblers winter in Central America and northwestern South America (Confer 1992). Migrants arrive in Maryland in early May and usually depart by mid-September. They nest on the ground with the nest entwined into and obscured by thick herbaceous vegetation. They are single brooded with a relatively brief three weeks from the start of incubation to fledging (Confer 1992). Nest dates for Maryland have been few, with eggs reported on 2 June and nestlings to 17 June (Stewart and Robbins 1958, MNRF). Fledglings have been seen from 10 June to 12 July. Egg dates from 33 upstate New York nests ranged from 15 May to 7 June (Confer 1992).

The Golden-winged Warbler has been known as a summer bird in western Maryland since the 1890s (Kirkwood 1895; Preble 1900). Eifrig (1904) considered it a fairly numerous nesting bird, citing breeders in Cumberland and Frostburg. Stewart and Robbins (1958) found them uncommon in Washington County west of the Hagerstown Valley and fairly common in Allegany and Garrett counties. From 1983 to 1987 Golden-winged Warblers were found in the same nesting range (D. Holmes 1996b) although they were not uniformly distributed as befitted a local species with specific habitat preferences. The two decades since 1987 have been unkind to this warbler; it was not refound in 72 blocks from the 1980s, a decline of 65%. It is now absent from Washington County and rare in Allegany County. Although still widespread in Garrett County, it was lost from 54% of previously occupied blocks. These losses have not been accompanied by a concomitant increase by the Blue-winged Warbler.

The Golden-winged Warbler has been declining on North American BBS routes since 1966 at 2.8% per year. They have been found on too few routes in Maryland for a reliable trend but they have been declining on these routes since 1980. It appeared the decline of this warbler was correlated with increases and range expansion by Blue-winged Warbler in the 1980s (F. Gill 1992), but the Blue-winged Warbler has actually declined in Maryland since the 1983–1987 Maryland-DC atlas. This tends to implicate some other causative agent, probably habitat loss, although cowbird brood parasitism may also be a factor (Confer 1992). Shrubland, which is prone to development and reversion to young forest, requires an ongoing regime of forest harvest or active maintenance of shrubby fields, a labor intensive management activity. Golden-winged Warbler needs to be actively monitored and its habitat preserved and managed for it to remain among Maryland's nesting birds.

WALTER G. ELLISON

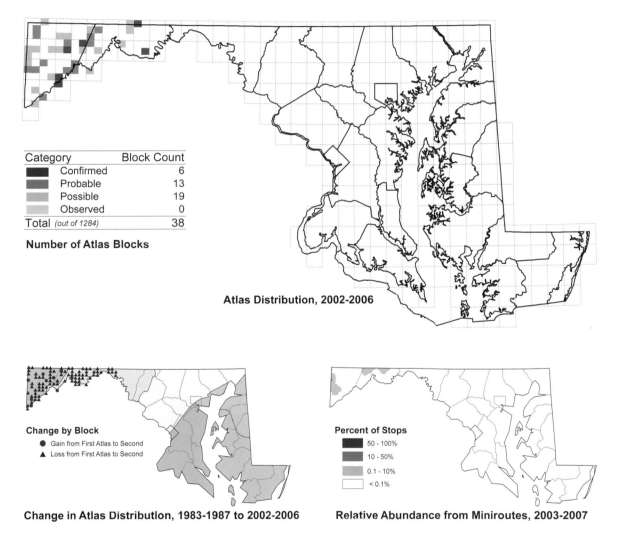

Category	Block Count
Confirmed	6
Probable	13
Possible	19
Observed	0
Total *(out of 1284)*	38

Number of Atlas Blocks

Atlas Distribution, 2002-2006

Change by Block
- ● Gain from First Atlas to Second
- ▲ Loss from First Atlas to Second

Change in Atlas Distribution, 1983-1987 to 2002-2006

Percent of Stops
- 50 - 100%
- 10 - 50%
- 0.1 - 10%
- < 0.1%

Relative Abundance from Miniroutes, 2003-2007

Atlas Region	1983-1987	2002-2006	Change No.	%
Allegheny Mountain	65	30	-35	-54%
Ridge and Valley	45	8	-37	-82%
Piedmont	0	0	0	-
Upper Chesapeake	0	0	0	-
Eastern Shore	0	0	0	-
Western Shore	0	0	0	-
Totals	110	38	-72	-65%

Change in Total Blocks between Atlases by Region

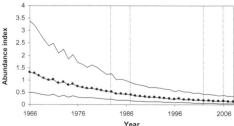

Breeding Bird Survey Results in Maryland

Nashville Warbler
Oreothlypis ruficapilla

The Nashville is a tiny, active wood warbler arrayed with bright yellow underparts, an olive back, and a gray half-hood featuring a prominent white eye-ring. Its bright accelerating trill calls attention to it in its haunts. These consist of combinations of low herbs and grassy plants adjacent to dense stands of saplings and shrubs, generally a mix of conifer and deciduous. It is at the southeastern edge of its broad boreal breeding distribution in the highest mountains of Maryland and of neighboring West Virginia. This species is now rare and declining in this region.

Nashville Warblers inhabit a wide array of regenerating short- and long-term shrubby habitats over their broad range. These include regenerating pastures; beaver ponds and meadows; the drier parts of shrub swamps; boggy meadows with numerous shrubs; burns; cutover boreal forest; and sandy scrub oak–pine barrens (J. Williams 1996). In Maryland and West Virginia most Nashville Warbler territories have been found in upland wet sphagnum meadows with scattered shrubs, often in the region's boggy "glades" (Hall 1983; D. Boone 1996b).

Nashville Warblers migrate to Mexico and northern Central America for the winter. They arrive in Maryland in late April and early May and are generally last seen in mid-

October. They are fairly common migrants in western Maryland becoming uncommon to the east. Only a few pairs have established nesting territories in Maryland and these were all west of the Allegheny Front. They nest on the ground in grassy and herbaceous plants. The nest is hard to find, and Nashville Warbler, as with most other Wood Warblers, is most readily confirmed by watching for parents carrying food to young or by finding begging fledglings. Maryland has few nesting records; two nests with eggs were found on dates ranging from 30 May to 15 June and a third nest with young was found on 16 June (Robbins and Stewart 1951; D. Boone 1981; MNRF). The range of egg dates for Ontario is from 24 May to 21 July (Peck and James 1987) suggesting later dates for renests could also be recorded here.

The Nashville Warbler was not known as a summer bird in Maryland's western mountains until 12 June 1949 when a pair was observed scolding and carrying food in Cranesville Swamp, Garrett County (Robbins 1949c). Stewart and Robbins (1958) called it uncommon and local in Garrett County and cited five locations with nesting season sightings. D. Boone (1981) found a nest near Sang Run and located several more singing males in Garrett County boglands in subsequent years (D. Boone 1996b). Nashville Warblers were recorded in a surprising 19 blocks during the 1983–1987 atlas, including reports from 6 blocks east of the Allegheny Front with territorial males in Green Ridge State Forest, Allegany County (2 blocks), and on the Blue Ridge at South Mountain north of Wolfsville, Frederick County (D. Boone

1996b). Almost half (47%) of the 19 reports were fleeting possible breeders that may have been unmated males or late migrants. They were found in only 2 blocks at traditional northeastern Garrett County sites in 2002–2006: territorial males at Finzel and Wolf swamps.

North American BBS trends over the past four decades suggest stability or slight annual increases, but U.S. route trends have been significantly negative. Although Maryland lacks sufficient data for Nashville Warbler, the species has declined at a steep 6.3% per year on 34 routes in the Allegheny Plateau region. Regional declines may reflect maturation of shrublands into forest on the Allegheny Plateau or degrada-

tion and loss of high-elevation shrub bogs. Although these warblers have occurred as far south as Cranberry Glades in Pocahontas County (Hall 1983), records for the West Virginia atlas were only from counties neighboring Garrett County (Buckelew and Hall 1994). With so few birds and nesting sites, Maryland could lose its nesting Nashville Warblers in less than a decade. The Maryland Department of Natural Resources has listed the Nashville Warbler as in need of conservation. It might be prudent to consider raising its status to threatened or endangered.

WALTER G. ELLISON

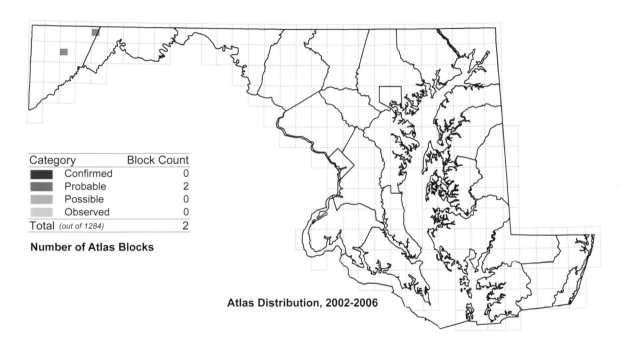

Category	Block Count
Confirmed	0
Probable	2
Possible	0
Observed	0
Total (out of 1284)	2

Number of Atlas Blocks

Atlas Distribution, 2002-2006

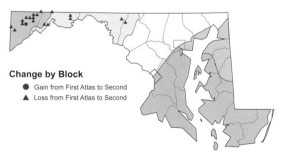

Change by Block
- ● Gain from First Atlas to Second
- ▲ Loss from First Atlas to Second

Change in Atlas Distribution, 1983-1987 to 2002-2006

	Atlas Region	1983-1987	2002-2006	Change No.	%
	Allegheny Mountain	13	2	-11	-85%
	Ridge and Valley	6	0	-6	-100%
	Piedmont	0	0	0	-
	Upper Chesapeake	0	0	0	-
	Eastern Shore	0	0	0	-
	Western Shore	0	0	0	-
Totals		19	2	-17	-89%

Change in Total Blocks between Atlases by Region

Northern Parula
Parula americana

The Northern Parula would be consistently overlooked if not for its distinctive rapidly rising buzzing song and its restless feeding behavior in branch tip leaf clusters. This is a small, short-tailed warbler, not much larger than a kinglet, with a yellow throat and breast, grayish blue upperparts, and a chartreuse patch on its upper back. It is a bird of moist mature forest ranging from deciduous to coniferous, often not far from water. These warblers are widespread in Maryland and DC, with a nesting range covering more than half of the state's atlas blocks, although they are scarce or lacking from Eastern Shore pinelands and from the most extensively farmed parts of the state.

The Northern Parula's varied forest breeding habitat is defined primarily by the availability of sites for its unusual pouchlike nest that is woven into thick hanging vegetation, often epiphytic plants or lichens. It prefers mature trees and fairly large woodlots, although these are sometimes linear corridors along streams. Wherever they nest in Maryland or DC, parulas require trees with pendulous branches and thick leaves on the ends of branches, places where flood debris collects on branch tips, or curtains of vines. To the north of Maryland they build their nests in luxuriant hanging clusters of lichens, largely in the genus *Usnea,* and in the Deep South they usually nest in the bromeliad Spanish moss, however they will use many other sites of similar structure where beard lichen and Spanish moss are unavailable (Petrides 1942; Bent 1953).

Northern Parulas are migratory, wintering in south Florida, the West Indies, southern Mexico, and northern Central America. They arrive in Maryland and DC in mid-April and are nest building by late April or early May. The fall migration is protracted, from late August to mid-October. The unusual nature of the nest site leads to highly variable nest height; 22 Maryland and DC nests ranged from 0.3 to 15 m (1–50 ft) with the mean in the middle of the range (Klimkiewicz 1996b; MNRF). The nests are very hard to find except when they are under construction (20 confirmations in this atlas); two-thirds of confirmations were of adults feeding young or fledglings. Maryland and DC egg dates range from 7 April to 14 June (Stewart and Robbins 1958; MNRF).

The Northern Parula has long been known as a summer bird in Maryland and DC, and for just as long its distribution has been somewhat localized. Coues and Prentiss (1883) suspected a few nested near DC; Kirkwood (1895) said that although they were most common as migrants, they nested in numbers. Kirkwood found them numerous near Berlin, Worcester County, and confirmed breeding there. Eifrig (1904) described them as rare in western Maryland. Stewart and Robbins (1958) recorded them statewide in variable numbers. They were rare in the Upper Chesapeake; uncommon in the Ridge and Valley, Allegheny Mountains, and most of the Eastern Shore; fairly common on the Piedmont; and common on the Western Shore and in the Pocomoke drainage on the lower Eastern Shore.

This warbler's status was similar during the 1983–1987 atlas project (Klimkiewicz 1996b) although it had apparently increased somewhat in the Ridge and Valley and Allegheny Mountains. It was scarcest in most of the state's heavily agricultural areas including the northern and western Piedmont, in lower Eastern Shore pine woods and wetlands, and in urbanized DC and Baltimore. Northern Parulas had increased by 26%, a net total of 134 blocks, during the 2002–2006 atlas. Most of this increase was in the northern and western Piedmont and in the Allegheny Mountains. Other local increases were in the DC area and on Elk Neck. On the other hand, there was a net loss of 8 blocks in the Pocomoke basin.

The Northern Parula has been stable or slowly increasing on North American BBS routes. Maryland BBS routes show a nonsignificant increasing trend. Recent range consolidations have been found in several northeastern states, for instance the second New York State atlas showed a 51% distributional increase (McGowan 2008h). Northeastern North American forests have increased in area and matured over the past two decades, although less so in Maryland than to the north. This warbler is capable of nesting in areas with little understory, so wooded suburbs are potential habitat and deer-browsing is less troublesome for them than for other migratory forest birds. At present the Northern Parula is doing better than several other forest interior birds, but its habitat remains vulnerable to development.

WALTER G. ELLISON

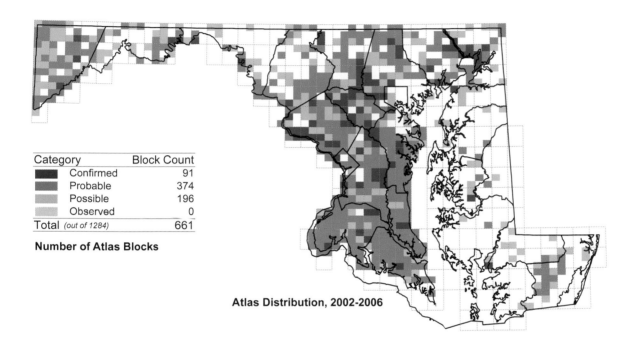

Category	Block Count
Confirmed	91
Probable	374
Possible	196
Observed	0
Total (out of 1284)	661

Number of Atlas Blocks

Atlas Distribution, 2002-2006

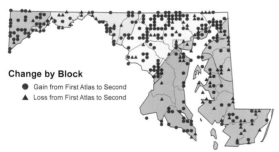

Change by Block
- ● Gain from First Atlas to Second
- ▲ Loss from First Atlas to Second

Change in Atlas Distribution, 1983-1987 to 2002-2006

Percent of Stops
- 50 - 100%
- 10 - 50%
- 0.1 - 10%
- < 0.1%

Relative Abundance from Miniroutes, 2003-2007

Atlas Region	1983-1987	2002-2006	Change No.	%
Allegheny Mountain	44	72	+28	+64%
Ridge and Valley	51	63	+12	+24%
Piedmont	160	226	+66	+41%
Upper Chesapeake	18	23	+5	+28%
Eastern Shore	52	48	-4	-8%
Western Shore	196	223	+27	+14%
Totals	521	655	+134	+26%

Change in Total Blocks between Atlases by Region

Breeding Bird Survey Results in Maryland

Charles Lentz

Yellow Warbler
Dendroica petechia

This brightly colored little bird is one of the most wide-spread of the warblers, breeding from west Mexico to Alaska, and from Newfoundland to Georgia. Often called summer yellow-bird or wild canary, its common occurrence along brushy edges, its bright yellow color, and the habit of singing from a visible perch have made it familiar to generations of country dwellers. Male and female are both yellow, brighter below and on the head and more greenish on the wings and back. The adult male in breeding condition is bright lemon yellow below and on the head, with obvious chestnut red streaking on the breast (the "petechia" of the Latin name); the fall plumage is slightly duller. The female is similar, but the yellow is paler, the greenish coloration extends to the sides and top of the head, and the streaking is faint or absent. The fairly short and simple song is variable. Usually it begins with three to five upslurred whistled *swee* notes, followed by a staccato warble. A common transcription is *sweet, sweet, sweet, I'm so sweet*.

Open scrubby fields and edges, brushy roadsides, and damp willow and alder thickets, frequently near water, are the preferred habitat of this species. It feeds mostly by glean-ing small insects and spiders at low to middle levels. Occa-sionally it will hover or sally after a flying insect. Caterpil-lars (including tent and gypsy moth caterpillars) are favored menu items; it will also eat a few berries (Bent 1953). A fork or crotch in a bush or small tree is the usual site of the nest; commonly it is 1 to 3 m (3–10 ft) above the ground but may be considerably higher. The nest is a small cup constructed from grass, shredded bark, and plant fibers, and lined with fine grass, hair, and plant fuzz. The Yellow Warbler is a pre-ferred host of the Brown-headed Cowbird. It commonly deals with the nest parasite by building a second floor over the intruding egg, leaving it abandoned. The cowbird may be persistent, in which case the warbler may repeat the pro-cedure until the nest is as many as six stories high—with a cowbird egg in every level (Bent 1953).

As one would expect from the extent of the breeding range, the Yellow Warbler is found across Maryland wher-ever there is suitable habitat. It is comparatively scarce in the Eastern and Western Shore regions of the state. The first migrants may trickle into the state during the first week in April, but most arrive during the last third of the month. The postbreeding exodus is among the earliest of all our migrants, beginning in July. By the end of August most of the birds have left although northern migrants may be found until the end of October. The species winters from eastern Mexico to the West Indies, Central America, and northern South America.

There have been significant changes in the distribution

of the Yellow Warbler since the previous atlas (Vaughn 1996b). Over most of the state, its abundance has decreased, as the blocks with breeding birds (possible, probable, and confirmed) have dropped from 765 to 687, a loss of 10%. However, that loss has not been uniform. The Coastal Plain regions have borne the brunt of the decrease: Upper Chesapeake, 67 blocks to 55 blocks (18%); Western Shore, 109 blocks to 89 blocks (18%); and Eastern Shore, 133 blocks to 78 blocks (41%). The western part of the state has suffered much smaller decreases (Allegheny Mountain region, 3%;

Ridge and Valley region, 8%), while the Piedmont has actually seen an 8% increase (251 blocks to 271 blocks). Breeding Bird Survey data do not show this decrease: across North America, the trend is neutral, and both Maryland and U.S. BBS data show a slight increase. However there was a strong decline on the Upper Coastal Plain from Maryland to North Carolina; the majority of block losses were on the Maryland Coastal Plain.

T. DENNIS COSKREN

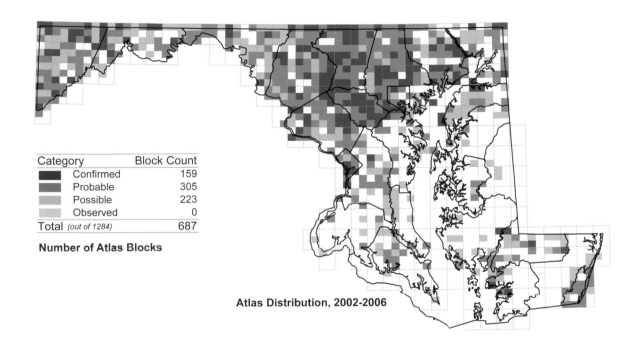

Category	Block Count
Confirmed	159
Probable	305
Possible	223
Observed	0
Total (out of 1284)	687

Number of Atlas Blocks

Atlas Distribution, 2002-2006

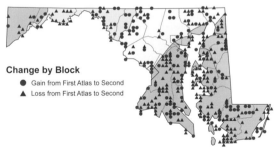

Change by Block
- ● Gain from First Atlas to Second
- ▲ Loss from First Atlas to Second

Change in Atlas Distribution, 1983-1987 to 2002-2006

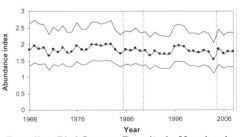

Percent of Stops
- 50 - 100%
- 10 - 50%
- 0.1 - 10%
- < 0.1%

Relative Abundance from Miniroutes, 2003-2007

Atlas Region	1983-1987	2002-2006	Change No.	%
Allegheny Mountain	87	84	-3	-3%
Ridge and Valley	118	109	-9	-8%
Piedmont	251	271	+20	+8%
Upper Chesapeake	67	55	-12	-18%
Eastern Shore	133	78	-55	-41%
Western Shore	109	89	-20	-18%
Totals	765	686	-79	-10%

Change in Total Blocks between Atlases by Region

Breeding Bird Survey Results in Maryland

Chestnut-sided Warbler
Dendroica pensylvanica

This striking wood warbler with its yellow crown, black and yellow streaked upperparts, rakish black streak through the eye, white underparts, and emblematic dark chestnut flanks is characteristic of forest edge and regrowing deciduous shrublands and stands of saplings of the northern hardwood forests of northeastern North America. This warbler's bright and snappy *pleased-pleased-pleased-ta-meetcha* song is a common feature of brushy pastures and logged forests in Garrett County. Small numbers nest eastward into the Allegany County hill country, and a few still breed at upper elevations on Catoctin and South Mountains.

These warblers feed and nest at low levels in shrubs and saplings, usually placing the nest under 3 m (10 ft). Maryland nesting records are surprisingly few for this bird of semiopen habitats. Egg dates range from 28 May to 26 June (MNRF). There were 13 reports of recently fledged juvenile Chestnut-sided Warblers during this atlas from 22 June to 7 August.

The historic distribution of the Chestnut-sided Warbler was larger in Maryland as this species regularly nested in northern Carroll County and northwestern Baltimore County (in the Gunpowder River drainage), with a few records farther south on the Piedmont (Stewart and Robbins

1958). This warbler ceased nesting in these areas prior to the 1983–1987 atlas (Robbins 1996c) and was still missing there in the current atlas. Possible reasons for this disappearance include habitat loss from increasingly intensive agricultural or suburban development. Comparisons between the two Maryland-DC atlases show that the breeding range of the Chestnut-sided Warbler is nearly stable, although it has been lost from 5 blocks, matching the net loss in the Catoctin and South Mountain Range. It seems the species continues to decline in the eastern part of its range in Maryland. Interestingly, atlas field-workers from 2002–2006 confirmed breeding for this species at twice the rate reported in the first atlas. Whether this increase was due to more intensive fieldwork or increased numbers of Chestnut-sided Warblers is unknown.

Recent BBS results for Chestnut-sided Warbler have shown declines, although the declines are not as strong as for forest interior nesting migrants to the neotropics (less than 1% per annum). Regional BBS trends in the Ridge and Valley and Allegheny Plateau provinces have been stable. One of the few states with an increasing BBS trend is neighboring Pennsylvania.

Given that the Chestnut-sided Warbler seldom occupies, or thrives in, urban or suburban edge and brush (Richardson and Brauning 1995), the encroachment of residential development in western Maryland could reduce its numbers in the foreseeable future. Currently, this handsome warbler of the hill country appears to be holding its own in its stronghold in the western hills and mountains of Maryland.

WALTER G. ELLISON

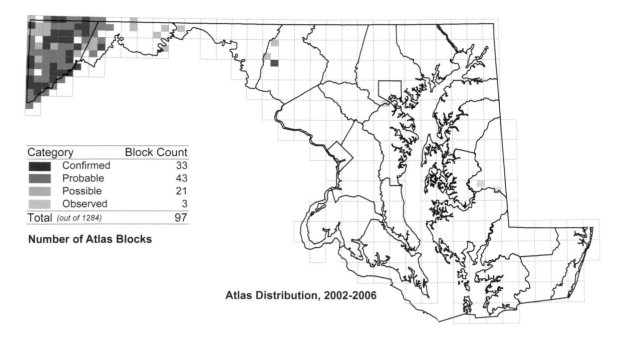

Category	Block Count
Confirmed	33
Probable	43
Possible	21
Observed	3
Total (out of 1284)	97

Number of Atlas Blocks

Atlas Distribution, 2002-2006

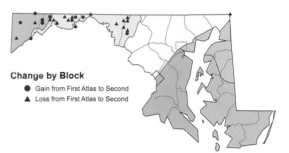

Change by Block
- ● Gain from First Atlas to Second
- ▲ Loss from First Atlas to Second

Change in Atlas Distribution, 1983-1987 to 2002-2006

Percent of Stops
- 50 - 100%
- 10 - 50%
- 0.1 - 10%
- < 0.1%

Relative Abundance from Miniroutes, 2003-2007

Atlas Region	1983-1987	2002-2006	Change No.	Change %
Allegheny Mountain	83	86	+3	+4%
Ridge and Valley	18	11	-7	-39%
Piedmont	1	0	-1	-100%
Upper Chesapeake	0	0	0	-
Eastern Shore	0	0	0	-
Western Shore	0	0	0	-
Totals	102	97	-5	-5%

Change in Total Blocks between Atlases by Region

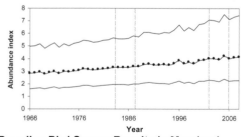

Breeding Bird Survey Results in Maryland

Magnolia Warbler
Dendroica magnolia

The American Wood Warblers (Parulidae) have been called "butterflies of the bird world" (R. Peterson 1934) and the Magnolia Warbler has some of the most complex butterfly-like plumage patterns among the many American warbler species. Magnolia Warblers have bright yellow breasts streaked with black, flashy, large white tail spots and wing bars (fused in males), and yellow rumps. The song sounds like an abbreviated Chestnut-sided Warbler, a bright and sweetly whistled *weeta-weeta-wetoo*.

Magnolia Warblers are birds of shrubs and small trees, only uncommonly rising above midstory to canopy height to feed and sing. Despite their name, these warblers are also birds of coniferous evergreen vegetation. They are particularly fond of short-needled small to medium-size conifers including hemlocks, fir, and spruce, particularly red spruce in Maryland. Lesser numbers of these colorful warblers also occupy young stands of pine including white and red pines. Magnolia Warblers are most common where young conifers are released by removal of the forest canopy, on forest edge, Christmas tree farms, and in logging openings and pastures regenerating to forest. This warbler is a high-elevation bird in Maryland, occurring only in the Allegheny Mountains at elevations 635 m (2,100 ft) and above, with most found above 760 m (2,500 ft; Stewart and Robbins 1958).

Magnolia Warblers are migratory, wintering from southern Mexico south to Costa Rica, rarely Panama (Hall 1994) and residing in Maryland from late April until late October. Magnolia Warbler nests are usually saddled on the branches of small evergreens about halfway to the trunk from the tips, generally at heights of less than 4.5 m (14.7 ft); mean nest height for 14 Maryland nests was 3.8 m (12.4 ft) (Robbins 1996d; MNRF). Most bird observers do not enjoy plowing through thickets of spruce and hemlock, so nest confirmations for Magnolia Warbler are rare. Nonetheless, the 20% breeding confirmation rate for this project doubled the rate for 1983–1987. Most Magnolia Warbler confirmations for 2002–2006 were of adults carrying food (7 records) and fledglings (4 records).

Magnolia Warbler was recorded in essentially the same nesting range delineated in Stewart and Robbins (1958) and in 1983–1987 (Robbins 1996d). Atlas workers from 2002 to 2006 located Magnolia Warblers in a net total of 6 more blocks, all of which had some land above 760 m (2,500 ft) elevation. Gaps in Garrett County are in the more heavily farmed and relatively low-elevation parts of the county, especially the northwestern part of the county and the area immediately surrounding Oakland. Observed records from Wicomico and Anne Arundel counties were of late spring migrants seen into atlas safe dates. An observed record from Dan's Rock, Allegany County, at an elevation above 760 m (2,500 ft) is more intriguing.

Although Maryland BBS trends are unreliable, as they arise from just seven survey routes and have shown great fluctuations in birds counted, the trend appears to be positive. North American BBS trends for Magnolia Warbler have been positive to stable. As birds of successional habitats in both the breeding and wintering ranges, Magnolia Warblers have suffered less from habitat loss than other neotropical migratory birds and remain among the most numerous of boreal wood warblers. Magnolia Warblers may still decline in Maryland if the landscape of Garrett County becomes suburbanized, with less hemlock and spruce at upper elevations.

WALTER G. ELLISON

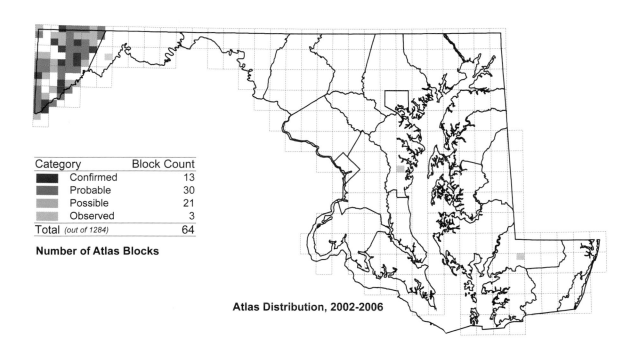

Category	Block Count
Confirmed	13
Probable	30
Possible	21
Observed	3
Total *(out of 1284)*	64

Number of Atlas Blocks

Atlas Distribution, 2002-2006

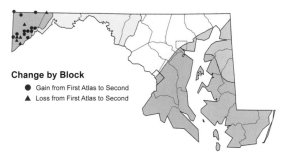

Change by Block
- ● Gain from First Atlas to Second
- ▲ Loss from First Atlas to Second

Change in Atlas Distribution, 1983-1987 to 2002-2006

Percent of Stops
- 50 - 100%
- 10 - 50%
- 0.1 - 10%
- < 0.1%

Relative Abundance from Miniroutes, 2003-2007

Atlas Region	1983-1987	2002-2006	Change No.	%
Allegheny Mountain	58	63	+5	+9%
Ridge and Valley	0	1	+1	-
Piedmont	0	0	0	-
Upper Chesapeake	0	0	0	-
Eastern Shore	0	0	0	-
Western Shore	0	0	0	-
Totals	58	64	+6	+10%

Change in Total Blocks between Atlases by Region

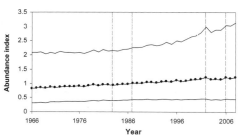

Breeding Bird Survey Results in Maryland

Mark Hoffman

Black-throated Blue Warbler

Dendroica caerulescens

The rising musical wheezy song of the Black-throated Blue Warbler is a characteristic sound of upper elevation northern hardwoods with a thick shrub layer in northeastern North America and in the Appalachian Mountains. Males and females are very different in general appearance. The male is elegantly attired in dark blue, black, and white and the female olive with a broken eye ring and dark cheeks; both share a patch of white like a vest pocket handkerchief at the base of the primaries. As a breeding bird in Maryland, this is a warbler of shrubs and the lower branches of northern hardwoods and mixed forests on mountain ridges and slopes of the Allegheny Mountains.

Black-throated Blue Warblers are most numerous in extensive, moist mature forest with abundant shrubs. In most of the range almost invariant components of the habitat are beech and sugar maple, usually in combination with short-needled conifers such as spruce and hemlock (R. Holmes 1994). The species of shrubs dominating the understory vary across the bird's range, but mountain laurel and great laurel (also known as rhododendron) are particularly frequent in Maryland (Robbins 1996e). These warblers also occur where the forest structure has changed in favor of a thick shrub layer. Such alterations include open, selectively cut woodlands with a flush of sapling growth, places where the can-opy has been thinned by caterpillar outbreaks, the shrubby open canopy woodlands around bogs and beaver ponds, and wind- and ice-damaged forests on ridgelines. This species is limited to extensive woodlands and is not found in relatively open landscapes or small woodlots (Robbins, Dawson, and Dowell 1989).

The Black-throated Blue Warbler winters in the Caribbean basin, with the greatest number in the Greater Antilles (R. Holmes 1994). They return to Maryland in late April and early May. The last seen in fall are usually noted in late October. These warblers build their nests in shrubs and do most of their feeding in shrubs as well (R. Holmes 1994). The nest, a fairly substantial cup of bark strips, bits of wood, and rootlets bound together with spider web, is seldom built higher than 1.2 m (4 ft). The nest is usually placed in almost impenetrable thickets and the parents are very secretive; so it was not surprising that no active nests were found during this atlas project. The two published Maryland egg dates were both recorded on 3 June 1925 (Kirkwood in Stewart and Robbins 1958); nests with young have been found from 11 June to 22 July (Robbins 1996e; MNRF). In New Hampshire egg dates range from mid-May to mid-August with frequent second broods (R. Holmes et al. 2005).

This warbler was first confirmed as breeding in Maryland when Preble (1900) found a used nest in Garrett County. Eifrig (1904) called it very abundant in summer on high ground in Garrett County. Stewart and Robbins (1958) found it common above 610 m (2,000 ft) in the Allegheny Mountains. The status of Black-throated Blue Warbler remained similar in both 1983–1987 (Robbins 1996e) and 2002–2006. In the second atlas period it was found in a net total of 8 more blocks in the Allegheny Mountains, occupying more blocks in the relatively lower elevations of northwestern Garrett County and the Potomac Valley. A bird seen during safe dates in Green Ridge State Forest in eastern Allegany County was intriguing but apparently not on territory. Eight other safe date reports from eastern Maryland were apparently of late migrants or birds summering out of range.

Black-throated Blue Warblers are recorded in only small numbers on BBS routes because much of their habitat is far from roadsides. BBS trends suggest this warbler is increasing slowly or has relatively stable numbers; they have shown a significant increase at 1.6% per year since 1980, including an increase of 2.9% per year on 58 Allegheny Plateau routes. They increased their atlas distribution by 10% between projects in New York State despite a declining BBS trend there (Collins 2008). At present it appears that the maturation and regeneration of forest in the Northeast favors them, but this habitat is still subject to development. Habitat loss in the limited suitable forest habitat in their West Indian winter range is also of concern, so their status requires continued scrutiny.

WALTER G. ELLISON

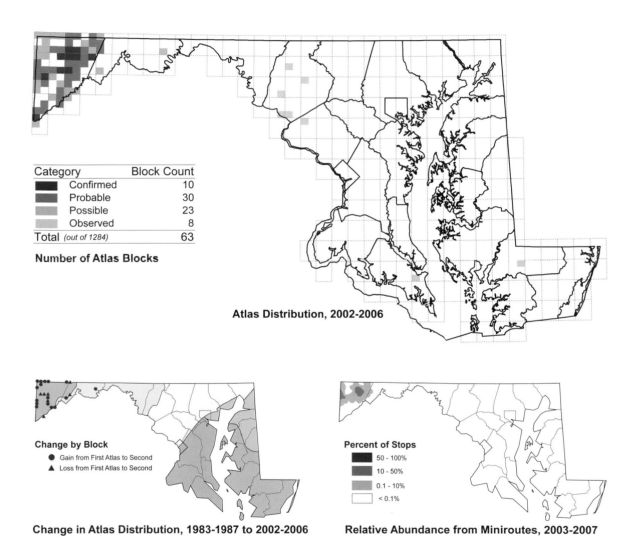

Category	Block Count
Confirmed	10
Probable	30
Possible	23
Observed	8
Total (out of 1284)	63

Number of Atlas Blocks

Atlas Distribution, 2002-2006

Change by Block
- ● Gain from First Atlas to Second
- ▲ Loss from First Atlas to Second

Change in Atlas Distribution, 1983-1987 to 2002-2006

Percent of Stops
- 50 - 100%
- 10 - 50%
- 0.1 - 10%
- < 0.1%

Relative Abundance from Miniroutes, 2003-2007

Atlas Region	1983-1987	2002-2006	Change No.	Change %
Allegheny Mountain	54	62	+8	+15%
Ridge and Valley	0	1	+1	-
Piedmont	0	0	0	-
Upper Chesapeake	0	0	0	-
Eastern Shore	0	0	0	-
Western Shore	0	0	0	-
Totals	54	63	+9	+17%

Change in Total Blocks between Atlases by Region

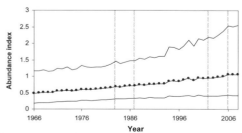

Breeding Bird Survey Results in Maryland

Eric Skrzypczak

Yellow-rumped Warbler
Dendroica coronata

Although hardly the dullest wood warbler, the Yellow-rumped Warbler, with its tasteful blue gray, black, and white plumage splashed with yellow, is a rather subdued member of the particolored warbler genus *Dendroica*. The Yellow-rumped Warbler is divided into two clear-cut subspecies groups. The western Audubon's (*auduboni*) Warbler is only a very rare fall and winter visitor to Maryland. The local nesting population is representative of the nominate *coronata*, long known as the Myrtle Warbler because of its predilection for the waxy fruits of the myrtle. The Myrtle Warbler is also distinctive in having the majority of its population winter in the temperate zone, largely in the southeastern United States but extending northward along the Atlantic coast to southern New England. This warbler winters in good numbers on Maryland's Coastal Plain, especially along tidewater.

For all practical purposes the Yellow-rumped Warbler is a recent addition to the regular nesting birdlife of Maryland. There was a single accidental record of nesting in 1879 at Havre de Grace, Harford County (Kumlien 1880), still a most unlikely location for this species to nest. After a 104-year hiatus the Yellow-rumped Warbler was found during the summer months on the high Allegheny Plateau at Rock Lodge, Garrett County, where Dan Boone found a male on territory and carrying food in June 1983 (Blom 1996m). Although the species' history dictated a cautious interpretation of the behavior, the bird was almost certainly feeding young warblers. There was one other summer report (within established safe dates) from Garrett County during the 1983–1987 atlas (Blom 1996m).

In the decades between the 1983–1987 Maryland-DC atlas and the current project, the Yellow-rumped Warbler was reported only five times during summer in Garrett County, including two instances of confirmed breeding. Maryland's first nest in 118 years was found at New Germany State Park on 14 June 1996 by Czaplak and Todd (Southworth 1997). On 23 July 1999, Fran Pope saw a male carrying food for young on Maple Glade Road, Garrett State Forest (Southworth and Ringler 2000). In Maryland, Myrtle Warblers tend to prefer relatively tall white pines, Norway spruces, and hemlocks; they have also occurred in red spruce associated with bogs. This warbler prefers fairly open coniferous and mixed woodlands to those with dense closed canopies.

The 2002–2006 Maryland-DC atlas fieldwork demonstrated that the Yellow-rumped Warbler is more widespread and better established than the handful of reports from the 1990s implied. Only one atlas report was of confirmed breeding. A female was observed by Joel Martin feeding young in Herrington Manor State Park on 18 June 2005. There were also eight reports of possible nesting and six reports of probable breeders, a substantial jump from two reports in 1983–1987.

Recent trends in neighboring states and in the BBS indicate a substantial increase by the Yellow-rumped Warbler on the Allegheny Plateau over the past 25 years (Buckelew and Hall 1994; D. Gross 1992a;). The current long-term trends for Yellow-rumped Warbler from the BBS are stable continentwide and increasing in the United States. Current trends suggest that, until global climate change affects the abundance of northern conifers in Maryland, the Yellow-rumped Warbler's numbers should remain stable or continue to increase in the Old Line State's western mountains.

WALTER G. ELLISON

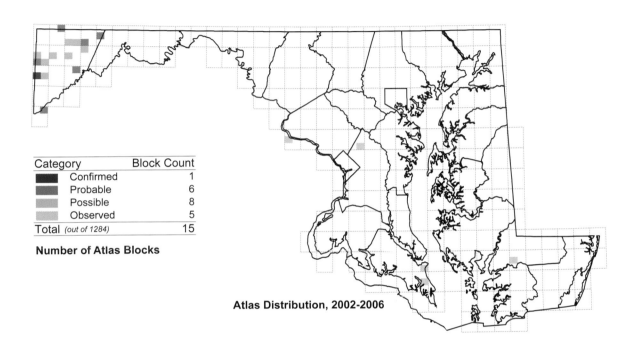

Category	Block Count
Confirmed	1
Probable	6
Possible	8
Observed	5
Total (out of 1284)	15

Number of Atlas Blocks

Atlas Distribution, 2002-2006

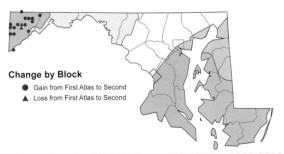

Change by Block
● Gain from First Atlas to Second
▲ Loss from First Atlas to Second

Change in Atlas Distribution, 1983-1987 to 2002-2006

Atlas Region	1983-1987	2002-2006	Change No.	%
Allegheny Mountain	2	15	+13	+650%
Ridge and Valley	0	0	0	-
Piedmont	0	0	0	-
Upper Chesapeake	0	0	0	-
Eastern Shore	0	0	0	-
Western Shore	0	0	0	-
Totals	2	15	+13	+650%

Change in Total Blocks between Atlases by Region

Black-throated Green Warbler
Dendroica virens

The Black-throated Green Warbler is another of those characteristically northern species that breeds mostly in the higher country of western Maryland. Its name is a fair description of the bird: the back to the top of the head is dull olive green, and (for the adult male, at least) the throat and uppermost breast are black. The sides of the head are yellow, with an olive green cheek patch. As is true for many of the *Dendroica* warblers, the wings are dark with two well-marked white wing bars. The underparts are mostly white (including the throat of females and young birds). The sides of the breast are black, and heavy black streaking extends down the flanks of adult birds. On first-winter birds, the streaking appears more washed out. The song is rather buzzy and usually distinctive, with two common patterns: *zhrrr, zheee, zhoo, zhoo, zheee,* or *zhrrr, zhrrr, zhrrr, zhoo, zheee.* The first is sung around the edge of the male's territory, to warn off other males; the second is sung near the middle of the territory, to attract a female (H. Harrison 1984; Morse 1993).

Usually this species prefers fairly open coniferous or mixed forest, but in Maryland it readily accepts deciduous forest. It feeds mostly at middle to upper levels in the canopy (and is therefore heard more often than seen) but sometimes descends to lower levels. Feeding is largely by gleaning, but it also flycatches and hovers to pick prey off foliage. As with most warblers, its primary food is insects, but it can also feed on berries (e.g., poison ivy), especially during migration. The breeding range in Maryland is mostly at high elevations in the Allegheny Mountains, but higher ridges of the Ridge and Valley region also support a sparse breeding population as far east as the Blue Ridge in Washington and Frederick counties (3 blocks, 1 confirmed). A single observed record in northern Baltimore County from the present atlas work, as well as possible and probable records from the previous atlas in the piney woods of Wicomico County (Robbins 1996f), suggest that the Black-throated Green Warbler may occasionally breed in the Piedmont and Coastal Plain regions.

Like most of our warblers, the Black-throated Green is a neotropical migrant. The wintering range extends from southern Florida through most of the Caribbean to the northernmost fringe of South America and from the Rio Grande Valley of southernmost Texas through the Gulf slope of Mexico to Central America, where it is a characteristic winter resident of highland pine forests. It generally arrives in April, when its insect prey is becoming abundant. A few birds may appear as early as the beginning of April, but the majority of migrants show up from the last week of April through the first half of May. Fall migrants first appear in late August and are mostly gone by mid-October. Strays may linger until the middle of November, and there are a few records in December and early January.

There is little information on nests in Maryland, but elsewhere the nest tends to be out on a horizontal branch at a height of 1 m (3 ft) to 24 m (80 ft) (H. Harrison 1975). Conifers are preferred, but hardwoods are acceptable. The bulk of the nest is fine bark, twigs, moss, grass, and spider webs; it is lined with fur, fine stems, rootlets, and feathers.

Black-throated Green Warblers are common throughout the Allegheny Mountain region, having been found in most blocks. Eastward, the species is uncommon on the higher ridges of the Ridge and Valley region, with a few on the east edge of the region at South Mountain. The population appears to be stable, or possibly even increasing slightly, as the number of blocks with possible, probable, or confirmed breeders increased from 87 to 92 in the second atlas.

T. DENNIS COSKREN

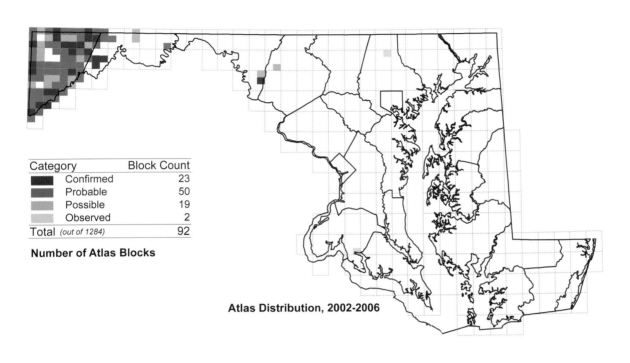

Category	Block Count
Confirmed	23
Probable	50
Possible	19
Observed	2
Total (out of 1284)	92

Number of Atlas Blocks

Atlas Distribution, 2002-2006

Change by Block
- ● Gain from First Atlas to Second
- ▲ Loss from First Atlas to Second

Change in Atlas Distribution, 1983-1987 to 2002-2006

Percent of Stops
- 50 - 100%
- 10 - 50%
- 0.1 - 10%
- < 0.1%

Relative Abundance from Miniroutes, 2003-2007

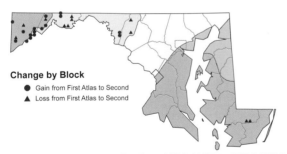

Atlas Region	1983-1987	2002-2006	Change No.	Change %
Allegheny Mountain	77	83	+6	+8%
Ridge and Valley	8	9	+1	+13%
Piedmont	0	0	0	-
Upper Chesapeake	0	0	0	-
Eastern Shore	2	0	-2	-100%
Western Shore	0	0	0	-
Totals	87	92	+5	+6%

Change in Total Blocks between Atlases by Region

Breeding Bird Survey Results in Maryland

Gary Van Velsir

Blackburnian Warbler
Dendroica fusca

The Blackburnian Warbler, a gemlike mite of a bird, is one of most colorful members of a notably colorful family. The breeding male, with its flaming orange throat and superciliary line set off by a black cheek patch and cap, is instantly recognizable. It is indeed (by sheer accident of nomenclature) "black and burning." A white-streaked black back, large white wing panels on dark wings, and black streaking on the yellowish flanks complete this bird's dapper ensemble. The female is easily recognized as a washed-out version of the stunning male, with paler yellow orange and dark gray replacing the flaming orange and black, respectively. Male and female young of the year have a similar pattern, but are paler and more washed-out than their elders. The song is one of the higher-pitched of the warblers, starting with a variable set of high squeaky *tsees* and *tsits* and characteristically finishing with a rising *seeeeeeee* that reaches near (or past) the human hearing range.

The habitat of this species is tall mature coniferous or mixed forest, particularly northern hemlock or red spruce. In mixed forest it shows a distinct preference for the conifers. Locally it may accept drier oak forest. It spends most of its time in the upper reaches of the trees, where it gleans the foliage for its insect prey and is one of the chief culprits in the incidence of "warbler neck" among birders. One of the spruce budworm warblers, it takes advantage of periodic budworm outbreaks to increase its numbers. It is more versatile in its requirements than some other warblers (such as the Cape May Warbler, *Dendroica tigrina*) and is not as strongly cyclical in its abundance. This species is associated with boreal forests and its breeding range in Maryland is at high elevations (most above 670 m [2,200 ft]), mostly in the Allegheny Mountains of the far west, but with a few on higher ridges of the Ridge and Valley region. Because of the dearth of mature red spruce in the state, hemlock is by far the most likely nest tree.

The Blackburnian Warbler is a neotropical migrant, with a wintering range farther to the south than most. The majority of the birds winter in humid montane and foothill forest of the Andes from Venezuela south to Peru, but it is fairly common in similar habitat as far northwest as the mountains of Costa Rica. Among the neotropical migrants, it is a later arrival in Maryland. The first appear in late April, but the majority of the birds are found during the middle two weeks of May. The fall migration is centered on the beginning of September, with most birds passing though from mid-August to mid-September.

The nest is typically high in a conifer (averaging 10 m [32 ft]), but it may be above 25 m (83 ft) and rarely as low as 2 m (6 ft). It is usually placed far out on a horizontal branch or in a small fork near the top of the tree. The Blackburnian Warbler is rarely confirmed by finding the nest; according to H. Harrison (1975) it is "one of the most difficult of warbler nests to discover."

The breeding range of the Blackburnian Warbler in Maryland seems to have remained virtually unchanged since previous work (Stewart and Robbins 1958; Robbins 1996g). The change in total blocks (confirmed, probable, and possible) between atlases is as follows: total blocks, up 2; Allegheny Mountain blocks, up 4; Ridge and Valley blocks, down 2. The species remains fairly common in proper habitat in Garrett County (Allegheny Mountains); is uncommon to rare (as before) on the higher ridges of Allegany County (Ridge and Valley); and is sporadic to the east, found in 1 block, confirmed, on the crest of the Blue Ridge in western Frederick County. These changes between atlases are statistically nonsignificant. The main threat to the species in Maryland is the loss of mature forest habitat. A potential decline of Blackburnian Warbler numbers due to loss of wintering habitat (or other threats) is not supported by Breeding Bird Survey data, which show a slight increase in numbers across the United States and the rest of North America.

T. DENNIS COSKREN

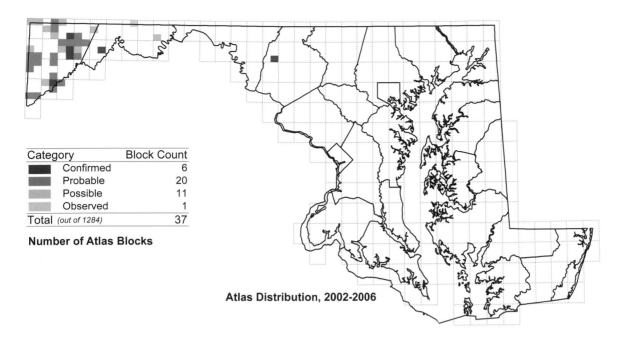

Category	Block Count
Confirmed	6
Probable	20
Possible	11
Observed	1
Total (out of 1284)	37

Number of Atlas Blocks

Atlas Distribution, 2002-2006

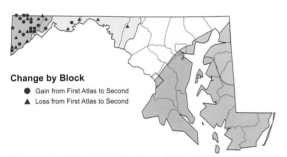

Change by Block
- ● Gain from First Atlas to Second
- ▲ Loss from First Atlas to Second

Change in Atlas Distribution, 1983-1987 to 2002-2006

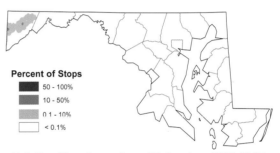

Percent of Stops
- 50 - 100%
- 10 - 50%
- 0.1 - 10%
- < 0.1%

Relative Abundance from Miniroutes, 2003-2007

| | | | Change | |
Atlas Region	1983-1987	2002-2006	No.	%
Allegheny Mountain	30	34	+4	+13%
Ridge and Valley	5	3	-2	-40%
Piedmont	0	0	0	-
Upper Chesapeake	0	0	0	-
Eastern Shore	0	0	0	-
Western Shore	0	0	0	-
Totals	35	37	+2	+6%

Change in Total Blocks between Atlases by Region

George M. Jett

Yellow-throated Warbler

Dendroica dominica

The Yellow-throated Warbler, neatly attired in gray, black, and white, has one spot of bright color, the yellow throat for which it is named. These warblers feed high in the canopy of the floodplain and pine woods that they haunt, gleaning prey from limbs, under bark, and between the scales of pine cones. Most observers first notice their presence by their sweet, loud, cascading song, somewhat similar in tone and pattern to the song of the Louisiana Waterthrush, another bird of stream sides. Although they were found in all of Maryland's counties and DC during the 2002–2006 atlas project they are generally localized and uncommon over much of the state.

The Yellow-throated Warbler has two major habitat preferences in Maryland that correspond roughly to the two nesting subspecies in the state. Nominate *dominica* prefers loblolly pine woods and mixed woods of loblolly pine, oak, and other pines and hardwoods. Midwestern *albilora* is most closely associated with mature floodplain forest usually dominated by sycamore. Most territories are not far from water. Other habitats that may host Yellow-throated Warblers include bald cypress swamps (Sprunt 1953) and dry ridge-top pine-oak woodlands (Hall 1983).

Yellow-throated Warblers are migratory, wintering in the southeastern United States as far north as coastal North Carolina, although the majority winter farther to the south in Mexico, northern Central America, and the Greater Antilles (Hall 1996). These are early migrants because of a northerly winter range and bark-gleaning food habits. The spring vanguard often arrives in Maryland in the last few days of March.

However, they also begin their autumn departure early, during July and August, and are gone by mid-September. These warblers usually build their nests in clusters of living or dead leaves near branch tips in the canopy, often higher than 12 m (40 ft). Most nests in a West Virginia study ranged from 25 to 30 m (83–100 ft) off the ground (R. H. Canterbury in Hall 1996). Nesting is best confirmed by finding nest-building females; 18 of 35 confirmations during this project were for nest-building. Few Yellow-throated Warbler nests have been found in Maryland. A nest with eggs was found on 16 May 1919 in Dorchester County (R. Jackson 1941), and an active nest was observed on 10 July 1954 in Anne Arundel County (Stewart and Robbins 1958; MNRF). Kyle Rambo reported banding females in full breeding condition in St. Mary's County from 10 to 20 May 2006.

During the late nineteenth and early twentieth century the Yellow-throated Warbler was apparently a rare vagrant from the south in Maryland and DC (Kirkwood 1895). It was found nesting in Sussex County, Delaware, in 1903 (Rhoads and Pennock 1905). The first confirmed Maryland nesting was in Dorchester County in 1919 (R. Jackson 1941). A slow northward range expansion that began during the 1940s has continued to the present for both subspecies that occur in Maryland and DC (Hall 1996). Southern *dominica* reached southern New Jersey in 1922 (Fables 1955) and it is now well established in three counties there (Walsh et al. 1999). Presumptive *albilora* was found nesting at Harrison Island in the Potomac River, Montgomery County, in 1953 (Stewart and Robbins 1958) and was first found on the Delaware River in New Jersey in 1954 (Walsh et al. 1999). By the late 1950s these warblers were common on the Eastern Shore and on the southern Western Shore, with smaller numbers northward into Anne Arundel and Prince George's counties, and along the Potomac to Montgomery County (Stewart and Robbins 1958).

From 1983 to 1987 Yellow-throated Warblers had spread

along the Potomac to western Allegany County and in other river systems on the Piedmont including the Patapsco, Monocacy, and Susquehanna; a few had spread into the Upper Chesapeake region (Jeschke 1996d). The distribution from 2002 to 2006 was similar in scope, with reports from a net total of 15 more blocks. Noteworthy increases were in Dorchester and Wicomico counties, northern Calvert and Anne Arundel counties, and along the Potomac in Washington County. Three reports in Garrett County were also noteworthy, especially territorial birds at Youghiogheny River Reservoir, which is part of the Ohio River drainage.

Very few Yellow-throated Warblers are observed on BBS routes because of their specialized habitats and an early attenuation in singing in early June. They have been stable or slowly increasing on BBS routes from 1966 to 2007. Habitat preservation is the key to retaining this elegant warbler among Maryland and DC's nesting birds. As long as floodplain forest is protected above the fall line and mature pine woodlands remain on the Eastern Shore, we will continue to enjoy Yellow-throated Warblers here.

WALTER G. ELLISON

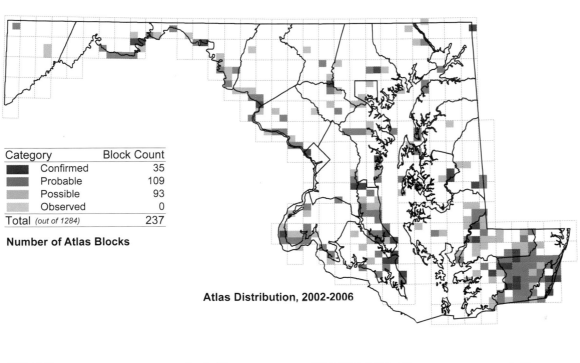

Category	Block Count
Confirmed	35
Probable	109
Possible	93
Observed	0
Total (out of 1284)	237

Number of Atlas Blocks

Atlas Distribution, 2002-2006

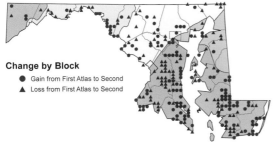

Change by Block
- ● Gain from First Atlas to Second
- ▲ Loss from First Atlas to Second

Change in Atlas Distribution, 1983-1987 to 2002-2006

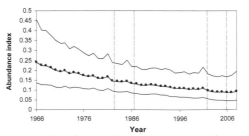

Percent of Stops
- 50 - 100%
- 10 - 50%
- 0.1 - 10%
- < 0.1%

Relative Abundance from Miniroutes, 2003-2007

Atlas Region	1983-1987	2002-2006	Change No.	%
Allegheny Mountain	2	3	+1	+50%
Ridge and Valley	24	25	+1	+4%
Piedmont	38	34	-4	-11%
Upper Chesapeake	9	5	-4	-44%
Eastern Shore	83	103	+20	+24%
Western Shore	64	65	+1	+2%
Totals	220	235	+15	+7%

Change in Total Blocks between Atlases by Region

Breeding Bird Survey Results in Maryland

Mark Hoffman

Pine Warbler
Dendroica pinus

Across the state, wherever there is a sizable stand of pines, the sweet trill of the Pine Warbler is a characteristic sound on a summer afternoon. Among the wood warblers, this species is one of the less strikingly marked. The adult male is greenish olive above, with darker wings and tail. His throat and breast are yellow, shading to white on the belly and undertail coverts. Two white wing bars are present, and there is a trace of a yellow superciliary stripe. Dark streaking, more or less obscure, is present along the sides of the breast. The adult female is similar, but duller. Immatures are slightly browner above and more uniform below (male yellowish and female whitish). The song is a simple trill, usually slower and sweeter than that of the Chipping Sparrow, which often shares its habitat. The pitch of the song may vary during its performance. Usually the variation is minor, no more than a slight unsteadiness of pitch, but some songs are distinctly on two separate pitches.

The name of this species is a good guide to its habitat, as it is almost invariably associated with pines during the breeding season. It will accept mixed woods, but its activities will be almost completely restricted to the pines. Usually this bird feeds at medium to high levels, but on occasion they will feed on the ground. It frequently searches the trunk and larger branches in the manner of a Black-and-white Warbler or Brown Creeper, but it is also an adept flycatcher. It also explores clumps of pine needles. Its prey is mostly insects and spiders, including beetles, moths, ants, bugs, flies, and scale

insects. It also takes plant matter, such as pine seeds, grapes, and the berries of dogwood, sumac, Virginia creeper, and poison ivy. The nest, which is 3 m (10 ft) or higher in a pine, is constructed of pine needles, weed stalks, and twigs. It is lined with pine needles, feathers, fern down, and hair. During the fieldwork for this atlas, nest building was observed as early as 6 April, and eggs have been found in Maryland as early as 19 April (Iliff, Ringler, and Stasz 1996), making this one of our earliest nesting warblers.

The Pine Warbler is present in all parts of Maryland, but its abundance varies greatly. The Western and (especially) Eastern Shore regions have large numbers, with most blocks recording the species. This, of course, reflects the extensive stands of pine woods in these regions. It is moderately common in the Piedmont and Upper Chesapeake regions. The Ridge and Valley region has few blocks with Pine Warblers, except in most of Allegany County and westernmost Washington County, where the species is common. The species is also scarce in the Allegheny Mountain region. A few Pine Warblers remain in the state during the winter, mostly in the southeast, where winters are milder. There is an increase in abundance in early March, and by the end of the month (when warblers of other species are just beginning to appear) Pine Warblers have reached full numbers. Late August sees the beginning of fall migration, but the loss is gradual through mid-December, by which time those remaining are mostly winter residents. Unlike most of our warblers, the Pine Warbler is not a neotropical migrant. It migrates only a short distance, to the southeastern United States.

The population trend of the Pine Warbler in Maryland is encouraging, although not entirely consistent. The total number of blocks reporting the species has risen by 44 since the previous atlas (Willoughby 1996b), an increase of 7%. However, within the Western Shore region (a part of its stronghold within the state), the number of blocks having breeding Pine Warblers has gone from 197 to 185, a drop of 6%. The Ridge and Valley region has also seen a decrease, but only by 2 blocks, not statistically significant. The Piedmont has seen the largest numerical increase, going from 45 blocks to 78, an increase of 73%. The Allegheny Mountain region has climbed from 4 blocks to 13, an increase of 225%.

The Eastern Shore increased from 287 blocks to 302, up 5%, and the Upper Chesapeake region increased by 1 block, not a significant change. Breeding Bird Survey data show a slight increase in abundance (ca. 1% annually) for both North America and the United States, and a slighter decrease (down 0.3% annually) for Maryland. Interestingly, early routes showed a 4% increase, while late routes had a 2% decrease.

T. DENNIS COSKREN

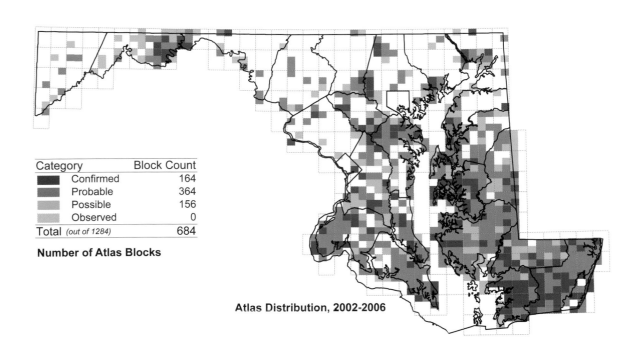

Category	Block Count
Confirmed	164
Probable	364
Possible	156
Observed	0
Total (out of 1284)	684

Number of Atlas Blocks

Atlas Distribution, 2002-2006

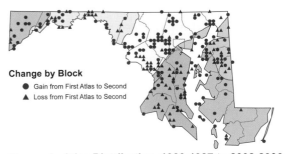

Change by Block
- ● Gain from First Atlas to Second
- ▲ Loss from First Atlas to Second

Change in Atlas Distribution, 1983-1987 to 2002-2006

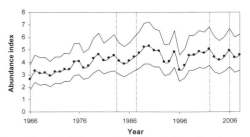

Percent of Stops
- 50 - 100%
- 10 - 50%
- 0.1 - 10%
- < 0.1%

Relative Abundance from Miniroutes, 2003-2007

Atlas Region	1983-1987	2002-2006	Change No.	%
Allegheny Mountain	4	13	+9	+225%
Ridge and Valley	50	48	-2	-4%
Piedmont	45	78	+33	+73%
Upper Chesapeake	55	56	+1	+2%
Eastern Shore	287	302	+15	+5%
Western Shore	197	185	-12	-6%
Totals	638	682	+44	+7%

Change in Total Blocks between Atlases by Region

Breeding Bird Survey Results in Maryland

Prairie Warbler
Dendroica discolor

Eric Skrzypczak

The Prairie Warbler is a denizen of open brushy areas and is not found in true prairie. The breeding male is bright yellow below, from the throat to the belly, and olive above, from crown to back. The back is marked with chestnut streaking. A yellow patch below the eye is outlined in black, and there is a yellow stripe over the eye. Crisp black streaking extends from the sides of the neck to the flanks. The wings and tail are dark. Female and immature birds are slightly duller; the black streaking is less distinct, the black and yellow cheek patch is missing, and there is no chestnut on the back. The species is usually first detected by its distinctive song, a series of 5 to 13 buzzy notes rising in pitch: *zhoo, zhoo, zho zho, zha, zhay, zhee, zhee*. The notes are evenly spaced or slightly accelerating, and the delivery may vary in speed, sometimes running together into a continuous song, but with each note on a clearly separate pitch.

To find this bird, search (especially by listening) in second-growth dry scrubby areas with shrubs and small trees. It is generally active, feeding on insects at low to medium levels by flycatching, gleaning, and hover-feeding. The song is often given from a higher perch in a larger tree. It shares with the Palm Warbler (*Dendroica palmarum*) and the waterthrushes the habit of tail-wagging, although to a lesser degree, and the motion may be from side to side. The female constructs a nest from shredded bark, plant down, and fine grass, interweaving it with supporting vegetation and lining it with hair, grass, and feathers. It is usually located 0.3 to 3 m (1–10 ft) above the ground. The nest is often a target of the Brown-headed Cowbird. The Prairie Warbler may deal with the intruder by abandoning the nest or, like the Yellow Warbler, by building a floor over the offending eggs (Bent 1953). Maryland egg dates extend from 12 May to 11 July (19 July in DC; MNRF). Nest building was observed between 8 May and 19 June for this atlas, and recently fledged young were seen between 20 June and 10 August.

The Prairie Warbler is found across Maryland, but its abundance varies greatly from region to region. In the present atlas, most blocks in the central Ridge and Valley, southern Piedmont, Western Shore, and southern Eastern Shore regions recorded the species. It was particularly scarce in the northern Eastern Shore, eastern portion of Upper Chesapeake, eastern and western Ridge and Valley, and Allegheny Mountain regions. In common with many of our warblers, the first few migrants appear near the beginning of April, but most arrive during the first week of May. Numbers begin to decrease at the end of July, but many remain through the beginning of September, and a few may hang on through the end of November or even later. In the winter, some are present in peninsular Florida and southern Louisiana, but many more go on to the West Indies; the species is particularly common in the Bahamas. A few winter in Central America and northernmost South America.

Throughout the state, populations of the Prairie Warbler have dropped since the previous atlas (Ringler 1996e). Overall, total blocks with Prairie Warbler have decreased by 23%. No region has seen an increase; the Allegheny Mountain region has fared the worst, with a 71% decline, and the Eastern Shore has done the least poorly, with only a 10% loss. The species is still fairly common, or even common, in some parts of the state, but the trends regionally range from poor to terrible. These discouraging data are in accord with BBS data, which show the Prairie Warbler to have statistically significant annual decreases across all of North America (2%), within the United States (2%), and especially in Maryland (4%). The reason for the decrease is probably loss of habitat from two directions: conversion of more agriculture to "clean" farming with the clearing of brushy areas and hedgerows, and the natural succession of formerly brushy areas to more mature forest unsuited to this species' requirements. According to Bent (1953) the Prairie Warbler greatly increased in abundance with the settlement of the continent and the opening up of the original forest, and we may be seeing a reversion to its previous less-common status.

T. DENNIS COSKREN

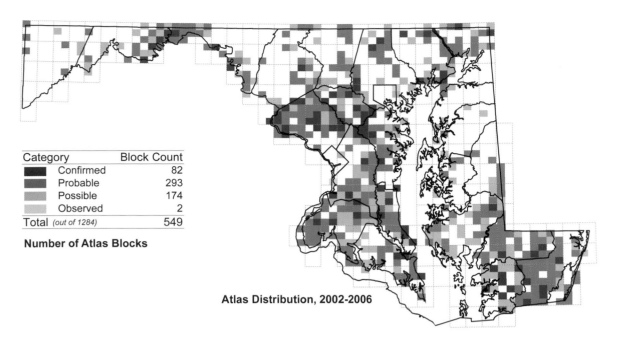

Category		Block Count
■	Confirmed	82
■	Probable	293
■	Possible	174
■	Observed	2
Total *(out of 1284)*		549

Number of Atlas Blocks

Atlas Distribution, 2002-2006

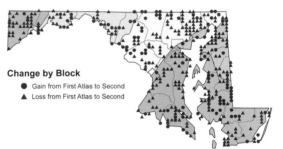

Change by Block

● Gain from First Atlas to Second
▲ Loss from First Atlas to Second

Change in Atlas Distribution, 1983-1987 to 2002-2006

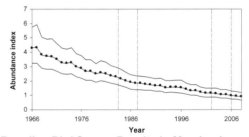

Percent of Stops

■ 50 - 100%
■ 10 - 50%
■ 0.1 - 10%
☐ < 0.1%

Relative Abundance from Miniroutes, 2003-2007

Atlas Region		1983-1987	2002-2006	Change No.	%
	Allegheny Mountain	34	10	-24	-71%
	Ridge and Valley	64	55	-9	-14%
	Piedmont	184	151	-33	-18%
	Upper Chesapeake	51	21	-30	-59%
	Eastern Shore	172	155	-17	-10%
	Western Shore	200	154	-46	-23%
Totals		705	546	-159	-23%

Change in Total Blocks between Atlases by Region

Breeding Bird Survey Results in Maryland

Cerulean Warbler
Dendroica cerulea

The Cerulean Warbler is a compact, short-tailed warbler of the treetops that would likely be overlooked if males were not persistent singers. This warbler's song is sufficiently similar to the song of the Northern Parula as to require some thought on the part of a listener to recognize it; the pattern of Cerulean Warbler song resembles the rapid ascending spiral of Blackburnian Warbler but it has the husky buzzing tone of the parula rising to a high pitch instead of the more abrupt termination of most parula songs, *zuree-zuree-zuree-zi-zi-zi-zi-zreeeee*.

Cerulean Warblers live in mature deciduous forests with large trees, often with breaks in the canopy or a pronounced super-canopy marked by occasional very tall trees breaking through the average canopy height (Hamel 2000). These warblers seldom come down to lower levels in the forest. In more than 250 sightings in a Tennessee study, the mean placed them at 77% of the height of the trees in which they were seen; that is, on average, a Cerulean Warbler would be seen at about 12 m (40 ft) in a 15-m (50-ft) tree (Robbins, Fitzpatrick, and Hamel 1992). Cerulean Warblers generally occur in two types of hardwood forest in Maryland, tall floodplain forest and mature ridge top oak-hickory forest; in both instances the warblers tend to avoid areas with midstory trees. These warblers are found almost exclusively in large forest tracts, the minimum forest area reported for a Maryland study was 138 ha (341 ac), and the maximum population density was predicted for forests of 3,000 ha (7,413 ac) and more (Robbins, Dawson, and Dowell 1989).

Cerulean Warblers reside in Maryland from the last days of April until late August and winter in subtropical humid forest on the east slope of the Andes from Colombia to Bolivia. These warblers saddle their compact lichen-studded nests well out on high limbs of hardwoods, seldom building below 9 m (30 ft). None of the 15 atlas confirmations (one duplicate confirmation in a single block) in this project were of nests. Dates for confirmed nesting ranged from a nest building pair on 5 May 2002 at McKeldin Area, Patapsco Valley State Park, Carroll County (C. Stirrat) to fledglings seen 12 July 2002 on Dans Mountain, Allegany County (S. Sires).

In the late nineteenth century, Cerulean Warblers were rare in eastern Maryland (Kirkwood 1895) and were not found nesting there until 1900 in Baltimore County (Kirkwood 1901). Eifrig (1904) reported that it was fairly numerous in western Maryland and stated they bred at Cumberland, Allegany County. Stewart and Robbins (1958) showed the nesting range included northwestern Garrett County (few records), the Savage River Valley, the western Ridge and Valley region from western Washington County through Allegany County, the Potomac Valley from DC to Montgomery County, the Patapsco Valley, the Loch Raven area of Baltimore County, and the Susquehanna Valley. In the 1983–1987 atlas project, Cerulean Warblers were found in a similar array of locations, adding Catoctin and South mountains, the Patuxent River Valley, lower Elk Neck, and the upper Big Elk Creek and Christina River valleys in northeastern Cecil County (Chestem 1996b). Over the past two decades the range has shrunk, especially in the Piedmont where it underwent a 66% decline in blocks during the present atlas project. Their favored nesting habitat, the upper limbs of mature trees along wooded stream valleys, was systematically destroyed during sewer line construction as rural communities were converted to residential (C. Robbins, pers. comm.). Other losses were from Elk Neck and parts of central Allegany County and northeastern Garrett County.

The Cerulean Warbler has shown a long-term decline on

North American BBS routes of 4% per year with the greatest losses in the areas of highest abundance in the bird's range such as the Cumberland Plateau (Robbins, Fitzpatrick, and Hamel 1992). Trends for Maryland have been more equivocal but also unreliable, as an average of only 0.27 Cerulean Warblers were reported on just 15 routes in the Free State. The nesting range of this warbler has expanded to the north and east as it has declined. Perhaps habitat loss has forced exploration on the part of some birds otherwise deprived of suitable nesting habitat. Gains in newly occupied areas are clearly insufficient to offset heavy losses in the core of the range. The 40% net decline in atlas blocks between atlas projects in the 1980s and 2000s is a clear warning that this warbler requires careful monitoring, habitat preservation, and management if it is to remain in Maryland and DC's avifauna.

WALTER G. ELLISON

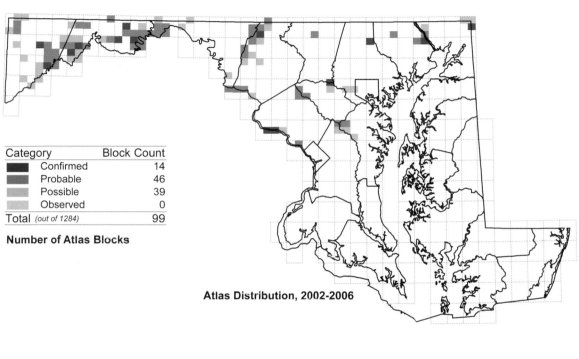

Category	Block Count
Confirmed	14
Probable	46
Possible	39
Observed	0
Total (out of 1284)	99

Number of Atlas Blocks

Atlas Distribution, 2002–2006

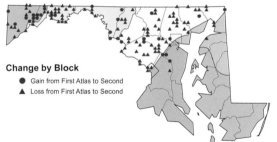

Change by Block
● Gain from First Atlas to Second
▲ Loss from First Atlas to Second

Change in Atlas Distribution, 1983–1987 to 2002–2006

Percent of Stops
- 50 - 100%
- 10 - 50%
- 0.1 - 10%
- < 0.1%

Relative Abundance from Miniroutes, 2003–2007

Atlas Region	1983-1987	2002-2006	Change No.	%
Allegheny Mountain	31	27	-4	-13%
Ridge and Valley	64	45	-19	-30%
Piedmont	64	22	-42	-66%
Upper Chesapeake	3	0	-3	-100%
Eastern Shore	0	0	0	-
Western Shore	4	5	+1	+25%
Totals	166	99	-67	-40%

Change in Total Blocks between Atlases by Region

Breeding Bird Survey Results in Maryland

Mark Hoffman

Black-and-white Warbler

Mniotilta varia

With its zebra-striped plumage and branch and trunk creeping habits, the Black-and-white Warbler is distinctive. The bark feeding habit has much to do with this warbler's early spring arrival, ahead of most other Wood Warblers, and with its broad wintering range that includes the southeastern U.S. coast north to North Carolina, Florida, and the Gulf Coast. The song is characteristic, thin and high-pitched with squeaky notes see-sawing up and down *wee-see, wee-see, wee-see*. It is often the first indication to observers that this warbler has returned in spring. Although these warblers occur widely in Maryland and DC, they have a spotty distribution with large uninhabited spaces between occupied areas.

Black-and-white Warblers are forest birds occupying a wide array of deciduous and mixed woodlands. These range from mature second-growth to old forests with varying moisture regimes and physiography. They are usually absent from small forest fragments and are often absent in regions lacking large woodlots (Whitcomb et al. 1977; Robbins, Dawson, and Dowell 1989).

The wintering range extends from the southeastern United States south through Central America and the Antilles at least sparingly south to Peru (Kricher 1995). Black-and-white Warblers generally reside in Maryland and DC from early April until mid-October. They nest on the ground and nestle the leafy cup into leaf litter next to tree roots, logs, or rocks and ledges. Kirkwood (1895) found a nest placed in a

rock crevice at the summit of Dans Mountain in Allegany County. Nests are very hard to find, just three confirmations in this project were for active nests. Over half of confirmations were of recent fledglings. Maryland egg dates range from 12 May to 19 June, with nestlings to 7 July (Stasz 1996b; MNRF). An active nest was reported to 12 July 2006 near Waterbury, Anne Arundel County during this atlas although nest contents were not detailed (D. Perry).

Numbers of Black-and-white Warblers have long been variable across Maryland and DC. Kirkwood (1895) found them notably less numerous in the nesting season than during migration, but C. Richmond (1888) and Eifrig (1904) considered them common in DC and western Maryland, respectively. Stewart and Robbins (1958) found them common east to Catoctin and South mountains, fairly common west of Chesapeake Bay, locally fairly common on the lower Eastern Shore, and rare in the Upper Chesapeake. During the 1983–1987 atlas project they had declined, particularly in the Ridge and Valley and Piedmont, where they had become patchily distributed (Stasz 1996b). It is likely habitat loss contributed to these losses in the Piedmont and Hagerstown Valley, but this explanation appears less tenable for western Washington and Allegany counties. On the other hand, modest increases in range seemed evident in Dorchester County and parts of the Upper Chesapeake.

This atlas project documented further losses in the Maryland nesting range of Black-and-white Warbler. The greatest losses were on the Western Shore, where they were lost from 45 blocks, and the Eastern Shore where they were not refound in 23 blocks. Notable losses there were in western Dorchester County and from Worcester County, especially in the urbanizing northeast. Although they were missed in only a net of 5 blocks in the Piedmont, these losses were concentrated in the DC to Baltimore I-95 corridor; gains were seen to the north and west. The preserves along the Patuxent and Patapsco rivers currently coincide with many of the blocks hosting this warbler in the eastern Piedmont and Western Shore.

The Black-and-white Warbler has shown modest declines of less than 1% per year on North American and eastern BBS routes. However, declines have been steeper since 1980. Maryland has evinced a larger decline of 4% per year since 1966. This warbler's patchy range and declines in Maryland

and DC are almost certainly related to habitat loss and degradation. Robbins, Dawson, and Dowell (1989) determined a minimum woodlot size of 202.5 ha (500 ac) was necessary to successfully support a population of these warblers. Areas that lack this warbler do indeed have small, often isolated woodlots. Preservation of forest habitat in sufficient quantity and extent is needed to prevent further declines in this unusual tree-climbing warbler.

WALTER G. ELLISON

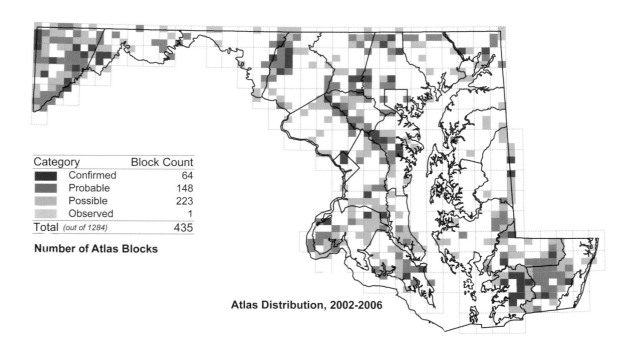

Category	Block Count
Confirmed	64
Probable	148
Possible	223
Observed	1
Total (out of 1284)	435

Number of Atlas Blocks

Atlas Distribution, 2002-2006

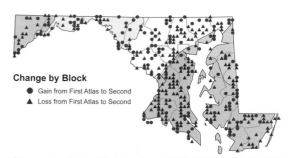

Change by Block
- ● Gain from First Atlas to Second
- ▲ Loss from First Atlas to Second

Change in Atlas Distribution, 1983-1987 to 2002-2006

Percent of Stops
- 50 - 100%
- 10 - 50%
- 0.1 - 10%
- < 0.1%

Relative Abundance from Miniroutes, 2003-2007

Atlas Region	1983-1987	2002-2006	Change No.	Change %
Allegheny Mountain	76	72	-4	-5%
Ridge and Valley	45	50	+5	+11%
Piedmont	101	96	-5	-5%
Upper Chesapeake	25	17	-8	-32%
Eastern Shore	117	94	-23	-20%
Western Shore	149	104	-45	-30%
Totals	513	433	-80	-16%

Change in Total Blocks between Atlases by Region

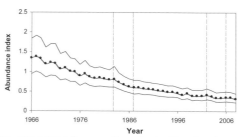

Breeding Bird Survey Results in Maryland

American Redstart
Setophaga ruticilla

A small, slender, and active wood warbler, the American Redstart frequently spreads its tail as it moves about, exposing bright patches of orange or yellow, thereby flushing prey, which are then energetically chased down. Our redstarts derive their group name from the redstarts of the Old World, chats that have a similar tail pattern and feeding behavior. This warbler inhabits deciduous forests, preferring those with thick understories of saplings and tall shrubs. The range of habitats occupied is broad, ranging from regenerating dense stands of saplings and shrub swamps to mature forest. Hunt (1996) found that New England redstarts had smaller territories and settled earlier in spring in young successional habitats than in mature forest. In Maryland these warblers are area-limited, requiring large blocks of forest to reach maximum population densities (Robbins, Dawson, and Dowell 1989). This may reflect the need for the variety of stand ages offered by larger blocks of forest.

American Redstarts are typically seen in Maryland and DC from mid-April until mid-October and winter broadly across the tropical Americas. They are single brooded and commence autumn migration as soon as their fledglings are independent. Many are on the move by late July. This redstart builds a compact woven nest constructed in three-way branch forks at a wide range of heights, although the average height is about 9.1 m (30 ft). Within a week of the arrival of territorial males, females return and quickly start nesting. Maryland and DC egg dates range from an extraordinarily early 25 April to 30 June (Robbins 1949c; Stewart and Robbins 1958). Early egg dates typically do not occur until mid-May (e.g., 12 May for Alabama; Imhof 1976).

Kirkwood (1895) called American Redstart fairly common but local in Maryland and DC, but Eifrig (1904) found them common to abundant in western Maryland. In the 1950s redstarts were common over much of Maryland but local on the Eastern Shore with most in the Pocomoke drainage, and uncommon in the Upper Chesapeake region (Stewart and Robbins 1958). In the 1983–1987 atlas project, the species was widespread but showed regions of absence or scarcity including the Hagerstown and Frederick valleys and much of the agricultural northern Piedmont; east of Chesapeake Bay they were largely limited to the Pocomoke basin (Fallon 1996). The distribution has changed little; 2002–2006 atlas results show a net loss of only 13 blocks. These warblers increased in the Allegheny Mountains by 10 blocks (13%) and were lost from 19 blocks on the Western Shore (18%), with most losses in suburban Prince George's County.

American Redstarts have significantly declined on North American BBS routes, at a rate of less than 1% per year over the past four decades. Maryland BBS routes show no significant trend because of small numbers per route and strong fluctuations among years, although their numbers appear to have increased. Redstarts have declined in northern New England apparently because much of the forest there has

matured and lost its well-developed understory (Hunt 1998), a process that could also occur in Maryland's protected forest lands. Deer browsing of the understory might also affect redstart habitat although this has not been documented. This warbler bears watching in Maryland because it depends on forest habitat, is scarce in areas with only small woodlots, and has declined in parts of the state experiencing rapid suburbanization.

WALTER G. ELLISON

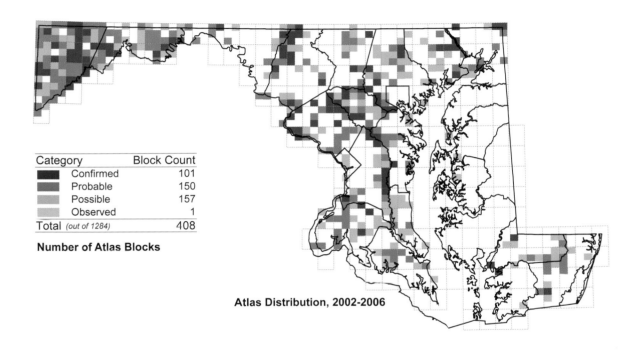

Category	Block Count
Confirmed	101
Probable	150
Possible	157
Observed	1
Total *(out of 1284)*	408

Number of Atlas Blocks

Atlas Distribution, 2002-2006

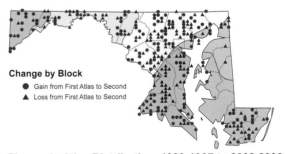

Change by Block
- ● Gain from First Atlas to Second
- ▲ Loss from First Atlas to Second

Change in Atlas Distribution, 1983-1987 to 2002-2006

Percent of Stops
- 50 - 100%
- 10 - 50%
- 0.1 - 10%
- < 0.1%

Relative Abundance from Miniroutes, 2003-2007

			Change	
Atlas Region	1983-1987	2002-2006	No.	%
Allegheny Mountain	75	85	+10	+13%
Ridge and Valley	73	69	-4	-5%
Piedmont	118	121	+3	+3%
Upper Chesapeake	13	9	-4	-31%
Eastern Shore	31	32	+1	+3%
Western Shore	107	88	-19	-18%
Totals	417	404	-13	-3%

Change in Total Blocks between Atlases by Region

Breeding Bird Survey Results in Maryland

Prothonotary Warbler

Protonotaria citrea

Although brightly colored, the Prothonotary Warbler can disappear in sun-dappled swamps, but when it is finally seen it glows so intensely its image burns itself into the memory. The combination of a solid bright yellow head and breast with cool blue gray wings and lower back is striking. It was enthusiasm over a sighting of this beautiful warbler on the Potomac River near DC that contributed to alleged spy Alger Hiss' downfall. He excitedly told his accuser Whittaker Chambers about the sighting and the information was used to catch him in contradictory testimony. This warbler, a bird of wet and shaded places, is Maryland and DC's only cavity nesting wood warbler. The males are persistent morning singers, producing a rapid monotonous *sweet-sweet-sweet-sweet*. Prothonotary Warblers are fairly widespread but localized along the Potomac River and on the Coastal Plain.

These are birds of places with shaded standing or slow-moving water such as swamps, wooded pond margins, wet riverbanks, and wet woods with large vernal pools. Birds nesting on vernal pools appear to move on to more permanent wetlands in late June during dry summers. Prothonotary Warblers require fairly extensive stands of floodplain forest around streams and do not occur in narrow strips of trees along streams running through farmland (C. Keller et al. 1993).

Prothonotary Warblers are migratory, wintering in coastal Central America and northern South America, often in mangrove swamps (Petit 1999). They return to Maryland in early to mid-April and leave by late August, with a few remaining until mid-September. They nest in natural cavities in stubs, logs, dead branches, and bald cypress knees, especially those dug by other birds such as Downy Woodpeckers, chickadees, and titmice. They will also accept nest boxes, including in one study boxes made from milk cartons (Fleming and Petit 1986; Petit et al. 1987). Nesting begins during April and may continue to August when the last fledglings are being fed. At least some females attempt second broods (Walkinshaw 1941); this may be especially true during wet summers. Maryland egg dates range from 28 April to 3 July with nestlings to 25 July (Wilmot 1996c; MNRF).

The Prothonotary Warbler was apparently a rare visitor to much of Maryland and DC during the late nineteenth century (Coues and Prentiss 1883, Kirkwood 1895). The first nesting on the Delmarva Peninsula was detected on the upper Choptank River near Marydel in Delaware in 1898 (W. Smith 1899; Rhoads and Pennock 1905). This suggests that this warbler may have already been nesting farther south on the Delmarva in the extensive swamplands in the Pocomoke drainage during the early twentieth century. They first nested in the DC area at Dyke Marsh, Virginia, in 1928 (Cooke 1929). Stewart and Robbins (1958) described a range that included much of the Coastal Plain, the lower Susquehanna valley, and the Potomac valley westward to Allegany County.

The atlas map from 1983 to 1987 revealed a fairly similar distribution and clearly illustrated the Prothonotary Warbler's association with major stream systems with reports westward on the Potomac along the Chesapeake and Ohio Canal to Cumberland and good numbers of blocks on most tidal streams on the Coastal Plain (Wilmot 1996c). This range has contracted slightly since the 1980s, with records from a net of 25 fewer blocks in 2002–2006 with the largest losses on the western Potomac in Allegany County, as well as on the lower Patuxent River and other Western Shore streams. These warblers have increased in DC, with reports from several blocks on the Anacostia River, and they have established a small population on Youghiogheny River Reservoir in northwestern Garrett County.

Prothonotary Warblers have declined slowly on North American BBS routes, with a significant 1.8% per year decline since 1980. Declines on Maryland BBS routes have not been statistically significant. This beautiful warbler's habitat is vulnerable to logging, changes in water quality, and development (especially at vernal pool sites). The declines detected between atlas projects, although relatively small, are worrisome. Closed wetland habitat must be preserved and monitored if we are to retain this little flame of the swamp in Maryland and DC.

WALTER G. ELLISON

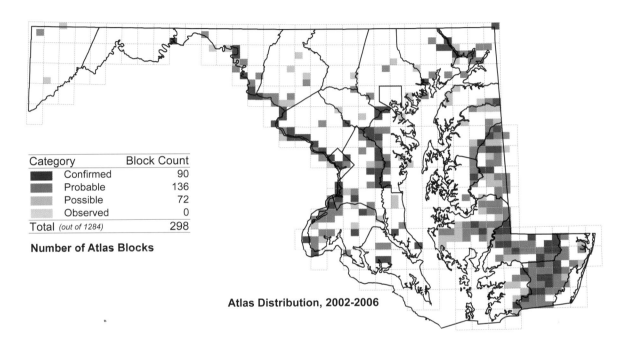

Category	Block Count
Confirmed	90
Probable	136
Possible	72
Observed	0
Total (out of 1284)	298

Number of Atlas Blocks

Atlas Distribution, 2002-2006

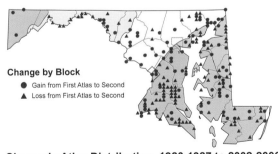

Change by Block
- ● Gain from First Atlas to Second
- ▲ Loss from First Atlas to Second

Change in Atlas Distribution, 1983-1987 to 2002-2006

Percent of Stops
- 50 - 100%
- 10 - 50%
- 0.1 - 10%
- < 0.1%

Relative Abundance from Miniroutes, 2003-2007

Atlas Region	1983-1987	2002-2006	Change No.	%
Allegheny Mountain	0	3	+3	-
Ridge and Valley	24	16	-8	-33%
Piedmont	32	46	+14	+44%
Upper Chesapeake	46	38	-8	-17%
Eastern Shore	122	121	-1	-1%
Western Shore	97	72	-25	-26%
Totals	321	296	-25	-8%

Change in Total Blocks between Atlases by Region

Breeding Bird Survey Results in Maryland

Worm-eating Warbler

Helmitheros vermivorum

A most subtle bird, the olive and warm buff Worm-eating Warbler is easily overlooked in its forest haunts. It is best detected by its long-winded rapidly trilled insectlike song. Even this clue may be overlooked when the superficially similar trilled songs of Chipping Sparrow and Pine Warbler are part of the local songbird chorus. Singing males perch at about 3 to 6 m (10–20 ft) on bare branches, but the song is so hard to place that the bird will often remain unseen. Adults feeding young are much easier to observe. This is the only Maryland bird that tends to specialize in hunting prey in hanging clusters of dead leaves (Greenberg 1987). Worm-eating Warblers do not eat earthworms, but they do eat large numbers of caterpillars, also traditionally called worms. Although they are widespread in Maryland their need for extensive tracts of forest leads to a patchy nesting range.

The Worm-eating Warbler inhabits dry to moist, but not wet, forest. Thick leaf litter is characteristic; therefore oaks, with their slowly decaying leaves, are frequently present. There is always a thick shrub layer from 1 to 3 m (3–10 ft) high, often composed of evergreen shrubs and small trees such as mountain laurel, rhododendron, blueberries, American holly, and sweet pepperbush. West of Chesapeake Bay they show a preference for steep slopes on hills, bluffs, and ravines, but east of the bay they often occur in fairly well-drained flatwoods of oak and Virginia pine with a thick shrub understory. These warblers do not occur in small isolated woodlots; the minimum woodlot holding a nesting pair

in a Maryland study was 21 ha (52 ac; Robbins, Dawson, and Dowell 1989).

Worm-eating Warblers are migratory, wintering in the Greater Antilles and from southeastern Mexico to central Panama (AOU 1998). They return to Maryland and DC in late April and early May. They raise a single brood and depart from late August to mid-September. The nest is a cup of leaf litter and leaf skeletons nestled into the forest floor against sloping ground and usually shaded by a shrub or sapling (Hanners and Patton 1998). Maryland and DC egg dates range from 23 May to 25 June (Stewart and Robbins 1958; MNRF), and nestlings have been found to 4 July (Stasz 1996c; MNRF). Clutch initiation over five years in a Connecticut study ranged from 13 to 22 May and late renests were commenced from 10 June to 1 July (Gale et al. 1997; Hanners and Patton 1998) implying that earlier and later dates are likely for Maryland and DC.

The historic status of the Worm-eating Warbler appears to have been fairly similar to its present status, with early authorities all assessing it as not common and localized (e.g., Kirkwood 1895; Eifrig 1904). Stewart and Robbins (1958) detailed the status and range, calling it fairly common in the western Ridge and Valley and on the Blue Ridge, locally so on the Piedmont, uncommon on the Western Shore and along the Pocomoke River and tributaries, and otherwise rare east of Chesapeake Bay and in the Alleghenies. By the 1980s the range had increased in the Alleghenies and on the Eastern Shore (Stasz 1996c). It was still scarce in the Upper Chesapeake, and essentially absent from the Frederick and Hagerstown valleys.

Worm-eating Warblers were found in 54 more blocks in 2002–2006 than in 1983–1987. There were some losses in three regions, the Piedmont, Western Shore, and the Allegheny Mountains. However these birds were found in far more blocks in the Upper Chesapeake, especially in Caroline County, and also increased on the Eastern Shore. There were notable losses of blocks from Prince George's County and from just above the fall line between DC and Baltimore.

There has been no clear trend for Worm-eating Warbler on North American BBS routes. It has been stable or slowly increasing since 1966. However, there is some evidence that this species is severely undersampled by the BBS. A banding study in Belize showed more Worm-eating Warblers wintering in that country (C. Robbins and B. Dowell, unpubl. data) than Rich et al. (2004) had calculated as breeding in all of

North America based on the BBS. Maryland trends are suspect because of fluctuations and small numbers of birds per route, but they have increased strongly since 1980. Although Worm-eating Warblers are sensitive to forest fragmentation and are ground-nesters, they have nonetheless held their own in Maryland and DC. However, given the losses from atlas blocks in the Baltimore and DC metropolitan area they should be closely watched, and large forest preserves with healthy shrub layers need to be maintained.

WALTER G. ELLISON

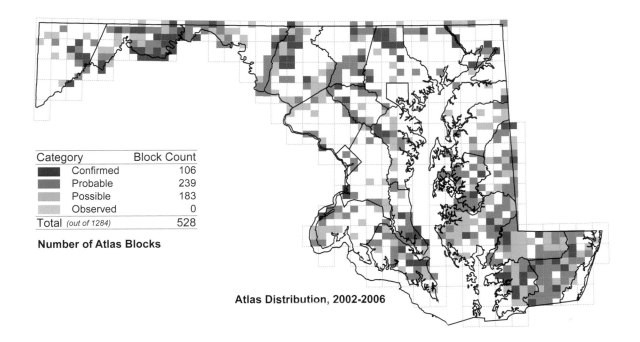

Category	Block Count
Confirmed	106
Probable	239
Possible	183
Observed	0
Total (out of 1284)	528

Number of Atlas Blocks

Atlas Distribution, 2002-2006

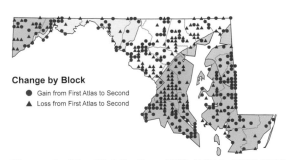

Change by Block
- ● Gain from First Atlas to Second
- ▲ Loss from First Atlas to Second

Change in Atlas Distribution, 1983-1987 to 2002-2006

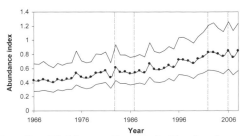

Percent of Stops
- 50 - 100%
- 10 - 50%
- 0.1 - 10%
- < 0.1%

Relative Abundance from Miniroutes, 2003-2007

Atlas Region	1983-1987	2002-2006	Change No.	%
Allegheny Mountain	24	23	-1	-4%
Ridge and Valley	82	95	+13	+16%
Piedmont	106	101	-5	-5%
Upper Chesapeake	12	31	+19	+158%
Eastern Shore	150	181	+31	+21%
Western Shore	96	93	-3	-3%
Totals	470	524	+54	+11%

Change in Total Blocks between Atlases by Region

Breeding Bird Survey Results in Maryland

Swainson's Warbler
Limnothlypis swainsonii

George M. Jett

Rare to uncommon throughout most of its southern U.S. breeding range (R. Brown and Dickson 1994), Swainson's Warbler is one of Maryland's rarest and most enigmatic breeding bird species. Its reclusive habits, dull leaf brown plumage, and dense swamp habitat make observations a challenge for even the most dedicated observer. Swainson's Warbler was not discovered in Maryland until 1942 when Joseph Cadbury recorded a territorial bird in Great Cypress Swamp in Wicomico County near the Maryland-Delaware border (Stewart and Robbins 1947). In 1947 four territorial males were documented in the Delaware portion of Great Cypress Swamp, establishing the first Delaware breeding records and the northernmost breeding locality for this species (Meanley 1950). Although sporadic, breeding season observations have since been reported for at least eight other Maryland localities, all from the Pocomoke River watershed. The most frequent and consistent observations have centered around three areas: Great Cypress Swamp, upper Nassawango Creek, and Hickory Point Swamp along the lower Pocomoke River (Heckscher and McCann 2006). Historically, Swainson's Warbler was probably more widespread on the lower Eastern Shore. However, forest clearing for agriculture, wetland loss and degradation because of extensive ditching and channelization, and past logging have greatly diminished this species' swamp habitat in Maryland (Meanley 1971; Vaughn and Robbins 1996; Heckscher and McCann 2006).

Atlantic and Gulf Coastal Plain Swainson's Warblers are associated with expansive bottomland hardwood swamps and floodplain forests (Meanley 1971; Eddleman et al. 1980; R. Brown and Dickson 1994; Graves 2001). It requires moist to seasonally wet but not inundated, deeply shaded forest with a dense understory of shrubs and vine tangles, and sparse ground vegetation. In areas to the south, canebrakes are often used. In Maryland, it is typically associated with dense thickets of sweet pepperbush, highbush blueberry, and greenbrier within mature to old growth forested wetlands dominated by black gum, red maple and bald cypress (Meanley 1950, 1971; Springer and Stewart 1948; Stewart and Robbins 1947, 1958; McCann and Heckscher, unpubl. data).

Swainson's Warbler is a neotropical migrant, wintering primarily in the western Caribbean Islands and Mexico's Yucatan Peninsula (R. Brown and Dickson 1994). Spring migrants typically arrive in Maryland in mid- to late April. Observations on Maryland's lower Eastern Shore extend from as early as 9 April to as late as 30 August (Meanley 1950; Iliff, Ringler, and Stasz 1996). Nests are typically placed 1 to 2 m (3.2–6.5 ft; range 0.4–3.1 m [1.3–10 ft]) above the ground along the edge of or close to dense patches of understory vegetation and tangled curtains of vines. Although relatively large compared to those of other warblers, nests are notoriously difficult to find. They resemble unorganized clumps of leaves suspended from vine tendrils or from the intersecting branches of shrubs or small trees.

During the first atlas, territorial birds were documented in 3 blocks at widely scattered locations in the Pocomoke River drainage: Great Cypress Swamp, near Shad Landing State Park, and Hickory Point Swamp (Vaughn and Robbins 1996). During the second atlas, there were also 3 records with breeding codes (2 possible, 1 confirmed) and 1 observed record of a singing male in Calvert County. The 2 possibles were located in the upper Nassawango Creek area; the confirmed record is not mapped to protect the birds from disturbance. Although these results show this species continues to be a rare breeder, it is unlikely that it has undergone a significant change in distribution within this region. Observations have occasionally been made in the past in the upper Nassawango Creek area (e.g., Ringler 1989, 1990; Southworth and Southworth 1995). The species may have been missed during the second atlas at Hickory Point Swamp. Observations there, although infrequent, date back to 1946 (Stewart and Robbins 1958; Bennett et al. 1998; Heckscher and McCann 2006). The first atlas block record near Shad Landing was probably of an isolated, unmated territorial male. The species' status at Great Cypress Swamp is uncertain. Although it has been observed repeatedly there since 1942, it seems to have slowly disappeared. The last known breeding record was of a pair with a nest in 1998 from the Maryland portion of the swamp (Heckscher 2000). Harassment by birders has been cited as one factor that may have contributed to the species' decline at this site (Shoch 1992; Vaughn and Robbins 1996). Perhaps more important though are the long-term cumulative effects of swamp clearing for agriculture, loss of old forest conditions and floristic diversity due to logging, and hydrological changes resulting from ditching, channelization, and groundwater withdrawal for irrigation (Heckscher 2000; Heckscher and McCann 2006).

BBS data show a significant ($P = 0.02$) rangewide population increase of 4.5% and an increase of 7.0% ($P = 0.04$) for the Coastal Plain physiographic region. Data are too scant for Maryland and the southern Appalachians to yield trend estimates here.

Despite recent accounts to the contrary (R. Brown and Dickson 1994; Graves 2001; Somershoe et al. 2004), Swainson's Warbler persists as a breeding species in Maryland. However, it remains highly rare and may be declining. It is currently listed as state endangered, a designation that is unlikely to change in the foreseeable future given its precarious status in the state. This species is so rare and isolated from other breeding localities (over 160 km [100 mi] to Dismal Swamp, Virginia) that site abandonment and extirpation may be unrecoverable. Conservation efforts should focus on the protection and large-scale restoration of palustrine forested wetlands in the Pocomoke watershed. Systematic, carefully designed surveys should be conducted to locate breeding sites, estimate breeding population size, and evaluate conservation needs and long-term viability.

JAMES M. MCCANN

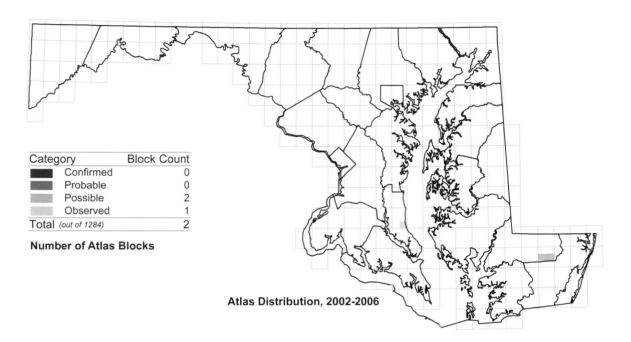

Category	Block Count
Confirmed	0
Probable	0
Possible	2
Observed	1
Total (out of 1284)	2

Number of Atlas Blocks

Atlas Distribution, 2002-2006

Change by Block
● Gain from First Atlas to Second
▲ Loss from First Atlas to Second

Change in Atlas Distribution, 1983-1987 to 2002-2006

Atlas Region	1983-1987	2002-2006	Change No.	Change %
Allegheny Mountain	0	0	0	-
Ridge and Valley	0	0	0	-
Piedmont	0	0	0	-
Upper Chesapeake	0	0	0	-
Eastern Shore	3	2	-1	-33%
Western Shore	0	0	0	-
Totals	3	2	-1	-33%

Change in Total Blocks between Atlases by Region

George M. Jett

Ovenbird
Seiurus aurocapilla

Like the two waterthrushes, the Ovenbird is an unobtrusively colored bird. The plumage of the back, wings. and tail is olive, lighter than that of the waterthrushes, without any streaking or wingbars. The underparts are white, heavily streaked with black on the breast and flanks but unmarked on the throat and belly. The crown is tawny orange, bordered with black. A bold white eye-ring sets off a black eye, giving the bird an alert appearance. There is no difference in plumage between the male and female. The song, usually given from a low branch, commonly consists of five or six loud and emphatic two-syllable phrases, delivered in an extended crescendo, usually represented as: *teacher, teacher, teacher, TEACHER, TEACHER*. Frequently the syllables are reversed: *chertea, chertea, chertea, CHERTEA, CHERTEA*. Sometimes a simpler one-syllable series is used: *teach, teach, teach, TEACH, TEACH*. Less well known is a flight song (sometimes given from a perch) made up of a complex jumble of whistles and chirps interspersed with a few of the more usual *teacher* phrases (Bent 1953). Usually given at dusk or during the night, the song is sung only after the bird has been on the nesting ground for a couple of weeks.

The Ovenbird's habitat is deep rich deciduous and mixed forest with a sparse to moderate understory. The bird may be found walking along the forest floor bobbing its head and flicking its tail; the gait gives a name to the genus (*Seiurus* = shake-tail). It is a ground feeder, gleaning such terrestrial fauna as small insects, snails, myriapods, spiders, and earthworms. It habitually turns over leaves to search for prey.

The nest is invariably on the ground, usually in a slight depression in a fairly open situation so the birds may approach it from any direction. It is built of grass, mosses, plant fibers, pine needles, and such materials, lined with fine rootlets, fibers, and hair, and is roofed over with a dome like an old-fashioned Dutch oven (whence "ovenbird"). The nest is often covered with leaf litter; it blends well with the forest floor and is invisible from above. The entrance is from one side, at ground level. The sitting female flushes very reluctantly, generally not until she is in danger of being stepped on. Maryland egg dates range from 29 April to 17 July and nestling dates from 15 May to 18 July (MNRF).

The breeding range of the Ovenbird extends from the Rocky Mountains to the Atlantic coast, and from Canada southward to South Carolina and Arkansas. Within Maryland, the species is found across the state, absent only from those areas that lack the requisite mature forest habitat (e.g., the open country of the Hagerstown Valley in Washington County).

Like most of the warblers, Ovenbirds generally spend winter in the neotropics, mostly in the West Indies and Central America. Some remain in Florida and south Texas, and rarely along the southern Atlantic and Gulf coasts. Spring migrants may arrive in Maryland as early as the first week of April, but most appear later in the month when the species becomes common. Numbers swell in mid-September with an influx of birds from the north, then dwindle into October, when most have left. Occasionally single birds will attempt to winter in Maryland.

The Ovenbird has held its own in Maryland over the past few decades. Reporting blocks in this atlas showed a 7% statewide increase over the first atlas. The three breeding bird atlases from Howard County show a consistent increase from the first (1973–1975; Klimkiewicz and Solem 1978), through the second (1983–1987; Cheevers 1996c), to the third (2002–2006); out of 132 quarterblocks, Ovenbirds were found in 67, 76, and 103 quarterblocks, respectively. This is in agreement with Breeding Bird Survey data across the country, which indicates a slight increase in population. The increase may reflect a maturation of second-growth forest from abandoned farmland.

T. DENNIS COSKREN

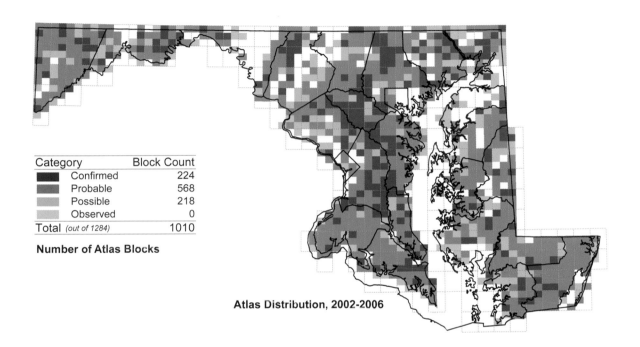

Category	Block Count
Confirmed	224
Probable	568
Possible	218
Observed	0
Total *(out of 1284)*	1010

Number of Atlas Blocks

Atlas Distribution, 2002-2006

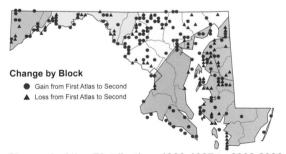

Change by Block
● Gain from First Atlas to Second
▲ Loss from First Atlas to Second

Change in Atlas Distribution, 1983-1987 to 2002-2006

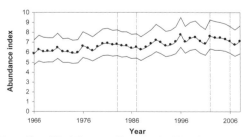

Percent of Stops
50 - 100%
10 - 50%
0.1 - 10%
< 0.1%

Relative Abundance from Miniroutes, 2003-2007

Atlas Region	1983-1987	2002-2006	Change No.	%
Allegheny Mountain	82	88	+6	+7%
Ridge and Valley	89	100	+11	+12%
Piedmont	228	265	+37	+16%
Upper Chesapeake	81	79	-2	-2%
Eastern Shore	243	240	-3	-1%
Western Shore	219	234	+15	+7%
Totals	942	1006	+64	+7%

Change in Total Blocks between Atlases by Region

Breeding Bird Survey Results in Maryland

Northern Waterthrush

Parkesia noveboracensis

Although the Northern Waterthrush is one of our less colorful warblers, its modest plumage is attractive enough and somewhat reminiscent of the spotted thrushes. The male and female share the same pattern of unmarked olive brown back, wings, and tail, and white to pale yellowish underparts marked with bold dark streaks on the flanks and breast, extending also onto the throat. A thin but obvious buffy to whitish supercilium completes the ensemble. Its plumage differs from that of its sibling species, the Louisiana Waterthrush, in the uniformity of color on the underparts, the streaked throat, and the narrower eyebrow stripe. Other, more obscure, distinguishing features are duller pinkish legs and a slightly smaller bill (Bent 1953). An easier identification is provided by the rapid descending song, which typically begins with three or four slurred whistles and ends with an emphatic *chup chup chup!*

As indicated by their names, both waterthrushes share an affinity for an aquatic habitat. However, they choose distinctly different versions: the Northern Waterthrush is usually found in high-elevation wooded swamps and bogs, while the Louisiana Waterthrush is associated with streams. Both species feed on the ground, poking into crevices and turning over leaves to hunt for small invertebrate prey. As they walk (not hop), they constantly teeter up and down, very much like several other waterside birds (e.g., American Dipper [*Cinclus mexicanus*], Spotted Sandpiper).

The nest of the Northern Waterthrush is on or near the ground, usually near or over water. Often it is among the roots of overturned or living trees, in a hollow stump, or on the side of an overhanging stream bank (H. Harrison 1975). It is built of mosses, twigs, bark strips, and skeletonized leaves, and is lined with mosses, hair, and fine grass.

Maryland is near the southern extremity of the breeding range of the Northern Waterthrush, and within the state the species is almost restricted to the Allegheny Mountain region of Garrett County. Two observed blocks for the present atlas suggest that there could be occasional nesting pairs at higher elevations of the Ridge and Valley region. However, high-elevation wetlands in Maryland are nearly restricted to the high plateau of Garrett County, so that the other counties offer poor prospects.

Like most other boreal warbler species, the Northern Waterthrush winters in the neotropics. Its normal winter range extends from northwestern Mexico to northern Peru, including the West Indies. As on its breeding territory, the wintering birds are usually near water and many forage in coastal mangroves. The first spring migrants may arrive in Maryland as early as mid-April, but most arrive during the first half of May. Fall migration extends from late July into October, with several records in November and December, but mid-September sees the largest southward flow.

Since work was completed on the first atlas in 1987 (Robbins 1996h), the number of blocks hosting breeding Northern Waterthrushes (possible, probable, or confirmed) dropped from 24 to 10, a decline of 58%. However, overall population trends from BBS results across its United States breeding range do not show such a drop, and the Maryland decline may reflect a withdrawal from the southern margin of its breeding range.

T. DENNIS COSKREN

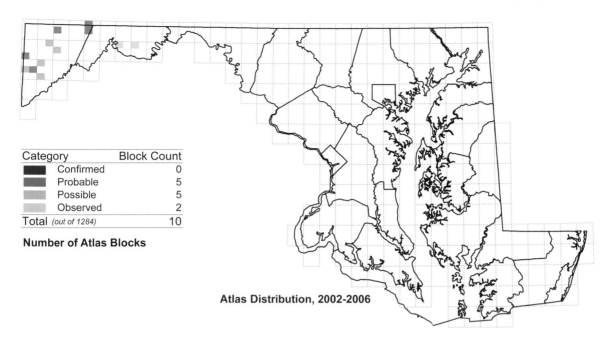

Category	Block Count
■ Confirmed	0
■ Probable	5
■ Possible	5
■ Observed	2
Total (out of 1284)	10

Number of Atlas Blocks

Atlas Distribution, 2002-2006

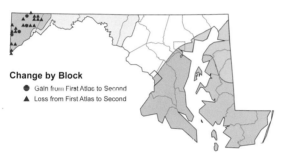

Change by Block

● Gain from First Atlas to Second
▲ Loss from First Atlas to Second

Change in Atlas Distribution, 1983-1987 to 2002-2006

Atlas Region	1983-1987	2002-2006	Change No.	%
Allegheny Mountain	24	10	-14	-58%
Ridge and Valley	0	0	0	-
Piedmont	0	0	0	-
Upper Chesapeake	0	0	0	-
Eastern Shore	0	0	0	-
Western Shore	0	0	0	-
Totals	24	10	-14	-58%

Change in Total Blocks between Atlases by Region

Mark Hoffman

Louisiana Waterthrush

Parkesia motacilla

The Louisiana Waterthrush is a rather large, brown, thrushlike wood warbler. It is a bird of wooded streams ranging from rushing brooks to swampy blackwater coastal rivers. Its descending loud ringing song rises above the rush and babble of rocky torrents but the song is frequent only until mid-May. The call, a distinctive loud *chik* that sounds like quartz pebbles being clashed together, provides a more consistent cue to this waterthrush's presence. It breeds earlier than most other wood warblers, perhaps because it requires spring high water and abundant aquatic insects to successfully nest (Hess et al. 2000). Although it is widespread with nesting records from all Maryland counties and DC, it occurred in fewer than half of the blocks surveyed in this project.

The Louisiana Waterthrush's basic habitat requirements are unpolluted headwater streams in mature extensive forest (Prosser and Brooks 1998). Within these strictures this warbler occurs along a wide array of running waters, including the wooded upper reaches of Coastal Plain rivers and creeks, ravines with permanent and ephemeral brooks, and streams with wet riparian forest. Forest cover can be almost pure hemlock, hardwoods, or swampy woods with bald cypress. C. Keller et al. (1993) demonstrated that narrow riparian corridors generally do not host these waterthrushes. Although their feeding behavior is similar to the closely related Northern Waterthrush, they have longer and heavier bills and ac-

cordingly take larger prey including many caddis flies and earthworms (Craig 1984, 1987).

The Louisiana Waterthrush is a long-distance migrant wintering in the Greater Antilles, Mexico, and northern Central America. It is among the earliest warblers to return to nest in Maryland and DC, with the vanguard appearing in the last week of March. After raising a single brood, it is early to depart, commencing its autumn migration in July with the last of the species usually gone by the first week of September. The nest is built at ground level in root clusters, under overhangs in stream cut-banks, or under fallen logs with the entrance decorated with dead leaves (W. Robinson 1995). The nest is hard to find, so most breeding confirmations in this project were based on family groups and adults seen with food. Maryland and DC nests with eggs have been found from 17 April to 14 June with nestlings to 19 June (Ricciardi 1996e; MNRF). Reports of waterhrushes feeding young into mid-July during this project imply later egg and nestling dates.

Kirkwood (1895) stated, "This species probably spends the summer with us in greater numbers than is generally supposed." Eifrig (1904) called the Louisiana Waterthrush rather common in "Carolinian" parts of western Maryland (i.e., warmer and lower elevation). Stewart and Robbins (1958) detailed its status as common on the lower Coastal Plain and fairly common from the upper Coastal Plain westward to relatively low elevations (up to 700 m [2,296 ft]) in the Allegheny Mountains. It was found in 46% of atlas blocks scattered across the length of Maryland in 1983–1987 (Ricciardi 1996e). It was largely absent east of Chesapeake Bay except in the Pocomoke basin and on the upper Choptank and Tuckahoe Creek. It was found in the vast majority of blocks from northern Cecil south to St. Mary's County and west to southern Carroll and southeastern Frederick County. It was present in the Blue Ridge and along the Potomac but scarce in the Frederick and Hagerstown valleys. It was also numerous in the western Ridge and Valley but largely confined to the lower elevation parts of Garrett County. In the 2002–2006 atlas it showed a very similar distribution marked by slight increases in the Frederick and Hagerstown valleys and slight losses in Caroline and Garrett counties.

Louisiana Waterthrushes are often overlooked on BBS routes because of their early peak of songfulness and the rush and mutter of their nesting streams. BBS trends are not significant and suggest stability or perhaps a slow increase over the whole range; this is also true for Maryland BBS routes. Improved distribution in the Hagerstown and Frederick valleys may reflect protection of riparian buffer woodlands to limit runoff from erosion. On the other hand, crop agriculture renders many streams turbid in spite of buffers, and urbanization removes contiguous forest and pollutes streams as well. This chocolate brown voice of cool forested streams appears to be holding its own but still bears careful watching because of its sensitivity to pollution and forest fragmentation.

WALTER G. ELLISON

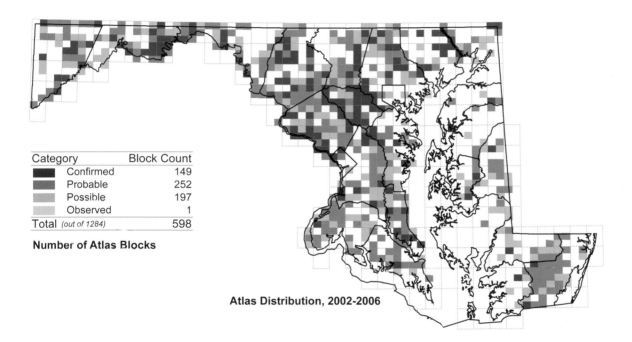

Category	Block Count
Confirmed	149
Probable	252
Possible	197
Observed	1
Total (out of 1284)	598

Number of Atlas Blocks

Atlas Distribution, 2002-2006

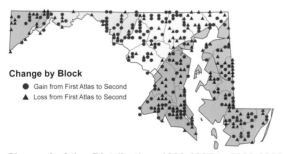

Change by Block
- ● Gain from First Atlas to Second
- ▲ Loss from First Atlas to Second

Change in Atlas Distribution, 1983-1987 to 2002-2006

Percent of Stops
- 50 - 100%
- 10 - 50%
- 0.1 - 10%
- < 0.1%

Relative Abundance from Miniroutes, 2003-2007

Atlas Region	1983-1987	2002-2006	Change No.	Change %
Allegheny Mountain	52	45	-7	-13%
Ridge and Valley	92	104	+12	+13%
Piedmont	188	205	+17	+9%
Upper Chesapeake	30	19	-11	-37%
Eastern Shore	87	83	-4	-5%
Western Shore	139	141	+2	+1%
Totals	588	597	+9	+2%

Change in Total Blocks between Atlases by Region

Breeding Bird Survey Results in Maryland

Mark Hoffman

Kentucky Warbler
Oporornis formosus

The Kentucky Warbler is far more often heard than seen. Although not overly furtive, this warbler is a bird of the ground in places with thick shrubs and herbs, seldom venturing higher than 6 m (20 ft) above the ground. The song of this warbler, however, is far-carrying and rich-toned, a loud husky *curry-curry-curry-curry*. The similarity of this song to another bird of thickets, the Carolina Wren, can lead to observers overlooking the Kentucky Warbler although the warbler's song has a hoarser quality and is more stereotyped in delivery. The Kentucky Warbler is also more a bird of the forest interior than is the Carolina Wren, occupying wet woods with rank herbage, abundant shrubs, and curtains of vines most often near running water and intermittent streams in ravines.

The Kentucky Warbler winters in Central America in humid lowland forests (McDonald 1998). Returning Kentucky Warblers arrive rather late for a southern warbler species; a few appear in late April, but most arrive in the first half of May. Autumn departure is early, with almost all gone by the end of August. These warblers nest on or just above the ground, often under ferns or at the bases of shrubs and robust herbs; the well-hidden nest is hard to find. There were only two nest reports among the 44 breeding confirmations on this project dated 23 May (ON) and 9 July (NE); published egg dates for Maryland and DC range from 16 May to 31 July

(MNRF). Most breeding confirmations for this project were for recently fledged young, with 21 records from 20 June to 24 July.

The breeding distribution of the Kentucky Warbler in Maryland has constricted considerably over the past two decades; this warbler now occurs in 284 fewer blocks than in 1983–1987 (a 38% decline). The general outline of the Kentucky Warbler's range is similar to the 1983–1987 range, but there are now large gaps in what was once a more uniform distribution. Historically, the Kentucky Warbler was rare in Western Maryland, where it increased in the 1960s and 1970s (Stewart and Robbins 1958; Davidson 1996g). The Allegheny Mountains and Ridge and Valley, where the species has lost ground in upper-elevation blocks, showed large losses by percentage of blocks between the two atlas projects (46% and 50% respectively). Similar losses in the rapidly suburbanizing Piedmont (46%) represented nearly 40% of the total losses for this species in the state. But losses of atlas blocks were registered in all Maryland physiographic regions with the fewest losses recorded in the Western Shore region (21%), where this warbler is still widespread in southern parts of the region (also experiencing rapid development).

Breeding Bird Survey trends for Kentucky Warbler show declines; the long-term annual rate of decline for Maryland is 3.1%, with the rate accelerating over the past 25 years. Possible causes for the current steep decline in the occupied range of the Kentucky Warbler include unsuitability of small habitat fragments for successful nesting attributable to edge predators and cowbird brood parasitism (Robbins, Dawson, and Dowell 1989; Gibbs and Faaborg 1990; S. Robinson et al.

1995); loss of understory herbs and shrubbery in forests because of heavy browsing by burgeoning numbers of white-tailed deer (D. Boone and Dowell 1986; McShea et al. 1995); and perhaps loss of lowland humid forest in the Central American wintering range (McDonald 1998). The Kentucky Warbler requires careful monitoring because of its decline. Large wooded preserves must be maintained to house viable nesting populations, and deer must be culled within or excluded from these preserves. Many places occupied by Kentucky Warblers are dominated by invasive plants unpalatable

to deer such as multiflora rose. Whether these habitats provide refuge from deer browsing or are sinks offering only poor nesting conditions for these warblers is unknown but could potentially complicate invasive plant control decisions in forest preserves. Although the Kentucky Warbler is still fairly common in some areas, it has been lost from much of the landscape over the past two decades in Maryland and its future is uncertain.

WALTER G. ELLISON

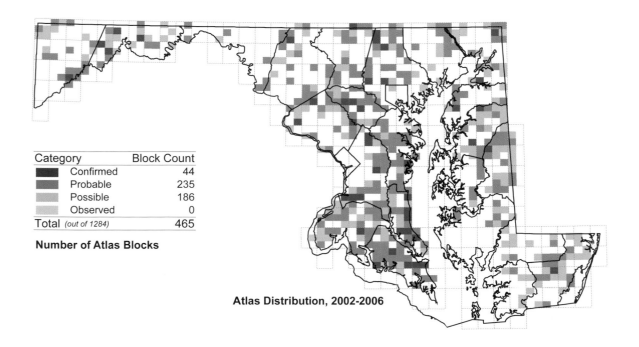

Category	Block Count
Confirmed	44
Probable	235
Possible	186
Observed	0
Total *(out of 1284)*	465

Number of Atlas Blocks

Atlas Distribution, 2002-2006

Change by Block

● Gain from First Atlas to Second
▲ Loss from First Atlas to Second

Change in Atlas Distribution, 1983-1987 to 2002-2006

Percent of Stops

■ 50 - 100%
■ 10 - 50%
■ 0.1 - 10%
□ < 0.1%

Relative Abundance from Miniroutes, 2003-2007

Atlas Region	1983-1987	2002-2006	Change No.	%
Allegheny Mountain	52	28	-24	-46%
Ridge and Valley	80	40	-40	-50%
Piedmont	236	127	-109	-46%
Upper Chesapeake	76	54	-22	-29%
Eastern Shore	129	77	-52	-40%
Western Shore	174	137	-37	-21%
Totals	747	463	-284	-38%

Change in Total Blocks between Atlases by Region

Breeding Bird Survey Results in Maryland

Mourning Warbler
Oporornis philadelphia

The Mourning Warbler is a subtly attractive wood warbler arrayed in yellow and olive with an ash gray hood. It is a bird of the thorny undergrowth of forest clearings in hilly and mountainous country, with a distinctive bright chortling song. Although it has a broad breeding range from northeastern British Columbia to Newfoundland ranging southward to the Great Lake states, New England, and in the Appalachians to West Virginia (Pitochelli 1993), it is only locally numerous and spottily distributed because of its ephemeral habitat. Maryland hosts a small, geographically isolated population presently confined to the state's highest elevations in southwestern Garrett County.

Mourning Warblers live among early successional shrubs in moist forests. They generally settle on sites where forest is regenerating two to three years after trees were lost because of fire, wind throw, insect damage, disease, flooding, or logging. These locations usually feature abundant deciduous saplings and brambles and sedge, often *Carex crinita* (fringed sedge). Males require at least a few relatively sizable saplings from 5 to 6 m (16–20 ft) tall for song perches. In Maryland they are mostly found in openings and along shrubby road shoulders with oak, maple, cherry, and hickory saplings. Historically they were also found in oak-chestnut sproutlands (Stewart and Robbins 1958).

These warblers winter in southern Central America and northwestern South America (Pitochelli 1993). They are late spring migrants, returning from mid- to late May, with a few still passing through Maryland and DC in the first week of June. The fall departure takes place largely during September with a few to early October. Mourning Warblers nest on or just above the ground in dense cover. They are very secretive near the nest, but are easily agitated when an observer is near nestlings or recent fledglings; they hop in the open giving loud *squick* calls and occasional rodent-run distraction displays. There are no published egg dates for Maryland. The first proof of nesting was provided by adults carrying food to hidden young on 26 June 1949 at Roth Rock Tower on Backbone Mountain, Garrett County (Robbins 1949a; W. Murphy and Robbins 1996). A fledgling was found on 1 July 2006 on Backbone Mountain north of Table Rock (F. Pope, this project). Egg dates for Pennsylvania range from 31 May to 3 July (Leberman 1992d).

The Mourning Warbler was considered a rare transient in Maryland and DC by Kirkwood (1895). It was not found in summer in Maryland until the 1930s by Maurice Brooks (1936). Chandler Robbins confirmed breeding at Roth Rock Tower in 1949 in a 12-year-old logging cut of oak, maple, and hickory saplings (Robbins 1949a). Stewart and Robbins (1958) stated it was local and uncommon on Backbone Mountain above 914 m (3,000 ft) with some as low as 805 m (2,640 ft) on the mountain's east slope. From 1983 to 1987 it was found in 7 blocks with only 3 records of probable and confirmed breeding (W. Murphy and Robbins 1996). Although there were 5 reports north of Backbone Mountain the only probable record was near Herrington Manor State Park. During the 2002–2006 atlas this warbler was found in only 4 blocks on Backbone Mountain in the Table Rock and Davis quadrangles.

The Mourning Warbler has declined on North American BBS routes, most steeply since 1980. There are no Maryland BBS data. The most likely reason for this decline appears to be the net trend toward reforestation over much of the bird's range. This is one of a handful of birds that benefit from forest clear-cutting, and the reduction of this practice on eastern forest lands may have lowered its numbers in recent decades. The Maryland population appears to be isolated, with the closest summering birds in Randolph County 40 km (25 mi) to the south during the West Virginia atlas (Buckelew and Hall 1994), although older records exist for counties neighboring Maryland (Hall 1983). The nearest records to the north, in Pennsylvania, are over 200 km (130 mi) away in Clearfield and Centre counties (McWilliams and Brauning 2000). The Maryland population is officially listed as endangered by the Maryland DNR and bears close watching because it is dependent on ongoing disturbance of mature forest for its continued existence.

WALTER G. ELLISON

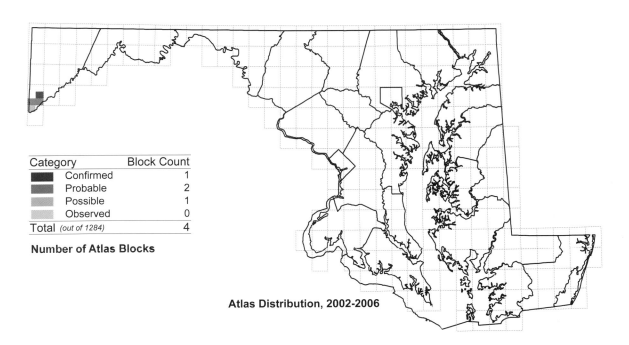

Category	Block Count
Confirmed	1
Probable	2
Possible	1
Observed	0
Total (out of 1284)	4

Number of Atlas Blocks

Atlas Distribution, 2002-2006

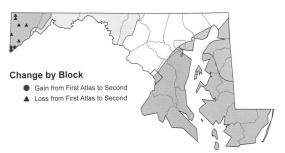

Change by Block
● Gain from First Atlas to Second
▲ Loss from First Atlas to Second

Change in Atlas Distribution, 1983-1987 to 2002-2006

Atlas Region	1983-1987	2002-2006	Change No.	%
Allegheny Mountain	7	4	-3	-43%
Ridge and Valley	0	0	0	-
Piedmont	0	0	0	-
Upper Chesapeake	0	0	0	-
Eastern Shore	0	0	0	-
Western Shore	0	0	0	-
Totals	7	4	-3	-43%

Change in Total Blocks between Atlases by Region

George M. Jett

Common Yellowthroat
Geothlypis trichas

The Common Yellowthroat is a wrenlike warbler, somewhat shy but active and aggressively curious. The first description of this species was based on a specimen collected in Maryland in the mid-eighteenth century. In *Gleanings of Natural History,* George Edwards called it the Maryland Yellowthroat, a name sometimes still applied to the subspecies breeding in Maryland and DC (AOU 1998). The male's song is a familiar, bright repetitive ditty usually transcribed as *witchety-witchety-witchety,* although it sometimes sounds more like *what's it to ya.* Yellowthroats may be found in wet thickets throughout Maryland and DC. This is the most widespread and numerous warbler in Maryland and it was one of the 10 most frequently recorded birds during both statewide atlas projects.

Common Yellowthroats breed in dense, rather coarse vegetation usually in wet places. As R. Stewart (1953) observed, the association with water is because the thick plant cover yellowthoats require is best developed where it is wet; they can occur in thick cover in relatively dry situations. The wide array of habitats occupied in Maryland and DC include marshes with tall emergent vegetation, shrub swamps, damp brushy meadows, overgrown abandoned farmland, power line rights of way, field ditches, stream banks, and lake shores.

Although a few yellowthroats winter on the Coastal Plain, the great majority are migratory, leaving Maryland and DC for the southeastern United States, Mexico, Central America, and the West Indies. They return by mid-April and most depart again by late October. Common Yellowthroats nest on or just above the ground in rank grasses or low shrubs (R. Stewart 1953). As with most ground and near-ground nesting warblers, the yellowthroat has a distraction display characterized by adults running rodentlike on the ground with quivering wings held over their backs, males appear more prone to giving this display (W. Ellison, pers. obs.). Because parents are very wary around nests, most confirmations were based on adults feeding young or observations of fledglings. Maryland and DC egg dates range from 4 May to 4 August (Stewart and Robbins 1958; MNRF), and nestlings have been reported to 22 August (Joyce 1996c; MNRF).

Early listings of birds in Maryland and DC all called the Common Yellowthroat common to abundant (C. Richmond 1888; Kirkwood 1895). Stewart and Robbins (1958) noted slight differences in numbers among regions, finding them abundant on the Coastal Plain, fairly common on the Piedmont and in the Ridge and Valley, and common in the Allegheny Mountains.

During the 1983–1987 atlas project, yellowthroats were reported from more than 1,200 blocks, with very few gaps in their distribution (Joyce 1996c). They were missed in a few blocks on the lower Eastern Shore in areas dominated by pine savanna and tidal marshland. They had essentially the same distribution in 2002–2006, with a small net increase in blocks, notably on the lower Eastern Shore. This seems to be related to improvements in atlas coverage in these largely uninhabited wetlands rather than to any actual change in distribution.

Continental BBS trends have been modestly declining at a rate of less than 1% per year. Declines have been more pronounced in Maryland at 1.7% per year. Loss of wetlands and

shrubland to suburbanization and reforestation have likely taken a toll on Common Yellowthroat populations. However, their habitat is still sufficiently widespread and their territory sizes so small that they evince little decline at the scale of standard atlas blocks. In spite of lower populations, yellowthroats will probably remain widely distributed and at least fairly common into the foreseeable future.

WALTER G. ELLISON

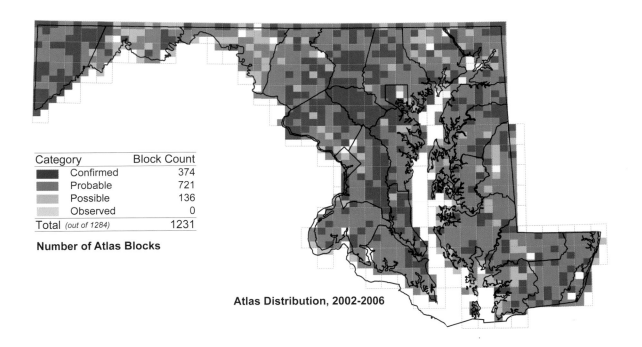

Category	Block Count
Confirmed	374
Probable	721
Possible	136
Observed	0
Total (out of 1284)	1231

Number of Atlas Blocks

Atlas Distribution, 2002-2006

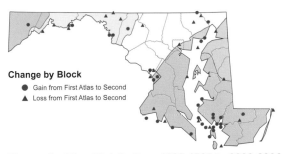

Change by Block
- ● Gain from First Atlas to Second
- ▲ Loss from First Atlas to Second

Change in Atlas Distribution, 1983-1987 to 2002-2006

Percent of Stops
- 50 - 100%
- 10 - 50%
- 0.1 - 10%
- < 0.1%

Relative Abundance from Miniroutes, 2003-2007

Atlas Region	1983-1987	2002-2006	Change No.	%
Allegheny Mountain	91	92	+1	+1%
Ridge and Valley	142	139	-3	-2%
Piedmont	312	309	-3	-1%
Upper Chesapeake	119	118	-1	-1%
Eastern Shore	312	322	+10	+3%
Western Shore	243	246	+3	+1%
Totals	1219	1226	+7	+1%

Change in Total Blocks between Atlases by Region

Breeding Bird Survey Results in Maryland

Hooded Warbler
Wilsonia citrina

White outer tail spots flash as the Hooded Warbler flits through the forest understory. These are often an observer's key for identifying the species, which can blend with the forest greens if seen from behind. On the underside the Hooded Warbler is bright yellow, and the male has a startlingly black hood. Older females often have a black border to the olive on the crown or a shadow of a hood, but are never as well marked as the male.

During the nesting season, Hooded Warblers inhabit mature forests that are generally larger than about 15 ha (37 ac) (L. Ogden and Stutchbury 1994) and have sufficient understory for nesting. These may include forests with swamps or seeps dominated by sweet pepperbush, greenbrier, arrowwood, or spicebush (Stewart and Robbins 1958), or upland forests that have tree fall gaps or patches of mountain laurel, or that have been lightly logged.

Hooded Warblers winter primarily on the Gulf-Caribbean slope from Mexico through Central America and show habitat segregation by sex (Lynch et al. 1985). Males generally winter in mature forest, while females use secondary forest, scrub, and disturbed habitats. There are increasing numbers of Hooded Warbler records in Maryland in mid-April, but most individuals arrive later in April or in early May. Hooded Warblers depart Maryland for the wintering grounds in late August or September, with some individuals staying into October.

Female Hooded Warblers build an open cup nest that is often wrapped with camouflaging dead leaves and leaf skeletons. Nests are placed low (0.3–1.4 m above the ground [1–4.7 ft]; L. Ogden and Stutchbury 1994) in young trees, shrubs, or herbs, and incorporate two or more upright or oblique stems as support. Hooded Warbler nests are frequently parasitized by Brown-headed Cowbirds in Maryland (Stasz 1996d) and rangewide (L. Ogden and Stutchbury 1994). Most confirmations of breeding for this atlas project (n = 58) were from observations of adult Hooded Warblers feeding young (n = 28; earliest date 31 May, latest date 9 August) or of fledglings (n = 19; earliest date 12 June, latest date 23 August). Few nests were located, with 5 reports of nest-building (earliest date 11 May, latest date 16 June), one of an occupied nest (14 June), and one nest with young (11 July); no nests with eggs were found, but published dates for eggs in Maryland range from 21 May to 30 July (Stewart and Robbins 1958; Stasz 1996d).

Hooded Warblers occur in all regions of Maryland; their current distribution is generally similar to that reported by Stewart and Robbins (1958). They were documented in 391 atlas blocks during 2002–2006, 10% fewer than during 1983–1987 (Stasz 1996d). The Ridge and Valley was the only region with an increase in Hooded Warbler occurrence (50 blocks in 2002–2006 versus 43 in 1983–1987), but the species remains quite common on the Western Shore, where it was found in 171 blocks (7% fewer than during 1983–1987), and on the Allegheny Plateau (67 blocks in 2002–2006 versus 73 in 1983–1987). The Piedmont showed the largest decline, with Hooded Warblers found in 28% fewer blocks than in 1983–1987 (74 versus 103 blocks). On the Eastern Shore, there were notable losses along the Pocomoke River system but several new occurrences in Dorchester County.

Estimates from Breeding Bird Survey data of rangewide

Hooded Warbler population trends are positive but nonsignificant. In Maryland, trends are negative but nonsignificant over the long term (1966–2007), with a slight positive trend prior to 1980 and a slight negative trend afterward. The species is still basically holding its own, owing in part to statewide forest conservation efforts. Forest fragmentation associated with suburban expansion and heavy browsing of the forest understory by white-tailed deer will likely result in further losses of Hooded Warblers in Maryland.

DEANNA K. DAWSON

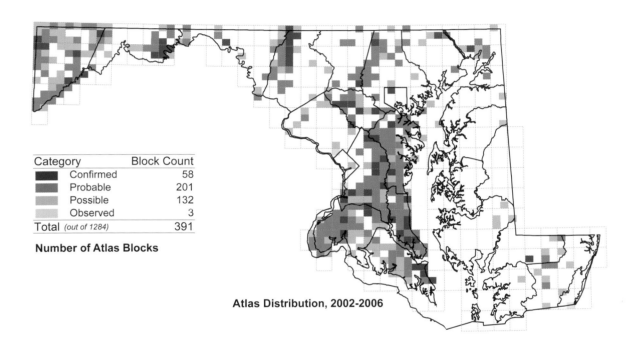

Category	Block Count
Confirmed	58
Probable	201
Possible	132
Observed	3
Total (out of 1284)	391

Number of Atlas Blocks

Atlas Distribution, 2002-2006

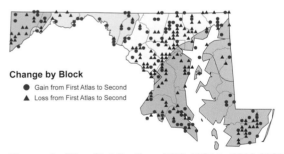

Change by Block
- ● Gain from First Atlas to Second
- ▲ Loss from First Atlas to Second

Change in Atlas Distribution, 1983-1987 to 2002-2006

Percent of Stops
- 50 - 100%
- 10 - 50%
- 0.1 - 10%
- < 0.1%

Relative Abundance from Miniroutes, 2003-2007

Atlas Region	1983-1987	2002-2006	Change No.	Change %
Allegheny Mountain	73	67	-6	-8%
Ridge and Valley	43	50	+7	+16%
Piedmont	103	74	-29	-28%
Upper Chesapeake	6	6	0	0%
Eastern Shore	23	21	-2	-9%
Western Shore	183	171	-12	-7%
Totals	431	389	-42	-10%

Change in Total Blocks between Atlases by Region

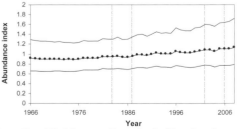

Breeding Bird Survey Results in Maryland

Canada Warbler
Wilsonia canadensis

One of our skulking species, the Canada Warbler is distinctive when it shows itself. The breeding male is mostly uniform medium gray above, shading to black along the front and sides of the head. Below, the throat, breast, and belly are lemon yellow, and the undertail coverts are white. A bold white eye-ring sets off a black eye. The most distinctive feature is a "necklace" of bold black streaks across the upper breast. The female has the same pattern but the black is replaced by gray, and immatures are similar but with paler and less well-marked plumage. The song is forceful and distinctive in its lack of a simple melodic pattern, consisting of a series of high whistled notes that skip rapidly about the scale with little or no repetition. If you enjoy the 12-tone school of classical music, then you will very much appreciate the compositions of the Canada Warbler.

This warbler is another of our boreal species. Its habitat of choice is thick leafy understory in mainly deciduous forest, commonly near water, similar to the haunts of its congeners Wilson's Warbler (*W. pusilla*) and Hooded Warbler. Density of foliage in the shrub layer and the availability of conspicuous singing perches appear to be prime factors in choosing territories (Hallworth et al. 2008). The Canada Warbler feeds mostly within the understory by gleaning insects and spiders, often by flycatching. It was once called the Canada flycatching warbler or Canada flycatcher (Bent 1953). Like other boreal species, the Canada Warbler is a breeding resident only in westernmost Maryland, with most of the nesters in Garrett County and a few spilling over onto the higher ridges of western Allegany County. Although it

is mostly a high-elevation species between 640 and 988 m (2,100–3,240 ft; Robbins 1950; Stewart and Robbins 1958), it has been found as low as 466 m (1,530 ft) in Maryland during the breeding season (Bridge 1966).

The Canada Warbler is a long-distance neotropical migrant, wintering mostly along the lower east slope of the Andes from Venezuela south to Peru, and rarely in Central America. In company with many other boreal migrants it begins arriving in Maryland in late April, but most of the population arrives during the middle two weeks of May. The fall migration begins in August, with most birds passing though from mid-August to late September.

The nest is usually on or near the ground, often on mossy stumps or logs and sometimes on steep embankments, in a site with dense cover. Commonly it is composed of dry (often skeletonized) leaves, grass, fern fronds, and weed stems, and is lined with fine plant fibers, fine rootlets, and hair.

Since the completion of fieldwork for the first atlas in 1987 (Robbins 1996i), the Canada Warbler has declined. The geographic extent of the breeding grounds has remained about the same, but within that area the number of blocks hosting breeding Canada Warblers dropped from 46 to 36. This is consistent with trends elsewhere within its range. Breeding Bird Survey results from 1966 to 2001 have shown an annual population decline averaging 1.9% per year over that period; the most severe declines have been in the northeastern United States (http://audubon2.org/watchlist/viewSpecies.jsp?id=61). The reasons for the decline are not well understood but may include habitat degradation within the breeding range (particularly the loss of a dense understory) and the loss or degradation of habitat in its tropical wintering grounds.

T. DENNIS COSKREN

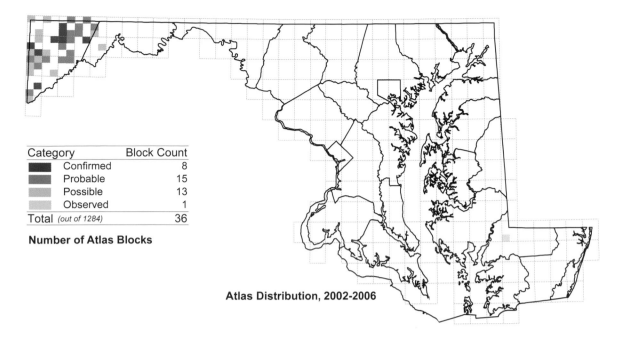

Category	Block Count
Confirmed	8
Probable	15
Possible	13
Observed	1
Total (out of 1284)	36

Number of Atlas Blocks

Atlas Distribution, 2002-2006

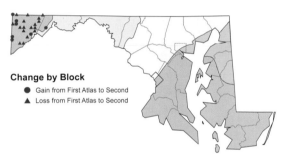

Change by Block
● Gain from First Atlas to Second
▲ Loss from First Atlas to Second

Change in Atlas Distribution, 1983-1987 to 2002-2006

Percent of Stops
- 50 - 100%
- 10 - 50%
- 0.1 - 10%
- < 0.1%

Relative Abundance from Miniroutes, 2003-2007

Atlas Region	1983-1987	2002-2006	Change No.	%
Allegheny Mountain	45	36	-9	-20%
Ridge and Valley	1	0	-1	-100%
Piedmont	0	0	0	-
Upper Chesapeake	0	0	0	-
Eastern Shore	0	0	0	-
Western Shore	0	0	0	-
Totals	46	36	-10	-22%

Change in Total Blocks between Atlases by Region

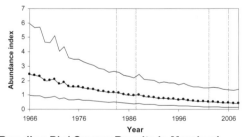

Breeding Bird Survey Results in Maryland

Charles Lentz

Yellow-breasted Chat
Icteria virens

Often the first indication that an observer has found a Yellow-breasted Chat is the sound of a distantly cawing crow from a meadow with scattered mounds of multiflora rose, which is followed after several breathless seconds by an odd coughing sound, and after another pause a rapid nasal chatter. At this point a visual scan may locate the long-tailed bird with the bright orange-yellow breast atop a bush as it produces its amazing array of sounds; it may even pop up into the air on shallowly rowing wings. Yellow-breasted Chats are shy birds, very hard to see and know outside of their rather short season of songfulness. Whether the Yellow-breasted Chat is actually a wood warbler or a member of another closely related family has long been a matter of controversy. The most recent DNA sequence analyses strongly indicate that this bird is not a warbler (Lovette and Bermingham 2002).

Yellow-breasted Chats prefer shrubby fields and leave them soon after the invading trees begin to form a closed canopy. Chats may be found in overgrown pastureland and fields, hedgerows adjacent to tall grass fields, the drier parts of shrub swamps, shrubby stream-bank thickets, and regenerating clear-cuts in forestry lands. Thorny vegetation is an almost constant feature of the haunts of Yellow-breasted Chats, including brambles, multiflora rose, greenbriers, and hawthorns. Yellow-breasted Chats nest low in shrubs and small trees, often amid thorns. Not surprisingly, given such well-protected nest sites, most atlas observers did not find actual nests when they recorded Yellow-breasted Chat, only five reports for this project involved finding a nest, and no reports recorded nest contents. Maryland egg dates for this species range from 18 May to 16 July, and nestlings have been seen as late as 26 July (Stewart and Robbins 1958; MNRF).

Yellow-breasted Chats were called common by early chroniclers of Maryland and DC's birdlife (Kirkwood 1895; Eifrig 1904). Stewart and Robbins (1958) found them common on the Coastal Plain, fairly common westward to the Allegheny Front, and uncommon beyond it. These chats were widespread in the 1983–1987 atlas, occurring in 79% of the blocks, and were among the 50 most prevalent bird species in the state and DC (Cheevers 1996d). Chats have lost ground over the past two decades as the number of blocks with records fell by almost 20%. There were losses in all regions; those in excess of 20% were the Allegheny Mountains (58%), the Ridge and Valley (25%), and the Piedmont (28%). The Yellow-breasted Chat occurred in less than 66% of blocks from 2002 to 2006.

Trends for Yellow-breasted Chat on the BBS have been mixed. Its continental population appears to be stable, but it has declined by 2% per year in Maryland although the trend has slowed over the past quarter century. The BBS trend agrees with the losses shown on the 2002–2006 atlas map. The primary reason for this chat's decline appears to be habitat loss from a variety of sources, including maturation of scrub habitat to young forest, losses to suburban development, and loss of hedgerows and fallow fields to crop fields on farms. Yellow-breasted Chats are still fairly numerous on the lower Eastern Shore in managed forest lands where they occupy cutover lands not long after timber harvest (ca.

3–5 years). This bird should be monitored carefully to maintain its numbers. Management techniques that would favor Yellow-breasted Chats include deliberately creating openings in managed forests, maintaining power line rights-of-way as shrublands, and retaining more shrubby open land on farms and around housing developments (Eckerle and Thompson 2001).

WALTER G. ELLISON

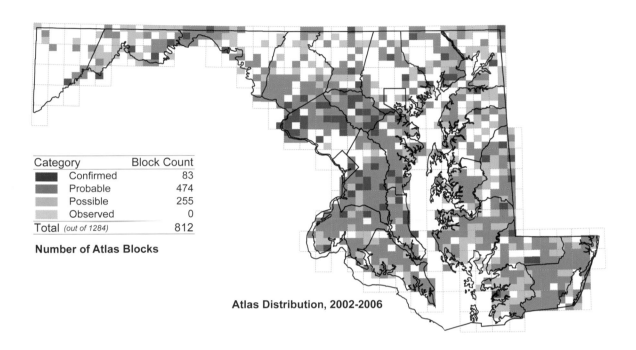

Category	Block Count
Confirmed	83
Probable	474
Possible	255
Observed	0
Total *(out of 1284)*	812

Number of Atlas Blocks

Atlas Distribution, 2002-2006

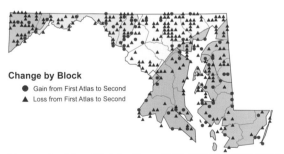

Change by Block
- ● Gain from First Atlas to Second
- ▲ Loss from First Atlas to Second

Change in Atlas Distribution, 1983-1987 to 2002-2006

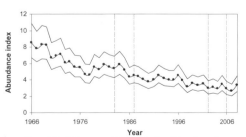

Percent of Stops
- 50 - 100%
- 10 - 50%
- 0.1 - 10%
- < 0.1%

Relative Abundance from Miniroutes, 2003-2007

Atlas Region	1983-1987	2002-2006	Change No.	Change %
Allegheny Mountain	48	20	-28	-58%
Ridge and Valley	118	88	-30	-25%
Piedmont	258	186	-72	-28%
Upper Chesapeake	101	84	-17	-17%
Eastern Shore	254	231	-23	-9%
Western Shore	225	200	-25	-11%
Totals	1004	809	-195	-19%

Change in Total Blocks between Atlases by Region

Breeding Bird Survey Results in Maryland

Fran Saunders

Eastern Towhee
Pipilo erythrophthalmus

The Eastern Towhee is a shy bird often heard before it is seen. The querulous calls that give the bird its common name emanate from thickets and brush piles. This call could be rendered as *shwee* although traditional interpretations include *towhee* and *chewink* (an earlier vernacular name). This towhee's energetic two-footed scratching in dry leaf litter and its bright brittle song usually rendered as *drink your teeee* also call attention to it.

Essentially a very large emberizid sparrow, the Eastern Towhee is almost as big as a robin, with shiny jet black or chocolate brown upperparts relieved by a touch of white in the wing and at the corners of the long slender tail; the breast and belly are white fringed by a broad open vest of cinnamon brown. Eastern Towhees inhabit a variety of habitats characterized by thick shrubby growth, including overgrown old pastures and fields, dry brushy forest with well-spaced trees, floodplain thickets and understory brush, pine savanna, and grasslands with luxuriant hedgerows and scattered shrubs.

This towhee regularly winters in small numbers on the Coastal Plain with a few on the Piedmont. Most local nest-

ers occupy their nesting territories during spring migration from March through early May. Towhees nest throughout spring and summer and often raise two broods. Maryland egg dates range from 22 April to 28 August (MNRF). First brood nests are usually built on the ground and later nests show a strong tendency to be built above the ground in small shrubs (Greenlaw 1996). Towhees showed very similar relative frequencies of possible, probable, and confirmed breeding between the 1983–1987 (Farrell 1996b) and 2002–2006 Maryland-DC atlas projects. The great majority of breeding confirmations in this atlas were for recently fledged young. Juvenile Eastern Towhees have brown rather than red eyes and a distinctive streaky yellow brown plumage that is held for almost a month, facilitating use of the FL code.

The distribution of the Eastern Towhee has changed little in recent years. Farrell (1996b) reported that this towhee's status was little different in the 1980s than in reports as far back as the nineteenth century. In 2002–2006 there was very little difference in the distribution and number of blocks recording Eastern Towhee versus 1983–1987. Many of the differences in block gains and losses relate to coverage differences between atlas projects, although gains recorded in north-central Carroll County and losses in Dorchester, Somerset, and Worcester counties may be genuine.

Although the Eastern Towhee's breeding range in Maryland and DC is essentially the same between the two state-wide atlas projects, the species has suffered a long-term downward population trend on BBS routes since 1966 that was most pronounced through the 1980s (Farrell 1996b). The trend from 1980 to 2006 indicates the towhee population may have stabilized at the low level reached in the 1980s. Possible reasons for the Eastern Towhee decline include loss of habitat through reforestation, destruction of understory vegetation by burgeoning deer populations, development of nesting habitat for housing, and removal of hedgerows and expanding cropland for intensive crop agriculture (mostly in parts of the Eastern Shore). In spite of population declines, the Eastern Towhee remains as widely distributed as it has ever been in the Old Line State, but it bears watching as a species of potential concern.

WALTER G. ELLISON

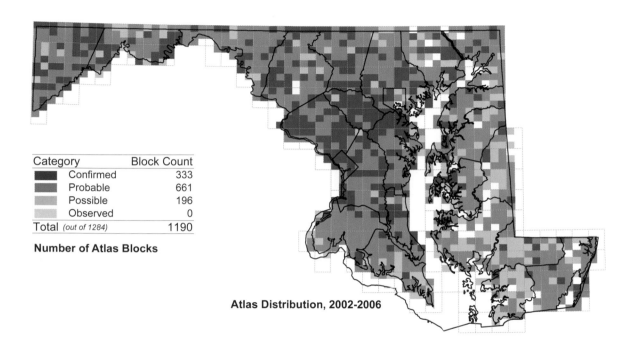

Category	Block Count
Confirmed	333
Probable	661
Possible	196
Observed	0
Total (out of 1284)	1190

Number of Atlas Blocks

Atlas Distribution, 2002-2006

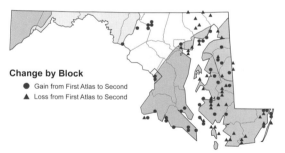

Change by Block
- ● Gain from First Atlas to Second
- ▲ Loss from First Atlas to Second

Change in Atlas Distribution, 1983-1987 to 2002-2006

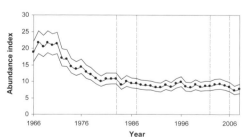

Percent of Stops
- 50 - 100%
- 10 - 50%
- 0.1 - 10%
- < 0.1%

Relative Abundance from Miniroutes, 2003-2007

Atlas Region	1983-1987	2002-2006	Change No.	Change %
Allegheny Mountain	93	93	0	0%
Ridge and Valley	147	148	+1	+1%
Piedmont	308	308	0	0%
Upper Chesapeake	113	111	-2	-2%
Eastern Shore	288	279	-9	-3%
Western Shore	237	247	+10	+4%
Totals	1186	1186	0	0%

Change in Total Blocks between Atlases by Region

Breeding Bird Survey Results in Maryland

Bill Sherman

Chipping Sparrow
Spizella passerina

The Chipping Sparrow is small and slender, with a jaunty air; it is a persistent if monotonous singer and a familiar presence on our lawns and road shoulders as well as in wilder haunts. Chipping Sparrows are most common in areas with a mix of short grass and scattered trees and shrubs; they are numerous on the lawns of homes, businesses, cemeteries, schools, and churches. "Chippies" also are found on grassy road shoulders in woodlands, and lawnlike openings in dry woods including pine-oak and oak-hickory. These rusty-capped, gray-breasted sparrows avoid wet swampy woods, marshes, woods with thick understory shrubbery, mature forest with only small openings, and urban areas with little greenery.

The Chipping Sparrow derives its name from its rapidly trilled machine-gun song. This sparrow nests in shrubs and trees favoring conifers, including pines and ornamental spruce and cedar. These sparrows are double brooded and are also persistent renesters; they have a long breeding season in Maryland and DC. Nesting dates reported for this atlas project ranged from nest-building on 10 April to late fledglings found on 14 September. Maryland egg dates range from 14 April to 2 September and nestlings have been reported as late as 2 October (MNRF). Atlas workers confirmed nesting by Chipping Sparrows in over 67% of the blocks in which it was recorded; 47% of these reports were of finely streaked noisy fledglings.

The Chipping Sparrow is the most widespread sparrow in Maryland and DC. It was found in 1,216 blocks from 2002–2006 and ranked among the top 20 most widely distributed of Maryland's birds. This sparrow has maintained this status here since the nineteenth century (Fletcher 1996b). Chipping Sparrows were absent from several blocks dominated by coastal wetlands (particularly on the Eastern Shore), some of the sparsely vegetated dunes on Assateague Island, and a few highly urbanized blocks (e.g., in Baltimore City). These sparrows declined in downtown Baltimore, but occurred in 4 new blocks in Washington, DC, which may reflect differences in coverage between atlas projects or changes in parks and lawns in the two cities. Other block additions were made along the lower Potomac, on the shores of the lower Chesapeake Bay, and on Atlantic barrier beach islands.

In the early years of the BBS, the Chipping Sparrow declined slightly but significantly in eastern North America (Robbins, Bystrak, and Geissler 1986), but recent trends have stabilized. The Chipping Sparrow remains almost ubiquitous in Maryland and DC and is likely to remain so. Unlike many of Maryland's woodland and grassland bird species, Chipping Sparrows actually benefit from suburban sprawl.

WALTER G. ELLISON

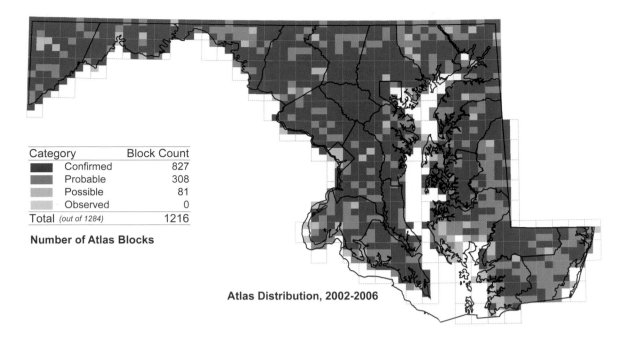

Category	Block Count
Confirmed	827
Probable	308
Possible	81
Observed	0
Total (out of 1284)	1216

Number of Atlas Blocks

Atlas Distribution, 2002-2006

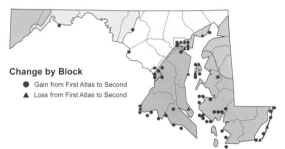

Change by Block
- ● Gain from First Atlas to Second
- ▲ Loss from First Atlas to Second

Change in Atlas Distribution, 1983-1987 to 2002-2006

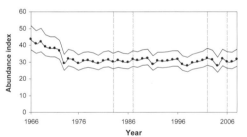

Percent of Stops
- 50 - 100%
- 10 - 50%
- 0.1 - 10%
- < 0.1%

Relative Abundance from Miniroutes, 2003-2007

			Change	
Atlas Region	1983-1987	2002-2006	No.	%
Allegheny Mountain	93	93	0	0%
Ridge and Valley	147	148	+1	+1%
Piedmont	312	315	+3	+1%
Upper Chesapeake	113	110	-3	-3%
Eastern Shore	276	297	+21	+8%
Western Shore	234	247	+13	+6%
Totals	1175	1210	+35	+3%

Change in Total Blocks between Atlases by Region

Breeding Bird Survey Results in Maryland

Field Sparrow
Spizella pusilla

This small brown and gray bird, one of the proverbial "little brown jobs," is fairly common across the state, but is not well known to most people. The back is brown with dark streaking, the breast and sides are buffy shading to a gray belly, and the brown wings have two white bars. The head is grayish with a rufous crown and an indistinct rufous streak on the cheek. The clearest distinguishing features are a white eye ring and a pink bill. The sexes are alike. Juvenile birds have a streaked breast and sides. During the breeding season the Field Sparrow is usually detected and identified by song rather than appearance. That song consists of a two- to four-second series of sweet high whistled notes, accelerating to a trill at the end. The pitch may be constant or not. To lay claim to his territory the singer performs in succession from four or five elevated perches, such as the tops of shrubs or saplings or high on weed stems. Singing commences at daybreak and, for an unmated male, continues throughout the day, but less so during the afternoon. Once paired, he sings much less until nesting begins, and then he is an early morning performer (Walkinshaw 1968).

The Field Sparrow is well named as a bird of grassland, but its habitat is not clean well-kept and weed-free fields. Rather, the bird prefers weedy fields with low brush and brambles, and brushy forest edges. Thus it is an inhabitant of early successional stages. Its basic diet is grass seeds and weed seeds (80%–90% in fall and winter); grass is dominant, but chickweed, sorrel, knotweed, lamb's quarters, pigweed, and ragweed are among the menu items. During the breeding season this vegetable fare drops to less than half, as it is augmented by high-protein insects: beetles, grasshoppers, and caterpillars are the most important, but, ants, flies, wasps, leafhoppers, and spiders are also taken. The nestlings are fed exclusively on animal food (Ehrlich et al. 1988). Food is taken on the ground, or close to it. The grass seeds are obtained by perching on a stem and "riding" it to the ground, where the seeds are picked off. The insects are taken by sitting for a few seconds on a low perch and seizing any prey item that shows itself. If none appear, the bird moves on to another low perch and tries again. The nest, which is built chiefly by the female, is constructed of grass, with some weed stems and leaves. It is lined with finer grasses, rootlets, and hair. Early nests, built before shrubs have leafed out, are on or very near the ground in a dense clump of grass or weeds. Later nests are far more often placed in a dense low shrub, as high as 58 cm (23 in) off the ground (Walkinshaw 1968). There are usually two broods in a season (sometimes three), and two to five eggs are laid for each brood; the early broods having more eggs. The Field Sparrow is a favored target of the Brown-headed Cowbird, and the later broods are less likely to be parasitized; the long nesting season "hedges the bet." A nest targeted by the cowbird is likely to be abandoned. Egg dates range from 21 April to 25 August (Iliff, Ringler, and Stasz 1996). Nestlings have been seen as late as 1 September (MNRF).

Within its chosen habitat the Field Sparrow is a common bird in the late spring and summer. After the breeding season finishes, it forms loose flocks, often mixed with other species. Numbers decrease when the weather becomes colder in mid-November and the majority migrate south, but many remain all winter, especially on the Coastal Plain.

The Field Sparrow is found all across Maryland, and it was present in most blocks in all regions. However, the change in the number of blocks between atlas projects is slightly negative. Blocks recording the species declined by 8% in the Upper Chesapeake and Western Shore and by 7% in the Piedmont region since the first atlas (Willoughby 1996d); and overall in Maryland by 4%. This decline is in agreement with BBS data, which show a 4.2% annual decrease in Maryland and a 2.8% decrease across the whole United States. As such, this species may be subject to further declines and will bear careful watching. Management of shrubby grasslands would benefit Field Sparrow and other species with similar habitat requirements.

T. DENNIS COSKREN

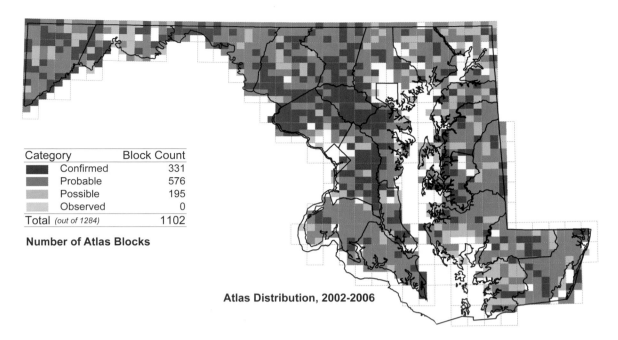

Category	Block Count
Confirmed	331
Probable	576
Possible	195
Observed	0
Total (out of 1284)	1102

Number of Atlas Blocks

Atlas Distribution, 2002-2006

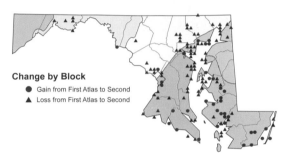

Change by Block
- ● Gain from First Atlas to Second
- ▲ Loss from First Atlas to Second

Change in Atlas Distribution, 1983-1987 to 2002-2006

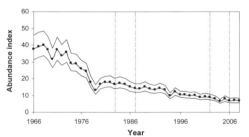

Percent of Stops
- 50 - 100%
- 10 - 50%
- 0.1 - 10%
- < 0.1%

Relative Abundance from Miniroutes, 2003-2007

Atlas Region	1983-1987	2002-2006	Change No.	%
Allegheny Mountain	92	92	0	0%
Ridge and Valley	147	143	-4	-3%
Piedmont	302	280	-22	-7%
Upper Chesapeake	110	101	-9	-8%
Eastern Shore	269	271	+2	+1%
Western Shore	230	212	-18	-8%
Totals	1150	1099	-51	-4%

Change in Total Blocks between Atlases by Region

Breeding Bird Survey Results in Maryland

Vesper Sparrow
Pooecetes gramineus

The Vesper Sparrow is named for its penchant for singing throughout the day from dawn, through midday heat, and well into the gloaming. They are often among the last birds to be heard for the day in the wide-open spaces they inhabit, hence the name promoted by such bird-writing luminaries as Thornton Burgess and John Burroughs. At one time this sparrow was named Bay-winged Bunting because it has an inconspicuous patch of chestnut on its lesser coverts. This is a nondescript pale brown and streaked bird with few prominent distinguishing marks; even its characteristic white outer tail feathers are difficult to see under most circumstances. The song, although an attractive combination of clear sweet notes and musical trills, is sufficiently similar to those of Song Sparrow or a distant Indigo Bunting that it is often overlooked by inexperienced observers.

The Vesper Sparrow was once found throughout Maryland and occurred as a nester in DC (Stewart and Robbins 1958); it is now largely restricted to four regions in the state: the central Delmarva Peninsula, the Frederick Valley, the Hagerstown Valley, and the Allegheny Mountains. Once common, it has declined precipitously over the past half century, and losses are clearly evident between the 1983–1987 (S. A. Smith 1996e) and 2002–2006 atlas projects. It is plausible that this sparrow may become endangered in Maryland by the time another breeding bird atlas project is undertaken.

Vesper Sparrows prefer areas with sparse, short vegetation, with large areas of bare ground and scattered elevated song perches such as fence posts, isolated trees and shrubs, and power lines. Maryland haunts include crop fields with vegetated buffers or hedgerows, overgrazed pastureland, and recently reclaimed strip mines. Most Vesper Sparrows winter south of Maryland, but some winter in small numbers on the lower Eastern Shore and casually elsewhere on the Coastal Plain. The usual period of occurrence in most of the state is from late March to mid-November. The nest is a small grass-lined cup on the ground under grass or weeds. Maryland egg dates range from 5 May to 1 August (Stewart and Robbins 1958; MNRF).

Kirkwood (1895) called the Vesper Sparrow an abundant migrant but a not very common breeding bird. Eifrig (1904) described it as a very common breeder above elevations of 605 m (2,000 ft) in western Maryland. By the 1950s Stewart and Robbins (1958) found it common east to the fall line, fairly common in the Upper Chesapeake, uncommon on the Eastern Shore and the northern Western Shore, and rare on the lower Western Shore. In the first statewide atlas, the species showed widespread range contraction. The four basic regions of occurrence were already in evidence, although they were found throughout Washington County, occurred much

George M. Jett

more widely on the Piedmont, and had a token presence in lower Patuxent Valley farmland. The range has become more restricted over the two decades between atlas projects. They are now found in fewer than 200 blocks, a 39% loss since 1987. The worst losses were on the Piedmont, a net total of 59 blocks, nearly 50% less than in the 1980s. Losses in excess of 33% also occurred in most other regions. They did show a modest increase on the Eastern Shore near the southeastern edge of their overall breeding range. They have been essentially lost from the Western Shore.

The Vesper Sparrow has declined on North American BBS routes at a rate of 1% per year and at a much higher 3.6% per year in eastern North America. The decline in Maryland has been very steep, at 6.7% per year. Reasons for the decline include habitat loss to development and reforestation (largely north of Maryland) and changes in agricultural practices, including less overgrazing of pastures, removal of hedgerows and grassy buffers between fields, mechanized field preparation, and harvest by heavy machinery. Iowa studies indicate that these sparrows have poor nesting success in crop fields and cannot sustain their numbers in them (Rodenhouse and Best 1983; Stallman and Best 1996). Maintaining a Maryland Vesper Sparrow population will require preserving large farms and encouraging farmers to maintain hedgerows and grassy field margins in dry cropland. Unfortunately, the maintenance of hay fields and Conservation Reserve Program tall grass buffers will not help preserve this species (D. Johnson and Igl 1995).

WALTER G. ELLISON

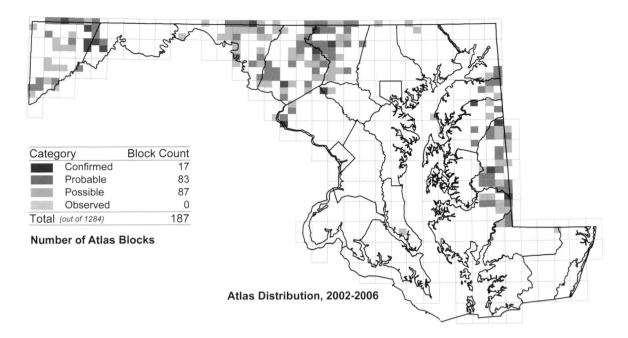

Category	Block Count
Confirmed	17
Probable	83
Possible	87
Observed	0
Total (out of 1284)	187

Number of Atlas Blocks

Atlas Distribution, 2002-2006

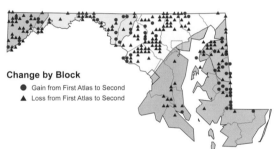

Change by Block
- ● Gain from First Atlas to Second
- ▲ Loss from First Atlas to Second

Change in Atlas Distribution, 1983-1987 to 2002-2006

Percent of Stops
- 50 - 100%
- 10 - 50%
- 0.1 - 10%
- < 0.1%

Relative Abundance from Miniroutes, 2003-2007

Atlas Region	1983-1987	2002-2006	Change No.	%
Allegheny Mountain	55	34	-21	-38%
Ridge and Valley	61	39	-22	-36%
Piedmont	120	61	-59	-49%
Upper Chesapeake	41	27	-14	-34%
Eastern Shore	19	25	+6	+32%
Western Shore	12	1	-11	-92%
Totals	308	187	-121	-39%

Change in Total Blocks between Atlases by Region

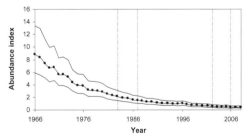

Breeding Bird Survey Results in Maryland

Savannah Sparrow
Passerculus sandwichensis

The Savannah Sparrow is an unobtrusive, small, streaked songbird of open grassy land that resembles the more common Song Sparrow but is paler and shorter tailed. The song is typical of many grassland sparrows in being rather weak and insectlike, a three-part ditty beginning with a few high ticks followed by two descending wiry trills. This sparrow reaches the southeastern limit of its breeding range in Maryland and is largely limited to western parts of the state from the Frederick Valley westward.

Across its broad North American range Savannah Sparrows occupy a wide array of grassy places including muskeg, tundra, and salt marsh. In Maryland they are most often found in luxuriant hay fields and lightly grazed pastures. In comparison to the more widely distributed Grasshopper Sparrow, this species prefers areas with fewer forbs, less bare ground, and deeper litter (Wiens 1973).

Savannah Sparrows occur year-round in Maryland, with wintering primarily on the lower Western Shore and east of Chesapeake Bay (Stewart and Robbins 1958). Migrants begin arriving on nesting grounds in late March and remain on them to September. They nest on the ground, placing the fine grass cup in thick grass with an arch of grass and litter. Maryland nest records are few; just one active nest was found during this project. Nest records include eggs in a nest found 5 June 2003 6.4 km (4 mi) north of Accident, Garrett County (C. Englar), a nest with eggs observed from 24 June to 2 July, and three nests with young found from 30 May to 5 June (S. A.

Smith 1996f; MNRF). Eighty-five percent of breeding confirmations were of adults with food and recent fledglings.

Savannah Sparrows were unreported, perhaps overlooked, as nesters in Maryland by early authorities and were not recorded in June or July before being confirmed breeding during a Garrett County breeding bird census (Stewart and Robbins 1951). Prior to that, the only published summer record was an August 1906 sighting from Mountain Lake Park, Garrett County (Eifrig 1909). Stewart and Robbins (1958) delineated the nesting range as being primarily in Garrett County where they were common; they also noted scattered records to the east in the Hagerstown, Frederick, and Worthington valleys in the Piedmont, and cited isolated populations near Fort Howard, Baltimore County; near Sandy Point in Anne Arundel County; and on Assateague Island.

By 1983–1987 Savannah Sparrows had expanded greatly in the Hagerstown and Frederick valleys and spread east of the Allegheny Front into Allegany County. However, they were not found in the handful of tidewater sites from the late 1950s (S. A. Smith 1996f). In 2002–2006 they were found in 14 additional blocks in the Ridge and Valley region and in 13 more blocks in the Piedmont, east to Howard and Harford counties. There also appears to be a small population inhabiting lightly grazed horse pastures in the Bohemia Creek valley in southern Cecil County.

Savannah Sparrows have declined on North American BBS routes at a rate of 1% per year, largely since 1980. They have declined at a more pronounced 2.1% per year in eastern North America. In Maryland they have been recorded on only 13 BBS routes with fewer than one bird per route. As with other grassland birds, habitat loss and changes in agricultural practices, such as increased frequency of haying and

greater emphasis on crop farming, have contributed to ongoing declines. Despite these trends, this sparrow has been able
to maintain and even increase its nesting range in Maryland.
The continued presence of Savannah Sparrows in Maryland
depends on the health of dairy and horse farming in the Old
Line State.

WALTER G. ELLISON

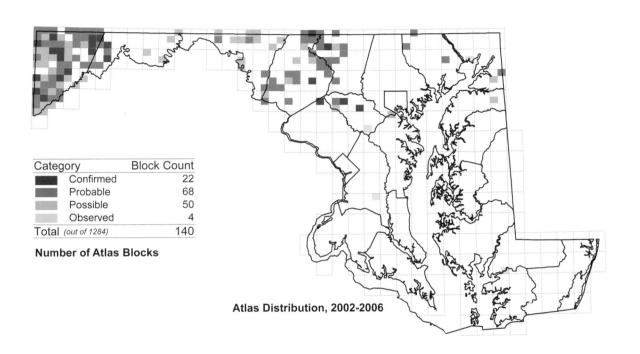

Category	Block Count
Confirmed	22
Probable	68
Possible	50
Observed	4
Total *(out of 1284)*	140

Number of Atlas Blocks

Atlas Distribution, 2002-2006

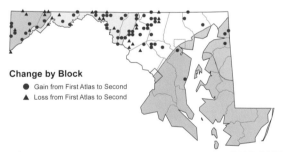

Change by Block

● Gain from First Atlas to Second
▲ Loss from First Atlas to Second

Change in Atlas Distribution, 1983-1987 to 2002-2006

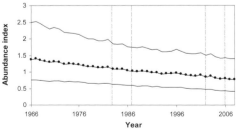

Percent of Stops

	50 - 100%
	10 - 50%
	0.1 - 10%
	< 0.1%

Relative Abundance from Miniroutes, 2003-2007

Atlas Region	1983-1987	2002-2006	Change No.	%
Allegheny Mountain	72	72	0	0%
Ridge and Valley	17	31	+14	+82%
Piedmont	19	32	+13	+68%
Upper Chesapeake	0	3	+3	-
Eastern Shore	0	0	0	-
Western Shore	0	2	+2	-
Totals	108	140	+32	+30%

Change in Total Blocks between Atlases by Region

Breeding Bird Survey Results in Maryland

Bill Sherman

Grasshopper Sparrow

Ammodramus savannarum

The Grasshopper Sparrow is a tiny warm-buff sparrow with two distinctive songs, one thin and buzzing like a grasshopper, the other complex and tinkling. This sparrow is secretive except when males are establishing and defending territories. At such times it often sings persistently from exposed perches such as tall weeds, fences, and power lines. Unfortunately the two songs are sufficiently high pitched that they are among the first lost to observers as their high-range hearing attenuates with age. Although this bird has declined, it remains the second most broadly distributed bird of open grasslands in Maryland behind the Eastern Meadowlark. As with other birds of open country, its range coincides with those parts of Maryland that still have extensive farmland.

Grasshopper Sparrows live in dry grassy meadows with low amounts of plant litter, some exposed ground, and few or no woody plants. Maryland haunts include dry hay fields, fallow weedy fields, airports, reclaimed strip mines, capped landfills, soybean fields planted in winter wheat stubble after harvest, and grassy margins and drainage ditches in crop fields. These sparrows are rare during the winter as far north as Maryland and DC. The majority return to nesting areas in early May. They are only uncommonly seen during fall migration, and most are gone by late September. The nest is a cup of woven fine grass with a grass dome and side entrance (Vickery 1996). Egg dates from an intensive study of Grasshopper Sparrow ecology at Bluestem (formerly Chino)

Farms in Queen Anne's County 5.3 km (3.3 mi) east of Chestertown range from 11 May to 31 August with nestlings from 22 May to 10 September; some pairs there attempt three broods in a season (D. E. Gill, pers. comm.; D. Small, pers. comm.).

Historical references all call the Grasshopper Sparrow a common and widespread nester in Maryland and DC (C. Richmond 1888; Kirkwood 1895; Stewart and Robbins 1958). By the 1980s the species had lost some ground to the decline of agriculture and the increase of suburban development as shown by the 1983–1987 atlas maps (D. Holmes 1996c); they were absent from forested blocks in Allegany and Garrett counties and holes surrounding the Washington and Baltimore metropolitan areas, and distribution was sparse on the lower Eastern Shore. The 2002–2006 atlas project shows further losses west of the Bay with a net loss of 100 blocks on the Piedmont and Western Shore. This was somewhat balanced east of the Bay Bridge by a net gain of 55 blocks in the Upper Chesapeake and Eastern Shore.

Although the Grasshopper Sparrow has retained a broad distribution in Maryland and DC, its population has fallen steeply on BBS routes, with declines of 5.6% per year in eastern North America and 6.1% per year in Maryland. Since 1980 declines have lessened but remain widespread and serious. Studies reviewed by Vickery (1996) imply that these sparrows often have low nesting success in small grasslands among crop fields. D. Gill et al. (2006) have found that Grasshopper Sparrows can thrive in conservation reserve grasslands managed with prescribed burning and careful applications of herbicide to control invasive noxious weeds. For them to remain widespread in Maryland, grassland habitats must be preserved and managed.

WALTER G. ELLISON

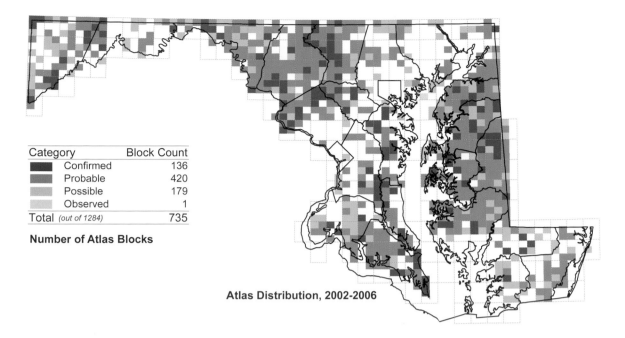

Category	Block Count
■ Confirmed	136
■ Probable	420
■ Possible	179
■ Observed	1
Total (out of 1284)	735

Number of Atlas Blocks

Atlas Distribution, 2002-2006

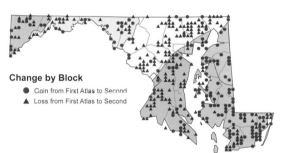

Change by Block
- ● Gain from First Atlas to Second
- ▲ Loss from First Atlas to Second

Change in Atlas Distribution, 1983-1987 to 2002-2006

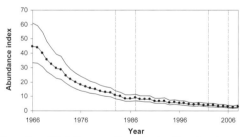

Percent of Stops
■	50 - 100%
■	10 - 50%
■	0.1 - 10%
□	< 0.1%

Relative Abundance from Miniroutes, 2003-2007

			Change	
Atlas Region	1983-1987	2002-2006	No.	%
▦ Allegheny Mountain	64	61	-3	-5%
▦ Ridge and Valley	102	95	-7	-7%
▦ Piedmont	237	176	-61	-26%
▦ Upper Chesapeake	77	92	+15	+19%
▦ Eastern Shore	137	177	+40	+29%
▦ Western Shore	171	132	-39	-23%
Totals	788	733	-55	-7%

Change in Total Blocks between Atlases by Region

Breeding Bird Survey Results in Maryland

Henslow's Sparrow
Ammodramus henslowii

George M. Jett

Perhaps the most secretive of the shy grassland sparrows, the Henslow's Sparrow is tiny, flat-headed, large-billed, and spike-tailed. They have a weak easily overlooked song, a high pitched katydid-like *tssip* or *ssslik*, uttered at fairly long intervals. It is hard to hear at a distance over the breezes that often sweep over the open weedy places they inhabit. The song is delivered from low perches or the ground and may be sung all night at the peak of territorial advertisement. These sparrows prefer to run away rather than flushing. When they flush they quickly drop back into cover and are seldom seen again by an observer. Once widespread in Maryland, if never very common, they are now limited to the Allegheny Mountains of far western Maryland.

The basic description of Henslow's Sparrow habitat is tall rank grasslands, often with scattered weedy forbs, thick litter buildup, and little or no woody growth. They usually avoid regularly harvested fields and recently burned grasslands lacking litter (Herkert, Vickery, and Kroodsma 2002). Favored locations include fallow fields, untidy infrequently cut hay fields, and reclaimed surface mines planted to grasses and forbs. Historically they nested in the relatively dry upland edges of tidal marshes on the lower Eastern Shore. Broomsedge is a frequent element in occupied fields, probably because it is typical of less heavily managed fields. This grass was highly correlated with occupied territories in an Indiana study (Bajema et al. 2001). Because their haunts are usually ephemeral, soon succumbing to plant succession, these sparrows show little tendency to return to locations in succeeding years. However, a minority of males will return to fields that retain suitable structure for nesting (Skipper 1998a).

Henslow's Sparrows winter in the southeastern United States north to coastal North Carolina in damp meadows with thick litter including bogs and open pine savanna (Plentovich et al. 1999). They are present in Maryland from mid-April at least to September, although their historic range of occurrence was from late March to early November (Stewart and Robbins 1958). They place nests slightly above or on the ground on a bed of plant litter and under a dome of grass (Winter 1999). As with other grassland sparrows they are probably double brooded. Maryland egg dates range from 10 May to 2 July, although they likely nest until late July or early August (D. Boone and Dowell 1996; MNRF).

Kirkwood (1895) considered the Henslow's Sparrow widespread on the Maryland Piedmont and upper Western Shore but only locally numerous. Stewart and Robbins (1958) described a statewide nesting range and called them fairly common on the Coastal Plain, uncommon on the Piedmont and in the Allegheny Mountains, and rare in the Ridge and Valley. By the 1983–1987 atlas they had declined drastically and were found in just 14 blocks, only 3 in eastern Maryland on the lower Eastern Shore. They have not been seen on the lower Eastern Shore since 1989 (H. Wierenga in D. Boone and Dowell 1996). They were also found in eastern Allegany County in the 1980s. Although they were found in a total of 8 more blocks during 2002–2006, they had become restricted to western Allegany and Garrett counties, primarily from Oakland to the north and east.

Henslow's Sparrow has always been uncommon and locally distributed, and BBS routes record few, 0.14 per route. They have declined rapidly on continental BBS routes, at a catastrophic rate of 8.1% per year. This species is not recorded on Maryland BBS routes. Henslow's Sparrow is currently listed as threatened by the DNR. Skipper (1998b) studied them on reclaimed surface mines in Garrett County from 1989 to 1997, locating 12 colonies. In spite of some site fidelity on the best habitat, they were lost from one-third of the sites found during the study, emphasizing the bird's tendency to abandon sites once they become unsuitable. To retain this sparrow in Maryland, we must maintain reclaimed surface mine grassland via mowing and prescribed burning of small portions of fields to prevent shrub invasion and maintain adequate amounts of litter. These techniques could be used on state-owned grasslands, and a subset of Conservation Reserve Program fields could also be managed similarly. This bird's status is sufficiently tenuous that it might disappear before another atlas project is started.

WALTER G. ELLISON

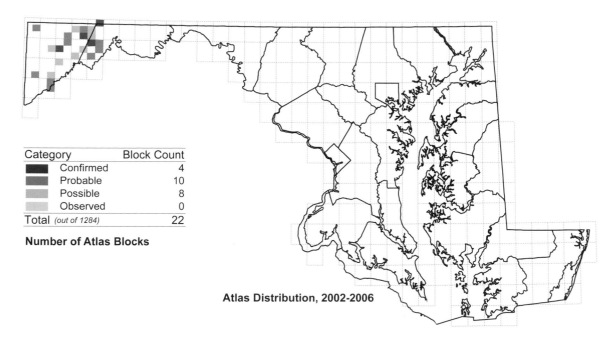

Category	Block Count
Confirmed	4
Probable	10
Possible	8
Observed	0
Total (out of 1284)	22

Number of Atlas Blocks

Atlas Distribution, 2002-2006

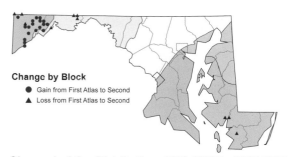

Change by Block

● Gain from First Atlas to Second
▲ Loss from First Atlas to Second

Change in Atlas Distribution, 1983-1987 to 2002-2006

Atlas Region		1983-1987	2002-2006	Change No.	Change %
	Allegheny Mountain	7	21	+14	+200%
	Ridge and Valley	4	1	-3	-75%
	Piedmont	0	0	0	-
	Upper Chesapeake	0	0	0	-
	Eastern Shore	3	0	-3	-100%
	Western Shore	0	0	0	-
Totals		14	22	+8	+57%

Change in Total Blocks between Atlases by Region

George M. Jett

Saltmarsh Sparrow
Ammodramus caudacutus

The Saltmarsh Sparrow is one of North America's most secretive and difficult to detect songbirds. Because it is barely audible more than a few meters away and the males of this promiscuous, colonial species do not establish territories, standard survey techniques are unreliable for this salt marsh endemic. Also, it breeds in expansive high marsh salt meadows that are mostly inaccessible by roads. The Chesapeake Bay area is the southern limit of this coastal sparrow's breeding range and the northern limit of its wintering range. Most individuals have migrated by the end of October to coastal marshes along the southeastern Atlantic and Gulf coasts for the winter and begin returning again in April (AOU 1983; Robbins and Bystrak 1977). However, it has been found throughout the winter on the lower Eastern Shore (Stewart and Robbins 1958).

In September 1862 the Saltmarsh Sparrow was first documented in this area from a DC specimen (Ridgway 1882). Its full distribution was described in the mid-twentieth century as the salt and brackish marshes of Worcester, Somerset, and Dorchester counties, with scattered locations north to Queen Anne's County as well as along the Western Shore north to Anne Arundel County and up the Potomac River to southern Charles County (Hampe and Kolb 1947; Stewart and Robbins 1947, 1958).

This sparrow prefers interior tracts of high marsh dominated by saltmeadow cordgrass and seashore saltgrass, but is also found in higher needlerush marshes not flooded by daily tides (Stewart and Robbins 1958; N. Hill 1968; Benoit and Askins 2002). However, flooding by high tides is a predictable major threat every 28 days, and the Saltmarsh Sparrow's breeding ecology has evolved as an adaptation to this threat; nesting is timed to lunar cycles and renesting occurs about 3 days after nest flooding (Gjerdrum et al. 2005; Shriver et al. 2007). Eggs are laid from 11 May to 21 August in nests built by females usually on or within 8 cm (3 in) of the ground (MNRF). Incubation lasts 11 days and chicks fledge about 10 days later (N. Hill 1968).

By the 1980s the Saltmarsh Sparrow had disappeared from most of the Western Shore, its northern Eastern Shore breeding areas in Queen Anne's County, and most marshes north of Assateague Island in coastal Worcester County (O'Brien 1996b). The disappearance and range contraction have continued in the past 20 years. This sparrow now occurs almost exclusively in the lower three Eastern Shore counties, with 1 occupied atlas block still found on the Western Shore in southern St. Mary's County. Even within its remaining stronghold on the lower Eastern Shore, the number of atlas blocks declined 23% between the first and second atlas, dropping from 47 to 36 blocks.

Some of this decline may be accounted for by differences in observer effort during the second atlas. In particular, canoes were used for atlasing in Dorchester County during the first atlas (S. Droege, pers. comm.) but rarely used during the second atlas (L. Davidson, pers. obs.). Canoe work allows access to some of the more inaccessible high marsh meadow habitat. However, it is also thought that this species' population is declining. The National Audubon Society and American Bird Conservancy include it on their 2007 WatchList (red list) as a species of global conservation concern. No BBS trend data are available for this species because this survey technique is insufficient for accurately assessing its population.

Salt marshes in the lower Chesapeake Bay and coastal bays of Worcester County are highly threatened by many factors, including fragmentation and hydrologic changes from ditching, inundation and erosion from subsidence and sea level rise, and invasion and conversion of short grass meadows by phragmites. Invasion by phragmites has been determined to negatively impact this species (Benoit and Askins 1999). Of particular interest and concern is the impact that prescribed fire and illegal burning of the lower Eastern Shore marshes may have on overwintering individuals and on early spring food availability and habitat for breeding individuals. This is especially true because they prefer to breed in areas with a deep layer of thatch (Gjerdrum et al. 2005). As sea levels continue to rise and the timing and intensity of flooding increases, this highly adapted species may no longer be able to successfully reproduce without the creation or restoration of large areas of high marsh meadows.

LYNN M. DAVIDSON

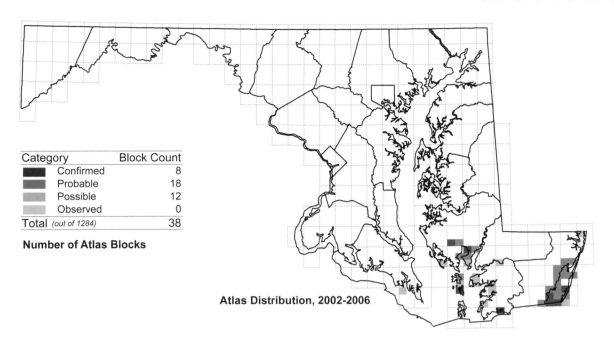

Category	Block Count
Confirmed	8
Probable	18
Possible	12
Observed	0
Total (out of 1284)	38

Number of Atlas Blocks

Atlas Distribution, 2002-2006

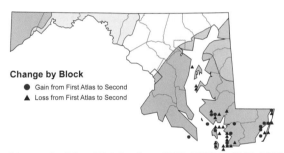

Change by Block

● Gain from First Atlas to Second
▲ Loss from First Atlas to Second

Change in Atlas Distribution, 1983-1987 to 2002-2006

Atlas Region	1983-1987	2002-2006	Change No.	%
Allegheny Mountain	0	0	0	-
Ridge and Valley	0	0	0	-
Piedmont	0	0	0	-
Upper Chesapeake	0	0	0	-
Eastern Shore	47	36	-11	-23%
Western Shore	1	1	0	0%
Totals	48	37	-11	-23%

Change in Total Blocks between Atlases by Region

Seaside Sparrow
Ammodramus maritimus

The Seaside Sparrow is one of the more obscure birds in the state, known by few outside the birding community. The adult male and female are grayish olive above with obscure streaks. Below, the bird is grayish, slightly darker on the breast, which is marked with faint darker gray streaks. The throat is white with dark gray malar streaks. The best field mark is a yellow patch on the lores. Immature birds are browner above and buffier below with more obvious dark streaking on the breast. The feet are comparatively large, which may be an adaptation for walking on the soft mud of the marsh. The song is as unremarkable as the plumage and is likely to be taken for the song of an insect rather than that of a bird: *tup, teetle-zhrrrrrrrrr* is a reasonable approximation. It lasts about 2 seconds. One male may sing different versions in succession. The performance is usually given from a slightly elevated perch, such as from the side of a tall grass stalk or the top of a marsh elder; sometimes it is given in flight, "the male fluttering upward 10 or 20 feet, and gliding back down into the marsh grass while buzzing" (G. Woolfenden 1968).

The habitat is given by the bird's common name: the species is restricted to salt marshes by the seaside. It shares this habitat with a close relative, the Saltmarsh Sparrow, but is commoner in the lower marsh and tends to feed in the more open areas. The Seaside Sparrow is usually found in the lower marsh dominated by smooth cordgrass or needlerush; it also uses the upper marsh, where saltmeadow cordgrass is common. Ideally there will be easy access to expanses of mud, either open or with sparse grass, where the bird commonly feeds. The diet of the Seaside Sparrow is much more insectivorous than that of most sparrows, consisting of about 80% animal matter. This fact is reflected in its bill, which is longer and more spikelike than that of other sparrows. Menu items include beetles, moths, small crabs, and snails. Vegetable food is mostly grass seed especially in the winter when insects are harder to come by (G. Woolfenden 1968).

The nest is located in the lower, wetter part of the marsh, just above the summer high tide mark. It may be located 20 to 30 cm (8–12 in) off the ground in a shrub (especially marsh elder), a tussock of *Juncus* (rushes), or a clump of smooth cordgrass, or it may be on the ground. The female constructs the nest of *Spartina* (cordgrass), *Juncus* stems, or both and lines it with finer grass. Normally only one brood is raised with four to five eggs in each clutch (sometimes three or six; H. Harrison 1975). Egg dates range from 16 May to 4 July and nestlings have been seen from 20 May to 14 June (MNRF). The late egg date probably represents an unusual second nesting. Most Seaside Sparrows in Maryland migrate to marshes farther south on the coast of the United States, but a few remain to winter, perhaps augmented by migrants from the north. They return in early spring and full numbers are present by the end of April.

The Seaside Sparrow is fairly common within its limited habitat and within a very limited range in the Eastern Shore region of Maryland. The distribution in the 1950s was limited to brackish or salty tidewater wetlands north to Kent Narrows, Queen Anne's County (historically to Kent County; Kirkwood 1895), on the east and Sandy Point, Anne Arundel County, on the west and west to southern Charles County on the Potomac (Stewart and Robbins 1958). In this atlas, it was found only in the salt marshes of the east side of Chesapeake Bay in Queen Anne's (1 block), Dorchester, Wicomico, and Somerset counties and the salt marshes of Sinepuxent Bay. This represents a significant range contraction since the first atlas when the species was still found in isolated blocks on the western side of Chesapeake Bay as far north as northern Anne Arundel County and in a single block along the lower Potomac River in Charles County (O'Brien 1996c). One block hosted the species at Miller Island in Baltimore County in the Upper Chesapeake region, but this site was significantly altered by the construction of the Hart-Miller Dredged Material Containment Facility. The most likely explanation for the contraction elsewhere is local degradation of habitat.

T. DENNIS COSKREN

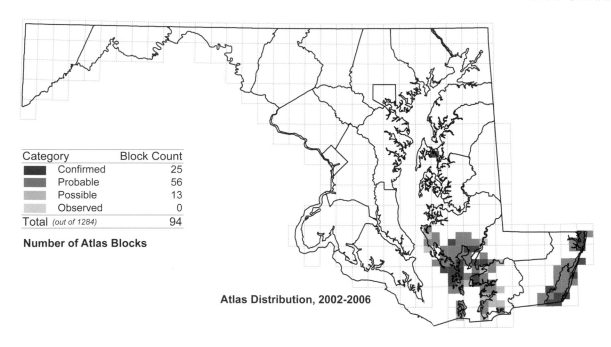

Category		Block Count
	Confirmed	25
	Probable	56
	Possible	13
	Observed	0
Total *(out of 1284)*		94

Number of Atlas Blocks

Atlas Distribution, 2002-2006

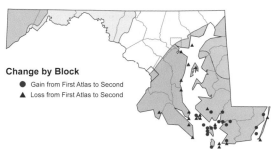

Change by Block

● Gain from First Atlas to Second
▲ Loss from First Atlas to Second

Change in Atlas Distribution, 1983-1987 to 2002-2006

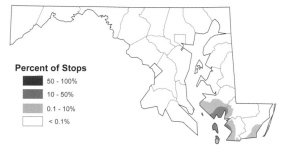

Percent of Stops

	50 - 100%
	10 - 50%
	0.1 - 10%
	< 0.1%

Relative Abundance from Miniroutes, 2003-2007

Atlas Region	1983-1987	2002-2006	Change No.	%
Allegheny Mountain	0	0	0	-
Ridge and Valley	0	0	0	-
Piedmont	0	0	0	-
Upper Chesapeake	1	0	-1	-100%
Eastern Shore	86	93	+7	+8%
Western Shore	9	0	-9	-100%
Totals	96	93	-3	-3%

Change in Total Blocks between Atlases by Region

Song Sparrow
Melospiza melodia

As befits its name, the Song Sparrow is a persistent singer with a spritely, if not very melodic, ditty that may be heard on sunny mornings during even the coldest months. Although it is not as elusive as some of its relatives, this sparrow is not overly conspicuous or showy, as both sexes are garbed in brown, gray, black, and creamy white. This is one of the most widespread and familiar birds in Maryland and DC; among its close relatives only the Eastern Towhee and Chipping Sparrow exceed its nesting range.

The Song Sparrow's habitat combines access to running or standing water, the presence of at least a few shrubs or small trees, and unmowed open grassy areas. In Maryland and DC it occupies grassy banks of streams, brushy fields, hedgerows in open land, the upland edges of tidal marshes, and parks and yards with patches of uncut grass. They are scarce or absent in open cropland or grassland, closed-canopy forests, and pine woods.

Song Sparrows are resident year-round in Maryland and DC, with wintering birds found in progressively lower numbers westward from the Coastal Plain, where they are common, to the Alleghenies, where they are uncommon (Stewart and Robbins 1958). The majority of wintering birds are migrants from the north, but some are resident throughout the year (Brackbill 1953). They commence nesting in late February and March, with many migrants arriving at this time. Song Sparrows build a cup nest of grass and weed stems in a variety of sites from ground level to more than 3 m (10 ft) in small trees and shrubs (Nice 1937). They generally nest several times in a breeding season, often three and sometimes four times (Nice 1937). As is typical for most small songbirds, the vast majority of atlas confirmations of nesting were of fledglings and of adults feeding young. Maryland egg dates range from 12 April to 21 August (Stewart and Robbins 1958; MNRF), and nestlings have been observed to 23 September (Harvey and Solem 1996; MNRF).

Early bird lists called the Song Sparrow common to abundant (C. Richmond 1888; Kirkwood 1895; Eifrig 1904). Stewart and Robbins (1958) called them common except inland on the southern Western Shore and on the Eastern Shore where they were uncommon. In the 1983–1987 atlas they showed a similar distribution, with notable absences from the Nanjemoy area and several blocks on Cobb Neck in Charles County, St. Mary's County, and much of inland Dorchester, Wicomico, Somerset, and Worcester counties (Harvey and Solem 1996). From 2002–2006 they were found in 13 fewer blocks, with most of these losses inland on the lower Eastern Shore and some gains on Cobb Neck.

The long term continental BBS trend has been modestly downward at 0.5% per year. In Maryland the recent trend has also been negative, but not significantly so. Declines on the lower Eastern Shore are probably related to habitat change, particularly loss of hedgerows in farmland and ditching of streams. The Song Sparrow appears sufficiently adaptable to continue to be a common and widespread bird in Maryland and DC for decades to come.

WALTER G. ELLISON

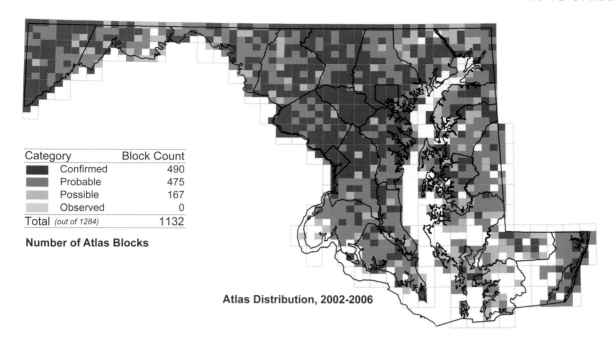

Category	Block Count
■ Confirmed	490
■ Probable	475
■ Possible	167
■ Observed	0
Total *(out of 1284)*	1132

Number of Atlas Blocks

Atlas Distribution, 2002-2006

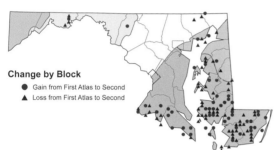

Change by Block
- ● Gain from First Atlas to Second
- ▲ Loss from First Atlas to Second

Change in Atlas Distribution, 1983-1987 to 2002-2006

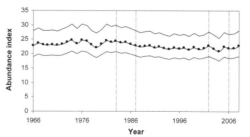

Percent of Stops
- ■ 50 - 100%
- ■ 10 - 50%
- ■ 0.1 - 10%
- □ < 0.1%

Relative Abundance from Miniroutes, 2003-2007

Atlas Region	1983-1987	2002-2006	Change No.	%
Allegheny Mountain	93	93	0	0%
Ridge and Valley	146	145	-1	-1%
Piedmont	314	313	-1	0%
Upper Chesapeake	115	112	-3	-3%
Eastern Shore	251	239	-12	-5%
Western Shore	222	226	+4	+2%
Totals	1141	1128	-13	-1%

Change in Total Blocks between Atlases by Region

Breeding Bird Survey Results in Maryland

Swamp Sparrow
Melospiza georgiana

The Swamp Sparrow is a dark, rather shy bird of open wet places with rank growth, both woody and grasslike. It is most easily noticed when it calls or sings as it keeps low and tends to stay out of sight unless it is curious or has mounted an elevated perch to sing. The song is a deliberate musical trill slower and sweeter than that of the more familiar Chipping Sparrow; the most common call is a sharp *cheep* whose quality lies between the alarm notes of Eastern Phoebe and White-throated Sparrow. This sparrow is at or very near the southeastern edge of its breeding range in Maryland. There are two nesting subspecies: nominate *georgiana* found in inland freshwater wetlands, principally in Garrett County, and *nigrescens* (Coastal Plain Swamp Sparrow) found near the shores of Chesapeake Bay and its tidewater rivers. The latter, first described in Maryland from the marshes of the upper Nanticoke River (Bond and Stewart 1951), ranges from southern New York to eastern Virginia (Greenberg et al. 2007). This Coastal Plain subspecies was listed by the DNR as "in need of conservation" in 2003.

Swamp Sparrows nest in wetlands, often where there is standing water and dense rank vegetation. Inland these range from open cattail marshes with few shrubs, to wet grass and sedge meadows with scattered shrubs, and shrub swamps including bogs and willow and alder swales. Coastal Plain Swamp Sparrows inhabit the upper edges of brackish (but not saline) tidal marshes where open saltmeadow cordgrass gives way to stands of marsh elder and groundsel tree.

These sparrows are resident in winter east of Maryland's mountains. They are especially numerous during the colder months on the Coastal Plain although the locally nesting *nigrescens* winter south of Maryland on the Atlantic coast (Greenberg et al. 2007). Spring migrants arrive on their nesting territories in late March and April inland (Stewart and Robbins 1958) and in mid-May on the Coastal Plain (Greenberg and Droege 1990). Swamp Sparrow nests are built off the ground in grassy plants and shrubs, usually less than 30 cm (12 in) high (Mowbray 1997). Maryland egg dates range from 24 May 2002 southeast of McComas Beach, Garrett County (C. Robbins, this project) to 22 June (Stewart and Robbins 1958; MNRF). Adults have been reported feeding fledglings as late as 9 August 2003 in Garrett County (W. Pope, this project) implying later laying dates in July.

Eifrig (1904) termed the Swamp Sparrow not rare in suitable habitat in western Maryland, but the species was not found nesting in eastern Maryland until Bond and Stewart (1951) described the Coastal Plain subspecies. Stewart and Robbins (1958) added the marshes at the mouth of Big Elk Creek in Cecil County to the Coastal Plain range and found these sparrows common at and above 725 m (2,400 ft) in Garrett County. The 1983–1987 atlas reaffirmed their status in the Allegheny Mountains but also recorded them in scattered blocks eastward to the Piedmont as far as Baltimore County (Droege and Blom 1996). Coastal Plain Swamp Sparrows were recorded from both historical locations and many other sites along Chesapeake Bay and its major tidal rivers including the Potomac and Patuxent south to western Wicomico County. The origin of Piedmont sparrows may be from either inland or the Coastal Plain, but given their freshwater marsh habitats an inland origin seems more likely.

The 2002–2006 atlas documented a net loss of 13 blocks between atlas projects with the majority (7) from the Western Shore, especially lower Potomac blocks. Although *nigrescens* were not found at the type locality on the upper Nanticoke or in the Elk Creek marshes, they had a net increase east of Chesapeake Bay, particularly on the lower Nanticoke River, Fishing Bay, and western Somerset County. Although Coastal Plain Swamp Sparrows are uncommon in Maryland, they are common on Delaware Bay (Walsh et al. 1999; Hess et al. 2000). Inland sites in the Piedmont and Ridge and Valley continue to host this species, including new locations in Washington, Frederick, Howard, and Harford counties.

Swamp Sparrows have increased on BBS routes in North America at a rate of 1.3% per year with much of the increase since 1980. There are insufficient data from Maryland BBS routes to determine a reliable trend. Although inland Swamp Sparrows appear to be stable, it seems likely Coastal Plain birds are declining (Beadell et al. 2003) as tidal wetlands are under threat from invasive plants, falling productivity, and rising sea levels. These sparrows seem to be tolerant of common reed invading their marshes (Benoit and Askins 1999) and are not limited to large wetlands (Benoit and Askins 2002), but they will decline if marshes and shrub swamps are not preserved in Maryland and DC.

WALTER G. ELLISON

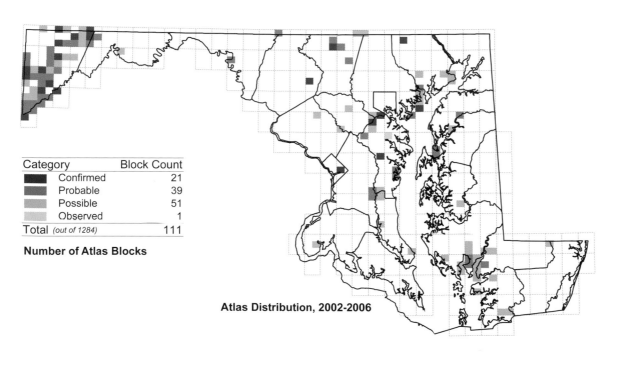

Category	Block Count
Confirmed	21
Probable	39
Possible	51
Observed	1
Total *(out of 1284)*	111

Number of Atlas Blocks

Atlas Distribution, 2002-2006

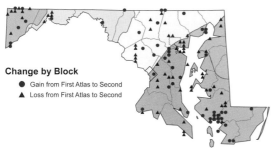

Change by Block
- ● Gain from First Atlas to Second
- ▲ Loss from First Atlas to Second

Change in Atlas Distribution, 1983-1987 to 2002-2006

Percent of Stops
- 50 - 100%
- 10 - 50%
- 0.1 - 10%
- < 0.1%

Relative Abundance from Miniroutes, 2003-2007

Atlas Region	1983-1987	2002-2006	Change No.	%
Allegheny Mountain	50	48	-2	-4%
Ridge and Valley	4	3	-1	-25%
Piedmont	16	12	-4	-25%
Upper Chesapeake	14	9	-5	-36%
Eastern Shore	17	23	+6	+35%
Western Shore	23	16	-7	-30%
Totals	124	111	-13	-10%

Change in Total Blocks between Atlases by Region

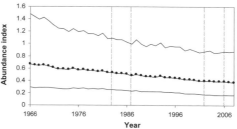

Breeding Bird Survey Results in Maryland

Dark-eyed Junco
Junco hyemalis

The Dark-eyed Junco, formerly called Slate-colored Junco, is most familiar to Marylanders and Washingtonians as a common winter denizen of wooded country with brushy edges and hedgerows and residential neighborhoods with an abundance of evergreen ornamentals and feeders. The dapper gray and white Slate-colored Junco is a member of a highly variable species found throughout northern North America and at high elevations far to the south both in the East and West. Maryland hosts the two subspecies of Dark-eyed Junco that comprise the Slate-colored Junco—the widespread wintering *hyemalis* and the locally nesting altitudinal migrant of the southern Appalachian Mountains *carolinensis*, also called the Carolina Junco. In Maryland, Carolina Juncos nest only in the Allegheny Mountain region in Garrett County; this subspecies also ranges northward into southwestern Pennsylvania (Mulvihill 1992d), and southward in the mountains to northern Georgia (AOU 1998).

During the breeding season, the Dark-eyed Junco is a bird of brushy openings in coniferous and mixed forest, usually in association with hemlock or spruce in areas with rough hilly terrain. Males require fairly high treetop perches for broadcasting their sweet trilling songs. Juncos nest along trails, by wood roads, at bog edges, on ledges, and along streams with high undercut banks. They are found at 760 m (2,500 ft) and higher on mountain ridges and upper slopes from the Allegheny Front westward.

Dark-eyed Juncos winter throughout Maryland and DC but they are uncommon at that season in Garrett County (Stewart and Robbins 1958). They have left wintering areas in eastern and central Maryland by late April in most years and begin arriving on nesting territory by late March and early April. Nesting birds probably depart their breeding sites by late September to early October, about the same time wintering birds begin to reappear to the east. Nests are usually placed on the ground although rare instances of nesting in artificial structures and trees exist (Sprunt 1968). They usually build a nest nestled into some sort of overhanging structure such as a bank along a road or stream, on a ledge, or in the soil and roots of wind-thrown trees (Nolan et al. 2002). Nests are hard to find, and the majority of atlas breeding confirmations involved locating the distinctive black-streaked brown fledglings. Maryland nest dates are sparse; the fewer than 10 records for eggs span 18 May to 20 July (Stewart and Robbins 1958; MNRF). Reports of fledglings extend the nesting season in Garrett County to 19 August (W. and F. Pope, this project). Mid- to late April egg dates from the Mountain Lake Biological Station in neighboring Virginia suggest that earlier egg dates could be found in Maryland in the future (Nolan et al. 2002).

The status of Dark-eyed Juncos in Maryland has changed little over the past hundred years. Eifrig (1904) said that Carolina Juncos bred "in numbers" at upper elevations, presumably in Garrett County, and wintered locally with Slate-colored Juncos. By 1918 Eifrig (1920) noted they appeared to be "growing less common" and they were found only above 760 m (2,500 ft). Stewart and Robbins (1958) reported they were fairly common above 900 m (3,000 ft) on Backbone Mountain in southwestern Garrett County and uncommon elsewhere in the county above 760 m (2,500 ft). Juncos were found in just 15 blocks in 1983–1987, 8 in the Backbone Mountain area and 7 others scattered to the northeast (Hobbs 1996).

In 2002–2006 Dark-eyed Juncos appear to have increased substantially although they were still restricted to the Allegheny Mountain region. They were found in 13 more blocks, an 87% increase in block occupancy. More than half the records were of confirmed nesting. They were confirmed in all 7 blocks in the Backbone Mountain area, with blocks running northeast to Finzel. There were also clusters of blocks on the upper Youghiogheny drainage including Cranesville Swamp and the western Casselman River drainage. Of interest were possible breeding records west of Youghiogheny River Reservoir and above 760 m (2,500 ft) on Dans Mountain in Allegany County.

Dark-eyed Juncos have been declining on North American BBS routes at 1.3% per year from 1966 to 2008. On U.S. routes the decline has been less steep and not statistically significant. On the 82 Allegheny Plateau routes, these juncos have been stable. The Carolina Junco's general status has changed rather little; they may have increased within their fairly narrow distributional limits in Maryland's Allegheny Mountains. Reasons for the recent range expansion are not obvious but may include a spate of mild winters, some improvement in atlas coverage, increases in winter bird feeding, or some combination of the three. At present juncos appear likely to continue to nest in Maryland's high western mountains.

WALTER G. ELLISON

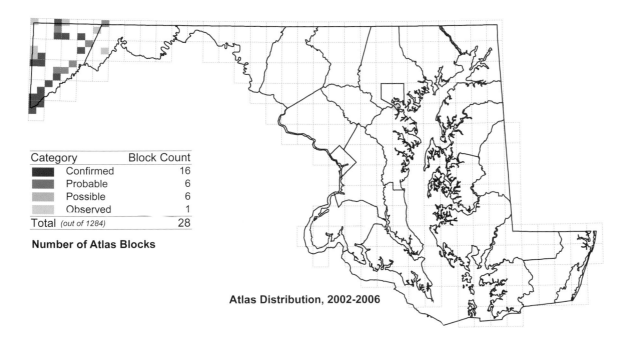

Category	Block Count
Confirmed	16
Probable	6
Possible	6
Observed	1
Total (out of 1284)	28

Number of Atlas Blocks

Atlas Distribution, 2002-2006

Change by Block
- Gain from First Atlas to Second
▲ Loss from First Atlas to Second

Change in Atlas Distribution, 1983-1987 to 2002-2006

Atlas Region	1983-1987	2002-2006	Change No.	Change %
Allegheny Mountain	15	28	+13	+87%
Ridge and Valley	0	0	0	-
Piedmont	0	0	0	-
Upper Chesapeake	0	0	0	-
Eastern Shore	0	0	0	-
Western Shore	0	0	0	-
Totals	15	28	+13	+87%

Change in Total Blocks between Atlases by Region

Summer Tanager
Piranga rubra

The "summer redbird" is near the limit of its range here and is common only in the southeastern part of the state. It favors open oak-hickory forest, loblolly and Virginia pine, and mixed woods. The soft whistled song and the staccato call are as distinctive as the red plumage. The adult male is easy to identify, with his solid rosy red plumage, less strident than the red of the Scarlet Tanager or the Cardinal. The wings and tail are a duskier red, and the bill is light colored. The female is yellow olive, darker above and lighter below; some females have a faint orange to reddish wash. First-spring males have variable amounts of red and olive, changing to completely red in August. The song is a clear robinlike series of whistled phrases; only a few notes may have the burry quality of the Scarlet Tanager's song. More distinctive is the call, a peremptory *pi-tuck* or *pick-it-up*.

The Summer Tanager is a characteristic bird of the open loblolly pine and mixed pine-oak forests of eastern Maryland. It also accepts open stands of oak-hickory forest. Usually the birds feed at medium to high levels, hawking and gleaning insects. Its prey are members of the order Hymenoptera (bees and wasps). Adult insects are taken, leading to an intense dislike of this species by beekeepers; wasp nests are also attacked and the larvae devoured. Other invertebrate prey include beetles, caterpillars, and spiders. A smaller proportion of the diet is made up of fruits. Blackberries, raspberries, blueberries, and huckleberries are on the menu (Bent 1958). The nest, constructed by the female, is often in an oak and is generally 3 to 11 m (10–35 ft) above the ground, away from the trunk, and on a horizontal limb. It is small, shallow, and flimsy. Grass, weed stems, leaves, and bark are used to build the nest, which is lined with finer grass. The male usually feeds the female as she incubates the eggs (H. Harrison 1975). The extreme egg dates are 24 May and 13 July (Stewart and Robbins, 1958). The incubation period is about 12 days. Both male and female feed the young birds. Recently fledged young have been found as late as 10 August (present atlas). The species is normally single brooded.

In Maryland the Summer Tanager is found mostly in the Eastern and Western Shore regions, where it is fairly common in the proper season and habitat. This is similar to its distribution during the 1983–1987 atlas (Wilmot 1996d). The present atlas found a scattering of possible breeding birds in the Upper Chesapeake and Piedmont regions, and the earlier atlas found a couple of possible breeders in the Ridge and Valley region as well. But the species is common only where the flora has a more southern aspect, for example, loblolly pine and American mistletoe. Migrants may arrive as early as mid-April; most arrive during the first week of May. The abundance decreases sharply between mid- and late August, but stragglers may linger until the end of October (Iliff, Ringler, and Stasz 1996). A few birds may winter in south Texas; most birds, however, continue south to winter in Mexico, Central America, and the northern half of South America.

Summer Tanagers have held their own in Maryland. Between atlas periods net losses from the Upper Chesapeake and Ridge and Valley regions were more than balanced by strong gains in the Eastern Shore region (up 34 blocks, or 20%) and Western Shore region (up 23 blocks, or 19%). The Piedmont held virtually steady. The net change for the state was up 49 blocks, a 15% gain. Breeding Bird Survey data show a slight increase in abundance (ca. 0.9% annually) for Maryland, slightly better than the overall U.S. increase of 0.3%.

T. DENNIS COSKREN

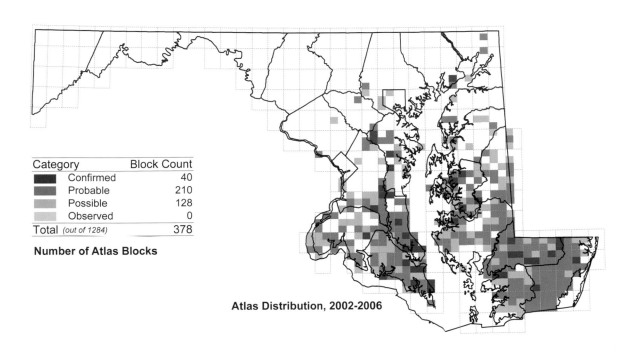

Category	Block Count
Confirmed	40
Probable	210
Possible	128
Observed	0
Total (out of 1284)	378

Number of Atlas Blocks

Atlas Distribution, 2002-2006

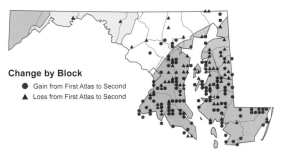

Change by Block
● Gain from First Atlas to Second
▲ Loss from First Atlas to Second

Change in Atlas Distribution, 1983-1987 to 2002-2006

Percent of Stops
■	50 - 100%
■	10 - 50%
■	0.1 - 10%
□	< 0.1%

Relative Abundance from Miniroutes, 2003-2007

Atlas Region	1983-1987	2002-2006	Change No.	Change %
Allegheny Mountain	0	0	0	-
Ridge and Valley	2	0	-2	-100%
Piedmont	9	10	+1	+11%
Upper Chesapeake	25	18	-7	-28%
Eastern Shore	170	204	+34	+20%
Western Shore	120	143	+23	+19%
Totals	326	375	+49	+15%

Change in Total Blocks between Atlases by Region

Breeding Bird Survey Results in Maryland

George M. Jett

Scarlet Tanager
Piranga olivacea

A lengthy full-length view of a male Scarlet Tanager is a much sought after prize among birders, not only because of the male's fire engine red plumage offset by velvet black wings and tail, but also because the bird lives high in leafy treetops and moves deliberately, sitting still for lengthy periods, making it surprisingly hard to see. If one learns this tanager's bright scratchy repetitive singsong: *hurry-worry-flurry-bury* or its querulous *chip-burr* low-level alarm call, one realizes that the bird is fairly common in forests throughout much of Maryland and DC. Scarlet Tanagers prefer reasonably mature woodlands, usually dominated by open-crowned tall deciduous trees, but they will occupy mixed forests dominated by hemlock or white pine. This tanager also prefers fairly extensively wooded country and is generally absent from small woodlots and landscapes where the forest is broken into small fragmented units (Robbins, Dawson, and Dowell 1989).

Scarlet Tanagers are migratory and spend the nonbreeding period at middle elevations along the east slopes of the Andes and their outlying foothills south to Peru and Bolivia. These tanagers return to Maryland in late April and early May; by early October most have departed. Nests are usually built sufficiently high in trees, well screened by foliage, to make finding one a challenge even for acute observers. Only a little over 20% of atlas records for the Scarlet Tanager were nesting confirmations, with most of these involving observation of recent fledglings (73 records) or adults with food for young (99 records). Egg dates for this tanager in Maryland and DC range from 12 May to 1 August (Stewart and Robbins 1958).

The Scarlet Tanager was reported from approximately the same distribution and the same number of blocks in 2002–2006 as in 1983–1987 (Cheevers 1996e). Blocks with Scarlet Tanagers decreased by 10% on the Eastern Shore, particularly in Dorchester and Somerset counties, and increased slightly on the Piedmont. The Scarlet Tanager is approaching the southern edge of its breeding range on the Coastal Plain (it drops out in the coastal Carolinas) and is uncommon on Maryland's lower Eastern Shore (Price et al. 1995). Holes in this tanager's Maryland and DC nesting range include parts of the agricultural Hagerstown and Frederick valleys, the pine savannas and marshes of the lower Eastern Shore, the dunes of Assateague, and the urbanized Baltimore and DC areas. All of these areas have few or fragmented forests, pointing up the Scarlet Tanager's difficulty in living in such landscapes (Robbins, Dawson, and Dowell 1989; Rosenberg et al. 1999).

National BBS trends suggest Scarlet Tanagers have a reasonably stable long-term population level, but from the 1980s to the present the species has declined significantly in Maryland. Trends from neighboring states with more extensive forested lands have been stable. The Scarlet Tanager's decline in Maryland may reflect Maryland's lower proportion of forest land and its rapidly suburbanizing landscape east of Allegany County. This tanager is still geographically widespread in spite of declines in Maryland and DC. There will likely be Scarlet Tanagers here two decades hence, but their range may have contracted.

WALTER G. ELLISON

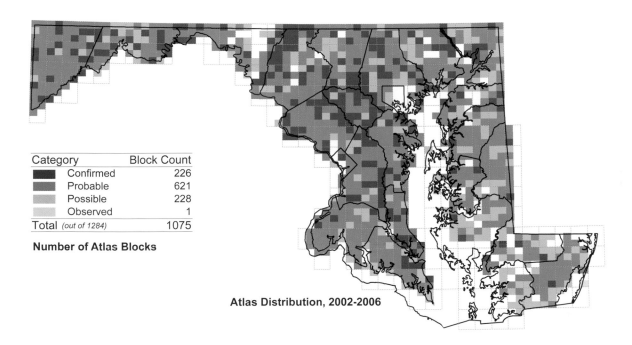

Category	Block Count
Confirmed	226
Probable	621
Possible	228
Observed	1
Total *(out of 1284)*	1075

Number of Atlas Blocks

Atlas Distribution, 2002-2006

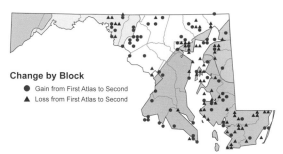

Change by Block
- ● Gain from First Atlas to Second
- ▲ Loss from First Atlas to Second

Change in Atlas Distribution, 1983-1987 to 2002-2006

Percent of Stops
- 50 - 100%
- 10 - 50%
- 0.1 - 10%
- < 0.1%

Relative Abundance from Miniroutes, 2003-2007

Atlas Region	1983-1987	2002-2006	Change No.	%
Allegheny Mountain	93	93	0	0%
Ridge and Valley	133	135	+2	+2%
Piedmont	292	300	+8	+3%
Upper Chesapeake	109	106	-3	-3%
Eastern Shore	221	198	-23	-10%
Western Shore	225	238	+13	+6%
Totals	1073	1070	-3	0%

Change in Total Blocks between Atlases by Region

Breeding Bird Survey Results in Maryland

Charles Lentz

Northern Cardinal
Cardinalis cardinalis

The Northern Cardinal, or more familiarly the redbird, is a deservedly popular bird known by nearly as many people as robins and crows. This cardinal combines bright plumage, fire engine red in males and warm brown, red, and pink in females, a sweet whistled far-carrying song, tolerance of humans and their works, and a jaunty air accentuated by its pert crest. It has been selected as the official state bird in seven states including neighboring Virginia and West Virginia. This was the most widely recorded bird during the 2002–2006 Maryland-DC atlas project, with reports from 1,256 blocks, and it was the second most widely distributed bird during the 1983–1987 atlas (Robbins and Blom 1996).

Northern Cardinals are primarily birds of shrubby thickets along forest edges, but they will often inhabit the interior of moist woodlands with thick shrubs and will haunt hedgerows and thickets in open land. Cardinals are very widespread, occurring from mature woods with small clearings to islands of trees in salt marshes to urban yards. They even occur on islands such as Bloodsworth and Smith that are untenanted by many other widespread songbirds.

Cardinals are resident year-round where they occur; only a few disperse more than 18 km (10 mi) from their birthplaces (Dow and Scott 1971), although some will disperse over 100 km (62 mi) including two individuals that went northward to Pennsylvania from Maryland banding stations (Stewart and Robbins 1958). Males commence territorial singing as early as late January. Females generally raise two broods; some raise three (Laskey 1944). Nest building is often under way by mid- to late March. The nest is a sturdy but untidy structure of sticks, weed stems, and bark strips built in a shrub or small tree. The mean height of 164 Maryland nests was 1.8 m (6 ft; Cullom 1996a; MNRF). Maryland and DC egg dates range from 5 April to 2 September, with nestlings reported to 5 September (Cullom 1996a; MNRF). Although cardinal nests are not very hard to find and are easy to identify as used nests, most atlas confirmations are of recent fledglings. The tittering high-pitched begging of baby cardinals is a characteristic sound of spring and summer in Maryland and DC.

The Northern Cardinal has been common in Maryland westward to Allegany County for at least a century (Kirkwood 1895; Eifrig 1904). It has been less common west of the Allegheny Front. Stewart and Robbins (1958) called it uncommon in Garrett County save at lower elevations. They were found throughout Maryland and DC in 1983–1987 with 62% of records referring to confirmed nesting. They were found throughout Garrett County and were recorded on more than 10% of relative abundance survey stops in the county (Cullom 1996a). The only notable absences were from the open tidal wetlands of southern Dorchester and northwestern Somerset counties. They were found in essentially the same distribution of atlas blocks in 2002–2006 with small increases in lower Eastern Shore marshlands. Nesting confirmations increased to 71% in the later atlas project.

Northern Cardinals have shown stable or slowly increasing numbers on North American BBS routes since 1966. They have declined in some southern states offset by increases in northern states. Indeed, cardinals have expanded their nesting range northward over the past 60 years to southwestern Quebec and central Maine (Beddall 1963; Halkin and Linville 1999). Maryland BBS routes show long-term stability, but they have been increasing at 1.1% per year since 1980. This increase seems to be in part a recovery from declines after harsh winters in the late 1970s. Northern Cardinals appear resilient and tolerant of most man-made landscapes and therefore should remain common well into the future.

WALTER G. ELLISON

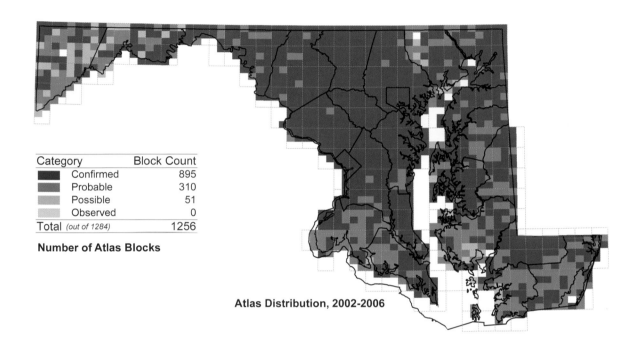

Category	Block Count
Confirmed	895
Probable	310
Possible	51
Observed	0
Total (out of 1284)	1256

Number of Atlas Blocks

Atlas Distribution, 2002-2006

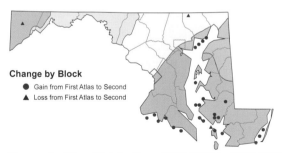

Change by Block
● Gain from First Atlas to Second
▲ Loss from First Atlas to Second

Change in Atlas Distribution, 1983-1987 to 2002-2006

Percent of Stops
50 - 100%
10 - 50%
0.1 - 10%
< 0.1%

Relative Abundance from Miniroutes, 2003-2007

Atlas Region	1983-1987	2002-2006	Change No.	%
Allegheny Mountain	93	92	-1	-1%
Ridge and Valley	148	148	0	0%
Piedmont	316	315	-1	0%
Upper Chesapeake	118	122	+4	+3%
Eastern Shore	307	321	+14	+5%
Western Shore	246	252	+6	+2%
Totals	1228	1250	+22	+2%

Change in Total Blocks between Atlases by Region

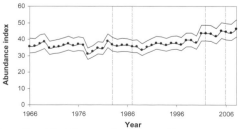

Breeding Bird Survey Results in Maryland

Rose-breasted Grosbeak
Pheucticus ludovicianus

A stocky, muscular, heavy-billed bird slightly smaller than a robin, the Rose-breasted Grosbeak shows a radical difference in the outward appearance of males and females. Males are black and white with the rose breast cited in the species' name, whereas females have a sparrowlike plumage, brown above and white below, with a strong brown and white head pattern. These grosbeaks are accomplished singers, producing a varied, mellow, robinlike warble, sometimes even uttered by birds on the nest. Rose-breasted Grosbeaks are essentially birds of northern hardwood forest dominated by sugar maple, beech, oak, and birch; in Maryland they nest largely west of the Allegheny Front and are uncommon and local east of it.

Rose-breasted Grosbeaks are most numerous in edge habitats and young to middle-aged deciduous woodlands; they also occupy mature forests with a generous understory of shrubs and small trees in lesser numbers. These grosbeaks often occur in regenerating timber cuttings, along roadsides, in old farmland reverting to forest, along streams, and in parks and wooded suburbs. Leberman (1992e) observed that Rose-breasted Grosbeaks in Pennsylvania were largely found in regions with a mean maximum July temperature of less than 29°C (84°F) and above 304 m (1,000 ft) and these restrictions generally apply in Maryland as well. Nonetheless, this species occurs in all of the counties along the southern tier of the Keystone State bordering Maryland.

Rose-breasted Grosbeaks summer in Maryland from late April until early October and spend the winter over a large part of tropical America from southern Mexico to northwestern South America. This grosbeak nests in trees, occasionally in large shrubs, generally at moderate heights of 9.2 m (30 ft) and below (Wyatt and Francis 2002), although nests may be higher in tall forests (Ellison 1985a). The flimsy nests are made of sticks and built by both sexes; males also help with incubation (Wyatt and Francis 2002). Maryland egg dates range from 22 May to 13 June (Dowell 1996b; MNRF), but egg dates have been reported into July to the north of Maryland (Peck and James 1987); adults with food for young were reported as late as 16 July during this project (I. Yoder).

Breeding season records of Rose-breasted Grosbeaks range back into the nineteenth century when Kirkwood (1895) reported a nest on Dans Mountain, Allegany County; Eifrig (1904) called the bird common at higher elevations. By the late 1950s they were described as fairly common in the Allegheny Mountains and rare to the east (Stewart and Robbins 1958). Rare eastern Maryland nestings have taken place from time to time, most remarkably on the Coastal Plain in Calvert County (Ball 1930; Stewart and Robbins 1958).

During the 1983–1987 atlas project, Rose-breasted Grosbeaks were widely recorded in the Alleghenies but were also found sporadically in the Ridge and Valley region, with most reports in central and eastern Allegheny County and 4 blocks on northern parts of South and Catoctin Mountains (Dowell 1996b). Two 1980s records of this grosbeak were from Baltimore County, including a confirmation near Phoenix (Dowell 1996b). The distribution was broadly similar to its 1980s status during the 2002–2006 survey. The range, if not the numbers, of this grosbeak has increased in the Ridge and Valley with more occupied blocks in central and eastern Allegany County, western Washington County, and on the Maryland Blue Ridge southeast to the vicinity of Braddock Heights, Frederick County. Although a few summering Rose-breasted Grosbeaks were encountered on the Piedmont and Coastal Plain, none could be given a higher code than observed (O). Such birds may occasionally attract a mate and become the source of the very few eastern Maryland breeding records.

North American BBS routes have shown a small, steady, but significant, declining trend for the Rose-breasted Grosbeak over the past four decades. The Maryland BBS trend is unreliable because of the small number of routes recording these grosbeaks and the small number of birds seen per route. Historic fluctuations in numbers have been noted to the north of Maryland (Forbush 1929; Leberman 1992e), perhaps because of fluctuations in the amount of young forest with shrubby understories. In general this species has expanded its range to the south over the past 50 years (Leberman 1992e). The Rose-breasted Grosbeak is well established in Garrett County and appears to have gained a firmer foothold in the Ridge and Valley where it remains uncommon. Its continued presence in Maryland depends on the extent and dynamic health of upland deciduous forests.

WALTER G. ELLISON

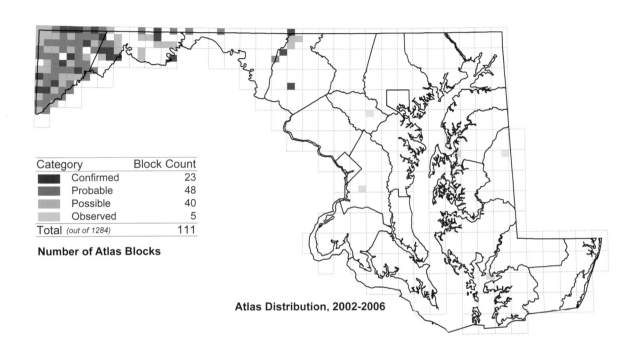

Category	Block Count
Confirmed	23
Probable	48
Possible	40
Observed	5
Total (out of 1284)	111

Number of Atlas Blocks

Atlas Distribution, 2002-2006

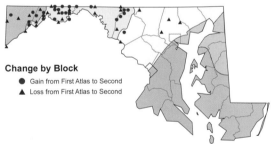

Change by Block
● Gain from First Atlas to Second
▲ Loss from First Atlas to Second

Change in Atlas Distribution, 1983-1987 to 2002-2006

Percent of Stops
■ 50 - 100%
■ 10 - 50%
▨ 0.1 - 10%
□ < 0.1%

Relative Abundance from Miniroutes, 2003-2007

Atlas Region	1983-1987	2002-2006	Change No.	Change %
Allegheny Mountain	87	84	-3	-3%
Ridge and Valley	17	26	+9	+53%
Piedmont	5	1	-4	-80%
Upper Chesapeake	0	0	0	-
Eastern Shore	0	0	0	-
Western Shore	0	0	0	-
Totals	109	111	+2	+2%

Change in Total Blocks between Atlases by Region

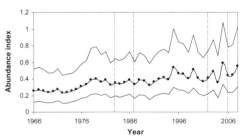

Breeding Bird Survey Results in Maryland

Blue Grosbeak
Passerina caerulea

Of the three widespread blue birds prone to perching on roadside wires in Maryland, the Blue Grosbeak is the bull-necked blue bird fanning and twitching its tail and throwing its head back to pour out its jumbled warble. Its peeved metallic *bink* call-note is distinctive and readily attracts attention. A southern bird, this grosbeak is uniformly distributed only below the fall line. Over the past century these birds have haltingly spread northward in the mid Atlantic states, mostly at low elevations near the coast.

The Blue Grosbeak is a bird of open grassy lands with scattered tall herbs, shrubs, and medium-size trees. The species' Maryland haunts include regenerating farm fields and pastures grown up to scattered shrubs and saplings, hedgerows, regenerating timber cuts, floodplain thickets, the margins of marshes, and field ditches in crop fields, especially corn.

Blue Grosbeaks summer in Maryland and DC from late April to early October; they winter from southern Mexico to Panama (J. Ingold 1993). Nests are placed in robust weedy herbs, thick shrubs, and small trees. The nest is a bulky structure of grass and weed stems well hidden by surrounding branches and leaves. Nests are so well hidden that they were seldom found by atlas observers; 85% of breeding confirmations were of adults with food (often large leggy insects) or of recent fledglings. The average nest height in Maryland is 1.3 m (4.3 ft; Willoughby 1996c; MNRF). Maryland egg dates range from 5 May to 3 September (Willoughby 1996c; MNRF). The early May date seems exceptional; atlas reports indicate most nests are not established until the third week of May or later. Blue Grosbeaks can be double brooded, with adults often seen feeding fledglings in Maryland until mid-September.

Kirkwood (1895) observed that the Blue Grosbeak was more or less regular in southern Maryland north to the vicinity of Washington, DC, and cited a nesting record at Laurel. By the 1950s the species had increased its range to include the southern Piedmont north to central Howard County and, although rare, extended westward along the Potomac to Cumberland; they were also uncommon in the Upper Chesapeake and northern Eastern Shore and were rare south of Talbot and Caroline counties (Stewart and Robbins 1958). By the 1983–1987 statewide atlas, these grosbeaks had spread into new areas and become ubiquitous in others (Willoughby 1996c). The species was found throughout the Eastern Shore and was reported throughout the Piedmont and Washington County although it was spottily distributed. Farther west, it was recorded in several blocks around Cumberland and was found in a few blocks in Garrett County.

Although Blue Grosbeaks occurred in essentially the

same number of blocks during 2002–2006, this masks a balance between losses and modest gains. They declined in the Ridge and Valley (net loss of 12 blocks) and are now almost absent around Cumberland, and they were lost from a net total of 24 blocks on the Piedmont. In spite of these losses they were still found in the Allegheny Mountains including two probable breeding records. There were gains east of the Bay Bridge with a net increase of 21 blocks.

The North American BBS trend for Blue Grosbeak has been positive at 1% per year since 1966 although the trend has slowed since 1980. Maryland BBS results have been similar. Trends across the range have been similar to other birds of successional habitats, showing broad areas of both increase and decrease. Maryland's pattern of atlas gains and losses seems to reflect similar shifting trends for shrublands and edges across the Free State. These grosbeaks are well established on Maryland's Coastal Plain, but they appear to be losing ground in the suburbanizing Piedmont and western Maryland. Rapid conversion of agricultural lands in southern Maryland and the Eastern Shore to housing and industry could threaten this species in its current stronghold.

WALTER G. ELLISON

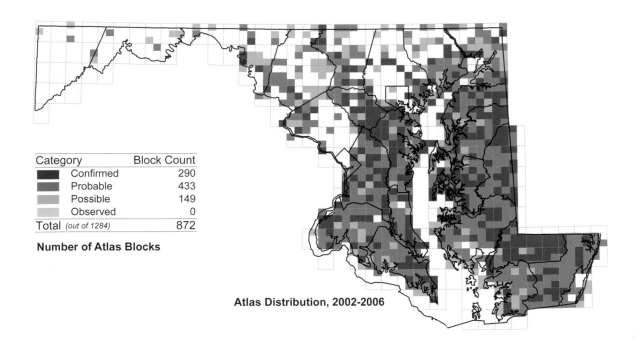

Category	Block Count
Confirmed	290
Probable	433
Possible	149
Observed	0
Total *(out of 1284)*	872

Number of Atlas Blocks

Atlas Distribution, 2002-2006

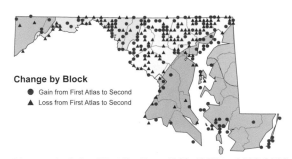

Change by Block
- ● Gain from First Atlas to Second
- ▲ Loss from First Atlas to Second

Change in Atlas Distribution, 1983-1987 to 2002-2006

Percent of Stops
- 50 - 100%
- 10 - 50%
- 0.1 - 10%
- < 0.1%

Relative Abundance from Miniroutes, 2003-2007

Atlas Region	1983-1987	2002-2006	Change No.	Change %
Allegheny Mountain	4	5	+1	+25%
Ridge and Valley	51	39	-12	-24%
Piedmont	192	168	-24	-13%
Upper Chesapeake	109	115	+6	+6%
Eastern Shore	285	305	+20	+7%
Western Shore	234	235	+1	0%
Totals	875	867	-8	-1%

Change in Total Blocks between Atlases by Region

Breeding Bird Survey Results in Maryland

Bill Sherman

Indigo Bunting
Passerina cyanea

One of the most common summer birds in Maryland, the Indigo Bunting is unknown to the average nonbirder in spite of its vivid color. The male in his alternate plumage is deep blue, darkest on the head. The winter-plumaged male (seen from late August until departure) is much duller. The blue is partly to mostly replaced by plain brown, but generally enough blue remains to make identification easy. The female is light brown with a whitish throat and belly. Her breast is usually streaked, but the streaking is at best faint and obscure. She is the plainest of plain brown birds; the lack of all field marks is the best field mark. Immature birds resemble the female but tend to be more heavily streaked. The loud song consists of a series of well-articulated strident finchlike notes that last between 1 and 6 seconds. Its easiest identifying feature is a characteristic doubling (sometimes tripling) of the notes: *This, this, is the song, song, song, of the, of the Indigo, Indigo, Bunting, Bunting.* It is usually given from an open perch, but sometimes is performed as a flight display: "The bird gives the song from an altitude of 75 to 100 feet, fans the air rather laboriously or stiffly, and propels the body rather slowly in a straight line" (G. M. Sutton in Taber and Johnston 1968). This bird is a persistent singer, starting at dawn and

continuing through the heat of the afternoon into dusk. It is also one of our last singers during late summer, usually singing well into August, and sometimes into September.

This species has been greatly assisted by the clearing of the forests that covered Maryland in colonial days as it nests in open brushy country, along the edges of roads, and in field hedgerows. It avoids the interior of mature forest as well as fields without at least some brush. The male seeks moderately high and open perches—often on wires—from which to sing, and the female looks for dense cover near the ground for a secure nest site. The Indigo Bunting feeds by gleaning at low to medium height. The diet is varied, including a wide array of insects; seeds and fruit are also important foods. The nest is usually placed in the crotch of a shrub or briar tangle within dense cover, from 0.6 m (2 ft) to 4 m (12 ft) above the ground. A woven cup of dry grass, fine twigs, weed stems, and bark strips, it is lined with softer material, such as finer grasses, rootlets, and sometimes hair or feathers. Occasionally the nest will include a snakeskin (H. Harrison 1975). The Indigo Bunting is a rather late breeder; nest building was observed between 12 May and 17 July for this atlas, and published egg dates range from 24 May to 16 August (Stewart and Robbins 1958).

The Indigo Bunting is one of the most common and widespread birds in Maryland, found in an overwhelming majority of the blocks in all regions. The previous atlas had only a few blocks without the species, and the present atlas increased the total by 13 (1%). Some overshooting migrant buntings can arrive in Maryland at the beginning of March, but the main influx comes during the first week in May. Numbers remain high until mid-September, then begin falling off through late October or early November. The bulk of the population winters from central Mexico through Central America to the Caribbean, but some remain in south Texas, the southern half of peninsular Florida, and southernmost Louisiana.

Although the number of blocks increased from the first atlas (Hitchner 1996c), BBS data indicate a slight drop in the population of Indigo Buntings. Maryland data are consistent with data from all of North America and from within the United States in showing an annual loss of about 0.5%. As with other early-successional species (e.g., Prairie Warbler, Eastern Towhee), the decline may be a result of loss of habitat from three directions: human development of scrubland habitats, maturation into forest, and the loss of brushy edge habitat through more intensive use of farmland.

T. DENNIS COSKREN

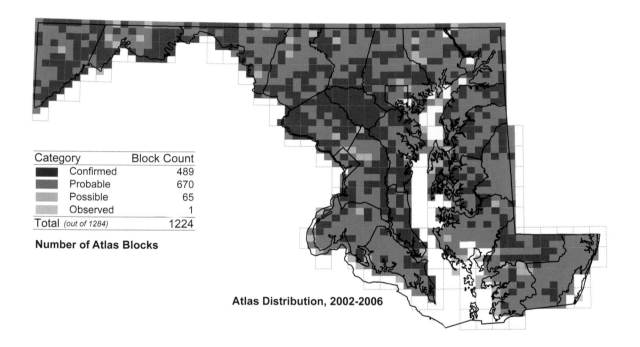

Category	Block Count
Confirmed	489
Probable	670
Possible	65
Observed	1
Total (out of 1284)	1224

Number of Atlas Blocks

Atlas Distribution, 2002-2006

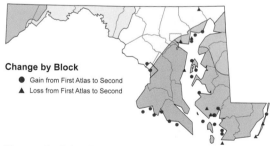

Change by Block
- Gain from First Atlas to Second
▲ Loss from First Atlas to Second

Change in Atlas Distribution, 1983-1987 to 2002-2006

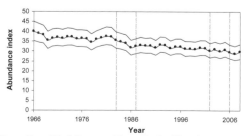

Percent of Stops
- 50 - 100%
- 10 - 50%
- 0.1 - 10%
- < 0.1%

Relative Abundance from Miniroutes, 2003-2007

Atlas Region	1983-1987	2002-2006	Change No.	%
Allegheny Mountain	93	93	0	0%
Ridge and Valley	148	148	0	0%
Piedmont	315	315	0	0%
Upper Chesapeake	118	119	+1	+1%
Eastern Shore	287	293	+6	+2%
Western Shore	244	250	+6	+2%
Totals	1205	1218	+13	+1%

Change in Total Blocks between Atlases by Region

Breeding Bird Survey Results in Maryland

Dickcissel
Spiza americana

For several fortunate atlas observers one of the highlights of their atlas fieldwork was hearing the staccato *dick-dick-siss-siss-siss* of the Dickcissel followed by a look at the singing male, with his yellow breast, black cravat, and rusty shoulder, perched on a tall weed, shrub, or roadside wire. If fortune had continued to smile, observers might also have found multiple males in a loose colony or seen the less brightly marked female carrying food to her offspring. This grassland bunting is a rare but increasingly regular nester in Maryland farmland, particularly on the Piedmont and Coastal Plain.

Because of the ephemeral nature of many of their favored breeding haunts Dickcissels are generally nomadic and vary in numbers from year to year on both a local and regional scale. Researchers studying grassland birds in Conservation Reserve Program fields at Bluestem (formerly Chino) Farms in northern Queen Anne's County have found that maintenance of these grasslands can lead to considerable site fidelity in this species, even far from the core of their breeding range (D. E. Gill, pers. comm.; D. Small, pers. comm.). Dickcissels nest in open grassland with a strong admixture of forbs, usually tall weedy ones. In Maryland they occur in fallow fields, unkempt hay fields with plenty of clover and tall weeds, alfalfa fields, old fields with a light scatter of small shrubs, weedy crop-field buffer strips, and field ditches.

The status of the Dickcissel in Maryland and DC is complex. They are scarce autumnal migrants, mostly along the coast, as they are along the entire eastern seaboard. A few may winter, usually at feeding stations with large flocks of ground-feeding birds. Most winter in rice and sorghum fields principally in central Venezuela, but also in southern Central America and other countries in northern South America (Temple 2002). Breeders arrive from mid-May to mid-June. The nest is usually built slightly above the ground in rank supporting vegetation with much overhead cover, usually less than half a meter (1.6 ft) above the ground (A. Gross 1921; M. Patterson and Best 1996; Zimmerman 1966). Maryland nesting dates have been thoroughly documented at Bluestem Farms 5.3 km (3.3 mi) east of Chestertown with egg sets found from 6 June (first laid egg 2 June) to 2 August, and nestlings from 17 June to 16 August (D. E. Gill, pers. comm.; D. Small, pers. comm.).

The Dickcissel was once a common nesting bird along the Atlantic seaboard from Massachusetts to South Carolina. They disappeared during the mid- to late nineteenth century (Rhoads 1903; A. Gross 1956). This loss took place in Maryland and DC during the 1870s (Coues and Prentiss 1883; Kirkwood 1895). They did not nest again in Maryland until 1928 in Montgomery County (Wetmore and Lincoln 1928). They have nested sporadically thereafter, especially in western Montgomery County and the Frederick and Hagerstown valleys (Stewart and Robbins 1958; S. A. Smith 1996d). They were reported from 26 atlas blocks from 1983–1987, with most in known regions of occurrence. But their presence in 10 blocks east of Chesapeake Bay implied a new nesting area, although none were confirmed breeding there (S. A. Smith 1996d).

In recent decades nesting Dickcissels have occasionally

irrupted eastward from their midwestern core range, presumably because of drought. Such Dickcissel flight-years occurred in 1988 and 1996 (Ringler 1988b; Southworth 1997). During 2002–2006 they were found in an impressive 66 blocks with another two records of birds simply observed. They increased in the Frederick Valley, continued in small numbers in the Hagerstown Valley, and have clearly become firmly established in the Upper Chesapeake and Eastern Shore. Of particular interest are several records from the eastern Piedmont, the Western Shore, and single reports from Allegany and Worcester counties. Analysis of the 85 atlas reports in the 2002–2006 database shows that 78% of them were made

in just two years 2004 (n = 25) and 2005 (n = 41), suggesting that this project coincided with two flight-years.

Dickcissels declined continentally on BBS routes from 1966 to 1979 at a rate of 5.5% per year; since 1980 the decline has leveled of and the overall trend over the past four decades is approximately stable. They have been found on only three Maryland BBS routes with all reports after 1980. This bird depends on the maintenance of grassland, especially well-managed Conservation Reserve lands, fallow fields, and broad weedy crop buffer strips.

WALTER G. ELLISON

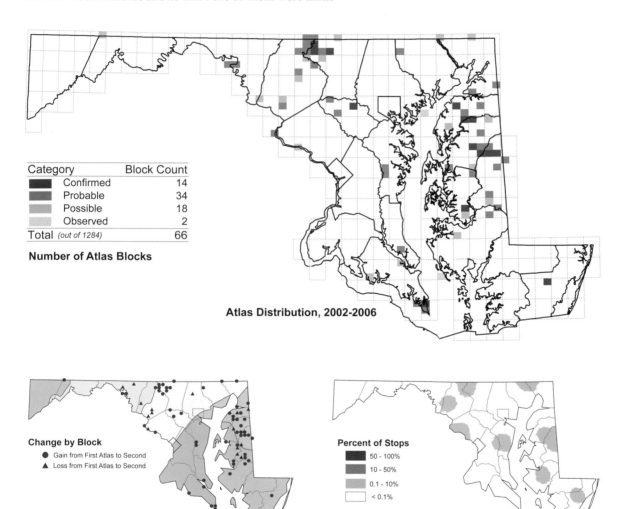

Category	Block Count
Confirmed	14
Probable	34
Possible	18
Observed	2
Total (out of 1284)	66

Number of Atlas Blocks

Atlas Distribution, 2002-2006

Change by Block
- ● Gain from First Atlas to Second
- ▲ Loss from First Atlas to Second

Change in Atlas Distribution, 1983-1987 to 2002-2006

Percent of Stops
- ■ 50 - 100%
- ■ 10 - 50%
- ■ 0.1 - 10%
- □ < 0.1%

Relative Abundance from Miniroutes, 2003-2007

			Change	
Atlas Region	1983-1987	2002-2006	No.	%
Allegheny Mountain	0	1	+1	-
Ridge and Valley	4	4	0	0%
Piedmont	12	24	+12	+100%
Upper Chesapeake	4	17	+13	+325%
Eastern Shore	6	12	+6	+100%
Western Shore	0	8	+8	-
Totals	26	66	+40	+154%

Change in Total Blocks between Atlases by Region

Bobolink
Dolichonyx oryzivorus

One of the avian pleasures of a sunny morning in upland dairy or horse-farming country is a hay field hosting a chorus of Bobolinks, with the males fluttering aloft like little black and white kites each pouring out a cascade of bubbling notes recalling a bluegrass breakdown. Loss of farmland and changes in the harvesting of hay threaten this small blackbird that is largely limited in Maryland to the Allegheny Mountain region.

Bobolinks breed in mesic to wet grasslands with grass of at least moderate height, little open ground, and few or no woody plants. Prior to European settlement they likely occupied floodplains, the drier parts of marshes, and prairies maintained by fires periodically set by Native Americans. At present they are almost exclusively found in hay meadows in Maryland. Upstate New York studies have demonstrated that Bobolinks prefer older and larger-than-average hayfields (Bollinger and Gavin 1992).

The Bobolink has the longest distance migration of any Western Hemisphere songbird, wintering as far south as Argentina. They reside in Maryland from late April to mid-October. The nest is a bowl of grass stems nestled into a hollow on the ground, often tucked beneath a weedy herb tuft (S. Martin and Gavin 1995). Few dates for active nests in Maryland have been recorded. During this project, nests with eggs were found north of Friendsville, Garrett County, on 15 June 2006 and 24 June 2006 (F. Ammer). Dates for confirmed breeding in this project ranged from nest building

in late May to fledged young in late July. Only one brood is usually attempted, although renesting is frequent. Second broods may be attempted under extraordinary conditions (Gavin 1984).

Kirkwood (1895) knew the Bobolink only as a common transient in Maryland. Eifrig (1904) alluded to nesting at upper elevations in the Allegheny Mountains but considered them more common in migration. Stewart and Robbins (1958, fig. 60) illustrated the main range in the Alleghenies as a swath running northeastward from just north of Backbone Mountain to northwest of Frostburg; they also cited two isolated breeding records from the Piedmont in Baltimore and Frederick counties. During the first statewide atlas they were found in more than three-quarters of the blocks in the Allegheny Mountains and had spread to lower elevations in the region (Dowell 1996c). There was also a scatter of records during atlas safe dates in the Frederick Valley from northwestern Carroll County to southeastern Frederick County, suggesting range expansion into the western Piedmont.

From 2002–2006 Bobolinks were similarly distributed in the Alleghenies, increased in the Frederick Valley including five breeding confirmations, and were found during safe dates eastward to the fall line with confirmations at Fair Hill Natural Resources Management Area northeast of Elkton, Cecil County, and west of Havre de Grace, Harford County. Three reports (two of them observed) from the Eastern Shore probably referred to nonbreeding birds.

Bobolinks have been declining over the past four decades on BBS routes at a rate of 1.8% per year with an accelerated rate of loss since 1980. Their status on Maryland BBS reports is uncertain because of the small number of routes recording them, but they have not obviously declined. The extensive declines in northeastern North America appear to be related to a decrease in dairy and horse farming and changes in farming practices with respect to hay cropping. Hay field acreage has fallen; fields are now often planted to alfalfa, less suitable for Bobolinks; and hay is now cut more frequently (up to four times in a season) and cropped more closely than in the past (Bollinger and Gavin 1992). These changes con-

tribute to declines via habitat loss and lowered reproductive success. At present these grassland blackbirds are holding their own in Maryland. However, changes in hayfield harvest and loss of hay meadows to development or conversion to other crops would quickly reduce their populations here.

WALTER G. ELLISON

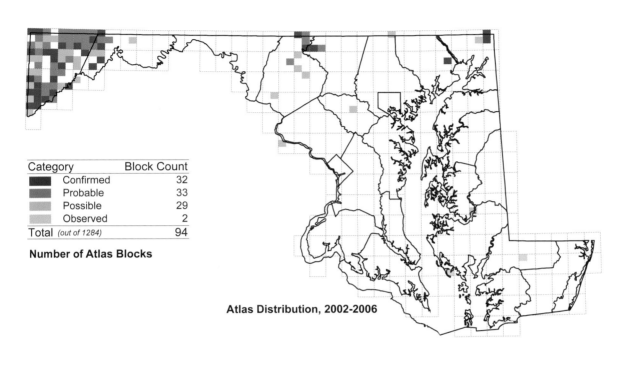

Category	Block Count
Confirmed	32
Probable	33
Possible	29
Observed	2
Total (out of 1284)	94

Number of Atlas Blocks

Atlas Distribution, 2002-2006

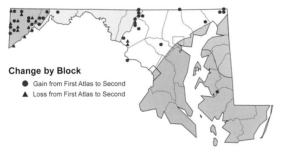

Change by Block
- ● Gain from First Atlas to Second
- ▲ Loss from First Atlas to Second

Change in Atlas Distribution, 1983-1987 to 2002-2006

Percent of Stops
- 50 - 100%
- 10 - 50%
- 0.1 - 10%
- < 0.1%

Relative Abundance from Miniroutes, 2003-2007

			Change	
Atlas Region	1983-1987	2002-2006	No.	%
Allegheny Mountain	71	72	+1	+1%
Ridge and Valley	2	5	+3	+150%
Piedmont	4	16	+12	+300%
Upper Chesapeake	0	0	0	-
Eastern Shore	0	1	+1	-
Western Shore	0	0	0	-
Totals	77	94	+17	+22%

Change in Total Blocks between Atlases by Region

Breeding Bird Survey Results in Maryland

Red-winged Blackbird
Agelaius phoeniceus

This striking bird is one of the more abundant in Maryland, and indeed across the continent. A common year-round resident in the eastern part of the state, it is one of the earliest spring arrivals in the remainder, where the male's song is a welcome reminder that winter's reign is drawing to a close. If the species were not so common, the adult male Red-winged Blackbird would be more appreciated for his elegant plumage: solid coal black accented by scarlet shoulder patches edged with buff (although the shoulder patch tends to be inconspicuous or invisible when the male is not in display). First-year males are variable: some are mostly black, some have buffy edges to the black feathers, black edges on the red feathers, and/or a suggestion of a superciliary line; others resemble the streaked females, but with a more pronounced reddish wing patch. The adult female is much more understated; she has sparrowlike plumage: dark brown above and heavily streaked brown and white below and on the head with an indistinct superciliary line. Sometimes she may be a little pinkish buff about the head and throat and her shoulders may be marked with reddish. The song of the male Red-winged Blackbird usually consists of one to three lower notes followed by a higher trill. A common and fairly accurate transcription is *o-ka-leeee*. The song is accompanied by an ostentatious display of the plumage: the tail is spread, the wings are half-spread, the feathers in the back are ruffled, and the shoulder epaulets are lifted so that they show to best advantage. The song is often given during a short flight.

The Red-wing is primarily a bird of the wetlands. Its first choice of habitat is cattail marshes, but wet hay meadows, wheat fields, wooded swamps, and riverside brush are acceptable. There must be dense vegetation in which to conceal the nest. The female constructs the nest, placing it in cattails, sedges, or bushes, preferably over or near water. Dense grass, weeds, or bushes can be used in dry situations. Occasionally the nest will be located in a tree. Leaves of sedges, cattails, and grasses are commonly used as building material. Fine grass is the usual lining and milkweed is often used to bind the structure (H. Harrison 1975). The height above the ground is quite variable, from 5 cm (2 in) to almost 7 m (22 ft; Bent 1958). The species is semicolonial. Territories are small and nests may be fairly close in prime habitat. The red epaulets are the badges of ownership of a territory. When a male Red-winged Blackbird trespasses on the land of another (a frequent occurrence), he will minimize the display of red and the resulting ire of the rightful proprietor (Hansen and Rohwer 1986). The male may have one mate, or he may have several; breeding females often outnumber males.

Red-winged Blackbirds are permanent residents within Maryland and are abundant throughout the state during the breeding season. But the wintering population is mostly east of the fall line, where it is common to abundant, and large flocks may be found in fields and marshes. Westward the species becomes uncommon to rare in the winter. The first few breeding males may arrive as early as late January, but displaying and singing males come in mid- to late February (Stewart and Robbins 1958). Females and immature males join them in March when territories have been established.

Present atlas results show that the state is nearly saturated, with only a very few blocks lacking the species. The earlier atlas had a few more blocks without Red-wings (C. Miller 1996), mostly in the Eastern and Western Shore regions,

so that the statewide number of blocks with Red-winged Blackbirds has increased by 2%, to 1,249. However, BBS data show a steady decrease, both nationwide and in Maryland. Over a 40-year period, the result has been a decrease of 50%. This may be the result of blackbird control programs in the southern states, a change in agricultural practices that reduce the amount of food available to winter flocks, or a combination of both. In spite of this decline, Red-winged Blackbird is likely to remain a prominent part of Maryland's avifauna.

T. DENNIS COSKREN

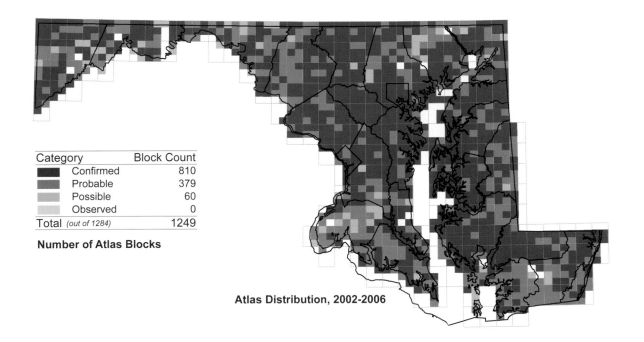

Category	Block Count
Confirmed	810
Probable	379
Possible	60
Observed	0
Total (out of 1284)	1249

Number of Atlas Blocks

Atlas Distribution, 2002-2006

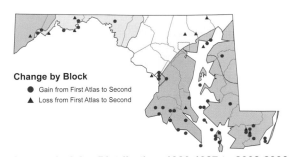

Change by Block
- ● Gain from First Atlas to Second
- ▲ Loss from First Atlas to Second

Change in Atlas Distribution, 1983-1987 to 2002-2006

Percent of Stops
- 50 - 100%
- 10 - 50%
- 0.1 - 10%
- < 0.1%

Relative Abundance from Miniroutes, 2003-2007

Atlas Region	1983-1987	2002-2006	Change No.	%
Allegheny Mountain	91	90	-1	-1%
Ridge and Valley	139	142	+3	+2%
Piedmont	314	312	-2	-1%
Upper Chesapeake	119	121	+2	+2%
Eastern Shore	316	332	+16	+5%
Western Shore	235	247	+12	+5%
Totals	1214	1244	+30	+2%

Change in Total Blocks between Atlases by Region

Breeding Bird Survey Results in Maryland

Eastern Meadowlark
Sturnella magna

The high lonesome gliding whistle of the Eastern Meadowlark is an early sign of spring in wide open lands, carrying far even on breezy days. This squat, hefty blackbird may be conspicuous when it is announcing territory from prominent perches that range from power poles and lines to fence posts to small trees and shrubs. At such times its bright yellow breast with black chevron stands out. When it is feeding on the ground, it disappears into tall grass or blends into brown stubble fields. Although this is the most widespread grassland bird in Maryland and DC, it has lost nearly 25% of its statewide nesting range since 1987.

The Eastern Meadowlark occupies a broad array of open grassy lands possessing a selection of elevated song perches. Large territories are characteristic for the polygynous males, generally ranging from 2.8 to 3.2 ha (7–8 ac; Lanyon 1957), so small blocks of habitat limit populations. For nesting they prefer lusher parts of grasslands with tall grassy vegetation and deep litter. Maryland haunts include extensive hayfields, lightly to moderately grazed pastures, the drier portions of salt marshes, airports, capped landfills, reclaimed surface mines, and winter wheat fields, with the lowest densities in the last environment.

These meadowlarks are resident year-round in Maryland.

They are rare during the winter in high elevation Garrett County, becoming more frequent to the east until they are fairly common on the Coastal Plain. Nesting birds return in early to mid-March and generally remain on territory into September. Males often have more than one mate, and it has been demonstrated that females with polygynous mates have higher average nesting success (Knapton 1988). The grass cup nests are built on the ground and are domed over with grass. They are very hard to find because sitting females flush only when almost trod upon. They are double brooded and Maryland egg dates span almost three and a half months, from 4 May to 9 August (Hilton 1996d; MNRF).

Eastern Meadowlarks were common nesting birds throughout Maryland at least until the 1960s (Kirkwood 1895; Stewart and Robbins 1958). They were still very widespread in 1983–1987, with records from over 1,000 blocks, 83% of the blocks receiving coverage (Hilton 1996d). From 2002–2006 they were not found in 247 of the blocks where they had occurred in the 1980s, a statewide loss of 23%. There were lost blocks in all physiographic regions. Three regions had losses in excess of 30%: the Piedmont with a net loss of 98 blocks, almost entirely in the eastern part of the region; the Western Shore, with a loss of 66 blocks, and the Upper Chesapeake, where they were not refound in 33 blocks.. The combined losses in these regions represent 80% of the overall net loss. The fewest losses were in the Ridge and Valley and Allegheny Mountain regions.

Declines of the Eastern Meadowlark over the past four decades have been precipitous on BBS routes at all geographic scales. The decline for all routes has been 2.8% per year; it has been 3.2% per year in eastern North America and an alarming 5.2% per year in Maryland. These declines, already obvious in the 1980s, were graphically illustrated in Hilton (1996d). The contrast between the 1980s range map and the trend is an excellent example of the tendency of geographic range loss to lag behind large population declines. The causes of these declines are complex and closely related to those listed in foregoing accounts of grassland birds: loss of pasture and hay acreage, changes in mowing schedules, short crop rotations with wheat occurring only once in a two- to three-year cycle, and conversion to row crops. Suburban sprawl is also clearly an issue in the eastern Piedmont and Western Shore. Publicly owned grasslands need to be managed with prescribed burning and less frequent and later mowing, and the Conservation Reserve Program should be promoted to enable the Eastern Meadowlark to continue in Maryland's avifauna.

WALTER G. ELLISON

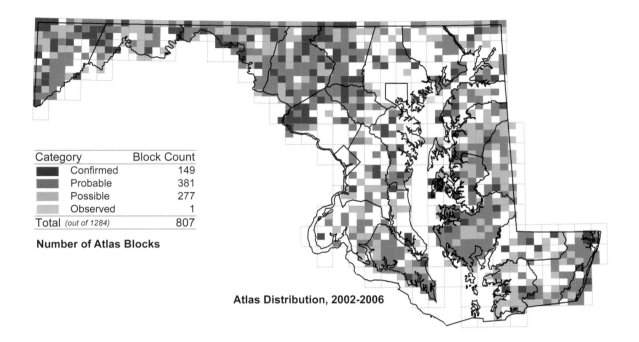

Category	Block Count
Confirmed	149
Probable	381
Possible	277
Observed	1
Total (out of 1284)	807

Number of Atlas Blocks

Atlas Distribution, 2002-2006

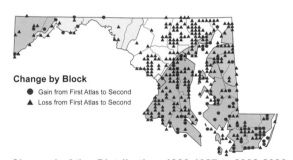

Change by Block
- ● Gain from First Atlas to Second
- ▲ Loss from First Atlas to Second

Change in Atlas Distribution, 1983-1987 to 2002-2006

Percent of Stops
- 50 - 100%
- 10 - 50%
- 0.1 - 10%
- < 0.1%

Relative Abundance from Miniroutes, 2003-2007

Atlas Region	1983-1987	2002-2006	Change No.	Change %
Allegheny Mountain	84	78	-6	-7%
Ridge and Valley	133	128	-5	-4%
Piedmont	288	190	-98	-34%
Upper Chesapeake	102	69	-33	-32%
Eastern Shore	255	216	-39	-15%
Western Shore	190	124	-66	-35%
Totals	1052	805	-247	-23%

Change in Total Blocks between Atlases by Region

Breeding Bird Survey Results in Maryland

Common Grackle
Quiscalus quiscula

For the greater part of the year the hefty, iridescent black, long-tailed Common Grackle forms large flocks that roost in woodlots, swamps, and marshes and wander the landscape in search of food in farm fields and wetlands, leaving much of the landscape untenanted by its kind. But during the nesting season, largely from April to July, grackles can seem omnipresent; they may be seen in every possible setting save deep woods. This grackle prefers a mixed landscape with scattered groves of trees and open ground. Although nest sites vary greatly, including such exotic locations as mailboxes and girders on highway overpasses, most nests are placed in trees and shrubs with thick, dense foliage. Common Grackles seem to be particularly fond of conifers, especially ornamentals planted as windbreaks, including pine, red cedar, Leland cypress, and Norway spruce, where they often form loose colonies of up to 20 pairs. Egg dates for Maryland and DC span a broad range of dates from 24 March to 16 August (MNRF); the latter date is extraordinary, the latest egg date for this project was in mid-June.

The nesting range of the Common Grackle in Maryland and the District of Columbia has changed little over the past half century (Stewart and Robbins 1958; Farrell 1996c). In

2002–2006 it was located in 1,235 atlas blocks, ranking eighth in prevalence among Maryland and DC's nesting birds, as compared to a rank of sixth in 1,222 blocks in 1983–1987 (Farrell 1996c). Common Grackles were not found in a handful of heavily wooded blocks in Garrett and Allegany counties, where this species had an uncharacteristically high number of reports of possible breeding; the statewide breeding confirmation rate was 83%. Common Grackles also proved scarce or absent on Assateague Island south of the campgrounds and picnic areas of North Assateague, went missing in 3 largely salt marsh and pine savanna blocks in southern Dorchester County, and went unreported on Bloodsworth and Smith islands (it nested in the 1980s on the latter island). It is generally assumed that all of Maryland's nesting grackles belong to the subspecies *stonei,* the Purple Grackle of the mid Atlantic and southeastern Coastal Plain. But it is possible that the grackles of the mountains of western Maryland might be intergrades with or pure *versicolor,* the Bronzed Grackle of the Midwest and inland Northeast. This requires further study.

Populations of the Common Grackle in North America were very large from the late 1960s to the early 1980s; southeastern Christmas Bird Counts north to Maryland occasionally recorded counts into the millions. This led to widespread attempts to control the numbers of birds at roosts in populated areas and to control crop depredation by these winter grain feeders. Breeding Bird Survey trends of the past 40 years have shown steady declines of 1.1% per year for North America and 1.7% per year in Maryland. Some of these declines may be attributable to the control measures that commenced in the late 1970s (Robbins, Bystrak, and Geissler 1986). Although grackles have been declining, the species remains very common, sometimes oppressively so, in Maryland and the District of Columbia and they should remain so into the foreseeable future.

WALTER G. ELLISON

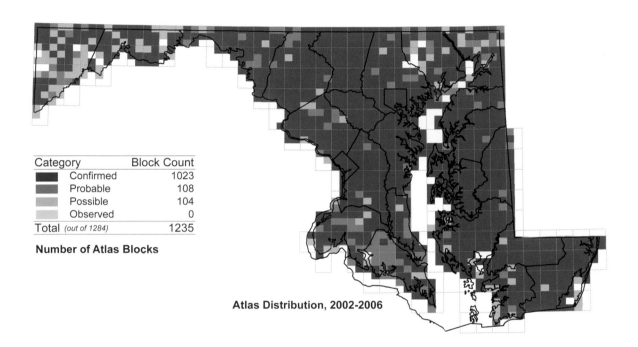

Category	Block Count
Confirmed	1023
Probable	108
Possible	104
Observed	0
Total (out of 1284)	1235

Number of Atlas Blocks

Atlas Distribution, 2002-2006

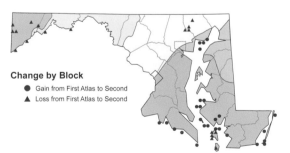

Change by Block
- ● Gain from First Atlas to Second
- ▲ Loss from First Atlas to Second

Change in Atlas Distribution, 1983-1987 to 2002-2006

Percent of Stops
- 50 - 100%
- 10 - 50%
- 0.1 - 10%
- < 0.1%

Relative Abundance from Miniroutes, 2003-2007

Atlas Region	1983-1987	2002-2006	Change No.	Change %
Allegheny Mountain	92	85	-7	-8%
Ridge and Valley	147	145	-2	-1%
Piedmont	316	312	-4	-1%
Upper Chesapeake	118	122	+4	+3%
Eastern Shore	305	317	+12	+4%
Western Shore	244	249	+5	+2%
Totals	1222	1230	+8	+1%

Change in Total Blocks between Atlases by Region

Breeding Bird Survey Results in Maryland

Boat-tailed Grackle
Quiscalus major

Unlike its ubiquitous relative the Common Grackle, the Boat-tailed Grackle is limited by habitat to a small nesting range in Maryland. These grackles nest colonially in salt marshes and usually feed along shores at low tide or on lawns near shorelines; they are seldom seen away from tidewater in Maryland. The iridescent glossy black males are much larger than Common Grackles, approaching the size of crows. The females are brown, far smaller than the males, and could easily be mistaken for another species. Colonies are composed of clusters of nesting females defended by dominant males (W. Post et al. 1996), a mating system seen more often in mammals than birds and unique among Maryland and DC birds.

Nesting colonies usually are found in stands of small trees or shrubs surrounded by or over water, or sometimes in tall rank grassy marsh vegetation such as common reeds or cattails. Boat-tailed Grackles feed away from their colonies on tidal flats, beaches, jetties, parking lots, or lawns. They do not usually feed in farm fields as do other blackbirds.

Boat-tailed Grackles are resident year-round, but because they usually form large roving flocks in the colder months, they can be easily overlooked at that time of year. Nesting commences in April, when the females build their large unkempt nests of grass, sedge, and cattail leaves. The nests,

which vary in height, are built in grassy vegetation and in trees and shrubs ranging from 1 to 3 m (3–10 ft) or more. In Maryland the majority of nests reported have been built in groundsel tree with an average height of 1.9 m (6.2 ft; O'Brien 1996d; MNRF). Maryland egg dates range from 24 April to 7 July with nestlings to 26 July (O'Brien 1996d; MNRF).

In the 1860s Boat-tailed Grackles appeared to occur northward only as far as North Carolina on the Atlantic seaboard (Coues 1870). By the 1890s they nested at least occasionally on Maryland's Atlantic coast (W. Wholey in Kirkwood 1895). They did not nest in Delaware until 1933 (Hess et al. 2000) and did not definitely nest in Chesapeake Bay salt marshes until the 1940s (Stewart and Robbins 1947). By 1958 Stewart and Robbins (1958) reported that these grackles nested in good numbers in Somerset and Worcester counties but were only rarely seen elsewhere. They continued to spread up the Atlantic coast, first nesting in New Jersey in 1952 (Fables 1955), and on Long Island, New York, in 1981 (Connor 1988). Boat-tailed Grackles apparently began nesting in Dorchester County in the 1970s and first nested on the Maryland Western Shore in St. Mary's County in 1980 (Ringler 1980).

From 1983 to 1987 the nesting range of the Boat-tailed Grackle included a few blocks on the southern Western Shore, and they appeared to be widespread in southern and western Dorchester County although no nesting confirmations were reported from there (O'Brien 1996d). They briefly nested north to Tilghman Island from 1991 to 1995 (Ringler 1991; Southworth and Southworth 1996). From 2002–2006 there was little change in the broad outlines of the nesting range, although they were found in a net total of 17 more blocks and appear to have declined slightly on the Western Shore. They were confirmed widely in Dorchester County and have also spread modestly up the lower Nanticoke, Wicomico, and Pocomoke rivers.

Numbers of Boat-tailed Grackles on BBS routes have been increasing significantly along the Atlantic coast. The condition of coastal wetlands and natural tidewater habitats could strongly influence the persistence and abundance of Boat-tailed Grackles in Maryland. This grackle's flexibility in feeding sites, especially its tendency to feed in urban settings, may insulate them slightly from degradation of natural habitats. At present they seem to be holding their own in Maryland, but they bear watching because of their limited and potentially fragile habitat.

WALTER G. ELLISON

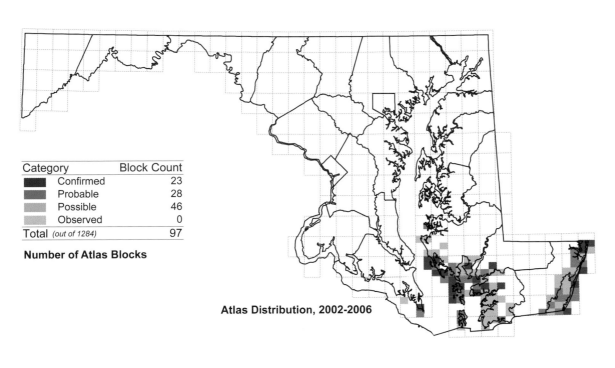

Category	Block Count
Confirmed	23
Probable	28
Possible	46
Observed	0
Total (out of 1284)	97

Number of Atlas Blocks

Atlas Distribution, 2002-2006

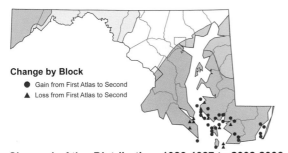

Change by Block
- ● Gain from First Atlas to Second
- ▲ Loss from First Atlas to Second

Change in Atlas Distribution, 1983-1987 to 2002-2006

Percent of Stops
- 50 - 100%
- 10 - 50%
- 0.1 - 10%
- < 0.1%

Relative Abundance from Miniroutes, 2003-2007

Atlas Region	1983-1987	2002-2006	Change No.	%
Allegheny Mountain	0	0	0	-
Ridge and Valley	0	0	0	-
Piedmont	0	0	0	-
Upper Chesapeake	0	0	0	-
Eastern Shore	73	93	+20	+27%
Western Shore	6	3	-3	-50%
Totals	79	96	+17	+22%

Change in Total Blocks between Atlases by Region

Breeding Bird Survey Results in Maryland

Brown-headed Cowbird
Molothrus ater

It is essential for the Brown-headed Cowbird to lay its eggs in the nests of other bird species. This adaptation may have originated from a nomadic life following herds of large grazing animals such as bison, making it hard to stay in one place long enough to lay eggs and raise nestlings (Friedmann 1929; Mayfield 1965). The exploitation of other birds' nesting efforts has made the parasitic cowbirds (five species in all) the target of human moral judgment. There is also concern over their effects on the populations of other birds, especially birds that naively accept cowbird eggs and raise the young. The Brown-headed Cowbird, which is most common in the Midwest and prefers open landscapes, may not be native to the eastern seaboard (Mayfield 1965). However, these cowbirds were first described from the Atlantic coast in North Carolina by Catesby in 1731. Whatever their native status, they certainly increased after cattle and horses arrived with European settlers and forests were cleared for farmland. This blackbird is found throughout Maryland and DC and is generally common except in heavily wooded and wetland landscapes.

Brown-headed Cowbirds inhabit open lands, although in the eastern United States they usually frequent mixed landscapes of forest, farms, and residential neighborhoods. They often associate with cattle and horses and feed in areas with bare ground and short grass. They seek the nests of their hosts in all habitats but most frequently parasitize birds in brushy areas, hedgerows, and in forests. In Maryland cowbirds appear to do most of their nest hunting in forests close to edges (Evans and Gates 1997).

Brown-headed Cowbirds are found year-round in Maryland and DC, but, as with other blackbirds, their flocking renders them local during the winter. Cowbirds are most numerous in eastern Maryland during the winter and are scarce in Allegany and Garrett counties (Hatfield et al. 1994). They arrive in nesting areas across Maryland and DC in March and April. Confirming nesting usually arises from detecting fledglings that beg persistently with loud buzzing calls. They are also often much larger than their caretakers. Unlike many other brood parasites, Brown-headed Cowbirds are not overly selective in choosing hosts and have parasitized more than two hundred species (Friedmann and Kiff 1985), although host abundance and quality affect parasitism rates (B. Woolfenden et al. 2004). Common Maryland and DC hosts include Red-eyed Vireo, Song Sparrow, Northern Cardinal, Wood Thrush, and Field Sparrow (Fletcher and Joyce 1996; MNRF). Cowbird eggs have been found in Maryland and DC from 17 April to 28 July (Fletcher and Joyce 1996; MNRF).

Brown-headed Cowbirds appear to be more common in Maryland and DC than they were a hundred years ago. Kirkwood (1895) and Eifrig (1904) noted that cowbirds were noticeably less numerous in the breeding season than during migration. Stewart and Robbins (1958) called them fairly common nesters in all regions save the Western Shore, where they were uncommon. They were reported throughout the state and District in 1983–1987 (Fletcher and Joyce 1996) and 2002–2006 in essentially the same number of blocks. There was a slight decline between atlas projects in the Allegheny Mountains and western Ridge and Valley in well-forested blocks. Cowbirds were missing from low-lying marsh islands on the lower Eastern Shore, but they were widespread on Assateague Island with its wild horses and scrub-nesting birds.

Brown-headed Cowbirds have been declining on BBS routes in North America at a rate of 1.1% per year and at 1.8% per year in the East. Declines appear to be in response to increases in mature forest in the north, as shown in the mountains of New York State between atlas projects (McGowan 2008i). Efforts to control blackbird numbers in winter roosts in the Gulf States may also be having an im-

pact. Maryland BBS trends are stable or declining slowly. Although many authors express concern that Brown-headed Cowbirds may be a cause of forest songbird declines, high cowbird parasitism is only symptomatic of a landscape pernicious to the survival of forest interior migratory songbirds. Humans have made the situation favorable for cowbirds, tilting the playing field against host species. Indeed cowbird declines may be partially attributable to declines of host species. Cowbirds will likely remain a fixture of Maryland and DC birdlife; how numerous they will be is less certain.

WALTER G. ELLISON

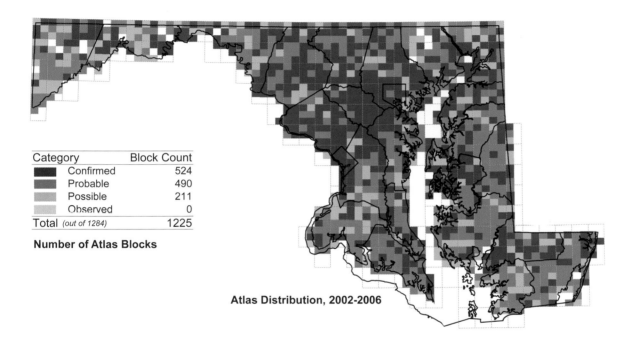

Category	Block Count
Confirmed	524
Probable	490
Possible	211
Observed	0
Total (out of 1284)	1225

Number of Atlas Blocks

Atlas Distribution, 2002-2006

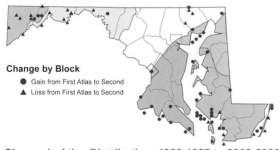

Change by Block
- ● Gain from First Atlas to Second
- ▲ Loss from First Atlas to Second

Change in Atlas Distribution, 1983-1987 to 2002-2006

Percent of Stops
- 50 - 100%
- 10 - 50%
- 0.1 - 10%
- < 0.1%

Relative Abundance from Miniroutes, 2003-2007

Atlas Region	1983-1987	2002-2006	Change No.	Change %
Allegheny Mountain	90	86	-4	-4%
Ridge and Valley	147	141	-6	-4%
Piedmont	312	312	0	0%
Upper Chesapeake	117	121	+4	+3%
Eastern Shore	301	311	+10	+3%
Western Shore	238	248	+10	+4%
Totals	1205	1219	+14	+1%

Change in Total Blocks between Atlases by Region

Breeding Bird Survey Results in Maryland

Charles Lentz

Orchard Oriole
Icterus spurius

The Orchard Oriole is a small, active oriole of open places with scattered trees. Males two years old and older are dark, black and chestnut; females are warblerlike, green above and yellow green below; one-year-old males are similar to females but have an extensively black throat. The Orchard Oriole's song differs from that of the Baltimore Oriole in being a complex, rapid ditty of fluty whistles, chatters, and nasal notes that generally lacks the slow delivery and mellowness of the larger bird. Among the places Orchard Orioles are found are open floodplain woodlands, stands of trees at the edge of marshes, hedgerows and wind breaks in open farmland, open subdivisions with scattered ornamentals, groves of trees around farm houses, and, of course, orchards.

Orchard Orioles are migratory and single brooded, taking only enough time in their breeding range to produce offspring and depart for Mexico, Central America, and northwestern South America as early as mid-July. These orioles are in Maryland from late April until early September. The Orchard Oriole's nest, a neatly woven purse of grasses, is green when first built and yellows with age; it is generally built farther in from branch tips and lower than the hanging nest of the Baltimore Oriole. Average height of Maryland nests for Orchard Oriole was 5.5 m (18 ft; Wilmot 1996e;

MNRF), and 9.1 m (30 ft; Cullom 1996b; MNRF) for Baltimore Oriole. Orchard Orioles often place nests in trees with active Eastern Kingbird nests (J. Dennis 1948; W. Ellison, pers. obs.). Maryland egg dates range from 13 May to 14 July, with nestlings to 14 August (Wilmot 1996e; MNRF); the latter date is exceptional as most nestlings have fledged by late July.

Traditionally, the Orchard Oriole has been rare to uncommon in western Maryland, and common to fairly common in central and eastern Maryland (Kirkwood 1895; Eifrig 1904; Stewart and Robbins 1958). In 1983–1987 the Orchard Oriole apparently had increased in the Ridge and Valley region but remained scarce in the Allegheny Mountains, with records in only 8 blocks and no confirmed nesting (Wilmot 1996e). The basic distributional limits for this oriole seen in 2002–2006 are essentially the same, but it occurred in many more blocks, a net increase of 123 (13%). By far the most impressive increase was in Garrett County where Orchard Orioles were found in 10 more blocks and were confirmed nesting in 2. The map shows some thinness in occurrence to the west and north, but Orchard Orioles were almost solidly distributed from central Frederick, and from southern Carroll, Baltimore, and Cecil counties southward.

Orchard Oriole numbers on BBS routes in their eastern North American breeding range have been stable over the past four decades. This oriole has increased by 2.6% per year in Maryland over the same time frame. The Orchard Oriole's population and range fluctuated notably in northeastern North America during the twentieth century. In the late nineteenth century and first decade of the twentieth century the Orchard Oriole was more widespread and numerous than in the mid-twentieth century; it has been increasing and expanding its range since the 1970s (Ellison 1985b; Walsh et al. 1999). The initial decline may have been in response to land use changes such as orchard clearing for crops in the

early twentieth century, with later increases following suburbanization in the late twentieth century, changes in pesticide selection and use, and perhaps long-term climate warming. At present the Orchard Oriole is doing well in Maryland and current trends seem to favor it, at least in the short term.

WALTER G. ELLISON

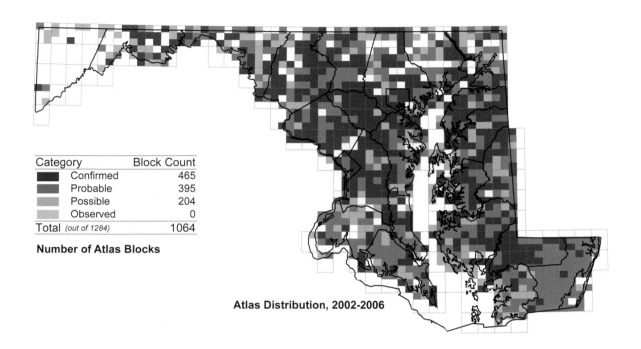

Category	Block Count
Confirmed	465
Probable	395
Possible	204
Observed	0
Total (out of 1284)	1064

Number of Atlas Blocks

Atlas Distribution, 2002-2006

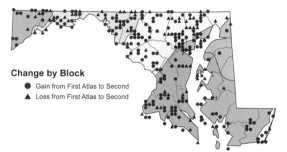

Change by Block
● Gain from First Atlas to Second
▲ Loss from First Atlas to Second

Change in Atlas Distribution, 1983-1987 to 2002-2006

Percent of Stops
- 50 - 100%
- 10 - 50%
- 0.1 - 10%
- < 0.1%

Relative Abundance from Miniroutes, 2003-2007

Atlas Region	1983-1987	2002-2006	Change No.	%
Allegheny Mountain	8	17	+9	+113%
Ridge and Valley	101	116	+15	+15%
Piedmont	236	276	+40	+17%
Upper Chesapeake	107	119	+12	+11%
Eastern Shore	282	303	+21	+7%
Western Shore	202	228	+26	+13%
Totals	936	1059	+123	+13%

Change in Total Blocks between Atlases by Region

Breeding Bird Survey Results in Maryland

David Brenneman

Baltimore Oriole
Icterus galbula

Mark Catesby was the first person to use the name Baltimore in connection with this colorful bird. In 1731 he called it the "Baltimore-Bird" in volume 1 of *Natural History of Carolina, Florida, and the Bahama Islands:* "This gold-colour'd bird I have only seen in Virginia and Maryland; there being none of them in Carolina. It is said to have its name from the Lord Baltimore's coat of arms . . . ; his Lordship being a proprietor of those countries." In the twenty-first century the Baltimore Oriole is still an important symbol to Marylanders: it was made the official state bird in 1947 and provides the name and colors for Baltimore's Major League Baseball team. This oriole was found in more than 60% of Maryland and DC's atlas blocks, but it is uncommon on the Coastal Plain.

Baltimore Orioles live in open woodlands with tall trees, especially floodplain forest along rivers, but also all manner of forest edges, and in small woodlots with tall shade trees including farmland, parks, and residential areas. They are migratory, wintering from Mesoamerica to northwestern South America. They are usually found in Maryland from late April until early October; a few may attempt to overwinter, generally at bird feeders offering suet or fruit. The nest is a remarkable and distinctive structure, a deep, pen-

dant pouch woven of bark strips and other tough plant fibers slung over the springy tips of the outer branches of tall trees, most often at heights from 6.1 to 12.2 m (20–40 ft). Because the nest usually survives well into the next season, this is one of the most likely species to be given used nest (UN) breeding confirmations. Eggs have been found in Maryland and DC from 18 May to 16 June (Stewart and Robbins 1958; Cullom 1996b; MNRF).

Kirkwood (1895) termed the Baltimore Oriole locally common in Maryland; he defined no regions of scarcity or absence. According to Eifrig (1904) they were uniformly common in western Maryland. Stewart and Robbins (1958) noted that these orioles were fairly common eastward to the fall line but were uncommon in the Upper Chesapeake and Eastern Shore, and rare, even absent, from the Western Shore. In the 1983–1987 atlas project they had clearly increased on the Western Shore, especially Prince George's and Anne Arundel counties, and they were also widespread in the Upper Chesapeake and on the Eastern Shore to just south of the Choptank River and the Salisbury area (Cullom 1996b). Over the past two decades the range has constricted on the Western Shore (15 fewer blocks), returning to prior low occupancy, continuing mostly along the valleys of the Potomac and Patuxent rivers. They occurred in 11 fewer blocks on the Eastern Shore, with losses concentrated in Talbot County and around Salisbury. These losses were balanced by gains in the Upper Chesapeake (14 more blocks) and scattered gains to the north and west.

Baltimore Orioles have been declining on North American BBS routes at a modest 0.6% per year over the past 40 years. This decline has been steeper since 1980, at 1.3% per year. They have declined at a greater rate in Maryland (1.6% per year), also showing a steeper decline since 1980. Trend patterns across eastern North America show strong declines along the eastern seaboard but a broad area of increase and stability through much of the Midwest. Reasons for declines may include loss of favored shade trees in many areas, especially elms to Dutch elm disease. The long-term population decline is not obvious on the 2002–2006 atlas map, although losses in southern Maryland may be related to it. Global warming could potentially cause a retreat from the southern edge of the species' range but this is speculative at present. Baltimore Orioles seem secure in their traditional Maryland nesting haunts, but losses from many places on the Coastal Plain over the past 20 years are cause for concern.

WALTER G. ELLISON

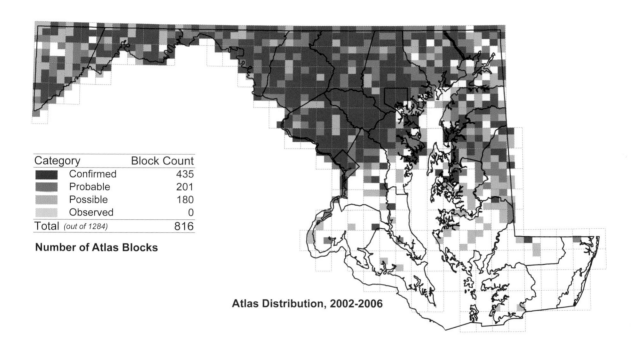

Category	Block Count
Confirmed	435
Probable	201
Possible	180
Observed	0
Total (out of 1284)	816

Number of Atlas Blocks

Atlas Distribution, 2002-2006

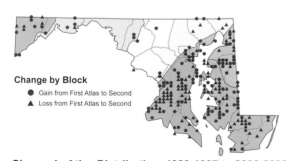

Change by Block
- ● Gain from First Atlas to Second
- ▲ Loss from First Atlas to Second

Change in Atlas Distribution, 1983-1987 to 2002-2006

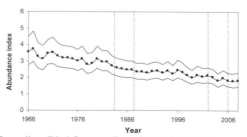

Percent of Stops
- 50 - 100%
- 10 - 50%
- 0.1 - 10%
- < 0.1%

Relative Abundance from Miniroutes, 2003-2007

Atlas Region	1983-1987	2002-2006	Change No.	Change %
Allegheny Mountain	84	87	+3	+4%
Ridge and Valley	143	147	+4	+3%
Piedmont	305	308	+3	+1%
Upper Chesapeake	87	101	+14	+16%
Eastern Shore	93	82	-11	-12%
Western Shore	103	88	-15	-15%
Totals	815	813	-2	0%

Change in Total Blocks between Atlases by Region

Breeding Bird Survey Results in Maryland

Purple Finch
Carpodacus purpureus

The Purple Finch is a chunky songbird with a large head, heavy bill, and short tail. This finch is more arboreal than the similar House Finch. When it does go to the ground it is most often to obtain grit or fallen seeds from road shoulders or to glean seeds beneath feeders. Purple Finches generally feed high in trees on their seeds and flowers; they are also fond of sunflower seeds and are inveterate visitors at bird feeders. Adult male Purple Finches are far less brown than House Finches and more pinkish red; young males (until the second autumn after hatching) and females have whiter underparts, with crisper brown streaking. They also have a more sharply defined head pattern with a darker ear-patch and broad whitish eyebrow compared to female House Finches. The song of the Purple Finch is a chortling, rich, varied warble given from a tall treetop in spring or in butterfly-like display flight on breeding territory.

Purple Finches may be seen throughout the year in one part of Maryland or another, but not in every year as their numbers fluctuate with the abundance of tree seeds in the North and in wintering areas. During the breeding season they are confined to the area west of the Allegheny Front. These finches nest in thick foliage on the side branches of generally short-needled conifers such as spruce, fir, and hemlock, so nesting habitat always includes some coniferous trees. Away from human habitation Purple Finches prefer forest edges with small to medium-size evergreens such as the margins of shrub swamps, bogs, and wetlands created by beavers. Over the past century this species has increasingly adjusted to nesting in man-made equivalent habitats, including rural and suburban yards with ornamental evergreens, Christmas tree farms, and conifer plantations.

Although early records are sketchy, it appears that the Purple Finch has been increasing its Maryland foothold in the Allegheny Mountains, perhaps for much of the past century. Summer records of Purple Finch in Maryland extend back to 1903 (Eifrig 1904), but nesting was not proven until 1949 (Robbins 1949b). Stewart and Robbins (1958) termed the Purple Finch uncommon, occasionally fairly common in Garrett County. As such, Blom (1996n) concluded that this finch's status during the 1983–1987 Maryland-DC atlas project was comparable to 1958. The results of the 2002–2006 atlas suggest a considerable expansion of the population and range of blocks occupied by the Purple Finch during the past two decades, with reports from 31% more blocks in the Allegheny Mountains. Purple Finch was also confirmed nesting in the mountains of Allegany County for the first time on Dans Mountain. It is possible that some of this increase in reports is attributable to better coverage by atlas workers than in the 1980s, but the increase in records was so widespread and large that a good part of it should be attributed to an increase in finch numbers. Considering that the introduction of the House Finch to eastern North America has apparently caused declines in Purple Finch populations (Wooton 1987), why has the Purple Finch done so well in western Maryland? Perhaps the climate of the Alleghenies is a bit rigorous for House Finches, allowing Purple Finches to better utilize dooryard ornamentals and evergreen plantings in addition to wilder sites. It is also possible that the advent of mycoplasmosis among House Finches contributed to low numbers in the Alleghenies, but since Purple Finches are also susceptible to the disease this seems less likely (Luttrell et al. 1996).

According to continental trends from BBS routes, the Purple Finch has declined by more than 60% since 1966. Although there are now more Purple Finches being recorded on the six Maryland BBS routes on which it occurs than in the past, no trend can be detected from so few routes. The Purple Finch has a wider range in Maryland than in the past and may be more numerous, but it is still not sufficiently common to offer it a certain future as one of Maryland's nesting birds. Given a strongly warming global climate, the species' status appears particularly uncertain.

WALTER G. ELLISON

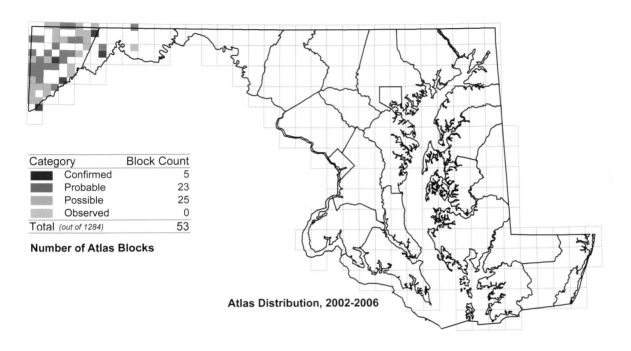

Category	Block Count
Confirmed	5
Probable	23
Possible	25
Observed	0
Total (out of 1284)	53

Number of Atlas Blocks

Atlas Distribution, 2002-2006

Change by Block
- ● Gain from First Atlas to Second
- ▲ Loss from First Atlas to Second

Change in Atlas Distribution, 1983-1987 to 2002-2006

Percent of Stops
- 50 - 100%
- 10 - 50%
- 0.1 - 10%
- < 0.1%

Relative Abundance from Miniroutes, 2003-2007

Atlas Region	1983-1987	2002-2006	Change No.	%
Allegheny Mountain	39	51	+12	+31%
Ridge and Valley	2	2	0	0%
Piedmont	0	0	0	-
Upper Chesapeake	0	0	0	-
Eastern Shore	0	0	0	-
Western Shore	0	0	0	-
Totals	41	53	+12	+29%

Change in Total Blocks between Atlases by Region

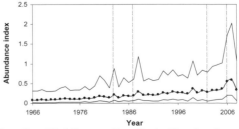

Breeding Bird Survey Results in Maryland

House Finch
Carpodacus mexicanus

The House Finch could be the generic songbird. The male is characterized by patches of strawberry red on its breast, head (notably above the eye), and rump, and its complex cheery song delivered from the highest available perch (or even from the air). Females, with their obscurely streaked uniformly pale brown plumage, are even more generic, rivaling the females of Brown-headed Cowbird, Indigo Bunting, and House Sparrow for blandness. The House Finch is well named; it is seldom seen very far from human habitation in much of its range. These finches are also inveterate feeder haunters and quickly become familiar to bird enthusiasts, although their similarity to the closely related Purple Finch can cause confusion when both are present until practice reveals the consistent differences between the two species. House Finches raise multiple broods over a long breeding season. Egg dates for Maryland extend from 26 March (Swift 1996) to 13 July 2004 (west of St. Mary's City, St. Mary's County; L. Lister, this project) and fledglings have been found as late as 7 September 2004 (Winfield, Carroll County; R. Ringler, this project). Female House Finches build nests in a variety of locations favoring ornamental conifers, vines on walls and trellises, hanging plants, and building ledges—even inside buildings if the windows are left open (G. Hill 1993). Such sites are easy to find and fledglings come to feeders soon after leaving their nests, contributing to a relatively high 50% rate of nesting confirmation.

More than 65 years ago House Finches were essentially unknown to eastern birders unless they had travelled widely in the American West, where the House Finch is native. Eastern U.S. pet dealers conducted an illegal cage bird trade in "Hollywood finches" or "California linnets" from the early twentieth century through the 1930s. In response to a crackdown by law enforcement officers, some dealers released their birds. One such release near New York City led to the establishment of a small population in western Long Island plant nurseries from 1941 to 1950 (Elliot and Arbib 1953).

House Finches were first identified in Maryland in April 1958 (Marshall 1958), and the first proven nesting came from Towson, Baltimore County, in 1963 (Garland 1963). The first DC record was established during 1962 (Pyle 1963). Many places in the mid Atlantic states first recorded the House Finch as a winter visitor (usually at feeders) with breeding following some years later (Hess et al. 2000; Swift 1996). This pattern of range expansion makes sense given that eastern House Finches have been well documented as partially migratory, with many northern individuals wintering in the southeastern United States (Belthoff and Gauthreaux 1991).

During the 1983–1987 Maryland-DC atlas project, the House Finch had become well established and widespread, with nesting in all of Maryland's physiographic regions and in all but one county, Somerset (Swift 1996). There were records of House Finch in 68% of 1983–1987 atlas blocks and nesting was confirmed in 44% of these (Swift 1996). Over the next two decades House Finches continued to claim new territory, adding 242 blocks in 2002–2006. Much of this increase occurred in Charles and St. Mary's counties and on the southern Eastern Shore (including Somerset County). Gaps are still evident in the atlas map in south central Dorchester County and in western Worcester and southeastern Wicomico counties. Small gaps were also evident in less populous, heavily forested blocks in western Maryland.

House Finch increases continued and persisted in spite of the advent of the often fatal, emergent respiratory disease mycoplasmosis, which was first noticed in 1994 (Luttrell et al. 1996). This disease led to widespread, significant declines in eastern North America (Hochachka and Dhondt 2000). It is easily recognized by feeder watchers because the advanced infection manifests as swelling around the eyes with attendant matted feathers. Most of the population decline has resulted in lower population densities rather than in loss of nesting range.

Breeding Bird Survey data continue to show a strong long-term increasing trend in Maryland (7% per year from 1966 to 2006). The rate of increase has decelerated to just over 2.5% per year over the past quarter century. The House Finch is likely here to stay in our yards, well-planted office parks, and school campuses.

WALTER G. ELLISON

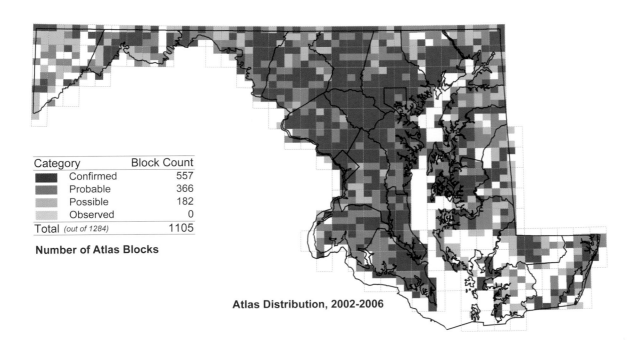

Category	Block Count
Confirmed	557
Probable	366
Possible	182
Observed	0
Total *(out of 1284)*	1105

Number of Atlas Blocks

Atlas Distribution, 2002-2006

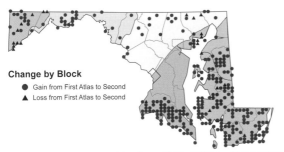

Change by Block
- ● Gain from First Atlas to Second
- ▲ Loss from First Atlas to Second

Change in Atlas Distribution, 1983-1987 to 2002-2006

Percent of Stops
- 50 - 100%
- 10 - 50%
- 0.1 - 10%
- < 0.1%

Relative Abundance from Miniroutes, 2003-2007

			Change	
Atlas Region	1983-1987	2002-2006	No.	%
Allegheny Mountain	61	65	+4	+7%
Ridge and Valley	121	142	+21	+17%
Piedmont	294	310	+16	+5%
Upper Chesapeake	85	105	+20	+24%
Eastern Shore	115	237	+122	+106%
Western Shore	183	242	+59	+32%
Totals	859	1101	+242	+28%

Change in Total Blocks between Atlases by Region

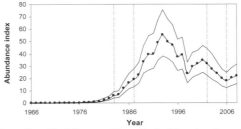

Breeding Bird Survey Results in Maryland

Pine Siskin
Spinus pinus

The Pine Siskin is most numerous in Maryland and DC in the winter. Siskins appear in Maryland in most winters, but their numbers vary enormously from one year to the next. Some winters see fewer than 10 reports; other winters find this state hosting thousands of these feisty little brown finches. Pine Siskins are primarily birds of the boreal forest, nesting in coniferous and mixed coniferous-birch forest. Until very recently Pine Siskins were uncertain breeders, lacking strong evidence of nesting in Maryland. The first nest in Maryland was found at Swallow Falls State Park in Garrett County on 23 April 1988, after the first statewide atlas project had ended (Ringler 1988a). Other more circumstantial inferences that strongly suggested nesting were drawn from observations in 1937, 1987, and 1988 (N. Stewart 1996b); the oldest record was also from Swallow Falls State Park. The reports from the late 1980s were of females in advanced breeding condition that were caught for banding in Laurel, at the fall line between the Piedmont and Coastal Plain. Between 1988 and the 2002–2006 atlas project there were several reports of siskins from May to late July in Garrett County but no reports of nesting (*Maryland Birdlife* 1989–2001). At the beginning of this atlas, adults were seen feeding young in a nest in a small white pine in a motel courtyard in McHenry, Garrett County, on 9 May 2002 (Ringler 2003a; W. Ellison, pers. obs.). Fledglings were subsequently seen near Swallow Falls State Park on 23 June 2002, in Mountain Lake Park on 28 June 2004 and 14 August 2005 (F. Pope), and on 23 June 2006 at Deep Creek Lake State Park (C. Skipper), all in Garrett County.

Pine Siskins eat a variety of weed and tree seeds, but their periodic irruptions imply they rely on abundant seed crops of spruce, fir, and birch (Dawson 1997; W. Ellison, pers. obs.). When these seed crops fail, siskins may pour southward into the mid Atlantic states and the Deep South. These irruptions may lead to nesting south of the boreal forest. The 1988 Garrett County nest and the breeding-condition females netted at Laurel were found after such an irruption (Farrand 1988). Siskins were reported from just 2 blocks during the first Maryland-DC atlas, including a possible breeding report from Garrett County and the probable nesting record (female with an incubation patch) from Laurel, Prince George's County, in May 1987 (N. Stewart 1996b).

In the 2002–2006 Maryland-DC atlas Pine Siskins were reported as possible to confirmed from 12 blocks. These included the 3 blocks with confirmations referenced above and another 3 with probable nesting. Most of these reports were from Garrett County, but there were two reports of possible breeders from Allegany County. Records of Pine Siskins

George M. Jett

were not concentrated into one, or a few, irruption years as was the case for prior nesting reports. There were reports of siskins for all survey years, although records were fewer in 2003 and 2006 (but the latter included a nesting record). The reasons for the great increase by Pine Siskin on Maryland's Allegheny Plateau over the last two decades are obscure. It is plausible that a small portion of the northeastern North American population of Pine Siskins may now rely on bird feeders as a breeding season food supply as they resort to nest sites in conifer stands nearby or in ornamental conifers. In adjacent Pennsylvania, the Pine Siskin has nested widely, if sporadically, in small numbers on the Allegheny Plateau, albeit mostly far north of Garrett County, but with one record as far south as Westmoreland County (D. Gross 1992b).

Given the Pine Siskin's nomadic habits and its predilection for nyjer thistle seed feeders, it is perhaps not surprising that it has become more numerous as a nester in the Maryland high country. An examination of continent-wide BBS results reveals declines over the past four decades. In light of these declines, the Pine Siskin's success in Maryland may be short-lived. This seems especially likely given predictions for continued worldwide climatic warming at the expense of this siskin's favored northern conifers.

WALTER G. ELLISON

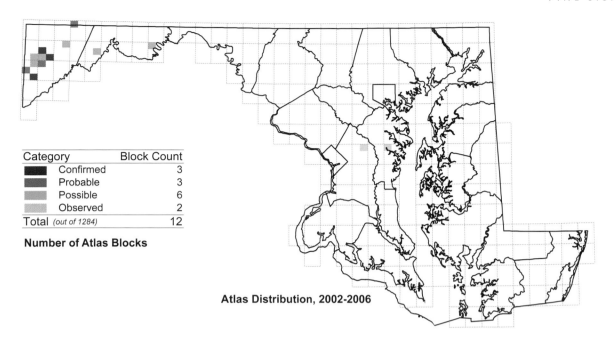

Atlas Distribution, 2002-2006

Category	Block Count
■ Confirmed	3
■ Probable	3
■ Possible	6
■ Observed	2
Total *(out of 1284)*	12

Number of Atlas Blocks

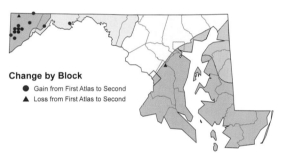

Change by Block
- ● Gain from First Atlas to Second
- ▲ Loss from First Atlas to Second

Change in Atlas Distribution, 1983-1987 to 2002-2006

Atlas Region	1983-1987	2002-2006	Change No.	Change %
Allegheny Mountain	1	11	+10	+1000%
Ridge and Valley	0	1	+1	-
Piedmont	0	0	0	-
Upper Chesapeake	0	0	0	-
Eastern Shore	0	0	0	-
Western Shore	1	0	-1	-100%
Totals	2	12	+10	+500%

Change in Total Blocks between Atlases by Region

Charles Lentz

American Goldfinch
Spinus tristis

A "charm of goldfinches" is the old English name for a flock of goldfinches. Although originally used to describe the European Goldfinch (*Carduelis carduelis*), the term applies equally well to the charming American Goldfinch, a much less aggressive bird than many of its relatives. In the breeding season the male American Goldfinch is smartly attired in bright yellow set off by a black forecrown, black and white wings, a white rump, and orange bill. Females are less striking but are nonetheless handsomely arrayed in moss green, dark brown, and pale yellow. Both sexes resemble breeding females after their autumn molt as they gather into nomadic winter flocks. This goldfinch is a generalist feeder on a wide array of tree and weed seeds; as such it is less prone to irruptive behavior than most other small finches, such as the closely related Pine Siskin. Much to the delight of homeowners, goldfinches readily frequent bird feeders during all four seasons, brightening backyards across the continent.

American Goldfinches occur throughout the year in all parts of Maryland. This species was the 11th most frequently reported bird in the 2002–2006 Maryland-DC atlas. It occurred in 50 more blocks in this atlas than in the first. This goldfinch is not particularly hard to confirm as a nester because its nests are often built at less than 6 m (20 ft) in small trees and shrubs at wood margins, in yards, and in brushy fields. In addition, pairs readily bring their fledglings to summer feeding stations. The confirmation rate for this bird was a relatively low 30%, which is largely attributable to a late

nesting schedule. Most goldfinches do not nest until July; nesting extends through August, and some nests may remain active until the third week of September. This schedule appears to be dictated by this goldfinch's dependence on the maturation of weed seeds to feed nestlings and to use as nest lining, especially those of thistles (Middleton 1993).

A feature of this atlas was a slight shift toward June nest initiation by Maryland and DC goldfinches: 8.7% of atlas confirmations for goldfinches were for nest building as early as 10 May 2003. This shift suggests that goldfinches may be responding to changes in flowering and seed maturation dates for plants during the warmer and drier springs of the past two decades, or to the abundance of earlier-blooming introduced species such as nodding thistle and Canada thistle. Nest dates for Maryland and DC range from 7 July 2002 northwest of Ellicott City, Howard County (R. Todd), to 21 September (MNRF) for eggs and 20 July 2003 near Bryantown, Queen Anne's County (G. Radcliffe), to 6 October (MNRF) for nestlings. Despite the rather low confirmation rate, the combined percentage for confirmed and probable nesting was 86%, compared to the combined 71% from the 1983–1987 atlas (Ricciardi 1996f).

During the first 20 years of the BBS, American Goldfinches declined across North America and in Maryland. Ricciardi (1996f) expressed concern over this decline and suggested that habitat management might be necessary to maintain this goldfinch's numbers. Over the next 20 years of the BBS, this decline reversed; American Goldfinch populations are increasing in Maryland and are stable continentwide. Most of this reversal in Maryland is because of strong increases on the Coastal Plain. Traditionally the American Goldfinch has been least numerous on the Coastal Plain, as it nears the southeastern edge of its nesting range (Price et al. 1995; Ricciardi 1996f). Forty-four (88%) of the 50 blocks added between the two statewide atlas projects were on the lower Eastern Shore and in southern Maryland. Why this goldfinch has increased so strongly on the Coastal Plain and held its own in the rest of Maryland and DC is not clear. Widespread bird feeding benefits this goldfinch; over the past 20 years many more bird enthusiasts have added specialty nyjer thistle seed feeders to their feeding stations in order to attract small finches into their yards. Several mild winters over the past two decades may also have increased survivorship for overwintering goldfinches. For now this popular bird appears to be doing well in the mid Atlantic states.

WALTER G. ELLISON

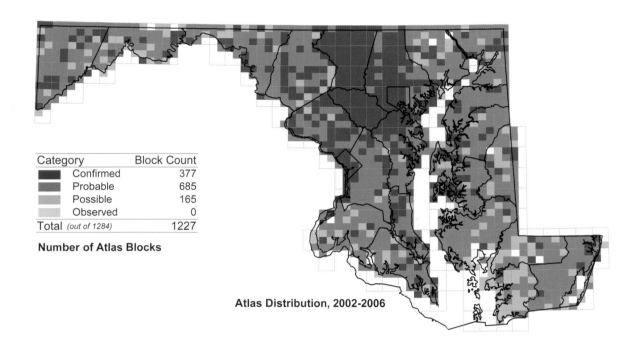

Category	Block Count
Confirmed	377
Probable	685
Possible	165
Observed	0
Total (out of 1284)	1227

Number of Atlas Blocks

Atlas Distribution, 2002-2006

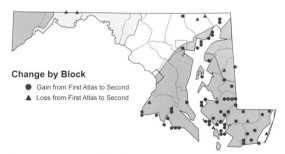

Change by Block
- ● Gain from First Atlas to Second
- ▲ Loss from First Atlas to Second

Change in Atlas Distribution, 1983-1987 to 2002-2006

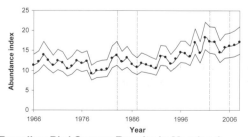

Percent of Stops
- 50 - 100%
- 10 - 50%
- 0.1 - 10%
- < 0.1%

Relative Abundance from Miniroutes, 2003-2007

			Change	
Atlas Region	1983-1987	2002-2006	No.	%
Allegheny Mountain	93	93	0	0%
Ridge and Valley	148	146	-2	-1%
Piedmont	313	314	+1	0%
Upper Chesapeake	112	119	+7	+6%
Eastern Shore	269	303	+34	+13%
Western Shore	237	247	+10	+4%
Totals	1172	1222	+50	+4%

Change in Total Blocks between Atlases by Region

Breeding Bird Survey Results in Maryland

Monroe Harden

House Sparrow
Passer domesticus

The House Sparrow is well named; it is seldom seen far from human habitation, especially here in its adopted American haunts. Hopping around on streets and sidewalks, buzzing rapidly across barnyards and fields, or lustily chirping in roosts in suburban hedges, this tough thick-necked urchin has inhabited North America since the mid-nineteenth century. Although this bird has been in decline ever since the early twentieth century, the "English" or House Sparrow remains common and is one of Maryland's most widespread nesting birds, with the 25th highest number of atlas blocks for a single species. They nest up to four times in a breeding season (McGillivray 1983). Maryland and DC egg dates range from 29 March to 17 August with nestlings to 6 September (MNRF).

The earliest attempts to add this familiar Old World bird of farm and city to the New World's bird fauna date back to 1850, when it was believed the bird would keep Old World insect pests (particularly cankerworm) in check in farm fields. This is one of the earliest, well-documented examples of ecological tinkering that had serious long-term negative consequences. The sparrow's spread across North America originated in New York but was abetted by many other local introductions, from both North American and European stock, over the next 30 years. House Sparrows were released

in Philadelphia in 1868 (Reid 1992b), at Washington, DC in 1871 (Cooke 1929), and in Baltimore in 1874 (Kirkwood 1895). Maryland's first report, however, dates from 1865 when they were seen in Hancock in western Washington County (Kirkwood 1895). By 1877 the House Sparrow had spread throughout the Old Line State (Dupree 1996d). As this sparrow became one of the most abundant birds in Maryland, it became evident that it was more of an agricultural nuisance than boon and a formidable competitor for the nest holes once used by more desirable native cavity-nesting birds such as martins and bluebirds.

By the late 1920s the House Sparrow had begun its long decline from its zenith as an ubiquitous and abundant bird to a merely common and localized one. The decline coincided with the rise of automotive transportation and the replacement of horses, first in cities and later on the farm. House Sparrows are most abundant around livestock where abundant food is available in fresh dung, waste grain in fields, and silage stored for animals. Even in cities and towns, sparrows picked seeds out of road apples in the streets and shared feedbags with horses. The land occupied by farms has also declined since the mid-twentieth century. As farming has become cleaner and more efficient and municipal sanitation improves, House Sparrows have fewer sources of food and nesting cavities in their favored haunts.

The 2002–2006 atlas map shows a slight decline in the number of blocks hosting House Sparrows. The small gaps in the sparrow's range continue to be in rural areas with extensive undeveloped land, including heavily wooded blocks in Garrett and Allegany counties, down range on Aberdeen Proving Ground, the pine woods and marshes of southeastern Dorchester County, the Nassawango and Pocomoke drainages, and the dunes and thickets of Assateague Island. Other less explicable gaps appear to relate to coverage lapses, although House Sparrows may be actively discouraged around some lower Eastern Shore chicken farms.

Data from the BBS show that the House Sparrow has been declining steeply in Maryland, and in North America as a whole, for the past five decades. The estimated annual

decline of 3.5% is exceeded by only nine other bird species in Maryland. These declines are evident for all USGS physiographic strata in Maryland, although the rate of decline appears greatest on the Upper Coastal Plain. But sparrows are unlikely to completely disappear as long as suitable nest cavities remain and grain is available as food. Indeed House Sparrows continue to do fairly well where poorly managed bird houses and bird feeding coexist.

WALTER G. ELLISON

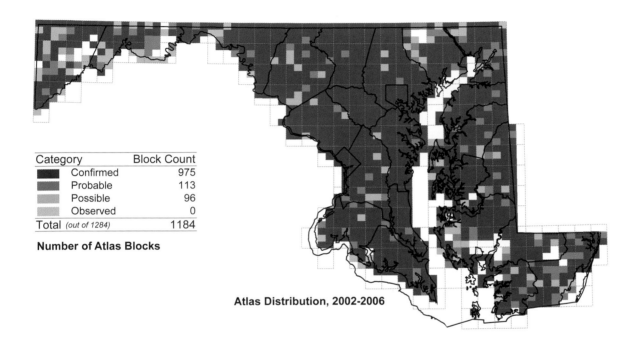

Category	Block Count
Confirmed	975
Probable	113
Possible	96
Observed	0
Total (out of 1284)	1184

Number of Atlas Blocks

Atlas Distribution, 2002-2006

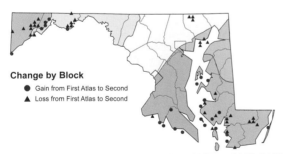

Change by Block
● Gain from First Atlas to Second
▲ Loss from First Atlas to Second

Change in Atlas Distribution, 1983-1987 to 2002-2006

Percent of Stops
50 - 100%
10 - 50%
0.1 - 10%
< 0.1%

Relative Abundance from Miniroutes, 2003-2007

Atlas Region	1983-1987	2002-2006	Change No.	%
Allegheny Mountain	87	83	-4	-5%
Ridge and Valley	146	141	-5	-3%
Piedmont	316	313	-3	-1%
Upper Chesapeake	118	114	-4	-3%
Eastern Shore	289	288	-1	0%
Western Shore	239	243	+4	+2%
Totals	1195	1182	-13	-1%

Change in Total Blocks between Atlases by Region

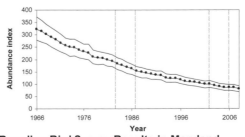

Breeding Bird Survey Results in Maryland

Additional Accounts of Potential Nesting Birds

The short accounts in this chapter present information on five birds that received breeding codes short of confirmation as proven breeding birds in Maryland and DC and one (Western Kingbird) that attempted a hybrid pairing with a closely related species. The Mississippi Kite has subsequently been proven to nest in Maryland.

Thirteen birds formerly nested in Maryland or DC on at least one occasion. Three species were recorded breeding from 1983 to 1987 and received full species accounts in Robbins and Blom: Northern Shoveler (Meritt 1996g), Wilson's Plover (Hoffman 1996a), and Bewick's Wren (Blom 1996h). Ten species were not found nesting during either the first atlas project or the current one. These were Green-winged Teal, the extinct eastern subspecies of the Greater Prairie-Chicken also known as the Heath Hen, Purple Gallinule, Roseate Tern, Sandwich Tern, the extinct Passenger Pigeon, Red-cockaded Woodpecker, Swainson's Thrush, Bachman's Sparrow, and Lark Sparrow. There were summer reports of Green-winged Teal, Purple Gallinule, Roseate and Sandwich terns, and Swainson's Thrush in 2002–2006, but none of the birds were deemed to be showing breeding behavior. Brief accounts of these 10 species are presented on pages 444 and 445 in Robbins and Blom (1996).

WHITE IBIS

Eudocimus albus
Long known to be a rare to uncommon breeding bird on Virginia's lower Eastern Shore in Northampton County, with a total of 74 counted on Cobb and Wreck islands in 2004 (B. Williams, Akers, et al. 2005), the White Ibis is primarily a rare postbreeding visitor in Maryland, largely on the lower Piedmont and Coastal Plain although reports exist westward to Allegany County (Ringler 2002c). Most reports for Maryland are of young birds, including the first record at PWRC in Laurel on 14 August 1961 (Robbins 1966). The very few spring reports of adults include 2 at Ocean City, 7 to 22 May 1966 (Robbins 1966), and an exceptionally early record at Blackwater NWR, 7 to 28 March 1988 (Ringler 1988a; Peterjohn and Davis 1996). On 27 May and 8 June 2005 at least one adult White Ibis was seen visiting a heron and ibis rookery on Holland Island, Dorchester County, by H. T. Armistead and J. Sparks (H. T. Armistead, pers. comm.; T. Day 2006).

Unfortunately keeping track of the bird was difficult and no further indication of nesting was obtained. This ibis should continue to be sought in heronries on the lower Eastern Shore.

MISSISSIPPI KITE

Ictinia mississippiensis
The Mississippi Kite has increased its breeding distribution to the north and west since the middle of the twentieth century, from enclaves in the lower Mississippi valley and along the southeastern coast in South Carolina, reoccupying what had formerly been a more extensive range during the early to mid-nineteenth century (Parker and Ogden 1979). The first sight report for Maryland was on 24 June 1978 at Owings Mills, Baltimore County; the first officially accepted record was for 11 May 1990 from Rockville, Montgomery County (Peterjohn and Davis 1996). These kites increased in Virginia during the 1980s and 1990s especially in Greenville County, along the North Carolina state line, where nesting evidence proved elusive. Nesting was first confirmed in Virginia in the DC suburbs at Woodbridge, Prince William County, in August 1996 (Quezon 1997). Nesting has subsequently spread into Fairfax County, Virginia (M. Adams and Hafner 2009). Late spring overshooting migrants have been annual in the mid Atlantic since the early 1980s. In Maryland nearly annual reports have been made from early May to late June since the late 1980s. Atlas field-workers submitted five reports of kites during the summer months although there were additional summer observations from 2002 to 2006. The atlas reports were one in Edgewater, Anne Arundel County (M. Braun, pers. comm.); two from Port Tobacco and Zekiah Swamp in Charles County (G. Jett, pers. comm.; D. and J. Coskren, pers. comm.); one in Harrisville, Dorchester County (L. Davidson, pers. comm.); and one far west on 16 June 2005 on Dans Mountain, Allegany County (S. Sires, pers. comm.). In early July 2008 an adult kite was seen feeding a recent fledgling in Salisbury, Wicomico County, establishing Maryland's first nesting record (M. Adams and Hafner 2009). Further nesting should be anticipated in Maryland. It is noteworthy that Mississippi Kites nested far to the north in Connecticut and New Hampshire (two pairs) in 2008 (Petersen 2009).

WILSON'S SNIPE

Gallinago delicata

If not for the spectacular flight display called winnowing—plunging dives accompanied by a far-carrying accelerating hooting sound produced by air rushing from the wingtips over stiffened outer tail feathers—and its repetitive *kek* calls, the Wilson's Snipe would often go undetected even where it is fairly common. This snipe reaches the southern edge of its nesting range in the Canaan Valley in Tucker County, West Virginia, adjacent to Garrett County (Hall 1983). In Pennsylvania most nesting snipes are found in the northern half of the state, but the Pennsylvania atlas documented displaying males in two blocks in Somerset County adjacent to Garrett County (Leberman 1992a). There were three atlas reports of Wilson's Snipe from 2002 to 2006. A persistently winnowing male was present at Patuxent River Park, near Aquasco, Prince George's County, from 1 to 7 June 2002 (J. Shenot in Ringler 2003b). A winnowing male descended to the ground and displayed to a female on Accident-Bittinger Road on Cherry Creek west of Bittinger, Garrett County, on 6 May 2005 (W. Ellison, pers. obs.), and two displaying males were seen in the Casselman River valley near Jennings, Garrett County, on 22 July 2005 (P. and E. Kreiss in T. Day 2006). The Coastal Plain bird may have been an optimistic late spring vagrant, but the Garrett County birds were close to the species' southernmost outpost in the Canaan Valley, indicating that this sandpiper may yet be confirmed nesting in the Free State.

EURASIAN COLLARED-DOVE

Streptopelia decaocto

The Eurasian Collared-Dove is a widespread Old World species noteworthy for its rapid twentieth-century westward expansion across Europe (Hengeveld 1993). It was introduced to Nassau in the Bahamas during the 1970s and spread westward to Florida during the 1980s (P. Smith 1987; Romagosa and McEneaney 1999). Since its arrival it has spread into a broad North American nesting range extending to the West Coast and northward through the Great Plains to the Canadian Prairie Provinces (Romagosa and McEneaney 1999). It has spread far less rapidly up the eastern seaboard, although nonbreeding records now exist as far north as upstate New York (McGowan 2008j) and southern New England (Perkins 2005; Ellison and Martin 2007). The first Maryland report was a bird at Port Deposit, Cecil County, from mid-April to mid-June 2002 (Iliff 2002). There were three atlas reports of this dove, including a small group resident in Selbyville, Sussex County, Delaware, just north of the state line in a Maryland atlas block in 2002 (M. Hoffman, pers. comm.). These birds were present in Selbyville from 1998 to at least 2002. Single calling and displaying doves were seen in Frederick County southwest of Frederick 20 to 24 June and near Thurmont 28

July 2006 (B. Gearhart, pers. comm.). In December 2006 and January 2007 two displaying and calling birds were present on Middle Hooper Island, Dorchester County (F. Jacobsen, pers. comm.; W. Ellison and N. Martin, pers. obs.). Subsequent observations, including persistent sightings at South Point, Worcester County, suggest that it is only a matter of time until evidence of breeding is found in Maryland.

WESTERN KINGBIRD

Tyrannus verticalis

Although the Western Kingbird is a well-known rare autumn visitor and even less frequent late spring and early summer vagrant along the Atlantic coastline of North America, it has not consistently nested east of the Mississippi River in the midwestern United States (Gamble and Bergin 1996). Maryland records of this attractive flycatcher date to a DC market specimen in 1874 (Coues and Prentiss 1883) with four-fifths of subsequent records from the autumn months. Spring and summer reports range from 15 May to 31 July, with an exceptional report for 17 March. A male Western Kingbird summered at Fort McHenry in Baltimore City for six years (2002–2007; J. Peters, pers. comm.). During its first two years in residence, the bird attempted to nest with local female Eastern Kingbirds. Two nests were subject to predation by Fish Crows in 2002, but a nest in 2003 fledged two young, both of which were apparently Eastern Kingbirds (T. Day and Iliff 2004; J. Peters, pers. comm.). Probably neither of the offspring were genetically related to the Western Kingbird that defended the nest and fed them, implying they were the result of extra pair matings by the female Eastern. This bird of the open spaces of the West is unlikely to attempt to nest in Maryland or DC again, but as long as a few stray spring migrants continue to appear here it is not out of the question.

RED CROSSBILL

Loxia curvirostra

Although a nest has yet to be found in Maryland or DC for this bird, the nomadic pine- and spruce-cone-feeding Red Crossbill has almost certainly nested in Maryland in the past and may yet sporadically breed in the region, most likely in the high mountains and valleys of Garrett County. Past records include a breeding condition female collected by George Marshall 23 May 1884 near Laurel (Ridgway 1884); a juvenile "barely able to fly" was seen a "short distance beyond [Washington] city limits" in 1885 (H. Smith 1885); summering birds were reported from the surprising Coastal Plain venue of Dorchester County in 1932 and 1933 (F. R. Smith in Stewart and Robbins 1958). Red Crossbills breed opportunistically in small numbers in Highland and Rockland counties in Virginia (Goetz 1981) and have nested as far south as the Great Smoky Mountains (Stupka 1963). During the spring of

2005, from March to early May, male and female Red Cross-bills visited the feeders of Irene Yoder just west of Gortner in Garrett County. Unfortunately, no juveniles were identified as such. Whenever these birds are seen at length in suitable habitat from early February to June in Maryland or DC, they should be watched carefully for nesting behavior, and flocks should be examined carefully for birds in juvenile plumage with uncrossed bills. Females often commence nesting in late winter and are fed at the nest by males for up to five days after the eggs have hatched (Adkisson 1996), thus nests are often very hard to find except at the building stage. There may be several cryptic species hidden within this highly variable taxon, which includes many forms that mate almost exclusively with others sharing call and song variants and details of bill shape and size (Groth 1993a, 1993b). There may be two of these cryptic species nesting in the Appalachians (Groth 1988).

WALTER G. ELLISON

APPENDIX A

Safe Dates for Maryland Breeding Birds

Canada Goose	10 May–31 July	Osprey	1 June–15 Aug.
Mute Swan	1 May–31 Aug.	Bald Eagle	15 Apr.–5 Aug.
Wood Duck	20 Apr.–15 Aug.	Northern Harrier	20 May–25 July
Gadwall	20 May–20 Aug.	Sharp-shinned Hawk	1 June–31 July
American Black Duck	20 May–31 July	Cooper's Hawk	20 May–31 July
Mallard	15 May–20 Aug.	Northern Goshawk	15 Apr.–31 Aug.
Blue-winged Teal	5 June–5 July	Red-shouldered Hawk	10 May–20 Aug.
Northern Shoveler	1 June–30 June	Broad-winged Hawk	15 June–10 Aug.
Green-winged Teal	1 June–30 June	Red-tailed Hawk	10 May–20 Aug.
Ring-necked Duck	20 May–15 Aug.	American Kestrel	15 May–15 July
Hooded Merganser	20 May–31 Aug.	Peregrine Falcon	20 May–15 Aug.
Ruddy Duck	10 June–25 Aug.	Black Rail	1 May–15 Aug.
Northern Bobwhite	15 Apr.–30 Sept.	Clapper Rail	1 May–31 Aug.
Ring-necked Pheasant	15 Apr.–30 Sept.	King Rail	1 May–31 Aug.
Ruffed Grouse	1 Apr.–31 July	Virginia Rail	20 May–15 Aug.
Wild Turkey	15 Apr.–30 Sept.	Sora	1 June–31 July
Pied-billed Grebe	20 May–10 July	Purple Gallinule	20 June–31 July
Brown Pelican	1 June–31 July	Common Moorhen	20 May–31 Aug.
Double-crested Cormorant	10 June–30 June	American Coot	10 Jun–25 Aug.
American Bittern	20 May–15 Aug.	Wilson's Plover	15 May–31 July
Least Bittern	20 May–31 July	Piping Plover	15 May–20 July
Great Blue Heron	15 May–30 June	Killdeer	20 Apr.–25 June
Great Egret	20 May–20 June	American Oystercatcher	15 May–25 July
Snowy Egret	20 May–30 June	Black-necked Stilt	15 May–15 July
Little Blue Heron	20 May–20 June	Spotted Sandpiper	10 June–30 June
Tricolored Heron	20 May–30 June	Willet	10 June–1 July
Cattle Egret	20 May–20 June	Upland Sandpiper	20 May–25 June
Green Heron	1 May–15 July	Wilson's Snipe	1 June–15 July
Black-crowned Night-Heron	10 May–30 June	American Woodcock	15 Apr.–31 Aug.
Yellow-crowned Night-Heron	20 Apr.–30 June	Laughing Gull	5 June–10 July
Glossy Ibis	20 May–30 June	Herring Gull	5 June–10 July
Black Vulture	1 May–31 July	Great Black-backed Gull	5 June–10 July
Turkey Vulture	15 May–20 Aug.	Least Tern	25 May–5 July

Gull-billed Tern	15 May–5 July	Northern Rough-winged Swallow	25 May–20 June
Roseate Tern	25 May–5 July	Bank Swallow	1 June–20 June
Common Tern	5 June–30 June	Cliff Swallow	1 June–25 June
Forster's Tern	15 May–25 June	Barn Swallow	25 May–25 June
Royal Tern	25 May–5 July	Carolina Chickadee	1 Mar.–31 Aug.
Sandwich Tern	25 May–5 July	Black-capped Chickadee	1 May–20 Sept.
Black Skimmer	25 May–5 July	Tufted Titmouse	1 Mar.–31 Aug.
Rock Pigeon	ALL YEAR	Red-breasted Nuthatch	1 June–31 July
Mourning Dove	15 Apr.–20 July	White-breasted Nuthatch	10 May–15 Aug.
Yellow-billed Cuckoo	15 June–31 July	Brown-headed Nuthatch	20 Mar.–15 Aug.
Black-billed Cuckoo	20 June–20 July	Brown Creeper	15 May–31 Aug.
Barn Owl	15 Apr.–30 Sept.	Carolina Wren	1 Mar.–30 Sept.
Eastern Screech-Owl	1 Apr.–15 Aug.	Bewick's Wren	10 May–31 Aug.
Great Horned Owl	15 Dec.–31 Aug.	House Wren	20 May–15 Aug.
Barred Owl	15 Jan.–31 Aug.	Winter Wren	20 May–31 Aug.
Long-eared Owl	1 May–30 Sept.	Sedge Wren	10 June–20 Sept.
Short-eared Owl	1 May–30 Sept.	Marsh Wren	25 May–25 Aug.
Northern Saw-whet Owl	5 May–10 Sept.	Golden-crowned Kinglet	20 May–10 Sept.
Common Nighthawk	5 June–15 July	Blue-gray Gnatcatcher	15 May–31 July
Chuck-will's-widow	1 May–10 Aug.	Eastern Bluebird	1 May–31 Aug.
Eastern Whip-poor-will	10 May–15 July	Veery	10 June–10 Aug.
Chimney Swift	15 May–10 Aug.	Swainson's Thrush	10 June–10 Aug.
Ruby-throated Hummingbird	15 May–31 July	Hermit Thrush	25 May–15 Sept.
Belted Kingfisher	10 Apr.–20 July	Wood Thrush	25 May–20 Aug.
Red-headed Woodpecker	25 May–20 Aug.	American Robin	1 May–31 July
Red-bellied Woodpecker	15 Mar.–31 Aug.	Gray Catbird	25 May–31 Aug.
Yellow-bellied Sapsucker	1 June–31 Aug.	Northern Mockingbird	1 Apr.–10 Sept.
Downy Woodpecker	15 Mar.–31 Aug.	Brown Thrasher	15 May–31 Aug.
Hairy Woodpecker	15 Mar.–31 Aug.	European Starling	1 Apr.–31 July
Red-cockaded Woodpecker	15 Mar.–31 Aug.	Cedar Waxwing	15 June–31 July
Northern Flicker	10 May–25 Aug.	Blue-winged Warbler	25 May–20 July
Pileated Woodpecker	15 Mar.–31 Aug.	Golden-winged Warbler	25 May–20 July
Olive-sided Flycatcher	15 June–31 July	Nashville Warbler	25 May–15 Aug.
Eastern Wood-Pewee	1 June–15 Aug.	Northern Parula	1 June–15 Aug.
Acadian Flycatcher	25 May–5 Aug.	Yellow Warbler	1 June–10 July
Alder Flycatcher	10 June–20 July	Chestnut-sided Warbler	1 June–10 Aug.
Willow Flycatcher	10 June–20 July	Magnolia Warbler	10 June–5 Aug.
Least Flycatcher	5 June–20 July	Black-throated Blue Warbler	5 June–5 Aug.
Eastern Phoebe	1 May–31 Aug.	Yellow-rumped Warbler	1 June–20 Aug.
Great Crested Flycatcher	25 May–31 July	Black-throated Green Warbler	10 June–5 Aug.
Eastern Kingbird	25 May–5 July	Blackburnian Warbler	10 June–31 July
Loggerhead Shrike	10 May–20 July	Yellow-throated Warbler	1 May–15 July
White-eyed Vireo	25 May–15 Aug.	Pine Warbler	25 Apr.–10 Aug.
Yellow-throated Vireo	25 May–15 Aug.	Prairie Warbler	25 May–20 July
Blue-headed Vireo	1 June–20 Aug.	Cerulean Warbler	25 May–5 Aug.
Warbling Vireo	10 June–10 Aug.	Black-and-white Warbler	15 May–25 July
Red-eyed Vireo	1 June–31 July	American Redstart	10 June–20 July
Blue Jay	10 June–5 Sept.	Prothonotary Warbler	10 May–20 July
American Crow	20 Apr.–31 Aug.	Worm-eating Warbler	20 May–20 July
Fish Crow	10 May–31 Aug.	Swainson's Warbler	20 Apr.–31 Aug.
Common Raven	1 Apr.–31 July	Ovenbird	20 May–5 Aug.
Horned Lark	10 Apr.–5 Sep.	Northern Waterthrush	5 June–15 July
Purple Martin	1 June–25 June	Louisiana Waterthrush	1 May–10 July
Tree Swallow	25 May–25 June	Kentucky Warbler	25 May–15 July

Mourning Warbler	15 June–31 July		Summer Tanager	5 June–10 Aug.
Common Yellowthroat	25 May–10 Aug.		Scarlet Tanager	25 May–10 Aug.
Hooded Warbler	25 May–25 July		Northern Cardinal	15 Mar.–30 Sept.
Canada Warbler	10 June–15 July		Rose-breasted Grosbeak	15 June–10 Aug.
Yellow-breasted Chat	25 May–5 Aug.		Blue Grosbeak	25 May–10 Aug.
Eastern Towhee	20 May–31 Aug.		Indigo Bunting	25 May–15 Aug.
Bachman's Sparrow	1 June–31 July		Dickcissel	1 June–31 Aug.
Chipping Sparrow	1 May–31 Aug.		Bobolink	15 June–30 June
Clay-colored Sparrow	1 June–31 July		Red-winged Blackbird	1 May–10 July
Field Sparrow	1 May–31 Aug.		Eastern Meadowlark	25 Apr.–10 Sept.
Vesper Sparrow	15 May–31 Aug.		Common Grackle	15 Apr.–30 June
Lark Sparrow	1 June–31 July		Boat-tailed Grackle	15 Apr.–31 Aug.
Savannah Sparrow	5 June–31 Aug.		Brown-headed Cowbird	1 May–10 July
Grasshopper Sparrow	25 May–31 Aug.		Orchard Oriole	1 June–5 July
Henslow's Sparrow	15 May–31 Aug.		Baltimore Oriole	1 June–25 July
Saltmarsh Sparrow	1 June–10 Aug.		Purple Finch	1 June–10 Aug.
Seaside Sparrow	1 June–10 Aug.		House Finch	1 May–15 July
Song Sparrow	1 May–31 July		Pine Siskin	15 June–31 Aug.
Swamp Sparrow	1 June–31 July		American Goldfinch	15 June–31 Aug.
White-throated Sparrow	10 June–31 July		House Sparrow	1 Feb.–31 Aug.
Dark-eyed Junco	1 June–31 July			

APPENDIX B

Total Species per Block, 1st and 2nd Atlases

			Species per block					
Quad no.	Quad name	Atlas	NW(1)	NE(2)	CW(3)	CE(4)	SW(5)	SE(6)
001	Friendsville	1st	69	76	77	70	76	90
		2nd	83	88	77	92	89	90
002	Accident	1st	97	74	70	62	68	72
		2nd	91	80	81	75	84	78
003	Grantsville	1st	79	94	69	83	75	108
		2nd	90	93	72	79	103	98
004	Avilton	1st	75	86	86	87	102	76
		2nd	81	80	82	90	99	86
005	Frostburg	1st	100	84	110	76	75	83
		2nd	95	66	121	79	105	72
006	Cumberland	1st	59	68	74	49	47	109
		2nd	55	68	79	56	66	57
007	Evitts Creek	1st	71	73	80	82	80	65
		2nd	64	85	63	82	58	64
008	Flintstone	1st	59	77	76	73	82	75
		2nd	57	56	72	74	64	80
009	Artemas	1st	65	67	75	72	60	64
		2nd	67	63	63	72	67	81
010	Bellegrove	1st	62	97	83	92	81	97
		2nd	68	86	77	83	87	93
011	Hancock	1st	91	95	93	91	95	—
		2nd	73	77	85	83	83	—
012	Cherry Run	1st	97	98	105	92	—	95
		2nd	83	87	80	70	—	75
013	Clear Spring	1st	78	52	90	69	79	75
		2nd	69	79	84	76	80	62
014	Mason Dixon	1st	70	77	80	75	81	62
		2nd	74	80	62	62	73	64
015	Hagerstown	1st	66	60	65	73	61	73
		2nd	65	60	61	66	57	68
016	Smithsburg	1st	66	88	67	81	75	91
		2nd	68	79	71	87	80	84
017	Blue Ridge Summit	1st	70	72	73	71	80	67
		2nd	78	77	84	82	77	71
018	Emmitsburg	1st	78	74	78	74	65	66
		2nd	84	78	83	81	76	71

Quad no.	Quad name	Atlas	Species per block					
			NW(1)	NE(2)	CW(3)	CE(4)	SW(5)	SE(6)
019	Taneytown	1st	73	59	71	63	71	64
		2nd	74	66	80	69	74	72
020	Littlestown	1st	61	55	69	66	60	88
		2nd	71	73	78	74	74	89
021	Manchester	1st	88	69	65	63	83	66
		2nd	88	77	88	78	89	85
022	Lineboro	1st	71	72	67	86	57	83
		2nd	80	73	79	82	62	84
023	New Freedom	1st	68	64	63	68	72	76
		2nd	75	76	78	75	67	74
024	Norrisville	1st	75	72	61	77	71	87
		2nd	78	56	79	70	78	82
025	Fawn Grove	1st	62	68	71	69	72	86
		2nd	62	52	60	50	48	68
026	Delta	1st	69	60	72	89	80	72
		2nd	55	60	72	81	37	77
027	Conowingo Dam	1st	73	63	82	78	81	81
		2nd	64	75	59	72	62	83
028	Rising Sun	1st	79	72	68	76	63	68
		2nd	68	73	70	64	66	62
029	Bay View	1st	82	72	68	68	67	65
		2nd	72	72	61	68	67	54
030	Newark West	1st	77	92	75	84	66	69
		2nd	79	81	70	90	67	63
031	Sang Run	1st	62	58	70	82	85	82
		2nd	68	66	78	78	95	83
032	McHenry	1st	68	107	77	90	75	84
		2nd	58	93	77	63	75	83
033	Bittinger	1st	75	75	89	77	85	75
		2nd	79	90	101	100	82	75
034	Barton	1st	87	75	81	63	64	63
		2nd	66	108	77	78	80	62
035	Lonaconing	1st	43	57	62	80	66	74
		2nd	78	74	69	112	71	59
036	Cresaptown	1st	58	54	98	—	—	—
		2nd	74	69	84	—	—	—
037	Patterson Creek	1st	82	75	71	68	—	58
		2nd	87	73	70	64	—	61
038	Oldtown	1st	72	63	75	82	86	70
		2nd	72	69	68	71	62	65
039	Paw Paw	1st	67	87	69	52	75	—
		2nd	70	83	65	68	59	—
040	Big Pool	1st	—	67	—	—	—	—
		2nd	—	71	—	—	—	—
041	Hedgesville	1st	80	73	—	77	—	—
		2nd	78	77	—	72	—	—
042	Williamsport	1st	67	63	77	70	72	73
		2nd	79	64	92	70	91	80
043	Funkstown	1st	72	67	70	70	66	77
		2nd	69	73	72	74	75	80
044	Myersville	1st	70	89	80	84	80	100
		2nd	73	89	81	88	90	89
045	Catoctin Furnace	1st	72	92	69	75	72	74
		2nd	69	82	72	81	78	77
046	Woodsboro	1st	73	79	88	80	70	69
		2nd	81	79	81	81	81	84
047	Union Bridge	1st	63	62	59	61	61	64
		2nd	71	75	72	72	69	77
048	New Windsor	1st	56	72	61	67	65	70
		2nd	64	71	81	74	79	81
049	Westminster	1st	72	67	73	67	76	73
		2nd	74	69	73	76	70	76

Note: — block not covered

Quad no.	Quad name	Atlas	NW(1)	NE(2)	CW(3)	CE(4)	SW(5)	SE(6)
					Species per block			
050	Hampstead	1st	61	66	68	77	65	76
		2nd	64	76	66	81	66	72
051	Hereford	1st	68	72	94	81	71	85
		2nd	78	74	86	83	73	79
052	Phoenix	1st	82	80	80	70	88	75
		2nd	79	54	76	61	82	73
053	Jarrettsville	1st	69	85	79	72	76	90
		2nd	55	71	41	70	51	50
054	Bel Air	1st	66	91	77	72	62	60
		2nd	65	76	54	57	62	71
055	Aberdeen	1st	73	88	79	71	68	73
		2nd	77	90	78	78	67	71
056	Havre de Grace	1st	65	70	59	71	69	—
		2nd	76	78	76	88	82	—
057	North East	1st	64	72	67	71	70	63
		2nd	56	84	81	67	63	61
058	Elkton	1st	75	70	63	75	60	76
		2nd	67	67	62	55	68	74
059	Oakland	1st	71	87	87	101	73	96
		2nd	83	79	98	100	83	80
060	Deer Park	1st	78	83	93	72	79	81
		2nd	89	81	81	80	85	91
061	Kitzmiller	1st	82	77	68	74	73	61
		2nd	80	81	61	73	82	78
062	Westernport	1st	58	76	87	—	—	—
		2nd	69	61	81	—	—	—
063	Keyser	1st	77	64	73	—	—	—
		2nd	68	53	61	—	—	—
064	Shepherdstown	1st	68	76	—	69	—	—
		2nd	80	71	—	66	—	—
065	Keedysville	1st	78	85	69	87	83	81
		2nd	68	82	71	90	82	85
066	Middletown	1st	91	76	93	72	72	66
		2nd	90	82	89	77	81	87
067	Frederick	1st	76	61	69	61	69	63
		2nd	79	77	74	75	88	63
068	Walkersville	1st	63	70	65	77	69	68
		2nd	79	83	77	81	76	81
069	Libertytown	1st	67	66	69	73	74	69
		2nd	80	82	87	81	87	83
070	Winfield	1st	77	89	86	85	84	80
		2nd	86	85	79	78	91	82
071	Finksburg	1st	78	89	83	77	89	72
		2nd	80	85	83	84	82	87
072	Reisterstown	1st	83	76	71	78	79	83
		2nd	80	72	77	67	87	73
073	Cockeysville	1st	77	68	72	71	82	73
		2nd	69	71	73	69	70	80
074	Towson	1st	75	72	52	81	58	55
		2nd	68	75	71	82	66	75
075	White Marsh	1st	78	70	82	67	64	76
		2nd	71	57	76	65	71	76
076	Edgewood	1st	90	83	37	67	80	53
		2nd	80	74	67	72	87	85
077	Perryman	1st	66	83	86	—	98	01
		2nd	78	85	76	86	64	48
078	Spesutie	1st	65	—	54	—	—	60
		2nd	86	33	82	68	—	65
079	Earleville	1st	78	65	75	60	62	59
		2nd	59	64	71	66	62	62
080	Cecilton	1st	53	76	58	66	56	61
		2nd	51	68	59	70	76	60

Quad no.	Quad name	Atlas	Species per block					
			NW(1)	NE(2)	CW(3)	CE(4)	SW(5)	SE(6)
081	Table Rock	1st	84	83	88	93	78	85
		2nd	86	78	81	86	92	90
082	Gorman	1st	89	81	84	81	83	—
		2nd	84	90	83	83	74	—
083	Mount Storm	1st	72	—	—	—	—	—
		2nd	64	—	—	—	—	—
084	Harpers Ferry	1st	70	86	68	78	—	—
		2nd	82	82	94	88	—	—
085	Point of Rocks	1st	68	75	65	76	—	68
		2nd	73	73	87	79	—	83
086	Buckeystown	1st	64	78	79	98	80	88
		2nd	75	76	79	86	89	84
087	Urbana	1st	74	70	81	77	84	93
		2nd	79	86	80	82	77	84
088	Damascus	1st	71	87	82	95	83	88
		2nd	74	88	86	92	79	87
089	Woodbine	1st	86	88	88	85	99	87
		2nd	88	84	84	88	98	95
090	Sykesville	1st	96	92	81	95	88	89
		2nd	86	88	87	96	83	91
091	Ellicott City	1st	76	67	86	86	82	89
		2nd	87	60	88	82	80	79
092	Baltimore West	1st	59	58	76	63	77	47
		2nd	65	71	79	70	76	67
093	Baltimore East	1st	50	59	40	61	38	31
		2nd	59	53	57	61	50	43
094	Middle River	1st	61	75	78	56	68	75
		2nd	66	62	64	61	59	70
095	Gunpowder Neck	1st	74	52	67	55	32	18
		2nd	67	54	70	32	60	23
096	Hanesville	1st	01	53	01	84	65	73
		2nd	49	63	01	83	73	84
097	Betterton	1st	80	92	85	78	71	73
		2nd	71	77	76	66	65	69
098	Galena	1st	73	66	69	60	63	66
		2nd	68	70	63	70	70	71
099	Millington	1st	72	84	65	84	77	82
		2nd	70	67	64	63	70	78
100	Davis	1st	86	71	78	—	—	—
		2nd	92	76	74	—	—	—
101	Waterford	1st	—	—	—	—	—	77
		2nd	—	—	—	—	—	88
102	Poolesville	1st	80	88	79	78	82	80
		2nd	84	87	96	84	82	75
103	Germantown	1st	84	83	79	91	78	92
		2nd	80	72	81	77	74	85
104	Gaithersburg	1st	85	77	73	62	79	71
		2nd	76	78	73	74	70	74
105	Sandy Spring	1st	83	95	70	76	78	79
		2nd	81	91	74	82	75	77
106	Clarksville	1st	81	89	89	85	95	97
		2nd	79	88	89	84	90	87
107	Savage	1st	85	89	85	81	89	82
		2nd	80	83	86	77	85	79
108	Relay	1st	92	77	83	68	71	63
		2nd	97	101	87	80	72	78
109	Curtis Bay	1st	55	50	58	74	63	63
		2nd	67	60	84	87	73	87
110	Sparrows Point	1st	60	83	58	—	76	—
		2nd	58	67	45	—	78	—
111	Swan Point	1st	—	—	—	69	—	78
		2nd	—	—	—	65	—	62

Note: — block not covered

Quad no.	Quad name	Atlas	Species per block NW(1)	NE(2)	CW(3)	CE(4)	SW(5)	SE(6)
112	Rock Hall	1st	84	68	86	82	76	67
		2nd	74	78	78	71	70	74
113	Chestertown	1st	65	81	91	61	60	56
		2nd	68	80	74	67	67	65
114	Church Hill	1st	72	78	54	56	62	64
		2nd	95	71	62	65	63	61
115	Sudlersville	1st	81	55	58	55	58	53
		2nd	81	84	72	69	67	72
116	Sterling	1st	79	84	67	91	—	—
		2nd	80	93	78	93	—	—
117	Seneca	1st	85	80	84	84	—	—
		2nd	78	74	87	83	—	—
118	Rockville	1st	75	93	86	74	83	80
		2nd	74	77	77	64	78	67
119	Kensington	1st	73	75	66	84	73	62
		2nd	70	81	69	75	63	76
120	Beltsville	1st	76	88	70	72	63	84
		2nd	86	75	66	72	67	78
121	Laurel	1st	90	79	101	88	86	88
		2nd	88	102	90	100	85	91
122	Odenton	1st	77	79	84	71	88	71
		2nd	83	76	96	83	82	70
123	Round Bay	1st	73	64	81	68	82	75
		2nd	80	70	76	78	80	72
124	Gibson Island	1st	70	61	64	61	64	99
		2nd	75	66	67	67	70	79
125	Love Point	1st	—	—	—	—	15	42
		2nd	—	—	—	—	36	53
126	Langford Creek	1st	61	35	79	64	99	58
		2nd	64	60	70	87	72	72
127	Centreville	1st	61	61	66	65	65	64
		2nd	69	62	67	72	72	78
128	Price	1st	41	61	56	53	54	60
		2nd	68	63	67	64	67	78
129	Goldsboro	1st	67	66	76	69	56	65
		2nd	64	70	72	73	65	67
130	Falls Church	1st	90	86	—	65	—	34
		2nd	76	82	—	60	—	—
131	Washington West	1st	52	61	82	76	73	42
		2nd	66	73	73	73	66	63
132	Washington East	1st	50	69	62	49	65	47
		2nd	61	82	73	65	92	58
133	Lanham	1st	87	91	85	87	75	87
		2nd	74	69	72	80	70	81
134	Bowie	1st	78	76	90	73	80	85
		2nd	71	82	85	85	81	81
135	South River	1st	91	76	77	72	69	81
		2nd	87	71	74	72	65	83
136	Annapolis	1st	75	65	77	—	78	—
		2nd	73	63	67	—	63	—
137	Kent Island	1st	71	71	65	63	64	31
		2nd	70	72	64	57	57	41
138	Queenstown	1st	59	62	70	64	51	68
		2nd	63	79	78	82	64	73
139	Wye Mills	1st	58	58	76	55	70	54
		2nd	64	71	71	71	69	67
140	Ridgely	1st	82	84	83	78	84	64
		2nd	73	67	78	71	83	64
141	Denton	1st	61	69	70	65	73	79
		2nd	69	80	66	76	71	79
142	Alexandria	1st	21	41	25	67	68	74
		2nd	—	56	—	74	—	83

Quad no.	Quad name	Atlas	NW(1)	NE(2)	CW(3)	CE(4)	SW(5)	SE(6)
					Species per block			
143	Anacostia	1st	70	50	52	84	82	93
		2nd	73	59	64	64	66	68
144	Upper Marlboro	1st	84	86	82	93	87	95
		2nd	75	89	70	83	75	81
145	Bristol	1st	97	91	94	84	115	89
		2nd	74	85	92	78	107	83
146	Deale	1st	75	70	72	64	76	63
		2nd	74	81	72	80	78	64
147	Claiborne	1st	50	67	01	76	76	66
		2nd	51	63	18	72	87	63
148	St. Michaels	1st	59	75	67	82	70	73
		2nd	45	77	73	86	68	79
149	Easton	1st	82	67	82	75	79	74
		2nd	73	66	78	74	79	86
150	Fowling Creek	1st	50	47	69	78	82	79
		2nd	82	74	87	60	84	64
151	Hobbs	1st	81	79	80	82	60	49
		2nd	73	66	77	76	71	74
152	Hickman	1st	71	—	83	—	66	—
		2nd	72	—	74	—	81	—
153	Mount Vernon	1st	—	83	83	90	88	86
		2nd	—	72	76	81	78	74
154	Piscataway	1st	80	89	87	88	84	82
		2nd	81	72	81	75	83	72
155	Brandywine	1st	87	93	86	83	101	92
		2nd	81	75	77	72	72	78
156	Lower Marlboro	1st	98	90	95	93	97	97
		2nd	92	87	87	80	91	85
157	North Beach	1st	69	73	78	82	86	49
		2nd	72	66	87	66	55	61
158	Tilghman	1st	75	65	69	01	—	—
		2nd	65	76	66	01	—	—
159	Oxford	1st	67	76	73	73	—	70
		2nd	68	75	73	65	—	64
160	Trappe	1st	73	84	76	78	71	70
		2nd	61	91	79	76	71	72
161	Preston	1st	82	50	81	68	71	62
		2nd	73	60	75	64	70	63
162	Federalsburg	1st	67	80	69	85	76	63
		2nd	74	73	66	84	75	83
163	Seaford West	1st	63	—	56	—	46	—
		2nd	72	—	77	—	73	—
164	Indian Head	1st	—	71	65	77	80	67
		2nd	35	75	80	77	72	66
165	Port Tobacco	1st	66	77	78	67	64	74
		2nd	76	67	72	66	64	83
166	La Plata	1st	65	85	91	91	67	61
		2nd	72	70	75	78	66	77
167	Hughesville	1st	73	79	70	76	73	71
		2nd	71	71	64	64	67	70
168	Benedict	1st	99	89	86	96	84	86
		2nd	78	90	85	88	72	73
169	Prince Frederick	1st	72	70	62	44	68	91
		2nd	72	80	74	73	70	92
170	Hudson	1st	—	49	—	53	01	01
		2nd	—	56	—	64	16	44
171	Church Creek	1st	06	75	78	79	77	61
		2nd	67	82	65	89	74	66
172	Cambridge	1st	57	80	63	53	65	70
		2nd	61	75	67	60	68	67
173	East New Market	1st	57	57	63	60	60	61
		2nd	77	67	70	66	64	72

Note: — block not covered

Quad no.	Quad name	Atlas	NW(1)	NE(2)	CW(3)	CE(4)	SW(5)	SE(6)
					Species per block			
174	Rhodesdale	1st	76	60	49	69	58	74
		2nd	82	83	65	86	71	79
175	Sharptown	1st	57	—	75	—	51	—
		2nd	75	—	90	—	80	—
176	Widewater	1st	—	68	—	59	—	44
		2nd	—	60	—	60	—	48
177	Nanjemoy	1st	62	70	79	88	66	76
		2nd	67	71	77	93	75	67
178	Mathias Point	1st	72	68	68	66	43	01
		2nd	73	61	67	62	58	15
179	Popes Creek	1st	69	64	63	68	77	69
		2nd	59	68	72	57	78	68
180	Charlotte Hall	1st	70	27	70	40	—	56
		2nd	70	71	61	73	76	76
181	Mechanicsville	1st	58	66	38	80	59	84
		2nd	76	69	78	75	74	80
182	Broomes Island	1st	75	58	53	73	65	69
		2nd	77	68	79	83	83	84
183	Cove Point	1st	—	—	89	—	67	93
		2nd	75	—	81	01	79	87
184	Taylors Island	1st	42	70	—	63	—	59
		2nd	69	86	—	78	—	79
185	Golden Hill	1st	69	78	87	89	83	74
		2nd	68	82	77	79	83	68
186	Blackwater River	1st	81	72	103	82	58	42
		2nd	69	75	86	82	62	39
187	Chicamacomico	1st	65	65	68	71	54	53
		2nd	72	70	85	82	63	73
188	Mardela Springs	1st	71	78	61	77	55	73
		2nd	76	91	67	79	69	77
189	Hebron	1st	77	—	69	63	70	63
		2nd	82	—	76	72	79	69
190	Delmar	1st	—	—	61	72	78	71
		2nd	—	—	71	70	76	71
191	Pittsville	1st	—	—	64	62	80	67
		2nd	—	—	68	78	74	73
192	Whaleysville	1st	—	—	87	69	81	58
		2nd	—	—	75	73	72	69
193	Selbyville	1st	—	—	63	53	60	63
		2nd	—	—	73	70	70	77
194	Assawoman Bay	1st	—	—	61	37	74	25
		2nd	—	—	78	34	74	34
195	King George	1st	42	01	—	—	—	—
		2nd	56	35	—	—	—	—
196	Colonial Beach North	1st	60	67	01	77	—	41
		2nd	71	69	15	84	—	61
197	Rock Point	1st	02	72	18	67	55	58
		2nd	30	85	36	73	70	74
198	Leonardtown	1st	64	60	66	83	63	80
		2nd	75	79	85	85	80	85
199	Hollywood	1st	63	69	76	65	76	85
		2nd	88	76	80	87	81	85
200	Solomons Island	1st	65	71	57	74	75	84
		2nd	69	68	64	79	83	93
201	Barren Island	1st	—	07	—	32	—	—
		2nd	—	40	—	24	—	—
202	Honga	1st	63	72	32	59	20	43
		2nd	62	78	45	69	40	50
203	Wingate	1st	60	40	73	38	55	42
		2nd	74	36	71	43	58	51
204	Nanticoke	1st	32	25	46	57	13	51
		2nd	39	27	65	76	12	86

Quad no.	Quad name	Atlas	Species per block					
			NW(1)	NE(2)	CW(3)	CE(4)	SW(5)	SE(6)
205	Wetipquin	1st	64	69	72	67	73	74
		2nd	64	80	75	83	79	80
206	Eden	1st	68	69	64	71	63	72
		2nd	73	71	77	76	73	68
207	Salisbury	1st	60	69	92	80	73	73
		2nd	71	83	76	78	69	72
208	Wango	1st	64	67	70	60	69	65
		2nd	78	72	86	78	84	77
209	Ninepin	1st	65	68	73	69	73	71
		2nd	71	78	82	71	78	75
210	Berlin	1st	61	74	68	75	82	72
		2nd	69	87	76	90	79	83
211	Ocean City	1st	67	—	65	—	34	—
		2nd	74	—	84	—	21	—
212	Stratford Hall	1st	—	52	—	—	—	—
		2nd	—	61	—	—	—	—
213	St. Clements Island	1st	01	51	—	—	—	—
		2nd	75	77	—	—	—	—
214	Piney Point	1st	72	61	01	61	—	55
		2nd	74	69	32	85	—	60
215	St. Marys City	1st	62	69	65	68	62	59
		2nd	69	69	72	76	64	73
216	Point No Point	1st	—	—	74	—	71	—
		2nd	—	—	87	—	79	—
217	Richland Point	1st	10	02	—	—	—	—
		2nd	—	19	—	—	—	—
218	Bloodsworth Island	1st	39	20	05	11	07	—
		2nd	33	38	16	27	36	15
219	Deal Island	1st	07	61	47	78	58	39
		2nd	08	66	48	77	53	37
220	Monie	1st	71	87	81	68	53	66
		2nd	80	77	73	69	64	75
221	Princess Anne	1st	75	70	71	75	71	73
		2nd	71	65	67	72	65	64
222	Dividing Creek	1st	79	72	77	73	88	71
		2nd	72	79	67	76	72	77
223	Snow Hill	1st	75	70	74	75	79	70
		2nd	81	79	81	78	80	76
224	Public Landing	1st	73	74	73	85	66	78
		2nd	81	81	73	82	71	81
225	Tingles Island	1st	51	60	19	53	33	45
		2nd	72	87	29	58	42	56
226	St. George Island	1st	53	61	—	—	—	—
		2nd	55	58	—	—	—	—
227	Point Lookout	1st	63	—	77	—	03	—
		2nd	77	—	87	—	24	—
228	Kedges Straits	1st	07	22	—	01	—	44
		2nd	39	18	—	06	—	30
229	Terrapin Sand Point	1st	—	—	01	—	01	—
		2nd	02	—	—	01	03	—
230	Marion	1st	67	79	51	64	70	74
		2nd	59	76	49	64	74	68
231	Kingston	1st	77	69	70	80	84	80
		2nd	68	65	64	68	81	85
232	Pocomoke City	1st	74	79	74	72	72	74
		2nd	68	75	82	75	77	73
233	Girdletree	1st	75	69	67	74	67	75
		2nd	76	75	74	76	76	79
234	Boxiron	1st	78	22	74	19	26	58
		2nd	97	59	69	24	43	63
235	Whittington Point	1st	64	11	48	—	32	—
		2nd	68	21	50	—	49	—

Note: — block not covered

Quad no.	Quad name	Atlas	Species per block					
			NW(1)	NE(2)	CW(3)	CE(4)	SW(5)	SE(6)
236	Ewell	1st	04	49	01	—	—	—
		2nd	—	41	—	19	—	—
237	Great Fox Island	1st	01	19	—	08	—	—
		2nd	10	03	—	02	—	—
238	Crisfield	1st	70	73	37	—	—	—
		2nd	63	56	43	—	—	—
239	Saxis	1st	85	67	—	—	—	—
		2nd	68	71	—	—	—	—
240	Marydel	1st	70	—	67	—	72	—
		2nd	70	—	69	—	66	—
241	Burrsville	1st	65	—	66	—	68	—
		2nd	63	—	62	—	67	—

Note: — block not covered

APPENDIX C

Examples of Distributional Changes Based on Quarterblocks

Quarterblocks provide an additional dimension to the atlas data by linking bird distribution to specific habitats, such as stream valleys or bogs, that would not be discernible at coarser scales. Examples from Howard County and from Garrett County show ways in which quarterblock maps increase our understanding of changes that are taking place.

Howard County

Howard County was the first county in the United States to have its species mapped in quarterblocks (Klimkiewicz and Solem 1978). The first statewide Maryland-DC atlas (Robbins and Blom 1996) used quarterblock maps to emphasize dramatic changes in the distribution of certain species between 1973–1975 and 1983–1987. Here we follow those very same species for another two decades. Each 7.5-minute USGS topographic map within the county is divided into six atlas blocks, and each black dot represents a quarterblock (6.2 sq km [2.3 sq mi]) where the species was found.

Canada Geese have moved into additional farm ponds as this semidomesticated population continues to expand. They were found in only 7 Howard County blocks in the 1970s, in 23 blocks in the 1980s, and in all 34 blocks in the latest atlas. Their quarterblock count increased from 8 to 49 to 130. Pileated Woodpecker distribution increased from 19 blocks in the 1970s to 33 blocks in the 1980s. They were in all 34 blocks in the latest atlas and continued to move into new quarterblocks, jumping 40% from the 1980s to the present. Tree Swallows continued to take advantage of bluebird nest boxes and moved into new areas; they were in only 3 blocks in the 1970s atlas; 6 blocks in the 1980s; and all 34 blocks in 2002–2006. Corresponding quarterblock counts increased from 5, to 8, to 106.

The 1983–1987 atlas came at a time of maximum expansion of Blue-winged Warblers, when much abandoned land was left idle during construction of Columbia in the central part of the county. The land is idle no more and Blue-winged Warblers are again very scarce at this southern limit of their range. Measured in whole blocks their numbers changed from 10 to 17 to 11. In quarterblocks they rose from 13 to 33 then fell to 17. Pine woods have been increasing in Howard County, especially on reservoir properties along the Patuxent River. Pine Warblers have responded by spreading from 6 to 8 to 22 blocks, and from 9 to 14 to 37 quarterblocks. House Finches, which were absent in the 1970s reached saturation in the 1980s when they were found in all 34 blocks and all but 7 quarterblocks. No longer tracked by whole blocks, they spread from 129 quarterblocks in the first atlas to 135 in the latest atlas.

Black-billed Cuckoos, which are at the southern limit of their breeding range, declined from 10 to 6 to 5 blocks and from 19 to 8 to 5 quarterblocks. As woodlands decline and the bright lights of suburbia spread westward through the county, Whip-poor-wills continue to decline. Block totals fell from 18 to 14 to 2, and quarterblocks from 30 to 22 to 3. Horned Larks were found in 27 of the 34 blocks in the 1970s, only 12 in the 1980s, and now 13; the quarterblock counts were 47, 17, and 24. The large fields that Horned Larks need for nesting are gradually disappearing, so further declines are expected.

Prothonotary Warblers, never common in Howard County during the breeding season, have found more suitable nesting habitat along the Potomac than along the Patuxent River. Their presence was registered in 6, then 2, and now 8 blocks, and in 9, 3, and 14 quarterblocks. Most streams as they cross the fall line flow too rapidly for these warblers. If small ponds were created along side streams, Prothonotary Warblers might use them. Vesper Sparrows are definitely on their way out as large farms are converted to residential communities. Their block totals fell from 19 to 12 to 2, and quarterblocks from 51 to 22 to 3. Grass-

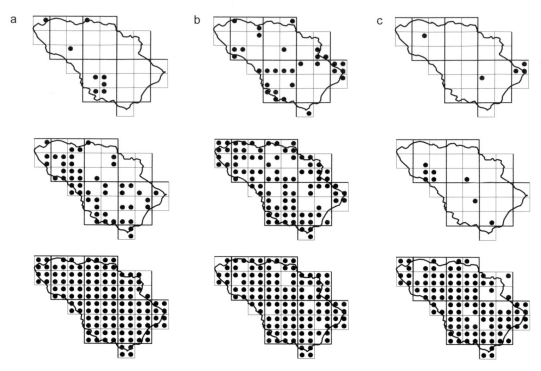

Comparison of quarterblock distribution in Howard County from (*top to bottom*) the 1970s, 1980s, and 2000s for (*a*) Canada Goose, (*b*) Pileated Woodpecker, (*c*) Tree Swallow.

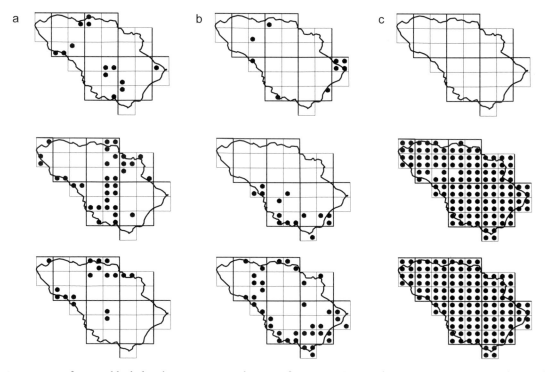

Comparison of quarterblock distribution in Howard County from (*top to bottom*) the 1970s, 1980s, and 2000s for (*a*) Blue-winged Warbler, (*b*) Pine Warbler, (*c*) House Finch.

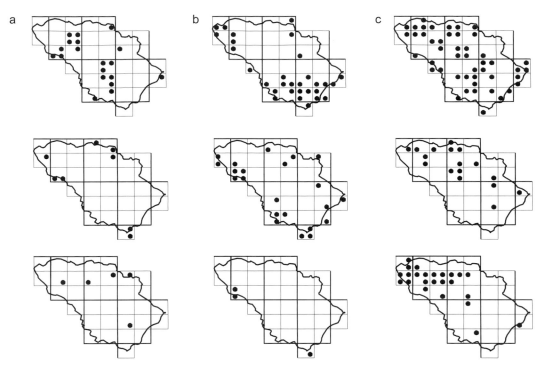

Comparison of quarterblock distribution in Howard County from (*top to bottom*) the 1970s, 1980s, and 2000s for (*a*) Black-billed Cuckoo, (*b*) Eastern Whip-poor-will, (*c*) Horned Lark.

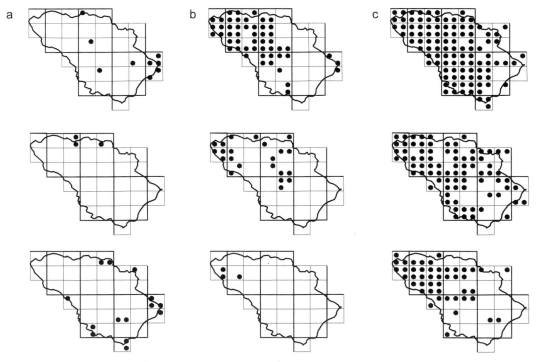

Comparison of quarterblock distribution in Howard County from (*top to bottom*) the 1970s, 1980s, and 2000s for (*a*) Prothonotary Warbler, (*b*) Vesper Sparrow, (*c*) Grasshopper Sparrow.

hopper Sparrows are also destined to disappear from Howard County unless remnants of the farming community can be maintained. Block totals were 32, 31, and 20; quarterblocks had detected the decline earlier with 103, 79, and 46.

Garrett County

As Maryland's temperatures creep upward, Garrett County, with its high elevation, has shown the most dramatic changes in bird distribution. During atlas work in the 1980s, some observers working in Garrett County failed to record the quarterblocks for their sightings. We have assigned 1 or 2 quarterblocks for each unspecified sighting, basing the selection on elevations, habitats, road access, subsequent observations, and other considerations. Thus comparisons between the first and second atlases for Garrett County are unbiased although some of the data may be assigned to the wrong quarterblock. A list of the assigned quarterblock records is available on the MOS website at www.mdbirds.org.

Garrett County is host to four closely related small flycatchers, all in the genus *Empidonax*. Three of the four species increased their distribution in Garrett County between the first atlas and the second, and for each of these species the increase was more apparent at the quarterblock level. Acadian Flycatcher, the southernmost species, was found in 64 of 83 blocks in the first atlas and 65 (2% increase) in the second; quarterblock coverage showed an increase of nearly 142%, from 65 to 157. Alder Flycatcher increased 38%, from 26 to 36 blocks; quarterblocks showed an increase of 88%, from 32 to 60. Willow Flycatcher remained at 45 blocks, with a 3% increase, from 67 to 69 quarterblocks. The Least Flycatcher also increased, from 54 to 58 blocks (7%) and 89 to 103 (16%) quarterblocks.

Carolina Wren and Northern Mockingbird are examples of southern species that have moved into the cooler climate of Garrett County in recent decades. In the first atlas, the Carolina Wren was found in 21 Garrett County blocks. In the second atlas it was found in 62, an increase of 195%. The increase at the quarterblock level is even more dramatic: from 24 to 120, or 400%. The mockingbird more than doubled its distribution; it was found in 14 Garrett County blocks on the first atlas and 31 on the second, an increase of 121%. At the quarterblock scale the increase was considerably greater: from 17 to 47, or 176%; most of this increase took place in the northwestern corner of the county. Hermit Thrush block distribution increased 61%, from 33 to 53 blocks, and 192% when measured in quarterblocks (39 to 114).

Several warblers of conservation concern are showing declines in Garrett County. Golden-winged Warbler distribution declined from 59 to 27 blocks (54%), and quarterblock coverage demonstrated a similar decline, from 81 to 39 (52%). Northern Waterthrush fell from 23 to10 blocks (57%), and from 33 to 16 quarterblocks (52%). Hooded Warbler declined from 67 to 61 blocks (9%), but quarterblocks showed an increase from 107 to 119 (11%). Canada Warbler fell from 46 to 35 blocks (24%), and from 63 to 57 quarterblocks (10%). Measured in blocks, Yellow-breasted Chat declined 71% from 41 to 12; quarterblocks also showed a serious drop, from 54 to 19, or 65%.

The maps from Garrett County raise a variety of questions: Why are Northern Waterthrushes declining so sharply when Alder Flycatchers are increasing in the same quarterblocks and human activities are not seriously impacting their nesting habitats? What could be causing Yellow-breasted Chats to disappear from so much of their former range in Garrett County?

This discussion has been limited to just a dozen species in 2 of Maryland's 23 counties. Quarterblock data for Washington, DC; Baltimore City; Maryland's central counties of Baltimore, southern Carroll, Howard, Montgomery, and Prince George's; and Garrett and Somerset counties are available on the MOS website (www.mdbirds.org), as are quarterblock data from the northwestern sixth of every 7.5-minute topographic map for the rest of Maryland.

Garrett County Quarterblocks

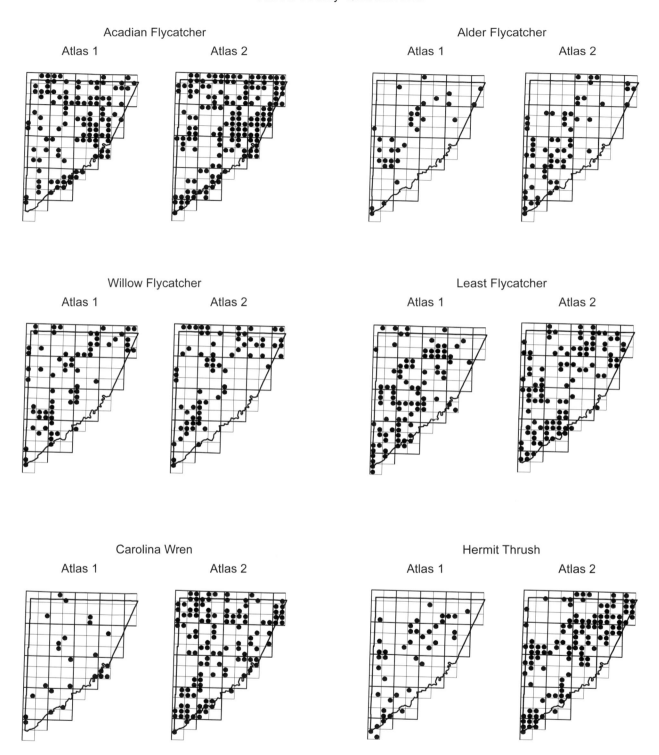

Comparison of quarterblock distribution in Garrett County from the 1980s (Atlas 1) and 2000s (Atlas 2) for four *Empidonax* flycatchers, Carolina Wren, and Hermit Thrush.

Garrett County Quarterblocks

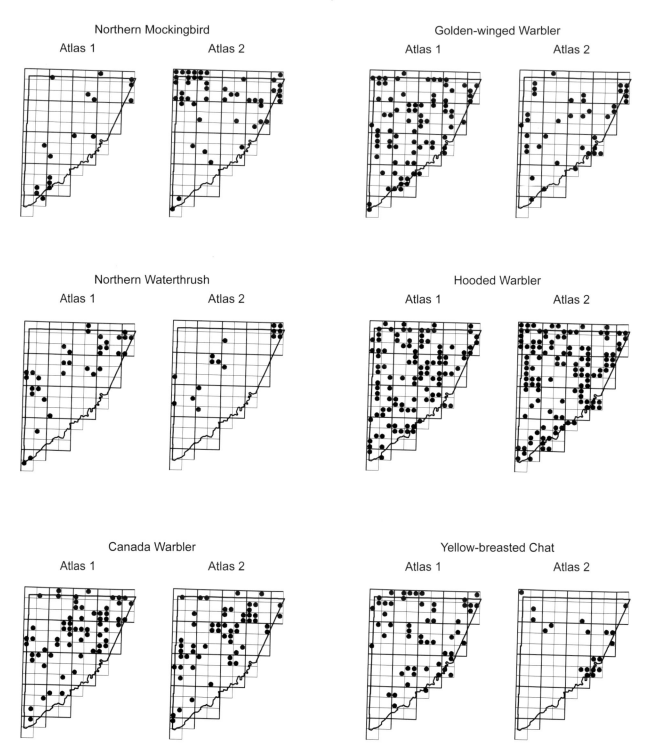

Comparison of quarterblock distribution in Garrett County from the 1980s (Atlas 1) and 2000s for Northern Mockingbird and five warbler species.

APPENDIX D

Abundance Definitions

The terms used to describe the relative abundance of birds are necessarily subjective. Atlas field-workers were not asked to record the actual number of each species they encountered in the course of their fieldwork, and miniroute data provide relative rather than actual abundance. In general, we tried to follow the guidelines for relative abundance terms established by Hall (1983), while taking into account our own experience and perceptions. In the species accounts, abundance terms cited from publications do not necessarily follow these guidelines. The following definitions are based on Hall (1983).

Very abundant. Species for which more than 1,000 individuals can be seen by a single observer in a full day's work in the field in suitable habitat. A species for which more than 300 individuals can be listed on a 39-km (24.5-mi) BBS.

Abundant. A species for which an observer can see between 201 and 1,000 individuals in a full day's work in suitable habitat. A species for which 101 to 300 individuals can be listed on a 39-km (24.5-mi) BBS. A species for which a singing male census has a density of more than 100 males per 100 ha (247 ac).

Very common. A species for which an observer can see between 51 and 100 individuals in a full day's work in suitable

habitat. A species for which between 31 and 100 individuals can be listed on a 39-km (24.5-mi) BBS. A species that has a density of between 51 and 100 males per 100 ha (247 ac) on a singing male census.

Common. A species for which an observer can see between 21 and 50 individuals in a day. A species for which between 11 and 30 can be listed on a 39-km (24.5-mi) BBS. A species that has a density of between 26 and 50 males per 100 ha (247 ac) on a singing male census.

Fairly common. A species for which an observer would list between 7 and 20 individuals in a full day in suitable habitat. A species for which between 4 and 10 can be listed on a 39-km (24.5-mi) BBS. A species with a density of between 11 and 25 males per 100 ha (247 ac) on a singing male census.

Uncommon. A species for which an observer will list between 1 and 6 individuals in a full day's work in suitable habitat. A species for which between 1 and 3 individuals will be listed on a 39-km (24.5-mi) BBS. A species with a density of between 6 and 10 males per 100 ha (247 ac) on a singing male census.

Rare. A species for which, in every appropriate season, an observer can expect to list between 1 and 6 birds.

APPENDIX E

Nonavian Fauna Cited

These taxonomic names are from Encyclopedia of Life, available at www.eol.org.

Invertebrates

ant	Formicidae
beetle	Coleoptera
blue crab	*Callinectes sapidus*
blue mussel	*Mytilus edulis*
caddisfly	Trichoptera
cankerworm	Geometridae
carpenter ant	*Camponotus* sp.
caterpillar	Lepidoptera (larval)
chigger	*Trombicula* sp.
crab	Decapoda
crayfish	Cambaridae
cricket	Gryllidae
dragonfly	Odonata
earthworm	Oligochaeta
fall webworm	*Hyphantria cunea*
fiddler crab	*Uca* sp.
fly	Diptera
ghost crab	*Ocypode quadrata*
grasshopper	Orthoptera
gypsy moth	*Lymantria dispar*
hemlock wooly adelgid	*Adelges tsugae*
katydid	Tettigoniidae
leafhopper	Cicadellidae
marine worms	Polychaeta
mole crab	*Emerita talpoida*
moth	Lepidoptera
mud crab	Panopeidae
myriapod	Protura
oyster	*Crassostrea virginica*
ribbed mussel	*Modiolus demissus*
17-year cicada	*Magicicada septendecim*
shrimp	Decapoda
snail	Gastropoda
spider	Araneae
spruce budworm	*Choristoneura fumiferana*
tent caterpillar	*Malacosoma* sp.
tick	Acarina
wasp	Hymenoptera
weevil	Curculionoidea

Fish

Atlantic salmon	*Salmo salar*
catfish (farmed)	*Ictalurus* sp.
menhaden	*Brevoortia tyrannus*
shad	*Alosa* sp.

Mammals

beaver	*Castor canadensis*
bison	*Bison bison*
cattle	*Bos taurus*
chipmunk	*Tamias striatus*
deer	*Odocoileus virginianus*
goat	*Capra hircus*
horse	*Equus caballus*
muskrat	*Ondatra zibethicus*
southern flying squirrel	*Glaucomys volans*
vole	*Microtus* sp.
white-tailed deer	*Odocoileus virginianus*

Amphibians

frog	Anura

APPENDIX F
Plants Cited

These taxonomic names are from Encyclopedia of Life, available at www.eol.org.

alder	*Alnus* sp.	corn	*Zea mays*
alfalfa	*Medicago sativa*	daisy fleabane	*Erigeron strigosus*
American holly	*Ilex opaca*	dogwood	*Cornus* sp.
arrowwood	*Viburnum* sp.	Dutch elm disease	*Ophiostoma ulmi*
aster	Asteraceae	eastern red cedar	*Juniperus virginiana*
bald cypress	*Taxodium distichum*	eelgrass	*Zostera marina*
barley	*Hordeum* sp.	elderberry	*Sambucus* sp.
beachgrass	*Ammophila breviligulata*	elm	*Ulmus* sp.
bee-balm	*Monarda didyma*	fescue	*Festuca* sp., *Lolium* sp.
beech	*Fagus grandifolia*	fir	*Abies* sp.
big cordgrass	*Spartina cynosuroides*	flowering dogwood	*Cornus florida*
bigtooth aspen	*Populus grandidentata*	goldenrod	*Solidago* sp.
birch	*Betula* sp.	grape	*Vitis* sp.
blackberry	*Rubus* sp.	grass	Poaceae
black birch	*Betula lenta*	greenbrier	*Smilax* sp.
black gum	*Nyssa sylvatica*	groundsel tree/shrub	*Baccharis halimifolia*
blueberry	*Vaccinium* sp.	hawthorn	*Crataegus* sp.
bramble	*Rubus* sp.	hemlock	*Tsuga canadensis*
broomsedge	*Andropogon virginicus*	hickory	*Carya* sp.
bur-reed	*Sparganium* sp.	highbush blueberry	*Vaccinium corymbosum*
buttonbush	*Cephalanthus occidentalis*	holly	*Ilex opaca*
Canada thistle	*Cirsium arvense*	huckleberry	*Gaylussaccia* sp.
cardinal flower	*Lobelia cardinalis*	iris	*Iris* sp.
cattail	*Typha* sp.	jewelweed	*Impatiens capensis, I. pallida*
cherry	*Prunus* sp.	joe-pye-weed	*Eupatorium* sp.
chestnut	*Castanea dentata*	knotweed	*Polygonum* sp.
chicory	*Cichorium intybus*	lamb's quarters	*Chenopodium album*
chickweed	*Stellaria* sp., *Cerastium* sp.	Leland cypress	*Cupressus × leylandii*
clover	*Trifolium* sp.	lizard's tail	*Saururus cernuus*
common reed	*Phragmites australis*	loblolly pine	*Pinus taeda*
cordgrass	*Spartina* sp.	locust	*Robinia* sp., *Gleditsia triacanthos*

marsh elder	*Iva frutescens*	sedge	*Carex* sp.
meadow fescue	*Lolium pratense*	shortleaf pine	*Pinus echinata*
milkweed	*Asclepias* sp.	skunk cabbage	*Symplocarpus foetida*
mimosa	*Albizzia julibrissin*	smooth brome	*Bromus inermis*
American mistletoe	*Phoradendron leucarpum*	smooth cordgrass	*Spartina alterniflora*
mountain laurel	*Kalmia latifolia*	sorghum	*Sorgum* sp.
multiflora rose	*Rosa fujisanensis*	sorrel	*Rumex* sp.
narrow-leaved cattail	*Typha angustifolia*	soybean	*Glycine max*
needlerush	*Juncus roemerianus*	Spanish moss	*Tillandsia usneoides*
nodding thistle	*Carduus nutans*	spicebush	*Lindera benzoin*
northern red oak	*Quercus rubra*	spiraea	*Spiraea* sp.
Norway spruce	*Picea abies*	spruce	*Picea* sp.
nyjer thistle	*Guizotia abyssinica*	sugar maple	*Acer saccharum*
oak	*Quercus* sp.	sumac	*Rhus* sp.
oats	*Avena sativa*	sunflower	*Helianthus* sp.
Olney's bulrush	*Schoenoplectus americanus*	sweetbay	*Magnolia virginiana*
orchard grass	*Dactylis glomerata*	sweet flag	*Acorus calamus*
pencil flower	*Stylosanthes biflora*	sweet gum	*Liquidambar styraciflua*
phragmites	*Phragmites australis*	sweet pepperbush	*Clethra alnifolia*
pickerelweed	*Pontederia cordata*	switchgrass	*Panicum virgatum*
pigweed	*Amaranthus* sp., *Chenopodium* sp.	sycamore	*Platanus occidentalis*
pin oak	*Quercus palustris*	Tatarian honeysuckle	*Lonicera tatarica*
pine	*Pinus* sp.	thistle	*Carduus* sp. & *Cirsium* sp.
poison ivy	*Toxicodendron radicans*	three-square	*Schoenoplectus pungens*
poverty grass	*Aristida dichotoma, Danthonia spicata, Sporobolus vaginaflorus*	tickseed sunflower	*Bidens aristosa*
		timothy	*Phleum pratense*
quaking aspen	*Populus tremuloides*	tobacco	*Nicotiana tabacum*
ragweed	*Ambrosia* sp.	trumpet vine / creeper	*Campsis radicans*
raspberry	*Rubus* sp.	tuckahoe	*Peltandra virginica*
red cedar	*Juniperus virginianus*	tulip-poplar	*Liriodendron tulipifera*
red maple	*Acer rubrum*	Virginia creeper	*Parthenocissus quinquefolia*
red pine	*Pinus resinosa*	Virginia pine	*Pinus virginiana*
red spruce	*Picea rubens*	water oak	*Quercus nigra*
redtop	*Agrostis gigantea*	wax myrtle	*Morella cerifera*
reed	*Phragmites australis*	wheat	*Triticum* sp.
reed canary grass	*Phalaris arundinacea*	white ash	*Fraxinus americana*
rhododendron	*Rhododendron maximum*	white oak	*Quercus alba*
rice	*Oryza sativa*	white pine	*Pinus strobus*
saltmeadow cordgrass	*Spartina patens*	wild columbine	*Aquilegia canadensis*
saltmeadow rush	*Juncus gerardi*	wild rice	*Zizania aquatica*
sassafras	*Sassafras albidum*	willow	*Salix* sp.
sea rocket	*Cakile edentula*	willow oak	*Quercus phellos*
seashore saltgrass	*Distichlis spicata*		

REFERENCES

Abbott, J. M. 1963. Bald Eagle survey for Chesapeake Bay, 1962. *Atlantic Naturalist* 18:22–27.

Adams, L. W., L. E. Dove, and T. M. Franklin. 1985. Mallard pair and brood use of urban stormwater-control impoundments. *Wildlife Society Bulletin* 13:46–51.

Adams, M. T., and M. Hafner. 2009. The nesting season: Middle Atlantic region. *North American Birds* 62:541–544.

Adamus, P. R. 1987. *Atlas of Breeding Birds in Maine, 1978–1983.* Augusta: Maine Department of Inland Fisheries and Wildlife.

Adkisson, C. S. 1996. Red Crossbill (*Loxia curvirostra*). In *The Birds of North America,* no. 256, ed. A. Poole and F. Gill. Philadelphia and Washington, DC: Academy of Natural Sciences and American Ornithologists' Union.

American Ornithologists' Union (AOU). 1973. Thirty-second supplement to the American Ornithologists' Union check-list of North American birds. *Auk* 90:411–419.

———. 1983. *Check-list of North American Birds.* 6th ed. Washington, DC: American Ornithologists' Union.

———. 1997. Forty-first supplement to the American Ornithologists' Union check-list of North American birds. *Auk* 114:542–552.

———. 1998. *Check-list of North American Birds.* 7th ed. Washington, DC: American Ornithologists' Union.

Anderson, D. W., and J. J. Hickey. 1972. Eggshell changes in certain North American birds. *Proceedings of the International Ornithological Congress* 15:514–540.

Andrle, R. F. 1971. Range extension of the Golden-crowned Kinglet in New York. *Wilson Bulletin* 83:313–316.

Andrle, R. F., and J. R. Carroll. 1988. *The Atlas of Breeding Birds in New York State.* Ithaca: Cornell University Press.

Armistead, H. T. 1970. The first Maryland breeding of American Coot at Deal Island. *Maryland Birdlife* 26:79–81.

———. 1987. The nesting season, June 1–July 31, 1987: Middle Atlantic coast region. *American Birds* 41:1418–1422.

Armistead, H. T., and W. C. Russell. 1967. First Black-necked Stilt and fourth Ruff. *Maryland Birdlife* 23:62–63.

Askins, R. A., M. J. Philbrick, and D. S. Sugeno. 1987. Relationship between the regional abundance of forest and the composition of forest bird communities. *Biological Conservation* 39:129–152.

Bailey, F. M. 1928. *Birds of New Mexico.* Washington, DC: Bureau of Biological Survey.

Bailey, H. H. 1913. *The Birds of Virginia.* Lynchburg, Va.: J. P. Bell.

Bailey, R. F., III. 2006. *Maryland's Forests and Parks: A Century of Progress.* Charleston, S.C.: Arcadia Publishing.

Baird, S. F., J. Cassin, and G. N. Lawrence. 1858. *Explorations and Surveys for a Railroad Route from the Mississippi River to the Pacific Ocean.* Washington, DC: A.O.P. Nicholson.

Bajema, R. A., T. L. DeVault, P. E. Scott, and S. L. Lima. 2001. Reclaimed coal mine grasslands and their significance for Henslow's Sparrows in the American Midwest. *Auk* 118:422–431.

Ball, W. H. 1930. Notes from eastern Maryland. *Auk* 47:94–95.

———. 1932. Notes from the Washington, DC region. *Auk* 49:362.

Ball, W. H., and R. B. Wallace. 1936. Further remarks on birds of Bolling Field, DC. *Auk* 53:345–346.

Bannor, B. K., and E. Kiviat. 2002. Common Moorhen (*Gallinula chloropus*). In *The Birds of North America,* no. 685a, ed. A. Poole and F. Gill. Philadelphia and Washington, DC: Academy of Natural Sciences and American Ornithologists' Union.

Beadell, J., R. Greenberg, S. Droege, and J. A. Royle. 2003. Distribution, abundance, and habitat affinities of the Coastal Plain Swamp Sparrow. *Wilson Bulletin* 115:38–44.

Beal, F. E. L. 1918. Food habits of the swallows, a family of valuable native birds. *U.S. Department of Agriculture Bulletin,* no. 619, Washington, DC.

Beddall, B. G. 1963. Range expansion of the Cardinal and other birds in the northeastern states. *Wilson Bulletin* 75:140–158.

Bedell, P. A. 1996. Evidence of dual breeding ranges for the Sedge Wren in the central Great Plains. *Wilson Bulletin* 108:115–122.

Bednarz, J. 1992. Red-shouldered Hawk (*Buteo lineatus*). Pages 102–103 in *Atlas of Breeding Birds in Pennsylvania,* ed. D. W. Brauning. Pittsburgh: University of Pittsburgh Press.

Behr, H. 1914. Some breeding birds of Garrett County, Maryland. *Auk* 31:548.

Bellrose, F. C. 1980. *Ducks, Geese, and Swans of North America*. 3rd ed. Harrisburg, Pa.: Stackpole Books.

Belthoff, J. R., and S. A. Gauthreaux Jr. 1991. Partial migration and differential winter distribution of House Finches in the eastern United States. *Condor* 93:374–382.

Belthoff, J. R., and G. Ritchison. 1989. Natal dispersal of Eastern Screech-Owls. *Condor* 91:254–265.

Bendel, P. R., and G. D. Therres. 1990. Nesting biology of Barn Owls from Eastern Shore marshes. *Maryland Birdlife* 46:119–123.

———. 1999. Proximity of waterbird colonies to development in Maryland. *Proceedings of the Annual Conference of the Southeastern Association of Fish and Wildlife Agencies* 53:425–433.

Bendire, C. E. 1895. The American Barn Owl breeding in Washington, DC, in winter. *Auk* 12:180–181.

Bennett, K. A., P. J. Bowman, C. M. Heckscher, W. A. McAvoy, and E. F. Zeulke. 1998. Ecological characterization of Delmarva's Great Cypress Swamp Conservation Area. Dover: Delaware Division of Fish and Wildlife. Unpublished report.

Benoit, L. K., and R. A. Askins. 1999. Impact of the spread of Phragmites on the distribution of birds in Connecticut tidal marshes. *Wetlands* 19:194–208.

———. 2002. Relationship between habitat area and the distribution of tidal marsh birds. *Wilson Bulletin* 114:314–323.

Bent, A. C. 1925. *Life Histories of North American Wild Fowl*. U.S. National Museum Bulletin, no.130.

———. 1926. *Life Histories of North American Marsh Birds*. U.S. National Museum Bulletin, no. 135.

———. 1929. *Life Histories of North American Shore Birds*. Part 2. U.S. National Museum Bulletin, no. 146.

———. 1939. *Life Histories of North American Woodpeckers*. U.S. National Museum Bulletin, no. 174.

———. 1948. *Life Histories of North American Nuthatches, Wrens, Thrashers, and Their Allies*. U.S. National Museum Bulletin, no. 195.

———. 1949. *Life Histories of North American Thrushes, Kinglets, and Their Allies*. U.S. National Museum Bulletin, no. 196.

———. 1953. *Life Histories of the North American Wood Warblers*. U.S. National Museum Bulletin, no. 203.

———. 1958. *Life Histories of North American Blackbirds, Orioles, Tanagers, and Allies*. U.S. National Museum Bulletin, no. 211.

Bergman, G., E. Kellomaki, K. Hyytia, and J. Koistinen. 1983. *Suomen Lintuatlas*. Helsingin yliopiston monistuspalvelu, Helsinki (Finland): Painatusjaos.

Bertin, R. I. 1977. Breeding habitats of the Wood Thrush and Veery. *Condor* 79:303–311.

Bevier, L. B. 1994. *Atlas of Breeding Birds of Connecticut, 1982–1986*. Hartford: State Geological and Natural History Survey of Connecticut.

Bildstein, K. L.1999. Racing with the sun: The forced migration of the Broad-winged Hawk. Pages 79–102 in *Gatherings of Angels: Migrating Birds and Their Ecology*, ed. K. P. Able. Ithaca: Cornell University Press.

Bildstein, K. L., and K. Meyer. 2000. Sharp-shinned Hawk (*Accipiter striatus*). In *The Birds of North America*, no. 482, ed. A. Poole and F. Gill. Philadelphia and Washington, DC: Academy of Natural Sciences and American Ornithologists' Union.

Bircham, P. M. M., J. C. A. Rathmell, and B. Jordan. 1994. *An Atlas of the Breeding Birds of Cambridgeshire*. Cambridge, UK: Cambridgeshire Bird Club.

Bird, D. M., and R. S. Palmer. 1988. American Kestrel (*Falco sparverius*). Pages 253–290 in *Handbook of North American Birds*. Vol. 5, *Diurnal Raptors*, part 2. Ed. R. S. Palmer. New Haven, Conn.: Yale University Press.

Black Duck Joint Venture. 2008. Bibliography of American Black Duck (*Anas rubripes*) ecology and management. Laurel, Md.: U.S. Fish and Wildlife Service.

Blom, E. A. T. 1996a. Clapper Rail (*Rallus longirostris*). Pages 124–125 in *Atlas of the Breeding Birds of Maryland and the District of Columbia*, ed. C. S. Robbins and E. A. T. Blom. Pittsburgh: University of Pittsburgh Press.

———. 1996b. King Rail (*Rallus elegans*). Pages 126–127 in *Atlas of the Breeding Birds of Maryland and the District of Columbia*, ed. C. S. Robbins and E. A. T. Blom. Pittsburgh: University of Pittsburgh Press.

———. 1996c. Rock Dove (*Columba livia*). Pages 172–173 in *Atlas of the Breeding Birds of Maryland and the District of Columbia*, ed. C. S. Robbins and E. A. T. Blom. Pittsburgh: University of Pittsburgh Press.

———. 1996d. Mourning Dove (*Zenaida macroura*). Pages 174–175 in *Atlas of the Breeding Birds of Maryland and the District of Columbia*, ed. C. S. Robbins and E. A. T. Blom. Pittsburgh: University of Pittsburgh Press.

———. 1996e. Acadian Flycatcher (*Empidonax virescens*). Pages 218–219 in *Atlas of the Breeding Birds of Maryland and the District of Columbia*, ed. C. S. Robbins and E. A. T. Blom. Pittsburgh: University of Pittsburgh Press.

———. 1996f. Least Flycatcher (*Empidonax minimus*). Pages 224–225 in *Atlas of the Breeding Birds of Maryland and the District of Columbia*, ed. C. S. Robbins and E. A. T. Blom. Pittsburgh: University of Pittsburgh Press.

———. 1996g. Red-breasted Nuthatch (*Sitta canadensis*). Pages 260–261 in *Atlas of the Breeding Birds of Maryland and the District of Columbia*, ed. C. S. Robbins and E. A. T. Blom. Pittsburgh: University of Pittsburgh Press.

———. 1996h. Bewick's Wren (*Thryomanes bewickii*). Pages 270–271 in *Atlas of the Breeding Birds of Maryland and the District of Columbia*, ed. C. S. Robbins and E. A. T. Blom. Pittsburgh: University of Pittsburgh Press.

———. 1996i. Sedge Wren (*Cistothorus platensis*). Pages 276–277 in *Atlas of the Breeding Birds of Maryland and the District of Columbia*, ed. C. S. Robbins and E. A. T. Blom. Pittsburgh: University of Pittsburgh Press.

———. 1996j. Golden-crowned Kinglet (*Regulus satrapa*). Pages 280–281 in *Atlas of the Breeding Birds of Maryland and the District of Columbia*, ed. C. S. Robbins and E. A. T. Blom. Pittsburgh: University of Pittsburgh Press.

———. 1996k. Blue-gray Gnatcatcher (*Polioptila caerulea*). Pages 282–283 in *Atlas of the Breeding Birds of Maryland and the District of Columbia*, ed. C. S. Robbins and E. A. T. Blom. Pittsburgh: University of Pittsburgh Press.

———. 1996l. Hermit Thrush (*Catharus guttatus*). Pages 288–289 in *Atlas of the Breeding Birds of Maryland and the District of Columbia*, ed. C. S. Robbins and E. A. T. Blom. Pittsburgh: University of Pittsburgh Press.

———. 1996m. Yellow-rumped Warbler (*Dendroica coronata*). Pages 332–333 in *Atlas of the Breeding Birds of Maryland and the District of Columbia*, ed. C. S. Robbins and E. A. T. Blom. Pittsburgh: University of Pittsburgh Press.

———. 1996n. Purple Finch (*Carpodacus purpureus*). Pages 428–429 in *Atlas of the Breeding Birds of Maryland and the District of Columbia*, ed. C. S. Robbins and E. A. T. Blom. Pittsburgh: University of Pittsburgh Press.

Blumton, A. K., J. D. Fraser, and K. Terwilliger. 1989. Loggerhead Shrike survey and census. Pages 116–118 in *Virginia Nongame and Endangered Wildlife Investigations: Annual Report, July 1, 1988 through June 30, 1989*. Richmond: Virginia Department of Game and Inland Fisheries.

Bogner, H. E., and G. A. Baldassarre. 2002. Home range, movement, and nesting of Least Bitterns in western New York. *Wilson Bulletin* 114:297–308.

Bolen, G. M., E. Morton, R. Greenberg, and S. Derrickson. 2002. Mallards replacing Black Ducks: Two views. Pages 16–22 in *Black Ducks and Their Chesapeake Bay Habitats: Proceedings of a Symposium*, ed. M. C. Perry. U.S. Geological Survey Information and Technology Report USGS/BRD ITR-2002-0005.

Bollinger, E. K., and T. A. Gavin. 1992. Eastern Bobolink populations: Ecology and conservation in an agricultural landscape. Pages 497–506 in *Ecology and Conservation of Neotropical Migrant Landbirds*, ed. J. M. Hagan III and D. W. Johnston. Washington, DC: Smithsonian Institution Press.

Bond, G., and R. E. Stewart. 1951. A new Swamp Sparrow from the Maryland Coastal Plain. *Wilson Bulletin* 63:38–40.

Bonney, R. E. 1988. Winter Wren (*Troglodytes troglodytes*). Pages 304–305 in *The Atlas of Breeding Birds in New York State*, ed. R. F. Andrle and J. R. Carroll. Ithaca: Cornell University Press.

Boone, D. D. 1975. Twenty-fourth annual nest card summary, 1972. *Maryland Birdlife* 34:46.

———. 1981. Nashville Warbler nest in Garrett County. *Maryland Birdlife* 37:2–4.

———. 1984. First confirmed nesting of a Goshawk in Maryland. *Wilson Bulletin* 96:129.

———. 1996a. Northern Goshawk (*Accipiter gentilis*). Pages 102–103 in *Atlas of the Breeding Birds of Maryland and the District of Columbia*, ed. C. S. Robbins and E. A. T. Blom. Pittsburgh: University of Pittsburgh Press.

———. 1996b. Nashville Warbler (*Vermivora ruficapilla*). Pages 320–321 in *Atlas of the Breeding Birds of Maryland and the District of Columbia*, ed. C. S. Robbins and E. A. T. Blom. Pittsburgh: University of Pittsburgh Press.

Boone, D. D., and B. A. Dowell. 1986. Catoctin Mountain Park bird study. National Park Service CX-3000-4-0152, Catoctin, Md.

———. 1996. Henslow's Sparrow (*Ammodramus henslowii*). Pages 400–401 in *Atlas of the Breeding Birds of Maryland and the District of Columbia*, ed. C. S. Robbins and E. A. T. Blom. Pittsburgh: University of Pittsburgh Press.

Boone, J. E. 1982. A Maryland nesting of the Golden-crowned Kinglet. *Maryland Birdlife* 38:79–85.

Brackbill, H. 1953. Migratory status of breeding Song Sparrows at Baltimore, Maryland. *Bird-Banding* 24:68.

———. 1957. Observations on a wintering flicker. *Bird-Banding* 28:40–41.

———. 1970. Tufted Titmouse breeding behavior. *Auk* 87:522–536.

Braun, E. L. 1950. *Deciduous Forests of Eastern North America*. Philadelphia: Blackiston.

Brauning, D. W., ed. 1992. *Atlas of Breeding Birds in Pennsylvania*. Pittsburgh: University of Pittsburgh Press.

Brennan, L. A. 1999. Northern Bobwhite (*Colinus virginianus*). In *The Birds of North America*, no. 397, ed. A. Poole and F. Gill. Philadelphia and Washington, DC: Academy of Natural Sciences and American Ornithologists' Union.

Brewer, D. 2001. *Wrens, Dippers, and Thrashers: A Guide to the Wrens, Dippers, and Thrashers of the World*. New Haven, CT: Yale University Press.

Brewer, R., G. A. McPeek, and R. J. Adams Jr. 1991. *The Atlas of Breeding Birds of Michigan*. East Lansing: Michigan State University Press.

Brichetti, P., and D. Cambi. 1985. Atlante degli uccelli nidificanti in provincia di Brescia (Lombardia), 1980–1984. *Monografie di Natura Bresciana*, no. 8.

Bridge, D. 1966. The birds of Big Run, Garrett County, Maryland. *Maryland Birdlife* 22:62.

Bridge, D., and M. Riedel. 1962. Maryland nest summary for 1961. *Maryland Birdlife* 18:64–70.

Brigham, R. M. 1989. Roost and nest sites of the Common Nighthawk: Are gravel roofs important? *Condor* 91:722–724.

Brinker, D. F. 1996a. Brown Pelican (*Pelecanus occidentalis*). Pages 42–43 in *Atlas of the Breeding Birds of Maryland and the District of Columbia*, ed. C. S. Robbins and E. A. T. Blom. Pittsburgh: University of Pittsburgh Press.

———. 1996b. American Oystercatcher (*Haematopus palliatus*). Pages 142–143 in *Atlas of the Breeding Birds of Maryland and the District of Columbia*, ed. C. S. Robbins and E. A. T. Blom. Pittsburgh: University of Pittsburgh Press.

———. 1996c. Laughing Gull (*Larus atricilla*). Pages 154–155 in *Atlas of the Breeding Birds of Maryland and the District of Columbia*, ed. C. S. Robbins and E. A. T. Blom. Pittsburgh: University of Pittsburgh Press.

———. 1996d. Herring Gull (*Larus argentatus*). Pages 156–157 in *Atlas of the Breeding Birds of Maryland and the District of Columbia*, ed. C. S. Robbins and E. A. T. Blom. Pittsburgh: University of Pittsburgh Press.

———. 1996e. Great Black-backed Gull (*Larus marinus*). Pages 158–159 in *Atlas of the Breeding Birds of Maryland and the District of Columbia*, ed. C. S. Robbins and E. A. T. Blom. Pittsburgh: University of Pittsburgh Press.

———. 1996f. Gull-billed Tern (*Sterna nilotica*). Pages 160–161 in *Atlas of the Breeding Birds of Maryland and the District of Columbia*, ed. C. S. Robbins and E. A. T. Blom. Pittsburgh: University of Pittsburgh Press.

———. 1996g. Royal Tern (*Sterna maxima*). Pages 162–163 in *Atlas of the Breeding Birds of Maryland and the District of Columbia*, ed. C. S. Robbins and E. A. T. Blom. Pittsburgh: University of Pittsburgh Press.

———. 1996h. Common Tern (*Sterna hirundo*). Pages 164–165 in *Atlas of the Breeding Birds of Maryland and the District of Columbia*, ed. C. S. Robbins and E. A. T. Blom. Pittsburgh: University of Pittsburgh Press.

———. 1996i. Forster's Tern (*Sterna forsteri*). Pages 166–167 in *Atlas of the Breeding Birds of Maryland and the District of Columbia*, ed. C. S. Robbins and E. A. T. Blom. Pittsburgh: University of Pittsburgh Press.

———. 1996j. Black Skimmer (*Rhynchops niger*). Pages 170–171 in *Atlas of the Breeding Birds of Maryland and the District of Columbia*, ed. C. S. Robbins and E. A. T. Blom. Pittsburgh: University of Pittsburgh Press.

Brinker, D. F., L. A. Byrne, P. J. Tango, and G. D. Therres. 1996. Population trends of colonial nesting waterbirds on Maryland's coastal plain. Final report. Annapolis: Maryland Department of Natural Resources.

Brinker, D. F., and K. M. Dodge. 1993. Breeding biology of the Northern Saw-whet Owl in Maryland: First nest record and associated observations. *Maryland Birdlife* 49:3–15.

Brinker, D. F., G. D. Therres, P. J. Tango, M. O'Brien, E. A. T. Blom, and H. L. Wierenga. 2002. Distribution and relative abundance of breeding rails and other marshbirds in Maryland's tidal marshes. *Maryland Birdlife* 58:3–17.

Brinker, D. F., B. Williams, and B. D. Watts. 2007. Colonial nesting seabirds in the Chesapeake Bay region: Where have we been and where are we going? *Waterbirds* 30 (Special Publication 1):93–104.

Brisbin, I. L., Jr., and T. B. Mowbray. 2002. American Coot (*Fulica americana*). In *The Birds of North America*, no. 697a, ed. A. Poole and F. Gill. Philadelphia and Washington, DC: Academy of Natural Sciences and American Ornithologists' Union.

Briskie, J. V., and S. G. Sealy. 1987. Polygyny and double-brooding in the Least Flycatcher. *Wilson Bulletin* 99:492–494.

Brooks, M. G. 1936. Notes on the land birds of Garrett County, Maryland. *Bulletin of the Natural History Society of Maryland* 7:6–14.

Brown, C. R. 1997. Purple Martin (*Progne subis*). In *The Birds of North America*, no. 287, ed. A. Poole and F. Gill. Philadelphia and Washington, DC: Academy of Natural Sciences and American Ornithologists' Union.

Brown, C. R., and M. B. Brown. 1995. Cliff Swallow (*Petrochelidon pyrrhonota*). In *The Birds of North America*, no. 149, ed. A. Poole and F. Gill. Philadelphia and Washington, DC: Academy of Natural Sciences and American Ornithologists' Union.

———. 1999. Barn Swallow (*Hirundo rustica*). In *The Birds of North America*, no. 452, ed. A. Poole and F. Gill. Philadelphia and Washington, DC: Academy of Natural Sciences and American Ornithologists' Union.

Brown, R. E., and J. G. Dickson. 1994. Swainson's Warbler (*Limnothlypis swainsonii*). In *The Birds of North America*, no. 126, ed. A. Poole and F. Gill. Philadelphia and Washington, DC: Academy of Natural Sciences and American Ornithologists' Union.

Bruch, A., H. Elvers, K. Luddecke, J. Schwarz, D. Westphal, and K. Witt. 1984. *Brutvogelatlas Berlin (West)*. Sonderheft (West Germany): Ornithologischer Bericht für Berlin (West).

Bruch, A., H. Elvers, Ch. Pohl, D. Westphal, and K. Witt. 1978. *Die Vögel in Berlin (West): Eine Übersicht*. Sonderheft (West Germany): Ornithologischer Bericht für Berlin (West).

Bryant, D. M., and A. K. Turner. 1982. Central place foraging by swallows (Hirundinidae): The question of load size. *Animal Behavior* 30:845–856.

Buckelew, A. R., Jr., and G. A. Hall. 1994. *The West Virginia Breeding Bird Atlas*. Pittsburgh: University of Pittsburgh Press.

Buckley, N. J. 1999. Black Vulture (*Coragyps atratus*). In *The Birds of North America*, no. 411, ed. A. Poole and F. Gill. Philadelphia and Washington, DC: Academy of Natural Sciences and American Ornithologists' Union.

Buckley, P. A., and F. G. Buckley. 2002. Royal Tern (*Sterna maxima*). In *The Birds of North America*, no. 700, ed. A. Poole and F. Gill. Philadelphia and Washington, DC: Academy of Natural Sciences and American Ornithologists' Union.

Buehler, D. A., J. D. Fraser, J. K. D. Seegar, G. D. Therres, and M. A. Byrd. 1991. Survival rates and population dynamics of Bald Eagles on Chesapeake Bay. *Journal of Wildlife Management* 55:608–613.

Bull, E. L., and J. A. Jackson. 1995. Pileated Woodpecker (*Dryocopus pileatus*). In *The Birds of North America*, no. 148, ed. A. Poole and F. Gill. Philadelphia and Washington, DC: Academy of Natural Sciences and American Ornithologists' Union.

Bull, J. 1974. *Birds of New York State*. Garden City, N.Y.: Doubleday Press.

Burger, J. 1978. Competition between Cattle Egrets and native North American herons, egrets, and ibises. *Condor* 80:15–23.

———. 1996. Laughing Gull (*Larus atricilla*). In *The Birds of North America*, no. 225, ed. A. Poole and F. Gill. Philadelphia and Washington, DC: Academy of Natural Sciences and American Ornithologists' Union.

Burger, L. W., M. R. Ryan, T. V. Dailey, and E. W. Kurzejeski. 1995. Reproductive strategies, success, and mating systems of Northern Bobwhites in Missouri. *Journal of Wildlife Management* 59:417–426.

Burns, J. T. 1982. Nests, territories, and reproduction of Sedge Wrens (*Cistothorus platensis*). *Wilson Bulletin* 94:338–349.

Butler, R. W. 1992. Great Blue Heron (*Ardea herodias*). In *The Birds of North America*, no. 25, ed. A. Poole, P. Stettenheim, and F. Gill. Philadelphia and Washington, DC: Academy of Natural Sciences and American Ornithologists' Union.

Bystrak, P. G. 1974. The Maryland Christmas counts of 1973. *Maryland Birdlife* 30:3–8.

Cadman, M. D., P. F. J. Eagles, and F. M. Helleiner. 1987. *Atlas of the Breeding Birds of Ontario*. Waterloo, Canada: University of Waterloo Press.

Cannings, R. J. 1993. Northern Saw-whet Owl (*Aegolius acadicus*). In *The Birds of North America*, no. 42, ed. A. Poole and F. Gill. Philadelphia and Washington, DC: Academy of Natural Sciences and American Ornithologists' Union.

Carnero, J. I., and S. J. Peris. 1988. *Atlas Ornitologico de la provincia de Salamanca*. Salamanca, España (Spain): Ediciones de la Diputacion de Salamanca.

Cartrale, J. S., E. M. Hopkins, and C. E. Keller. 1998. *Atlas of Breeding Birds of Indiana*. Indianapolis: Indiana Department of Natural Resources.

Cavitt, J. F., and C. A. Haas. 2000. Brown Thrasher (*Toxostoma rufum*). In *The Birds of North America*, no. 557, ed. A. Poole and F. Gill. Philadelphia and Washington, DC: Academy of Natural Sciences and American Ornithologists' Union.

Cercle de Ornithologique de Fribourg. 1993. *Atlas des Oiseaux nicheurs du canton de Fribourg*. Fribourg (Switzerland).

Cheevers, J. W. 1996a. Eastern Phoebe (*Sayornis phoebe*). Pages 226–227 in *Atlas of the Breeding Birds of Maryland and the District of Columbia*, ed. C. S. Robbins and E. A. T. Blom. Pittsburgh: University of Pittsburgh Press.

———. 1996b. White-eyed Vireo (*Vireo griseus*). Pages 306–307 in *Atlas of the Breeding Birds of Maryland and the District of Columbia*, ed. C. S. Robbins and E. A. T. Blom. Pittsburgh: University of Pittsburgh Press.

———. 1996c. Ovenbird (*Seiurus aurocapillus*). Pages 356–357 in *Atlas of the Breeding Birds of Maryland and the District of Columbia*, ed. C. S. Robbins and E. A. T. Blom. Pittsburgh: University of Pittsburgh Press.

———. 1996d. Yellow-breasted Chat (*Icteria virens*). Pages 372–373 in *Atlas of the Breeding Birds of Maryland and the District of Columbia*, ed. C. S. Robbins and E. A. T. Blom. Pittsburgh: University of Pittsburgh Press.

———. 1996e. Scarlet Tanager (*Piranga olivacea*). Pages 376–377 in *Atlas of the Breeding Birds of Maryland and the District of Columbia*, ed. C. S. Robbins and E. A. T. Blom. Pittsburgh: University of Pittsburgh Press.

Chestem, M. 1996a. American Robin (*Turdus migratorius*). Pages 292–293 in *Atlas of the Breeding Birds of Maryland and the District of Columbia*, ed. C. S. Robbins and E. A. T. Blom. Pittsburgh: University of Pittsburgh Press.

———. 1996b. Cerulean Warbler (*Dendroica cerulea*). Pages 344–345 in *Atlas of the Breeding Birds of Maryland and the District of Colum-*

bia, ed. C. S. Robbins and E. A. T. Blom. Pittsburgh: University of Pittsburgh Press.

Ciaranca, M. A., C. C. Allin, and G. S. Jones. 1997. Mute Swan (*Cygnus olor*). In *The Birds of North America,* no. 273, ed. A. Poole and F. Gill. Philadelphia and Washington, DC: Academy of Natural Sciences and American Ornithologists' Union.

Cink, C. L. 2002. Whip-poor-will (*Caprimulgus vociferus*). In *The Birds of North America,* no. 620, ed. A. Poole and F. Gill. Philadelphia and Washington, DC: Academy of Natural Sciences and American Ornithologists' Union.

Cink, C. L., and C. T. Collins. 2002. Chimney Swift (*Chaetura pelagica*). In *The Birds of North America,* no. 646, ed. A. Poole and F. Gill. Philadelphia and Washington, DC: Academy of Natural Sciences and American Ornithologists' Union.

Clark, J. M., and J. A. Eyre. 1993. *The Birds of Hampshire.* Hampshire, UK: Hampshire Ornithological Society.

Clark, R. J. 1975. A field study of the Short-eared Owl (*Asio flammeus*) (Pontoppidan) in North America. *Wildlife Monographs,* no. 47.

Clarkson, M. F. 1996. Cedar Waxwing (*Bombycilla cedrorum*). Pages 300–301 in *Atlas of the Breeding Birds of Maryland and the District of Columbia,* ed. C. S. Robbins and E. A. T. Blom. Pittsburgh: University of Pittsburgh Press.

Clarkson, S. 1996a. Tree Swallow (*Tachycineta bicolor*). Pages 236–237 in *Atlas of the Breeding Birds of Maryland and the District of Columbia,* ed. C. S. Robbins and E. A. T. Blom. Pittsburgh: University of Pittsburgh Press.

———. 1996b. Barn Swallow (*Hirundo rustica*). Pages 244–245 in *Atlas of the Breeding Birds of Maryland and the District of Columbia,* ed. C. S. Robbins and E. A. T. Blom. Pittsburgh: University of Pittsburgh Press.

———. 1996c. Warbling Vireo (*Vireo gilvus*). Pages 312–313 in *Atlas of the Breeding Birds of Maryland and the District of Columbia,* ed. C. S. Robbins and E. A. T. Blom. Pittsburgh: University of Pittsburgh Press.

Cline, M. L., B. D. Dugger, C. R. Paine, J. D. Thompson, R. A. Montgomery, and K. M. Dugger. 2004. Factors influencing nest survival of Giant Canada Geese in northeastern Illinois. Page 84 in *Proceedings of the 2003 International Canada Goose Symposium,* ed. T. J. Moser, R. D. Lien, K. C. VerCauteren, K. F. Abraham, D. E. Andersen, J. G. Bruggink, J. M. Coluccy, D. A. Graber, J. O. Leafloor, D. R. Luukkonen, and R. E. Trost. Madison, Wisc.

Coleman, J. S., and J. D. Fraser. 1989. Habitat use and home ranges of Black and Turkey Vultures. *Journal of Wildlife Management* 53:782–792.

Collins, J. E. 2008. Black-throated Blue Warbler (*Dendroica caerulescens*). Pages 488–489 in *The Second Atlas of Breeding Birds in New York State.* ed. K. J. McGowan and K. Corwin. Ithaca: Cornell University Press.

Colvin, B. A. 1985. Common Barn Owl population decline in Ohio and the relationship to agricultural trends. *Journal of Field Ornithology* 56:224–235.

Confer, J. L. 1992. Golden-winged Warbler (*Vermivora chrysoptera*). In *The Birds of North America,* no. 20, ed. A. Poole and F. Gill. Philadelphia and Washington, DC: Academy of Natural Sciences and American Ornithologists' Union.

———. 1998. Golden-winged Warbler (*Vermivora chrysoptera*). Pages 453–455 in *Bull's Birds of New York State,* ed. E. Levine. Ithaca: Cornell University Press.

———. 2008. Blue-winged Warbler (*Vermivora pinus*). Pages 466–467 in *The Second Atlas of Breeding Birds in New York State,* ed. K. J. McGowan and K. Corwin. Ithaca: Cornell University Press.

Confer, J. L., and K. Knapp. 1981. Golden-winged Warblers and Blue-winged Warblers: The relative success of a habitat specialist and a generalist. *Auk* 98:108–114.

Conner, R. N. 1976. Nesting habitat for Red-headed Woodpeckers in southwestern Virginia. *Bird-Banding* 47:40–43.

Conner, R. N., and C. S. Adkisson. 1977. Principal component analysis of woodpecker nesting habitat. *Wilson Bulletin* 89:122–129.

Connor, P. F. 1988. Boat-tailed Grackle (*Quiscalus major*). Pages 476–477 in *The Atlas of Breeding Birds in New York State,* ed. R. F. Andrle and J. R. Carroll. Ithaca: Cornell University Press.

Conover, M. R. 1998. Reproductive biology of an urban population of Canada Geese. Pages 67–70 in *Biology and Management of Canada Geese,* ed. D. H. Rusch, M. D. Samuel, D. D. Humburg, and B. D. Sullivan. *Proceedings of the International Canada Goose Symposium.* Milwaukee, Wisc.

Conroy, M. J., M. W. Miller, and J. E. Hines. 2002. Identification and synthetic modeling of factors affecting American Black Duck populations. *Wildlife Monographs,* no. 150.

Conway, C. J., and W. R. Eddleman. 1994. Virginia Rail. Pages 193–206 in *Migratory Shore and Upland Game Bird Management in North America,* ed. T. C. Tacha and C. E. Braun. Lawrence, Kans: Allen Press.

Cooke, M. T. 1929. Birds of the Washington, DC region. *Proceedings of the Biological Society of Washington* 42:1–80.

Cooper, R. J. 1981. Relative abundance of Georgia caprimulgids based on call-counts. *Wilson Bulletin* 93:363–371.

Cooper, T. R. 2008. King Rail (*Rallus elegans*) conservation plan. Fort Snelling, Minn.: U.S. Fish and Wildlife Service.

Cooper, T. R., K. Parker, and R. D. Rau. 2008. American Woodcock population status, 2008. Laurel, MD: U.S. Fish and Wildlife Service,

Costanzo, G. R., and T. F. Bidrowski. 2004. Resident Canada Goose colonization of Chesapeake Bay islands and implication of habitat degradation. Pages 112–113 in *Proceedings of the 2003 International Canada Goose Symposium,* ed. T. J. Moser, R. D. Lien, K. C. VerCauteren, K. F. Abraham, D. E. Andersen, J. G. Bruggink, J. M. Coluccy, D. A. Graber, J. O. Leafloor, D. R. Luukkonen, and R. E. Trost. Madison, Wisc.

Costanzo, G. R., and L. J. Hindman. 2007. Chesapeake Bay breeding waterfowl populations. *Waterbirds* 30 (Special Publication 1):17–24.

Coues, E. 1870. The natural history of *Quiscalus major. Ibis* 6 n.s.: 367–378.

Coues, E., and D. W. Prentiss. 1862. List of birds ascertained to inhabit the District of Columbia with times of arrival and departure of such as are non-resident, and brief notices of habits, etc. Annual Report of the Board of Regents, Smithsonian Institution, Washington, DC.

———. 1883. Avifauna Columbiana: Being a list of Birds ascertained to inhabit the District of Columbia, with the times of arrival and departure of such as are non-residents, and brief notices of habits, etc. 2nd ed. *U.S. National Museum Bulletin,* no. 26.

Court, E. J. 1921. Some records of breeding birds for the vicinity of Washington, DC. *Auk* 38:281–282.

———. 1924. Black Vulture (*Coragyps urubu*) nesting in Maryland. *Auk* 41:475–476.

———. 1936. Four rare nesting records for Maryland. *Auk* 53:95–96.

Craig, R. J. 1984. Comparative foraging ecology of Louisiana and Northern Waterthrushes. *Wilson Bulletin* 96:173–183.

———. 1987. Divergent prey selection in two species of water-thrushes (*Seiurus*). *Auk* 104:180–187.

Cramp, S., D. J. Brooks, E. Dunn, R. Gillmor, and J. Hall-Craggs. 1988. *The Birds of the Western Palearctic.* Vol. 5, *Tyrant Flycatchers to Thrushes.* Oxford: Oxford University Press.

Crocoll, S. 2008. Northern Goshawk (*Accipiter gentilis*). Pages 196–197 in *The Second Atlas of Breeding Birds in New York State,* ed. K. J. McGowan and K. Corwin. Ithaca: Cornell University Press.

Cullom, J. 1996a. Northern Cardinal (*Cardinalis cardinalis*). Pages 378–379 in *Atlas of the Breeding Birds of Maryland and the District of Columbia,* ed. C. S. Robbins and E. A. T. Blom. Pittsburgh: University of Pittsburgh Press.

———. 1996b. Baltimore Oriole (*Icterus galbula*). Pages 426–427 in *Atlas of the Breeding Birds of Maryland and the District of Columbia,* ed. C. S. Robbins and E. A. T. Blom. Pittsburgh: University of Pittsburgh Press.

Czaplak, D. 1996. Veery (*Catharus fuscescens*). Pages 286–287 in *Atlas of the Breeding Birds of Maryland and the District of Columbia,* ed. C. S. Robbins and E. A. T. Blom. Pittsburgh: University of Pittsburgh Press.

Davidson, L. M. 1996a. American Bittern (*Botaurus lentiginosus*). Pages 46–47 in *Atlas of the Breeding Birds of Maryland and the District of Columbia,* ed. C. S. Robbins and E. A. T. Blom. Pittsburgh: University of Pittsburgh Press.

———. 1996b. Least Bittern (*Ixobrychus exilis*). Pages 48–49 in *Atlas of the Breeding Birds of Maryland and the District of Columbia,* ed. C. S. Robbins and E. A. T. Blom. Pittsburgh: University of Pittsburgh Press.

———. 1996c. Sora (*Porzana carolina*). Pages 130–131 in *Atlas of the Breeding Birds of Maryland and the District of Columbia,* ed. C. S. Robbins and E. A. T. Blom. Pittsburgh: University of Pittsburgh Press.

———. 1996d. Black-necked Stilt (*Himantopus mexicanus*). Pages 144–145 in *Atlas of the Breeding Birds of Maryland and the District of Columbia,* ed. C. S. Robbins and E. A. T. Blom. Pittsburgh: University of Pittsburgh Press.

———. 1996e. Northern Rough-winged Swallow (*Stelgidopteryx serripennis*). Pages 238–239 in *Atlas of the Breeding Birds of Maryland and the District of Columbia,* ed. C. S. Robbins and E. A. T. Blom. Pittsburgh: University of Pittsburgh Press.

———. 1996f. Loggerhead Shrike (*Lanius ludovicianus*). Pages 302–303 in *Atlas of the Breeding Birds of Maryland and the District of Columbia,* ed. C. S. Robbins and E. A. T. Blom. Pittsburgh: University of Pittsburgh Press.

———. 1996g. Kentucky Warbler (*Oporornis formosus*). Pages 362–363 in *Atlas of the Breeding Birds of Maryland and the District of Columbia,* ed. C. S. Robbins and E. A. T. Blom. Pittsburgh: University of Pittsburgh Press.

Davis, D. E. 1960. The spread of the Cattle Egret in the United States. *Auk* 77:421–424.

Davis, M. B., T. R. Simmons, M. J. Groom, J. L. Weaver, and J. R. Cordes. 2001. The breeding status of American Oystercatcher on the east coast of North America and breeding success in North Carolina. *Waterbirds* 24:195–202.

Davis, W. E., Jr. 1993. Black-crowned Night-Heron (*Nycticorax nycticorax*). In *The Birds of North America,* no. 74, ed. A. Poole and F. Gill. Philadelphia and Washington, DC: Academy of Natural Sciences and American Ornithologists' Union.

Davis, W. E., Jr., and J. Kricher. 2000. Glossy Ibis (*Plegadis falcinellus*). In *The Birds of North America,* no. 545, ed. A. Poole and F. Gill. Philadelphia and Washington, DC: Academy of Natural Sciences and American Ornithologists' Union.

Davis, W. E., Jr., and J. A. Kushlan. 1994. Green Heron (*Butorides virescens*). In *The Birds of North America,* no. 129, ed. A. Poole and F. Gill. Philadelphia and Washington, DC: Academy of Natural Sciences and American Ornithologists' Union.

Dawson, W. R. 1997. Pine Siskin (*Carduelis pinus*). In *The Birds of North America,* no. 280, ed. A. Poole and F. Gill. Philadelphia and Washington, DC: Academy of Natural Sciences and American Ornithologists' Union.

Day, J. C., M. S. Hodgson, and N. Rossiter, eds. 1995. *The Atlas of Breeding Birds in Northumbria.* Morpeth, UK: Northumbria and Tyneside Bird Club.

Day, T. M. 2006. The nesting season: Middle Atlantic region. *North American Birds* 59:577–581.

Day, T. M., and M. J. Iliff. 2004. The nesting season: Middle Atlantic region. *North American Birds* 57:475–478.

Deans, P., and J. Sankey. 1992. *An Atlas of the Breeding Birds of Shropshire.* Shrewsbury, UK: Shropshire Ornithological Society.

DeCalesta, D. S. 1994. Effect of white-tailed deer on songbirds within managed forests in Pennsylvania. *Journal of Wildlife Management* 58:711–718.

de Juana, E. A. 1980. *Atlas Ornitologico de La Rioja.* Logroño, España (Spain): Instituto de Estudios Riojanos.

Dennis, J. V. 1948. Observations on the Orchard Oriole in the lower Mississippi Delta. *Bird-Banding* 19:12–21.

Dennis, M. K. 1996. *Tetrad Atlas of the Breeding Birds of Essex.* Colchester, UK: Essex Birdwatching Society.

Devilliers, P., W. Roggeman, J. Tricot, P. del Marmol, C. Kerwijn, J. P. Jacob, and A. Anselin. 1988. *Atlas des Oiseaux Nicheurs de Belgique.* Bruxelles (Belguim): Institut Royal des Sciences Naturelles de Belgique.

Dewey, S. R., P. L. Kennedy, and R. M. Stephens. 2003. Are dawn vocalization surveys effective for monitoring Goshawk nest-area occupancy? *Journal of Wildlife Management* 67:390–397.

Dingle, E. von S. 1942. Rough-winged Swallow. Pages 424–33 in *Life Histories of North American Flycatchers, Larks, Swallows, and Their Allies,* ed. A. C. Bent. U.S. National Museum Bulletin, no. 179.

Dinsmore, S. J., and A. Farnsworth. 2006. The changing seasons: Weather birds. *North American Birds* 60:14–26.

Dow, D. D., and D. M. Scott. 1971. Dispersal and range expansion by the Cardinal: An analysis of banding records. *Canadian Journal of Zoology* 49:185–198.

Dowell, B. A. 1996a. Cliff Swallow (*Hirundo pyrrhonota*). Pages 242–243 in *Atlas of the Breeding Birds of Maryland and the District of Columbia,* ed. C. S. Robbins and E. A. T. Blom. Pittsburgh: University of Pittsburgh Press.

———. 1996b. Rose-breasted Grosbeak (*Pheucticus ludovicianus*). Pages 380–381 in *Atlas of the Breeding Birds of Maryland and the District of Columbia,* ed. C. S. Robbins and E. A. T. Blom. Pittsburgh: University of Pittsburgh Press.

———. 1996c. Bobolink (*Dolichonyx oryzivorus*). Pages 412–413 in *Atlas of the Breeding Birds of Maryland and the District of Columbia,* ed. C. S. Robbins and E. A. T. Blom. Pittsburgh: University of Pittsburgh Press.

Drilling, N., R. Titman, and F. McKinney. 2002. Mallard (*Anas platyrhynchos*). In *The Birds of North America,* no. 658, ed. A. Poole and F. Gill. Philadelphia and Washington, DC: Academy of Natural Sciences and American Ornithologists' Union.

Droege, S., and E. A. T. Blom. 1996. Swamp Sparrow (*Melospiza georgiana*). Pages 408–409 in *Atlas of the Breeding Birds of Maryland and the District of Columbia,* ed. C. S. Robbins and E. A. T. Blom. Pittsburgh: University of Pittsburgh Press.

Duffy, K., and P. Kerlinger. 1992. Autumn owl migration at Cape May Point, New Jersey. *Wilson Bulletin* 104:312–320.

Dugger, B. D., K. M. Dugger, and L. H. Fredrickson. 1994. Hooded Merganser (*Lophodytes cucullatus*). In *The Birds of North America*, no. 98, ed. A. Poole and F. Gill. Philadelphia and Washington, DC: Academy of Natural Sciences and American Ornithologists' Union.

Duncan, C. D. 1996. Changes in the winter abundance of Sharp-shinned Hawks in New England. *Journal of Field Ornithology* 67:254–262.

Dunn, E. H., and D. L. Tessaglia-Hymes. 1999. *Birds at Your Feeder: A Guide to Feeding Habits, Behavior, Distribution, and Abundance.* New York: Norton.

Dunn, J. L., and E. A. T. Blom. 1983. *Field Guide to the Birds of North America.* Washington, DC: National Geographic Society.

Dupree, D. C. 1996a. Belted Kingfisher (*Ceryle alcyon*). Pages 200–201 in *Atlas of the Breeding Birds of Maryland and the District of Columbia*, ed. C. S. Robbins and E. A. T. Blom. Pittsburgh: University of Pittsburgh Press.

———. 1996b. House Wren (*Troglodytes aedon*). Pages 272–273 in *Atlas of the Breeding Birds of Maryland and the District of Columbia*, ed. C. S. Robbins and E. A. T. Blom. Pittsburgh: University of Pittsburgh Press.

———. 1996c. Eastern Bluebird (*Sialia sialis*). Pages 284–285 in *Atlas of the Breeding Birds of Maryland and the District of Columbia*, ed. C. S. Robbins and E. A. T. Blom. Pittsburgh: University of Pittsburgh Press.

———. 1996d. House Sparrow (*Passer domesticus*). Pages 436–437 in *Atlas of the Breeding Birds of Maryland and the District of Columbia*, ed. C. S. Robbins and E. A. T. Blom. Pittsburgh: University of Pittsburgh Press.

Dvorak, M., A. Ranner, and H-M. Berg, eds. 1993. *Atlas der Brutvogel Osterreichs.* Wien (Austria): Umweltbundesamt.

Dybbro, T. 1976. *De Danske Ynglefugles Udbredelse.* Copenhagen: Dansk Ornitologisk Forening.

Dykstra, C. R., J. L. Hays, and S. T. Crocoll. 2008. Red-shouldered Hawk (*Buteo lineatus*). In *The Birds of North America Online*, no. 107, ed. A. Poole. Ithaca: Cornell Lab of Ornithology. Available at http://bna.birds.cornell.edu/bna/species/107.

Eaton, S. W. 1992. Wild Turkey (*Meleagris gallopavo*). In *The Birds of North America*, no. 22, ed. A. Poole, P. Stettenheim, and F. Gill. Philadelphia nad Washington, DC: Academy of Natural Sciences and American Ornithologists' Union.

Eckerle, K. P., and C. F. Thompson. 2001. Yellow-breasted Chat (*Icteria virens*). In *The Birds of North America*, no. 575, ed. A. Poole and F. Gill. Philadelphia and Washington, DC: Academy of Natural Sciences and American Ornithologists' Union.

Eddleman, W. R., and C. J. Conway. 1994. Clapper Rail. Pages 167–179 in *Migratory Shore and Upland Game Bird Management in North America*, ed. T. C. Tacha and C. E. Braun. Lawrence, Kans.: Allen Press.

Eddleman, W. R., K. E. Evans, and W. H. Elder. 1980. Habitat characteristics and management of Swainson's Warbler in southern Illinois. *Wildlife Society Bulletin* 8:228–233.

Ehrlich, P. R., E. S. Dobkin, and D. Wheye. 1988. *The Birder's Handbook: A Guide to the Natural History of North American Birds.* New York: Simon and Schuster.

Eifrig, C. W. G. 1904. Birds of Allegany and Garrett counties, western Maryland. *Auk* 21:234–250.

———. 1905. Nesting of the Raven (*Corvus corax principalis*) at Cumberland, Maryland. *Auk* 22:312.

———. 1909. Additions to the List of Birds of Allegany and Garrett Counties, western Maryland. *Auk* 26:437–438.

———. 1915. Notes on some birds of the Maryland Alleghenies: An anomaly in the check-list. *Auk* 32:108–110.

———. 1920. In the haunts of Cairns' Warbler. *Auk* 37:551–558.

———. 1923. Prairie Horned Lark (*Otocoris alpestris praticola*) in Maryland in summer. *Auk* 40:126.

Elliot, J. J., and R. S. Arbib Jr. 1953. Origin and status of the House Finch in the eastern United States. *Auk* 70:31–37.

Ellison, W. G. 1985a. Rose-breasted Grosbeak (*Pheucticus ludovicianus*). Pages 332–333 in *The Atlas of Breeding Birds of Vermont*, ed. S. B. Laughlin and D. P. Kibbe. Hanover, N.H.: University Press of New England.

———. 1985b. Orchard Oriole (*Icterus spurius*). Pages 370–371 in *The Atlas of Breeding Birds of Vermont*, ed. S. B. Laughlin and D. P. Kibbe. Hanover, N.H.: University Press of New England.

———. 1992a. Different drummers: Identifying the rhythms of northeastern woodpeckers. *Birding* 24:350–355.

———. 1992b. Blue-gray Gnatcatcher (*Polioptila caerulea*). In *The Birds of North America*, no. 23, ed. A. Poole and F. Gill. Philadelphia and Washington, DC: Academy of Natural Sciences and American Ornithologists' Union.

———. 1993. Historical patterns of vagrancy by Blue-gray Gnatcatchers in New England. *Journal of Field Ornithology* 64:358–366.

Ellison, W. G., and N. L. Martin. 2007. Fall migration: New England region. *North American Birds* 61:36–41.

Elosegui, J. 1985. *Atlas de las aves nidificantes de Navarra.* Navarra, España (Spain): Caja de Ahorros de Navarra.

Emlen, J. T., Jr. 1954. Territory, nest building, and pair formation in the Cliff Swallow. *Auk* 71:16–35.

Emslie, S. D., J. S. Weske, M. M. Browne, S. Cameron, R. Boettcher, D. F. Brinker, and W. Golder. 2009. Population trends in Royal and Sandwich Terns along the Mid-Atlantic seaboard, USA, 1975–2005. *Waterbirds* 32:54–63.

Enser, R. W. 1992. *The Atlas of Breeding Birds in Rhode Island.* Providence: Rhode Island Department of Environmental Management.

Erskine, A. J. 1971. Growth and annual cycles in weights, plumages, and reproductive organs in goosanders in eastern Canada. *Ibis* 113:42–58.

———. 1992. *Atlas of Breeding Birds of the Maritime Provinces.* Halifax, Nova Scotia: Nimbus Publishing.

Erwin, R. M., J. G. Haig, D. B. Stotts, and J. S. Hatfield 1996. Reproductive success, growth and survival of Black-crowned Night Heron (*Nycticorax nycticorax*) and Snowy Egret (*Egretta thula*) chicks in coastal Virginia. *Wilson Bulletin* 108:342–356.

Erwin, R. M., and C. E. Korschgen 1979. Coastal waterbird colonies: Maine to Virginia, 1977. U.S. Fish and Wildlife Service FWS/OBS-79/08.

Erwin, R. M., G. M. Sanders, D. J. Prosser, and D. R. Cahoon. 2006. High tides and rising seas: Potential effects on estuarine waterbirds. *Studies in Avian Biology* 32:214–228.

Erwin, R. M., B. R. Truit, and J. E. Jimenez. 2001. Ground nesting waterbirds and mammalian carnivores in the Virginia barrier island region: Running out of options. *Journal of Coastal Research* 17:292–296.

Evans, D. R., and J. E. Gates. 1997. Cowbird selection of breeding areas: The role of habitat and bird species abundance. *Wilson Bulletin* 109:470–480.

Evarts, S. 2005. Blue-winged Teal (*Anas discors*). Pages 545–549 in

Bird Families of the World: Ducks, Geese, and Swans. Vol. 2. Ed. J. Kear. Oxford: Oxford University Press.

Evens, J., and G. W. Page. 1986. Predation on Black Rails during high tides in salt marshes. *Condor* 88:107–109.

Fables, D., Jr. 1955. *Annotated List of New Jersey Birds.* Roselle Park, N.J.: Urner Ornithological Club.

Fallon, F. W. 1996. American Redstart (*Setophaga ruticilla*). Pages 348–349 in *Atlas of the Breeding Birds of Maryland and the District of Columbia,* ed. C. S. Robbins and E. A. T. Blom. Pittsburgh: University of Pittsburgh Press.

Farmer, G. C., K. McCarty, B. Robertson, S. Robertson, and K. L. Bildstein. 2006. Suspected predation by accipiters on radio-tracked American Kestrels (*Falco sparverius*) in eastern Pennsylvania, U.S.A. *Journal of Raptor Research* 40:294–297.

Farrand, J., Jr. 1988. The changing seasons. *American Birds* 42:221–226.

Farrell, J. H. 1996a. Solitary Vireo (*Vireo solitarius*). Pages 308–309 in *Atlas of the Breeding Birds of Maryland and the District of Columbia,* ed. C. S. Robbins and E. A. T. Blom. Pittsburgh: University of Pittsburgh Press.

———. 1996b. Eastern Towhee (*Pipilo erythrophthalmus*). Pages 388–389 in *Atlas of the Breeding Birds of Maryland and the District of Columbia,* ed. C. S. Robbins and E. A. T. Blom. Pittsburgh: University of Pittsburgh Press.

———. 1996c. Common Grackle (*Quiscalus quiscula*). Pages 420–421 in *Atlas of the Breeding Birds of Maryland and the District of Columbia,* ed. C. S. Robbins and E. A. T. Blom. Pittsburgh: University of Pittsburgh Press.

Figley, W. K., and L. W. VanDruff. 1982. Ecology of urban Mallards. *Wildlife Monographs,* no. 81.

Fisk, E. J. 1978. Roof-nesting terns, skimmers, and plovers in Florida. *Florida Field Naturalist* 6:1–8.

Fitzpatrick, J. W. 1980. Wintering of North American Tyrant Flycatchers in the neotropics. Pages 67–78 in *Migrant Birds in the Neotropics,* ed. A. Keast and E. S. Morton. Washington, DC: Smithsonian Institution Press.

Fleming, W. J., and D. R. Petit. 1986. Modified milk carton nest box for studies of Prothonotary Warblers. *Journal of Field Ornithology* 57:313–315.

Fletcher, A. J. 1996a. Horned Lark (*Eremophila alpestris*). Pages 232–233 in *Atlas of the Breeding Birds of Maryland and the District of Columbia,* ed. C. S. Robbins and E. A. T. Blom. Pittsburgh: University of Pittsburgh Press.

———. 1996b. Chipping Sparrow (*Spizella passerina*). Pages 390–391 in *Atlas of the Breeding Birds of Maryland and the District of Columbia,* ed. C. S. Robbins and E. A. T. Blom. Pittsburgh: University of Pittsburgh Press.

Fletcher, A. J., and J. H. Farrell. 1996. Killdeer (*Charadrius vociferus*). Pages 140–141 in *Atlas of the Breeding Birds of Maryland and the District of Columbia,* ed. C. S. Robbins and E. A. T. Blom. Pittsburgh: University of Pittsburgh Press.

Fletcher, A. J., and R. B. Fletcher. 1959. Maryland nest summary for 1958. *Maryland Birdlife* 15:5–12.

Fletcher, A. J., and E. D. Joyce 1996. Brown-headed Cowbird (*Molothrus ater*). Pages 422–423 in *Atlas of the Breeding Birds of Maryland and the District of Columbia,* ed. C. S. Robbins and E. A. T. Blom. Pittsburgh: University of Pittsburgh Press.

Forbush, E. H. 1925. *Birds of Massachusetts and Other New England States.* Vol. 1, *Water Birds, Marsh Birds, and Shore Birds.* Boston: Massachusetts Department of Agriculture.

———. 1927. *Birds of Massachusetts and Other New England States.* Vol. 2, *Land Birds from Bobwhites to Grackles.* Boston: Massachusetts Department of Agriculture.

———. 1929. *Birds of Massachusetts and other New England States.* Vol. 3, *Land Birds from Sparrows to Thrushes.* Boston: Massachusetts Department of Agriculture.

Foss, C. R., ed. 1994. *Atlas of Breeding Birds in New Hampshire.* Concord: Audubon Society of New Hampshire.

Fox, T. 2005. Gadwall (*Anas strepera*). Pages 491–494 in *Bird Families of the World: Ducks, Geese, and Swans.* Vol. 2. Ed. J. Kear. Oxford: Oxford University Press.

Fraissinet, M., and E. Caputo. 1995. Atlante ornitologico degli uccelli nidificanti e svernanti nella città di Napoli. Salerno, Italia. Monografia 4 dell'Associazione Studi Ornitologici Italia Meridionota, Electa, Napoli, Italia (Italy).

Fraissinet, M., and M. Kalby, eds. 1989. Atlante degli uccelli nidificanti in Campania, 1983–1987. Assessorato agricoltura, caccia, pesca e le foreste, Regione Campania, Italia (Italy).

Frederick, P. C. 1997. Tricolored Heron (*Egretta tricolor*). In *The Birds of North America,* no. 306, ed. A. Poole and F. Gill. Philadelphia and Washington, DC: Academy of Natural Sciences and American Ornithologists' Union.

Frederickson, L. H. 1971. Common Gallinule breeding biology and development. *Auk* 88:914–919.

Friedmann, H. 1929. *The Cowbirds: A Study in the Biology of Social Parasitism.* Springfield, Ill.: C. Thomas Publishers.

Friedmann, H., and L. F. Kiff. 1985. The parasitic cowbirds and their hosts. *Proceedings of the Western Foundation of Vertebrate Zoology* 2:225–304.

Frohling, R. C. 1965. American Oystercatcher and Black Skimmer nesting on salt marsh. *Wilson Bulletin* 77:193–194.

Galati, R. 1991. *Golden-crowned Kinglets: Treetop Nesters of the North Woods.* Ames: Iowa State University Press.

Galati, R., and C. Galati. 1985. Breeding of the Golden-crowned Kinglet in northern Minnesota. *Journal of Field Ornithology* 56:28–40.

Gale, G. A., L. A. Hanners, and S. R. Patton. 1997. Reproductive success of Worm-eating Warblers in a forest landscape. *Conservation Biology* 11:246–250.

Gamble, L. R., and T. M. Bergin. 1996. Western Kingbird (*Tyrannus verticalis*). In *The Birds of North America,* no. 227, ed. A. Poole and F. Gill. Philadelphia and Washington, DC: Academy of Natural Sciences and American Ornithologists' Union.

Garland, M. 1963. First Maryland nest of House Finch. *Maryland Birdlife* 19:78.

Gates, J. E., D. F. Brinker, and J. E. McKearnan. 1992. *Maryland Waterbird Study. Final Report.* Frostburg, Md.: Appalachian Environmental Laboratory.

Gavin, T. A. 1984. Broodedness in Bobolinks. *Auk* 101:179–181.

Geister, I. 1995. *Ornitoloski atlas Slovenija.* Ljubljana (Slovenia): DZS.

Ghalambor, C. K., and T. E. Martin. 1999. Red-breasted Nuthatch (*Sitta canadensis*). In *The Birds of North America,* no. 459, ed. A. Poole and F. Gill. Philadelphia and Washington, DC: Academy of Natural Sciences and American Ornithologists' Union.

Gibbons, D. W., J. B. Reid, and R. A. Chapman. 1993. *The New Atlas of Breeding Birds in Britain and Ireland, 1988–1991.* London: T. and A. D. Poyser.

Gibbs, J. P., and J. Faaborg. 1990. Estimating the viability of Ovenbird and Kentucky Warbler populations in forest fragments. *Conservation Biology* 4:193–196.

Gill, D. E., P. Blank, J. Parks, J. B. Guerard, B. Lohr, E. Schwartzman, J. G. Gruber, G. Dodge, C. A. Rewa, and H. F. Sears. 2006.

Plants and breeding bird response on a managed Conservation Reserve Program grassland in Maryland. *Wildlife Society Bulletin* 34:944–956.

Gill, F. B. 1980. Historical aspects of hybridization between Blue-winged and Golden-winged Warblers. *Auk* 97:1–18.

———. 1992. Golden-winged Warbler (*Vermivora chrysoptera*). Pages 300–301 in *Atlas of Breeding Birds in Pennsylvania*, ed. D. W. Brauning. Pittsburgh: University of Pittsburgh Press.

Gill, F. B., R. A. Canterbury, and J. L. Confer. 2001. Blue-winged Warbler (*Vermivora pinus*). In *The Birds of North America*, no. 584, ed. A. Poole and F. Gill. Philadelphia and Washington, DC: Academy of Natural Sciences and American Ornithologists' Union.

Giudice, J. H., and J. T. Ratti. 2001. Ring-necked Pheasant (*Phasianus colchicus*). In *The Birds of North America*, no. 572, ed. A. Poole and F. Gill. Philadelphia and Washington, DC: Academy of Natural Sciences and American Ornithologists' Union.

Gjerdrum, C., C. S. Elphick, and M. Rubega. 2005. Nest site selection and nesting success in saltmarsh breeding sparrows: The importance of nest habitat, timing, and study site differences. *Condor* 107:849–862.

Gjershaug, J. O., P. G. Thingstad, S. Eldoy, and S. Byrkjeland, eds. 1994. *Norsk Fugleatlas*. Klaebu (Norway): Norsk Ornitologisk Forening.

Glayre, D., and D. Magnenat. 1984. Oiseaux de la haute vallée de l'Orbe. Nos Oiseaux, no. 398. Prangins (Switzerland).

Glue, D. E. 1993. Short-eared Owl (*Asio flammeus*). Pages 254–255 in *The New Atlas of Breeding Birds in Britain and Ireland, 1988–1991*, ed. D. Wingfield Gibbons, J. B. Reid, and R. A. Chapman. Staffordshire, UK: T. and A. D. Poyser.

Godfrey, W. E. 1966. *The Birds of Canada*. 1st ed. Ottawa: National Museums of Canada.

———. 1986. *The Birds of Canada*. Rev. ed. Ottawa: National Museum of Natural Sciences, National Museums of Canada.

Goetz, W. J. 1981. Nesting Red Crossbills in Rockingham County, Virginia. *Redstart* 48:90–92.

Good, T. P. 1998. Great Black-backed Gull (*Larus marinus*). In *The Birds of North America*, no. 330, ed. A. Poole and F. Gill. Philadelphia and Washington, DC: Academy of Natural Sciences and American Ornithologists' Union.

Goodrich, L. 1992. Sharp-shinned Hawk (*Accipiter striatus*). Pages 96–97 in *Atlas of Breeding Birds in Pennsylvania*, ed. D. W. Brauning. Pittsburgh: University of Pittsburgh Press.

Gorman, G. 1996. *The Birds of Hungary*. London: Christopher Helm.

Graves, G. R. 2001. Factors governing the distribution of Swainson's Warbler along a hydrological gradient in Great Dismal Swamp. *Auk* 118:650–664.

Greenberg, R. 1987. Seasonal foraging specialization in the Worm-eating Warbler. *Condor* 89:158–168.

Greenberg, R., P. Bichier, A. Cruz Angon, and R. Reitsma. 1997. Bird populations in shade and sun coffee plantations in central Guatemala. *Conservation Biology* 11:448–459.

Greenberg, R., and S. Droege. 1990. Adaptations to tidal marshes in breeding populations of the Swamp Sparrow. *Condor* 92:393–404.

Greenberg, R., P. P. Marra, and M. J. Wooler. 2007. Stable-isotope (C, N, H) analyses help locate the winter range of the Coastal Plain Swamp Sparrow (*Melospiza georgiana nigrescens*). *Auk* 124:1137–1148.

Greenlaw, J. S. 1996. Eastern Towhee (*Pipilo erythrophthalmus*). In *The Birds of North America*, no. 262, ed. A. Poole and F. Gill. Phil-adelphia and Washington, DC: Academy of Natural Sciences and American Ornithologists' Union.

Gregoire, J. A. 1996. Red-tailed Hawk (*Buteo jamaicensis*). Pages 108–109 in *Atlas of the Breeding Birds of Maryland and the District of Columbia*, ed. C. S. Robbins and E. A. T. Blom. Pittsburgh: University of Pittsburgh Press.

Gregoire, J. A., and D. F. Brinker. 1996. Broad-winged Hawk (*Buteo platypterus*). Pages 106–107 in *Atlas of the Breeding Birds of Maryland and the District of Columbia*, ed. C. S. Robbins and E. A. T. Blom. Pittsburgh: University of Pittsburgh Press.

Gregoire, J. A., and G. D. Therres. 1996a. Black Vulture (*Coragyps atratus*). Pages 88–89 in *Atlas of the Breeding Birds of Maryland and the District of Columbia*, ed. C. S. Robbins and E. A. T. Blom. Pittsburgh: University of Pittsburgh Press.

———. 1996b. Turkey Vulture (*Cathartes aura*). Pages 90–91 in *Atlas of the Breeding Birds of Maryland and the District of Columbia*, ed. C. S. Robbins and E. A. T. Blom. Pittsburgh: University of Pittsburgh Press.

Greij, E. D. 1994. Common Moorhen. Pages 145–157 in *Migratory Shore and Upland Game Bird Management in North America*, ed. T. C. Tacha and C. E. Braun. Lawrence, Kans: Allen Press.

Grinnell, J., and A. H. Miller. 1944. The distribution of the birds of California. *Pacific Coast Avifauna*, no. 27.

Griscom, L. 1923. *Birds of the New York City Region*. New York: American Museum of Natural History.

Griscom, L., and D. E. Snyder. 1955. *The Birds of Massachusetts: An Annotated and Revised Check List*. Salem, Mass.: Peabody Museum.

Gross, A. O. 1921. The Dickcissel (*Spiza americana*) of the Illinois prairies, part 2. *Auk* 38:163–184.

———. 1940. Eastern Nighthawk. Pages 206–234 in *Life Histories of North American Cuckoos, Goatsuckers, Hummingbirds, and Their Allies*, ed. A. C. Bent. U.S. National Museum Bulletin, no. 176.

———. 1942. Northern Cliff Swallow. Pages 463–484 in *Life Histories of North American Flycatchers, Larks, Swallows, and Their Allies*, ed. A. C. Bent. U.S. National Museum Bulletin, no. 179.

———. 1956. The recent reappearance of the Dickcissel (*Spiza americana*) in eastern North America. *Auk* 73:66–70.

Gross, D. A. 1992a. Yellow-rumped Warbler (*Dendroica coronata*). Pages 316–317 in *Atlas of Breeding Birds in Pennsylvania*, ed. D. W. Brauning. Pittsburgh: University of Pittsburgh Press.

———. 1992b. Pine Siskin (*Carduelis pinus*). Pages 416–417 in *Atlas of Breeding Birds in Pennsylvania*, ed. D. W. Brauning. Pittsburgh: University of Pittsburgh Press.

Groth, J. G. 1988. Resolution of cryptic species of Appalachian Red Crossbills. *Condor* 90:745–760.

———. 1993a. Evolutionary differentiation in morphology, vocalizations, and allozymes among nomadic sibling species in the North American Red Crossbill (*Loxia curvirostra*) complex. *University of California Publication in Zoology* 127:1–143.

———. 1993b. Call matching and positive assortative mating in Red Crossbills. *Auk* 110:398–401.

Guenzani, W., and F. Saporetti, eds. 1988. *Atlante degli uccelli nidificanti in Provincia di Varese (Lombardia), 1983–1987*. Varese, Italia (Italy): Edizioni Lativa.

Guest, J. P., D. Elphick, J. S. A. Hunter, and D. Norman. 1992. *The Breeding Bird Atlas of Cheshire and Wirral*. Cheshire, UK: Cheshire and Wirral Ornithological Society.

Guthery, F. S., and W. P. Kuvlesky Jr. 1998. The effect of multiple brooding on age ratios of quail. *Journal of Wildlife Management* 62:540–549.

Hader, R. J. 1975. Goshawks in Avery County, N.C. *Chat* 39:18–19.

Hagemeijer, W. J. M., and M. J. Blair, eds. 1997. *The EBCC Atlas of European Breeding Birds: Their Distribution and Abundance.* London: T. and A. D. Poyser.

Haggerty, T. M., and E. S. Morton. 1995. Carolina Wren (*Thryothorus ludovicianus*). In *The Birds of North America,* no. 188, ed. A. Poole and F. Gill. Philadelphia and Washington, DC: Academy of Natural Sciences and American Ornithologists' Union.

Haggerty, T. M., E. S. Morton, and R. C. Fleischer. 2001. Genetic monogamy in Carolina Wrens (*Thryothorus ludovicianus*). *Auk* 118:215–219

Haig, S. M. 1992. Piping Plover (*Charadrius melodus*). In *The Birds of North America,* no. 2, ed. A. Poole, P. Stettenheim, and F. Gill. Philadelphia and Washington, DC: Academy of Natural Sciences and American Ornithologists' Union.

Haig S. M., and L. W. Oring. 1985. Distribution and status of the Piping Plover throughout the annual cycle. *Journal of Field Ornithology* 56:334–345.

Halkin, S. L., and S. U. Linville. 1999. Northern Cardinal (*Cardinalis cardinalis*). In *The Birds of North America,* no. 440, ed. A. Poole and F. Gill. Philadelphia and Washington, DC: Academy of Natural Sciences and American Ornithologists' Union.

Hall, G. A. 1983. *West Virginia Birds: Distribution and Ecology.* Pittsburgh: Carnegie Museum of Natural History Special Publication No. 7.

———. 1994. Magnolia Warbler (*Dendroica magnolia*). In *The Birds of North America,* no. 136, ed. A. Poole and F. Gill. Philadelphia and Washington, DC: Academy of Natural Sciences and American Ornithologists' Union.

———. 1996. Yellow-throated Warbler (*Dendroica dominica*). In *The Birds of North America,* no. 223, ed. A. Poole and F. Gill. Philadelphia and Washington, DC: Academy of Natural Sciences and American Ornithologists' Union.

Halle, L. J. 1943. The Veery breeding in Washington, DC. *Auk* 60:104.

———. 1947. *Spring in Washington.* Baltimore: Johns Hopkins University Press.

Hallworth, M., A. Ueland, E. Anderson, D. J. Lambert, and L. Reitsma. 2008. Habitat selection and site fidelity of Canada Warblers (*Wilsonia canadensis*) in central New Hampshire. *Auk* 125:880–888.

Hamas, M. J. 1974. Human incursion and nesting sites of the Belted Kingfisher. *Auk* 91:835–836.

Hamel. P. B. 2000. Cerulean Warbler (*Dendroica cerulea*). In *The Birds of North America,* no. 511, ed. A. Poole and F. Gill. Philadelphia and Washington, DC: Academy of Natural Sciences and American Ornithologists' Union.

Hamilton, R. B. 1975. Comparative behavior of the American Avocet and the Black-necked Stilt (Recurvirostridae). *Ornithological Monographs,* no. 17.

Hamilton, W. J., Jr. 1933. A late nesting waxwing in central New York. *Auk* 50:114–115.

Hampe, I. E., and H. Kolb. 1947. *Preliminary List of the Birds of Maryland and the District of Columbia.* Baltimore: Natural History Society of Maryland.

Hanners, L. A., and S. R. Patton. 1998. Worm-eating Warbler (*Helmitheros vermivorum*). In *The Birds of North America,* no. 367, ed. A. Poole and F. Gill. Philadelphia and Washington, DC: Academy of Natural Sciences and American Ornithologists' Union.

Hansen, A. J., and S. Rohwer. 1986. Coverable badges and resource defense in birds. *Animal Behavior* 34:69–76.

Haramis, G. M. 1991. Wood Duck (*Aix sponsa*). Pages 15-1–15-11 in *Habitat Requirements for Chesapeake Bay Living Resources,* ed, S. L. Funderburk, S. J. Jordan, J. A. Milhursky, and D. Riley. 2nd ed. Solomons, Md.: Chesapeake Research Consortium.

Haramis, G. M., D. G. Jorde, G. H. Olsen, D. B. Stotts, and M. K. Harrison. 2002. Breeding productivity of Smith Island Black Ducks. Pages 22–30 in *Black Ducks and Their Chesapeake Bay Habitats: Proceedings of a Symposium,* ed. M. C. Perry. U.S. Geological Survey Information and Technology Report USGS/BRD ITR-2002-0005.

Haramis, G. M., and G. D. Kearns. 2000. A radio transmitter attachment technique for Soras. *Journal of Field Ornithology* 71:135–139.

———. 2007a. Soras in tidal marsh: Banding and telemetry studies on the Patuxent River, Maryland. *Waterbirds* 30 (Special Publication 1):105–121.

———. 2007b. Herbivory by resident geese: The loss and recovery of wild rice along the tidal Patuxent River. *Journal of Wildlife Management* 71:788–794.

Harding, B. D. 1979. *Bedfordshire Bird Atlas, 1968 to 1977.* Luton, UK: Bedfordshire Natural History Society.

Harrison, H. H. 1975. *A Field Guide to Birds' Nests in the United States East of the Mississippi River.* Boston: Houghton Mifflin.

———. 1984. *Wood Warblers' World.* New York: Simon and Schuster.

Harrison, K., ed. 1995. *An Atlas of Breeding Birds of Lancaster and District.* Preston, UK: Lancaster and District Birdwatching Society.

Harvey, D. A., and J. K. Solem. 1996. Song Sparrow (*Melospiza melodia*). Pages 406–407 in *Atlas of the Breeding Birds of Maryland and the District of Columbia,* ed. C. S. Robbins and E. A. T. Blom. Pittsburgh: University of Pittsburgh Press.

Hatch, J. W., and D. V. Weseloh. 1999. Double-crested Cormorant (*Phalacrocorax auritus*). In *The Birds of North America,* no. 441, ed. A. Poole and F. Gill. Philadelphia and Washington, DC: Academy of Natural Sciences and American Ornithologists' Union.

Hatfield, J. S., S. A. Ricciardi, G. A. Gough, D. Bystrak, S. Droege, and C. S. Robbins. 1994. Distribution and abundance of birds wintering in Maryland, 1988–1993. *Maryland Birdlife* 50:3–83.

Heckenroth, H. 1985. *Atlas der Brutvögel Niedersachsens, 1980.* Hannover (West Germany): Naturschutz und Laudschaftspflege in Niedersachsen.

Heckscher, C. M. 2000. Forest-dependent birds of the Great Cypress (North Pocomoke) Swamp: Species composition and implications for conservation. *Northeastern Naturalist* 7:113–130.

Heckscher, C. M., and J. M. McCann. 2006. Status of Swainson's Warbler on the Delmarva Peninsula. *Northeastern Naturalist* 13:521–530.

Hejl, S. J., J. A. Holmes, and D. E. Kroodsma. 2002. Winter Wren (*Troglodytes troglodytes*). In *The Birds of North America,* no. 623, ed. A. Poole and F. Gill. Philadelphia and Washington, DC: Academy of Natural Sciences and American Ornithologists' Union.

Hejl, S. J., K. R. Newlon, M. E. Mcfadzen, J. S. Young, and C. K. Ghalambor. 2002. Brown Creeper (*Certhia americana*). In *The Birds of North America,* no. 669, ed. A. Poole and F. Gill. Philadelphia and Washington, DC: Academy of Natural Sciences and American Ornithologists' Union.

Helm, R. N., D. N. Pashley, and P. J. Zwank. 1987. Notes on the nesting of the Common Moorhen and Purple Gallinule in southwestern Louisiana. *Journal of Field Ornithology* 58:55–61.

Hengeveld, R. 1993. What to do about the North American invasion by the Collared-Dove. *Journal of Field Ornithology* 64:477–489.

Henny, C. J., and N. E. Holgerson. 1974. Range expansion and population increase of the Gadwall in eastern North America. *Wildfowl* 25:95–101.

Hepp, G. R., and F. C. Bellrose. 1995. Wood Duck (*Aix sponsa*). In *The Birds of North America,* no. 169, ed. A. Poole and F. Gill. Philadelphia and Washington, DC: Academy of Natural Sciences and American Ornithologists' Union.

Herkert, J. R. 1995. Status and habitat requirements of the Veery in Illinois. *Auk* 112:794–797.

Herkert, J. R., D. E. Kroodsma, and J. P. Gibbs. 2001. Sedge Wren (*Cistothorus platensis*). In *The Birds of North America,* no. 582, ed. A. Poole and F. Gill. Philadelphia and Washington, DC: Academy of Natural Sciences and American Ornithologists' Union.

Herkert, J. R., P. D. Vickery, and D. E. Kroodsma. 2002. Henslow's Sparrow (*Ammodramus henslowii*). In *The Birds of North America,* no. 672, ed. A. Poole and F. Gill. Philadelphia and Washington, DC: Academy of Natural Sciences and American Ornithologists' Union.

Hess, G. K., R. L. West, M. V. Barnhill III, and L. M. Fleming. 2000. *Birds of Delaware.* Pittsburgh: University of Pittsburgh Press.

Hickey, J. J., ed. 1969. *Peregrine Falcon Populations: Their Biology and Decline.* Madison: University of Wisconsin Press.

Hill, G. E. 1993. House Finch (*Carpodacus mexicanus*). In *The Birds of North America,* no. 46, ed. A. Poole, P. Stettenheim, and F. Gill. Academy of Natural Sciences, Philadelphia, and American Ornithologists' Union, Washington, DC.

Hill, J. R., III. 1988. Nest-depth preference in pipe-nesting Northern Rough-winged Swallows. *Journal of Field Ornithology* 59:334–336.

Hill, N. P. 1968. Eastern Sharp-tailed Sparrow. Pages 795–812 in *Life Histories of North American Cardinals, Grosbeaks, Buntings, Towhees, Finches, Sparrow, and Allies,* ed. O. L. Austin Jr. U.S. National Museum Bulletin, no. 237.

Hilton, R. 1996a. Yellow-billed Cuckoo (*Coccyzus americanus*). Pages 178–179 in *Atlas of the Breeding Birds of Maryland and the District of Columbia,* ed. C. S. Robbins and E. A. T. Blom. Pittsburgh: University of Pittsburgh Press.

———. 1996b. Yellow-bellied Sapsucker (*Sphyrapicus varius*). Pages 206–207 in *Atlas of the Breeding Birds of Maryland and the District of Columbia,* ed. C. S. Robbins and E. A. T. Blom. Pittsburgh: University of Pittsburgh Press.

———. 1996c. Common Raven (*Corvus corax*). Pages 252–253 in *Atlas of the Breeding Birds of Maryland and the District of Columbia,* ed. C. S. Robbins and E. A. T. Blom. Pittsburgh: University of Pittsburgh Press.

———. 1996d. Eastern Meadowlark (*Sturnella magna*). Pages 416–417 in *Atlas of the Breeding Birds of Maryland and the District of Columbia,* ed. C. S. Robbins and E. A. T. Blom. Pittsburgh: University of Pittsburgh Press.

Hindman, L. J., and W. F. Harvey IV. 2004. Status and management of feral Mute Swans in Maryland. Pages 11–17 in *Mute Swans and Their Chesapeake Bay Habitats: Proceedings of a Symposium,* ed. M.C. Perry. U.S. Geological Survey, Biological Resources Discipline Information and Technology Report USGS/BRD/ITR-2004-0005.

Hindman, L. J., W. F. Harvey IV, and V. D. Stotts. 1992. Harvest and band recovery of captive-reared Mallards released by the State of Maryland, 1974–1987. *Proceedings of the Annual Conference of Southeastern Association of Fish and Wildlife Agencies* 46:215–222.

Hitchner, S. B. 1996a. Gray Catbird (*Dumetella carolinensis*). Pages 294–295 in *Atlas of the Breeding Birds of Maryland and the District of Columbia,* ed. C. S. Robbins and E. A. T. Blom. Pittsburgh: University of Pittsburgh Press.

———. 1996b. Brown Thrasher (*Toxostoma rufum*). Pages 298–299 in *Atlas of the Breeding Birds of Maryland and the District of Columbia,* ed. C. S. Robbins and E. A. T. Blom. Pittsburgh: University of Pittsburgh Press.

———. 1996c. Indigo Bunting (*Passerina cyanea*). Pages 384–385 in *Atlas of the Breeding Birds of Maryland and the District of Columbia,* ed. C. S. Robbins and E. A. T. Blom. Pittsburgh: University of Pittsburgh Press.

Hobbs, A. S. 1996. Dark-eyed Junco (*Junco hyemalis*). Pages 410–411 in *Atlas of the Breeding Birds of Maryland and the District of Columbia,* ed. C. S. Robbins and E. A. T. Blom. Pittsburgh: University of Pittsburgh Press.

Hochachka W. M., and A. A. Dhondt. 2000. Density-dependent decline of host abundance resulting from a new infectious disease. *Proceedings of the National Academy of Sciences USA* 97:5303–5306.

Hoffman, M. L. 1996a. Wilson's Plover (*Charadrius wilsonia*). Pages 136–137 In *Atlas of the Breeding Birds of Maryland and the District of Columbia,* ed. C. S. Robbins and E. A. T. Blom. Pittsburgh: University of Pittsburgh Press.

———. 1996b. Piping Plover (*Charadrius melodus*). Pages 138–139 in *Atlas of the Breeding Birds of Maryland and the District of Columbia,* ed. C. S. Robbins and E. A. T. Blom. Pittsburgh: University of Pittsburgh Press.

Holmes, D. W. 1996a. Blue-winged Warbler (*Vermivora pinus*). Pages 316–317 in *Atlas of the Breeding Birds of Maryland and the District of Columbia,* ed. C. S. Robbins and E. A. T. Blom. Pittsburgh: University of Pittsburgh Press.

———. 1996b. Golden-winged Warbler (*Vermivora chrysoptera*). Pages 318–319 in *Atlas of the Breeding Birds of Maryland and the District of Columbia,* ed. C. S. Robbins and E. A. T. Blom. Pittsburgh: University of Pittsburgh Press.

———. 1996c. Grasshopper Sparrow (*Ammodramus savannarum*). Pages 398–399 in *Atlas of the Breeding Birds of Maryland and the District of Columbia,* ed. C. S. Robbins and E. A. T. Blom. Pittsburgh: University of Pittsburgh Press.

Holmes, R. T. 1994. Black-throated Blue Warbler (*Dendroica caerulescens*). In *The Birds of North America,* no. 87, ed. A. Poole and F. Gill. Philadelphia and Washington, D.C.: Academy of Natural Sciences and American Ornithologists' Union.

Holmes, R. T., N. L. Rodenhouse, and T. S. Sillett. 2005. Black-throated Blue Warbler (*Dendroica caerulescens*). In *The Birds of North America Online,* no. 87, ed. A. Poole. Ithaca: Cornell Lab of Ornithology. Available at http://bna.birds.cornell.edu/bna/species/087.

Holmes, R. T., and T. W. Sherry. 2001. Thirty-year bird population trends in an unfragmented temperate deciduous forest: Importance of habitat change. *Auk* 118:589–609.

Holt, D. W., and S. M. Leasure. 1993. Short-eared Owl (*Asio flammeus*). In *The Birds of North America,* no. 62, ed. A. Poole and F. Gill. Philadelphia and Washington, DC: Academy of Natural Sciences and American Ornithologists' Union.

Hopp, S. L., A. Kirby, and C. A. Boone. 1995. White-eyed Vireo (*Vireo griseus*). In *The Birds of North America,* no. 168, ed. A. Poole and F. Gill. Philadelphia and Washington, DC: Academy of Natural Sciences and American Ornithologists' Union.

Houston, C. S., and D. E. Bowen Jr. 2001. Upland Sandpiper (*Bartramia longicauda*). In *The Birds of North America,* no. 557, ed. A. Poole and F. Gill. Philadelphia and Washington, DC: Academy of Natural Sciences and American Ornithologists' Union.

Houston, C. S., D. G. Smith, and C. Rohner. 1998. Great Horned Owl (*Bubo virginianus*). In *The Birds of North America*, no. 372, ed. A. Poole and F. Gill. Philadelphia and Washington, DC: Academy of Natural Sciences and American Ornithologists' Union.

Howe, M. A. 1982. Social organization in a nesting population of eastern Willets (*Catoptrophorus semipalmatus*). *Auk* 99:88–102.

Hughes, J. M. 1999. Yellow-billed Cuckoo (*Coccyzus americanus*). In *The Birds of North America*, no. 418, ed. A. Poole and F. Gill. Philadelphia and Washington, DC: Academy of Natural Sciences and American Ornithologists' Union.

Hunt, P. D. 1996. Habitat selection by American Redstarts along a successional gradient in northern hardwoods forest: Evaluation of habitat quality. *Auk* 113:875–888.

———. 1998. Evidence from a landscape population model of the importance of early successional habitat to the American Redstart. *Conservation Biology* 12:377–1389.

Hurley, R. J., and E. C. Franks. 1976. Changes in the breeding ranges of two grassland birds. *Auk* 93:108–115.

Ickes, R. 1992a. Ruffed Grouse (*Bonasa umbellus*). Pages 114–115 in *Atlas of Breeding Birds in Pennsylvania*, ed. D. W. Brauning. Pittsburgh: University of Pittsburgh Press.

———. 1992b. Black-billed Cuckoo (*Coccyzus erythropthalmus*). Pages 150–151 in *Atlas of Breeding Birds in Pennsylvania*, ed. D. W. Brauning. Pittsburgh: University of Pittsburgh Press.

———. 1992c. Great Crested Flycatcher (*Myiarchus crinitus*). Pages 210–211 in *Atlas of Breeding Birds in Pennsylvania*, ed. D. W. Brauning. Pittsburgh: University of Pittsburgh Press.

Iliff, M. J. 2002. The nesting season: Middle Atlantic Coast region. *North American Birds* 56:423–426.

Iliff, M. J., M. Hafner, and G. L. Armistead. 2000. First Maryland nest of Long-eared Owl (*Asio otus*) since 1950. *Maryland Birdlife* 56:3–10.

Iliff, M. J., R. F. Ringler, and J. L. Stasz. 1996. *Field List of the Birds of Maryland*. 3rd ed. Maryland Avifauna No. 2. Baltimore: Maryland Ornithological Society.

Imhof, T. A. 1976. *Alabama Birds*. 2nd ed. Tuscaloosa: University of Alabama Press.

Ingold, D. J. 1994. Influence of nest-site competition between European Starlings and woodpeckers. *Wilson Bulletin* 106:227–241.

Ingold, J. L. 1993. Blue Grosbeak (*Passerina caerulea*). In *The Birds of North America*, no. 79, ed. A. Poole and F. Gill. Philadelphia and Washington, DC: Academy of Natural Sciences and American Ornithologists' Union.

Jackson, B. J., and J. A. Jackson. 2000. Killdeer (*Charadrius vociferus*). In *The Birds of North America*, no. 517, ed. A. Poole and F. Gill. Philadelphia and Washington, DC: Academy of Natural Sciences and American Ornithologists' Union.

Jackson, J. A. 1983. Nesting phenology, nest site selection, and reproductive success of Black and Turkey Vultures. Pages 245–270 in *Vulture Biology and Management*, ed. S. R. Wilbur and J. A. Jackson. Berkeley: University of California Press.

Jackson, J. A., and W. E. Davis Jr. 1998. Range expansion of the Red-bellied Woodpecker. *Bird Observer* 26:4–12.

Jackson, R. W. 1941. Breeding birds of the Cambridge area, Maryland. *Bulletin of the Natural History Society of Maryland* 11:65–74.

Jacob, J. P., and M. Parquay. 1992. *Oiseaux nicheurs de Famenne, Atlas de Lesse et Lomme*. Liège (Belgium): Aves.

James, R. D. 1976. Foraging behavior and habitat selection of three species of vireos in southern Ontario. *Wilson Bulletin* 88:62–75.

———. 1979. The comparative foraging behavior of Yellow-throated and Solitary Vireos: The effect of habitat and sympatry.

Pages 137–163 in *The Role of Insectivorous Birds in Forest Ecosystems*, ed. J. G. Dickson, R. N. Connor, R. R. Fleet, J. C. Kroll, and J. A. Jackson. New York: Academic Press.

———. 1998. Blue-headed Vireo (*Vireo solitarius*). In *The Birds of North America*, no. 379, ed. A. Poole and F. Gill. Philadelphia and Washington, DC: Academy of Natural Sciences and American Ornithologists' Union.

Jeschke, C. F. 1996a. Eastern Screech-Owl (*Otus asio*). Pages 182–1833 in *Atlas of the Breeding Birds of Maryland and the District of Columbia*, ed. C. S. Robbins and E. A. T. Blom. Pittsburgh: University of Pittsburgh Press.

———. 1996b. Great Horned Owl (*Bubo viginianus*). Pages 184–185 in *Atlas of the Breeding Birds of Maryland and the District of Columbia*, ed. C. S. Robbins and E. A. T. Blom. Pittsburgh: University of Pittsburgh Press.

———. 1996c. Barred Owl (*Strix varia*). Pages 186–187 in *Atlas of the Breeding Birds of Maryland and the District of Columbia*, ed. C. S. Robbins and E. A. T. Blom. Pittsburgh: University of Pittsburgh Press.

———. 1996d. Yellow-throated Warbler (*Dendroica dominica*). Pages 338–339 in *Atlas of the Breeding Birds of Maryland and the District of Columbia*, ed. C. S. Robbins and E. A. T. Blom. Pittsburgh: University of Pittsburgh Press.

Jeschke, C. F., and D. F. Brinker. 1996. Northern Saw-whet Owl (*Aegolius acadicus*). Pages 188–189 in *Atlas of the Breeding Birds of Maryland and the District of Columbia*, ed. C. S. Robbins and E. A. T. Blom. Pittsburgh: University of Pittsburgh Press.

Jeschke, C. F., and G. D. Therres. 1996. Barn Owl (*Tyto alba*). Pages 180–181 in *Atlas of the Breeding Birds of Maryland and the District of Columbia*, ed. C. S. Robbins and E. A. T. Blom. Pittsburgh: University of Pittsburgh Press.

Johnsgard, P. A. 1983. *The Grouse of the World*. Lincoln: University of Nebraska Press.

Johnson, D. H., and L. D. Igl. 1995. Contributions of the Conservation Reserve Program to populations of breeding birds in North Dakota. *Wilson Bulletin* 107:709–718.

Johnson, D. H., J. D Nichols, and M. D. Schwartz. 1992. Pages 446–485 in *Population Dynamics of Breeding Waterfowl*, ed. B. D. J. Batt, A. D. Afton, M. G. Anderson, C. D. Ankney, D. H. Johnson, J. A. Kadlec, and G. L. Krapu. Minneapolis: University of Minnesota Press.

Johnson, L. S. 1998. House Wren (*Troglodytes aedon*). In *The Birds of North America*, no. 380, ed. A. Poole and F. Gill. Philadelphia and Washington, DC: Academy of Natural Sciences and American Ornithologists' Union.

Johnson, N. K., R. M. Zink, and J. A. Marten. 1988. Genetic evidence for relationships in the avian family Vireonidae. *Condor* 90:428–445.

Johnston, R. F. 1992. Rock Pigeon (*Columba livia*). In *The Birds of North America*, no. 13, ed. A. Poole and F. Gill. Philadelphia and Washington, DC: Academy of Natural Sciences and American Ornithologists' Union.

Jones, P. W., and T. M. Donovan. 1996. Hermit Thrush (*Catharus guttatus*). In *The Birds of North America*, no. 261, ed. A. Poole and F. Gill. Philadelphia and Washington, DC: Academy of Natural Sciences and American Ornithologists' Union.

Jorde, D. G., and D. B. Stotts. 2002. The midwinter survey of Black Ducks, locally and regionally. Pages 31–35 in *Black Ducks and Their Chesapeake Bay Habitats: Proceedings of a Symposium*, ed. M. C. Perry. U.S. Geological Survey Information and Technology Report USGS/BRD ITR-2002-0005.

Joyce, E. D. 1996a. Carolina Wren (*Thryothorus ludovicianus*). Pages 268–269 in *Atlas of the Breeding Birds of Maryland and the District of Columbia,* ed. C. S. Robbins and E. A. T. Blom. Pittsburgh: University of Pittsburgh Press.

———. 1996b. Marsh Wren (*Cistothorus palustris*). Pages 278–279 in *Atlas of the Breeding Birds of Maryland and the District of Columbia,* ed. C. S. Robbins and E. A. T. Blom. Pittsburgh: University of Pittsburgh Press.

———. 1996c. Common Yellowthroat (*Geothlypis trichas*). Pages 366–367 in *Atlas of the Breeding Birds of Maryland and the District of Columbia,* ed. C. S. Robbins and E. A. T. Blom. Pittsburgh: University of Pittsburgh Press.

Joyner P. H., S. Kelly, A. A. Shreve, S. E. Snead, J. M. Sleeman, and D. A. Pettit. 2006. West Nile virus in raptors from Virginia during 2003: Clinical, diagnostic, and epidemiologic findings. *Journal of Wildlife Diseases* 42:335–344.

Kale, H. W., II. 1965. Ecology and bioenergetics of the Long-billed Marsh Wren *Telmatodytes palustris griseus* (Brewster) in Georgia salt marshes. *Publication of the Nuttall Ornithological Club,* no. 5.

Kaufmann, G. W. 1989. Breeding ecology of the Sora, *Porzana carolina,* and the Virginia Rail, *Rallus limicola. Canadian Field-Naturalist* 103:270–282.

Kearns, G. D, N. B. Kwartin, D. F. Brinker, and G. M. Haramis. 1998. Digital playback and improved trap design enhances capture of migrant Soras and Virginia Rails. *Journal of Field Ornithology* 69:466–473.

Keller, C. M. E., C. S. Robbins, and J. S. Hatfield. 1993. Avian communities in riparian forests of different widths in Maryland and Delaware. *Wetlands* 13:137–144.

Keller, R. 1997. First nesting record of Ruddy Duck (*Oxyura jamaicensis*) in Berks County, Pennsylvania. *Pennsylvania Birds* 11:142–143.

———. 1998. Local notes, Berks County. *Pennsylvania Birds* 12:133–134.

———. 2000. Local notes, Berks County. *Pennsylvania Birds* 14:157–158.

Kennedy, E. D., and D. W. White. 1996. Interference competition from House Wrens as a factor in the decline of Bewick's Wrens. *Conservation Biology* 10:281–284.

Keppie, D. M., and R. M. Whiting Jr. 1994. American Woodcock (*Scolopax minor*). In *The Birds of North America,* no. 100, ed. A. Poole and F. Gill. Philadelphia and Washington, DC: Academy of Natural Sciences and American Ornithologists' Union.

Kiff, L. F. 1989. Historical breeding records of the Common Merganser in southeastern United States. *Wilson Bulletin* 101:141–143.

Kingery, H. E., and W. D. Graul, eds. 1978. *Colorado Bird Distribution Latilong Study.* Denver: Colorado Division of Wildlife.

Kirk, D. A., and M. J. Mossman. 1998. Turkey Vulture (*Cathartes aura*). In *The Birds of North America,* no. 339, ed. A. Poole and F. Gill. Philadelphia and Washington, DC: Academy of Natural Sciences and American Ornithologists' Union.

Kirkwood, F. C. 1895. A list of the birds of Maryland. *Transactions of the Maryland Academy of Science* 1895:241–281.

———. 1901. The Cerulean Warbler (*Dendroica cerulea*) as a summer resident in Baltimore County, Maryland. *Auk* 18:137–142.

———. 1930. A Raven in Baltimore County, Maryland. *Auk* 47:255.

Kleen, R. L. 1956. A trip to Sharps Island. *Maryland Birdlife* 12:3–5.

Klimkiewicz, M. K. 1972. Breeding bird atlas of Montgomery County, Maryland. *Maryland Birdlife* 28:130–141.

———. 1996a. Purple Martin (*Progne subis*). Pages 234–235 in *Atlas of the Breeding Birds of Maryland and the District of Columbia,* ed. C. S. Robbins and E. A. T. Blom. Pittsburgh: University of Pittsburgh Press.

———. 1996b. Northern Parula (*Parula americana*). Pages 322–323 in *Atlas of the Breeding Birds of Maryland and the District of Columbia,* ed. C. S. Robbins and E. A. T. Blom. Pittsburgh: University of Pittsburgh Press.

Klimkiewicz, M. K., and J. K. Solem. 1978. The breeding bird atlas of Montgomery and Howard counties, Maryland. *Maryland Birdlife* 34:3–39.

Knapton, R. W. 1988. Nesting success is higher for polygynously mated females than for monogamously mated females in the Eastern Meadowlark. *Auk* 105:325–338.

Komar N., S. Langevin, S. Hinten, N. Nemeth, E. Edwards, D. Hettler, B. Davis, R. Bowen, and M. Bunning. 2003. Experimental infection of North American birds with the New York 1999 strain of West Nile virus. *Emerging Infectious Diseases* 9:311–322.

Krementz, D. G. 1991. American Black Duck (*Anas rubripes*). Pages 16-1–16-7 in *Habitat Requirements for Chesapeake Bay Living Resources,* ed. S. L. Funderburk, S. J. Jordan, J. A. Mihursky, and D. Riley. 2nd ed. Solomons, MD: Chesapeake Research Consortium.

Kricher, J. C. 1995. Black-and-white Warbler (*Mniotilta varia*). In *The Birds of North America,* no. 158, ed. A. Poole and F. Gill. Philadelphia and Washington, DC: Academy of Natural Sciences and American Ornithologists' Union.

Kroodsma, D. E., P. A. Bedell, W. C. Liu, and E. Goodwin. 1999. The ecology of song improvisation as illustrated by North American Sedge Wrens. *Auk* 116:373–386.

Kroodsma, D. E., and J. Verner. 1987. Use of song repertoires among Marsh Wren populations. *Auk* 104:63–72.

———. 1997. Marsh Wren (*Cistothorus palustris*). In *The Birds of North America,* no. 308, ed. A. Poole and F. Gill. Philadelphia and Washington, DC: Academy of Natural Sciences and American Ornithologists' Union.

Kumer, J. 2004. Status of the endangered Piping Plover, *Charadrius melodus,* population in the Maryland coastal bays. Pages 8-94–8-99 in *Maryland's Coastal Bays: Ecosystem Health Assessment 2004,* ed. C. E. Wazniak and M. R. Hall. Annapolis: Maryland Department of Natural Resources.

Kumlien, L. 1880. The Yellow-rumped Warbler (*Dendroica coronata*) nesting in eastern Maryland. *Bulletin of the Nuttall Ornithological Club* 5:182–183.

LaDeau S. L., A. Marm Kilpatricik, and P. P. Marra. 2007. West Nile virus emergence and large-scale declines of North American bird populations. *Nature* 447:710–714.

Lanyon, W. E. 1957. The comparative biology of the meadowlarks (*Sturnella*) in Wisconsin. *Publication of the Nuttall Ornithological Club,* no. 1.

———. 1997. Great Crested Flycatcher (*Myiarchus crinitus*). In *The Birds of North America,* no. 300, ed. A. Poole and F. Gill. Philadelphia and Washington, DC: Academy of Natural Sciences and American Ornithologists' Union.

Laskey, A. R. 1944. A study of the Cardinal in Tennessee. *Wilson Bulletin* 56:27–44.

———. 1948. Some nesting data on the Carolina Wren at Nashville, Tennessee. *Bird-Banding* 19:101–121.

Laughlin, S. B., ed. 1982. *Proceedings of the Northeastern Breeding Bird Atlas Conference.* Woodstock: Vermont Institute of Natural Science.

Laughlin, S. B., and D. P. Kibbe. 1985. *Atlas of the Breeding Birds of Vermont*. Hanover, N.H.: University Press of New England.

Lauro, B., and J. Burger. 1989. Nest-site selection of American Oystercatchers (*Haematopus palliatus*) in salt marshes. *Auk* 106:185–192.

Leberman, R. C. 1992a. Common Snipe (*Gallinago gallinago*). Pages 140–141 in *Atlas of Breeding Birds in Pennsylvania*, ed. D. W. Brauning. Pittsburgh: University of Pittsburgh Press.

———. 1992b. Sedge Wren (*Cistothorus platensis*). Pages 256–257 in *Atlas of Breeding Birds in Pennsylvania*, ed. D. W. Brauning. Pittsburgh: University of Pittsburgh Press.

———. 1992c. Blue-gray Gnatcatcher (*Polioptila caerulea*). Pages 262–263 in *Atlas of Breeding Birds in Pennsylvania*, ed. D. W. Brauning. Pittsburgh: University of Pittsburgh Press.

———. 1992d. Mourning Warbler (*Oporornis philadelphia*). Pages 348–349 in *Atlas of Breeding Birds in Pennsylvania*, ed. D. W. Brauning. Pittsburgh: University of Pittsburgh Press.

———. 1992e. Rose-breasted Grosbeak (*Pheucticus ludovicianus*). Pages 364–365 in *Atlas of Breeding Birds in Pennsylvania*, ed. D. W. Brauning. Pittsburgh: University of Pittsburgh Press.

Legare, M. L., and W. R. Eddleman. 2001. Home range size, nest-site selection, and nesting success of Black Rails in Florida. *Journal of Field Ornithology* 72:170–177.

LeSchack, C. R., S. K. McKnight, and G. R. Hepp. 1997. Gadwall (*Anas strepera*). In *The Birds of North America*, no. 283, ed. A. Poole and F. Gill. Philadelphia and Washington, DC: Academy of Natural Sciences and American Ornithologists' Union.

Lesser, F. H., and A. R. Stickley Jr. 1967. Occurrence of the Saw-whet Owl in Florida. *Auk* 84:425.

Levine, E. 1998. *Bull's Birds of New York State*. Ithaca: Cornell University Press.

Lippens, L., and H. Wille. 1972. *Atlas des oiseaux de Belgique et d'Europe occidentale*. Tielt (Belgium).

Livezey, B. C. 1981. Duck nesting in retired croplands at Horicon National Wildlife Refuge, Wisconsin. *Journal of Wildlife Management* 45:27–37.

Long, J. 1981. *Introduced Birds of the World*. New York: Universe Books.

Longcore, J. R., D. G. McAuley, G. R. Hepp, and J. M. Rhymer. 2000. American Black Duck (*Anas rubripes*). In *The Birds of North America*, no. 481, ed. A. Poole and F. Gill. Philadelphia and Washington, DC: Academy of Natural Sciences and American Ornithologists' Union.

Lord, J., and D. J. Munns. 1970. *Atlas of Breeding Birds of the West Midlands*. London: Collins.

Lovette, I. J., and E. Bermingham. 2002. What is a wood-warbler? A molecular characterization of a monophyletic Parulidae. *Auk* 119:695–714.

Ludwig, G. V., P. P. Calle, J. A. Mangiafico, B. L. Raphael, D. K. Danner, J. A. Hile, T. L. Clippinger, J. F. Smith, R. A. Cook, and T. McNamara. 2002. An outbreak of West Nile virus in a New York City captive wildlife population. *American Journal of Tropical Medicine and Hygiene* 67:67–75.

Lunk, W. A. 1962. The Rough-winged Swallow *Stelgidopteryx ruficollis* (Vieillot): A study based on its breeding biology in Michigan. *Nuttall Ornithological Club Publication*, no. 4.

Luttrell, M. P., J. R. Fischer, D. E. Stallknecht, and S. H. Kleven. 1996. Field investigation of *Mycoplasma gallisepticum* infections in House Finches (*Carpodacus mexicanus*) from Maryland and Georgia. *Avian Diseases* 40:335–341.

Luukkonen, D. R., and J. D. Fraser. 1987. Loggerhead Shrike status

and breeding biology in Virginia. Master's thesis, Virginia Polytechnic Institute and State University, Blacksburg.

Lynch, J. F., E. S. Morton, and M. E. Van der Voort. 1985. Habitat segregation between the sexes of wintering hooded warblers (*Wilsonia citrina*). *Auk* 102:714–721.

Mallory, M., and K. Metz. 1999. Common Merganser (*Mergus merganser*). In *The Birds of North America*, no. 442, ed. A. Poole and F. Gill. Philadelphia and Washington, DC: Academy of Natural Sciences and American Ornithologists' Union.

Manke J. E., J. E. Carlson, and M. C. Brittingham. 2004. A century of hybridization: Decreasing genetic distance between American Black Ducks and Mallards. *Conservation Genetics* 5:395–403.

Marks, J. S., D. L. Evans, and D. W. Holt. 1994. Long-eared Owl (*Asio otus*). In *The Birds of North America*, no. 133, ed. A. Poole and F. Gill. Philadelphia and Washington, DC: Academy of Natural Sciences and American Ornithologists' Union.

Marshall, E. W. 1958. Banding recovery adds House Finch to Maryland list. *Maryland Birdlife* 14:96.

Marti, C. D., A. F. Poole, and L. R. Bevier. 2005. Barn Owl (*Tyto alba*). In *The Birds of North America Online*, no. 1, ed, A. Poole. Ithaca: Cornell Lab of Ornithology. Available at bna.birds.cornell.edu/bna/species/001.

Martin, A. C., H. S. Zim, and A. L. Nelson. 1951. *American Wildlife and Plants: A Guide to Wildlife Food Habits*. Washington, DC: U.S. Department of the Interior.

Martin, A. H. 1987. Atlas de las aves nidificantes en la Isla de Atenerife. Instituto de Estudios Canarios, Monografia, no. 32. Tenerife, Islas Canarias (Canary Islands).

Martin, E. M. 1996. Red-shouldered Hawk (*Buteo lineatus*). Pages 104–105 in *Atlas of the Breeding Birds of Maryland and the District of Columbia*, ed. C. S. Robbins and E. A. T. Blom. Pittsburgh: University of Pittsburgh Press.

———. 2004. Decreases in a population of Red-shouldered Hawks nesting in central Maryland. *Journal of Raptor Research* 38:312–319.

Martin, S. G., and T. A. Gavin. 1995. Bobolink (*Dolichonyx oryzivorus*). In *The Birds of North America*, no. 176, ed. A. Poole and F. Gill. Philadelphia and Washington, DC: Academy of Natural Sciences and American Ornithologists' Union.

Maryland Department of Natural Resources. 2005. Maryland wildlife diversity conservation plan. Maryland Department of Natural Resources. Available at www.dnr.maryland.gov/wildlife/divplan_wdcp.asp.

———. 2007. *Rare, Threatened, and Endangered Animals of Maryland*. Annapolis: Maryland Department of Natural Resources.

———. 2008. DNR owned lands acreage report. Maryland Department of Natural Resources. Available at http://dnr.maryland.gov/land/lps/pdfs/current.acreage.report.pdf.

Maryland Ornithological Society. 2001. Second Maryland/DC breeding bird atlas project handbook. N.p.: Maryland Ornithological Society.

Maryland Wood Duck Initiative. 2006. Public lands survey: 2006 Hooded Merganser summary. Available at www.mwdi.net/mwdi/Reports/Hooded_Mergansers/2006_Hooded_Meganser_Summary.asp.

Marzilli, V. 1989. Up on the roof. *Maine Fish and Wildlife* 31:25–29.

Master, T. L. 1992a. Short-eared Owl (*Asio flammeus*). Pages 164–165 in *Atlas of Breeding Birds in Pennsylvania*, ed. D. W. Brauning. Pittsburgh: University of Pittsburgh Press.

———. 1992b. Northern Flicker (*Colaptes auratus*). Pages 190–191 in *Atlas of Breeding Birds in Pennsylvania*, ed. D. W. Brauning. Pittsburgh: University of Pittsburgh Press.

————. 1992c. Yellow-throated Vireo (*Vireo flavifrons*). Pages 292–293 in *Atlas of Breeding Birds in Pennsylvania*, ed. D. W. Brauning. Pittsburgh: University of Pittsburgh Press.

Mathews, T. P. 1996a. Wild Turkey (*Meleagris gallopavo*). Pages 118–119 in *Atlas of the Breeding Birds of Maryland and the District of Columbia*, ed. C. S. Robbins and E. A. T. Blom. Pittsburgh: University of Pittsburgh Press.

————. 1996b. Northern Bobwhite (*Colinus virginianus*). Pages 120–121 in *Atlas of the Breeding Birds of Maryland and the District of Columbia*, ed. C. S. Robbins and E. A. T. Blom. Pittsburgh: University of Pittsburgh Press.

————. 1996c. American Woodcock (*Scolopax minor*). Pages 152–153 in *Atlas of the Breeding Birds of Maryland and the District of Columbia*, ed. C. S. Robbins and E. A. T. Blom. Pittsburgh: University of Pittsburgh Press.

Mayfield, H. F. 1965. The Brown-headed Cowbird, with old and new hosts. *Living Bird* 4:13–28.

Mazur, K. M., and P. C. James. 2000. Barred Owl (*Strix varia*). In *The Birds of North America*, no. 508, ed. A. Poole and F. Gill. Philadelphia and Washington, DC: Academy of Natural Sciences and American Ornithologists' Union.

McAuley, D. G., D. A. Clugston, and J. R. Longcore. 2004. Dynamic use of wetlands by Black Ducks and Mallards: Evidence against competitive exclusion. *Wildlife Society Bulletin* 32:465–473.

McAuley, D. G., J. R. Longcore, and G. F. Sepik. 1993. Behavior of radio-marked breeding American Woodcocks. Pages 115–124 in *Proceedings of the Eighth Woodcock Symposium*, ed. J. R. Longcore and G. F. Sepik. U.S. Fish and Wildlife Service Biological Report, no. 16, Washington, DC.

McCarty, J. P. 1996. Eastern Wood-Pewee (*Contopus virens*). In *The Birds of North America*, no. 245, ed. A. Poole and F. Gill. Philadelphia and Washington, DC: Academy of Natural Sciences and American Ornithologists' Union.

McCrimmon, D. A., Jr., J. C. Ogden, and G. T. Bancroft. 2001. Great Egret (*Ardea alba*). In *The Birds of North America*, no. 570, ed. A. Poole and F. Gill. Philadelphia and Washington, DC: Academy of Natural Sciences and American Ornithologists' Union.

McDonald, M. V. 1998. Kentucky Warbler (*Oporornis formosus*). In *The Birds of North America*, no. 324, ed. A. Poole and F. Gill. Philadelphia and Washington, DC: Academy of Natural Sciences and American Ornithologists' Union.

McGillivray, W. B. 1983. Intraseasonal reproductive costs for the House Sparrow (*Passer domesticus*). *Auk* 100:25–32.

McGilvrey, F. B. 1969. Survival in Wood Duck broods. *Journal of Wildlife Management* 33:73–76.

McGowan, K. J. 2008a. Barn Owl (*Tyto alba*). Pages 290–291 in *The Second Atlas of Breeding Birds in New York State*, ed. K. J. McGowan and K. Corwin. Ithaca: Cornell University Press.

————. 2008b. Red-headed Woodpecker (*Melanerpes erythrocephalus*). Pages 320–321 in *The Second Atlas of Breeding Birds in New York State*, ed. K. J. McGowan and K. Corwin. Ithaca: Cornell University Press.

————. 2008c. Yellow-bellied Sapsucker (*Sphyrapicus varius*). Pages 324–325 in *The Second Atlas of Breeding Birds in New York State*, ed. K. J. McGowan and K. Corwin. Ithaca: Cornell University Press.

————. 2008d. Great Crested Flycatcher (*Myiarchus crinitus*). Pages 356–357 in *The Second Atlas of Breeding Birds in New York State*, ed. K. J. McGowan and K. Corwin. Ithaca: Cornell University Press.

————. 2008e. Brown Creeper (*Certhia americana*). Pages 416–417 in *The Second Atlas of Breeding Birds in New York State*, ed. K. J. McGowan and K. Corwin. Ithaca: Cornell University Press.

————. 2008f. Winter Wren (*Troglodytes troglodytes*). Pages 424–425 in *The Second Atlas of Breeding Birds in New York State*, ed. K. J. McGowan and K. Corwin. Ithaca: Cornell University Press.

————. 2008g. Sedge Wren (*Cistothorus platensis*). Pages 426–427 in *The Second Atlas of Breeding Birds in New York State*, ed. K. J. McGowan and K. Corwin. Ithaca: Cornell University Press.

————. 2008h. Northern Parula (*Parula americana*). Pages 478–479 in *The Second Atlas of Breeding Birds in New York State*, ed. K. J. McGowan and K. Corwin. Ithaca: Cornell University Press.

————. 2008i. Brown-headed Cowbird (*Molothrus ater*). Pages 600–601 in *The Second Atlas of Breeding Birds in New York State*, ed. K. J. McGowan and K. Corwin. Ithaca: Cornell University Press.

————. 2008j. Rare, improbable, and historic breeders: Eurasian Collared-Dove (*Streptopelia decaocto*). Pages 628–629 in *The Second Atlas of Breeding Birds in New York State*, ed. K. J. McGowan and K. Corwin. Ithaca: Cornell University Press.

McKearnan, J. 1996a. Great Blue Heron (*Ardea herodias*). Pages 50–51 in *Atlas of the Breeding Birds of Maryland and the District of Columbia*, ed. C. S. Robbins and E. A. T. Blom. Pittsburgh: University of Pittsburgh Press.

————. 1996b. Great Egret (*Ardea alba*). Pages 52–53 in *Atlas of the Breeding Birds of Maryland and the District of Columbia*, ed. C. S. Robbins and E. A. T. Blom. Pittsburgh: University of Pittsburgh Press.

————. 1996c. Snowy Egret (*Egretta thula*). Pages 54–55 in *Atlas of the Breeding Birds of Maryland and the District of Columbia*, ed. C. S. Robbins and E. A. T. Blom. Pittsburgh: University of Pittsburgh Press.

————. 1996d. Little Blue Heron (*Egretta caerulea*). Pages 56–57 in *Atlas of the Breeding Birds of Maryland and the District of Columbia*, ed. C. S. Robbins and E. A. T. Blom. Pittsburgh: University of Pittsburgh Press.

————. 1996e. Tricolored Heron (*Egretta tricolor*). Pages 58–59 in *Atlas of the Breeding Birds of Maryland and the District of Columbia*, ed. C. S. Robbins and E. A. T. Blom. Pittsburgh: University of Pittsburgh Press.

————. 1996f. Cattle Egret (*Bubulcus ibis*). Pages 60–61 in *Atlas of the Breeding Birds of Maryland and the District of Columbia*, ed. C. S. Robbins and E. A. T. Blom. Pittsburgh: University of Pittsburgh Press.

————. 1996g. Black-crowned Night-Heron (*Nycticorax nycticorax*). Pages 64–65 in *Atlas of the Breeding Birds of Maryland and the District of Columbia*, ed. C. S. Robbins and E. A. T. Blom. Pittsburgh: University of Pittsburgh Press.

————. 1996h. Yellow-crowned Night-Heron (*Nyctanassa violacea*). Pages 66–67 in *Atlas of the Breeding Birds of Maryland and the District of Columbia*, ed. C. S. Robbins and E. A. T. Blom. Pittsburgh: University of Pittsburgh Press.

————. 1996i. Glossy Ibis (*Plegadis falcinellus*). Pages 68–69 in *Atlas of the Breeding Birds of Maryland and the District of Columbia*, ed. C. S. Robbins and E. A. T. Blom. Pittsburgh: University of Pittsburgh Press.

————. 1996j. Least Tern (*Sterna antillarum*). Pages 168–169 in *Atlas of the Breeding Birds of Maryland and the District of Columbia*, ed. C. S. Robbins and E. A. T. Blom. Pittsburgh: University of Pittsburgh Press.

McNair, D. B., and W. Post. 1993. Supplement to status and distribution of South Carolina birds. *Charleston Museum Ornithological Contribution*, no. 8.

McNicholl, M. K., P. E. Lowther, and J. A. Hall. 2001. Forster's Tern (*Sterna forsteri*). In *The Birds of North America*, no. 595, ed.

A. Poole and F. Gill. Philadelphia and Washington, DC: Academy of Natural Sciences and American Ornithologists' Union.

McShea, W. J., M. V. McDonald, E. S. Morton, R. Meier, and J. H. Rappole. 1995. Long-term trends in habitat selection by Kentucky Warblers. *Auk* 112:375–381.

McWilliams, G. M., and D. W. Brauning. 2000. *The Birds of Pennsylvania*. Ithaca: Cornell University Press.

Mead, C. J., and K. W. Smith. 1982. *The Hertfordshire Breeding Bird Atlas*. Tring Herts., UK: H.B.B.A.

Meade, G. M. 1988. Tufted Titmouse (*Parus bicolor*). Pages 292–293 in *The Atlas of Breeding Birds in New York State*, ed. R. F. Andrle and J. R. Carroll. Ithaca: Cornell University Press.

Meanley, B. 1943. Nesting of the Upland Plover in Baltimore County, Maryland. *Auk* 60:603.

———. 1950. Swainson's Warbler on the Coastal Plain of Maryland. *Wilson Bulletin* 62:93–94.

———. 1969. Natural history of the King Rail. *North American Fauna*, no. 67.

———. 1971. Natural history of the Swainson's Warbler. *North American Fauna*, no. 69.

———. 1975. *Birds and Marshes of the Chesapeake Bay Country*. Cambridge, Md.: Tidewater Publishers.

———. 1985. *The Marsh Hen: A Natural History of the Clapper Rail of the Atlantic Coast Salt Marsh*. Centreville, Md.: Tidewater Publishers.

———. 1992. King Rail (*Rallus elegans*). In *The Birds of North America*, no. 3, ed. A. Poole and F. Gill. Philadelphia and Washington, DC: Academy of Natural Sciences and American Ornithologists' Union.

Medler, M. D. 2008. Whip-poor-will (*Caprimulgus vociferus*). Pages 310–311 in *The Second Atlas of Breeding Birds in New York State*, ed. K. J. McGowan and K. Corwin. Ithaca: Cornell University Press.

Mehner, J. F., and G. J. Wallace. 1959. Robin populations and insecticides. *Atlantic Naturalist* 14:4–9.

Meritt, D. W. 1996a. Double-crested Cormorant (*Phalacrocorax auritus*). Pages 44–45 in *Atlas of the Breeding Birds of Maryland and the District of Columbia*, ed. C. S. Robbins and E. A. T. Blom. Pittsburgh: University of Pittsburgh Press.

———. 1996b. Canada Goose (*Branta canadensis*). Pages 72–73 in *Atlas of the Breeding Birds of Maryland and the District of Columbia*, ed. C. S. Robbins and E. A. T. Blom. Pittsburgh: University of Pittsburgh Press.

———. 1996c. Wood Duck (*Aix sponsa*). Pages 74–75 in *Atlas of the Breeding Birds of Maryland and the District of Columbia*, ed. C. S. Robbins and E. A. T. Blom. Pittsburgh: University of Pittsburgh Press.

———. 1996d. American Black Duck (*Anas rubripes*). Pages 76–77 in *Atlas of the Breeding Birds of Maryland and the District of Columbia*, ed. C. S. Robbins and E. A. T. Blom. Pittsburgh: University of Pittsburgh Press.

———. 1996e. Mallard (*Anas platyrhynchos*). Pages 78–79 in *Atlas of the Breeding Birds of Maryland and the District of Columbia*, ed. C. S. Robbins and E. A. T. Blom. Pittsburgh: University of Pittsburgh Press.

———. 1996f. Blue-winged Teal (*Anas discors*). Pages 80–81 in *Atlas of the Breeding Birds of Maryland and the District of Columbia*, ed. C. S. Robbins and E. A. T. Blom. Pittsburgh: University of Pittsburgh Press.

———. 1996g. Northern Shoveler (*Anas clypeata*). Pages 82–83 in *Atlas of the Breeding Birds of Maryland and the District of Columbia*,

ed. C. S. Robbins and E. A. T. Blom. Pittsburgh: University of Pittsburgh Press.

———. 1996h. Gadwall (*Anas strepera*). Pages 84–85 in *Atlas of the Breeding Birds of Maryland and the District of Columbia*, ed. C. S. Robbins and E. A. T. Blom. Pittsburgh: University of Pittsburgh Press.

———. 1996i. American Coot (*Fulica americana*). Pages 134–135 in *Atlas of the Breeding Birds of Maryland and the District of Columbia*, ed. C. S. Robbins and E. A. T. Blom. Pittsburgh: University of Pittsburgh Press.

Middleton, A. L. A. 1993. American Goldfinch (*Carduelis tristis*). In *The Birds of North America*, no. 80, ed. A. Poole and F. Gill. Philadelphia and Washington, DC: Academy of Natural Sciences and American Ornithologists' Union.

Miller, C. E. 1996. Red-winged Blackbird (*Agelaius phoeniceus*). Pages 414–415 in *Atlas of the Breeding Birds of Maryland and the District of Columbia*, ed. C. S. Robbins and E. A. T. Blom. Pittsburgh: University of Pittsburgh Press.

Miller, G. E. 1959. First nesting of the Cattle Egret in Maryland. *Maryland Birdlife* 15:22.

Mills, A. M. 1986. The influence of moonlight on the behavior of goatsuckers (Caprimulgidae). *Auk* 103:370–378.

Mirarchi, R. E., and T. S. Baskett. 1994. Mourning Dove (*Zenaida macroura*). In *The Birds of North America*, no. 117, ed. A. Poole and F. Gill. Philadelphia and Washington, DC: Academy of Natural Sciences and American Ornithologists' Union.

Mitra, S. S. 2008a. Gull-billed Tern (*Gelochelidon nilotica*). Pages 262–263 in *The Second Atlas of Breeding Birds in New York State*, ed. K. J. McGowan and K. Corwin. Ithaca: Cornell University Press.

———. 2008b. Chuck-will's-widow (*Caprimulgus carolinensis*). Pages 308–309 in *The Second Atlas of Breeding Birds in New York State*, ed. K. J. McGowan and K. Corwin. Ithaca: Cornell University Press.

Molina, K. C., and R. M. Erwin. 2006. The distribution and conservation status of the Gull-billed Tern (*Gelochelidon nilotica*) in North America. *Waterbirds* 29:271–295.

Monroe, B. L., Jr. 1974. Summary of highest counts of individuals for Canada and the United States. *American Birds* 28:568–576.

Montier, D. J. 1977. *Atlas of the Breeding Birds of the London Area*. London: London Natural History Society.

Moore, W. S. 1995. Northern Flicker (*Colaptes auratus*). In *The Birds of North America*, no. 166, ed. A. Poole and F. Gill. Philadelphia and Washington, DC: Academy of Natural Sciences and American Ornithologists' Union.

Morse, D. H. 1993. Black-throated Green Warbler (*Dendroica virens*). In *The Birds of North America*, no. 55, ed. A. Poole and F. Gill. Philadelphia and Washington, DC: Academy of Natural Sciences and American Ornithologists' Union.

Mossman, M. J. 1991. Black and Turkey Vultures. Pages 3–22 in *Proceedings of the Midwest Raptor Management Symposium and Workshop*, ed. M. N. Lefranc Jr. Washington, DC: National Wildlife Federation.

Mostrom, A. M., R. L. Curry, and B. Lohr. 2002. Carolina Chickadee (*Poecile carolinensis*). In *The Birds of North America*, no. 636, ed. A. Poole and F. Gill. Philadelphia and Washington, DC: Academy of Natural Sciences and American Ornithologists' Union.

Mowbray, T. B. 1997. Swamp Sparrow (*Melospiza georgiana*). In *The Birds of North America*, no. 279, ed. A. Poole and F. Gill. Philadelphia and Washington, DC: Academy of Natural Sciences and American Ornithologists' Union.

Mowbray, T. B., C. R. Ely, J. S. Sedinger, and R. E. Trost. 2002. Canada Goose (Branta canadensis). In *The Birds of North America,* no. 682, ed. A. Poole and F. Gill. Philadelphia and Washington, DC: Academy of Natural Sciences and American Ornithologists' Union.

Muller, M. J., and R. W. Storer. 1999. Pied-billed Grebe (*Podilymbus podiceps*). In *The Birds of North America,* no. 410, ed. A. Poole and F. Gill. Philadelphia and Washington, DC: Academy of Natural Sciences and American Ornithologists' Union.

Mulvihill, R. S. 1992a. Common Raven (*Corvus corax*). Pages 236–237 in *Atlas of Breeding Birds in Pennsylvania,* ed. D. W. Brauning. Pittsburgh: University of Pittsburgh Press.

———. 1992b. Golden-crowned Kinglet (*Regulus satrapa*). Pages 260–261 in *Atlas of Breeding Birds in Pennsylvania,* ed. D. W. Brauning. Pittsburgh: University of Pittsburgh Press.

———. 1992c. Hermit Thrush (*Catharus guttatus*). Pages 270–271 in *Atlas of Breeding Birds in Pennsylvania,* ed. D. W. Brauning. Pittsburgh: University of Pittsburgh Press.

———. 1992d. Dark-eyed Junco (*Junco hyemalis*). Pages 394–395 in *Atlas of Breeding Birds in Pennsylvania,* ed. D. W. Brauning. Pittsburgh: University of Pittsburgh Press.

Muntaner, J., X. Ferrer, and A. Martinez-Vilalta. 1984. *Atlas dels Ocells nidificants de Catalunya I Andorra.* Barcelona, España (Spain): Ketres.

Murphy, M. T. 1996. Eastern Kingbird (*Tyrannus tyrannus*). In *The Birds of North America,* no. 253, ed. A. Poole and F. Gill. Philadelphia and Washington, DC: Academy of Natural Sciences and American Ornithologists' Union.

Murphy, W. L. 1996. Alder Flycatcher (*Empidonax alnorum*). Pages 220–221 in *Atlas of the Breeding Birds of Maryland and the District of Columbia,* ed. C. S. Robbins and E. A. T. Blom. Pittsburgh: University of Pittsburgh Press.

Murphy, W. L., and C. S. Robbins. 1996. Mourning Warbler (*Oporornis philadelphia*). Pages 364–365 in *Atlas of the Breeding Birds of Maryland and the District of Columbia,* ed. C. S. Robbins and E. A. T. Blom. Pittsburgh: University of Pittsburgh Press.

Murray, R. D., M. Holling, H. E. M. Dott, and P. Vandome. 1998. *The Breeding Birds of South-east Scotland: A Tetrad Atlas, 1988–1994.* Edinburgh: Scottish Ornithologists' Club.

Nemeth, N., D. Gould, R. Bowen, and N. Komar. 2006. Natural and experimental West Nile virus infection in five raptor species. *Journal of Wildlife Diseases* 42:1–13.

Nice, M. M. 1937. Studies in the life history of the Song Sparrow, part 1. *Transactions of the Linnean Society of New York* 4:1–247.

Nichols, W. D. 1985. Tufted Titmouse (*Parus bicolor*). Pages 214–215 in *The Atlas of Breeding Birds of Vermont,* ed. S. B. Laughlin and D. P. Kibbe. Hanover, N.H.: University Press of New England.

Nicholson, C. P. 1997. *Atlas of the Breeding Birds of Tennessee.* Knoxville: University of Tennessee Press.

Nicolai, B. 1993. *Atlas der Brutvogel Ostdeutschlands.* Stuttgart (East Germany): Fischer.

Nisbet, I. C. T. 2002. Common Tern (*Sterna hirundo*). In *The Birds of North America,* no. 618, ed. A. Poole and F. Gill. Philadelphia and Washington, DC: Academy of Natural Sciences and American Ornithologists' Union.

Nol, E., and R. C. Humphrey. 1994. American Oystercatcher (*Haematopus palliatus*). In *The Birds of North America,* no. 82, ed. A. Poole and F. Gill. Philadelphia and Washington, DC: Academy of Natural Sciences and American Ornithologists' Union.

Nolan, V., Jr., and C. F. Thompson. 1975. The occurrence and significance of anomalous reproductive activities in two North American nonparasitic cuckoos *Coccyzus* spp. *Ibis* 117:496–503.

Nolan, V., Jr., E. D. Ketterson, D. A. Cristol, C. M. Rogers, E. D. Clotfelter, R. C. Titus, S. J. Schoech, and E. Snajdr. 2002. Dark-eyed Junco (*Junco hyemalis*). In *The Birds of North America,* no. 716, ed. A. Poole and F. Gill. Philadelphia and Washington, DC: Academy of Natural Sciences and American Ornithologists' Union.

Oakleaf, B., H. Downing, B. M. Rayner, and O. Scott. 1979. *Wyoming avian atlas.* Lander: Wyoming Game and Fish Department.

Oatman, G. F. 1985. Common Raven (*Corvus corax*). Pages 208–209 in *The Atlas of Breeding Birds of Vermont,* ed. S. B. Laughlin and D. P. Kibbe. Hanover, N.H.: University Press of New England.

O'Brien, M. 1996a. Willet (*Catoptrophorus semipalmatus*). Pages 146–147 in *Atlas of the Breeding Birds of Maryland and the District of Columbia,* ed. C. S. Robbins and E. A. T. Blom. Pittsburgh: University of Pittsburgh Press.

———. 1996b. Saltmarsh Sharp-tailed Sparrow (*Ammodramus caudacutus*). Pages 402–403 in *Atlas of the Breeding Birds of Maryland and the District of Columbia,* ed. C. S. Robbins and E. A. T. Blom. Pittsburgh: University of Pittsburgh Press.

———. 1996c. Seaside Sparrow (*Ammodramus maritimus*). Pages 404–405 in *Atlas of the Breeding Birds of Maryland and the District of Columbia,* ed. C. S. Robbins and E. A. T. Blom. Pittsburgh: University of Pittsburgh Press.

———. 1996d. Boat-tailed Grackle (*Quiscalus major*). Pages 418–419 in *Atlas of the Breeding Birds of Maryland and the District of Columbia,* ed. C. S. Robbins and E. A. T. Blom. Pittsburgh: University of Pittsburgh Press.

Ogden, J. C. 1978. Population trends of colonial wading birds on the Atlantic and Gulf coasts. Pages 137–153 in *Wading Birds,* ed. A. Sprunt IV, J. C. Ogden, and S. Winckler. New York: National Audubon Society. Research Report no. 7.

Ogden, L. J., and B. J. Stutchbury. 1994. Hooded Warbler (*Wilsonia citrina*). In *The Birds of North America,* no. 110, ed. A. Poole and F. Gill. Philadelphia and Washington, DC: Academy of Natural Sciences and American Ornithologists' Union.

Oring, L. W., E. M. Gray, and J. M. Reed. 1997. Spotted Sandpiper (*Actitis macularius*). In *The Birds of North America,* no. 289, ed. A. Poole and F. Gill. Philadelphia and Washington, DC: Academy of Natural Sciences and American Ornithologists' Union.

Oring, L. W., D. B. Lank, and S. J. Maxson. 1983. Population studies of the polyandrous Spotted Sandpiper. *Auk* 100:272–285.

Osborn, R. G., and T. W. Custer. 1978. *Herons and Their Allies: Atlas of Atlantic Coast Colonies, 1975 and 1976.* U.S. Fish and Wildlife Service, FWS/OBS-77/08.

Palmer, R. S., ed. 1962. *Handbook of North American Birds.* Vol.1, *Loons through Flamingos.* New Haven, CT: Yale University Press.

Parker, J. W., and J. C. Ogden. 1979. The recent history and status of the Mississippi Kite. *American Birds* 33:119–129.

Parnell, J. F., R. M. Erwin, and K. C. Molina. 1995. Gull-billed Tern (*Sterna nilotica*). In *The Birds of North America,* no. 140, ed. A. Poole and F. Gill. Philadelphia and Washington, DC: Academy of Natural Sciences and American Ornithologists' Union.

Parsons, K. C. 2003. Reproductive success of wading birds using phragmites marsh and upland nesting habitats. *Estuaries* 26:596–601.

Parsons, K. C., and T. L. Master. 2000. Snowy Egret (*Egretta thula*). In *The Birds of North America,* no. 489, ed. A. Poole and F. Gill. Philadelphia and Washington, DC: Academy of Natural Sciences and American Ornithologists' Union.

Patterson, M. P., and L. B. Best. 1996. Bird abundance and nesting success in Iowa CRP fields: The importance of vegetation structure and composition. *American Midland Naturalist* 135:153–167.

Patterson, R. M. 1981. Range expansion of Cliff Swallow into Maryland coastal plain. *Maryland Birdlife* 37:43–44.

Paxton, R. O. 1998. Piping Plover (*Charadrius melodus*). Pages 232–233 in *Bull's Birds of New York State*, ed. E. Levine. Ithaca: Cornell University Press.

Peakall, D. B. 1976. The Peregrine Falcon (*Falco peregrinus*) and pesticides. *Canadian Field-Naturalist* 90:301–307.

Peck, G. K., and R. D. James. 1983. *Breeding Birds of Ontario: Nidiology and Distribution*. Vol. 1, *Non-Passerines*. Toronto: Royal Ontario Museum of Life Sciences.

———. 1987. *Breeding Birds of Ontario: Nidiology and Distribution*. Vol. 2, *Passerines*. Toronto: Royal Ontario Museum of Life Sciences.

———. 1997. Breeding birds of Ontario: Nidiology and distribution. Vol. 2, Passerines, 1st rev., part A: Flycatchers to gnatcatchers. *Ontario Birds* 15:95–107.

Perkins, S. 2005. Spring migration: New England region. *North American Birds* 59:402–408.

Perring, F. H., and S. M. Walters. 1962. *Atlas of the British Flora*. London: T. Nelson.

Peterjohn, B. G. 2001. *The Birds of Ohio*. Wooster, Ohio: Wooster Book Co.

Peterjohn, B. G., and P. Davis. 1996. The first report of the Maryland/DC Records Committee. *Maryland Birdlife* 52:3–43.

Peterjohn, B. G., and D. L. Rice. 1991. *The Ohio Breeding Bird Atlas*. Columbus, Ohio: Department of Natural Resources.

Petersen, W. R. 2009. The nesting season: New England region. *North American Birds* 62:532–537.

Petersen, W. R., and W. R. Meservey. 2003. *Massachusetts Breeding Bird Atlas*. Amherst: University of Massachusetts Press.

Peterson, J. M. C. 1988. Common Raven (*Corvus corax*). Pages 286–287 in *The Atlas of Breeding Birds in New York State*, ed. R. F. Andrle and J. R. Carroll. Ithaca: Cornell University Press.

Peterson, R. T. 1934. *A Field Guide to the Birds*. 1st ed. Boston: Houghton Mifflin.

Petit, L. J. 1999. Prothonotary Warbler (*Protonotaria citrea*). In *The Birds of North America*, no. 408, ed. A. Poole and F. Gill. Philadelphia and Washington, DC: Academy of Natural Sciences and American Ornithologists' Union.

Petit, L. J., W. J. Fleming, K. E. Petit, and D. R. Petit. 1987. Nest-box use by Prothonotary Warblers (*Prothonotaria citrea*) in riverine habitat. *Wilson Bulletin* 99:485–488.

Petrides, G. A. 1942. Variable nesting habits of the Parula Warbler. *Wilson Bulletin* 54:252–253.

Phillips, A. R. 1991. *The Known Birds of North and Middle America*. Part 2, *Bombycillidae; Sylviidae to Sturnidae; Vireonidae*. Denver: A. R. Phillips.

Pickwell, G. B. 1931. The Prairie Horned Lark. *St. Louis Academy of Science Transactions* 27:1–153.

Pierotti, R. J., and T. P. Good. 1994. Herring Gull (*Larus argentatus*). In *The Birds of North America*, no. 124, ed. A. Poole and F. Gill. Philadelphia and Washington, DC: Academy of Natural Sciences and American Ornithologists' Union.

Pitocchelli, J. 1993. Mourning Warbler (*Oporornis philadelphia*). In *The Birds of North America*, no. 72, ed. A. Poole and F. Gill. Philadelphia and Washington, DC: Academy of Natural Sciences and American Ornithologists' Union.

Plentovich, S., N. R. Holler, and G. E. Hill. 1999. Habitat requirements of Henslow's Sparrows wintering in silvacultural lands of the gulf coastal plain. *Auk* 116:109–115.

Poole, A. F., R. O. Bierregaard, and M. S. Martell. 2002. Osprey (*Pandion haliaetus*). In *The Birds of North America*, no. 683, ed. A. Poole and F. Gill. Philadelphia and Washington, DC: Academy of Natural Sciences and American Ornithologists' Union.

Post, T. J. 2008. Ruffed Grouse (*Bonasa umbellus*). Pages 138–139 in *The Second Atlas of Breeding Birds in New York State*, ed. K. J. McGowan and K. Corwin. Ithaca: Cornell University Press.

Post, W., J. P. Poston, and G. T. Bancroft. 1996. Boat-tailed Grackle (*Quiscalus major*). In *The Birds of North America*, no. 207, ed. A. Poole and F. Gill. Philadelphia and Washington, DC: Academy of Natural Sciences and American Ornithologists' Union.

Pough, R. H. 1951. *Audubon Water Bird Guide*. Garden City, NY: Doubleday.

Poulin, R. G., S. D. Grindal, and R. M. Brigham. 1996. Common Nighthawk (*Chordeiles minor*). In *The Birds of North America*, no. 213, ed. A. Poole and F. Gill. Philadelphia and Washington, DC: Academy of Natural Sciences and American Ornithologists' Union.

Preble, E. A. 1900. List of summer birds of western Maryland. Pages 294–307 in *Allegany County*, ed. Maryland Geological Survey. Baltimore: Johns Hopkins Press.

Price, J., S. Droege, and A. Price. 1995. *The Summer Atlas of North American Birds*. New York: Academic Press.

Priednieks, J., M. Strazds, A. Strazds, and A. Petrins. 1989. *Latvijas Ligzdojoso Putnu Atlants, 1980–1984*. Riga (Latvia): Zinatne.

Prosser D. J., and R. P. Brooks. 1998. A verified habitat suitability index for the Louisiana Waterthrush. *Journal of Field Ornithology* 69:288–298.

Purroy, F. J. 1997. *Atlas de las Aves de España, 1975–1995*. Barcelona: Sociedad Española de Ornitologia and Lynx Edicions.

Pyle, R. L. 1963. House Finch reaches District of Columbia and Virginia. *Atlantic Naturalist* 18:32–33.

Quezon, A. J. 1997. First breeding record for Mississippi Kite in Virginia. *Raven*. 68:85–88.

Rasberry, D. A. 1996. Green Heron (*Butorides virescens*). Pages 62–63 in *Atlas of the Breeding Birds of Maryland and the District of Columbia*, ed. C. S. Robbins and E. A. T. Blom. Pittsburgh: University of Pittsburgh Press.

Reese, J. G. 1972. A Chesapeake Barn Owl population. *Auk* 89:106–114.

———. 1975. Diurnal vocalization by a wintering Black Rail. *Maryland Birdlife* 31:13–14.

———. 1980. Demography of European Mute Swans in Chesapeake Bay. *Auk* 97:449–464.

———. 1996a. Mute Swan (*Cygnus olor*). Pages 70–71 in *Atlas of the Breeding Birds of Maryland and the District of Columbia*, ed. C. S. Robbins and E. A. T. Blom. Pittsburgh: University of Pittsburgh Press.

———. 1996b. Osprey (*Pandion haliaetus*). Pages 92–93 in *Atlas of the Breeding Birds of Maryland and the District of Columbia*, ed. C. S. Robbins and E. A. T. Blom. Pittsburgh: University of Pittsburgh Press.

———. 1996c. Chuck-will's-widow (*Caprimulgus carolinensis*). Pages 192–193 in *Atlas of the Breeding Birds of Maryland and the District of Columbia*, ed. C. S. Robbins and E. A. T. Blom. Pittsburgh: University of Pittsburgh Press.

———. 1996d. Whip-poor-will (*Caprimulgus vociferus*). Pages 194–195 in *Atlas of the Breeding Birds of Maryland and the District of Columbia*, ed. C. S. Robbins and E. A. T. Blom. Pittsburgh: University of Pittsburgh Press.

Reid, W. 1992a. Common Merganser (*Mergus merganser*). Pages 84–

85 in *Atlas of Breeding Birds in Pennsylvania,* ed. D. W. Brauning. Pittsburgh: University of Pittsburgh Press.

———. 1992b. House Sparrow (*Passer domesticus*). Pages 420–421 in *Atlas of Breeding Birds in Pennsylvania,* ed. D. W. Brauning. Pittsburgh: University of Pittsburgh Press.

Remsen, J. V., Jr. 2001. True winter range of the Veery (*Catharus fuscescens*): Lessons for determining winter ranges of species that winter in the tropics. *Auk* 118:838–848.

Renno, O. 1993. *Eesti linnuatlas: Eesti haudelindude levikuatlas.* Estonia: Eesti Looduseuurijate Selts.

Reudink, M. W., S. G. Mech, S. P. Mullen, and R. L. Curry. 2007. Structure and dynamics of the hybrid zone between Black-capped Chickadee (*Poecile atricapillus*) and Carolina Chickadee (*P. carolinensis*) in southeastern Pennsylvania. *Auk* 124:463–478.

Rheinwald, G. 1977. *Atlas der Brutverbreitung westdeutscher Vogelarten 1975.* Bonn (West Germany): K. Druck.

———. 1982. *Brutvogelatlas der Bundesrepublik Deutschland 1980.* Lengede (West Germany): Dachverband Deutscher Avifaunisten.

———. 1993. *Atlas der Verbreitung und Häufigkeit der Brutvögel Deutschlands.* Bonn (Germany): Dachverband Deutscher Avifaunisten.

Rhoads, S. N. 1903. Exit the Dickcissel: A remarkable case of local extinction. *Cassinia* 7:17–28.

Rhoads, S. N., and C. J. Pennock. 1905. Birds of Delaware: A preliminary list. *Auk* 22:194–205.

Ricciardi, S. A. 1995. Sighting of Chuck-will's-widow fledgling. *Maryland Birdlife* 51:61–62

———. 1996a. Ruby-throated Hummingbird (*Archilochus colubris*). Pages 198–199 in *Atlas of the Breeding Birds of Maryland and the District of Columbia,* ed. C. S. Robbins and E. A. T. Blom. Pittsburgh: University of Pittsburgh Press.

———. 1996b. Downy Woodpecker (*Picoides pubescens*). Pages 208–209 in *Atlas of the Breeding Birds of Maryland and the District of Columbia,* ed. C. S. Robbins and E. A. T. Blom. Pittsburgh: University of Pittsburgh Press.

———. 1996c. Hairy Woodpecker (*Picoides villosus*). Pages 210–211 in *Atlas of the Breeding Birds of Maryland and the District of Columbia,* ed. C. S. Robbins and E. A. T. Blom. Pittsburgh: University of Pittsburgh Press.

———. 1996d. Great Crested Flycatcher (*Myiarchus crinitus*). Pages 228–229 in *Atlas of the Breeding Birds of Maryland and the District of Columbia,* ed. C. S. Robbins and E. A. T. Blom. Pittsburgh: University of Pittsburgh Press.

———. 1996e. Louisiana Waterthrush (*Seiurus motacilla*). Pages 360–361 in *Atlas of the Breeding Birds of Maryland and the District of Columbia,* ed. C. S. Robbins and E. A. T. Blom. Pittsburgh: University of Pittsburgh Press.

———. 1996f. American Goldfinch (*Carduelis tristis*). Pages 434–435 in *Atlas of the Breeding Birds of Maryland and the District of Columbia,* ed. C. S. Robbins and E. A. T. Blom. Pittsburgh: University of Pittsburgh Press.

Rich, T. D., C. J. Beardmore, H. Berlanga, P. J. Blancher, M. S. W. Bradstreet, G. S. Butcher, D. W. Demarest, E. H. Dunn, W. C. Hunter, E. E. Inigo-Elias, J. A. Kennedy, A. M. Martell, A. O. Panjabi, D. N. Pashley, K. V. Roosenberg, C. M. Rustay, J. S. Wendt, and T. C. Will. 2004. Partners in flight: North American landbird conservation plan. Ithaca: Cornell Lab of Ornithology.

Richardson, M., and D. W. Brauning. 1995. Chestnut-sided Warbler (*Dendroica pensylvanica*). In *The Birds of North America,* no. 190, ed. A. Poole and F. Gill. Philadelphia and Washington, DC:

Academy of Natural Sciences and American Ornithologists' Union.

Richmond, C. W. 1888. An annotated list of birds breeding in the District of Columbia. *Auk* 5:18–25.

Richmond, M. E. 2008. Herring Gull (*Larus argentatus*). Pages 256–257 in *The Second Atlas of Breeding Birds in New York State,* ed. K. J. McGowan and K. Corwin. Ithaca: Cornell University Press.

Ridgway, R. 1882. Birds new to or rare in the District of Columbia. *Nuttall Ornithological Club Bulletin* 7:253.

———. 1884. Probable breeding of the Red Crossbill in central Maryland. *Auk* 1:292.

Rines, M. W. 1998. Red-bellied Woodpeckers raising second brood. *Bird Observer* 26:14–15.

Ringler, R. F. 1977. The season: Spring migration, 1977. *Maryland Birdlife* 33:116–124.

———. 1980. The season: Spring migration, March 1–May 31, 1980. *Maryland Birdlife* 36:113–128.

———. 1983. The season: Breeding season, June 1–July 31, 1983. *Maryland Birdlife* 39:98–105.

———. 1985. The season: Breeding season, June 1–July 31, 1985. *Maryland Birdlife* 41:109–115.

———. 1986. The season: Spring migration, March 1–May 31,1986. *Maryland Birdlife* 42:60–79.

———. 1987. The season: Winter season, December 1, 1986–February 28, 1987. *Maryland Birdlife* 43:50–59.

———. 1988a. The season: Spring migration, March 1, 1988–May 31, 1988. *Maryland Birdlife* 44:84–106.

———. 1988b. The season: Breeding season, June 1–July 31, 1988. *Maryland Birdlife* 44:122–130.

———. 1989. The season: Spring migration, March 1–May 31, 1989. *Maryland Birdlife* 45:94–117.

———. 1990. The season: Spring migration, March 1–May 31, 1990. *Maryland Birdlife* 46:86–108.

———. 1991. The season: Spring migration, March 1–May 31, 1991. *Maryland Birdlife* 47:106–126.

———. 1996a. Pied-billed Grebe (*Podilymbus podiceps*). Pages 40–41 in *Atlas of the Breeding Birds of Maryland and the District of Columbia,* ed. C. S. Robbins and E. A. T. Blom. Pittsburgh: University of Pittsburgh Press.

———. 1996b. Spotted Sandpiper (*Actitis macularia*). Pages 148–149 in *Atlas of the Breeding Birds of Maryland and the District of Columbia,* ed. C. S. Robbins and E. A. T. Blom. Pittsburgh: University of Pittsburgh Press.

———. 1996c. Black-capped Chickadee (*Parus atricapillus*). Pages 254–255. in *Atlas of the Breeding Birds of Maryland and the District of Columbia,* ed. C. S. Robbins and E. A. T. Blom. Pittsburgh: University of Pittsburgh Press.

———. 1996d. Carolina Chickadee (*Parus carolinensis*). Pages 256–257 in *Atlas of the Breeding Birds of Maryland and the District of Columbia,* ed. C. S. Robbins and E. A. T. Blom. Pittsburgh: University of Pittsburgh Press.

———. 1996e. Prairie Warbler (*Dendroica discolor*). Pages 342–343 in *Atlas of the Breeding Birds of Maryland and the District of Columbia,* ed. C. S. Robbins and E. A. T. Blom. Pittsburgh: University of Pittsburgh Press.

———. 2001. The season: Breeding season, June 1–July 31, 2000. *Maryland Birdlife* 57:40–47.

———. 2002a. The season: Spring migration, March 1–May 31, 2001. *Maryland Birdlife* 58:30–43.

———. 2002b. The season: Breeding season, June 1–July 31, 2001. *Maryland Birdlife* 58:44–50.

———. 2002c. The season: Fall migration, August 1–November 30, 2001. *Maryland Birdlife* 58:50–67.

———. 2003a. The season: Spring migration, March 1, 2002–May 31, 2002. *Maryland Birdlife* 59:26–43.

———. 2003b. The season: Breeding season, June 1, 2002–July 31, 2002. *Maryland Birdlife* 59(3–4):8–12.

———. 2004a. The season: Spring migration, March 1–May 31, 2003. *Maryland Birdlife* 60:16–31.

———. 2004b. The season: Breeding season, June 1–July 31, 2003. *Maryland Birdlife* 60:43–48.

Ripley, S. D. 1977. *Rails of the World*. Boston: Godine.

Risberg, L., G. Aulen, K. Bylin, and T. Tyrberg. 1990. Sveriges fåglar. Vår Fågelvärld Supplement, no. 14. Stockholm (Sweden).

Robbins, C. S. 1949a. Breeding bird census: Mature and lumbered oak-maple ridge forest. *Audubon Field Notes* 3:259–261.

———. 1949b. Breeding bird census: Open hemlock-spruce bog. *Audubon Field Notes* 3:269.

———. 1949c. Summary of Maryland nest records, 1949. *Maryland Birdlife* 5:41–48.

———. 1950. Ecological distribution of the breeding Parulidae of Maryland. Master's thesis, George Washington University, Washington, DC.

———. 1966. The season: March, April, May 1966. *Maryland Birdlife* 22:77–85.

———. 1967. The season: April, May, June 1967. *Maryland Birdlife* 23:70–79.

———. 1973. The season: April, May, June 1973. *Maryland Birdlife* 29:114–124.

———. 1996a. Brown Creeper (*Certhia americana*). Pages 266–267 in *Atlas of the Breeding Birds of Maryland and the District of Columbia*, ed. C. S. Robbins and E. A. T. Blom. Pittsburgh: University of Pittsburgh Press.

———. 1996b. Northern Mockingbird (*Mimus polyglottos*). Pages 296–297 in *Atlas of the Breeding Birds of Maryland and the District of Columbia*, ed. C. S. Robbins and E. A. T. Blom. Pittsburgh: University of Pittsburgh Press.

———. 1996c. Chestnut-sided Warbler (*Dendroica pensylvanica*). Pages 326–327 in *Atlas of the Breeding Birds of Maryland and the District of Columbia*, ed. C. S. Robbins and E. A. T. Blom. Pittsburgh: University of Pittsburgh Press.

———. 1996d. Magnolia Warbler (*Dendroica magnolia*). Pages 328–329 in *Atlas of the Breeding Birds of Maryland and the District of Columbia*, ed. C. S. Robbins and E. A. T. Blom. Pittsburgh: University of Pittsburgh Press.

———. 1996e. Black-throated Blue Warbler (*Dendroica caerulescens*). Pages 330–331 in *Atlas of the Breeding Birds of Maryland and the District of Columbia*, ed. C. S. Robbins and E. A. T. Blom. Pittsburgh: University of Pittsburgh Press.

———. 1996f. Black-throated Green Warbler (*Dendroica virens*). Pages 334–335 in *Atlas of the Breeding Birds of Maryland and the District of Columbia*, ed. C. S. Robbins and E. A. T. Blom. Pittsburgh: University of Pittsburgh Press.

———. 1996g. Blackburnian Warbler (*Dendroica fusca*). Pages 336–337 in *Atlas of the Breeding Birds of Maryland and the District of Columbia*, ed. C. S. Robbins and E. A. T. Blom. Pittsburgh: University of Pittsburgh Press.

———. 1996h. Northern Waterthrush (*Seiurus noveboracensis*). Pages 358–359 in *Atlas of the Breeding Birds of Maryland and the District of Columbia*, ed. C. S. Robbins and E. A. T. Blom. Pittsburgh: University of Pittsburgh Press.

———. 1996i. Canada Warbler (*Wilsonia Canadensis*). Pages 370–371

in *Atlas of the Breeding Birds of Maryland and the District of Columbia*, ed. C. S. Robbins and E. A. T. Blom. Pittsburgh: University of Pittsburgh Press.

Robbins C. S., and E. A. T. Blom, eds. 1996. *Atlas of the Breeding Birds of Maryland and the District of Columbia*. Pittsburgh: University of Pittsburgh Press.

Robbins, C. S., and D. D. Boone. 1985. Threatened breeding birds of Maryland. *Maryland Birdlife* 41:87–108.

Robbins, C. S., and D. Bystrak. 1977. *Field List of the Birds of Maryland*. 2nd ed. Maryland Avifauna No. 2. Baltimore: Maryland Ornithological Society.

Robbins, C. S., D. Bystrak, and P. H. Geissler. 1986. *The Breeding Bird Survey: Its First Fifteen Years, 1965–1979*. Washington, DC: U.S. Fish and Wildlife Service. Resource Publication 157.

Robbins, C. S., D. K. Dawson, and B. A. Dowell. 1989. Habitat area requirements of breeding forest birds of the middle Atlantic states. *Wildlife Monographs*, no. 103.

Robbins, C. S., J. W. Fitzpatrick, and P. B. Hamel. 1992. A warbler in trouble: *Dendroica cerulea*. Pages 549–562 in *Ecology and Conservation of Neotropical Migrant Landbirds*, ed. J. M. Hagan III, and D. W. Johnston. Washington, DC: Smithsonian Institution Press.

Robbins, C. S., and R. E. Stewart. 1951. Breeding bird census: Scrub spruce bog. *Audubon Field Notes* 5:325.

Robertson, P. A. 1993. Pheasant (*Phasianus colchicus*). Pages 140–141 in *The New Atlas of Breeding Birds in Britain and Ireland, 1988–1991*, ed. D. Wingfield Gibbons, J. B. Reid, and R. A. Chapman. Staffordshire, UK: T. and A. D. Poyser.

Robertson, R. J., B. J. Stutchbury, and R. R. Cohen. 1992. Tree Swallow (*Tachycineta bicolor*). In *The Birds of North America*, no. 11, ed. A. Poole and F. Gill. Philadelphia and Washington, DC: Academy of Natural Sciences and American Ornithologists' Union.

Robinson, S. K. 1994. *Nesting Success of Forest Songbirds in Northwestern Illinois*. Final report, Project W-115-R-3. Urbana: Illinois Natural History Survey, Center of Wildlife Ecology.

Robinson, S. K., and R. T. Holmes. 1982. Foraging behavior of forest birds: The relationships among search tactics, diet, and habitat structure. *Ecology* 63:1918–1931.

Robinson, S. K., F. R. Thompson III, T. M. Donovan, D. R. Whitehead, and J. Faaborg. 1995. Regional forest fragmentation and the nesting success of migratory birds. *Science* 267:1987–1990.

Robinson, W. D. 1995. Louisiana Waterthrush (*Seiurus motacilla*). In *The Birds of North America*, no. 151, ed. A. Poole and F. Gill. Philadelphia and Washington, DC: Academy of Natural Sciences and American Ornithologists' Union.

Rodenhouse, N. L., and L. B. Best. 1983. Breeding ecology of Vesper Sparrow in corn and soybean fields. *American Midland Naturalist* 110:265–275.

Rodewald, P. G., and R. D. James. 1996. Yellow-throated Vireo (*Vireo flavifrons*). In *The Birds of North America*, no. 247, ed. A. Poole and F. Gill. Philadelphia and Washington, DC: Academy of Natural Sciences and American Ornithologists' Union.

Rodgers, J. A., Jr., and H. T. Smith. 1995. Little Blue Heron (*Egretta caerulea*). In *The Birds of North America*, no. 145, ed. A. Poole and F. Gill. Philadelphia and Washington, DC: Academy of Natural Sciences and American Ornithologists' Union.

Rohwer, F. C., W. P. Johnson, and E. R. Loos. 2002. Blue-winged Teal (*Anas discors*). In *The Birds of North America*, no. 625, ed. A. Poole and F. Gill. Philadelphia and Washington, DC: Academy of Natural Sciences and American Ornithologists' Union.

Romagosa, C. M., and T. McEneaney. 1999. Eurasian Collared-

Dove in North America and the Caribbean. *North American Birds* 53:348–353.

Root, T. 1988. *Atlas of Wintering North American Birds: An Analysis of Christmas Bird Count Data*. Chicago: University of Chicago Press.

Roseberry, J. L., D. L. Ellsworth, and W. D. Klimstra. 1987. Comparative post-release behavior and survival of wild, semi-wild, and game farm bobwhites. *Wildlife Society Bulletin* 15:449–455.

Roseberry, J. L., and W. D. Klimstra. 1984. *Population Ecology of the Bobwhite*. Carbondale: Southern Illinois University Press.

Rosenberg, K. V., J. D. Lowe, and A. A. Dhondt. 1999. Effects of forest fragmentation on breeding tanagers: A continental perspective. *Conservation Biology* 13:568–583.

Roth, R. R., and R. K. Johnson. 1993. Long-term dynamics of a Wood Thrush population breeding in a forest fragment. *Auk* 110:37–48.

Rufino, R., ed. 1989. *Atlas das aves que nidificam em Portugal Continental*. Lisboa, Portugal: CEMPA.

Rusch, D. H., S. Destefano, M. C. Reynolds, and D. Lauten. 2000. Ruffed Grouse (*Bonasa umbellus*). In *The Birds of North America*, no. 515, ed. A. Poole and F. Gill. Philadelphia and Washington, DC: Academy of Natural Sciences and American Ornithologists' Union.

Russell, W. A., Jr. 1996a. Blue Jay (*Cyanocitta cristata*). Pages 246–247 in *Atlas of the Breeding Birds of Maryland and the District of Columbia*, ed. C. S. Robbins and E. A. T. Blom. Pittsburgh: University of Pittsburgh Press.

———. 1996b. American Crow (*Corvus brachyrhynchos*). Pages 248–249 in *Atlas of the Breeding Birds of Maryland and the District of Columbia*, ed. C. S. Robbins and E. A. T. Blom. Pittsburgh: University of Pittsburgh Press.

———. 1996c. Fish Crow (*Corvus ossifragus*). Pages 250–251 in *Atlas of the Breeding Birds of Maryland and the District of Columbia*, ed. C. S. Robbins and E. A. T. Blom. Pittsburgh: University of Pittsburgh Press.

———. 1996d. Red-eyed Vireo (*Vireo olivaceus*). Pages 314–315 in *Atlas of the Breeding Birds of Maryland and the District of Columbia*, ed. C. S. Robbins and E. A. T. Blom. Pittsburgh: University of Pittsburgh Press.

Ryman, L. 1992. Osprey (*Pandion haliaetus*). Pages 90–91 in *Atlas of Breeding Birds in Pennsylvania*, ed. D. W. Brauning. Pittsburgh: University of Pittsburgh Press.

Sallabanks, R., and F. C. James. 1999. American Robin (*Turdus migratorius*). In *The Birds of North America*, no. 462, ed. A. Poole and F. Gill. Philadelphia and Washington, DC: Academy of Natural Sciences and American Ornithologists' Union.

Santner, S. 1992a. Long-eared Owl (*Asio otus*). Pages 162–163 in *Atlas of Breeding Birds in Pennsylvania*, ed. D. W. Brauning. Pittsburgh: University of Pittsburgh Press.

———. 1992b. Red-breasted Nuthatch (*Sitta canadensis*). Pages 244–245 in *Atlas of Breeding Birds in Pennsylvania*, ed. D. W. Brauning. Pittsburgh: University of Pittsburgh Press.

———. 1992c. Brown Creeper (*Certhia americana*). Pages 248–249 in *Atlas of Breeding Birds in Pennsylvania*, ed. D. W. Brauning. Pittsburgh: University of Pittsburgh Press.

Sattler, G. D., and M. J. Braun. 2000. Morphometric variation as an indicator of genetic interactions between Black-capped and Carolina Chickadees at a contact zone in the Appalachian Mountains. *Auk* 117:427–444.

Sauer, J. R. 1996. Common Nighthawk (*Chordeiles minor*). Pages 190–191 in *Atlas of the Breeding Birds of Maryland and the District*

of Columbia, ed. C. S. Robbins and E. A. T. Blom. Pittsburgh: University of Pittsburgh Press.

Saunders, N. C., and F. C. Saunders. 1996. Bank Swallow (*Riparia riparia*). Pages 240–241 in *Atlas of the Breeding Birds of Maryland and the District of Columbia*, ed. C. S. Robbins and E. A. T. Blom. Pittsburgh: University of Pittsburgh Press.

Schifferli, A., P. Geroudet, and R. Winkler, eds. 1980. *Verbrreitungsatlas der Brutvogel der Schweiz*. Sempach (Switzerland): Schweizerische Vogelwarte.

Schneider, K. J. 2008. Short-eared Owl (*Asio flammeus*). Pages 300–301 in *The Second Atlas of Breeding Birds in New York State*, ed. K. J. McGowan and K. Corwin. Ithaca: Cornell University Press.

Sealy, S. G. 1978. Possible influence of food on egg-laying and clutch size in the Black-billed Cuckoo. *Condor* 80:103–104.

Sedgwick, J. A. 2000. Willow Flycatcher (*Empidonax traillii*). In *The Birds of North America*, no. 30, ed. A. Poole and F. Gill. Philadelphia and Washington, DC: Academy of Natural Sciences and American Ornithologists' Union.

Semenchuk, G. P. 1992. *The Atlas of Breeding Birds in Alberta*. Edmonton: Federation of Alberta Naturalists.

Servello, F. A., and R. L. Kirkpatrick. 1987. Regional variation in the nutritional ecology of Ruffed Grouse. *Journal of Wildlife Management* 51:749–770.

Sharrock, J. T. R. 1976. *The Atlas of Breeding Birds in Great Britain and Ireland*. Staffordshire, UK: T. and A. D. Poyser.

Shields, M. 2002. Brown Pelican (*Pelecanus occidentalis*). In *The Birds of North America*, no. 609, ed. A. Poole and F. Gill. Philadelphia and Washington, DC: Academy of Natural Sciences and American Ornithologists' Union.

Shoch, D. T. 1992. Are there Swainson's Warblers in Delaware's Pocomoke Swamp? *Delmarva Ornithologist* 24:11–15.

Shriver, W. G., P. D. Vickery , T. P. Hodgman, and J. P. Gibbs. 2007. Flood tides affect breeding ecology of two sympatric sharp-tailed sparrows. *Auk* 124:552–560.

Sibley, D. A. 2000. *The Sibley Guide to Birds*. New York: Knopf.

Simpson, M. B., and P. G. Range. 1974. Evidence of breeding of Saw-whet Owls in western North Carolina. *Wilson Bulletin* 86:173–174.

Sitters, H. P. 1988. *Tetrad Atlas of the Breeding Birds of Devon*. Yelverton, UK: Devon Bird Watching and Preservation Society.

Skaar, P. D. 1975. *Montana Bird Distribution: Preliminary Mapping by Latilong*. Bozeman, Mont.: P. D. Skaar.

Skarphédinsson, K. H., G. Pétursson, and J. Hilmarsson. 1994. *Utbreidsia varpfugla á Sudvesturlandi, 1987–1992*. Natturufraedistofnun Islands, Reykjavik (Iceland).

Skipper, C. S. 1998a. Henslow's Sparrows return to previous nest site in western Maryland. *North American Bird Bander* 23:36–41.

———. 1998b. Monitoring and banding the Henslow's Sparrow in Garrett County, Maryland. *Maryland Birdlife* 54:106–117.

Smallwood, J. A., and D. M. Bird. 2002. American Kestrel (*Falco sparverius*). In *The Birds of North America*, no. 602, ed. A. Poole and F. Gill. Philadelphia and Washington, DC: Academy of Natural Sciences and American Ornithologists' Union.

Smith, D. B. 1999. Survival, behavior, and movements of captive-reared Mallards released in Dorchester County, Maryland. PhD diss., Louisiana State University, Baton Rouge.

Smith, D. G., A. Devine, and D. Walsh. 1987. Censusing screech owls in southern Connecticut. Pages 255–257 in *Biology and Conservation of Northern Forest Owls*, ed. R. W. Nero, R. J. Clark, R. J. Knapton, and R. H. Hamre. General Technical Report RM-142, Fort Collins, Colo.: U.S. Forest Service.

Smith, D. R. 1996. American Kestrel (*Falco sparverius*). Pages 110–111 in *Atlas of the Breeding Birds of Maryland and the District of Columbia*, ed. C. S. Robbins and E. A. T. Blom. Pittsburgh: University of Pittsburgh Press.

Smith, H. M. 1885. Breeding of *Loxia americana* in the District of Columbia. *Auk* 2:379–380.

Smith, H. M., and W. Palmer. 1888. Additions to the avifauna of Washington and vicinity. *Auk* 5:147–148.

Smith, K. G. 1986. Winter population dynamics of three species of mast-eating birds in the eastern United States. *Wilson Bulletin* 98:407–418.

Smith, K. G., J. H. Withgott, and P. G. Rodewald. 2000. Red-headed Woodpecker (*Melanerpes erythrocephalus*). In *The Birds of North America*, no. 518, ed. A. Poole and F. Gill. Philadelphia and Washington, DC: Academy of Natural Sciences and American Ornithologists' Union.

Smith, K. W., C. D. Dee, J. D. Fearnside, E. W. Fletcher, and R. N. Smith. 1993. *The Breeding Birds of Hertfordshire*. Boxmoor, UK: Hertfordshire Bird Club.

Smith, P. W. 1987. The Eurasian Collared-Dove arrives in the Americas. *American Birds* 41:1370–1379.

Smith, S. A. 1996a. Northern Harrier (*Circus cyaneus*). Pages 96–97 in *Atlas of the Breeding Birds of Maryland and the District of Columbia*, ed. C. S. Robbins and E. A. T. Blom. Pittsburgh: University of Pittsburgh Press.

———. 1996b. Ring-necked Pheasant (*Phasianus colchicus*). Pages 114–115 in *Atlas of the Breeding Birds of Maryland and the District of Columbia*, ed. C. S. Robbins and E. A. T. Blom. Pittsburgh: University of Pittsburgh Press.

———. 1996c. Upland Sandpiper (*Bartramia longicauda*). Pages 150–151 in *Atlas of the Breeding Birds of Maryland and the District of Columbia*, ed. C. S. Robbins and E. A. T. Blom. Pittsburgh: University of Pittsburgh Press.

———. 1996d. Dickcissel (*Spiza americana*). Pages 386–387 in *Atlas of the Breeding Birds of Maryland and the District of Columbia*, ed. C. S. Robbins and E. A. T. Blom. Pittsburgh: University of Pittsburgh Press.

———. 1996e. Vesper Sparrow (*Pooecetes gramineus*). Pages 394–395 in *Atlas of the Breeding Birds of Maryland and the District of Columbia*, ed. C. S. Robbins and E. A. T. Blom. Pittsburgh: University of Pittsburgh Press.

———.1996f. Savannah Sparrow (*Passerculus sandwichensis*). Pages 396–397 in *Atlas of the Breeding Birds of Maryland and the District of Columbia*, ed. C. S. Robbins and E. A. T. Blom. Pittsburgh: University of Pittsburgh Press.

Smith, S. M. 1993. Black-capped Chickadee (*Parus atricapillus*). In *The Birds of North America*, no. 39, ed. A. Poole and F. Gill. Philadelphia and Washington, DC: Academy of Natural Sciences and American Ornithologists' Union.

Smith, W. G. 1899. Abstracts of the Proceedings of the Delaware Valley Ornithological Club of Philadelphia 3:9.

Snyder, N. F. R., H. A. Snyder, J. L. Lincer, and R. T. Reynolds. 1973. Organochlorines, heavy metals, and the biology of North American accipiters. *Bioscience* 23:300–305.

Solem, J. K. 1996a. Tufted Titmouse (*Parus bicolor*). Pages 258–259 in *Atlas of the Breeding Birds of Maryland and the District of Columbia*, ed. C. S. Robbins and E. A. T. Blom. Pittsburgh: University of Pittsburgh Press.

———. 1996b. White-breasted Nuthatch (*Sitta carolinensis*). Pages 262–263 in *Atlas of the Breeding Birds of Maryland and the District of Columbia*, ed. C. S. Robbins and E. A. T. Blom. Pittsburgh: University of Pittsburgh Press.

———. 1996c. Wood Thrush (*Hylocichla mustelina*). Pages 290–291 in *Atlas of the Breeding Birds of Maryland and the District of Columbia*, ed. C. S. Robbins and E. A. T. Blom. Pittsburgh: University of Pittsburgh Press.

Somershoe, S. G., S. P. Hudman, and C. R. Chandler. 2004. Habitat use by Swainson's Warbler in a managed bottomland forest. *Wilson Bulletin* 115:148–154.

Soukup, S. S., and C. F. Thompson. 1997. Social mating system affects the frequency of extra-pair paternity in House Wrens. *Animal Behavior* 54:1089–1105.

Southworth, D. R. 1997. The season: Breeding season, June 1–July 31, 1996. *Maryland Birdlife* 53:86–95.

———. 1999. The season: Breeding season, June 1–July 31, 1998. *Maryland Birdlife* 55:74–79.

Southworth, D. R., and R. F. Ringler. 2000. The season: Breeding season, June 1, 1999–July 31, 1999. *Maryland Birdlife* 56:43–49.

Southworth, D. R., and L. Southworth. 1995. The season: Spring migration, March 1–May 31, 1995. *Maryland Birdlife* 52:80–91.

———. 1996. The season: Breeding season, June 1–July 31, 1995. *Maryland Birdlife* 52:91–96.

SOVON. 1987. *Atlas van de Nederlandse Vogels*. Arnhem (The Netherlands): SOVON.

Speiser, R., and T. Bosakowski. 1987. Nest site selection by Northern Goshawk in northern New Jersey and southeastern New York. *Condor* 89:387–394.

Spencer, O. R. 1943. Nesting habits of the Black-billed Cuckoo. *Wilson Bulletin* 55:11–22.

Spendelow, J. A., and S. R. Patton. 1988. *National Atlas of Coastal Waterbird Colonies in the Contiguous United States, 1976–82*. Biological Report 88(5). Washington, DC: U.S. Fish and Wildlife Service

Springer, P. F., and R. E. Stewart. 1948. Twelfth breeding bird census: Tidal marshes. *Audubon Field Notes* 2:223–226.

———. 1950. Gadwall nesting in Maryland. *Auk* 67:234–235.

Sprunt, A., Jr. 1942. Purple Martin (*Progne subis*). Pages 489–508 in *Life Histories of North American Flycatchers, Larks, Swallows, and Their Allies*, ed. A. C. Bent. U.S. National Museum Bulletin, no. 179.

———. 1953. Eastern Yellow-throated Warbler (*Dendroica dominica dominica*). Pages 349–359 in *Life Histories of North American Wood Warblers*, ed. A. C. Bent. U.S. National Museum Bulletin, no. 203.

———. 1968. Carolina Slate-colored Junco (*Junco hyemalis carolinensis*). Pages 1043–1049 in *Life Histories of North American Cardinals, Grosbeaks, Buntings, Towhees, Finches, Sparrow, and Allies*, ed. O. L. Austin Jr. U.S. National Museum Bulletin, no. 237.

Squires, J. R., and R. T. Reynolds. 1993. Northern Goshawk (*Accipiter gentilis*). In *The Birds of North America*, no. 298, ed. A. Poole and F. Gill. Philadelphia and Washington, DC: Academy of Natural Sciences and American Ornithologists' Union.

Stallman, H. R., and L. B. Best. 1996. Bird use of an experimental strip intercropping system in northeast Iowa. *Journal of Wildlife Management* 60:354–362.

Standley, P., and A. Swash. 1996. *The Birds of Berkshire*. Reading, UK.: Berkshire Atlas Group.

Stastny, K., A. Randik, and K. Hudec. 1987. *Atlas Hnizdniho Rozsireni Ptaku V CSSR, 1973–77*. Praha (Czech.): Academia.

Stastny K. A., V. Bejcek, and K. Hudec. 1995. *Atlas hnizdního rozšíření ptáků v České republice, 1985–1989*. Praha (Czech Republic): Academia.

Stasz, J. L. 1996a. Black-billed Cuckoo (*Coccyzus erythropthalmus*). Pages 176–177 in *Atlas of the Breeding Birds of Maryland and the District of Columbia,* ed. C. S. Robbins and E. A. T. Blom. Pittsburgh: University of Pittsburgh Press.

———. 1996b. Black-and-white Warbler (*Mniotilta varia*). Pages 346–347 in *Atlas of the Breeding Birds of Maryland and the District of Columbia,* ed. C. S. Robbins and E. A. T. Blom. Pittsburgh: University of Pittsburgh Press.

———. 1996c. Worm-eating Warbler (*Helmitheros vermivorus*). Pages 352–353 in *Atlas of the Breeding Birds of Maryland and the District of Columbia,* ed. C. S. Robbins and E. A. T. Blom. Pittsburgh: University of Pittsburgh Press.

———. 1996d. Hooded Warbler (*Wilsonia citrina*). Pages 368–369 in *Atlas of the Breeding Birds of Maryland and the District of Columbia,* ed. C. S. Robbins and E. A. T. Blom. Pittsburgh: University of Pittsburgh Press.

Stephens, D. A., and S. H. Sturts. 1991. *Idaho Bird Distribution: Mapping by Latilong.* Pocatello: Idaho Museum of Natural History Special Publication No. 11.

Stewart, N. J. 1996a. Common Moorhen (*Gallinula chloropus*). Pages 132–133 in *Atlas of the Breeding Birds of Maryland and the District of Columbia,* ed. C. S. Robbins and E. A. T. Blom. Pittsburgh: University of Pittsburgh Press.

———. 1996b. Pine Siskin (*Carduelis pinus*). Pages 432–433 in *Atlas of the Breeding Birds of Maryland and the District of Columbia,* ed. C. S. Robbins and E. A. T. Blom. Pittsburgh: University of Pittsburgh Press.

Stewart, P. A. 1962. *Waterfowl Populations in the Upper Chesapeake Region.* U.S. Fish and Wildlife Service Special Scientific Report– Wildlife, no. 65.

Stewart, R. E. 1951. Clapper Rail populations of the middle Atlantic states. *Transactions of the North American Wildlife Conference* 16:421–430.

———. 1953. A life history study of the Yellowthroat. *Wilson Bulletin* 65:99–115.

———. 1957. Eastern Glossy Ibis nesting in southeastern Maryland. *Auk* 74:509.

———. 1975. *Breeding Birds of North Dakota.* Fargo, N.D.: Tri-College Center for Environmental Studies.

Stewart, R. E., J. B. Cope, C. S. Robbins, and J. W. Brainerd. 1952. Seasonal distribution of bird populations at the Patuxent Research Refuge. *American Midland Naturalist* 47:257–363.

Stewart, R. E., and C. S. Robbins. 1947. Recent observations on Maryland birds. *Auk* 64:266–274.

———. 1951. Breeding bird census: Lightly grazed meadow. *Audubon Field Notes* 5:326–327.

———. 1958. The Birds of Maryland and the District of Columbia. *North American Fauna,* no. 62.

Stewart, R. E., and R. D. Titman. 1980. Territorial behaviour by prairie pothole Blue-winged Teal. *Canadian Journal of Zoology* 58:639–649.

Stickel, L. F., N. J. Chura, P. A. Stewart, C. M. Menzie, R. M. Prouty, and W. L. Reichel. 1966. Bald Eagle pesticide relations. *Transactions of the North American Wildlife and Natural Resources Conference* 31:190–200.

Stotts, V. D. 1987. A survey of breeding American Black Ducks in the Eastern Bay Region of Maryland in 1986. Report for Contract no. 14–16-005–86-017. Annapolis, Md.: U.S. Fish and Wildlife Service.

Stotts, V. D., and D. E. Davis. 1960. The Black Duck in the Chesapeake Bay of Maryland: Breeding behavior and biology. *Chesapeake Science* 1:127–154.

Stotz, D. F., R. O. Bierregaard, M. Cohn-Haft, P. Petermann, J. Smith, A. Whittaker, and S. V. Wilson. 1992. The status of North American migrants in central Amazonian Brazil. *Condor* 94:608–621.

Straight, C. A., and R. J. Cooper. 2000. Chuck-will's-widow (*Caprimulgus carolinensis*). In *The Birds of North America,* no. 499, ed. A. Poole and F. Gill. Philadelphia and Washington, DC: Academy of Natural Sciences and American Ornithologists' Union.

Stupka, A. 1963. *Notes on the Birds of Great Smoky Mountains National Park.* Knoxville: University of Tennessee Press.

Sullivan, E. 1992. Red-bellied Woodpecker raises two broods in Connecticut. *Connecticut Warbler* 12:24–25.

Swales, B. H. 1922. Prairie Horned Lark (*Otocoris alpestris praticola*) in Maryland in summer. *Auk* 39:568–569.

Swift, B. 1996. House Finch (*Carpodacus mexicanus*). Pages 430–431 in *Atlas of the Breeding Birds of Maryland and the District of Columbia,* ed. C. S. Robbins and E. A. T. Blom. Pittsburgh: University of Pittsburgh Press.

Taber, W., and D. W. Johnston. 1968. Indigo Bunting (*Passerina cyanea*). Pages 80–111 in *Life Histories of North American Cardinals, Grosbeaks, Buntings, Towhees, Finches, Sparrow, and Allies,* ed. O. L. Austin Jr. U.S. National Museum Bulletin, no. 237.

Tarof, S. A., L. M. Ratcliffe, M. M. Kasumovic, and P. T. Boag. 2005. Are Least Flycatcher (*Empidonax minimus*) clusters hidden leks? *Behavioral Ecology* 16:207–217.

Tate, J., Jr. 1986. The blue list for 1986. *American Birds* 40:227–236.

Tatu, K. S., J. T. Anderson, L. J. Hindman, and G. Seidel. 2007. Mute Swans' impact on submerged aquatic vegetation in Chesapeake Bay. *Journal of Wildlife Management* 71:1431–1439.

Taylor, B., and B. van Perlo. 1998. *Rails: A Guide to the Rails, Crakes, Gallinules and Coots of the World.* New Haven, Conn.: Yale University Press.

Taylor, D. W., D. L. Davenport, and J. Flegg. 1981. *The Birds of Kent: A Review of Their Status and Distribution.* Meopham, UK: Kent Ornithological Society.

Teixeira, R. M. 1979. *Atlas van de Nederlandse Broedvogels. Vereniging tot Behoud van Naturmonumenten in Nederland.* Nederland (The Netherlands): S-Graveland.

Telfair, R. C., II. 1994. Cattle Egret (*Bubulcus ibis*). In *The Birds of North America,* no. 113, ed. A. Poole and F. Gill. Philadelphia and Washington, DC: Academy of Natural Sciences and American Ornithologists' Union.

Temple, S. A. 2002. Dickcissel (*Spiza americana*). In *The Birds of North America,* no. 703, ed. A. Poole and F. Gill. Philadelphia and Washington, DC: Academy of Natural Sciences and American Ornithologists' Union.

Terres, J. K. 1980. *The Audubon Society Encyclopedia of North American Birds.* New York: Knopf.

Tessen, D. D. 1994. Western Great Lakes region. *National Audubon Society Field Notes* 48:945.

Therres, G. D. 1996a. Bald Eagle (*Haliaeetus leucocephalus*). Pages 94–95 in *Atlas of the Breeding Birds of Maryland and the District of Columbia,* ed. C. S. Robbins and E. A. T. Blom. Pittsburgh: University of Pittsburgh Press.

———. 1996b. Peregrine Falcon (*Falco peregrinus*). Pages 112–113 in *Atlas of the Breeding Birds of Maryland and the District of Columbia,* ed. C. S. Robbins and E. A. T. Blom. Pittsburgh: University of Pittsburgh Press.

———. 1996c. Ruffed Grouse (*Bonasa umbellus*). Pages 116–117 in *Atlas of the Breeding Birds of Maryland and the District of Columbia*, ed. C. S. Robbins and E. A. T. Blom. Pittsburgh: University of Pittsburgh Press.

———. 1996d. Breeding biology of Peregrine Falcons nesting on Maryland's coastal plain. *Maryland Birdlife* 52:47–51.

———. 1998. Maryland's endangered species law as a tool for biodiversity conservation. Pages 133–137 in *Conservation of Biological Diversity: A Key to the Restoration of the Chesapeake Bay Ecosystem and Beyond*, ed. G. D. Therres. Annapolis: Maryland Department of Natural Resources.

———. 2005. Recovery of Maryland's Bald Eagle nesting population. *Maryland Birdlife* 61: 35–45.

Therres, G. D., and D. F. Brinker. 2004. Mute Swan interactions with other birds in Chesapeake Bay. Pages 43–46 in *Mute Swans and Their Chesapeake Bay Habitats: Proceedings of a Symposium*, ed. M. C. Perry. U.S. Geological Survey, Biological Resources Discipline Information and Technology Report USGS/BRD/ITR-2004-0005.

Therres, G. D., S. Dawson, and J. C. Barber. 1993. Peregrine Falcon restoration in Maryland. Wildlife Technical Publication, no. 93-1. Annapolis: Department of Natural Resources, Fish, Heritage, and Wildlife Administration.

Therres, G. D., J. S. McKegg, and R. L. Miller. 1988. Maryland's Chesapeake Bay Critical Area Program: Implications for wildlife. *Transactions of the North American Wildlife and Natural Resources Conference* 53:392–400.

Therres, G. D., J. S. Weske, and M. A. Byrd. 1978. Breeding status of Royal Tern, Gull-billed Tern, and Black Skimmer in Maryland. *Maryland Birdlife* 34:75–77.

Thompson, B. C., J. A. Jackson, J. Burger, L. A. Hill, E. M. Kirsch, and J. L. Atwood. 1997. Least Tern (*Sterna antillarum*). In *The Birds of North America*, no. 290, ed. A. Poole and F. Gill. Philadelphia and Washington, DC: Academy of Natural Sciences and American Ornithologists' Union.

Titus, K. 1996. Cooper's Hawk (*Accipiter cooperii*). Pages 100–101 in *Atlas of the Breeding Birds of Maryland and the District of Columbia*, ed. C. S. Robbins and E. A. T. Blom. Pittsburgh: University of Pittsburgh Press.

Titus, K., and D. F. Brinker. 1996. Sharp-shinned Hawk (*Accipiter striatus*). Pages 98–99 in *Atlas of the Breeding Birds of Maryland and the District of Columbia*, ed. C. S. Robbins and E. A. T. Blom. Pittsburgh: University of Pittsburgh Press.

Titus, K., and J. A. Mosher. 1981. Nest-site habitat selected by woodland hawks in the central Appalachians. *Auk* 98:270–281.

———. 1987. Selection of nest tree species by Red-shouldered and Broad-winged hawks in two temperate forest regions. *Journal of Field Ornithology* 58:274–283.

Tomkins, I. R. 1955. The summer schedule of the Eastern Willet in Georgia. *Auk* 67:291–296.

———. 1965. The Willets of Georgia and South Carolina. *Wilson Bulletin* 77:151–167.

Traut, A. H., J. M. McCann, and D. F. Brinker. 2006. Breeding status and distribution of American Oystercatchers in Maryland. *Waterbirds* 29:302–307.

Trollinger, J. B., and K. K. Reay. 2001. *Breeding Bird Atlas of Virginia, 1985–1989*. Richmond: Virginia Department of Game and Inland Fisheries.

Turnbull, W. P. 1869. *The Birds of Eastern Pennsylvania and New Jersey*. Philadelphia: Henry Grambo.

Tyler, W. M. 1948. Brown Creeper (*Certhia familiaris*). Pages 56–79 in *Life Histories of North American Nuthatches, Wrens, Thrashers, and Their Allies*, ed. A. C. Bent. U.S. National Museum Bulletin, no. 195.

Tyrrell, W. B. 1936. *Bald Eagle Nest Survey of the Chesapeake Bay Region*. Washington, DC: National Audubon Society.

Ulke, T. 1935. Rare birds in the District of Columbia. *Auk* 51:461.

U.S. Fish and Wildlife Service (USFWS). 1987. *Migratory Nongame Birds of Management Concern in the United States: The 1987 List*. Washington, DC: U.S. Fish and Wildlife Service.

———. 2004. *U.S. Atlantic Coast Piping Plover Population: 2002–2003 Status Update*. Sudbury, Mass.: U.S. Fish and Wildlife Service.

———. 2008. *Waterfowl Population Status, 2008*. Washington, DC: U.S. Department of the Interior.

Van Ness, K. D., Jr. 1996a. Red-bellied Woodpecker (*Melanerpes carolinus*). Pages 204–205 in *Atlas of the Breeding Birds of Maryland and the District of Columbia*, ed. C. S. Robbins and E. A. T. Blom. Pittsburgh: University of Pittsburgh Press.

———. 1996b. Northern Flicker (*Colaptes auratus*). Pages 212–213 in *Atlas of the Breeding Birds of Maryland and the District of Columbia*, ed. C. S. Robbins and E. A. T. Blom. Pittsburgh: University of Pittsburgh Press.

———. 1996c. Pileated Woodpecker (*Dryocopus pileatus*). Pages 214–215 in *Atlas of the Breeding Birds of Maryland and the District of Columbia*, ed. C. S. Robbins and E. A. T. Blom. Pittsburgh: University of Pittsburgh Press.

———. 1996d. Willow Flycatcher (*Empidonax traillii*). Pages 222–223 in *Atlas of the Breeding Birds of Maryland and the District of Columbia*, ed. C. S. Robbins and E. A. T. Blom. Pittsburgh: University of Pittsburgh Press.

———. 1996e. Yellow-throated Vireo (*Vireo flavifrons*). Pages 310–311 in *Atlas of the Breeding Birds of Maryland and the District of Columbia*, ed. C. S. Robbins and E. A. T. Blom. Pittsburgh: University of Pittsburgh Press.

Van Velzen, W. T. 1967. First observed Brown Creeper nest in Maryland. *Maryland Birdlife* 23:68–69.

Vaughn, C. R. 1996a. Eastern Wood-Pewee (*Contopus virens*). Pages 216–217 in *Atlas of the Breeding Birds of Maryland and the District of Columbia*, ed. C. S. Robbins and E. A. T. Blom. Pittsburgh: University of Pittsburgh Press.

———. 1996b. Yellow Warbler (*Dendroica petechia*). Pages 324–325 in *Atlas of the Breeding Birds of Maryland and the District of Columbia*, ed. C. S. Robbins and E. A. T. Blom. Pittsburgh: University of Pittsburgh Press.

Vaughn, C. R., and C. S. Robbins. 1996. Swainson's Warbler (*Limnothlypis swainsonii*). Pages 354–355 in *Atlas of the Breeding Birds of Maryland and the District of Columbia*, ed. C. S. Robbins and E. A. T. Blom. Pittsburgh: University of Pittsburgh Press.

Veit, R. R., and W. R. Petersen. 1993. *Birds of Massachusetts*. Lincoln: Massachusetts Audubon Society.

Vickery, P. D. 1996. Grasshopper Sparrow (*Ammodramus savannarum*). In *The Birds of North America*, no. 239, ed. A. Poole and F. Gill. Philadelphia and Washington, DC: Academy of Natural Sciences and American Ornithologists' Union.

Viverette, C. B., S. Struve, L. J. Goodrich, and K. L. Bildstein. 1996. Decreases in migrating Sharp-shinned Hawks (*Accipiter striatus*) at traditional raptor-migration watch sites in eastern North America. *Auk* 113:32–40.

Walbeck, D. E. 1989. Observations of roof-nesting Killdeer and Common Nighthawks in Frostburg, Maryland. *Maryland Birdlife* 45:3–9.

Walkinshaw, L. H. 1941. The Prothonotary Warbler: A comparison of nesting conditions in Tennessee and Michigan. *Wilson Bulletin* 53:3–21.

———. 1966. Summer biology of Traill's Flycatcher. *Wilson Bulletin* 78:31–46.

———. 1968. Eastern Field Sparrow (*Spizella pusilla pusilla*). Pages 1217–1235 in *Life Histories of North American Cardinals, Grosbeaks, Buntings, Towhees, Finches, Sparrow, and Allies,* ed. O. L. Austin Jr. U.S. National Museum Bulletin, no. 237.

Walsh, J., V. Elia, R. Kane, and T. Halliwell. 1999. *Birds of New Jersey.* Bernardsville: New Jersey Audubon Society.

Walters, R. E., E. Sorensen, and S. Casjens. 1983. *Utah Bird Distribution: Latilong Study, 1983.* Salt Lake City: Utah Division of Wildlife Resources.

Warrilow, G. J. 1996. *Atlas of the Breeding Birds of Leicestershire and Rutland.* Privately published.

Wasilco, M. R. 2008a. American Oystercatcher (*Haematopus palliates*). Pages 236–237 in *The Second Atlas of Breeding Birds in New York State,* ed. K. J. McGowan and K. Corwin. Ithaca: Cornell University Press.

———. 2008b. Laughing Gull (*Larus atricilla*). Pages 252–253 in *The Second Atlas of Breeding Birds in New York State,* ed. K. J. McGowan and K. Corwin. Ithaca: Cornell University Press.

———. 2008c. Forster's Tern (*Sterna forsteri*). Pages 272–273 in *The Second Atlas of Breeding Birds in New York State,* ed. K. J. McGowan and K. Corwin. Ithaca: Cornell University Press.

Watts, B. D. 1995. Yellow-crowned Night-Heron (*Nyctanassa violacea*). In *The Birds of North America,* no. 161, ed. A. Poole and F. Gill. Philadelphia and Washington, DC: Academy of Natural Sciences and American Ornithologists' Union.

Watts, B. D., and M. A. Byrd. 2006. Status and distribution of colonial waterbirds in coastal Virginia: The 2003 breeding season. *Raven* 77:3–22.

Watts, B. D., A. C. Markham, and M. A. Byrd. 2006. Salinity and population parameters of Bald Eagles (*Haliaeetus leucocephalus*) in the lower Chesapeake Bay. *Auk* 123:393–404.

Watts, B. D., G. D. Therres, and M. A. Byrd. 2007. Status, distribution, and the future of Bald Eagles in the Chesapeake Bay area. *Waterbirds* 30 (Special Publication 1):25–38.

———. 2008. Recovery of the Chesapeake Bay Bald Eagle population. *Journal of Wildlife Management* 72:152–158.

Weber, P., D. Munteanu, and A. Papadopol, eds. 1994. *Atlasul provisoriu al pasarilor clocitoare din Romania.* Medias, Romania: Societatea Ornitologica.

Weeks, H. P., Jr. 1994. Eastern Phoebe (*Sayornis phoebe*). In *The Birds of North America,* no. 94, ed. A. Poole and F. Gill. Philadelphia and Washington, DC: Academy of Natural Sciences and American Ornithologists' Union.

Weller, M. W. 1961. Breeding biology of the Least Bittern. *Wilson Bulletin* 73:11–35.

Wennerstrom, J. 1996. Winter Wren (*Troglodytes troglodytes*). Pages 274–275 in *Atlas of the Breeding Birds of Maryland and the District of Columbia,* ed. C. S. Robbins and E. A. T. Blom. Pittsburgh: University of Pittsburgh Press.

Weske, J. S., and H. Fessenden. 1963. Glossy Ibis nesting in tidewater Maryland away from the ocean. *Bird-Banding* 34:161.

Wetmore, A. 1935. The Short-billed Marsh Wren breeding in Maryland. *Auk* 52:455.

Wetmore, A., and F. C. Lincoln. 1928. The Dickcissel in Maryland. *Auk* 45:508–509.

Whitcomb, B. L., R. F. Whitcomb, and D. Bystrak. 1977. Long-term turnover and effects of selective logging on the avifauna of forest fragments. *American Birds* 31:17–23.

Whitehead, D. R., and T. Taylor. 2002. Acadian Flycatcher (*Empidonax virescens*). In *The Birds of North America,* no. 614, ed. A. Poole and F. Gill. Philadelphia and Washington, DC: Academy of Natural Sciences and American Ornithologists' Union.

Whitmore, R. C. 1980. Reclaimed surface mines as avian habitat islands in the eastern forest. *American Birds* 34:13–14.

Wiemeyer, S. N., P. R. Spitzer, W. C. Krantz, T. G. Lamont, and E. Cromartie. 1975. Effects of environmental pollutants on Connecticut and Maryland ospreys. *Journal of Wildlife Management* 39:124–139.

Wiens, J. A. 1973. Interterritorial habitat variation in Grasshopper and Savannah Sparrows. *Ecology* 54:877–884.

Wierenga, H. L. 1996a. Black Rail (*Laterallus jamaicensis*). Pages 122–123 in *Atlas of the Breeding Birds of Maryland and the District of Columbia,* ed. C. S. Robbins and E. A. T. Blom. Pittsburgh: University of Pittsburgh Press.

———. 1996b. Virginia Rail (*Rallus limicola*). Pages 128–129 in *Atlas of the Breeding Birds of Maryland and the District of Columbia,* ed. C. S. Robbins and E. A. T. Blom. Pittsburgh: University of Pittsburgh Press.

Wiggins, D. A. 2004. Short-eared Owl (*Asio flammeus*): A technical conservation assessment. USDA Forest Service, Rocky Mountain Region. Available at www.fs.fed.us/r2/projects/scp/assessments/shortearedowl.pdf.

Wilkinson, J. R. 1996. Eastern Kingbird (*Tyrannus tyrannus*). Pages 230–231 in *Atlas of the Breeding Birds of Maryland and the District of Columbia,* ed. C. S. Robbins and E. A. T. Blom. Pittsburgh: University of Pittsburgh Press.

Williams, B., B. Akers, M. Beck, R. Beck, and J. Via. 2005. The thirtieth annual beach-nesting and colonial waterbirds survey of the Virginia barrier islands, 2004. *Raven* 76:14–20.

Williams, B., D. F. Brinker, and B. D. Watts. 2007. The status of colonial nesting wading bird populations within the Chesapeake Bay and Atlantic barrier island–lagoon system. *Waterbirds* 30 (Special Publication 1):82–92.

Williams, J. M. 1996. Nashville Warbler (*Vermivora ruficapilla*). In *The Birds of North America,* no. 205, ed. A. Poole and F. Gill. Philadelphia and Washington, DC: Academy of Natural Sciences and American Ornithologists' Union.

Willoughby, E. J. 1996a. Brown-headed Nuthatch (*Sitta pusilla*). Pages 264–265 in *Atlas of the Breeding Birds of Maryland and the District of Columbia,* ed. C. S. Robbins and E. A. T. Blom. Pittsburgh: University of Pittsburgh Press.

———. 1996b. Pine Warbler (*Dendroica pinus*). Pages 340–341 in *Atlas of the Breeding Birds of Maryland and the District of Columbia,* ed. C. S. Robbins and E. A. T. Blom. Pittsburgh: University of Pittsburgh Press.

———. 1996c. Blue Grosbeak (*Guiraca caerulea*). Pages 382–383 in *Atlas of the Breeding Birds of Maryland and the District of Columbia,* ed. C. S. Robbins and E. A. T. Blom. Pittsburgh: University of Pittsburgh Press.

———. 1996d. Field Sparrow (*Spizella pusilla*). Pages 392–393 in *Atlas of the Breeding Birds of Maryland and the District of Columbia,* ed. C. S. Robbins and E. A. T. Blom. Pittsburgh: University of Pittsburgh Press.

Wilmot, G. B. 1996a. Hooded Merganser (*Lophodytes cucullatus*). Pages 86–87 in *Atlas of the Breeding Birds of Maryland and the Dis-*

trict of Columbia, ed. C. S. Robbins and E. A. T. Blom. Pittsburgh: University of Pittsburgh Press.

———. 1996b. Red-headed Woodpecker (*Melanerpes erythrocephalus*). Pages 202–203 in *Atlas of the Breeding Birds of Maryland and the District of Columbia,* ed. C. S. Robbins and E. A. T. Blom. Pittsburgh: University of Pittsburgh Press.

———. 1996c. Prothonotary Warbler (*Protonotaria citrea*). Pages 350–351 in *Atlas of the Breeding Birds of Maryland and the District of Columbia,* ed. C. S. Robbins and E. A. T. Blom. Pittsburgh: University of Pittsburgh Press.

———. 1996d. Summer Tanager (*Piranga rubra*). Pages 374–375 in *Atlas of the Breeding Birds of Maryland and the District of Columbia,* ed. C. S. Robbins and E. A. T. Blom. Pittsburgh: University of Pittsburgh Press.

———. 1996e. Orchard Oriole (*Icterus spurius*). Pages 424–425 in *Atlas of the Breeding Birds of Maryland and the District of Columbia,* ed. C. S. Robbins and E. A. T. Blom. Pittsburgh: University of Pittsburgh Press.

Wilson, M. D., and B. D. Watts. 2006. Effects of moonlight on detection of Whip-poor-wills: Implications for long-term monitoring strategies. *Journal of Field Ornithology* 77:207–211.

Wilson, M. D., B. D. Watts, and D. F. Brinker. 2007. Status review of Chesapeake Bay marsh lands and breeding marsh birds. *Waterbirds* 30 (Special Publication 1):122–137.

Winter, M. 1999. Nesting biology of Dickcissels and Henslow's Sparrows in southwestern Missouri prairie fragments. *Wilson Bulletin* 111:515–526.

Wise-Gervais, C. 2005. American Kestrel (*Falco sparverius*). Pages 154–155 in *Arizona Breeding Bird Atlas,* ed. T. E. Corman and C. Wise-Gervais. Albuquerque: University of New Mexico Press.

Withers, P. C., and P. L. Timko. 1977. The significance of ground effect to the aerodynamic cost of flight and energetics of the Black Skimmer (*Rhynchops niger*). *Journal of Experimental Biology* 70:13–26.

Withgott, J. H., and K. G. Smith. 1998. Brown-headed Nuthatch (*Sitta pusilla*). In *The Birds of North America,* no. 349, ed. A. Poole and F. Gill. Philadelphia and Washington, DC: Academy of Natural Sciences and American Ornithologists' Union.

Witmer, M. C. 1996. Annual diet of Cedar Waxwing based on U.S. Biological Survey records (1885–1950) compared to diet of American Robin: Contrasts in dietary patterns and natural history. *Auk* 113:414–430.

Witmer, M. C., D. J. Mountjoy, and L. Elliot. 1997. Cedar Waxwing (*Bombycilla cedrorum*). In *The Birds of North America,* no. 309, ed. A. Poole and F. Gill. Philadelphia and Washington, DC: Academy of Natural Sciences and American Ornithologists' Union.

Woolfenden, B. E., H. L. Gibbs, C. M. McLaren, and S. G. Sealy. 2004. Community-level patterns of parasitism: Use of three common hosts by a brood parasitic bird, the Brown-headed Cowbird. *Écoscience* 11:238–248.

Woolfenden, G. E. 1968. Northern Seaside Sparrow (*Ammospiza maritima maritima*). Pages 819–831 in *Life Histories of North American Cardinals, Grosbeaks, Buntings, Towhees, Finches, Sparrow, and Allies,* ed. O. L. Austin Jr. U.S. National Museum Bulletin, no. 237.

Wooton, J. T. 1987. Interspecific competition between introduced House Finch populations and two associated passerine species. *Oecologia* 71:325–331.

Woutersen, K., and M. Platteeuw. 1998. *Atlas of the Birds of Huesca.* Huesca, España (Spain): Kees Woutersen Publications.

Wunz, G. A., and D. W. Brauning. 1992. Wild Turkey (*Meleagris gallopavo*). Pages 116–117 in *Atlas of Breeding Birds in Pennsylvania,* ed. D. W. Brauning. Pittsburgh: University of Pittsburgh Press.

Wurster, D. H., C. F. Wurster Jr., and W. N. Strickland. 1965. Bird mortality following DDT spray for Dutch elm disease. *Ecology* 46:488–499.

Wyatt, V. E., and C. M. Francis. 2002. Rose-breasted Grosbeak (*Pheucticus ludovicianus*). In *The Birds of North America,* no. 692, ed. A. Poole and F. Gill. Philadelphia and Washington, DC: Academy of Natural Sciences and American Ornithologists' Union.

Yeatman, L. 1976. *Atlas des oiseaux nicheurs de France de 1970 à 1975.* Paris: Ministère de la Qualité de la Vie Environement.

Yeatman-Berthelot, D., and G. Jarry, eds. 1994. *Nouvel Atlas des Oiseaux Nicheurs de France, 1985–1988.* Paris: Société Ornithologique de France.

Yingling, S. 1996. European Starling (*Sturnus vulgaris*). Pages 304–305 in *Atlas of the Breeding Birds of Maryland and the District of Columbia,* ed. C. S. Robbins and E. A. T. Blom. Pittsburgh: University of Pittsburgh Press.

Zeleny, L. 1976. *The Bluebird: How You Can Help Its Fight for Survival.* Bloomington: Indiana University Press.

Zicus, M. C. 1990. Nesting biology of Hooded Mergansers using nest boxes. *Journal of Wildlife Management* 54:637–643.

Zimmerman, J. L. 1966. Polygyny in the Dickcissel. *Auk* 83:534–546.

Zimpfer, N. L., G. S. Zimmerman, E. D. Silverman, and M. D. Koneff. 2008. Trends in duck breeding populations, 1955–2008. Administrative Report. Laurel, Md.: U.S. Fish and Wildlife Service, Division of Migratory Bird Management.

Zucker, E. E. 1996. Chimney Swift (*Chaetura pelagica*). Pages 196–197 in *Atlas of the Breeding Birds of Maryland and the District of Columbia,* ed. C. S. Robbins and E. A. T. Blom. Pittsburgh: University of Pittsburgh Press.

Zusi, R. L. 1962. Structural adaptations of the head and neck in the Black Skimmer. *Proceedings of the Nuttall Ornithological Club,* no. 3.

Zusi, R. L., and D. Bridge. 1981. On the slit pupil of the Black Skimmer (*Rhynchops niger*). *Journal of Field Ornithology* 51:338–340.

INDEX

Common names are indexed under the last word (or words, if hyphenated), and scientific names are indexed under the generic name. Each species account for a breeding species can easily be identified because it is indicated by a range of two pages (e.g., 212–213); other references to a species are indexed only when a significant point is stated or mapped.